Modern Heterogeneous Catalysis

Modern Heterogeneous Catalysis

An Introduction

Rutger A. van Santen

The Author

Prof. Dr. Rutger A. van Santen
Eindhoven University of Technology
Faculty of Chemistry and Chemical Engineering
Institute of Complex Molecular Systems
Den Dolech 2
5600MB Eindhoven
The Netherlands

Cover
The cover artwork was kindly provided by the author.

All books published by **Wiley-VCH** are carefully produced. Nevertheless, authors, editors, and publisher do not warrant the information contained in these books, including this book, to be free of errors. Readers are advised to keep in mind that statements, data, illustrations, procedural details or other items may inadvertently be inaccurate.

Library of Congress Card No.: applied for

British Library Cataloguing-in-Publication Data
A catalogue record for this book is available from the British Library.

Bibliographic information published by the Deutsche Nationalbibliothek
The Deutsche Nationalbibliothek lists this publication in the Deutsche Nationalbibliografie; detailed bibliographic data are available on the Internet at <http://dnb.d-nb.de>.

© 2017 Wiley-VCH Verlag GmbH & Co. KGaA, Boschstr. 12, 69469 Weinheim, Germany

All rights reserved (including those of translation into other languages). No part of this book may be reproduced in any form – by photoprinting, microfilm, or any other means – nor transmitted or translated into a machine language without written permission from the publishers. Registered names, trademarks, etc. used in this book, even when not specifically marked as such, are not to be considered unprotected by law.

Print ISBN: 978-3-527-33961-7
ePDF ISBN: 978-3-527-69452-5
ePub ISBN: 978-3-527-69450-1
Mobi ISBN: 978-3-527-69451-8

Cover Design Adam Design, Weinheim, Germany
Typesetting SPi Global Private Limited, Chennai, India
Printing and Binding
CPI Group (UK) Ltd, Croydon, CR0 4YY

Printed on acid-free paper

C9783527339617_030924

To Edith

Contents

Preface *xv*
Acknowledgments *xix*
Arrangement of This Book *xxi*

Part I Physical Chemistry and Kinetics *1*

1   Heterogeneous Catalysis *3*
1.1   What is Heterogeneous Catalysis? *3*
1.2   Early Developments *4*
1.2.1   Early Nineteenth Century Discoveries *5*
1.2.2   Later Nineteenth Century Discoveries *7*
1.3   The Three Basic Laws of Catalysis *7*
1.3.1   Berzelius' Catalysis Law *7*
1.3.2   Ostwald's Catalysis Law *8*
1.3.3   Sabatier's Catalysis Law *10*
         References *12*

2   Heterogeneous Catalytic Processes *15*
2.1   Introduction *15*
2.2   Important Heterogeneous Catalytic Reactions and Processes *19*
2.2.1   Hydrogenation and Dehydrogenation Reactions *19*
2.2.1.1   Hydrogenation Reactions: Transition Metal Catalysts *19*
2.2.2   Hydrocarbon Transformation Reactions *22*
2.2.2.1   Brønsted Acid Catalysis *22*
2.2.3   Oligomerization and Polymerization Catalysis *27*
2.2.3.1   Methanol to Ethylene and Aromatics: The Aufbau Reaction *27*
2.2.3.2   Fischer–Tropsch Catalysis *29*
2.2.3.3   Disproportionation and Metathesis Reaction: Single Site Catalysis *32*
2.2.3.4   Polymerization: Surface Coordination Complex Catalyst *35*
2.2.4   Hydrodesulfurization and Related Hydrotreating Reactions *36*
2.2.4.1   Hydrodesulfurization *36*
2.2.4.2   The Biomass Refinery *37*
2.2.5   Oxidation and Reduction Reactions *38*

2.2.5.1	Steam Reforming	*38*
2.2.5.2	NH$_3$ to NO$_x$ Oxidation	*39*
2.2.5.3	NO Reduction Catalysis	*45*
2.3	Summary	*48*
	References	*51*
3	**Physical Chemistry, Elementary Kinetics**	*59*
3.1	Introduction	*59*
3.2	Catalyst Characterization	*63*
3.2.1	Langmuir Adsorption Isotherm	*64*
3.2.2	Measurement of Pore Volume	*65*
3.2.3	Porosity	*66*
3.2.4	Temperature-Programmed Reactivity Measurements	*67*
3.2.5	Spectroscopic Techniques	*68*
3.3	Elementary Kinetics	*68*
3.3.1	Lumped Kinetics Expressions: Kinetic Determination of Key Reaction Intermediates	*68*
3.3.1.1	The Rate Constant of an Elementary Reaction	*68*
3.3.1.2	Elementary Catalytic Reaction Kinetics	*71*
3.3.2	Sabatier Principle and Volcano Curves: Brønsted–Evans–Polanyi Relations	*78*
3.3.2.1	Brønsted–Evans–Polanyi Relations of Elementary Surface Reaction Rate Constants	*79*
3.3.2.2	The Sabatier Volcano Curve	*84*
3.3.2.3	The Reaction Energy Diagram of the Catalytic Reaction Cycle	*90*
3.3.2.4	Electrocatalysis and Sabatier Principle Optimum	*94*
3.3.2.5	Temperature Dependence of Catalytic Reaction Rate	*96*
3.3.2.6	Summary: The Order of Reaction Rate	*101*
3.4	Transient Kinetics: The Determination of Site Concentration	*104*
3.5	Diffusion	*106*
3.5.1	Concentration Profiles	*106*
3.5.2	Effectiveness Factor	*107*
3.5.3	Diffusion in Zeolitic Micropores	*110*
	References	*114*
4	**The State of the Working Catalyst**	*117*
4.1	Introduction	*117*
4.2	Surface Reconstruction	*119*
4.3	Compound Formation: Activation and Deactivation	*123*
4.4	Supported Small Metal Particles	*124*
4.4.1	Nature of the Support Material	*125*
4.4.2	Reactivity and Stability	*127*
4.4.3	Summary	*128*
4.5	Structure Sensitivity of Transition Metal Particle Catalysts	*129*
4.5.1	Particle Size and Structure Dependence of Heterogeneous Catalytic Reactions	*129*
4.5.2	Site Generation	*133*

4.6	Alloys and Other Promotors	*133*
4.7	The Working Zeolite Catalysts	*136*
4.8	The State of the Mixed Oxide Surface	*139*
4.9	Summary	*139*
	References	*140*

5	**Advanced Kinetics: Breakdown of Mean Field Approximation**	***145***
5.1	Introduction	*145*
5.2	The Kinetic Monte Carlo Method: RuO_2 Catalyzed Oxidation	*146*
5.3	Single Molecule Spectroscopy	*149*
5.4	Catalytic Self-Organizing Systems	*153*
5.4.1	Introduction	*153*
5.4.2	Heterogeneous Catalytic Self-Organizing Systems	*154*
5.4.2.1	Alternation Between Two Different Surface Phases: Spiral Wave Formation	*155*
5.4.2.2	Competitive Reactive Steps Prevent Reactive Phase Formation	*165*
	References	*165*

	Part II Molecular Heterogeneous Catalysis	***167***
	Introduction	*167*
	References	*171*

6	**Basic Quantum-Chemical Concepts, The Chemical Bond Revisited (Jointly Written with I. Tranca)**	***173***
6.1	Introduction	*173*
6.2	The Definitions of Partial Density of States and Bond Order Overlap Population	*174*
6.3	Diatomic Molecules that Have σ Bonds	*176*
6.4	Diatomic Molecules with π Bonds	*182*
6.5	Comparison of the Electronic Structure of Molecules and Solids	*186*
6.6	Chemical Bonding in Transition Metals	*190*
6.6.1	The Electronic Structure of the Transition Metals	*190*
6.6.2	The Relative Stability of Transition Metal Structures	*199*
	Appendix	*205*
	References	*205*

7	**Chemical Bonding and Reactivity of Transition Metal Surfaces**	***209***
7.1	Introduction	*209*
7.2	The Nature of the Surface Chemical Bond	*210*
7.2.1	The Electronic Structure of the Transition Metal Surface	*210*
7.2.2	Chemisorption of Atoms and Molecules (This Section has been Jointly Written with I. Tranca.)	*213*
7.2.2.1	The H Adatom: The Covalent Surface Bond	*214*

7.2.2.2	Adsorption of the Carbon Atom: The Surface Molecular Complex *220*
7.2.2.3	The Oxygen Adatom: The Polar Surface Bond *228*
7.2.3	Adsorption Site Preference as a Function of Accessible Free Valence *233*
7.2.3.1	Chemisorption of Molecular Fragments CH_x, NH_x, and OH_x Species: Coordination Preference as a Function of Accessible Free Valence *233*
7.2.3.2	CH_3 and NH_3 Chemisorption: The Agostic Interaction *235*
7.2.4	Adsorption as a Function of Coordinative Unsaturation of Surface Atoms: Relation with d Valence Band Energy Shift *243*
7.2.5	Chemisorption of CO: Donative and Backdonative Interactions *249*
7.2.6	Lateral Interactions *258*
7.2.7	Scaling Laws *259*
7.2.8	In Summary: The Adsorbate Chemical Bond *262*
7.3	The Transition States of Elementary Surface Reactions *264*
7.3.1	Adsorbate σ-Bond Activation *265*
7.3.1.1	Activation of Methane *265*
7.3.1.2	The Oxidative Addition and Reductive Elimination Model *268*
7.3.1.3	The Umbrella Effect *269*
7.3.1.4	Activation Entropy *270*
7.3.1.5	σ-Bond Activation of Molecules that Bind Through their Lone Pair Orbital *270*
7.3.2	Dissociation of Diatomic Molecules with π-Bonds *273*
7.3.2.1	Principle of Non-Shared Bonding with the Same Surface Metal Atom *274*
7.4	Reactivity of Surfaces at High Coverage *274*
7.4.1	Decreased Surface Reactivity and Site Blocking *275*
7.4.2	Adatom Co-Assisted Activation *276*
7.4.2.1	Hydrogen Activated Dissociation *276*
7.4.2.2	Oxygen Assisted X–H Bond Cleavage *277*
7.4.2.3	Reactivity of the Oxide Overlayers *281*
	References *286*

8	**Mechanisms of Transition Metal Catalyzed Reactions** *293*
8.1	Introduction *293*
8.2	Hydrogenation Reactions *293*
8.2.1	Ammonia Synthesis *293*
8.2.1.1	Heterogeneous Catalytic Reaction *293*
8.2.1.2	Enzyme Catalysis *296*
8.2.2	Synthesis Gas Conversion to Methane and Liquid Hydrocarbons *297*
8.2.3	Hydroconversion of Hydrocarbons *306*
8.2.3.1	Ethylene *306*
8.2.3.2	Acetylene *309*
8.2.3.3	Hydrogenolysis and Isomerization *310*
8.2.4	NH_3 and CH_4 to HCN *314*
8.2.5	Electrocatalysis; H_2 Evolution *316*

8.3	Oxidation Reactions *321*
8.3.1	Synthesis Gas from Methane *321*
8.3.1.1	Steam Reforming *321*
8.3.1.2	Methane Oxidation *322*
8.3.1.3	NH_3 Oxidation to NO and N_2 *323*
8.3.1.4	Selective Oxidation of Ethylene *325*
8.4	Uniqueness of a Metal for a Particular Selective Reaction *337*
	References *338*
9	**Solid Acid Catalysis, Theory and Reaction Mechanisms** *345*
9.1	Introduction *345*
9.2	Elementary Theory of Surface Acidity and Basicity *345*
9.2.1	The Pauling Charge Excess *345*
9.2.2	The Chemistry of the Zeolitic Proton *351*
9.2.2.1	Vibrational Spectroscopy of the OH Bond *352*
9.2.2.2	Quantum Chemistry of the Zeolite Acidic OH Chemical Bond *358*
9.2.2.3	The Proton Transfer Reaction *364*
9.2.2.4	Chemical Reactivity Probes of Proton Donation Affinity: H/D Exchange Reactions *367*
9.3	Mechanism of Reactions Catalyzed by Zeolite Protons *371*
9.3.1	Introduction to Acid-Catalyzed Reactions and Their Mechanism *371*
9.3.2	Elementary Reactions in Acid Catalysis *374*
9.3.2.1	Alkene Protonation *374*
9.3.2.2	Alkane Activation *378*
9.3.2.3	Alkene Isomerization *383*
9.3.2.4	*n*-Butene Isomerization *386*
9.3.2.5	β-C–C Bond Cleavage *388*
9.3.2.6	The Hydride Transfer Reaction *390*
9.3.3	Catalytic Reaction Cycles and Kinetics *394*
9.3.3.1	Physical Chemistry of Zeolite Catalysis *394*
9.3.3.2	Catalytic Cracking *396*
9.3.3.3	Bifunctional Catalysis *402*
9.3.3.4	Methanol Aufbau Chemistry: Alkylation by Methanol *413*
9.4	Acid Catalysis and Hydride Transfer by Enzyme Catalysts *420*
	References *421*
10	**Zeolitic Non-Redox and Redox Catalysis, Lewis Acid Catalysis** *429*
10.1	Introduction *429*
10.2	Non-Reducible Cations; The Electrostatic Field *429*
10.3	Catalysis with Non-Framework Non-Reducible Cations *433*
10.3.1	Alkali and Earth Alkali Ions *433*
10.3.2	Non Redox Oxycationic Clusters *437*
10.4	Catalysis by Non-Framework Redox Complexes *440*
10.4.1	NO Reduction Catalysis: Selective Catalytic Reduction *440*
10.4.2	N_2O Decomposition Catalysis *441*

10.4.3	Selective Oxidation of Benzene and Methane: The Panov Reaction *443*	
10.5	Related Homogeneous and Enzyme Oxidation Catalysts *446*	
10.6	Lewis Acid Catalysis by Non-Reducible Cations Located in the Zeolitic Framework *453*	
10.6.1	Bayer–Villiger Oxidation *454*	
10.6.2	Meerwein–Ponndorf–Verley Reduction and Oppenauer Oxidation *456*	
10.6.3	Homogeneous and Biocatalyst Analogs *461*	
10.6.4	Propylene Epoxidation *461*	
10.7	Catalysis by Redox Cations located in the Zeolitic Framework: The Thomas Oxidation Catalysts *463*	
10.7.1	Bayer–Villiger Oxidation with Molecular Oxygen *464*	
10.7.2	Zeolite Catalysts for Caprolactam Synthesis *464*	
10.7.3	Alkane Oxidation *467*	
10.8	Summary of Zeolite Catalysis *468*	
	References *469*	

11 Reducible Solid State Catalysts *475*
11.1 Introduction *475*
11.2 Chemical Bonding of Transition Metal Oxides and Their Surfaces *475*
11.2.1 Electronic Structure of the Metal Oxide Chemical Bond *475*
11.2.2 The Electronic Structure of the Transition Metal Oxides *478*
11.2.3 The Electronic Structure of the Transition Metal Oxide Surface *488*
11.2.4 Trends in Adsorption Energies of O Adatoms to Transition Metal Oxide Surfaces *490*
11.2.5 Reconstruction of Polar Surfaces *492*
11.3 Mechanism of Oxidation Catalysis by Group V, VI Metal Oxides *494*
11.3.1 Reactivity Trends *494*
11.3.2 Selective Oxidation of Propylene and Propane *500*
11.3.2.1 Propylene to Acrolein Conversion *500*
11.3.2.2 Propane Ammoxidation *502*
11.3.3 Methane Conversion to Higher Hydrocarbons *503*
11.3.3.1 Direct CH_4 Conversion to Aromatics *503*
11.3.3.2 The Mechanism of Oxidative Methane Coupling, the LiO–MgO System *504*
11.4 Metathesis and Polymerization Catalysis: Surface Coordination Complexes *507*
11.4.1 Alkene Disproportionation and Metathesis *507*
11.4.2 Polymerization of Propylene *511*
11.4.3 Ziegler–Natta Polymerization versus Metathesis Reaction *514*
11.5 Sulfide Catalysts *515*
11.6 Electrocatalysis: The Oxygen Evolution Reaction (OER) *520*
11.6.1 Trends in OER Reactivity *520*
11.6.2 Reaction Mechanism of OER Reaction *522*
11.6.3 Summary Mechanism of OER Reaction *531*

11.6.4	Comparison with the OER in Enzyme Catalysis	*532*
11.7	Photocatalytic Water Splitting	*538*
11.7.1	Device Considerations	*538*
11.7.2	Mechanism of Photoactivation of Water	*540*
	References	*545*

Index *553*

Preface

The science of heterogeneous catalysis has flourished for over more than a century. Its processes have contributed significantly to our modern way of life by enabling the production of indispensable commodities such as fuel and fertilizer as well as materials with previously unknown properties such as the polymers.

Both serendipity and empirical chemical discovery have played a part in the development of many of the catalytic systems now used in large chemical processes. The early observations of catalysis and their scientific formulation by Ostwald, Sabatier, and Haber provided the insights and tools necessary for later developments in this field.

While the nineteenth century was the century characterized by categorizing scientific phenomena, the twentieth century can be seen as the century of developing the physical sciences at the molecular level. The formulation of a molecular theory and the development of predictive tools for heterogeneous catalysis has been a long and tortuous process that is now reaching a reasonable level of completion.

This book provides an introduction and overview of the molecular basis of heterogeneous catalysis. It took almost an entire century to arrive at this understanding because heterogeneous catalytic systems are extremely complex. The two main themes discussed in this text are the reaction mechanisms of the various currently known heterogeneous catalytic reactions and the relationship of catalytic reactivity to catalyst structure and composition.

The catalyst is, in itself, a complex material. It is usually a porous inorganic solid composed of various compounds that are often distributed as nano-sized particles over a heterogeneous surface. The catalytic reactions are also complex because they include many reaction steps that often compete. The reaction product is often a multicomponent mixture that requires the availability of advanced analytical tools for its characterization. The overall catalytic system is not easy to study because its performance often depends on reaction conditions and because catalytic activity may change over time. In addition, a catalyst may eventually deactivate during the reaction process.

Clearly, catalyst research requires an integration of many skills. Knowledge of applied inorganic chemistry is needed for the preparation of catalytic materials, training in spectroscopic measurement is essential for characterizing the catalyst as synthesized and at working conditions, and reactor engineering skills are required for the proper measurement and interpretation of catalyst performance data.

The simulation of kinetics based on a mechanistic modeling of the reaction has always been an important tool in catalyst science. More recently, applied quantum-chemical calculations have been used as an aid in determining the mechanism of a particular reaction. These calculations have become relevant to heterogeneous catalysis only in the past decades with the development of computer hardware which is sufficiently powerful for calculating catalyst site models for a solid surface. Since the accurate calculation of elementary reaction rates is also now possible, detailed molecular information on many catalytic systems has become available. This information complements and enriches the chemical reactivity studies on well-defined scientific surface models, providing a powerful predictive tool for heterogeneous catalysis.

This detailed molecular information has enabled reinterpretation of many mechanistic questions that are fundamental to heterogeneous catalytic reactions. A correct mechanistic molecular model complemented by quantitative reactivity data is indispensable for predicting the activity and selectivity of a reaction catalyzed by a particular catalyst. This information can then be used to determine the optimum catalyst surface composition and structure for best catalytic performance. In this book, the concise and comprehensive descriptions of various catalytic systems and reactions can be considered as case studies that exemplify this approach.

One of the most important changes in theoretical chemistry has been in the development of quantitative methods to allow predictive catalysis. In the early nineties of the last century, theoretical concepts had already been formulated, but computational hardware was not yet available to allow their quantitative application to specific systems. Theoretical catalysis was mainly qualitative; conceptually useful but of limited direct practical use. At this time, the mechanism of catalytic reactions could be described in a kinetics context, but could only be applied using parametrized models. This changed dramatically when the necessary computer software and hardware became available for quantitative applications. This computational ability gave rise to the quantitative theories of catalyst activity based on both the molecular computational and spectroscopic data described in this book. The theoretical developments are based not only on chemical bonding data, but more importantly on information about elementary reaction transition states that is sufficiently accurate to allow discrimination between the different mechanisms that determine catalyst activity and selectivity.

In the experimental realm, contributions were largely due to the important advances in the development of sophisticated spectroscopic techniques, in the development of model catalytic systems, and in the discovery of practical catalysts that are atomistically well defined.

These developments are important because there is now quantitative molecular information that can be used to define a catalyst's performance–structure relationship for most catalytic systems of interest. This relationship provides the scientific framework to the reaction mechanisms that are the main focus of the book.

Eindhoven
January 2016

Acknowledgments

Writing this book has been both enjoyable and challenging. Over the past three decades I have co-authored several monographs on related topics with my colleagues and friends. The first book on theoretical heterogeneous catalysis was published in 1991, a second book on chemical kinetics and catalysis written with Hans Niemantsverdriet appeared in 1995, and the book "Molecular Heterogeneous Catalysis" that I wrote with Matthew Neurock was completed in 2006. These books document the gradual progress in our understanding of the molecular foundations of heterogeneous catalysis over the past 30 years. This book could not have been written without the experience of writing the previous books and the insights derived by communication with numerous contacts in the scientific community.

Since my first exposure to heterogeneous catalysis at Shell Research and later when I joined the faculty of Chemistry and Chemical Engineering of the Technical University Eindhoven, I have been fortunate to be surrounded by great colleagues and interested and sometimes excellent students and postdoctoral fellows. A scientific career also requires travel and the cultivation of contacts within the international scientific community. My communications with these numerous contacts (too many to mention here individually) have been instrumental in collecting the information and insights brought together in this book. An account of the importance of these contacts for my scientific work can be found in the book "40 Years Catalysis Research" which appeared on my 65th birthday. Here, I would like to mention a few notable colleagues from the past. Wolfgang Sachtler, who introduced me to heterogeneous catalysis at Shell, was also one of the early leaders of the "Dutch School of Catalysis." One of its main research themes was the mechanism of heterogeneous catalytic reaction. The debates of that period provided the origin for many of the ideas on reaction mechanisms that I describe in this book. A second person I would like to mention is Vladimir Kazansky from Moscow, who visited our laboratories many times in the 1990s. Through him I became familiar with the Russian school of thinking, and I credit him with many of the insights on solid acid catalysis that are presented in this book.

The Laboratory of Inorganic Chemistry that I joined in 1988 had already been the leading academic center of molecular approach to heterogeneous catalysis for many years. This approach was started by George Schuit who occupied the first chair of this laboratory at the end of the 1950s. With Bruce Gates and James Katzer he co-authored the early textbook "Chemistry of Catalytic Processes" that was published in 1979, and which I consulted frequently. It is interesting to compare the mechanistic discussions included in that earlier book with the discussions presented in the current book, which now contains a detailed molecular foundation for several of the early qualitative ideas.

Tonek Jansen and I joined the laboratory at around the same time, and together we started the computational catalysis program. He became the leading expert in microkinetics and kinetics Monte Carlo methods, and I am grateful for his help and inspirational ideas. The sections of this book dealing with electrocatalysis could not have been written without my acquaintance with Marc Koper, who joined our laboratory for a few years as a Royal Netherlands Academy of Arts and Sciences Fellow and taught us the molecular aspects of modern electrocatalysis.

I am very grateful for the hospitality of the Institute of Complex Molecular Sciences, and its director Bert Meijer, which gave me the opportunity to concentrate on the writing of this book for the past 3 years. I would also like to thank Emiel Hensen, my successor and currently head of the Laboratory of Inorganic Materials Chemistry. My communications with him and with faculty members, PhD students, and postdocs were invaluable and created an inspiring intellectual climate for writing. There are a few persons I would like to thank individually. The contributions of Ivo Filot to the microkinetics illustrations and results included in the book can be found in Chapters 3, 7, and 8. My work with Ionuth Tranca on chemical bonding contributed to most of Chapter 6 and important parts of Chapters 7 and 11. Over the years, I have enjoyed collaborative work with Evgeny Pidko, who helped me with sections on Brønsted and Lewis acid catalysis in Chapters 9 and 10. His student Chong Liu contributed the calculations on zeolitic protons. Discussions during many visits with Matthew Neurock continue until today. Matthew agreed to the use of many of the ideas that have their origin in our joint 2006 book, which are presented in this book within renovated context. I was fortunate to work with one of his students Craig Plaisance, a post-doctoral fellow at the TU/e, who contributed to the material discussed in Chapter 11, especially through our discussions that provided important insights on computational electrocatalysis.

Generous assistance is required to transform a book like this into a suitable format and to get the text and figures prepared and organized. I sincerely thank Floris Hieselaar, Tiana Plaisance, Freke Sens and Tom van den Berg for practical support in compiling the text properly, Koen Pieterse for producing some of the high quality figures and Bram Vermeer and Jeanne Daniele for editorial support.

Arrangement of This Book

This book consists of two parts. Part I is an introduction to the physical chemistry of heterogeneous catalysis. It aims to teach basic and well-established scientific theories to the student who is unfamiliar with heterogeneous catalysis. It starts with two chapters on the history of the discovery and development of catalytic systems, followed by an introduction to the chemistry and implementation of the major modern heterogeneous catalytic processes. Various classes of catalytic materials and the different types of known reactions are also described.

The following three chapters are essentially an introduction to the kinetics and reactivity of heterogeneous catalysis. The development of the chemical insights that form the basis of catalysis science is presented in a systematic way. The material is organized to provide an introduction to currently understood microkinetics and reactivity–structure relationships, and is illustrated by examples of recent theoretical applications to working catalyst systems and relevant experiments that reveal the molecular details of a working catalyst.

Part II of the book can be considered to be an advanced course of the molecular approach to heterogeneous catalysis. Four chapters deal with the mechanism of heterogeneous catalytic reactions on the molecular level from the perspective of a physical organic chemist. The chapters include the details of bond breaking, bond formation, and the skeletal organization of molecules. The complete spectrum of heterogeneous catalytic systems is covered. Relevant catalytic activity, spectroscopic data, and kinetics are discussed. Information on organometallic and coordination chemistry is also included, which is important in heterogeneous catalysis because elementary reactions occur in contact with an inorganic surface. These reactions also display unique aspects that relate to the chemistry of surfaces, requiring insights from inorganic chemistry.

The four mechanistic chapters of Part II are organized by the types of various catalytic systems: transition metal catalysts, solid acid catalysts, Lewis acidic catalysts, and reducible oxide catalysts and related inorganic materials such as the sulfides. Each chapter contains descriptions of the catalytic reactions, their mechanisms, and their relation to catalyst composition. The main heterogeneous catalytic reactions that are currently used or that appear to be promising candidates for further exploration are discussed. In addition, the physical chemistry and molecular basis of electrocatalytic hydrolysis are presented. Recent advances in this field have significantly contributed to our general understanding of heterogeneous catalytic reactivity. These descriptions are based mainly on experimental and computational data providing detailed molecular information on the systems. The relationship between catalytic activity, selectivity, and catalyst stability with the structure and composition of the catalyst is essential to these discussions.

Because of the large variation in system types and reactions it is also important to present the general aspects that unify the different methods for reactant activation and the various reaction mechanisms. This information is provided in summary sections. In addition, sections are included to describe several reactions and systems such as enzymatic reactions that often share common features with the heterogeneous catalytic systems. These specialized sections feature the cross-referencing of descriptions of reactions on different systems to highlight their similarities.

The molecular basis for the reactivity of heterogeneous catalysts cannot be properly understood without a solid understanding of the surface chemical bond. The electronic structure of the catalyst surface determines the trends in chemical reactivity across different systems. For this reason, Part II of the book contains one chapter that provides an introduction to the quantum chemistry of chemical bonding in molecules and solids. The computational electronic structure and stability data that provide a basis for this chapter are now readily available from state-of-the-art density functional theory (DFT) calculations. These have become a useful tool to the chemist who wishes to analyze surface reactivity as a function of surface structure or composition. The following chapter contains a discussion of the surface chemical bonding on transition metals and the use of these theoretical concepts to analyze transition states of elementary reactions catalyzed by transition metal surfaces. The chemical bonding and reactivity concepts introduced in these two chapters are often referred to in the four later mechanistic chapters that focus more on the chemical aspects of catalysis.

This book has been designed to be accessible and interesting for students and researchers with very different backgrounds, from those who are new to the field to those with advanced research experience. Thus, the book is divided into two parts: an introductory part with basic information on heterogeneous catalysis and a more specialized part that covers the molecular aspects of catalyst reactivity. The book is tailored to suit both types of readers by presenting general information complemented by inserts. These inserts serve to include material that is useful, but not always of general interest or necessary in order to follow the main text. The inserts allow for the addition of relevant up-to-date information that is of more specific interest to the advanced reader, and they provide a rich source of illustrative material that can be used in lectures.

Catalysis is about reactivity, catalyst synthesis, and spectroscopy. This book focuses on reactivity and provides the reader with the tools to use computational and experimental data to select materials with the desired catalytic properties. While this text covers many topics (catalytic systems and their reactions, physical chemical processes, quantum-chemistry theory) that together present a complete picture of catalytic science, the topical sections have been written to be read independently so that the book is suitable for browsing. References are also provided to complementary materials concerning computational quantum-chemical methods and approaches as well as catalytic measurement and spectroscopic techniques.

Part I

Physical Chemistry and Kinetics

1

Heterogeneous Catalysis

1.1 What is Heterogeneous Catalysis?

The phenomenon of catalysis applies to a wide range of chemical reactions. By using a catalyst, chemical reactions that are non-selective can be made selective and those that require a high temperature can be conducted at a lower temperature (see Figure 1.1). In this and the next chapter, we will describe the discovery of this phenomenon in the nineteenth century, the understanding of catalysis that followed, and the main chemical industrial processes that developed in the twentieth century.

There are many chemical compounds that show catalytic action. In this book, we will mainly discuss heterogeneous catalysts. These are solid materials that are positioned in a reactor where they are exposed to reacting gases or liquids. The reagents flow over the catalyst and are converted by it into products. These products and the unconverted reagents are continuously removed from the reactor, while the catalyst remains in the reactor and becomes exposed to fresh reactants. This continuous process operation is possible because the catalyst is in a separate phase.

We will focus primarily on inorganic materials, which were also the main focus of the early exploratory catalytic studies. When the molecular principles of heterogeneous catalyst activity were gradually understood, it became apparent that similarities exist between biochemical enzymatic processes and molecular organometallic catalysts. The knowledge of the molecular chemistry of the catalytic action of enzymes in biochemical systems as well as information about the action of molecular organic and metal–organic complexes has been of great importance to the further development of the molecular scientific basis of heterogeneous catalytic systems discussed later in this book.

Heterogeneous inorganic catalysts are often quite robust, which makes them especially useful for chemical reactions that operate under hostile conditions.

A specific advantage of the heterogeneous catalyst is that it can be used in a continuous process operation, instead of in a batch-type process where it must be separated in an individual step from the reaction product. Continuous process operation made the development of the large-scale chemical process industry possible once heterogeneous catalysis was discovered early in the previous

Modern Heterogeneous Catalysis: An Introduction, First Edition. Rutger A. van Santen.
© 2017 Wiley-VCH Verlag GmbH & Co. KGaA. Published 2017 by Wiley-VCH Verlag GmbH & Co. KGaA.

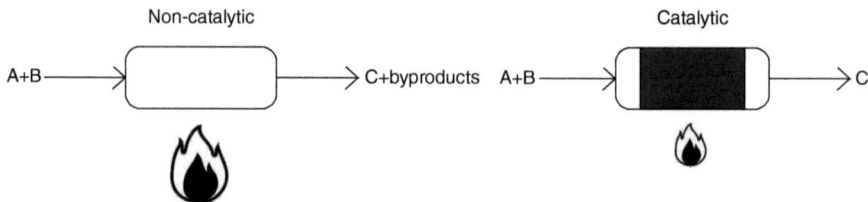

Figure 1.1 Non-catalytic versus catalytic processes. Non-catalytic processes are usually non-selective and require a high temperature of reaction. Catalytic processes are selective and can be conducted at lower temperatures. They save energy, reduce waste, and produce materials that are otherwise not available.

century. Heterogeneous catalysis has become basic to more than 80% of current bulk chemical processes used in the chemical and petrochemical industries [1].

The modern science of heterogeneous catalytic chemistry developed as the understanding of its chemical basis at a molecular level increased. The current understanding of the catalytic reaction mechanism at the physical and chemical levels and its relation to the structure and composition of heterogeneous catalysts is based to a large extent on the computational catalytic results obtained during the past decades. The ultimate aim of catalysis science is to predict the proper catalytic material for a given chemical conversion reaction. This is still far off due to the complexity of experimental catalytic systems, but the large body of empirical chemical information currently available about the reactivity of different catalytic materials is helpful. Computational methods provide tools for estimating the rate and selectivity of a catalytic reaction as a function of catalyst composition. In addition to spectroscopic characterization of a catalyst, this has become indispensable for research.

In this chapter, we will describe the discovery of the basic principles of heterogeneous catalysis. The next chapters will provide an introduction to current catalytic processes, followed by chapters on the principles of physical chemistry and inorganic chemistry of heterogeneous catalysis that are the basis of modern catalysis science.

The discovery of quantum mechanics provided the chemical sciences a predictive foundation, which led to the development of computational catalysis. Combined with the application of advanced spectroscopic techniques and advances in material synthesis, this discovery has provided the molecular foundation of heterogeneous catalysis that is the main topic of part II of this book.

1.2 Early Developments

In its infancy, the science of chemistry was highly exploratory and focused primarily on the categorization of chemical materials and their properties. Heterogeneous catalysis science started in that early period of modern chemistry, before the understanding of the molecular nature of matter. It has its origin in the early part of the nineteenth century, after the founding of modern chemistry by

Lavoisier and colleagues at the end of the eighteenth century. They introduced the law of conservation of mass, and with it an accurate understanding of the nature of oxygen. This knowledge sparked the development of the significant heterogeneous catalytic oxidation processes of the nineteenth century.

The discovery by Davy and Döbereiner of catalytic oxidation contributed to the earlier recognition by Berzelius of catalysis as a separate phenomenon, and to the development of two important heterogeneous catalytic processes: oxidation of HCl to produce chlorine and oxidation of SO_2 to yield oleum, a highly concentrated sulfuric acid.

A second period at the end of the nineteenth century was marked by the development of chemical thermodynamics, which aided in the definition of catalysis by Ostwald and provided its physical chemical foundation. During this period, the rise of coal gasification for domestic lighting and heating made hydrogen readily available. This allowed Sabatier's discovery of catalytic hydrogenation to develop into a practical technology.

Sabatier also discovered an important law of catalysis that defines the condition of maximum activity of a catalyst. The three catalysis laws of Berzelius, Ostwald, and Sabatier are the main topics of this chapter.

This chapter closes with the invention of the Haber–Bosch ammonia synthesis process that is the culminating technological success of this period. As the source for artificial fertilizer, it made a significant impact on the world's growing need for fertilizers. It contributed to chemical technology by creating the continuous heterogeneous catalytic process, which can be operated at high pressure. The development of this process provided the basis for many other modern heterogeneous catalytic processes.

1.2.1 Early Nineteenth Century Discoveries

The early discoveries of heterogeneous catalytic reactions were concurrent with electrocatalytic exploration, which significantly improved the understanding of catalytic reactivity [2]. We will occasionally refer back to this parallel development in the physical chemistry of heterogeneous catalysis and electrocatalysis.

It took some time to recognize the chemical nature of the effect of metals on decomposition reactions as displayed by the decomposition of ammonia or alcohols. These were previously attributed to the effect of heat or electrochemical action, as the latter had been demonstrated in the electrolysis of water [3–5]. A crucial observation by H. Davy was published in 1817. During the course of his studies on the miner's safety lamp, he discovered that a Pt wire remained hot in the presence of coal gas and air. In this experiment, the flame heating the wire was extinguished by exposure to gas enriched with coal gas that is essentially methane. He concluded that the Pt wire assisted in the burning of the gas by air without a flame [3, 6]. In 1823, Döbereiner, another famous chemist from that age, discovered that Pt sponge would react at room temperature with hydrogen to produce a flame. As previously shown by Thénard in the decomposition of ammonia, the effect is material-dependent and could only be observed for particular metals [7].

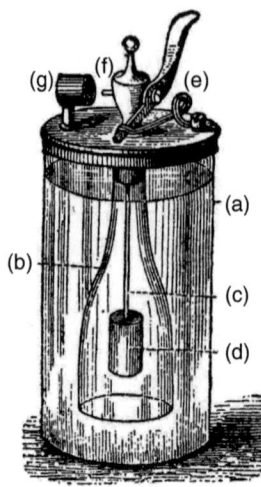

Figure 1.2 An early design of Döbereiner's lamp. This lamp contains hydrogen gas that is ignited using platinum as a catalyst. In a bottle filled with sulfuric acid, zinc metal (e) reacts with sulfuric acid to produce hydrogen gas. $Zn + H_2SO_4 \rightarrow Zn^{2+} + SO_4^{2-} + H_2$. When the stopcock (e) is opened, the hydrogen is directed by a thin tube onto a platinum sponge (g), and then a flame lights instantaneously. The flame goes out when the stopcock is closed. Hydrogen production ceases as gas pressure builds. (a) Glass cylinder, (b) open bottle, (c) wire, (d) zinc, (e) stopcock, (f) nozzle, (g) platinum sponge [7, 8].

Based on his discovery, Döbereiner designed a lamp (Figure 1.2) that generates a flame. The hydrogen is produced by the reaction of sulfuric acid with zinc, which over time is converted into $ZnSO_4$. This lamp became widely used in industry and even in households, until it was replaced by matches and other lighters. Döbereiner's invention can be considered a first practical application of catalysis [7].

Twenty years after Davy's initial experiment, the electro-catalytic analogue of the Döbereiner reaction was discovered by Grove who designed a fuel cell in 1839. Grove ultimately proved that electrical energy, instead of heat, could be produced from an electrochemical reaction between hydrogen and oxygen over a platinum electrode. He called his device a gas voltaic battery. Electricity was stored by the electrochemical decomposition of water and released by the reaction of the products with platinum (Figure 1.3).

The fuel cell is currently attracting renewed interest as a source of electrical energy in the context of a hydrogen-based energy economy, in combination with electricity produced by the electrolysis of water and generated by renewable energy sources such as wind or solar.

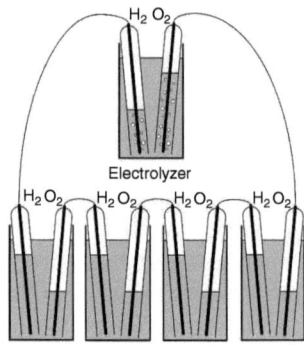

Figure 1.3 Grove cells (gas voltaic battery). This fuel cell consists of five sealed containers (cells) charged with hydrogen (H) and oxygen (O), connected by a voltmeter. Two platinum electrodes each have one end immersed in a container of sulfuric acid and the other end connected to a hydrogen–oxygen cell. The constant flow of energy between the electrodes causes the water level to rise in the cells holding the water and gases [1, 9, 10].

Several other important heterogeneous catalytic reactions were discovered during this same period. The discovery of catalytic oxidation of ammonia to nitrous oxides over Pt by Kuhlmann in 1838 was critical for the development of the ammonia-based fertilizer industry half a century later [3, 6, 11, 12].

Since the Middle Ages, the production of sulfuric acid by burning sulfur was widespread because of its use as an agent to prepare wool for dying. A major innovation in this process was introduced in the eighteenth century. The Lead Chamber Process [13, 14] oxidizes SO_2 to SO_3 by NO_2 yielding NO, which is then re-oxidized by air to NO_2. In this process, NO is the homogeneous gas phase catalyst that drew the attention of Davys and Berzelius.

1.2.2 Later Nineteenth Century Discoveries

At the end of the nineteenth century, the developing synthetic organic dye industry required oleum (fuming sulfuric acid) of substantially higher concentration than the product of the lead chamber process. In 1875, Messel developed the contact process based on the use of Pt as a catalyst, which was later replaced by vanadium oxide [13, 14]. The vanadium-based process is more stable since unlike Pt, vanadium oxide is not deactivated by the arsenic present in the sulfur.

The other large-scale oxidation process that developed during the same period was the catalytic Deacon process (1868) that oxidizes HCl to chlorine. It replaced the commonly used stoichiometric oxidation process by MnO_2. This is an early example of replacing a stoichiometric reaction that consumes expensive chemicals and produces harmful waste with an environmentally friendly catalytic reaction [15–19].

Chloric acid is a co-product of the soda production process. Chlorine is converted with hydroxide into hypochlorite, which is a bleaching chemical. The Deacon process was based on a Cu catalyst of CuO and $CuCl_2$ that reacts to produce a chlorine. The process had several drawbacks, such as the lack of stability of the catalyst under harsh reaction conditions and insufficient activity [20, 21]. This process has been replaced by the electrocatalytic production of chlorine that was discovered in 1800 and developed into a commercial process by Griesheim in 1888. Initially, Fe or graphite electrodes were used, but modern large-scale electrochemical chlorine production facilities use mercury as the cathode and TiO_2 as the anode materials respectively.

Alternative improved catalysts were developed for the Deacon process only recently in the last decade of the twentieth century. RuO_2-based catalysts have been developed by Sumitomo and Bayer showing outstanding stability and activity while producing Cl_2 of high chemical purity [20–22].

1.3 The Three Basic Laws of Catalysis

1.3.1 Berzelius' Catalysis Law

In 1835, Berzelius summarized the chemical reactivity phenomena that he recognized as catalytic for chemical systems: the acid-enhanced conversion of starch into sugar, the oxidation of SO_2 assisted by NO in the lead chamber process, and the many observations of oxidation reactions made possible only by contact with particular materials. He generalized his observation in the following sentences [5, 11, 23, 24]:

> Substances, both simple and compound, in solid form as well as in solution, have the property of exerting an effect on compound bodies, which is quite different from ordinary chemical affinity, in that they promote the conversion of the component parts of the body they influence into other states, without necessarily participating in the process with their own component parts; the body effecting the change does not take part in the reaction and remains unaltered through the reaction.

The definition of a catalyst by Berzelius as a material that enhances a chemical reaction, but itself remains unchanged by reaction, became readily accepted [5, 11, 23, 24]. He proposed the term "catalysis" derived from the Greek words "kata" meaning down and "lyein" meaning loosen. He understood catalysis as the decomposition of a substance by a mysterious force induced by contact with the catalytically active material. Later in 1843, he mentions: "The catalytic force manifests itself by the excitation of the electrical relations that have so far evaded our researches" [3, 4] (Figure 1.4).

This was not an unreasonable suggestion in view of the frequently observed relationship between chemical reactivity and electrochemical activity. Its elucidation had to wait for the next century, once the proper nature of atoms and molecules became understood.

This theory created a great controversy between those who attributed catalysis to a physical force like Ostwald's thermodynamics law, and those who proposed that the interaction was chemical, as was the view of the inorganic chemist Sabatier. In fact, the catalysis laws of Ostwald and of Sabatier essentially complement each other.

Both Ostwald and Sabatier are Nobel laureates that received awards for their contributions to catalysis. Ostwald was awarded for his work on catalysis and fundamental principles governing equilibria in kinetics in 1909, and Sabatier for his work on hydrogenation of organic molecules by finely dispersed metal particles in 1912 [1, 6, 25, 26].

1.3.2 Ostwald's Catalysis Law

Ostwald is the founder of physical chemistry. His main interest was to formulate the chemical laws that regulate chemical transformations. He defined catalysis

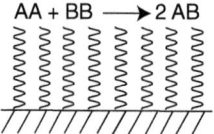

Figure 1.4 Catalytic force as defined by Berzelius. A catalyst influences the course of a reaction, but is not changed by it.

in chemical thermodynamics terms: the catalyst does not alter the equilibrium between the reaction and product molecules, but it enhances or suppresses the rate of their reaction [1, 6, 24, 27, 28].

This so-called Ostwald catalysis law was a breakthrough in catalysis science, because thermodynamics predicts the conditions required by a reaction and provides an essential scientific basis to catalytic investigation. This innovative application of the science of chemical thermodynamics enables selection of the process conditions for reactor experiments to test catalytic materials.

The selection of an appropriate catalyst material could not be determined by prediction and thus required empirical investigation. It is with this strategy that the ammonia synthesis process was discovered (see Insert 1). The process required the discovery of catalytic hydrogenation, the subject of Sabatier's Nobel award.

The two basic physical chemical relationships that were known at that time to govern the rate and equilibrium of a catalytic reaction were the van Hoff equation for the equilibrium constant K_{eq} (Eq. (1.1)) (with $\Delta G°$ the Gibbs free energy at standard conditions, R gas constant, and T temperature):

$$K_{eq} = e^{\left(-\frac{\Delta G°}{RT}\right)} \tag{1.1}$$

and r_A, the Arrhenius reaction rate expression (Eq. (1.2)) (with A the pre-exponential factor, and E the activation energy)

$$r_A = Ae^{\left(-\frac{E}{RT}\right)} \tag{1.2}$$

The constants in the Arrhenius rate expression are dependent on catalyst material and had to be empirically determined, as there was no method available to deduce them using thermodynamics. It was known that the most active catalyst would have a lower Arrhenius activation energy, leading to the general understanding that a catalyst decreases the activation energy of a reaction. The Sabatier principle to be discussed next makes clear that this activation energy is the result of the interplay within the complex reaction network on the catalyst surface.

The character of this relationship within specific catalytic systems is an important topic of this book. As we will describe in the later chapters, a major theoretical advance made at the end of the last century was the development of techniques used to relate the kinetic parameters of the elementary reactions within the catalytic reaction network to the thermodynamic data of the corresponding reaction intermediates.

At the time of the invention of the Haber–Bosch ammonia synthesis process, only the thermodynamics of the overall reaction could be determined. This data, combined with a measurement of the overall reaction rate of the catalytic reaction (which is catalyst-dependent), was used to determine the optimum reaction condition for the process [11, 14, 25].

> **Insert 1: The Haber–Bosch Ammonia Synthesis Process: The Haber and Nernst Controversy**
>
> Haber and Nernst disagreed initially on the proper conditions for ammonia synthesis from nitrogen and hydrogen. Nernst disagreed on Haber's data obtained at atmospheric conditions and concluded that the endothermic reaction can be conducted at higher pressures. However, improved data from Haber indicated that the exothermic reaction ($\Delta H = -92$ kJ mol^{-1}) should produce an acceptable yield at a pressure of 200 atmosphere a temperature of around 800 K. He identified Os and Ru as proper catalysts.
>
> The process was purchased by the German company BASF and a commercial plant started in 1913. Carl Bosch solved the substantial engineering problems of the process, such as the required use of high pressures. Nearly 20 000 tests were conducted with reactors designed by Bosch before the optimum, economic catalyst formulation was found. The catalyst was a complex composition of Fe promoted with potassium oxide and alumina. The mechanism and operation of this catalyst only became understood after a 100 years, aided by the studies of Ertl!
>
> This process was integrated in the production of hydrogen from coal and in the selective oxidation of ammonia which became the Ostwald process of high-temperature catalytic oxidation of NH_3.
>
> These revolutionary scientific discoveries were recognized by Nobel awards to all four of these scientists.

Ostwald believed that a catalyst did not induce a reaction, but rather accelerated it without formation of an intermediate. He theorized that catalyzed gas reactions resulted from the absorption of gases in the cavities of the porous metal where compression and local temperature elevation led to chemical combination.

1.3.3 Sabatier's Catalysis Law

Ostwald's view of chemistry of the working catalyst was rejected by Sabatier. He studied reactions such as the hydrogenation of ethylene to ethane and the conversion of CO_2 to methane through the use of transition metal powders, prepared by reducing their corresponding oxides. Based on these studies, Sabatier formulated a chemical theory of catalysis that involves the formation of unstable chemical compounds as intermediates. These determine the product selectivity and the rate of the catalytic reaction. He assumed that a hydrogenation reaction catalyzed by Ni involved various nickel hydrides with concentrations that relate to the reactivity of the nickel [6]. He argued that the formation and decomposition of intermediate compounds formed between reagent and catalyst corresponded to a lowering of the Gibbs energy of the system. This view is completely validated by modern molecular insights into the action of a catalyst [29].

Sabatier's law can be considered the third basic law of catalysis [1, 6, 14, 25, 28, 30]. It is a rule that formulates the condition of optimum catalytic reactivity. In modern terms it can be worded as: *In a catalytic reaction, reacting molecules form intermediate complexes with the catalyst (surface). These complexes should be of intermediate stability. If they are too stable they will not decompose to achieve product formation. If they are too unstable reagent molecules will not be activated and surface reaction intermediate complexes will not be formed.*

As we now know, the catalytic reaction is composed of a sequence of elementary reaction steps.

- Molecules adsorb onto the catalyst surface where they become activated and form intermediate reaction complexes.
- These complexes then rearrange and recombine.
- Final product formation occurs by desorption of the product molecules that regenerates the free catalyst surface (Figure 1.5a).

Sabatier's rule is a direct consequence of the cyclic nature of the catalytic reaction sequence. Reacting molecules interact with a catalyst. This chemical reaction

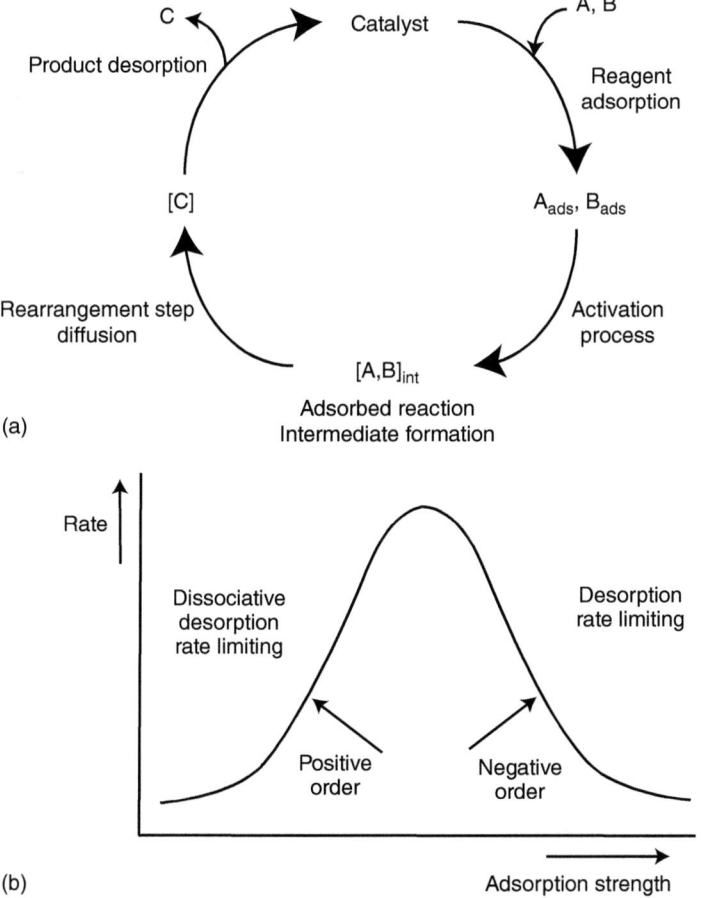

Figure 1.5 (a) Schematic representation of the cyclic nature of the catalytic heterogeneous reaction. (b) Volcano plot illustrating the Sabatier principle. The rate of reaction is shown as a function of the interaction energy of the substrate. The rate is maximized at optimum adsorption strength. To the left of the Sabatier maximum, the rate has a positive order in reactant concentration, and to the right it has a negative order [25].

leads to the formation of the intermediate complexes between reagent molecules and catalyst surface atoms. The strong interaction between catalyst and adsorbed reaction intermediates causes dissociation reactions as chemical bonds in reactants break, and association reactions as new chemical bonds form from the production of different reaction intermediates. The latter leads to the product molecules that in subsequent steps desorb from the surface.

The reaction activation rates of the reagent molecules, the formation of intermediate product molecules, and the desorption of product molecules all compete. Chemical bond activation of the adsorbed molecules requires a stronger interaction with the catalyst surface than the formation of product molecule sand product by association reactions. Because cleavage of the chemical bonds between reaction intermediate and catalyst surface is necessary, the maximum overall rate of the catalytic reaction is determined by the optimum interaction energy of reaction intermediates with the surface. This leads to the volcano-type dependence of the overall catalytic reaction rate on catalyst reactivity as shown in Figure 1.5b.

The kinetics of the full network of interacting elementary reactions must be considered in order to determine the catalyst's maximum performance. As we will discuss in more detail in Chapter 3, at the catalyst's operational maximum other reaction steps compete and there is no longer a single reaction step controlling the reaction rate.

The Sabatier principle is basic to our understanding of the physical chemistry of catalyst activity. The understanding of catalyst selectivity would come later, with the molecular theory of catalysis that is discussed at length in Chapter 5.

References

1 Knozinger, H. (2009) *Heterogeneous Catalysis and Solid Catalysts*, Wiley-VCH Verlag GmbH.

2 Lupskowski, J. and Ross, P.N. (eds) (1998) *Electrocatalysis*, Wiley-VCH Verlag GmbH.

3 Robertson, A.J.B. (1975) The early history of catalysis. *Platinum Met. Rev.*, **19**, 64–69.

4 Robertson, A.J.B. (1983) The development of ideas on heterogeneous catalysis. *Platinum Met. Rev.*, **27**, 31–39.

5 Roberts, M.W. (2000) Birth of the catalytic concept (1800–1900). *Catal. Lett.*, **67**, 1–4. doi: 10.1023/A:1016622806065.

6 Wisniak, J. (2010) The history of catalysis. From the beginning to Nobel Prizes. *Educ. Química*, **21**, 60–69.

7 Hoffmann, R. (1998). Dobereiner's lighter *Am. Sci.* **86**, 326, http://www.americanscientist.org/issues/pub/1998/4/d-bereiners-lighter (accessed 5 October 3016).

8 Kauffman, G.B. (1999) Johann Wolfgang Döbereiner's Feuerzeug. *Platinum Met. Rev.*, **43**, 122–128.

9 Wendt, H., Gotz, M., and Linardi, M. (2000) Fuel cell technology. *Quim. Nova*, **23**, 538–546.

10 Bossel, U. (2000) *The Birth of the Fuel Cell 1835–1845*, European Fuel Cell Forum, p. 7.
11 Califano, S. (2012) *Pathways to Modern Chemical Physics*, Springer, Berlin.
12 Hunt, L.B. (1958) The ammonia oxidation process for nitric acid manufacture. *Platinum Met. Rev.*, **2**, 129–134.
13 Jones, E.M. (1916) Chamber process manufacture of sulfuric acid. *Ind. Eng. Chem.*, **42**, 1–3.
14 Lloyd, L. (2011) *Handbook of Industrial Catalysts*, Springer, Boston, MA.
15 Itoh, H., Kono, Y., Ajioka, M., Takezaka, S., and Katzita, M. (1989) US Patent 4,803,065.
16 Arnold, C. and Kobe, K.A. (1952) Thermodynamics of the Deacon process. *Chem. Eng. Prog.*, **48**, 293–296.
17 Hisham, M.W.M. and Benson, S.W. (1995) Thermochemistry of the Deacon process. *J. Phys. Chem.*, **99**, 6194–6198.
18 Deacon, H. (1868) Improvement in the manufacture of chlorine. US Patent, p. 370.
19 Schmittinger, P. (2008) *Chlorine: Principles and Industrial Practice*, Wiley VCH Verlag GmbH.
20 Over, H. (2012) Atomic-scale understanding of the HCl oxidation over RuO_2, a novel Deacon process. *J. Phys. Chem. C*, **116**, 6779–6792.
21 Over, H. and Schomäcker, R. (2013) What makes a good catalyst for the Deacon process? *ACS Catal.*, **3**, 1034–1046.
22 Busca, G. (2014) *Heterogeneous Catalytic Materials*, Elsevier, pp. 429–446, http://linkinghub.elsevier.com/retrieve/pii/B9780444595249000134 (accessed 5 October 2016).
23 Lindström, B. and Pettersson, L.J. (2003) A brief history of catalysis. *CATTECH*, **7**, 130–138.
24 Ertl, G. and Gloyna, T. (2003) Vom Stein der Weisen zu Wilhelm Ostwald. *Z. Phys. Chem.*, **217**, 1207.
25 van Santen, R.A. (2010) in *Novel Concepts in Catalysis and Chemical Reactors: Improving the Efficiency for the Future* (eds A. Cybulski, J.A. Moulijn, and A. Stankiewicz), John Wiley & Sons, Inc., Hoboken, NJ, pp. 1–30.
26 Niemantsverdriet, J.W. (2003) *Organic Chemistry Principles and Industrial Practice*, Wiley-VCH Verlag GmbH.
27 Ostwald, W. (1902) Catalysis. *Nature*, **65**, 522–526. doi: 10.1038/065522a0.
28 Davis, B.H. (1997) in *Handbook of Heterogeneous Catalysis*, vol. 1 (eds G. Ertl, H. Knozinger, and J. Weitkamp), Wiley-VCH Verlag GmbH, Weinheim, p. 13.
29 Sabatier, P. (1897) Action du nickel sur l'ethylene. Synthese de l'ethane. *C.R. Acad. Sci.*, **124**, 1358–1361.
30 Sabatier, P. (1913) *La Catalyse en Chimie Organique*, Librarie Polytechnique, Paris.

2

Heterogeneous Catalytic Processes

2.1 Introduction

This chapter provides an overview of the main catalytic reactions and their chemistry as an introduction to the detailed discussion of various catalytic reactions and their corresponding catalysts in the following chapters.

In this section, we will continue to follow the development of catalysis in the twentieth century as a function of time. In the next section, we will briefly discuss the chemistry of several common catalytic processes.

The development of new heterogeneous catalytic processes is primarily driven by the need to provide chemical solutions for an evolving society. This process innovation has often led to the early exploitation of recent scientific discoveries and has given rise to new scientific advances.

An example of this type of process development is the previously mentioned discovery of ammonia synthesis from nitrogen and hydrogen at the beginning of the nineteenth century. This process was developed before World War I when the public realized that the main source of nitrates for explosives and fertilizers (guano from Chile) would become scarce. Direct production of NO from the oxidation of N_2 in air is not practical because of the reaction's high endothermicity ($\Delta H = 90\,\text{kJ}\,\text{mol}^{-1}$) [1]. The commercial process that was developed involved initial ammonia synthesis and consecutive oxidation of ammonia to NO_x.

The discovery of the ammonia synthesis process was based on the recent development of chemical thermodynamics, Sabatier's discovery of catalytic hydrogenation, and Bosch's development of hydrogen production by the steam reforming reaction of coal.

These founding studies provided new insights into thermochemistry, and made continuous catalytic process technology possible through the development of modern heterogeneous catalysis.

As oil replaced coal and the consumption of vehicle and airplane fuel increased, the need for oil conversion processes also increased. The end of the twentieth century saw a growing interest in catalysis to reduce environmentally harmful emissions, and more recently, there is increased focus on developing alternative energy conversion processes based on renewable resources such as biomass and solar energy.

Modern Heterogeneous Catalysis: An Introduction, First Edition. Rutger A. van Santen.
© 2017 Wiley-VCH Verlag GmbH & Co. KGaA. Published 2017 by Wiley-VCH Verlag GmbH & Co. KGaA.

Table 2.1 The development of new catalytic processes and the introduction of new catalytic materials in response to changing societal needs.

	Catalyst	Process
1900	Nobel metal	Hydrogenation
1910 World War I	Promoted iron	Nitrogen to ammonia Synthesis gas to methanol and liquid fuel
1920	Sulfides	Desulfurization–denitrogenation
1930 Automobiles	Solid acids	Catalytic cracking
1940 World War II	Super acids Anionic catalysis	Synthetic kerosene Alkylates Synthetic rubber
1950	Coordination catalysis	Polymers
1960 Petrochemical industry	Bifunctional catalysis Zeolites Reducible oxidic systems	Hydrocracking Catalytic cracking Selective oxidation
1970 Energy crisis	Novel synthetic acidic zeolites Organometallic complexes	Methanol to gasoline Synthesis gas to chemicals
1980 Environment	Noble metal alloys Mixed oxides Immobilized complexes Redox zeolite systems	Exhaust catalysis Stack gas treating Bulk and fine chemicals
1990 Environment Raw materials: natural gas and coal	Organometallic complexes in nano/mesoporous materials Supported reducible oxides Zeolitic redox systems Reducible mixed oxides Ga, Zn promoted zeolites $Si_xAl_{1-x}(PO_4)_2$ zeolitic systems Co nanoparticles	Enantiomeric catalysis NO_x, SO_2 reduction N_2O utilization (Panov reaction) Selective alkane oxidation and ammoxidation Alkane dehydrogenation: methanol to olefins Fine chemicals Synthesis gas to hydrocarbons
2000 Climate	Electrocatalysts Cr/molten salts, Lewis acidic zeolites Early transition metals Metal organic framework Microporous systems Hybrid organic/inorganic systems	Fuel cell Glucose to diesel Hydrogen storage CO_2 storage and activation Photocatalysis

Table 2.1 summarizes the development of important processes and catalysts over time, and the corresponding needs of society that gave rise to the exploitation of these processes. The invention of new processes is often related to the discovery of new catalytically active materials. Transition metals are frequently employed, and even the early hydrogenation process innovations were based on their use. The discovery of sulfide catalysts based on transition metals, stable in the presence of H_2S, led to the development of a process to remove sulfur and nitrogen from crude oil and liquefied coal-based process streams.

In the second part of the nineteenth century, the selective oxidation catalysis process using reducible mixed oxide catalysts was introduced for the large-scale production of base chemicals required by the new and rapidly developing polymer industry.

High surface area non-reducible oxides such as alumina or silica materials comprise another important class of catalytic materials that were developed at this time. These function primarily as catalyst supports, but they may also have some acidic or basic properties. Their ability to support small catalytically active particles and to optimally tune the porous structure substantially improved catalyst process technology over time.

A new catalytic material introduced around 1960 had a large impact on oil refining: the microporous solid acid zeolite Y catalyst. Zeolite Y has the crystal structure of the faujasite structure shown in Figure 2.1.

Zeolites are a large class of mainly crystalline silica–alumina compounds containing microporous materials with a dimension comparable to that of organic molecules. Zeolitic materials have been produced in a large variety of compositions resulting in many important catalytic applications [2]. Before their application as catalysts, they were extensively used as drying agents in organic solvents or as boiling stones to control the boiling of liquids.

The zeolite structure is based on tetrahedral building units that contain a cation surrounded by four oxygen atoms. They form a three-dimensional network by sharing cation–oxygen bonds, so that the oxygen atoms form bridging connections between the cations. There are approximately 200 different polymorphs in existence based on this building principle, and those that contain micropores of a size accessible by organic molecules have potential as catalysts. Polymorphs of SiO_2 or $AlPO_4$ fit this profile, although non-zeolitic materials of this composition have no catalytic activity because they do not contain catalytically activating sites.

Changes in the framework composition can be used to create catalytically active sites. In the siliceous framework, when Si^{4+} is replaced by Al^{3+} the framework gets a negative charge. The natural zeolites are alumina-silicates

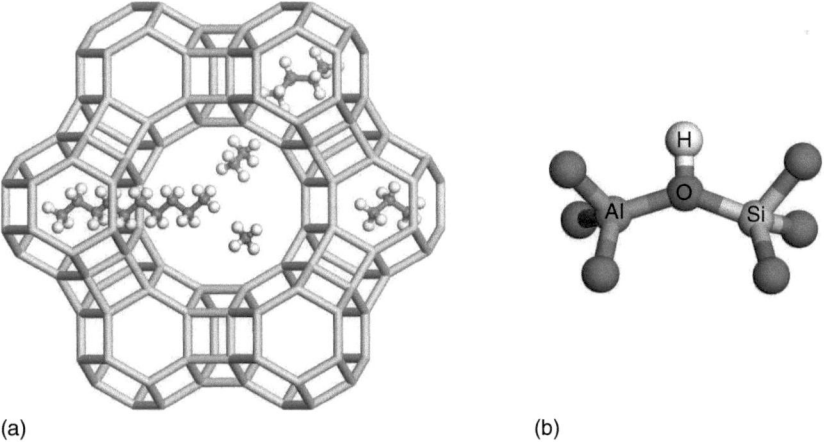

(a) (b)

Figure 2.1 (a) A representation of the micro-cavity structure of the zeolite–faujasite interacting with a few hydrocarbon molecules. (b) The protonic site in the zeolite. The proton attached to the Si and Al bridging framework O atom is shown. The lines connect the framework of Si or Al cations. These are tetrahedrally surrounded by four oxygen atoms. The tetrahedra are connected through the oxygen atoms. At the center of each connecting line there is a bridging oxygen atom. An acidic proton is located next to the Al on an oxygen atom that bridges with Si.

in which this negative charge is compensated for by alkali cations located in the zeolite micropores. An acidic material useful for catalysis is generated by ion exchange of these cations by ammonium ions. When ammonia desorbs by heating this leaves a proton located on an O atom that bridges a Si and Al cation as is shown in Figure 2.1. The presence of these protons converts the material into a solid acid. Other cations such as Fe^{3+} or Ga^{3+} can also be used instead of Al^{3+}.

Similarly, in the $AlPO_4$ structure, the P^{5+} cation can be replaced by a Si^{4+} cation. This can also be used to generate a solid acid catalyst and again the proton will bridge a Si and Al cation. These materials are called *SAPO's*. Application of this material is discussed in Section 2.2.3.1. A more complete discussion of the use of zeolite catalysts including their application as oxidation catalysts is given in Chapters 10 and 11.

As catalysts, zeolites are especially useful as solid acids that are stable at high temperatures where conventional acids cannot be used, such as in the conversion of heavy oil-derived liquids to produce lighter product fractions for use in gasoline and other products. Their beneficial application in the catalytic cracking process is due to their ability to reduce deposits of deactivating coke. When hydrocarbons react in the micropores of zeolite material the limited micropore size excludes the large aromatic oligomer molecules, thus inhibiting deposition of aromatic coke (see Insert 1).

Insert 1: The Invention of the Catalytic Cracking Process

In the early 1960s, C.J. Plank and E.J. Rosinski of Socony Mobil Oil Company invented zeolite Y, a rare earth metal containing a solid acid zeolite catalyst exhibiting a faujasite structure (see Figure 2.1). This material was used to crack the hydrocarbons of heavy petroleum fractions to produce lighter materials such as gasoline. The discovery of this robust catalyst considerably reduced the non-selective deposition of coke residue. The useful product yield was increased from 70% to 90% by switching from a process using non-microporous clay and amorphous silica-alumina catalysts to a process using zeolite-based catalysts. The use of zeolite catalysts has already saved the United States an estimated 200 000 barrels of imported crude oil per year. Plank and Rosinski were elected to the U.S. National Inventors Hall of Fame in 1979 [3, 4].

Since the end of the twentieth century, there has been increasing interest in developing catalytic processes to reduce environmental impact and increase resource utilization. The Shell SHOP process (see Insert 5) that produces synthetic linear biodegradable detergents with high efficiency from ethylene is an example of a process that has attracted this interest. The desire to produce chemicals directly from natural gas has driven a search to replace the process using olefin in selective oxidation by one using alkane activation. The development of catalysts to reduce exhaust emissions from automotive exhaust typifies catalysis designed to clean end-of-pipe emissions. Other important developments were the replacement of stoichiometric organic reactions by catalytic processes. An example that will be discussed in the next section is the replacement of the stoichiometric production of propylene epoxide through chlorohydrin and CaCl$_2$ production by catalytic epoxidation with hydroperoxide.

The current demand for efficient renewable energy conversion systems has led to the invention of new processes to convert biomass to liquid fuels. Likewise, the use of renewable resources like solar or wind energy to generate electricity has spurred an interest in improved hydrogen storage systems combined with efficient H$_2$ generation and even the production of liquid fuels.

The increased scientific understanding of the operation of catalytically reactive systems and of the preparation methods of practical catalysts are major factors in improving existing processes and adapting them to the changing conditions and needs of modern society.

2.2 Important Heterogeneous Catalytic Reactions and Processes

The chemistry of some important catalytic processes is summarized in this section. Detailed discussions of their mechanisms and a molecular-level description of the activation of molecules by their corresponding catalysts can be found later in part II.

We will describe the following classes of reactions:

- Hydrogenation and dehydrogenation reactions.
- Hydrocarbon transformation reactions such as cracking or isomerization reactions.
- Association reactions such as oligomerization or polymerization reactions.
- Hydrodesulfurization and related hydrodenitrogenation and hydrodeoxygenation reactions.
- Selective oxidation and reduction reactions.

In the following sections, a short introduction of individual processes and their corresponding catalysts will be presented. For an extensive overview of catalytic reactions, refer to K. Weissermel and H.-J.Arpe [5]; an excellent introduction to process aspects is presented in the book A. Moulijn *et al.* [6]; an important reference for catalytic reactions and processes is G. Ertl *et al.* [7].

2.2.1 Hydrogenation and Dehydrogenation Reactions

2.2.1.1 Hydrogenation Reactions: Transition Metal Catalysts

2.2.1.1.1 Ammonia Synthesis

The ammonia synthesis reaction requires cleavage of a strong N$_2$ bond as well as the dissociation of H$_2$. A catalyst that strongly binds N atoms is desired since this will favor the dissociation of N$_2$. Surface science experiments have shown that the reaction is highly structure-sensitive, so it is important to provide the proper sites for the activation of N$_2$. The nitrogen atoms generated by the dissociation of N$_2$ are hydrogenated by coadsorbed H atoms to yield ammonia (see Figure 2.2).

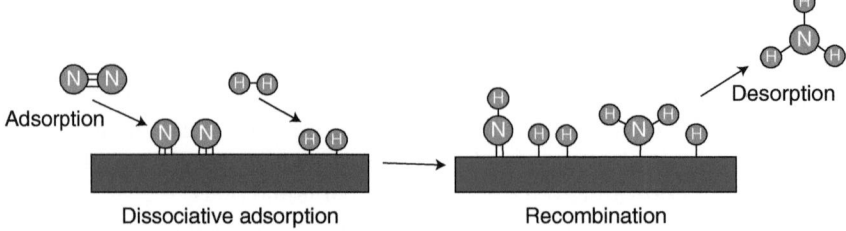

Figure 2.2 Molecular events on the ammonia synthesis catalyst.

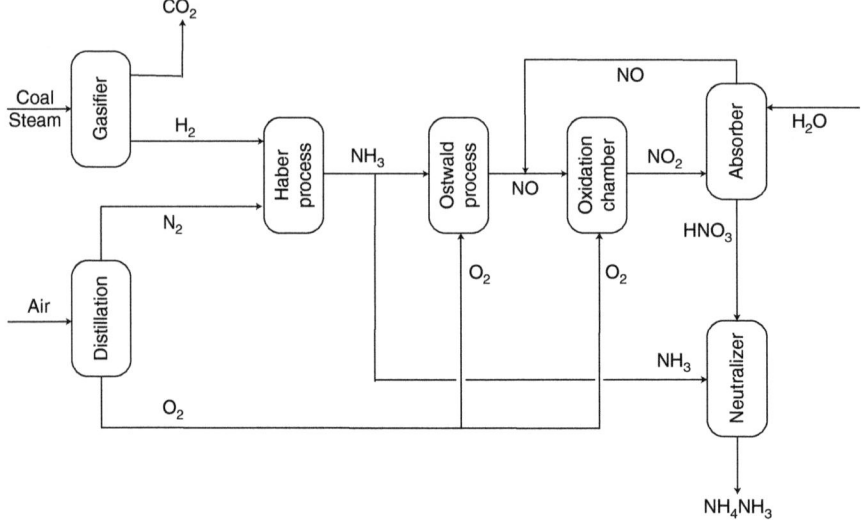

Figure 2.3 Process scheme of a fertilizer plant [6].

The most commonly used catalyst for this process is Fe-based, but some more recent processes are based on Ru. Transition metals are sometimes activated by promoting substances such as potassium or alumina. When using an Fe catalyst, promotion by potassium increases the bond cleavage rate of N_2 to produce dissociated adsorbed N atoms. Promotion with alumina is used to strengthen the catalyst, and in this case alumina can be considered as a structural promoter.

Figure 2.3 shows the scheme of the integrated process of fertilizer production and its component reaction steps. The synthesis gas, a mixture of CO and H_2, is produced by steam reforming of coal or natural gas. (Coal gasification is an endothermic reaction that requires a temperature between 1200 and 2000 K, depending on the reactor technology used. Natural gas steam reforming is also endothermic and requires temperatures between 1000 and 1300 K.) The addition of alkali to coal facilitates steam reforming, while steam reforming of natural gas is catalyzed by a Ni-based catalyst stabilized by Mg on alumina.

The hydrogen process stream is generated by the catalytic water–gas shift reaction of H_2O and CO to yield CO_2 and H_2 ($\Delta H = -41.1\,\text{kJ mol}^{-1}$). Cu/ZnO catalysts can be applied at 500–700 K [8–11]. Nitrogen is cryogenically separated from air in the distillation step, and the Haber process then converts N_2 and H_2 into ammonia at a temperature of 750 K and pressure of 100–250 bar. Ammonia is converted in a consecutive step at high temperature to NO_x by the Ostwald process (using a Pt/Rh gauze catalyst). The NO_x is further oxidized to NO_2. Finally, ammonium nitrate is produced by recombination of HNO_3 with ammonia.

2.2.1.1.2 Synthesis Gas to Methanol

Methanol is an important bulk chemical also produced from synthesis gas. The preferred catalyst in this process is Cu-based because it can dissociate H_2, making hydrogen atoms available to combine with CO.

The first catalyst for methanol production was introduced by BASF in 1923 and based on a mixture of Zn–chromite. Cu is a more active hydrogenation catalyst than chromite, and thus this initial catalyst was replaced by a CuO/ZnO-based catalyst with higher activity and the ability to operate at a lower temperature and pressure. The CuO/ZnO catalyst became widely used once the synthesis gas could be purified to prevent Cu poisoning of the process. Today, methanol synthesis represents a major chemical production process with a worldwide yield of 65 million tons per year.

Cu is only moderately reactive, therefore it will not dissociate CO and the C—O bond remains intact. If the Cu catalyst in this process is replaced by a more reactive Ni catalyst, methane will be produced instead of methanol. The more reactive Ni will dissociate the C—O bond and the adsorbed C atoms will become hydrogenated to methane. The dominant mechanism of methanol formation is thought to proceed by CO_2 activation and the formation of intermediate formate species, possibly stabilized by ZnO. Note the similarity of this catalyst with the low temperature water–gas shift catalyst. This indicates that H_2 and CO_2 will be formed by decomposition of the adsorbed intermediate formate species.

2.2.1.1.3 Olefin Hydrogenation

An early application of the hydrogenation of double bonds in organic molecules is the fat hardening process. The hydrogenation of the double bonds in fat produces more saturated triglyceride fats with a higher melting point, useful in the production of margarine.

The Ni catalyst used to hydrogenate the unsaturated C=C bonds in this process is based on Sabatier's discoveries. Normann developed the process that had its first commercial application in 1910. Hydrogenation is an exothermic reaction that can be conducted at low temperatures with a high partial pressure of H_2 applied to prevent coke formation. Kieselgur (a siliceous high surface area natural mineral) is used as a catalytic support [12, 13].

The transition metal catalyst Ni dissociates H_2 and adsorbs the alkene, which is strengthened to activate the C=C double bonds. The reactivity of Ni is low enough to avoid cleaving the C=C bonds, which would produce methane and other light products.

2.2.2 Hydrocarbon Transformation Reactions

2.2.2.1 Brønsted Acid Catalysis

2.2.2.1.1 Catalytic Cracking

Microporous solid acidic zeolites are widely used as cracking catalysts that convert heavy oil fractions into lighter oil fractions. The essential overall reaction involves converting longer chain linear alkanes into monomeric or dimeric aromatics and shorter chain linear and branched alkane and alkene molecules. In catalytic cracking, zeolite Y is used (see Figure 2.1). This zeolite contains microchannels that connect its so-called supercages through 12-ring openings [2, 14–18].

Protonation of the hydrocarbons induces C—C bond cleavage. At the higher reaction temperatures required by this endothermic reaction (770 K), light alkene and alkane molecules are initially formed. Consecutive reactions lead to oligomerization of the olefins and the formation of aromatics. This complex network of reactions leads to a mixture of products highly useful as components of gasoline. The elementary reaction steps and the mechanism of the catalytic cracking process are discussed in detail in Chapter 9.

As discussed in Insert 1, the use of microporous zeolites is essential in this process because it prevents the formation of large aromatic molecules that are the precursors to coke.

Because the catalyst deactivates rapidly, a unique process has been designed that continuously removes the spent catalyst then regenerates and replenishes it in the reactor. This is illustrated in Figure 2.4.

2.2.2.1.2 Alkane to Aromatics: Bifunctional Catalysis

The dehydrogenation of alkanes is important because the olefin produced is an essential intermediate for many different processes. Due to thermodynamic constraints, dehydrogenation reactions require high temperatures, which often lead to adverse reactions in which C—C bonds are broken and deactivating coke is formed.

In the oil industry, dehydrogenation must often occur in combination with additional conversion steps under the same reaction conditions. To achieve this, a catalyst must have multiple functions. A bifunctional zeolitic catalyst that consists of transition metal particles distributed on a solid acidic material can be used for hydrogenation and dehydrogenation. Unlike zeolitic systems used for catalytic cracking that do not contain a transition metal, bifunctional catalytic systems can be used in hydroconversion processes that employ high-pressure hydrogen to suppress coke deposition.

At this point, we will introduce catalysis that is basic to the catalytic platforming process. Insert 2 provides background information on this process. The catalytic platforming process is an important refinery process that converts linear alkanes into aromatics. In this process, bifunctional catalysts with Brønsted acidic reactivity are used, which consist of transition metal particles distributed over a high surface are a alumina support.

The function of the transition metal in a bifunctional catalytic system is to cause dehydrogenation of the alkane so that alkenes are formed. These are converted in consecutive reaction steps by acidic protons into other alkenes, and in turn

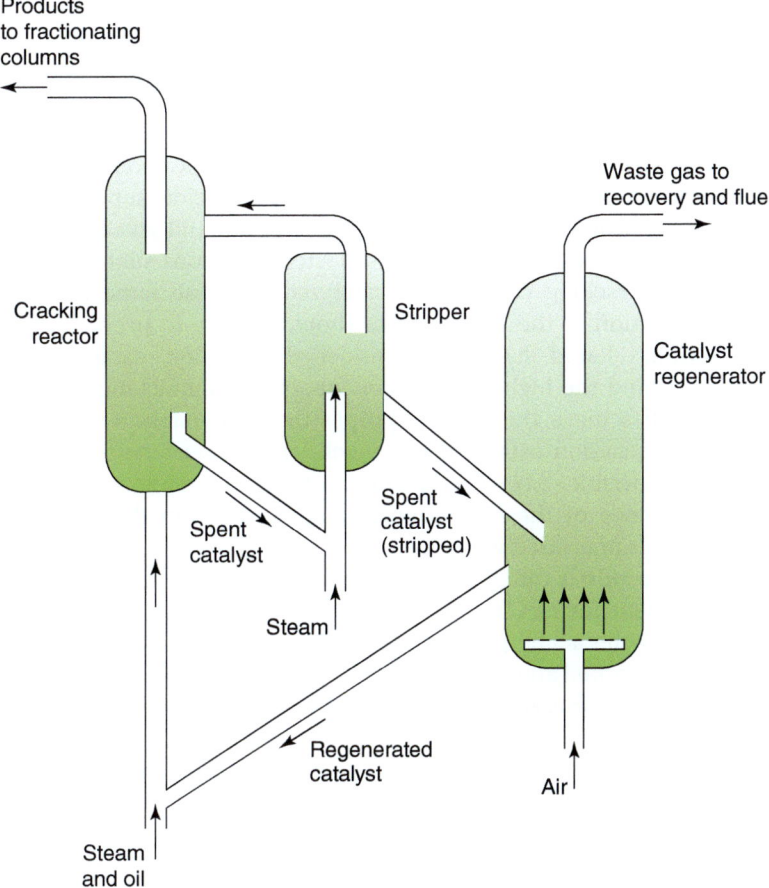

Figure 2.4 The fluid catalytic cracking process reactor. A mixture of steam and oil is fed to the cracking reactor, where it is exposed for a short time (2–4 s) to the cracking catalyst. Then, the catalyst and product are separated, and the catalyst is cleared of deposited coke by air and steam treatment. After regeneration, it is added back to the cracking reactor. After several cycles, the catalyst becomes permanently deactivated, so spent catalyst is continuously removed from the system and replenished by fresh catalyst. (The Essential Chemical Industry: Cracking and related refinery processes (2014). Available at: http://www.essential chemicalindustry.org/processes/cracking-isomerisation-and-reforming.html, with kind permission from *The Essential Chemical Industry – online* [19].)

these can be hydrogenated to product alkane molecules at low temperature. In catalytic platforming, higher temperatures are used so that aromatics will be the main product instead of alkanes. The lower temperature reactions are used in the hydroisomerization and hydrocracking process that is discussed in the next section.

When alkanes are activated by transition metals, the C—H bonds are the first to be cleaved. This occurs despite the lower bond energy of the alkane C—C bonds because the hydrogen atoms surround the carbon atoms of the molecule, and thus they are the first atoms to touch the metal surface. This is known as the umbrella effect. Once a C—H bond is broken, alkenes can be formed when a second C—H bond breaks, but C—C bond cleavage may also occur. Since this cleavage leads to the production of light gasses, it must be suppressed. The use of Pt offers an advantage in this regard: other metals like Ni will produce undesirable methane as a non-selective product.

Acid catalysis is required for the reaction that transforms the alkenes (which were initially formed by the transition metal) into aromatics. The endothermicity of the reaction prescribes a high temperature so that a solid acidic support must be used. The zeolitic protons catalyze olefin isomerization and alkadiene ring closure reactions. However, in addition, non-selective reactions such as olefin oligomerization and cracking can also be catalyzed by these same protons. The alkene concentration at the protonic site should be low to prevent these extraneous reactions. Rapid hydrogenation–dehydrogenation catalyzed by the transition metal and the higher activation energy of the cracking reaction and ring closure will achieve this. At the proper thermodynamic conditions, consecutive dehydrogenation catalyzed by Pt will lead to benzene formation, which is highly endothermic ($\Delta H = 260\,\text{kJ}\,\text{mol}^{-1}$; $n\text{-}C_6$ to benzene).

The reaction sequence of the bifunctional catalytic reaction that converts n-hexane to benzene is presented in Figure 2.5.

The ring closure reaction is catalyzed by the acidic protons of the high surface area alumina. Hexadiene formed from the alkane by dehydrogenation catalyzed by Pt is converted by the acidic protons through an intermediate cyclopentyl cation that remains adsorbed and isomerizes to a cyclohexyl cation. Upon deprotonation, cyclohexene desorbs and consecutive dehydrogenation catalyzed by Pt leads to benzene formation.

Initially, in the development of the platforming process, Cr_2O_3 was used as a dehydrogenation catalyst, but was later replaced by the more active and selective Pt-based catalyst.

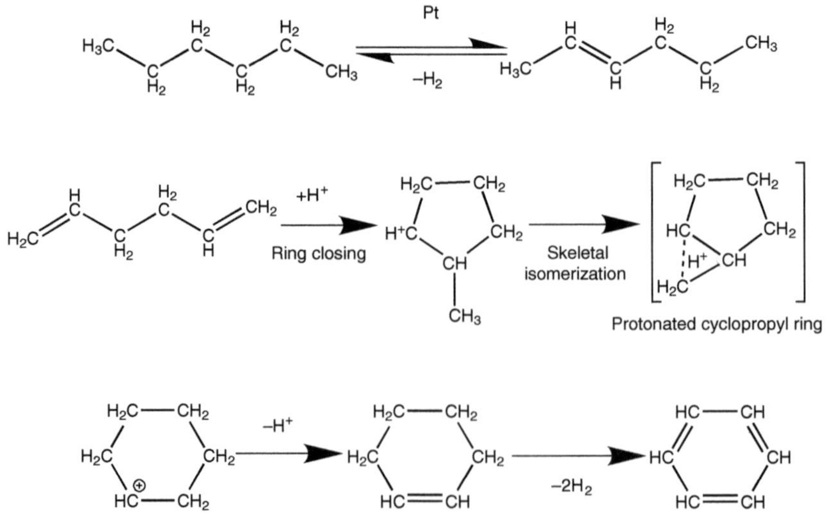

Figure 2.5 Bifunctional reaction scheme to form aromatics from n-hexane. The transition metal achieves C—H bond activation and formation. The acidic proton catalyzes hydrocarbon skeleton isomerization. This reaction will be explained in detail in Section 9.3.3.3.2.

Insert 2: The Catalytic Platforming Process

The platforming process was developed at Universal Oil Products Company in 1949 by the Russian chemist Vladimir Haensel [20, 21]. This process converts light petroleum distillate (naphtha) into higher octane fuels (reformates) with a high content of aromatic compounds. The process uses platinum-based catalysts and operates at high temperatures (800–1000 K) [22, 23] with hydrogen pressures ranging from 345 to 3450 kPa (1 kPa = 10^{-2} atm).

Several generations of processes were developed to further reduce hydrogen process pressure and catalyst regeneration cycle intensity.

High H_2 pressure is necessary to hydrogenate coke precursors before they can form catalyst-deactivating coke, and it also increases the selectivity for the octane-reducing hydrogenolysis reactions. Initially, monometallic catalysts were used, but these were later replaced by bimetallic catalysts containing either Re or Ir. These newer catalysts allowed the use of lower process pressures (1380–2070 kPa), offered a longer cycling length of a year, and produced a gasoline of 95–98 octane [20, 21].

In 1971, the CCR platforming process that featured Continuous Catalyst Regeneration was introduced (Figure 2i.1). This process can be conducted at an ultra-low pressure of 345 kPa, and produces gasoline with a 108 octane rating. The catalyst is removed from the last reactor, reconditioned by burning off the coke to restore its performance, and sent back to the first reactor. During this process catalyst deactivation is also caused by Pt particle sintering. During catalyst reconditioning Pt is redispersed as small metal particles on the catalyst support [20, 21].

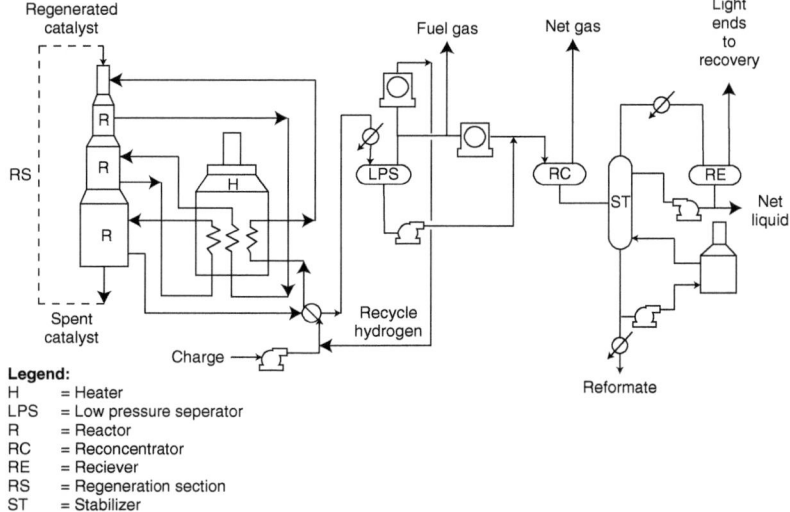

Legend:
H = Heater
LPS = Low pressure seperator
R = Reactor
RC = Reconcentrator
RE = Reciever
RS = Regeneration section
ST = Stabilizer

Figure 2i.1 The UOP CCR (continuous catalytic reactor) platforming process. Continuous regeneration of the catalyst is combined with the platforming reaction. A portion of gas from the separator is compressed and recycled to the reactor section, and the net hydrogen produced is sent to the hydrogen users in the refinery complex. The separator liquid is pumped to a product stabilizer where the more volatile light hydrocarbons are fractionated from the high-octane liquid product [20, 21, 23].

2.2.2.1.3 Hydroisomerization, Hydrocracking: Bifunctional Catalysis

Branched alkanes are desirable substitutes for aromatics or lead that are no longer allowed as components in gasoline. Branched molecules can be made by isomerization of linear C_5–C_7 alkanes or by hydrocracking of linear long-chain or cyclic alkanes to lighter oil product fractions.

The isomerization reaction of a linear alkane to a branched alkane is slightly exothermic ($\Delta H = -7$ kJ mol^{-1}). The hydrocracking reaction that consumes hydrogen is endothermic ($\Delta H = 72$ kJ mol^{-1} C_{10} to C_4, C_6), while the alkane to aromatics reaction is even more endothermic ($\Delta H = 288$ kJ mol^{-1} C_7 to toluene). The optimum temperatures of these reactions are respectively 520, 710, and 820 K.

The reactions are also executed at high partial hydrogen pressures to reduce the deactivating oligomerization reactions of the intermediate olefins. This shifts the equilibrium in the first equation of Figure 2.6 to the hydrogenated alkanes.

Due to the high temperature of the platforming reaction, the protons of the bifunctional catalyst applied in this reaction must not be too acidic so that non-selective cracking reactions that lower productivity can be avoided. For this reason, alumina, which is less active, is used as a support. In contrast, at the lower temperature of the hydroisomerization reactions, strongly acidic materials such as the solid acid zeolites must be used.

Hydrocracking catalysis must allow some amount of cracking of the intermediate alkenes instead of allowing only skeletal isomerization. Therefore, the bifunctional hydrocracking catalysts should contain a hydro-dehydrogenation function that competes less efficiently with proton activation than the functions of the hydroisomerization catalysts. The hydro-dehydrogenation function of a hydrocracking catalyst is usually based on WS_2 or MoS_2. The acidity of the support must also be tuned to a slightly higher temperature by using a less acidic zeolite or an

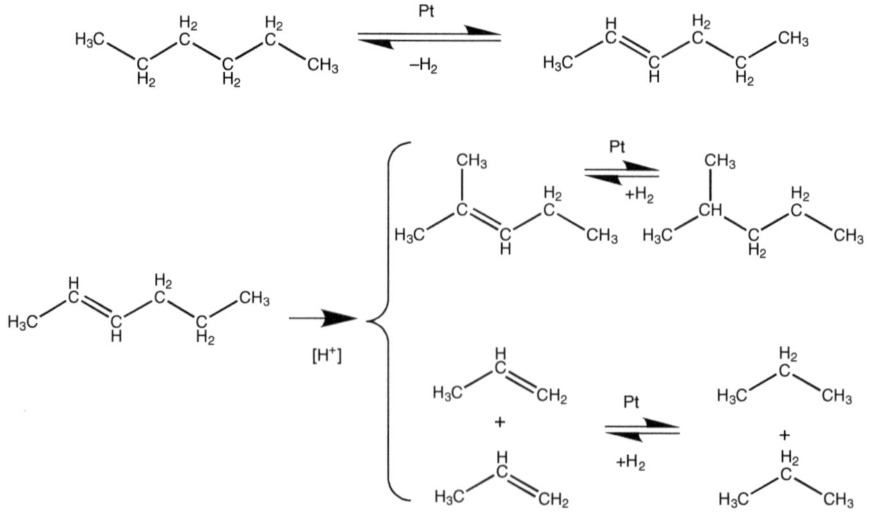

Figure 2.6 Reaction scheme of bifunctional hydroisomerization and hydrocracking.

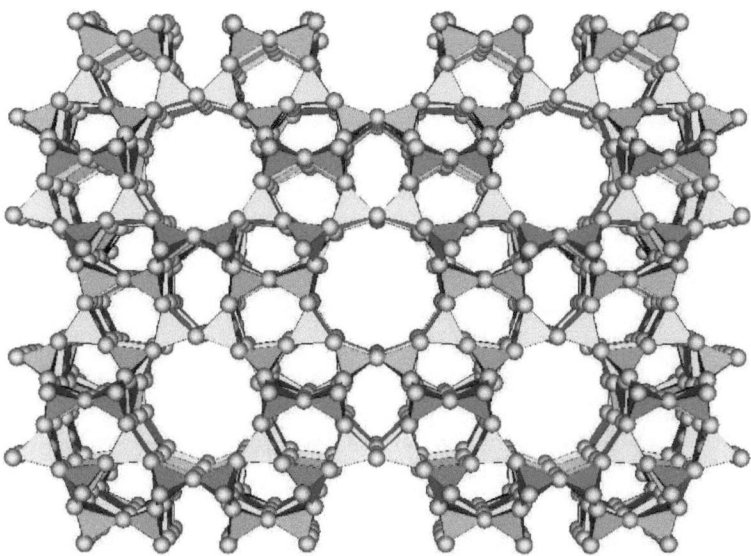

Figure 2.7 Structure of ZSM-5, a zeolite with 10-ring channel structure of low Al/Si ratio. The tetrahedral form a three-dimensional connected framework of 10-ring circumference channels.

amorphous SiO_2/Al_2O_3 support. This mechanism of the reactions is schematically illustrated in Figure 2.7 and discussed extensively in Chapter 9.

2.2.3 Oligomerization and Polymerization Catalysis

2.2.3.1 Methanol to Ethylene and Aromatics: The Aufbau Reaction

The Aufbau reaction based on methanol conversion is an important reaction that produces liquid fuels. Methanol is produced on a large scale from synthesis gas derived from coal or natural gas. Around 1970, Mobil researchers discovered that the ZSM-5 zeolite (Figure 2.7) could convert methanol to aromatics (see Figure 2.8). This provided an important new selective method for liquid fuel production from coal or natural gas (see also Insert 3).

Ethylene and propylene are important bulk chemicals used in the chemical process industry. By using a zeolitic system with narrower pores, ethylene or propylene can be selectively produced instead of aromatics. In practice, catalysts with an $AlPO_4$ lattice framework of the chabazite structure are used (see Figure 2.9). The $AlPO_4$ material is converted into a solid acid by the partial substitution of P^{5+} by Si^{4+}.

While protonic zeolitic materials must be used in these processes, the actual catalysis does not occur by direct proton activation, but by reaction with an

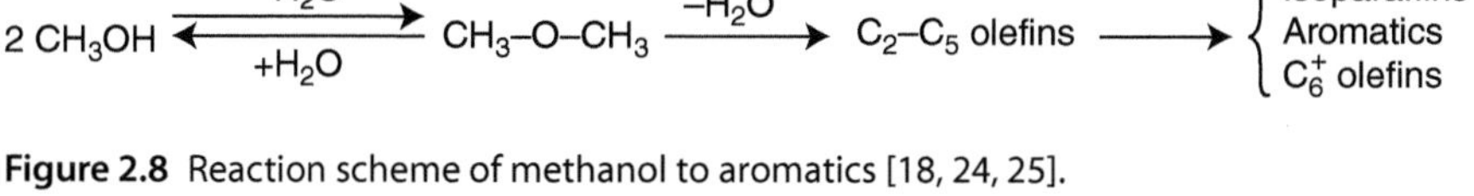

$$2\ CH_3OH \underset{+H_2O}{\overset{-H_2O}{\rightleftarrows}} CH_3\text{–}O\text{–}CH_3 \xrightarrow{-H_2O} C_2\text{–}C_5 \text{ olefins} \longrightarrow \begin{cases} \text{Isoparaffins} \\ \text{Aromatics} \\ C_6^+ \text{ olefins} \end{cases}$$

Figure 2.8 Reaction scheme of methanol to aromatics [18, 24, 25].

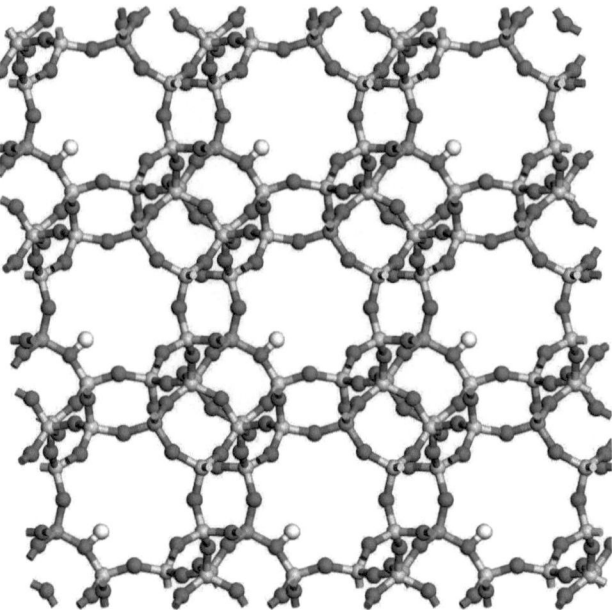

Figure 2.9 Structure of H-SAPO-34. This $AlPO_4$-based material is converted into a solid acid by substitution of P^{5+} with Si^{4+} and adsorption of a proton on a bridging O atom, that connects a Si^{4+} and Al^{3+} cation. This catalyst is especially suitable for ethylene and propylene production from methanol because its cavities are connected through 8-ring openings that only enable diffusion by small molecules.

aromatic organic intermediate that is generated *in situ*. This intermediate acts as an organocatalyst. The reaction proceeds through the alkylation and subsequent transformation of this cationic intermediate (see Figure 4.10).

Insert 3: The Methanol to Gasoline Process and the Methanol to Olefin Process

Chang and Silvestry [18, 26, 27], researchers at Mobil Oil Company in the 1970s, accidently discovered the conversion of methanol into higher hydrocarbons (MTH) over a new synthetic acidic aluminosilicate zeolite catalyst H-ZSM5 (Zeolite Socony Mobil 5). This material with a low framework Al/Si ratio was created using a newly developed synthesis route that employed an organic basic template molecule. In response to the second oil crises in 1987, a natural gas to gasoline plant began operation in New Zealand with a production capacity of 600 000 tons of gasoline per year [28, 29].

This process was replaced in the 1990s by the more efficient Fischer–Tropsch process (see Section 2.2.3.2) that can selectively produce linear alkanes, and when combined with the hydrocracking process, produces branched alkanes for aromatic-free gasoline.

This MTG (methanol to gasoline) technology is currently used to selectively produce lower olefins such as ethylene and propylene. This reaction is catalyzed by the H-SAPO-34 catalyst discovered by Union Carbide (see Figure 2.9) and can achieve selectivity of up to 80% to produce ethylene and propylene at an operating temperature of 670 K. Coking of the H-SAPO-34 catalyst in this process is rapid, necessitating catalyst regeneration [25, 30].

In the 1990s, the MTO process was developed by the joint UOP/Norsk Hydro companies (now INEOS, see Figure 3i.1) [31], using an H-SAPO-34-based catalyst in a fluidized bed reactor that included continuous regeneration of the catalyst combined with an olefin cracking process (OCP).

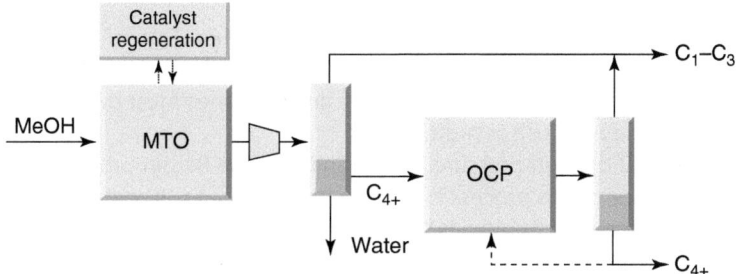

Figure 3i.1 INEOS MTO process scheme. (Olsbye *et al.* 2012 [25]. Reproduced with permission of John Wiley and Sons.)

In 2009, a second generation MTG plant converting coal to liquid fuels began operation in Shanxi Coal Institute, Shanxi Province of China. The plant is operated by the Jincheng Anthracite Mining Group (JAMG) with gasoline production of 100 kilotons per year [25].

2.2.3.2 Fischer–Tropsch Catalysis

The other important hydrocarbon Aufbau reaction is the Fischer–Tropsch (FT) reaction (see also Insert 4). In this reaction, CO and H_2 conversion is catalyzed by metals such as Fe, Co, or Ru to produce a mixture of gasoline and diesel-range molecules. Low-temperature processes mainly produce linear alkanes. At higher temperatures, aromatics will also be produced, especially in reactions catalyzed by Fe. The Fe is converted by contact with the reacting gas into a carbide phase, while Co and Ru remain metallic and produce mostly waxes. Typical reaction conditions for the FT process are temperatures in the range of 450–580 K, and pressures of 20–40 bar. At higher CO pressures, oxygenates may also be produced.

The main non-selective reaction in this process is the undesirable formation of light gases such as methane. The activation of CO controls the reaction rate as well as its selectivity, while the relative rates of methane formation and CO bond dissociation determine non-selective methane formation. A fast rate of C—O bond cleavage is beneficial because it creates a high activation barrier for CO activation.

Ni is usually the preferred catalyst for methane formation, but not for the Fischer–Tropsch reaction. The active metals used have lower activation energies for C—O bond cleavage. The chain growth probability is high due to the production of partially hydrogenated C adatoms necessary for chain growth polymerization. Competitive methane formation is inhibited, so that the growing hydrocarbon chains have a high probability of remaining adsorbed to the surface.

Insert 4: Fischer–Tropsch and Coal Liquefaction Processes

Early processes to produce liquid fuels were based on coal, while more recent processes are based on natural gas conversion. Related technologies can also be developed to convert biomass to liquid fuels (see Section 2.2.4 on hydrotreating).

There are two competing process methods for liquid fuels:

a. Intermediate synthesis gas production and the subsequent Fischer–Tropsch reaction. Synthesis gas is not produced by steam reforming of coal, but by auto-thermal oxidation of natural gas that is already rich in H_2. The linear alkanes must be converted into branched alkanes for use as motor fuel.
b. Direct liquefaction of coal by high pressure hydrogenation (the Bergin process). The technical disadvantage of this process is that the oil produced still contains a substantial level of impurities. Technologies to upgrade synthetic oil produced by this process have provided the basis for many modern hydrotreating and hydrocracking refinery processes (see Section 2.2.4 on hydrotreating).

Process (a) **Synthesis Gas Production and Fischer–Tropsch Reaction** The FT process was first developed in the 1920s by Frans Fischer and Hanz Tropsch at the Kaiser-Wilhelm-Institut für Kohlenforschung in Mülheiman der Ruhr in Germany (now the Max Plank Institute). They used an iron catalyst at an elevated temperature of 400–450 °C and a pressure of 150 bar to convert carbon monoxide and hydrogen into "synthol," a mixture of oxygen-containing hydrocarbons.

In 1936, the first FT synthesis plant began operation by Ruhrchemie AG in Oberhausen, Germany. The plant had a production capacity of 70 000 tons per year. Later, in 1938, nine plants were operational, with a combined production of 660 000 tons per year [32, 33].

In the 1950s, in order to reduce dependence on imported oil, the South African Coal, Oil and Gas Corporation (SASOL I) began operating the first large-scale coal-based FT plant in Sasolburg, South Africa. This plant employed a process using an alkalized iron catalyst supported on silica and promoted by copper. In 1980 and 1983, two more SASOL plants (SASOL II and SASOL III) were brought online. Currently, the three SASOL plants are the only coal liquefaction plants in the world producing liquid fuels from the FT process [34] (Figure 4i.1).

More recently, FT processes for the conversion of natural gas into liquid fuels have been commercialized and used in plants worldwide. In 1993, the first FT plant for the conversion of natural gas from remote fields into synthetic hydrocarbons was installed by the Shell company at Bintulu, Malaysia. This plant employs a process using a cobalt catalyst, with a production capacity as high as 12 000 barrels per day [36]. In 2005, the Oryx GTL plant, a joint venture between SASOL and Qatar Petroleum, was launched with a production capacity of about 32 000 barrels per day [36]. And in 2011, the Shell Pearl GTL facility was launched in Qatar. It employs an upgraded and expanded version of the Fischer–Tropsch process used at Bintulu, which can operate at low temperatures, and it is the largest GTL plant in the world (10 times the size of Bintulu) with the highest hydrocracking capacity in a single location.

Process (b) **The Bergius Direct Coal Liquefaction Process** In 1913, Bergius invented a process to produce liquid fuel product at an H_2 pressure of 100 bar and at a temperature of 450 °C. In 1925, a two-stage process was developed by BASF that included a hydrogenation stage to reduce sulfur (hydrotreating) and a hydrocracking stage. At the end of World War II, it was the major fuel-producing technology used in Germany [37, 38].

Related technology was used for the Shell Hycon process developed in 1989 that converts heavy oil residue at high H_2 pressure into lighter oil fractions [6, 39–43]. This process is followed by hydrotreating to remove heavy metals, sulfur, and nitrogen in order to produce oil fractions that can be further refined. The flexicoking process of Exxon/Mobil [6, 44, 45], an analogous process based on the Fischer–Tropsch reaction, also gasifies heavy oil residue to synthesis gas for use in refineries.

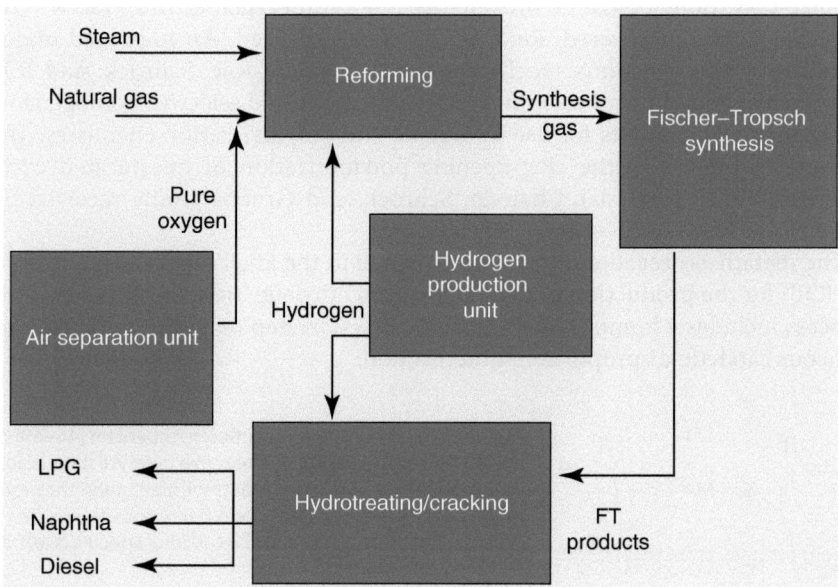

Figure 4i.1 Overall process scheme of the Sasol Fischer–Tropsch plant that converts natural gas into diesel and gasoline components [35].

2.2.3.3 Disproportionation and Metathesis Reaction: Single Site Catalysis

In the metathesis and disproportionation reaction, molecules with a C=C double bond react with each other through cleavage of their respective double bonds. In this process, new alkene molecules are formed by recombination of the reaction products.

The petrochemical process used to convert propylene to ethylene and butene was an early example of this type of reaction. The term metathesis, derived from the Greek words "meta" which means change, and "thesis," meaning position, is the exchange of the parts of two substances. For example, in the reaction AB + CD → AC + BD, B changes position with C.

The reaction is catalyzed by a reactive single-site oxidic cluster immobilized on a high surface area alumina support, such as the catalyst prepared by the reaction of $Mo(CO)_6$ to form a reactive MoO_2 oxy cation immobilized on a high surface area Υ-Al_2O_3 support. When C=C bond cleavage occurs, the reactive Mo cation forms intermediate carbene species that react with a second alkene molecule through intermediate formation of a metallacycle as illustrated in Figure 2.10.

The disproportionation reaction was discovered at Phillips Petroleum by R.L. Banks and G.C. Bayley in 1964. They used propylene heated with a catalyst of $Mo(CO)_6$ or $W(CO)_6$ on a high surface area alumina support to yield ethylene and 2-butenes. This manner of catalyst preparation is an early example of using homogeneous coordination complexes to create highly dispersed reaction centers on a heterogeneous catalyst.

Since C=C bonds must be broken, reactive cations such as Mo^{6+} or W^{6+}, or as subsequently discovered, Re^{7+} or Ta^{5+} must be used. An improved understanding of this reaction's mechanics by Y. Chauvin, R.R. Schrock, and R.H. Grubbs resulted in the development of highly active and selective homogeneous organometallic catalysts for use in organic and polymerization chemistry. This process is applied in the ring-opening polymerization of unsaturated cyclic alkenes (ROMP catalysis). Chauvin, Schrock, and Grubbs jointly received the 2005 Nobel Prize in Chemistry [47–50].

The metathesis reaction is also fundamental in the Shell higher olefin process (SHOP) for the production of detergents from ethylene. Insert 5 shows that this process includes a homogeneous catalytic reaction step in addition to a heterogeneous catalytic disproportionation reaction.

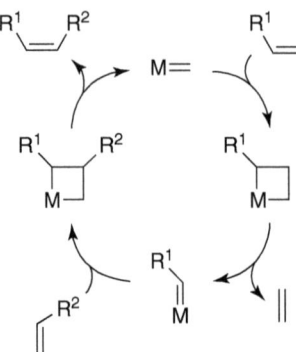

Figure 2.10 The Chauvin reaction mechanism [46–49] of the metathesis reaction. The reactive catalyst state is an intermediate carbene. This reacts with an olefin through a metallocycle intermediate. Upon decomposition, a new olefin molecule is formed and a carbene species is left on the catalyst cation.

Insert 5: The Shell Higher Olefin Process (SHOP)

The SHOP process, invented in 1968 at Shell Oil Company by W. Keim and colleagues, produces linear alpha alcohols used in the production of detergent from ethylene and CO and hydrogen. It resulted from discoveries in the rapidly developing field of coordination chemistry and homogeneous organometallic catalysis, as well as new developments in heterogeneous catalysis as a disproportionation reaction [50]. It was commercialized by Shell Company in 1977 [51–53], and by 2002 the annual yield from this process was 1.2 million tons [50, 51, 54, 55].

The process consists of the following steps:

A. Oligomerization of ethylene using a homogeneous Ni catalyst produces a mixture of linear α-olefins with a maximum chain length of 10 ethylene units (Figure 5i.1).

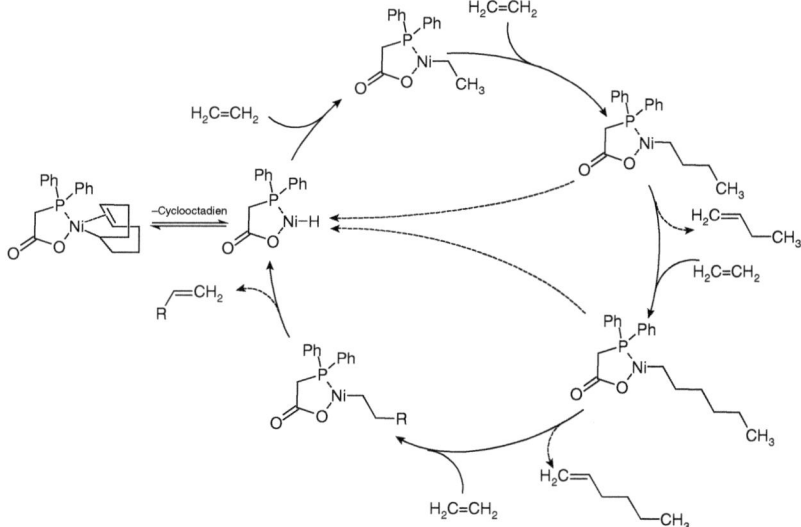

Figure 5i.1 The catalytic reaction cycle of the homogeneous ethylene oligomerization process that yields a range of α-olefin. (Moulijn et al. 2013 [56]. Reproduced with permission of John Wiley and Sons.)

B. The second step in this process is the homogeneous hydroformylation reaction of 1-alkene with CO to create the end-on aldehyde for conversion to the detergent molecules.

The mechanism of the hydroformylation reaction is shown in Figure 5i.2. In this process, the catalyst is a homogeneous Co carbonyl complex $HCo(CO)_4$. The olefin replaces a CO molecule in this complex and reacts with the hydride to give an alkyl species. In subsequent steps, CO and H_2 are added. Several improved versions of this process later replaced the Co carbonyl catalyst with more selective catalysts based on metal organic complexes of Rh [57].

The range of olefins useful for detergent production is restricted to C_{12}–C_{18}, so the metathesis reaction rearranges the olefins not used in the carbonylation

(Continued)

Insert 5: (Continued)

reaction with ethylene to produce olefins in the desired range. Since the double bond must be redistributed over the alkene chain, a double bond shift catalyst must first be used to redistribute the π bonds statistically along the linear hydrocarbon chain.

C. Double bond shift by potassium using an alumina-based heterogeneous base catalyst.

The role of potassium is to abstract an allylic hydrogen atom from the alkene. C=C bond shift occurs by subsequent hydride atom backdonation from potassium to the C atom that was initially part of the double bond.

D. Disproportionation by a Re VII cation on an alumina catalyst using ethylene to yield olefins in the proper product range.

The steps of this complex process are presented in Figure 5i.3.

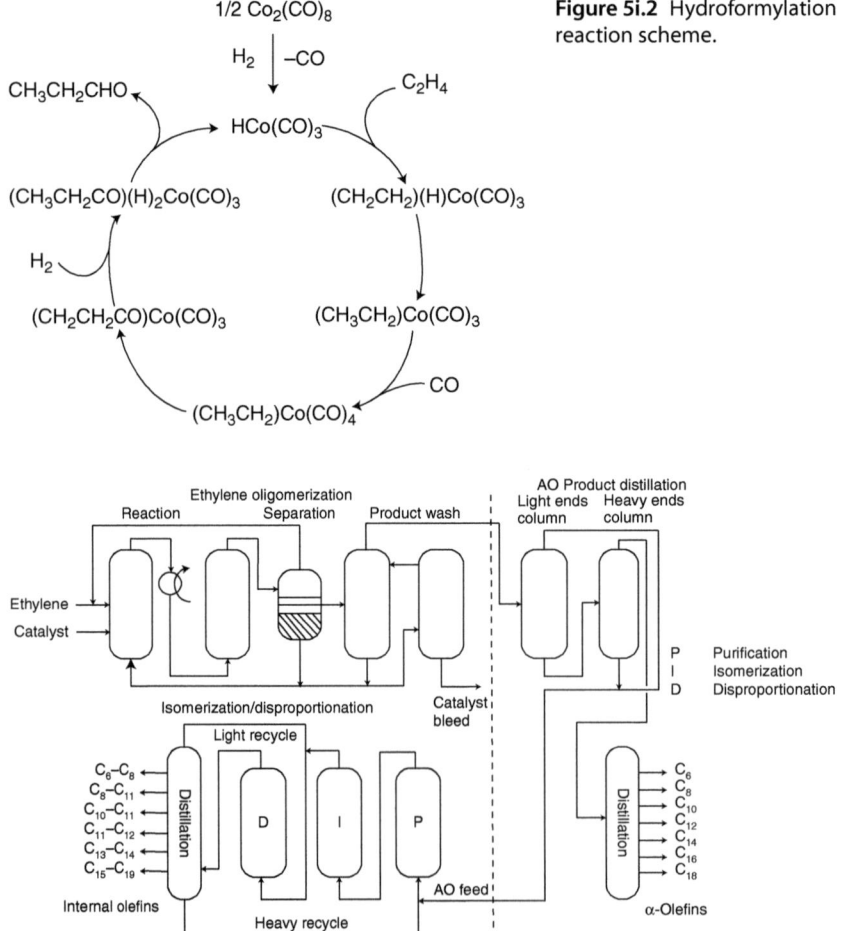

Figure 5i.2 Hydroformylation reaction scheme.

Figure 5i.3 Flow schematic of the Shell higher olefin process (SHOP); AO is α-olefin. (Vogt 2006 [58]. Reproduced with permission of John Wiley & Sons.)

2.2.3.4 Polymerization: Surface Coordination Complex Catalyst

Ethylene and propylene can be polymerized to long chains with a typical molecular weight between 25 000 and 500 000 by using Ziegler catalysts [59]. Ziegler discovered the original catalysts based on $TiCl_3$ by chance in 1954 [60–65].

This reaction is initiated by alkylation of a Ti^{3+} reaction center, which produces a Ti-alkyl intermediate. $TiCl_3$ is then activated using diethyl aluminum chloride as a cocatalyst. Chain growth occurs by subsequent insertion of the olefin into the developing alkyl chain. The insertion reaction is analogous to the homogeneous Ni^{2+} oligomerization process of the SHOP reaction discussed earlier in Insert 5. The reaction is terminated by hydride transfer to a coadsorbed olefin.

Figure 2.11 shows the schematic of the chain growth cycle and termination steps of the olefin polymerization reaction. The key intermediate reaction step of the olefin polymerization mechanism is the formation of metal cation-alkyl, metallacycle intermediate. In contrast to the metathesis reaction, the C=C bond no longer cleaves, but instead the olefin inserts into attached alkyl.

The growing polymer chain is σ-coordinated to one site on the titanium with one adjacent coordination site vacant (i). The chain growth occurs through coordination of the next olefin monomer to the cis coordination site of the polymer chain (ii), followed by insertion of the coordinated olefin into the Ti—C bond. This cis-migration of the σ-bond polymer chain to the coordinated olefin occurs through a four-center transition state (iii). The chain is terminated by ß-hydride elimination, in which a ß-hydride from the polymer chain is transferred to the titanium or the coordinated olefin (iv). The chemistry is a prototype of many reactions catalyzed by coordination complexes (e.g., see also Figure 5i.2).

When propylene is used in this process, a unique stereochemistry develops because chiral centers develop within the alkyl chain. Natta discovered that these can be ordered in a regular way along the backbone of the polymer, yielding polymers with unique and varying properties. Ziegler and Natta received the Nobel Prize in Chemistry in 1963 [49, 64, 65].

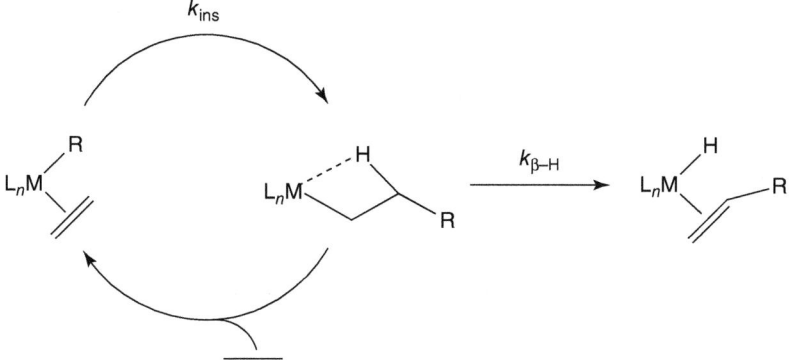

Figure 2.11 Polymerization chain growth of olefin [66]. The olefin inserts into the growing olefin chain. The reaction is terminated by hydrogen transfer from the alkyl chain to the Ti reaction center. (Busico *et al.* 2006 [66]. Reproduced with permission of National Academy of Sciences.)

Table 2.2 The development of Ziegler–Natta catalysts.

Generation	Catalyst	Yield (kg PP per gram catalyst)	Isotacticity index (%)[a]	Process steps
First	δ-TiCl$_3$/DEAC	2–4	90–94	Deashing and atactic polypropylene extraction
Second	δ-TiCl$_3$/isoamylether/ AlCl$_3$/DEAC	10–15	94–97	Deashing
Third	MgCl$_2$/ester/TiCl$_4$/ TEA/ester	15–30	95–97	No purification
Fourth	MgCl$_2$/ester/TiCl$_4$/ TEA/PhSi(OEt)$_3$	>100[b]	>98	No purification, no extrusion/pelletization

Subsequent generations of catalysts increased the relative product yield. As a result, in modern polymerization processes, the catalyst content in the polymers is so low that it is no longer necessary to separate them from the product [69].
a) Measured as the weight percentage of polymer soluble in boiling heptanes.
b) In liquid (bulk) polypropylene.

Supported systems such as TiCl$_3$/MgCl$_2$ developed later, which maximized the number of active sites on the surface. This led to significantly higher activity and selectivity, and eliminated the need to remove catalyst from the product (Table 2.2). More recently, homogeneous Ti and Zr organometallic complexes that use this same mechanism have been discovered, but these deliver superior activity [63, 65, 67, 68].

2.2.4 Hydrodesulfurization and Related Hydrotreating Reactions

2.2.4.1 Hydrodesulfurization

In fuel-producing processes, product streams of crude oil or liquefied coal contain aromatic molecules that include sulfur and nitrogen and thus require hydrodesulphurization and hydrodenitrogenation. Likewise, the conversion of biomass to fuel requires selective deoxygenation to remove water and oxygen. Active hydrodesulphurization, hydrodenitrogenation, or hydrodeoxygenation catalysts can be sulfides such as MoS$_2$ or WS$_2$ promoted by third-row metal ions such as Co or Ni. The preferred catalysts for hydrodesulphurization of oil fractions are promoted by Co, while those used in hydrodenitrogenation are promoted by Ni [70, 71].

The promoting ions weaken the interaction with adsorbed H$_2$S and ammonia, and hence increase the number of surface vacancy positions. Molecules that are to be reacted can adsorb on these surface vacancies so that C—S and C—N bonds are activated by the transition metal cations. Catalysis consists of a sequence of hydrogen additions and C—S, C—N, or C—O bond cleavage reactions.

In the petrochemical industry, these catalysts are used in the hydrotreating step of process streams that can have complex compositions and may contain deactivating metals (residue from heavy oil) as well as heavy asphaltene molecules. There is rapid deactivation of the catalysts unless the sulfide is supported on particles with appropriate pore size distributions that are adapted

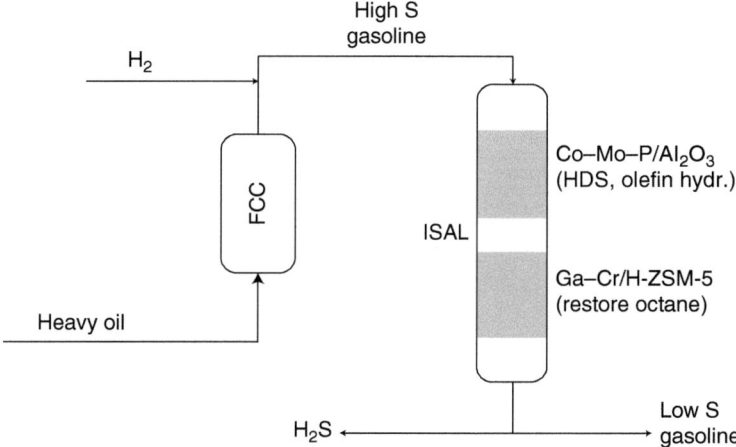

Figure 2.12 Hydrotreating process scheme of upgrading heavy oil to gasoline. The ISAL™ process was jointly developed by UOP and Intevep, S.A.

to the particular feed stream. In the Shell Hycon process, a variety of reactors containing a series of catalysts with varying pore size distribution and metal sulfide content are used to purify heavy feedstock, yielding feedstreams that are more easily processed.

The hydrotreating process is often used in combination with the hydrocracking process. This is illustrated in Figure 2.12 which shows how heavy oil is upgraded to gasoline. The harsh conditions required to remove sulfur by hydrogenation from the gasoline product of fluid catalytic cracking (FCC) produces the undesirable side effect of hydrogenating the olefins, thus reducing the gasoline's octane. The ISAL process addresses this problem in the first part of its reactor by hydrotreating the oil to remove sulfur using a Co—Mo—P/Al_2O_3 catalyst. In a second step, the octane is restored by passing the oil through a bed containing a Ga—Cr/H-ZSM-5 catalyst that achieves isomerization and cracking [72–75].

2.2.4.2 The Biomass Refinery

The use of biomass as raw material presents an alternative to producing liquid fuels from fossil energy sources. Biomass is considered a promising renewable energy source [76–80] because it is sustainable and carbon-neutral. The conversion processes used in a biomass refinery are similar to the various oil conversion processes described in earlier subsections. We will illustrate this for the conversion of wood to energy.

The two competing process routes for converting wood to energy are by intermediate synthesis gas production, to be discussed in the next subsection, or by fast pyrolysis, which we will discuss here. Processes that use fast pyrolysis produce bio-fuels with high water (15–30 wt%) and oxygen (10–50 wt%) content. In order to reduce the oxygen content of the bio-fuel, oxygen is removed as water. Hydrodeoxygenation (HDO) based on hydrotreating (HT) can be conducted at a temperature of less than 400 °C under high hydrogen pressure, using sulfur-containing NiMo and CoMo catalysts [81–84]. These processes can

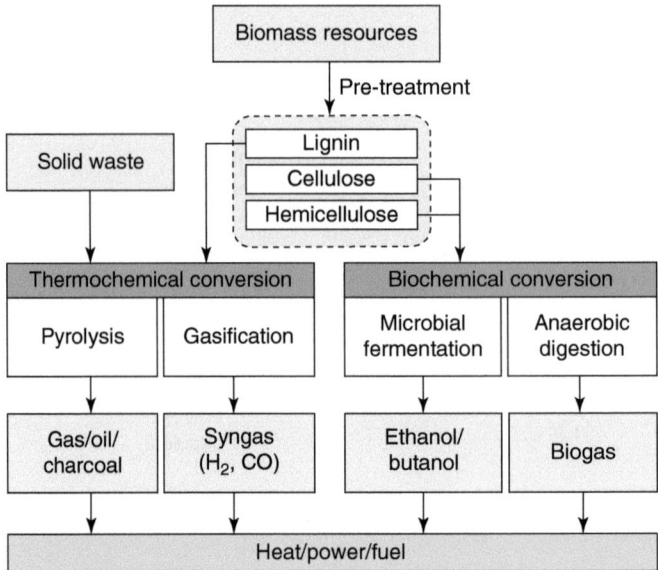

Figure 2.13 The various approaches for converting biomass to biofuel [88].

be combined with catalytic cracking of the heavy oil components of the pyrolysis product or with high pressure heavy feed treating processes [85–87] to further refine the biofuel.

The options for converting biomass to biofuel chemically or biotechnically are presented in Figure 2.13.

2.2.5 Oxidation and Reduction Reactions

2.2.5.1 Steam Reforming

The steam reforming reaction converts natural gas, coal, or biomass that is combined with water into CO and H_2. It is an important reaction to produce the synthesis gas with high hydrogen content that is required for ammonia synthesis. Synthesis gas with lower hydrogen content for the Fischer–Tropsch process is made by autothermal oxidation with oxygen [56, 89].

In the steam reforming reaction of methane derived from natural gas, the transition metal catalyst must activate both CH_4 and H_2O, but must not be deactivated by non-reactive oxide formation or carbon overlayer formation [90–92].

The most active transition metal that can be used as a catalyst in this reaction is Rh, but it is expensive. In practice, a promoted $Ni/MgAl_2O_3$ catalyst is used [56, 93, 94] as a more economical alternative. The main cause of catalyst deactivation at the high reaction temperature (800 K) is graphite formed by the decomposition of methane. This can be suppressed by using Ni metal with a small particle size, and by enhancing removal of carbon deposits with a reaction stream that uses parallel gasification catalyzed by promoters such as K or La.

The steam reforming reaction is strongly dependent on the metal's particle size. The activation of methane in this reaction is enhanced on the edge or corner sites. (See Section 4.5.1.)

2.2.5.2 NH_3 to NO_x Oxidation

In the production of nitric acid, NH_3 is first formed by the Bosch–Haber process, and NO is produced from NH_3 by the Ostwald process. This is an essential step in the process that transforms N_2 into nitrates, which are mainly used in fertilizers [95–97]. The intermediate formation of ammonia from N_2 in this process is required because N_2 to NO oxidation is an endothermic reaction that prevents conversion at acceptable temperatures [98].

Catalytic oxidation of ammonia occurs selectively at a temperature of 1100 K and is catalyzed by a Pt/Rh gauze. The high temperature prevents non-selective formation of N_2 or N_2O that are dominant products at lower temperatures [95, 99, 100].

A schematic of the Ostwald reaction process is shown in Figure 2.14. Ammonia is mixed with dry air at a 1 : 10 ratio, and the mixture is introduced into the top of a cylindrical catalyst chamber. The catalyst chamber contains a Pt or Pt/Rh gauze catalyst at a temperature of 1100 K. The NO and NO_2 is then passed through an oxidation chamber as gases and quenched with water to produce the nitric acid [100].

2.2.5.2.1 Selective Oxidation of Alkanes and Olefins

Butane to Maleic Acid Oxidation The selective oxidation of light alkanes is a desirable reaction because these molecules are readily available from natural gas. Here, we will discuss the selective oxidation of butane to maleic acid that replaces the classical processes based on benzene oxidation.

The catalyst used for butane oxidation is based on reducible vanadium oxides. In the oxidation reaction of butane, the oxygen atoms of the reducible oxide abstract H atoms from butane to yield H_2O and an alkene as the initial products. The reduced catalyst is regenerated by reoxidation with gas phase oxygen.

A selectivity problem arises when other alkanes are used instead of butane since the alkene intermediates produced in this oxidative dehydrogenation process are more reactive than the alkanes initially converted. This may readily lead to total oxidation, because selective oxidation can only occur when the final product molecule is less prone to oxidation than the initial alkane.

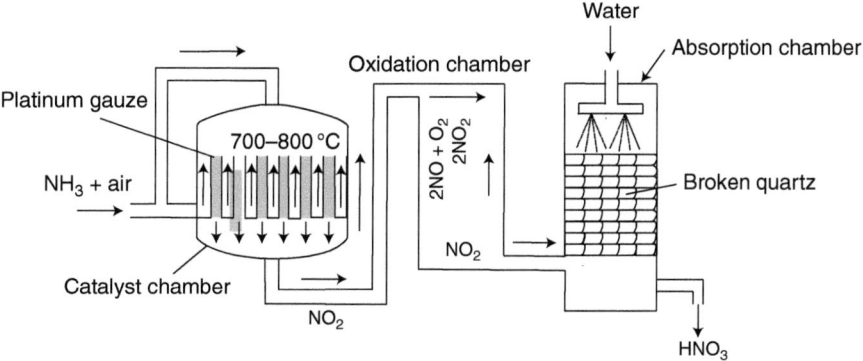

Figure 2.14 Ostwald process for nitric acid production [100].

2 Heterogeneous Catalytic Processes

In the butane oxidation reaction, the final maleic anhydride product is more resilient with respect to oxygen activation than the butane molecule. Hence, this conversion reaction can be conducted selectively. The preferred catalyst used for this process is a vanadium phosphorus oxide (VPO) catalyst [101–103]. (See also Section 4.7.) A circulating fluidized bed process for this reaction was developed by Dupont in the 1990s [6]. (See Figure 2.15.)

The CFB reactor increases selectivity by carrying out the two parts of the catalytic cycle (oxidation of butane/reduction and reoxidation of the catalyst) in two separate reactors. The oxidized catalyst and butane mixture is fed into the riser section of the reactor in the absence of oxygen. There, it reacts to form maleic anhydride leading to a reduction of vanadium in the catalyst. The reduced catalyst is then separated from the product stream and fed into a regenerator where it is reoxidized by air. The Dupont tetrahydrofuran (THF) plant in Asturias, Spain

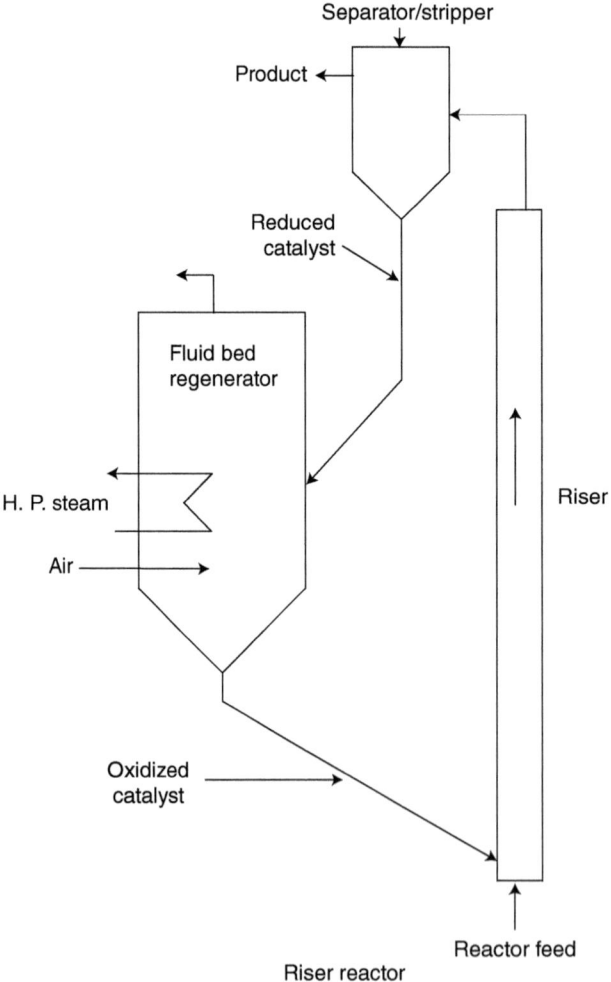

Figure 2.15 The circulating fluid bed reaction of the butane oxidation process [6, 104, 105].

oxidizes butane to maleic anhydride, with a production capacity of about 100 million pounds of THF per year. In a subsequent step, hydrogenate crude aqueous maleic acid is catalytically hydrogenated to THF. Special attrition-resistant catalyst particles have been developed to enhance catalytic life [104–106].

Propylene or Propane Conversion to Acrolein and Acrylonitrile Two essential building blocks of synthetic fibers and plastics are acrylonitrile and acrylic acid. Conventionally, they are produced in consecutive reaction steps from acrolein. A major breakthrough in the 1950s was the discovery of mixed oxides such as Bi_2O_3/MoO_3 that selectively convert propylene to acrolein [107]. Direct conversion of propylene to acrylonitrile became possible with the discovery of the ammoxidation reaction. The reaction is catalyzed by related mixed oxides based on Mo and also containing Sb, Fe, or U. When propane is converted instead of propene, mixed oxides with higher reactivity than Mo must be used, such as V.

These oxidation reactions follow a complex cycle of steps that take place on different reactive centers of the catalyst, and all the steps must occur under the same reaction conditions. The catalyst is tuned to avoid total combustion.

When propylene is converted, Bi or Sb is added to MoO_3 to activate the allylic CH_3 group of propylene, since propylene cannot react with Mo oxide alone. Oxygen from acrolein production or NH from the partial oxidation of ammonia in the ammoxidation reactions can be inserted by Mo or Fe into the allyl intermediate.

The Sohio process is used for manufacturing acrylonitrile from propylene. It was discovered and commercialized in 1957 by Standard Oil of Ohio (which became BP in 1987). This discovery led to an exponential increase in the supply of acrylonitrile that was in high demand for the production of a wide range of polymer products [108, 109]. The ammoxidation reaction is fast, highly selective, and efficiently yields acrylonitrile without additional recycling steps [110, 111]. It is summarized by:

$$CH_2 = CH - CH_3 + NH_3 + 3/2 O_2 \rightarrow CH_2 = CH - CN + 3H_2O \Delta H$$

The world production capacity for acrylonitrile has grown to more than 6 million tons per year [110].

Recently, the alternative method of producing acrylonitrile from propane instead of propene has attracted more interest due to the rising market price of propene and the falling market price (and greater availability) of propane derived from natural gas. This type of process was commercialized in Thailand in 2012 in a joint venture of PTT Asahi Kasei Chemical and Marubeni (PTT Asahi Kasei Chemical Co. Ltd) with a productivity of about 200 000 tons per year [110].

Ethylene Epoxidation: Ag Catalysis Ethylene epoxide is an important intermediate molecule in the production of glycol which is used as antifreeze. Selective oxidation of ethylene is catalyzed by a Ag catalyst promoted with alkali and chlorine [112]. Epoxide formation competes with non-selective total combustion when acetaldehyde is formed rapidly as the intermediate product instead of epoxide.

There are two reaction paths for non-selective acetaldehyde formation. These must be suppressed to allow the selective formation of ethylene epoxide.

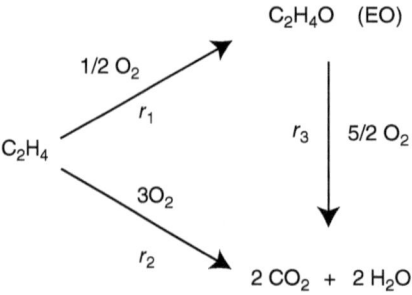

Figure 2.16 Non-selective total combustion of ethylene with parallel and consecutive reaction paths. (Reproduced with permission of Nett Technologies Inc.)

Acetaldehyde formation is catalyzed by Ag and can be formed in parallel to the epoxide. Selective formation of the epoxide is promoted by modification of Ag with coadsorbed Cl. A second, non-selective reaction path to acetaldehyde is by a consecutive reaction of the epoxide. This second route involves total combustion of ethylene epoxide, and is catalyzed by acidic protons of the α alumina support, which aid in the isomerization of the epoxide to acetaldehyde. For this reason, a support with low surface area must be used. Modification of the alumina surface by alkali promotion is used to decrease the concentration of the harmful acidic protons on the catalyst support (see also Insert 6).

The kinetic reaction scheme of Figure 2.16 [113] illustrates the parallel and consecutive reaction paths that can lead to the non-selective total combustion of ethylene.

Insert 6: The Ethylene Epoxidation Process

Catalytic ethylene epoxidation replaced the classical stoechiometric chlorohydrin process that was invented in 1859 by the French chemist Adolph Wurtz and used commercially beginning in 1914 [114, 115]. In the chlorohydrin process, the epoxide is produced by the reaction of Cl_2, H_2O, and $Ca(OH)_2$. For every molecule of ethylene oxide that is formed, a salt molecule is produced, and this waste problem has rendered the process environmentally unacceptable. As such, there was an incentive to develop catalytic ethylene epoxidation, and to replace the chlorohydrin route method used to convert propylene to olefin epoxide. We will discuss propylene epoxide production in the next section.

In 1931, the French chemist Theodore Lefort discovered the direct oxidation of ethylene with air over a Ag catalyst [116]. This method was first commercialized in 1937 by Carbide and Carbon Chemicals (later Union Carbide) [117]. At the time, its main advantage was the elimination of the Cl_2 catalyst previously used in the process [117, 118]. Currently, ethene oxide is commercially produced directly from ethene and O_2 at 90% selectivity. This high selectivity can only be achieved when a compound containing Cl such as vinyl chloride is added to the oxygen as a moderator, although only at the ppm level. The vinyl chloride readily combusts on Ag. The Cl becomes coadsorbed to the Ag catalyst and increases the selectivity of the chlorine-free catalyst from 40% to 60%. The selectivity of the reaction was further improved by adding alkali promoters to the catalyst support to suppress its acidity.

In 1957, Shell modified the direct oxidation process that used air as the oxidant to instead use high purity oxygen. The annual production capacity of this modified process is 15 million tons [118].

Propylene Epoxidation: Lewis Acid Catalysis Propylene epoxide is an important component in the production of polyurethane plastics. Selective oxidation of propylene with oxygen catalyzed by Ag to yield epoxide is not possible because propylene's reactive CH_3 group causes efficient combustion. However, selective epoxidation of propylene is possible in a reaction with hydrogen peroxide or a hydroperoxide compound.

The preferred heterogeneous catalyst for these reactions contain single-center Ti-oxide reaction centers dispersed on a high surface area silica support or included in a siliceous zeolite framework (see also Insert 7). The latter catalyst is called Ti-silicalite (TS-1). The silica-supported catalyst can be used to epoxidize propylene using the hydroperoxide compound, while the zeolitic system can epoxidize propylene using hydrogen peroxide.

The Lewis acidic Ti center activates the RO–OH group of the peroxide part of the molecule to generate an electrophilic O atom that will insert into the C=C bond of propylene. This activation mode of oxygen will not lead to competitive intermediate allyl formation.

Precautions must be taken to suppress non-selective decomposition reactions of the peroxide, which are catalyzed by small anatase or rutile clusters that may be present due to poor catalyst preparation. In the hydroperoxide reaction, silylation of the catalyst is used to eliminate the non-selective sites.

Insert 7: Propylene Epoxidation Processes

Shell has developed a heterogeneous route to produce propylene oxide catalyzed by Ti/SiO_2. The SMPO (styrene monomer–propylene oxide) process converts propylene and ethylbenzene into propylene oxide and styrene monomer.

The SMPO process produces styrene (SM) at a rate of about 800 000 tons per year (worldwide total: 10 million tons per year), and propylene oxide at a rate of 400 000 tons per year (worldwide total: 3.5 million tons per year) [119–121]. There are now five SMPO plants in existence world-wide, with the latest plant start-up in China in 2006 [121].

The process involves four main reaction steps [121]:

1. The air auto-oxidation of ethylbenzene EB to the corresponding hydroperoxide (EBHP) in liquid phase at 400 K and 30 mbar. EB can be obtained from alkylation of benzene with ethene over H-ZSM5 catalyst at 700 K, 15 bar (the Mobil–Badger process).
2. The catalytic epoxidation of propylene with EBHP over Ti/SiO_2 (or over homogeneous Ti or Mo) at 400 K that yields propylene oxide (PO) and methyl-phenyl carbinol (MPC).
3. The dehydration of MPC to SM over alumina.
4. The hydrogenation of methyl phenyl ketone (MPK), a byproduct of the first two reactions, to MPC for subsequent conversion over titanium at 440–550 K to styrene monomer (SM).

(Continued)

Insert 7: (Continued)

The reaction steps are illustrated in Figure 7i.1.

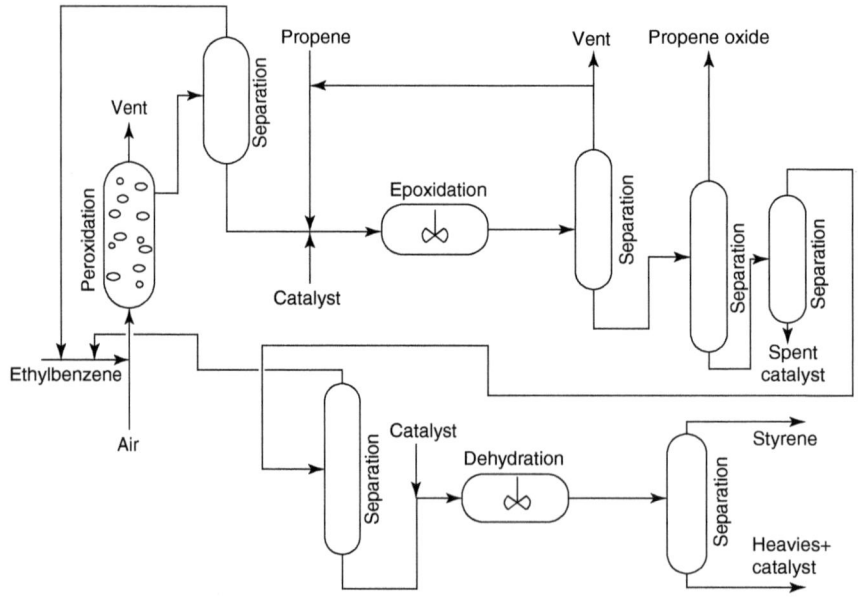

Figure 7i.1 The sequence of reactions that constitute the SMPO process. (Buijink et al. 2011 [121]. Reproduced with permission of Elsevier.)

A simplified flow scheme for the SMPO process is illustrated in Figure 7i.2.

Figure 7i.2 Simplified SMPO process scheme. (Buijink et al. 2011 [121]. Reproduced with permission of Elsevier.)

> In 2000, a new process for the continuous production of propylene oxide based on hydrogen peroxide (direct catalytic oxidation of an olefin with aqueous hydrogen peroxide, or HPPO) was developed by the Dow Company leading to a simple and highly selective process that produces only water as a by-product. The reaction is catalyzed by a proprietary titanium silicalite TS-1 zeolite. The process involves a sequence of steps: reaction, distillation, decomposition, phase separation, condensation, and further distillation; with recycling of the various streams to lengthen the life of the catalyst and increase reaction selectivity [122].
>
> The HPPO process has major environmental benefits compared to the conventional processes. Wastewater production is decreased by up to 80%, energy use is decreased by 35%, and the plant's infrastructure and physical footprint are significantly reduced by the use of simple raw materials (H_2O_2 as a green oxidant and propylene) [121, 122].
>
> A plant using the hydrogen peroxide-based propylene oxide HPPO process was built by a joint venture between the BASF and Dow companies in Antwerp, Belgium. It became operational in 2008, and with an annual production of 300 000 metric tons of propylene oxide it is the world's largest HPPO plant [122].
>
> A mega-sized plant (three times larger than any existing plant) that can synthesize 230 000 tons of hydrogen peroxide per year using Solvay's anthraquinone technology was built close to the BASF/Dow HPPO plant, allowing a significant reduction of transportation and production costs [122].
>
> Dow and BASF received two awards for their jointly developed HPPO technology: the Institution of Chemical Engineers' Innovation and Excellence Award in Core Engineering in 2009, and the Presidential Green Chemistry Challenge Award in 2010 [122].

2.2.5.3 NO Reduction Catalysis

2.2.5.3.1 Automotive Exhaust Catalysis: NO Reduction by CO

Catalytic mufflers (catalytic converters) were introduced in the 1970s to control and reduce the pollutants in the emissions of gasoline-powered vehicles. The three major components of exhaust are CO, unburned HC, and NO_x.

Early catalytic converters employed a catalyst consisting of a Pt/Pd/Rh alloy on a reducible oxide such as CeO_2 to reduce the NO content of the exhaust. The catalyst consisted of a ceramic monolithic honeycomb-shaped structure with small channels. Within these channels the active components (composed mainly of noble metals such as Pt, Pd, and Pt, rare-earth oxides such as CeO_2, or alkaline-earth oxides such as BaO) were dispersed with alumina as a thin layer or wash coat of 10–150 µm. The reducible CeO_2 functioned as an oxygen source, and BaO could then take up that oxygen to become peroxide [123, 124].

Because automotive exhaust contains excess oxygen and the NO reduction reaction to N_2 requires close to stoichiometric reductive conditions, a sensor/computer system is used to regulate the exhaust composition (see Figure 2.17). Because this was not completely efficient, the CeO_2 and BaO acted as a buffer to moderate oxygen excess or shortage.

In this reaction, NO decomposes on the transition metal surface, and the nitrogen atoms recombine to yield N_2. Non-selective production of NH_3 or N_2O must be prevented, so NO dissociation and N_2 formation must be relatively fast. O

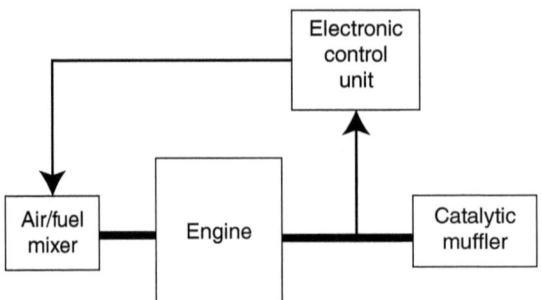

Figure 2.17 Three-way catalyst (TWC) control system [125].

atoms generated by the decomposition of NO are removed by CO to produce carbon dioxide, or by H_2 to produce water.

In 1979, the three-way catalyst (TWC) was developed to control and remove the three primary pollutants (CO, unburned HC, and NO_x) by the oxidation of CO and HC to CO_2 and water, and the reduction of NO_x to molecular nitrogen [126].

The TWC is based on three elements [126, 127]: electronic fuel injection (EFI) providing a stoichiometric air/fuel ratio λ in the gas mixture (if $\lambda < 1$ there is high activity for NO reduction, if $\lambda > 1$ there is high activity for CO and HC oxidation) [123]; an oxygen sensor-governor in the exhaust system; and an electronic system to control a feedback loop using the oxygen sensor signal to determine the amount of fuel to inject to maintain the exhaust gas close to the stoichiometric point under the current conditions [123].

Pt, Pd, and Rh catalysts are preferred because they remain metallic under typical operating conditions and do not produce volatile oxides that are the precursors to metal loss. They are robust enough to resist poisoning by lead and halide, and by sulfur dioxides from additives in the fuel [128–130].

2.2.5.3.2 Diesel Engines: NO Reduction by NH_3

The exhaust of diesel engines cannot be regulated to eliminate excess oxygen. This O_2 excess prevents the use of noble metal catalysts and instead requires the addition of a reducing agent to the exhaust gas. Urea is used for this purpose, decomposing when it reacts with water to produce NH_3 and CO_2. The ammonia reduces NO to N_2.

Catalysts used for this reaction are reducible oxide catalysts such as V_2O_5/TiO_2 or zeolites that have been ion exchanged with Cu^{2+} or Pt^{2+} ions. On these catalysts NO or NO_2 will recombine with NH_4^+ to produce reactive NH_4NO_2. This decomposes to H_2O and N_2. NH_4NO_3 (ammonium nitrate) formation must be prevented because it will decompose to N_2O, which is a prominent greenhouse gas [131].

Two processes that have been used to reduce NO_x are selective catalytic reduction (SCR) patented in the United States in 1957 by the Engelhard Corporation, and Thermal DeNox (TDN) developed in 1975 in the United States by R.K. Lyon at Exxon (see also Insert 8).

In the SCR process, NH_3 is adsorbed on the catalyst substrate, where it catalytically reacts with NO_x and is converted into nitrogen molecules and water

[131, 132]. This can take any of the following forms:

$$\text{Standard reaction}: NH_3 + NO + 1/4O_2 \rightarrow N_2 + 3/2H_2O \quad (2.1)$$

$$\text{Fast-SCR}: NH_3 + 1/2NO + 1/2NO_2 \rightarrow N_2 + 3/2H_2O \quad (2.2)$$

$$NO_2 \text{ SCR}: NH_3 + 3/4NO_2 \rightarrow 7/8N_2 + 3/2H_2O \quad (2.3)$$

Insert 8: Diesel Exhaust Treatment

The first generation SCR process was introduced and tested on trucks in the mid-1990s, with a deNO$_x$ efficiency of about 50–60%, and an SCR catalyst to engine displacement ratio of about 6 : 1 (swept volume ratio) (Figure 8i.1).

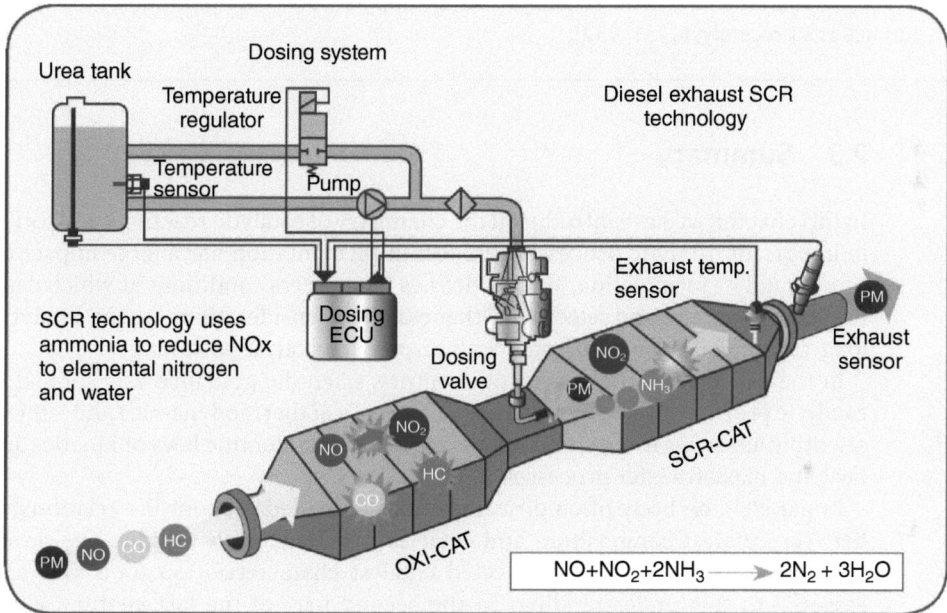

Figure 8i.1 Large scale diesel exhaust-Urea-SCR system [133].

Urea-SCR was commercialized by Daimler Chrysler (now Daimler AG) in 2005 under the trademark Blue TEC. It was designed for use in both heavy-duty and passenger vehicles, and based on the use of extruded honeycomb monolith catalysts consisting of V_2O_5–WO_3–TiO_2 [131].

In 2013, cycle-average de-NO$_x$ efficiencies approaching 95% were realized. Today, with the fourth generation of commercial SCR technology, NO$_x$ emissions can be reduced by 90% using a system half the size of the original.

- **Soot Removal** Due to the nature of the diesel combustion process, high amounts of soot or particulate matter (PM) are formed [132]. An SCR catalyst is often placed downstream of a soot trap called the diesel particulate filter (DPF)

(Continued)

> **Insert 8: (Continued)**
>
> in order to meet the requirements for reducing soot particle emissions. This exposes the SCR catalyst to very high temperatures during DPF regeneration, so that its hydrothermal stability is a major concern. The DPF mechanically separates the particles by forcing the exhaust gas to diffuse through porous walls, leading to high filtration efficiency (85% or more). Soot deposits can produce a substantial backpressure, so the DPF must employ regeneration (oxidation of the stored soot particles) to avoid increased fuel consumption and decreased engine efficiency. Continuously regenerating trap (CRT) technology is the preferred method for DPF regeneration. Regeneration is achieved through CRT functions by initiating soot oxidation by NO_2, using oxidation of NO on the Pt catalysts such as in SCR. The Pt catalyst can be applied as a coating on the DPF to act as a precatalyst [131, 132].

2.3 Summary

In this chapter, we have introduced the chemistry of catalytic reactivity and its use in important catalytic processes. Process implementation has a large impact on actual catalyst formulation, since it defines the process conditions at which catalysts must be active and selective. Other essential considerations are the catalysts' long-term stability or robustness with respect to catalyst poisoning.

In the first part of the nineteenth century, scientific guidance was applied to catalytic process development initially by chemical thermodynamics and later by scientific advances in reactor engineering based on scientific laws of kinetics and heat and mass-transfer processes.

In parallel, the body of empirical chemical knowledge about the relationship between catalyst composition and catalytic reactivity grew rapidly. The development of increasingly sophisticated catalyst characterization tools was also essential to this understanding. In the second part of the last century, along with the advancements in heterogeneous catalysis, the science of molecular catalysis progressed based on coordination chemistry and later on reactions using organometallic complexes, providing the foundation for the modern molecular scientific basis of catalysis.

A predictive approach to catalysis requires knowledge of the mechanism of the targeted reaction. As we will see in the chapters devoted to the mechanism of catalytic reactions, more than one mechanism has been proposed for many catalytic reactions. While there is consensus on the mechanism for some reactions, the mechanistic debate still continues for several other important reactions. One reason for this uncertainty is that the mechanism of a reaction may change with catalyst composition or when reaction conditions change. In addition, mechanistic information is often only indirectly deduced by experiment and is not uniquely related to kinetic expressions. One of the great challenges to catalytic theory is the development of tools that can determine, with reasonable effort, the reaction mechanism for a given catalytic composition and hence its activity and selectivity.

Theoretical advancements have made it possible to identify the optimum composition of a catalyst once its mechanism is known.

Catalyst discovery is driven by the desire for low-cost large-scale methods for producing useful molecules, by the need to replace scarce but essential molecules with synthetic versions and by the demand for commodity products derived from alternative feedstocks.

Most catalytic processes have been developed empirically by testing many materials with a particular reaction in mind, for example, the processes for ammonia synthesis or hydrocarbon cracking. But serendipitous discoveries have also proved to be technological game changers; for example, the discovery of unexpected new product molecules such as the polymers resulting from olefin polymerization. Accidental discovery often plays a role in finding new catalysts or novel catalytic reactions, as well. Examples include the discovery of the Ti catalyst for propylene epoxidation, the Ag catalysts for ethylene epoxidation and the discovery of catalytic cracking by a zeolite.

New pathways for particular products can often be designed based on empirical information that suggests a material composition as a candidate for a particular elementary reaction step. This empirical information is presented in this book, complemented by theories and tools of computational catalysis that help to refine it.

A pragmatic method of discovery is to hypothesize a mechanism for the reaction by comparing the stoichiometric compositions of the reaction product and the reagents. For example, oxygen can be added to a hydrocarbon from O_2, from H_2O, from a peroxide or from CO insertion. Oxygen can be removed through decomposition by dehydration or through reaction with H_2. In addition, in NO reduction oxygen can be removed by CO, and in biomolecules decarbonylation of CO_2 is also possible. Catalyst formulations for most of these necessary reaction steps are known from existing processes, and this information aids in the design of exploratory experiments.

The comparison of the hydrogen content of the product versus its reagent can also be used for pragmatic discovery. When relative hydrogen content increase is required, a hydrogen donor such as H_2 must be used and the reaction is exothermic. Hydrogen decrease can be achieved either by an oxidative reaction or by a non-oxidative reaction which is endothermic.

And finally, pragmatic discovery can be achieved by comparing the number of carbon atoms in the product molecules versus the number in the reagent molecules. This comparison distinguishes reactions with very different temperature requirements – decomposition reactions are endothermic while isomerization and oligomerization reactions are exothermic.

Chemical processes often combine several conversion steps in the same reactor or in successive process steps. Heterogeneous catalytic and homogeneous catalytic process steps sometimes complement each other in the overall production process, as in the SHOP process.

The sciences of molecular homogenous catalysis and surface-related heterogeneous catalysis have become close partners not only in practice but also in the scientific sense. As we will discuss in depth later in Part II of this book, when considered at the molecular level, the activation and adsorption of molecules to surface atoms and in a molecular coordination complex are essentially similar. Increasing knowledge about each of the two systems has led to cross-fertilization, which has significantly aided the advances of the past 20 years in the molecular-level understanding of heterogeneous catalytic reactions.

In the second part of the twentieth century, several important homogeneous catalytic reactions based on organometallic complexes were discovered that are analogous to previously discovered heterogeneous catalytic reactions. Half a century after the early discoveries of heterogeneous catalytic hydrogenation catalyzed by transition metal particles by Sabatier, Wilkinson discovered the $RhCl(PPh_3)_3$ complex as an organometallic hydrogenation catalyst in 1965 [134]. This was followed by the discovery of molecular metathesis catalysis with the use of Mo, Ta, or Re complexes as catalysts by Schrock and Grubbs. The reaction had been discovered earlier in the context of refining processes that employed catalytically active components supported on a high surface area alumina. A final example of molecular-level catalyst discovery occurred in olefin polymerization catalysis, where the original discovery of heterogeneous $TiCl_3$-based catalysts was followed by the discovery of very active organo-metallic complexes based on Ti, V, or Zr by Kaminsky [135].

A major advantage of molecular catalytic systems is that they can be tuned by the choice of organic ligands around the catalyst's center to control activity and selectivity. (For an excellent introduction to homogeneous catalysis we refer to van Leeuwen [57].)

The use of heterogeneous catalysts at low temperatures is sometimes beneficial for separating the product from the catalyst. The catalytically reactive site can still be an "immobilized" organo-metallic complex, which can be usefully exploited in association- or insertion-type catalytic reactions.

Heterogeneous catalysts are uniquely suited to reactions that require the dissociation of a strong molecular bond such as N_2 or CO; for example, ammonia synthesis and the Fischer–Tropsch reaction. In these cases, reaction centers that consist of (at a minimum) a small cluster of active metal atoms are needed. Single-site catalysts such as the organometallic catalysts cannot be applied in these reactions. Instead, the catalyst surface must provide the required ensemble size. (We describe this in greater detail in Chapters 4 and 8.) In addition, organometallic complexes are not stable at the high temperatures required for these reactions, but the solid state matrix of the heterogeneous catalytic system gives the catalyst the required robustness.

A single-site catalyst may be beneficially applied as a heterogeneous catalyst by virtue of the unique ligand environment provided by the lattice in which it is embedded. This is especially true for heterogeneous catalytic zeolitic systems, where the interplay between microchannel structure and catalyst reactivity provides unique opportunities (see Chapter 9).

Another characteristic of heterogeneous catalysts is that they often require the presence of reaction centers of different catalytic reactivity so that chemically diverse reaction steps can be combined in coupled catalytic reaction cycles conducted at the same reaction conditions. This is possible due to the unique design of the catalyst. These types of multifunctional catalysts include the mixed oxides used for selective oxidation, and the bifunctional solid acid catalysts promoted by a transition metal used for hydroisomerization or hydrocracking.

References

1 Thiemann, M. and Scheibler, K.W. (2000) Nitric acid, nitrous acid, and nitrogen oxides, in *Ullmann's Encyclopedia of Industrial Chemistry*, Wiley-VCH Verlag GmbH & Co. KGaA.
2 Perego, C. and Carati, A. (2008) Zeolites and Zeolite-Like Materials in Industrial Catalysis, in Cejka, J. *Zeolites: from model systems to industrial catalysis*, Chapter 14, pp. 357–380 Transworld Research Network.
3 Plank, C.J. (1983) The invention of zeolite cracking catalysts. *ACS Symp. Ser.*, **222**, 253–271, http://cat.inist.fr/?aModele=afficheN&cpsidt=9627522 (accessed 21 September 2016).
4 Van Santen, R. (2012) *Catalysis in perspective; Historic review* in (Beller, M., Renken, A., Van Santen, R.A., eds.) Chapter 1, *From Principles to Applications*, Wiley-VCH Verlag GmbH, pp. 3–19.
5 Weissermel, K. and Arpe, H.J. (1978) *Industrielle Organische Chemie*, Verlag Chemie, Weinheim.
6 Moulijn, J.A., Makkee, M., and van Diepen, A. (2001) *Chemical Process Technology*, John Wiley & Sons, Ltd..
7 Ertl, G., Knözinger, H., Schüth, F., and Weitkamp, J. (eds) (2008) *Handbook of Heterogeneous Catalysis*, 2nd edn, Wiley-VCH Verlag GmbH.
8 Newsome, D.S. (1980) The water-gas shift reaction. *Catal. Rev. Sci. Eng.*, **21**, 275–318.
9 Turco, M., Bagnasco, G., Costantino, U., Marmottini, F., Montanari, T., Ramis, G., and Busca, G. (2004) Production of hydrogen from oxidative steam reforming of methanol. II. Catalytic activity and reaction mechanism on Cu/ZnO/Al2 3 hydrotalcite-derived catalysts. *J. Catal.*, **228**, 56–65.
10 Turco, M., Bagnasco, G., Costantino, U., Marmottini, F., Montanari, T., Ramis, G., and Busca, G. (2004) Production of hydrogen from oxidative steam reforming of methanol. I. Preparation and characterization of Cu/ZnO/Al2O3 catalysts from a hydrotalcite-like LDH precursor. *J. Catal.*, **228**, 43–55.
11 Fiorot, S., Galletti, C., Specchia, S., Saracco, G., and Specchia, V. (2007) Development of water gas shift supported catalysts for fuel processor units. *Int. J. Chem. React. Eng.*, **5** (1), A(110), pp. 1–13. doi: 10.2202/1542-6580.1536
12 Chakrabarty, M.M. (2003) *Chemistry and Technology of Oils & Fats*, Allied Publishers.
13 Almqvist, E. (2003) *History of Industrial Gases Springer*, Kluwer Academic/Plenum Publishers, New York, https://books.google.com/books?id=OI0fTJhydh4C&pgis=1 (accessed 21 September 2016).
14 Guisnet, M. and Gilson, J.-P. (2002) *Zeolites for Cleaner Technologies*, World Scientific.

15 Weitkamp, J. and Hunger, M. (2007) Acid and base catalysis on zeolites. *Stud. Surf. Sci. Catal.*, **168**, 787–835, http://www.sciencedirect.com/science/article/pii/S016729910780810X (accessed 21 September 2016).
16 Wielers, A.F.H., Vaarkamp, M., and Post, M.F.M. (1991) Relation between properties and performance of zeolites in paraffin cracking. *J. Catal.*, **127**, 51–66.
17 Degnan, T.F., Chitnis, G.K., and Schipper, P.H. (2000) History of ZSM-5 fluid catalytic cracking additive development at Mobil. *Microporous Mesoporous Mater.*, **35–36**, 245–252.
18 Chang, C.D. (1983) Hydrocarbons from methanol. *Catal. Rev. Sci. Eng.*, **25**, 1–118.
19 The Essential Chemical Industry (2014) Cracking and Related Refinery Processes, http://www.essentialchemicalindustry.org/processes/cracking-isomerisation-and-reforming.html (accessed 21 September 2016).
20 Lapinski, M., Baird, L., and James, R. (2004) in *Handbook of Petroleum Refining Processes*, 3rd edn (ed. R.A. Meyers), McGraw-Hill, New York, pp. 4.01–4.31.
21 Stine, M. (2006) *Encyclopedia of Hydrocarbons*, Istituto della Encioclopedia Italiana Fondata da Giovanni Treccana S.p.A.
22 Turaga, U.T. and Ramanathan, R. (2003) Catalytic naphtha reforming: revisiting its importance in the modern refinery. *J. Sci. Ind. Res. (India)*, **62**, 963–978.
23 Pujado, P.R. and Moser, M. (2006) in *Handbook of Petroleum Processing* (eds D.S.J.S. Jones and P.R. Pujado), Springer, Dordrecht, p. 1349.
24 Stöcker, M. (1999) Methanol-to-hydrocarbons: catalytic materials and their behavior. *Microporous Mesoporous Mater.*, **29**, 3–48.
25 Olsbye, U., Svelle, S., Bjrgen, M., Beato, P., Janssens, T.V.W., Joensen, F., Bordiga, S., Lillerud, K.P., Bjørgen, M., Beato, P. et al. (2012) Conversion of methanol to hydrocarbons: how zeolite cavity and pore size controls product selectivity. *Angew. Chem. Int. Ed.*, **51**, 5810–5831, http://www.ncbi.nlm.nih.gov/pubmed/22511469 (accessed 18 September 2014).
26 Haw, J.F., Song, W., Marcus, D.M., and Nicholas, J.B. (2003) The mechanism of methanol to hydrocarbon catalysis. *Acc. Chem. Res.*, **36**, 317–326.
27 Chang, C.D. and Silvestri, A.J. (1977) The conversion of methanol and other O-compounds to hydrocarbons over zeolite catalysts. *J. Catal.*, **47**, 249–259.
28 Kvenvolden, K.A. and Cooper, C.K. (2003) Natural seepage of crude oil into the marine environment. *Geo-Marine Lett.*, **23**, 140–146.
29 Olah, G.A., Goeppert, A., and Prakash, G.K.S. (2011) *Beyond Oil and Gas: The Methanol Economy*, 2nd edn, John Wiley & Sons, Inc.
30 Svelle, S., Olsbye, U., Joensen, F., and Bjørgen, M. (2007) Conversion of methanol to alkenes over medium- and large-pore acidic zeolites: steric manipulation of the reaction intermediates governs the ethene/propene product selectivity. *J. Phys. Chem. C*, **111**, 17981–17984.
31 Seddon, D., (2010) *Petrochemical Economics*, Catalytic Science Series, vol. **8**, Imperial College Press.
32 Falbe, J. (1977) *Fischer–Tropsch-Synthese aus Kohle*, Thieme, Stuttgart.
33 Anderson, R.B. (1956) *Catalysis*, Von Nostrand-Reinhold, New York.

34 Maitlis, P.M. and de Klerk, A. (2013) *Greener Fischer–Tropsch Processes for Fuels and Feedstocks*, John Wiley & Sons, Inc..
35 ORYX GTL www.oryxgtl.com.qa (accessed 21 September 2016).
36 Perego, C., Bortolo, R., and Zennaro, R. (2009) Gas to liquids technologies for natural gas reserves valorization: the Eni experience. *Catal. Today*, **142**, 9–16.
37 Fischer, F. (1925) *The Conversion of Coal Into Oils*, Van Nostrand Company, New York.
38 Robinson, K.K. (2009) Reaction engineering of direct coal liquefaction. *Energies*, **2**, 976–1006.
39 Ancheyta, J. and Speight, J.G. (eds) (2007) *Hydroprocessing of Heavy Oils and Residuals*, CRC Press.
40 Van Dongen, R.H. and Groeneveld, K.J.W. (1987) *Heavy Residue Conversion by Shell Hycon Process*, Hydrocarbon Technology International.
41 Scherzer, J. and Gruia, A.J. (1996) *Hydrocracking Science and Technology*, CRC Press.
42 Kwant, P.B. and Van Zijll Langhout, W.C. (1985) The development of Shell's residue hydroconversion process. Proceedings of the Conference on the Complete Upgrading of Crude Oil, Siofolk, Hungary, 25–27 September 1985.
43 Scheffer, B., van Koten, M.A., Röbschläger, K.W., and de Boks, F.C. (1998) The shell residue hydroconversion process: development and achievements. *Catal. Today*, **43**, 217–224.
44 Meyers, R.A. (2004) *Handbook of Petroleum Refining Processes*, McGraw-Hill, New York.
45 Gray, M.R. (1996) *Upgrading Petroleum Residues and Heavy Oils*, Marcel Dekker, New York.
46 Sutthasupa, S., Shiotsuki, M., and Sanda, F. (2010) Recent advances in ring-opening metathesis polymerization, and application to synthesis of functional materials. *Polym. J.*, **42**, 905–915.
47 Chauvin, Y. (2006) Olefin metathesis: the early days (Nobel Lecture). *Angew. Chem. Int. Ed.*, **45**, 3740–3747.
48 Jean-Louis Hérisson, P. and Chauvin, Y. (1971) Catalyse de transformation des oléfines par les complexes du tungstène. II. Télomérisation des oléfines cycliques en présence d'oléfines acycliques. *Die Makromol. Chem.*, **141**, 161–176. doi: 10.1002/macp.1971.021410112
49 Grubbs, R.H. and Schrock, R.R. (2005) Development of the metathesis method in organic synthesis. Nobel Prize Background Description, pp. 1–12.
50 Keim, W. (2013) Oligomerization of ethylene to α-olefins: discovery and development of the shell higher olefin process (SHOP). *Angew. Chem. Int. Ed.*, **52**, 12492–12496.
51 Behr, A. and Neubert, P. (2012) *Applied Homogeneous Catalysis*, John Wiley & Sons, Inc..
52 Lutz, E.F. (1986) Shell higher olefins process. *J. Chem. Educ.*, **63**, 202–203.
53 Singh, O.M. (2006) Metathesis catalysts: historical perspective, recent developments and practical applications. *J. Sci. Ind. Res.*, **65**, 957–965.
54 Vogt, D. (2005) in *Multiphase Homogeneous Catalysis* (ed. B. Cornils), Wiley-VCH Verlag GmbH, Weinheim, p. 330.

55 Belgiorno, V., De Feo, G., Della Rocca, C., and Napoli, R.M.A. (2003) Energy from gasification of solid wastes. *Waste Manage.*, **23**, 1–15.
56 Moulijn, J.A., Makkee, M., and van Diepen, A.E. (2013) *Chemical Process Technology*, 2nd edn, John Wiley & Sons, Inc..
57 van Leeuwen, P.W.N.M. (2004) *Homogeneous Catalysis*, Kluwer Academic Publishers.
58 Vogt, D. (2006) in *Aqueous-Phase Organometallic Catalysis: Concepts and Applications* (eds B. Cornils and W.A. Hermann), John Wiley & Sons, Inc., p. 780.
59 Salamone, J.C. (1996) *Polymeric Materials Encyclopedia, Twelve Volume Set*, CRC Press.
60 Cerruti, L. (1999) Historical and philosophical remarks on Ziegler–Natta catalysts: a discourse on industral catalysis. *Hyle*, **5**, 3–41, http://www.hyle.org/journal/issues/5/cerruti.htm (accessed 21 September 2016).
61 Hagemeyer, A., Strasser, P., Anthony, F., and Volpe, J. (2006) *High-Throughput Screening in Chemical: Technologies, Strategies and Applications*, John Wiley & Sons, Inc..
62 Boor, J.J. (2012) *Ziegler–Natta Catalysts Polymerizations*, Elsevier.
63 Steinborn, D. (2011) *Fundamentals of Organometallic Catalysis*, John Wiley & Sons, Inc..
64 Hoff, R. and Mathers, R.T. (2010) *Handbook of Transition Metal Polymerization Catalysts*, John Wiley & Sons, Inc..
65 Ugbolue, S.C.O. (2009) *Polyolefin Fibres: Industrial and Medical Applications*, Elsevier.
66 Busico, V., Cipullo, R., Pellecchia, R., Ronca, S., Roviello, G., and Talarico, G. (2006) Design of stereoselective Ziegler–Natta propene polymerization catalysts. *Proc. Natl. Acad. Sci. U.S.A.*, **103**, 15321–15326, http://www.pnas.org/cgi/doi/10.1073/pnas.0602856103 (accessed 21 September 2016).
67 Swaddle, T.W. (1997) *Inorganic Chemistry: An Industrial and Environmental Perspective*, Academic Press.
68 Kissin, Y.V., Nowlin, T.E., and Mink, R.I. (1993) Ethylene oligomerization and chain growth mechanisms with Ziegler–Natta catalysts. *Macromolecules*, **26** (9), 2151–2158.
69 Soares, J.B.P. and McKenna, T.F.L. (2013) *Polyolefin Reaction Engineering*, John Wiley & Sons, Inc..
70 Topsoe, H., Clausen, B.S., and Massoth, F.E. (1996) *A Review of: "Hydrotreating Catalysis Science and Technology"*, Springer-Verlag, New York.
71 Kabe, T., Qian, W., and Ishihara, A. (2000) *Hydrodesulfurization and Hydrodenitrogenation: Chemistry and Engineering*, Wiley-VCH Verlag GmbH.
72 Brunet, S., Mey, D., Pérot, G., Bouchy, C., and Diehl, F. (2005) On the hydrodesulfurization of FCC gasoline: a review. *Appl. Catal., A*, **278**, 143–172.
73 Liu, K., Song, C., and Subramani, V. (2009) *Hydrogen and Syngas Production and Purification Technologies*, John Wiley & Sons, Inc..
74 Song, C. (2003) An overview of new approaches to deep desulfurization for ultra-clean gasoline, diesel fuel and jet fuel. *Catal. Today*, **86**, 211–263.

References

75 Delmon, B., Froment, G.F., and Grange, P. (1997) *Hydrotreatment and Hydrocracking of Oil Fractions: Studies in Surface Science and Catalysis*, Elsevier.
76 Klass, D.L. (1998) *Biomass for Renewable Energy, Fuels and Chemicals*, Academic Press.
77 Gullón, P., Conde, E., Moure, A., Domínguez, H., and Parajó, J.C. (2010) Selected process alternatives for biomass refining: a review. *Open Agric. J.*, **4**, 135–144.
78 Johansson, T.B. and Burnham, L. (1993) *Renewable Energy: Sources for Fuels and Electricity*, Island Press.
79 Rosillo-Calle, F. and Rosillo-Calle, F. (2012) *The Biomass Assessment Handbook*, Earthscan.
80 de Jong, W. and van Ommen, J.R. (2014) *Biomass as a Sustainable Energy Source for the Future: Fundamentals of Conversion Processes*, John Wiley & Sons, Inc..
81 Furimsky, E. (2000) Catalytic hydrodeoxygenation. *Appl. Catal., A*, **199**, 147–190.
82 Fogassy, G., Thegarid, N., Toussaint, G., van Veen, A.C., Schuurman, Y., and Mirodatos, C. (2010) Biomass derived feedstock co-processing with vacuum gas oil for second-generation fuel production in FCC units. *Appl. Catal., B*, **96**, 476–485.
83 Gunawardena, D.A. and Fernando, S.D. (2013) Methods and applications of deoxygenation for the conversion of biomass to petrochemical products, in *Biomass Now – Cultivation and Utilization*, InTech, http://www.intechopen.com/books/biomass-now-cultivation-and-utilization/methods-and-applications-of-deoxygenation-for-the-conversion-of-biomass-to-petrochemical-products (accessed 21 September 2016).
84 Pinheiro, A., Hudebine, D., Dupassieux, N., and Geantet, C. (2009) Impact of oxygenated compounds from lignocellulosic biomass pyrolysis oils on gas oil hydrotreatment. *Energy Fuels*, **23**, 1007–1014.
85 Vispute, T.P., Zhang, H., Sanna, A., Xiao, R., and Huber, G.W. (2010) Renewable chemical commodity feedstocks from integrated catalytic processing of pyrolysis oils. *Science*, **330** (6008), 1222–1227.
86 Dickerson, T. and Soria, J. (2013) Catalytic fast pyrolysis: a review. *Energies*, **6**, 514–538.
87 de Miguel Mercader, F., Groeneveld, M.J., Kersten, S.R.A., Way, N.W.J., Schaverien, C.J., and Hogendoorn, J.A. (2010) Production of advanced biofuels: co-processing of upgraded pyrolysis oil in standard refinery units. *Appl. Catal., B*, **96**, 57–66.
88 Chung, J.N. (2013) Grand challenges in bioenergy and biofuel research: engineering and technology development, environmental impact, and sustainability. *Front. Energy Res.*, **1**, 1–4.
89 de Klerk, A. (2012) *Fischer–Tropsch Refining*, John Wiley & Sons, Inc..
90 Demirbas, A. (2009) *Biohydrogen: For Future Engine Fuel Demands. Green Energy and Technology*, Springer Science & Business Media.

91 Hadden, R.A., Howe, J.C., and Waugh, K.C. (1991) in *Catalyst Deactivation. Studies in Surface Science and Catalysis* (eds C.H. Bartholomew and J.B. Butt), Elsevier, p. 177.
92 Anthony, D., Rand, J., and Dell, R. (2008) *Hydrogen Energy: Challenges and Prospects*, Royal Society of Chemistry.
93 Kolb, G. (2008) *Fuel Processing: For Fuel Cells*, John Wiley & Sons, Inc..
94 Kolanski, K.W. (2002) *Surface Science: Foundations of Catalysis and Nanoscience*, John Wiley & Sons, Inc..
95 Ross, J.R.H. (2011) *Heterogeneous Catalysis: Fundamentals and Applications*, Elsevier.
96 Greenwood, N.N. and Earnshaw, A. (1997) *Chemistry of the Elements*, 2nd edn, Butterworth-Heinemann.
97 Jennings, J.R. (1991) *Catalytic Ammonia Synthesis: Fundamentals and Practice*, Springer Science & Business Media.
98 Wiberg, E. and Wiberg, N. (2001) *Inorganic Chemistry*, Academic Press.
99 Bartholomew, C.H. and Farrauto, R.J. (2006) *Fundamentals of Industrial Catalytic Processes*, 2nd edn, John Wiley & Sons, Inc..
100 Srivastava, A.K. and Jain, P.C. (2008) *Chemistry* Vols. **1 and 2**, V. K. (India) Enterprises.
101 Abon, M. and Volta, J.-C. (1997) Vanadium phosphorus oxides for n-Butane oxidation to maleic anhydride. *Appl. Catal., A*, **157**, 173–193.
102 Jackson, S.D., Hargreaves, J.S.J (2009) *Metal Oxide Catalysis*, Vol **2**, Wiley-VCH.
103 Guliants, V.V. and Carreon, M.A. (2005) in *Catalysis* (ed. J.J. Spivey), Royal Society of Chemistry, p. 208.
104 Contractor, R.M. (1999) Dupont's CFB technology for maleic anhydride. *Chem. Eng. Sci.*, **54**, 5627–5632.
105 Lintz, H. and Reitzmann, A. (2007) Alternative reaction engineering concepts in partial oxidations on oxidic catalysts. *Catal. Rev.*, **49**, 1–32.
106 Cortés Corberán, V. and Vic Bellon, S. (1994) *New Developments in Selective Oxidation II*, Elsevier.
107 Sawyer, D.T. and Martell, A.E. (1992) *Industrial Environmental Chemistry: Waste Minimization in Industrial Processes and Remidiation of Hazardous Waste*, Springer Science & Business Media.
108 Grasselli, R.K. (2002) Fundamental principles of selective heterogeneous oxidation catalysis. *Top. Catal.*, **21**, 79–88.
109 Grasselli, R.K., Burrington, J.D., and Brazdil, J.F. (1981) Mechanistic features of selective oxidation and ammoxidation catalysis. *Faraday Discuss. Chem. Soc.*, **72**, 203.
110 Cespi, D., Passarini, F., Neri, E., Vassura, I., Ciacci, L., and Cavani, F. (2014) Life cycle assessment comparison of two ways for acrylonitrile production: the SOHIO process and an alternative route using propane. *J. Clean. Prod.*, **69**, 17–25.
111 Langvardt, P.W. (2011) Acrylonitrile, in *Ullmann's Encyclopedia of Industrial Chemistry*, John Wiley & Sons, Inc..

112 Serafin, J.G., Liu, A.C., and Seyedmonir, S.R. (1998) Surface science and the silver-catalyzed epoxidation of ethylene: an industrial perspective. *J. Mol. Catal. A: Chem.*, **131**, 157–168.

113 Voge, H.H. and Adams, C.R. (1967) Catalytic oxidation of olefins. *Adv. Catal.*, **17**, 151–221, http://linkinghub.elsevier.com/retrieve/pii/S0360056408606872 (accessed 21 September 2016).

114 Wurtz, A. (1859) Ueber das aethylenoxyd. *Ann. Chem. Pharm.*, **110**, 125–128. doi: 10.1002/jlac.18591100116

115 Rebsdat, S.R. and Mayer, D. (2002) *Ethylene Oxide*, Wiley-VCH Verlag GmbH & Co. KGaA, Weinheim.

116 Lefort, T.E. (1931) Procédé d'obtention de l'éthylèneglycol. FR 729 952 739, p. 562.

117 McKetta, J.J. Jr., (1984) *Encyclopedia of Chemical Processing and Design: Ethanol as Fuel: Options: Advantages, and Disadvantages to Exhaust Stacks: Cost. Chemical Processing and Design Encyclopedia*, vol. **20**, CRC Press.

118 Dever, J.P., George, K.F., Hoffman, W.C., and Soo, H. (1994) Ethylene oxide, in *Kirk-Othmer Encyclopedia of Chemical Technology*, John Wiley & Sons, Inc., New York.

119 Moulijn, J.A., van Leeuwen, P.W.N.M., and van Santen, R.A. (1993) *Catalysis: An Integrated Approach to Homogeneous, Heterogeneous and Industrial Catalysis*, Studies in Surface Science and Catalysis, Elsevier.

120 Van Santen, R.A., van Leeuwen, P.W.N.M., Moulijn, J.A., and Averill, B.A. (2000) *Catalysis: An Integrated Approach*, Studies in Surface Science and Catalysis Series, vol. **123**.

121 Buijink, J.K.F., Lange, J.-P., Bas, A.N.R., Horton, A.D., and Niele, F.G.M. (2011) in *Mechanism in Homogeneous and Heterogeneous Epoxidation Catalysis*, vol. **45** (ed. S. Ted Oyama), Elsevier, p. 528.

122 Jentoft, F.C. (2014) Preface. *Adv. Catal.*, **57**, ix–x, http://www.sciencedirect.com/science/article/pii/B9780128001271099884 (accessed 21 September 2016).

123 Thomas, J.M. and Thomas, W.J. (2005) *Principles and Practice of Heterogeneous Catalysis*, VCH Verlagsgesellschaft mbH, Weinheim.

124 Farrauto, R.J. and Heck, R.M. (1999) Catalytic converters: state of the art and perspectives. *Catal. Today*, **51**, 351–360.

125 NETT Technologies Inc http://www.nettinc.com/t-series/itemid-735 (accessed 21 September 2016).

126 Guillén-Hurtado, N., Rico-Pérez, V., García-García, A., Lozano-Castelló, D., and Bueno-López, A. (2012) Three-way catalysts: past, present and future. *DYNA*, **79**, 114–121.

127 Twigg, M.V. (2007) Progress and future challenges in controlling automotive exhaust gas emissions. *Appl. Catal., B*, **70**, 2–15.

128 Taylor, K.C. (1993) Nitric oxide catalysis in automotive exhaust systems. *Catal. Rev.*, **35**, 457–481.

129 Shelef, M. and McCabe, R. (2000) Twenty-five years after introduction of automotive catalysts: what next? *Catal. Today*, **62**, 35–50.

130 Gandhi, H. (2003) Automotive exhaust catalysis. *J. Catal.*, **216**, 433–442.

131 Nova, I. and Tronconi, E. (2014) *Urea-SCR Technology for deNOx After Treatment of Diesel Exhausts*, Springer, New York.

132 Ozinger, N., Deutschmann, O., Knozinger, H., Kochloefl, K., and Turek, T. (2009) *Heterogeneous Catalysis and Solid Catalysts*, Wiley-VCH Verlag GmbH & Co. KGaA.

133 Nova, I. and Tronconi, E. (2014) *Urea-SCR Technology for deNOX. After Treatment of Diesel Exhaust. Fundamental and Applied Catalysts*, Springer Science & Business Media.

134 Young, J.F., Osborn, J.A., Jardine, F.H., and Wilkinson, G. (1965) Hydride intermediates in homogeneous hydrogenation reactions of olefins and acetylenes using rhodium catalysts. *Chem. Commun.*, 131.

135 Kaminsky, W. (2001) Olefin polymerization catalyzed by metallocenes. *Adv. Catal.*, **46**, 89–159, http://linkinghub.elsevier.com/retrieve/pii/S0360056402460221 (accessed 21 September 2016).

3

Physical Chemistry, Elementary Kinetics

3.1 Introduction

A heterogeneous catalyst typically consists of millimeter-sized particles that are prepared in a specific shape that controls their strength. These particles are packed or circulated in a reactor bed with a dimension of centimeters in laboratory experiments, or with a dimension of a meter or more in industrial systems. Small particles of metal or other catalytically active materials are distributed on high surface area porous supports, so that the maximum amount of material is available for reaction.

The different characteristic dimensions of the catalytic reactor bed are illustrated in Figure 3.1.

Hydrodynamics determines the flow properties in the reactor that has a process time that can vary from minutes to hours. Within the catalyst particles, the flow is subject to diffusional resistance, which depends on the length and distribution of the particle pore sizes. The dimensions of these micropores can vary in size from a micron to several nanometers. The catalyst particle can also be an aggregate of crystallites, such as a zeolitic system with smaller micropores on the order of a nanometer. These crystallites are often embedded in an amorphous material to give the large catalyst particles their strength.

The catalytically reactive nanoparticles distributed on the surface of the porous catalyst particle can vary significantly in size as well as location and this distribution can have a large impact on the overall catalyst reactivity. The reactivity of the active particles depends on their size and shape, but due to the effects of diffusion, their contribution to overall catalysis will also depend on their position in the pore structure.

Ultimately, the chemical reactivity of the catalytic particle is determined by the atomistic structure of the catalytically reactive centers on that particle.

The early discoveries in heterogeneous catalysis during the previous century were accomplished without a scientific framework of the understanding of surface chemistry and reactivity at a molecular level. The content of this book, which is written a century later, describes the current deep understanding of the molecular basis of heterogeneous catalyst reactivity. This is described in detail in Part II. This chapter serves as an introduction to the basic physical chemical and kinetic

Modern Heterogeneous Catalysis: An Introduction, First Edition. Rutger A. van Santen.
© 2017 Wiley-VCH Verlag GmbH & Co. KGaA. Published 2017 by Wiley-VCH Verlag GmbH & Co. KGaA.

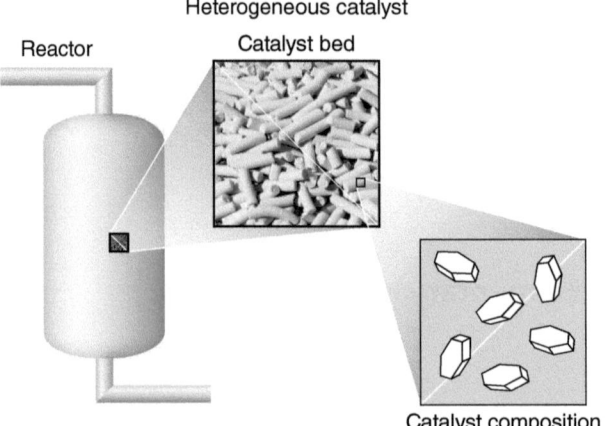

Figure 3.1 The relative dimensions of catalytic process components; reactor(meters), catalyst particle(millimeters), catalytically active components(microns or less).

concepts that will provide a context for the later chapters on mechanism and catalytic modeling.

Operation of a catalytic reactor depends not only on the chemical process of catalysis, but also on the processes of mass and heat transfer. These processes are determined by physical properties such as catalyst particle size, pore volume and distribution, and internal surface area as well as the stoichiometry of the catalyst's composition. Since it has been possible to measure these properties since the beginning of catalytic science in the previous century, they were the earliest parameters to be related to catalyst performance.

The scientific basis of mass or heat transfer limitations of the catalyst particles and the consequences of these limitations for reactor design was formulated in the middle of the last century, and this has evolved into the important discipline of chemical reaction engineering [1, 2]. In parallel, kinetics also developed into an important branch of physical chemistry. The measurement of the rate of reaction as a function of partial pressures of the reactant and product became an important tool for deducing information on rate-controlling steps, and for correlating catalyst performance with catalyst composition and physical structure [3].

A reaction mechanism can never be firmly established based solely on overall kinetics measurements because too many molecular reactivity parameters are unknown. This process has been enhanced by the recent availability of molecular reactivity information from computation and surface science data. The phenomenological development of catalyst kinetics became an important tool for analyzing catalyst performance, for estimating reactor requirements, and for optimizing process design. Complemented by chemical thermodynamics, the determination of reaction mechanisms became essential to reactor modeling and also provided important inputs for the improvement of catalytically active materials.

In order to predict which material is optimum for a catalytic reaction, we need to know how catalyst performance relates to catalyst composition and structure. Therefore, the characterization of the structure of a catalyst at various scales relevant to catalytic operations is a core activity of heterogeneous catalysis science.

In the early twentieth century, quantitative product determination was difficult due to the complexity of product analysis. Now, with new tools for physical analysis and with modern test facilities, this can be done almost automatically on a laboratory scale.

Determining catalyst structure and composition on a molecular level still presents a significant challenge because the complexity of the catalytic system makes complete molecular characterization difficult. Synthesis control on a molecularly defined level presents a similar challenge. Therefore, empirical exploration remains essential to catalyst research. There is ample scope for further refinement of the techniques for catalyst characterization and synthesis.

The ultimate difficulty in the search for an optimum catalyst is that reactivity is controlled on a molecular level by the topology of the atoms that constitute a catalyst's surface site. Catalyst preparation affords full molecular control and complete atomistic characterization only in the rare case when model molecular catalytic systems are used. Figure 3.2 shows an example of such a catalyst with an Ir_4 cluster prepared by immobilizing the corresponding carbonyl complex on the internal aluminosilicate surface of a zeolite.

Conventional heterogeneous catalysts are inhomogeneous, and the molecular character of the optimum reaction center is often not known. An additional complication is that the catalyst surface often reorganizes during reaction. As we will see in Chapter 4, this reorganization may initially activate the catalyst, but will ultimately lead to its deactivation. Advanced spectroscopic techniques that can be applied *ex situ* as well as *in situ* (*in operando*) are available to unravel the complex chemical constitution of the working catalyst.

Due to the inhomogeneous nature of the catalysts, model catalyst studies based on the techniques of surface science [5, 6] have been very useful, because model catalytic systems are more homogeneous and allow for detailed atomistic characterization of the catalytically active system. Examples of the single crystal surfaces of transition metals used in such experiments are shown in Figure 3.3.

Computational quantum-chemical approaches such as density functional theory (DFT) [8] have reached a high level of maturity and are now the basis of computational catalysis. This approach is very useful for simulating the reactivity of realistic models of catalytically reactive centers at a molecular level [8–10]. These types of calculations are used to study the relative stability of reaction intermediates and are indispensable for microkinetic simulations. The resulting detailed scientific understanding of the relationship of molecular

Figure 3.2 Ir_4 cluster supported on the zeolite fragment $Al_3Si_3O_6H_{12}{}^{3-}$. (Ferrari *et al.* 2008 [4]. Reproduced with permission of American Chemical Society.)

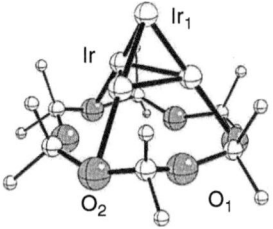

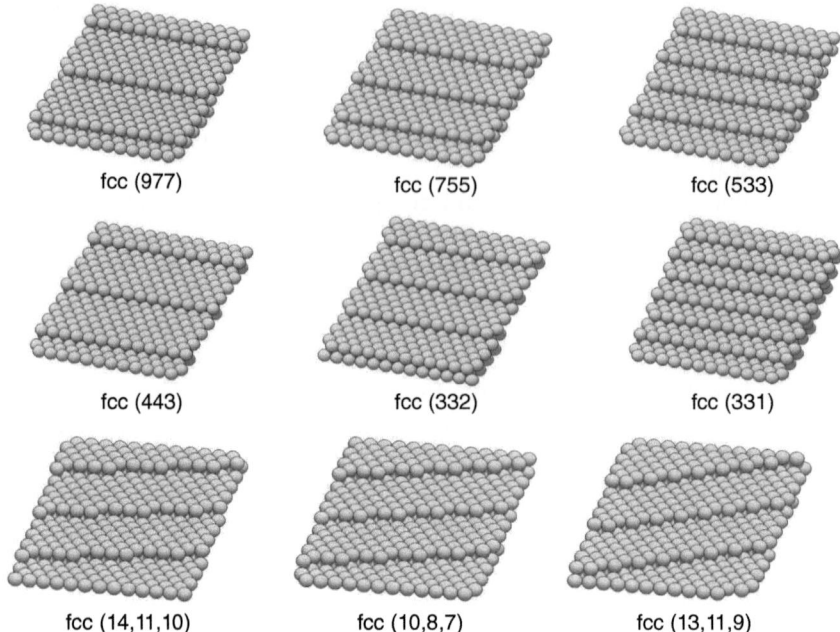

Figure 3.3 Schematic representation of the surface structures of several stepped (first two rows) and kinked (bottom row) crystal faces deduced from the bulk unit cell of a face-centered cubic crystal. The numbers within the brackets denote crystallographic directions. The non-stepped surface is the (111) surface, with each surface atom having six neighbor atoms in the plane and three below the plane. Contraction of interlayer spacing and other modes of restructuring that are commonly observed are not shown. (Somorjai and Li 2010 [7]. Reproduced with permission of John Wiley & Sons.)

surface chemical reactivity to catalyst surface composition and structure has a great influence on our predictive abilities.

Computational chemistry and physics provide important tools to analyze the atomistic structure of catalytically reactive sites by comparing calculated spectra to measured spectra [10, 11].

Section 3.2 briefly describes some of the most important techniques for characterizing heterogeneous catalysts. In Section 3.3, we will present the fundamental kinetics equations and elementary rate equations used in heterogeneous catalytic kinetics, and explain the design of the Sabatier volcano curve. Section 3.4 will introduce transient kinetics methods that are useful to unravel the elementary rate constants of the overall catalytic reaction cycle, and also to determine the fraction of reactive surface centers. The final section considers diffusion and its interplay with kinetics.

3.2 Catalyst Characterization

A heterogeneous catalyst typically consists of a high surface area inert support on which small catalytically active particles have been distributed. The smaller the catalytically reactive particles, the more efficient the process, because small size maximizes the number of particle atoms that are exposed. The high surface area of the inert support allows for a high density of the small catalytically active particles in a finite volume. The support is also needed to stabilize the particles against particle growth, so that they remain small. Most often the "inert" high surface area support is an oxide that has some innate reactivity. If this reactivity is not beneficial, promotors or moderating compounds are used to suppress it. A challenge in catalyst synthesis lies in the ability to prepare catalysts with uniform particle size and well-defined pore structure. Since this is often not achieved, a significant goal of catalyst research is to understand this inhomogeneity and the actual composition of the heterogeneous catalyst.

Catalyst performance characterization involves determining the kinetics properties of the catalytic reaction as well as the physical characterization of the catalyst.

In the laboratory, catalyst performance is measured in microreactors as shown in Figure 3.4. (For an introduction to the chemical reactor engineering of such reactors refer to Levenspiel [1].) These catalyst test reactors require only a few

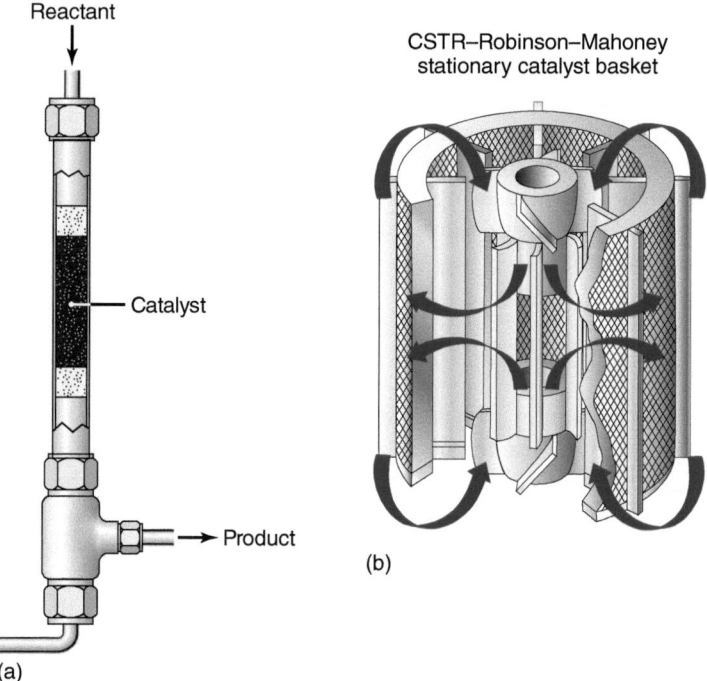

Figure 3.4 (a) Plug flow micro reactor. (b) Continuous stirred tank reactor (CSTR) [12].

grams of catalyst, and can be operated isothermally, without internal pressure or flow gradients when the catalyst particle dimensions, flow conditions, and reactor size are chosen properly [13]. Most of the kinetics data discussed in this book have been obtained from such measurements.

The four essential questions for catalyst characterization are as follows:

1. What is the surface area and porosity of the catalyst?
2. What is the composition of the catalytically reactive particles, and what is their size and shape distribution?
3. What fraction of the exposed surface contains the catalytically reactive centers?
4. What is the state of the catalyst at reaction conditions?

While the first two questions can be answered by studying the catalyst before and after the reaction, the third and fourth require a characterization study of the catalyst while it is in operation.

Volumetric adsorption methods are fundamental to heterogeneous catalyst characterization. They are based on the chemisorption of molecules on a reactive surface or on the physical adsorption of molecules on the entire catalyst surface. Chemisorbing molecules form chemical bonds with a surface which is material-specific. Physical adsorption is due to weak dispersion or van der Waals interaction between adsorbate and surface, and is not material-specific. This property can also be used to determine the pore volume of the catalyst.

3.2.1 Langmuir Adsorption Isotherm

A classical expression used in chemisorption is the Langmuir's adsorption isotherm (Eq. (3.1)), which computes the surface coverage as a function of gas pressure and size of adsorbing molecule. This expression is also important in kinetics modeling studies. By measuring the change in gas pressure, the amount of gas adsorbed can be deduced, and the surface area and equilibrium adsorption constant can be determined:

$$\theta = \frac{K_{eq}\frac{P}{P_0}}{1 + K_{eq}\frac{P}{P_0}} \tag{3.1}$$

In Eq. (3.1), θ is the fraction of surface sites occupied by the adsorbing gas, P is gas pressure, P_0 is the pressure of gas at standard temperature and pressure conditions, and K_{eq} is the equilibrium constant of adsorption defined as the ratio of the rate constant of adsorption k_{ads}, and the rate constant of desorption k_{des}. θ is determined by using the surface area of the adsorbing gas as a unit measure of material.

Langmuir's view of a surface is analogous to a checkerboard surface like the terraced surfaces in Figure 3.3 without the step-edges. Equation (3.1) is valid as long as one molecule adsorbs per site, the number of sites is finite, the sites are identical, and the adsorbing molecules do not interact. The characteristic of Langmuir adsorption is its initial linear dependencies of surface coverage on gas pressure (Henry's law) and on the saturation of surface coverage at higher pressure,

Figure 3.5 Langmuir adsorption isotherm; Coverage θ as a function of pressure. θ saturates at high pressure.

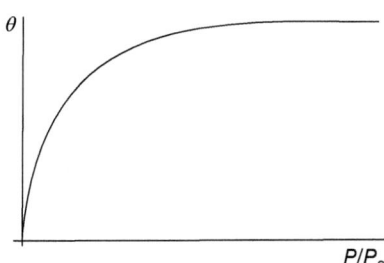

which is due to the finite number of adsorption sites (see Figure 3.5). The Langmuir adsorption isotherm for adsorption of a non-dissociating molecule is deduced from equilibration of rates of adsorption $\overrightarrow{v_{ads}}$ and desorption $\overleftarrow{v_{des}}$ respectively:

$$\overrightarrow{v_{ads}} = \overleftarrow{v_{des}} \Rightarrow k_{ads}\frac{P}{P_0}(1-\theta)N_s = k_{des}\theta N_s$$

$$\Rightarrow \frac{\theta}{1-\theta} = \frac{k_{ads}\frac{P}{P_0}}{k_{des}} = K_{eq}\frac{P}{P_0}$$

N_s is the total number of sites, θ is surface coverage, k is the rate of adsorption and desorption, P is pressure, and P_0 is the pressure at STP (Standard Temperature and Pressure).

The volumetric adsorption of reactive gasses that dissociate, such as CO or H_2, can be used to measure the surface area of catalytically reactive particles. In later chapters we will discuss the questionable nature of the assumptions of identical sites and the absence of lateral interactions. Nonetheless, the expression in Eq. (3.1) is very useful and is widely used to deduce free energies of adsorption.

3.2.2 Measurement of Pore Volume

At the time of its discovery, the BET (Brunauer, Emmet, Teller) [14] expression based on physical adsorption represented a breakthrough for determining the total surface area and pore volume of a catalyst. The BET equation uses the change in pressure at a temperature close to gas condensation.

As long as the pressure of the gas is below the condensation pressure, Langmuir-type adsorption will occur over the entire surface, but when pressure increases, condensation will occur instead of adsorption saturation.

The BET adsorption isotherm is given by Eq. (3.2).

$$\frac{P}{V_{ads}(P_s - P)} = \frac{1}{CV_m} + \frac{C-1}{CV_m}\frac{P}{P_s} \qquad (3.2)$$

V_{ads} is the volume of gas adsorbed and V_m is the volume per molecule. P_s is the saturation pressure at the condensation temperature of gas. From the plot of volume of gas adsorbed V_{ads} versus P/P_s from Eq. (3.2), the monolayer volume of gas V_m that can be adsorbed can be deduced and, since the surface area of the molecule is known, will give the total surface area of the entire catalyst particle. C is defined in Insert 1, that gives the derivation of the BET equation in Insert 1.

Insert 1: Multilayer Adsorption; the BET Equation

As in the derivation of the Langmuir adsorption, we will equilibrate the rates of adsorption and desorption, but since we have to consider condensation, subsequent layers on top of each other have to be considered (after [3]):

$$\text{Layer 1} \rightarrow r_0^{ads}\theta_0 = r_0^{des}\theta_1; r_0^{ads} = k_0^{ads}\frac{P}{P_s}$$

$$\text{Layer 2} \rightarrow r_1^{ads}\theta_1 = r_1^{des}\theta_2; r_1^{ads} = k_1^{ads}\frac{P}{P_s}$$

In the second and following layers, molecules only absorb on previously absorbed molecules. Now, assume that all $r_i(i \neq 0) = r_1$. At condensation pressure $P = P_0$:

$$r_1^{des} = r_{evap} = r_{cond} = k_1^{ads} \text{ and } r_1^{ads} = r_{cond}\frac{P}{P_s}$$

Hence

$$\frac{\theta_2}{\theta_1} = \frac{r_1^{ads}}{r_1^{des}} = \frac{r_{cond}\frac{P}{P_s}}{r_{1,evap}}$$

and

$$\frac{\theta_2}{\theta_1} = \frac{P}{P_s} = x \Rightarrow \theta_i = x^{i-1}\theta_1$$

$$\frac{\theta_1}{\theta_0} = \frac{k_0^{ads}\frac{P}{P_s}}{r_0^{des}} \Rightarrow \theta_i = Cx^i\theta_0 \left(C = \frac{k_0^{ads}}{r_0^{des}} = K_{ads}^0\right)$$

The total number of molecules adsorbed per unit surface area n_t is $n_t = \sum_{i=1}^{\infty} i\theta_i = C\theta_0 \sum_{i=1}^{\infty} ix^i$.

Conservation of number of sites gives: $1 = \theta_0 + \sum_{i=1}^{\infty}\theta_i = \theta_0\left(1 + C\sum_{i=1}^{\infty}x^i\right)$ with $\sum_{i=1}^{\infty}x^i \approx \frac{x}{1-x}$ $(x < 1)$ and $\sum_{i=1}^{\infty}ix^i = \frac{x}{1-x^2}$ gives $n_t = \frac{Cx}{(1-x)(1-x+Cx)}$ that can be written as Eq. (3.2).

3.2.3 Porosity

In Figure 3.6, a characteristic adsorption isotherm is shown for N_2 adsorption on a high surface area η-alumina.

In addition to the increasing multilayer adsorption of N_2, Figure 3.6 illustrates that at even higher temperatures hysteresis occurs in the adsorption behavior so that the amount of liquid adsorbed varies when pressure increases or decreases. This hysteresis relates to the capillary forces that arise when liquid condenses into the mesopores ($2\,\text{nm} < d < 50\,\text{nm}$) of the catalyst particles. Hysteresis

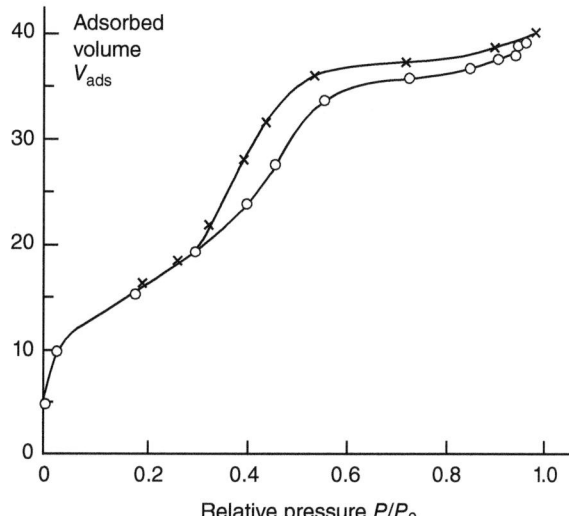

Figure 3.6 Adsorption of N_2 at 77.3 K on η-alumina. O = increasing pressures and × = decreasing pressures. (After [3].)

occurs due to the non-uniformity of the pores. In this case, we refer to the Kelvin equation:

$$P_K = P \exp\left(\frac{-2\gamma V}{r_b RT}\right) \quad (3.3)$$

P_K is the equilibrium pressure of the liquid in the pore, P is the equilibrium pressure of the free liquid, γ is the surface tension of the fluid, r_b is the capillary radius, R is the gas constant, and T and V are the temperature and the volume of the liquid.

A pore with a small neck radius will fill at relatively low pressure, but it will only completely fill at the higher pressure consistent with its larger body radius. For desorption to occur, the pressure must initially be low enough to cause evaporation of the liquid at the bottleneck.

The hysteresis loops of physical adsorption isotherms can be used to deduce the dimensions and shapes of the capillaries in the catalyst particles. These are called *T-plots*, and their values can also be obtained by measuring the physical adsorption of non-reactive gas molecules such as N_2 or noble gas molecules [11].

3.2.4 Temperature-Programmed Reactivity Measurements

Temperature-programmed reduction (TPR) or temperature-programmed oxidation (TPO) experiments can be used to determine the weight loss or increase of the total amount of reactive material. Reduction uses H_2 uptake as a function of temperature, while oxidation uses oxygen consumption as a function of temperature. Instead of measuring chemisorption, which is a surface measurement, TPR experiments measure the bulk reactivity properties of a particular reactive material. These measurements are element-specific because the temperatures of oxidation or reduction will depend on the material studied. Additionally, temperatures can be varied with particle size and can thus be used as an indicator of reactive particle sizes. As we will discuss in Chapter 11, smaller metal-oxide particles tend to reduce at higher temperatures.

3.2.5 Spectroscopic Techniques

The evolution of catalyst characterization has involved the use of spectroscopy for collecting data of increasing detail at the molecular level. Some important spectroscopic techniques are listed:

- *Electron microscopy* can determine the particle size distribution of the catalytically reactive particles on the catalyst at the nano- and micron levels. High resolution studies reveal atomistic-level details of the catalyst.
- *X-Ray photo emission spectroscopy* (*XPS*) enables identification of low concentration elements on the surface of catalyst materials based on the energy of emitted electrons.
- *Extended X-ray absorption fine structure* (*EXAFS*) *spectroscopy* is used at synchrotron facilities to measure the local coordination of atoms that are part of the partially disordered materials of a reactive catalyst.
- *Infrared and Raman spectroscopies* are vibrational spectroscopies useful for probing the interaction of adsorbed molecules with catalyst surfaces. Changes in bond frequencies provide information on the activation of molecules by reactive substrates.

For a detailed discussion of these techniques, see [11, 15].

Methods for bulk solid-state characterization such as X-ray diffraction and solid-state nuclear magnetic resonance (NMR) are also widely used. X-ray diffraction can not only determine the structure of compounds that constitute the catalyst, but can also be used to measure particle size. In addition, to assist structure determination, solid-state NMR spectroscopy can also be used to probe the state and mobility of adsorbed molecules in microporous systems such as the zeolites [16].

While most physical characterization is conducted on catalyst particles before or after a reaction, there is an increasing interest in the development of methods to study the catalyst surface during the catalytic reaction. This is achieved by combining spectroscopies that can be applied in an ambient environment with catalyst reaction measurements [17].

3.3 Elementary Kinetics

3.3.1 Lumped Kinetics Expressions: Kinetic Determination of Key Reaction Intermediates

Knowledge of the kinetics of a reaction enables to calculate the rate of reaction as a function of reaction concentration and temperature. The kinetics of a reaction is based on a proposed mechanistic model. For a catalytic reaction, this will be a cyclic sequence of sometimes competing elementary reaction steps.

3.3.1.1 The Rate Constant of an Elementary Reaction

The elementary reaction is defined as a single chemical bond-breaking or formation event that connects the local equilibrium state of the initial reacting intermediate(s) with the next new reaction intermediate that is again in a local equilibrium state. The probability of progressing from one equilibrium state to the next for this particular transformation is the rate constant of the elementary reaction. It is given by Eyrings' transition state reaction rate expression [18]:

$$r_T = \frac{kT}{h} \exp\left(-\frac{\Delta G^{\#}}{kT}\right) \tag{3.4}$$

r_T is the rate constant, k is the Boltzmann constant, and h is the Planck constant. $\Delta G^{\#}$ is the activation-free energy of the reaction. This is an Arrhenius-type expression that depends exponentially on an energy divided by the temperature T. kT/h is a frequency of the order of 10^{13} per second. The reaction rate expression in Eq. (3.4) relates the reaction rate of the elementary reaction step to the free energy difference between the transition state and the ground state. The transition state is defined as the minimum free energy saddle point on the potential energy surface that connects the two reaction intermediate equilibrium states, which are the initial and final states of the elementary reaction (see Figure 3.7).

The Eyring transition state rate expression of an elementary reaction is based on the essential assumption that reaction probability is slow compared to the rates of energy exchange between the reacting system and its environment. This is valid as long as (and because) the activation energies of a bond formation or bond cleavage are usually higher than 5 kT. The transition state can be considered in equilibrium with the equilibrium ground states. If this condition is not satisfied, as can be the case for surface diffusion, molecular dynamics approaches based on the solution of the Newton's equations must be used.

The second assumption is that once the system is in the transition state it moves to the other equilibrium state without probability of reflection. It implies that there is no friction due to high viscosity of a reaction medium.

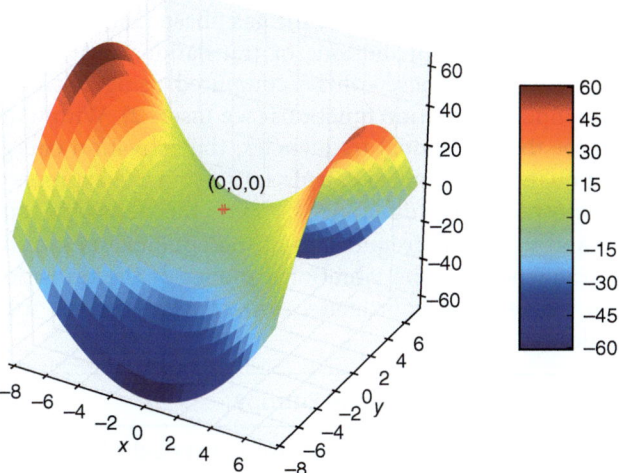

Figure 3.7 Potential energy saddle point diagram illustrating the relationship between transition state, reactant, and product state. The local energy minima (x = 0, y = −8 and x = 0, y = 6) indicate reactant and product state respectively. The potential energy saddle point (x = 0, y = 0) defines the transition state.

Transition state theory requires the definition of a reaction coordinate q of the reacting system. This is one of the degrees of freedom of the system along which the bond cleavage or formation reaction takes place, and it is defined uniquely in the transition state. At the transition state, the second derivative of the energy with respect to the reaction coordinate is negative, whereas in the local energy minima it is always positive. In the transition state, all the other free energies of the reacting system are at their local energy minimum, except for the translational degrees or rotational degrees of freedom of a freely moving molecular system.

Eyring's reaction rate expression can be rewritten as Eq. (3.5).

$$r_T = v_T \exp\left(-\frac{E_{act}}{kT}\right) \tag{3.5}$$

With:

$$v_T = e\frac{kT}{h}\exp\left(\frac{\Delta S^{\#}}{k}\right) \tag{3.6a}$$

$$E_{act} = \Delta H^{\#} + kT \tag{3.6b}$$

$$\Delta S^{\#} = \Delta S^{\#}_{transl} + \Delta S^{\#}_{rot} + \Delta S^{\#}_{vibr} \tag{3.6c}$$

$$E_{act} = E^{\#}_{electr} + \frac{1}{2}h\left(\sum_{i=0}^{N} v_i^0 - \sum_{i=1}^{N} v_i\right) + kT \tag{3.6d}$$

The activation energy of the reaction E_{act} (Eq. (3.6b)) is deduced from the expression in Eq. (3.4) by using its definition, which is the derivative of its logarithm with respect to $1/kT$ (for a more detailed discussion see [19], p. 138 and following). The pre-exponent v_T depends on the transition state entropy $\Delta S^{\#}$, and the activation energy E_{act} relates to the activation enthalpy $\Delta H^{\#}$, all of which are assumed to be independent of temperature. For a molecule that desorbs from the surface from an immobile adsorbed state, a typical value of v_T is 10^{17}; for an adsorbing molecule it is 10^9. This reflects the difference in mobility of the adsorbed molecule with respect to the gas phase. The activation entropy is a sum of the respective contributions of translational, rotational, or vibrational entropies (Eq. (3.6c)). These can be computed using statistical mechanics from the corresponding partition functions (see Insert 2). Within the harmonic approximation for the vibrational frequencies, the activation energy E_{act} can be deduced from quantum-chemical calculations using the expression Eq. (3.6d). $E^{\#}_{electr}$ is the electronic energy difference between the transition state and the ground state. v_i^0 and v_i are the frequencies of the equilibrium ground state vibrations and the frequencies of the vibrational modes of the transition state complex other than the reaction coordinate, respectively. N is the total number of vibrational nodes.

Insert 2: Statistical Mechanical Relations for Energy and Entropy

Once the partition function Q is known, thermodynamic properties can be calculated. The partition function Q is given by (after [20]):

$$Q = \sum_i \exp\left(-\frac{E_i}{kT}\right)$$

The energies E_i are the eigen energies of the system. From Q, the energy and entropy can be calculated from these relations:

$$E = kT^2 \frac{\partial}{\partial T} \ln Q$$

$$S = \frac{\partial}{\partial T}(kT \ln Q)$$

The partition function is the product of the respective partition functions of the degrees of freedom of the system.

$$Q = Q_{trans} \times Q_{rot} \times Q_{vib}$$

The expression of Q_{trans} for one degree of freedom is:

$$Q_{trans} = l\frac{(2\pi m kT)^{1/2}}{h}$$

m is the mass of the particle and l is the length of translation.

The expression for Q_{rot} of a linear molecule is:

$$Q_{rot} = \frac{8\pi^2 I kT}{h^2}$$

I is the momentum of inertia.

The expression for the contribution of one vibrational degree of freedom within the harmonic approximation is:

$$Q_{vib} = \frac{e^{-\frac{1}{2}\frac{h\nu}{kT}}}{1 - e^{-\frac{1}{2}\frac{h\nu}{kT}}}$$

When activation entropies are calculated instead, the following expression is used:

$$Q'_{vib} = \frac{1}{1 - \exp\left(-\frac{h\nu}{kT}\right)}$$

The zero-point vibrational energy correction $\frac{1}{2}h\nu$ has been incorporated into the activation energy expression calculated from the electronic energy.

The electronically calculated saddle point energy is to be corrected with the zero-point frequencies of all ground state degrees of freedom, except for the transition state vibrational frequencies of the non-reaction coordinate degrees of freedom. There is also a small correction to the activation energy barrier due to kT. Equations (3.5) is fundamental to the catalytic kinetics equations discussed in the next subsection.

3.3.1.2 Elementary Catalytic Reaction Kinetics

Because the catalytic reaction is a cycle that consists of a combination of elementary reaction steps, the energy reaction path and apparent activation energy have a complex relation with those same attributes of the elementary reactions that constitute the catalytic reaction cycle. This is the topic of this section.

The reaction step sequence that comprises a catalytic reaction cycle is called the *mechanism of the reaction*. Since this sequence of elementary reactions is often unknown, one must propose a plausible mechanism.

A comparison of predicted kinetics with the measured kinetics of the reaction can be used to evaluate potential candidates. However, alternative mechanistic proposals can result in very similar kinetics, when elementary rate constant parameters are not known and are adjusted to correspond with the experiment. Without additional information, kinetic parameters obtained in this way are very helpful for catalytic reactor design, but are not sufficiently accurate to establish a conclusive relation between kinetic performance and catalyst chemical reactivity indicators.

This additional information can be deduced from *in situ* spectroscopic information about the surface composition of the working catalyst or from model chemical experiments, for instance those using isotope-labeled molecules (as discussed in Section 3.4). Quantum-chemical calculations on model systems can provide information on the relative stability of proposed reaction intermediates and activation energies of elementary reaction steps, and this aids considerably in the validation of a proposed reaction mechanism.

To calculate kinetics one must solve the set of kinetic equations for the rate of change of the reaction intermediates of the reaction. These are not only gas phase intermediate molecules that can readsorb, but are dominantly reaction intermediates adsorbed on the catalyst surface and only indirectly related to the reaction product. The kinetics equations are coupled ordinary differential equations (ODEs) that calculate product formation from the surface concentrations of these reaction intermediates. The equations are usually solved making the steady state assumption, so that the stationary rate of conversion to product is calculated.

In the elementary surface kinetics that we consider in this section, the Langmuir checkerboard surface model is used to build a kinetic model of a reaction. As in the Langmuir adsorption model, one type of surface site is considered and the adsorbing intermediates have no lateral interactions. The concentration of intermediates can be described by an average coverage θ_i. This corresponds to the mean field equation that assumes that the rate of surface diffusion is fast compared to the rate of the individual elementary reaction steps, so surface concentrations can be considered to be uniformly distributed.

3.3.1.2.1 Site Number Conservation Law

The law of site number conservation states that the sum of the surface vacancy concentration and the surface coverages of reaction intermediates is equal to a constant.

$$\sum_i \theta_i + \theta_v = 1 \tag{3.7}$$

θ_i is the surface concentration of intermediate i and θ_v is the surface vacancy concentration. It implies, as we will see, a fundamental difference between catalytic kinetics and non-catalytic kinetics. The latter assumes mass-action law kinetics, according to which the rate of a reaction is proportional to the product of the concentrations of reacting molecules.

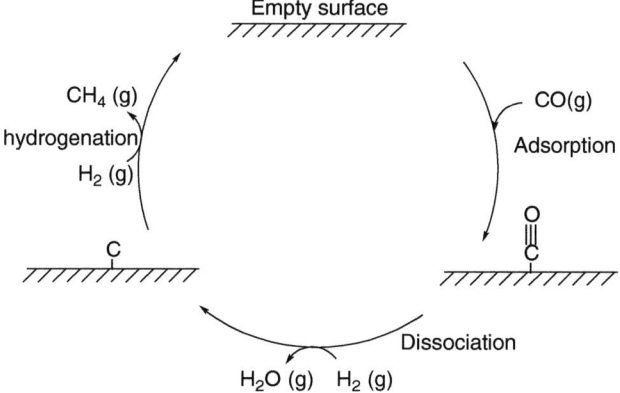

Figure 3.8 Lumped kinetics reaction cycle of methane formation from synthesis gas (schematic).

Lumped kinetic expressions are often used in elementary kinetics. Every individual reaction step is not explicitly considered, but instead several steps may be contracted into one rate expression. This simplifies the equations to be solved as well as the interpretation of their results.

As an initial example, we will provide the solutions for the elementary lumped kinetics model of the formation of methane from synthesis gas. The mechanism of this reaction is shown by the reaction cycle in Figure 3.8.

This lumped kinetics reaction cycle does not explicitly consider the dissociative adsorption of H_2. Reaction is initiated by adsorption of CO and H_2 to empty surface sites, and the cycle closes with regeneration of empty surface sites by product desorption. θ_H is ignored because this lumped kinetics reaction cycle does not explicitly consider dissociative adsorption of H_2.

The corresponding lumped kinetics equations are as follows:

$$R_{CH_4} = \frac{1}{N_s}\frac{d[P]}{dt} = k_H \theta_C \tag{3.8}$$

Equation (3.8) gives the rate of product formation R from adsorbed C atoms. θ_C is the surface coverage of C. The rate constant k_H is the lumped reaction rate constant of C atom hydrogenation to methane. It implicitly depends on H_2 partial pressure.

$$\frac{d\theta_C}{dt} = k_{diss}\theta_{CO}(1-\theta) - k_H \theta_C \tag{3.9a}$$

$$\frac{d\theta_{CO}}{dt} = k_{ads}^{CO}\frac{P^{CO}}{P_0^{CO}}(1-\theta) - k_{des}^{CO}\theta_{CO} - k_{diss}\theta_{CO}(1-\theta) \tag{3.9b}$$

$$\theta = \theta_C + \theta_{CO} \tag{3.9c}$$

P^{CO} is the partial pressure of CO, P_0^{CO} is the partial pressure of CO at STP, and θ is the total surface coverage.

Equations (3.9a)–(3.9c) assume that the surface coverage by O atoms can also be ignored, which implies that the rate constant of their removal is fast compared to that of the other reactions. This assumption is useful for our illustrated example. It simplifies the ODEs since now only two surface species, CO and C, must be considered. The rate of C-atom formation is proportional to the reaction rate constant of CO dissociation k_{diss} and the product of θ_{CO} and surface vacancy concentration, since two atoms are generated by dissociation of CO and hence a vacant surface position next to the adsorbed position is required.

Conventionally, one proceeds to identify the rate-controlling step of the reaction. We will compare the cases where the rate constant of dissociation of adsorbed CO, k_{diss}, is large and the rate of C atom hydrogenation is fast, with the opposite case where k_{diss} is rate-controlling. By comparing the predicted and experimentally demonstrated partial pressure dependence, it can be established which of the two cases applies. This also indicates which reaction intermediate dominates the surface coverage.

Equations (3.9) can be solved within the steady state approximation ($d\theta_i/dt = 0$):

$$R_{CH_4} = \frac{k_H(1 - \theta_{CO})}{1 + \frac{k_H}{k_{diss}\theta_{CO}}} \tag{3.10a}$$

$$\frac{\theta_{CO}}{1 - \theta_{CO}} = K_{ads}^{CO} \frac{p^{CO}}{p_0^{CO}} \frac{1}{1 + \frac{k_{diss}\theta_{CO}}{k_H} + \frac{k_{diss}(1-\theta_{CO})}{k_{des}^{CO}}} \tag{3.10b}$$

An analytical solution is not possible because of the higher-order equation for θ_{CO}.

However, the kinetic expressions for R_{CH_4} that correspond to the two different cases can be determined analytically:

$$R_{CH_4} = k_H(1 - \theta_{CO}), \quad k_{diss}\theta_{CO} \gg k_H \tag{3.11a}$$

$$\theta_{CO} \approx \frac{K_{ads}^{CO}}{1 + \frac{K_{diss}\theta_{CO}}{k_H}} \frac{p^{CO}}{p_0^{CO}} \ll 1, \quad k_{diss}\theta_{CO} \gg k_H \tag{3.11b}$$

$$R_{CH_4} = k_{diss}\theta_{CO}(1 - \theta_{CO}), \quad k_{diss}\theta_{CO} \ll k_H \tag{3.12a}$$

$$\theta_{CO} = \frac{K_{ads}^{CO}\frac{p^{CO}}{p_0^{CO}}}{1 + K_{ads}^{CO}\frac{p^{CO}}{p_0^{CO}}}, \quad k_{diss}\theta_{CO} \ll k_H; k_{diss}\theta_{CO} \ll k_{des}^{CO} \tag{3.12b}$$

$$R_{CH_4} = k_{diss}\frac{K_{ads}^{CO}\frac{p^{CO}}{p_0^{CO}}}{\left(1 + K_{ads}^{CO}\frac{p^{CO}}{p_0^{CO}}\right)^2} \tag{3.12c}$$

These equations will be analyzed in the subsection that follows.

3.3.1.2.2 The Apparent Activation Energy

In the case described by Eqs. (3.11), the apparent activation energy of the reaction is determined by the rate of surface carbon hydrogenation and the surface is mainly covered with "C" species. In the case described by Eqs. (3.12), the apparent activation energies depend on the CO coverage, which can vary.

The apparent activation energy E_{app} of the rate of methane formation now depends strongly on the coverage of CO:

$$E_{app} = E_{diss} + (1 - 2\theta_{CO})E_{ads}^{CO} \qquad (3.13)$$

In Eq. (3.13), E_{diss} is the activation energy of CO dissociation and E_{ads}^{CO} is the adsorption energy of CO. At low CO coverage E_{app} is lower than E_{diss} because E_{ads}^{CO} is subtracted. As can be seen in equation (3.12c), the order of reaction is first order in partial CO pressure. At high CO coverage E_{app} is higher than E_{diss}, because E_{ads}^{CO} must be added to E_{diss}. The order of reaction becomes negative order in the partial pressure of CO.

There is a clear physical interpretation of these different behaviors as shown by the rate of product formation as a function of partial pressure of CO. It is typical for heterogeneous catalytic reactions (Figure 3.9).

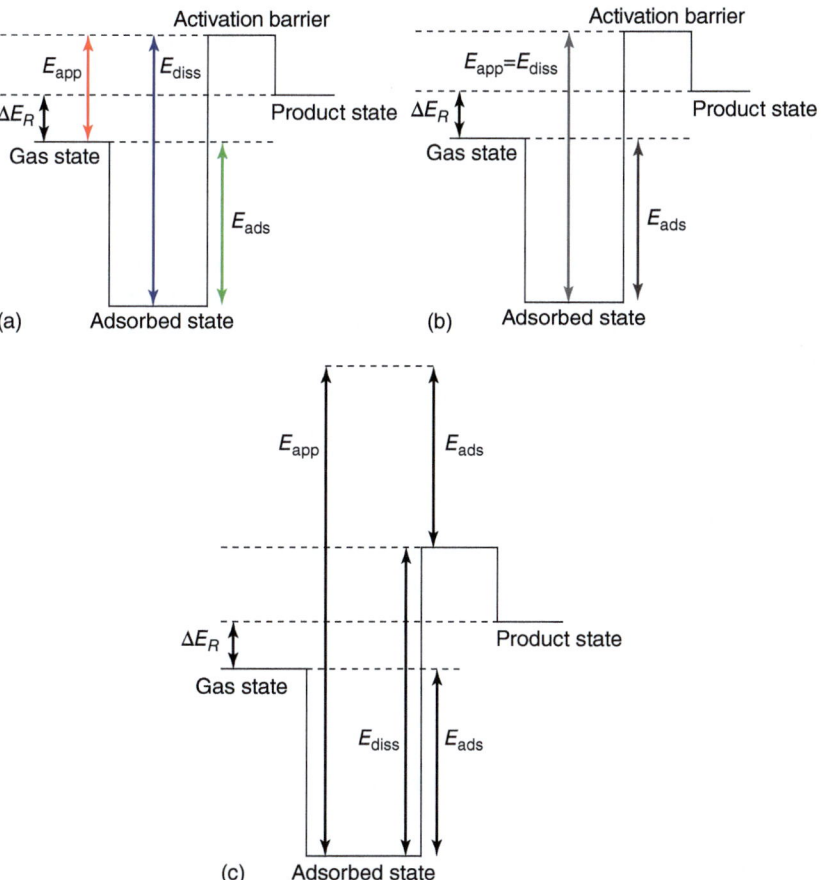

Figure 3.9 (a) $\theta_{CO} \ll 1$ and $E_{app} = E_{diss} + E_{ads}^{CO}$. (b) $\theta = 1/2$ and $E_{app} = E_{diss}$. (c) $\theta_{CO} \gg 1$ and $E_{app} = E_{diss} + E_{ads}^{CO}$. Schematic reaction energy diagram corresponding to elementary kinetics of methane formation from CO. The rate constant of CO dissociation is assumed to be rate-controlling ($E_{diss} > 0$; $E_{ads} < 0$).

In the case where θ_{CO} is substantially less than 1, the apparent activation energy is E_{diss} measured with respect to the gas phase. When $\theta_{CO} = 1/2$, the apparent activation energy is equal to E_{diss} measured with respect to its adsorbed state. And when θ_{CO} is nearly 1, the apparent is E_{diss} to which $|E_{ads}{}^{CO}|$ is added. In the third case, a CO molecule must desorb in order to create a surface vacancy for the adsorption of the additional adatom generated upon CO dissociation.

As a second example, in alkene hydrogenation, when the hydrocarbon surface coverage is low the adsorption energy of the alkene must be subtracted from the activation energy of the molecule. In contrast at high coverage the apparent activation energy is equal to the molecular activation energy.

This subsection has illustrated the close relationship between surface coverage, order of reaction, and apparent activation energy for the kinetics of a heterogeneous catalytic reaction.

3.3.1.2.3 Michaelis–Menten Kinetics

We will conclude this section by showing the similarity between the kinetics of enzyme catalysis, or homogeneous catalysis by organometallic complex catalysts, and the kinetics of heterogeneous catalysis.

The reason for this equivalence is that in both types of catalysis the number of catalytic sites is a constant, and in both cases the analog of the surface coverage is the number of occupied molecular catalytic complexes. The reaction rate expression used in enzyme catalysis assumes monomolecular reaction and is called the *Michaelis–Menten equation*.

This comparison is most easily demonstrated by considering the lumped kinetics expression of heterogeneous catalytic hydrogenation of an olefin, considered as a mono molecular reaction and to compare this with the analog expression for a mono molecular reaction when catalyzed by a molecular complex or enzyme.

The Michaelis–Menten type expression for the rate of reaction is deduced if one assumes rapid equilibration between the catalyst center and free reagent and that the reaction rate is limited by the hydrogenation reaction step. In the case of hydrogenation, this would be the reaction of adsorbed hydrogen atoms with adsorbed olefin.

The lumped kinetic equation according to heterogeneous catalysis is (again we ignore explicit dependence on hydrogen pressure and concentration):

$$R_s = N_s k_r \theta \qquad (3.14a)$$

$$\theta = \frac{K_{ads} \frac{P}{P_0}}{1 + K_{ads} \frac{P}{P_0}} \qquad (3.14b)$$

The equivalent Michaelis–Menten expression in the notation conventionally used in enzyme catalysis is:

$$R_e = k_2 E_0 \frac{S}{K_m + S} \qquad (3.14c)$$

The overall catalytic rate is proportional to the maximum number of available sites: N_s (Eq. (3.14a)) or enzyme molecules E_0 (Eq. (3.14c)). The Michaelis constant K_m is the equivalent of $1/K_{ads}$ and the partial pressure is the equivalent of the substrate concentration S. k_r is the lumped elementary rate constant of the rate-controlling surface reaction, and k_2 is the symbol corresponding to the enzyme-catalyzed reaction.

Both expressions show that at high reactant concentration, the rate of reaction becomes a constant, independent of reactant concentration. This is due to site saturation by adsorption, represented by the term in the denominator of expression (3.14b) or (3.14c).

These expressions do not allow us to predict N_s in the surface reaction for the activation energies or orders of reaction. This is a very important issue in heterogeneous catalysis because the distribution and reactivity of reactive surface centers is not uniform, and as a result the Langmuir assumption is no longer valid.

Spectroscopically, it is still a great challenge to determine the number of active sites on a working catalyst, which is often low compared to the total number of surface atoms. In Section 3.4, we will discuss the transient kinetics approach that can be used to derive this information.

It is important to realize that the kinetic equations represented thus far do not account for the reverse reaction of the product where readsorption and reaction back to reactant leads to equilibrium between the reactant and product at long contact times of gas or liquid flow.

3.3.1.2.4 Fundamental Catalytic Kinetics Equation

Generally, the overall rate of a catalytic reaction is determined by the elementary reaction rate constants, the surface coverage of reaction sites, and a chemical driving force that is expressed as affinity A_{ff}.

The chemical affinity of the reaction is zero when reaction is at equilibrium.

For the simple case of reactant A and product B, A_{ff} is defined as:

$$A_{\text{ff}} = R_g T \ln\left(\frac{R_f\,[A]}{R_b\,[B]}\right); \quad K_{eq} = \frac{R_f}{R_b} \tag{3.15}$$

R_f and R_b are the respective forward and backward rates of reactant conversion, respectively. [A] and [B] are the respective concentrations in the reaction mixture. R_g is the gas constant.

For a two-component system, assuming a reactant activation to be rate controlling, the rate expression that includes the back reaction becomes:

$$R = \frac{k_a K_{\text{ads}}^A [A] - k_b K_{\text{ads}}^B [B]}{1 + K_{\text{ads}}^A [A] + K_{\text{ads}}^B [B]} \tag{3.16}$$

In Eq. (3.16), K_{ads}^A and K_{ads}^B are the respective adsorption constants of reagent and product. k_a and k_b are the elementary reaction rate constants due to the surface reactions of adsorbed A and B. We have normalized the rate per surface site. As in Eqs. (3.14), we assume rapid equilibration between the surface and the reactant and product phase.

Equation (3.16) can be rewritten so that it explicitly shows that the rate R is zero when reactant and product are at equilibrium:

$$R = \frac{k_a K_{ads}^A [A] \left(1 - \exp\left(-\frac{A_{ff}}{R_g T}\right)\right)}{1 + K_{ads}^A [A] + K_{ads}^B [B]} \quad (3.17)$$

This leads to the Langmuir–Hinshelwood–Hougen–Watson (LHHW) expression:

$$R = k_a \theta_A \left(1 - \exp\left(-\frac{A_{ff}}{R_g T}\right)\right) \quad (3.18)$$

with:

$$\theta_A = \frac{K_{ads}^A [A]}{1 + K_{ads}^A [A] + K_{ads}^B [B]} \quad (3.19)$$

Equation (3.18) illustrates the three contributions to the overall rate of a catalytic reaction. These are the elementary reaction rate constant of the rate-controlling step (k_a), the fraction of the surface covered by the adsorbed species to be converted (θ_A), and the driving force of the reaction that is determined by the affinity. The expression can be readily extended to represent a case with many components. The essential assumptions are that a rate-controlling step exists and that all other reaction steps are equilibrated. In the section dealing with the simulation of the Sabatier volcano curve, we will see that at the maximum of the volcano curve this condition breaks down and a different approach must be followed.

3.3.1.2.5 Eley–Rideal versus Langmuir–Hinshelwood Kinetics

The rate expressions shown in Eqs. (3.14) and (3.18) assume that the adsorption equilibrium of the reactant and product molecules is rapidly established and that the reaction proceeds by a slow elementary reaction step in which the adsorbed molecular complex is activated by the surface.

The initial adsorption of reactants is not always necessary for the reaction. Sometimes, there is a direct reaction between gas phase molecules and adsorbed molecules. When a molecule reacts directly with a preadsorbed molecule or surface adatom without an adsorption step, the reaction is occurring through an Eley–Rideal mechanism. Examples of this type of reaction are seen in the activation of alkane molecules by surface oxygen atoms.

3.3.2 Sabatier Principle and Volcano Curves: Brønsted–Evans–Polanyi Relations

Computational catalysis has devised methods to determine the reactivity performance indicators of a catalyst at its optimum performance. In the Sabatier volcano optimum of catalytic reaction rate plotted against surface reactivity, the rates of competing elementary reaction steps balance. There is no rate-controlling step. Kinetic expressions such as Eqs. (3.9a)–(3.9c) now must be solved without assuming one rate-controlling reaction step.

In order to construct the Sabatier volcano curve, one needs to know how the reaction rate constants of the elementary reaction rates that constitute the catalytic reaction cycle depend on surface reactivity indicators. We will discuss this in the following subsection. The Brønsted–Evans–Polanyi (BEP) relation is very useful for evaluating trends in elementary reaction rate constants as a function of catalyst composition. The BEP relation correlates the activation energies of elementary reactions with equilibrium properties such as the corresponding reaction energies. Since equilibrium properties (such as adsorption energy, which is a kinetic property) are used to calculate a reaction rate, the relationship between the rate constant and equilibrium properties and the adsorption energies can only be estimated. While no rigorous statistical mechanical theory can be used to determine this relationship, useful approximation theories have been developed for a limited class of materials and reactions.

These theories involve the use of linear activation energy–reaction energy relations between the activation energy and adsorption energies of reaction products or reaction reagents. We will explain the concept of BEP as it applies to elementary surface reaction steps in Section 3.3.2.1. In Section 3.3.2.2, its use will be illustrated by a simulation of the Sabatier optimum using the kinetics model of the CO to methane reaction of Eqs. (3.10).

In Section 3.3.2.3, we will return to a discussion of the full catalytic reaction cycle. The concept of a reaction energy diagram and its relationship to the overall reaction rate will be introduced. The kinetics definition of the Sabatier reaction rate optimum will be given.

3.3.2.1 Brønsted–Evans–Polanyi Relations of Elementary Surface Reaction Rate Constants

The Brønsted–Evans equation relates the change in activation energy of an elementary reaction step with the change in reaction energies when the composition of a catalyst surface changes, but the structure of the reaction site remains unaltered. An example of the BEP relationship is shown in Figure 3.10.

In this figure, the DFT-calculated barriers for N_2 dissociation (measured with respect to the gas phase) for two surfaces of different topologies are plotted for a variety of transition metals as a function of the adsorption energy of the N atoms. A linear dependence is observed when dissociation is compared on surfaces with the same structure. The notation of the crystallographic orientations of the different surfaces can be different when the bulk structure is varied, while the surface topology can still be similar [22].

The BEP relation applies as long as the elementary reactions are studied at reaction centers of the same structure and the reaction path of the elementary reaction steps is similar.

The BEP equations can be written as:

$$E_f^{\#} = E_{f,0}^{\#} + \alpha_b \Delta E_r \tag{3.20a}$$

α_b is the BEP proportionality constant. $E_f^{\#}$ and $E_{f,0}^{\#}$ are the activation energies on two surfaces with different reactivity. ΔE_r is the change in reaction energy. When Eq. (3.20a) represents the activation energy change for the forward reaction, the

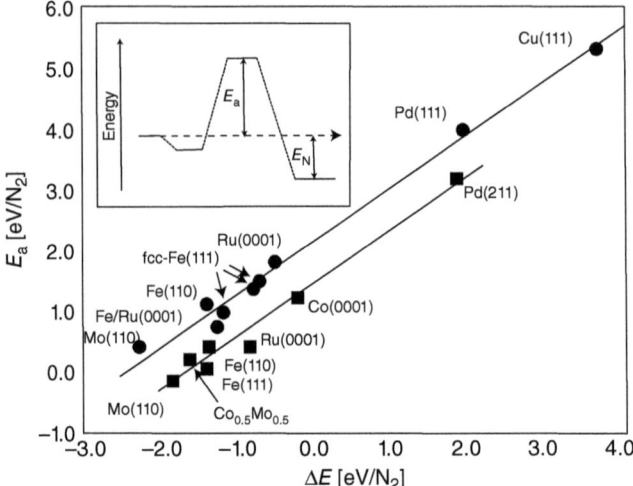

Figure 3.10 The calculated activation energies (transition state potential energies) for N_2 dissociation (E_a) on a range of metal surfaces plotted as a function of the adsorption energy for two nitrogen atoms (E_N^*). All energies are relative to N_2 (g). Results for both closely packed surfaces (filled circles) and more open surfaces (filled squares) are shown. The crystallographic orientations of the different surfaces are indicated within brackets. The inset shows a sketch of the energies for the N_2 dissociation reaction. (Dahl et al. 2001 [21]. Reproduced with permission of Elsevier.)

equation for the activation energy of the reverse reaction is:

$$E_r^{\#} = E_{r,0}^{\#} - (1 - \alpha_b)\Delta E_r \tag{3.20b}$$

$E_r^{\#}$ and $E_{r,0}^{\#}$ are the respective values of the activation energies of the reverse reaction. The changes in activation energies are illustrated by Figure 3.11.

The proportionality constant of forward and reverse reaction are related due to microscopic reversibility; the sum of the changes in the rate constants for the forward and reverse reactions must be equal to the change in reaction energies.

As Figure 3.11 illustrates, the value of α_B is close to 1 for a dissociation reaction. As a consequence, the proportionality for the reverse association reaction of two N atoms is nearly zero. This shows that surface reactions in which adsorbate chemical bonds are broken are sensitive to a change in surface reactivity. Conversely, the activation energy of a recombination reaction is independent of surface reactivity.

The BEP relation is an example of the Brønsted relation that is widely used to correlate activation energy changes of chemically related elementary reactions with a physical parameter that changes with reaction energy [23]. For the case of the surface reactions, it is the linear relation between activation energy change ΔE_{act} and reaction energy change ΔE_r:

$$\Delta E_{act} = \alpha_b \cdot \Delta E_r, \quad 0 < \alpha_b < 1 \tag{3.20c}$$

In gas phase kinetics, often a value of ½ is chosen for α_b. The value of α close to 1 for surface dissociation reactions implies that the structures of the transition

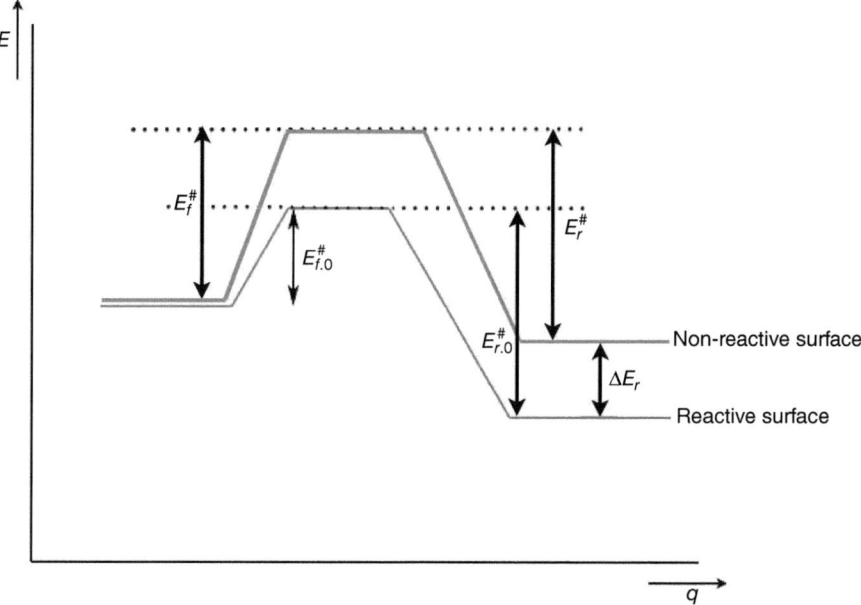

Figure 3.11 Sketch of activation energies and reaction energy according to Eqs. (3.20a) and (3.20b). Symbols are explained in the text.

state complex and that of the final dissociated state are close. This is because to activate a chemical bond the adsorbate–surface interaction has to be so strong that in the transition state the cleaving adsorbate can be considered to be nearly broken.

Knowing the α_b values, one can predict the activation energy of an elementary reaction step for a new system once the adsorption energies of the dissociation fragments are known for one system. The structure of the reaction center must be identical to that of another metal with known activation energy and fragment energies. The advantage of this procedure is that it eliminates the elaborate process of calculating the activation energy for each new system.

Note that in Figure 3.10 the least reactive transition metals are located to the right of the center in the periodic system, while metals of highest reactivity are in the center. This is caused by the nature of the bond between surface chemical and transition metal surface, which we will discuss in Chapter 8.

The strong dependence of the activation energy of dissociation on changes in surface reactivity as measured by the adsorption energies of the corresponding adatoms is due to the strong interaction that is needed between surface atoms and dissociating molecular chemical bond to achieve dissociation.

The high value of α_b implies that the changes in chemical bonding interaction in the transition state complex are very similar to changes in bonding interactions for the adatoms resulting from dissociation. An important result of this strong interaction is that the molecule in the transition state becomes immobile.

The activation entropy of an elementary surface reaction is small. Although there is a large change in entropy of a molecule between the gas phase and the adsorbed state, such changes are relatively small for reactions between fragments on the surface.

The higher reactivity of the more open stepped (211) surface compared to the more dense (111) surface is due to the higher reactivity of the surface metal atoms. The surface atoms on the (211) surface are more reactive than on the (111) surface because they have a lower coordination with neighbor metal atoms. When their coordinative unsaturation increases, the strength of the atomic chemical bond with a surface adatom also increases. In addition, the stepped configuration of the (211) surface has a favorable topology for low activation energy of N_2. This will be discussed in more detail in Chapter 4.

The changes in adsorption energy of an adatom when surface atom coordination varies are illustrated by the calculated adsorption energies of a C atom to different metal surfaces of Ru as shown in Table 3.1. Such trends are very similar to atoms such as C, N, or O.

It can be seen that the adsorption energy of the C atom increases with its coordination number of surface atoms. As long as the M—C interaction is compared for atoms adsorbed to the same number of surface atoms, the interaction with surface metal atoms with lower coordination to the nearest neighbor metal surface atoms shows an increase of the adsorption energy. This inverse relation between the strength of the adatom chemical bond and the strength of interaction of the metal atom in contact with the adsorbed nearest neighbor metal atoms will be discussed in detail in Chapter 7, where we will also describe its relation to the electronic structure of the surface chemical bond. A

Table 3.1 Adsorption energy of a C atom as a function of its coordination and surface atom coordinative unsaturation as calculated by DFT on different Ru surfaces (the crystallographic orientation of the surface of the hexagonal close-packed Ru metal are indicated within brackets; see [7] for notation).

Adsorption energies with respect to C gas phase and Ru(0001) bare surface			
Surface	Site	Adsorption energy (kJ mol^{-1})	Number of Ru neighbors for C (and for Ru)
Ru(0001)	Top	497	1 (9)
	Bridge	631	2 (9)
	Hollow hcp	688	3 (9)
	Hollow fcc	648	3 (9)
Ru(11$\bar{2}$0)	Top up	549	1 (7)
	Top down	675	3 (7) + 1 (11)
	Bridge short	666	2 (7)
	Bridge long	579	2 (7) + 2 (11)
Ru(10$\bar{1}$0)	Hollow	678	2 (11) + 1 (7)
Ru(1015)	Hcp	714	2 (7) + 1 (9)

The numbers in brackets in the column that shows the number of Ru neighbors for C (and for Ru) show the number of metal atom neighbors of the metal surface atoms and the number before it shows the number of metal atoms to which the C atom is attached.
Source: van Santen and Neurock 2006 [9]. Reproduced with permission of John Wiley & Sons.

more intuitive comprehension of this relation can be obtained by understanding the consequences of bond order conservation in chemical bonding when coordination changes. This will be discussed in the next subsection since this is very useful for understanding the differences in the reactivity of surfaces as a function of their degree of coordinative unsaturation of the surface atoms.

3.3.2.1.1 Bond Order Conservation

The concept of bond order conservation is based on the idea that for a single atom, its total binding capability with the surrounding atoms is a constant. This means that when more atoms bind to a central atom, the binding power must be redistributed over more chemical bonds and, hence, the binding power for each bond decreases. The sum of the bond orders of the chemical bonds attached to a surface atom then is approximately constant.

When a chemical bond is broken, the energy of the other remaining bonds increases to maintain a constant total bond order. This provides a qualitative explanation for the increase in reactivity of a surface metal atom when coordinative unsaturation increases, and rationalizes the weakening of internal chemical bonds of molecules adsorbed to a surface compared to the internal bonds of the free molecules (see Insert 3).

Insert 3: Bond Order Conservation Expression

Shustorovich [24] deduced the following expressions for the bond strength Q_n of a bond to be shared with similar neighbors:

$$Q_n = \frac{1}{n} Q_0 \left(2 - \frac{1}{n}\right) \tag{3i.1}$$

Q_0 is the bond strength when bonds do not have to be shared ($n = 1$).

The consequences of this expression on adsorbate strength are shown in Figure 3i.1.

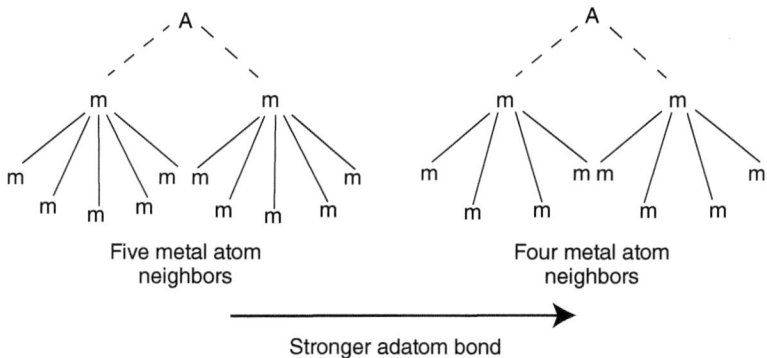

Figure 3i.1 Adatom A chemisorption as a function of metal atom coordination number m. Surfaces are more reactive when surface metal atoms have fewer neighbors.

(Continued)

Insert 3: (Continued)

The derivation is based on the assumption of a Morse potential between two particles ($n = 1$).

$$Q(x) = Q_0(2x - x^2)$$

$$x = e^{-\frac{r-r_0}{a}} \tag{3i.2}$$

x is the bond order ($x = 1$ when the distance r between the particles is equal to equilibrium distance, and a defines the frequencies of oscillations).

Conservation of bond order implies:

$$\sum_{i=1}^{n} x_i = x_0 = 1 \tag{3i.3}$$

3.3.2.2 The Sabatier Volcano Curve

The BEP relation can be used to deduce the optimum surface reactivity parameters that determine the maximum reaction rate. The surface reactivity is characterized by particular values of the adatom adsorption energies. The BEP approach is used to relate activation energies of elementary reaction rates with adsorption energies of the adatoms. This is done for a specific topology of the reaction site. Different reaction centers will have a different Sabatier reaction rate optimum. When this procedure is followed, a prediction can be made for the combination of surface topology and transition metal that will produce an optimum overall reaction rate.

We will illustrate this using the example of the lumped sum kinetics of the CO to methane reaction discussed in Section 3.3.1.2. In this case, surface reactivity indicators are the strength of the M—C or M—O bonds.

Insert 4 summarizes the simulation results based on the solutions of the corresponding steady state lumped kinetic equations (Eq. (3.10)). The BEP

Insert 4: Simulation of Sabatier Volcano Curve Maximum

1. **Brønsted–Evans–Polanyi Relations for Lumped Elementary rate Constants of the CO to CH_4 Reaction**

$$k_H = k_H^0 \exp\left(-\beta_H^C \Delta E_{C,ads}\right) \tag{4i.1}$$

$$k_{diss} = k_{diss}^0 \exp\left(\beta_d^C \Delta E_{C,ads} + \beta_d^O \Delta E_{O,ads}\right) \tag{4i.2}$$

$$\beta = \frac{\alpha_B}{kT} \tag{4i.3}$$

$\Delta E_{C,ads}$ and $\Delta E_{O,ads}$ are the differences in adsorption energy of a C adatom and O adatom, versus those on the reference surface. k_H^0 and k_{diss}^0 are the respective lumped and elementary rate constants for hydrogenation of C adatom to methane and CO dissociation on the reference surface, respectively.

β_H^C, β_d^C, and β_d^O are the BEP proportionality-related constants for the elementary reaction steps. Because k_H and k_{diss} refer to a reaction with a chemical bond breaking step, the BEP proportionality parameters α_B will be close to 1.

2. **Simulations of Volcano Curves (assume $\Delta E_0 = \Delta E_c$)** Figures 4i.1–4i.3 show Sabatier volcano curves for three values of K_{ads}^{CO}: $K_{ads}^{CO} = 100$, $K_{ads}^{CO} = 1$, and $K_{ads}^{CO} = 0.2$, respectively. The figures compare the rate of production of methane, R_{CH_4} and surface coverages of C and CO as a function of ΔE_C. Figure 4i.4 compares simulations for different values of k_{diss}^0 with $K_{ads}^{CO} = 1$.

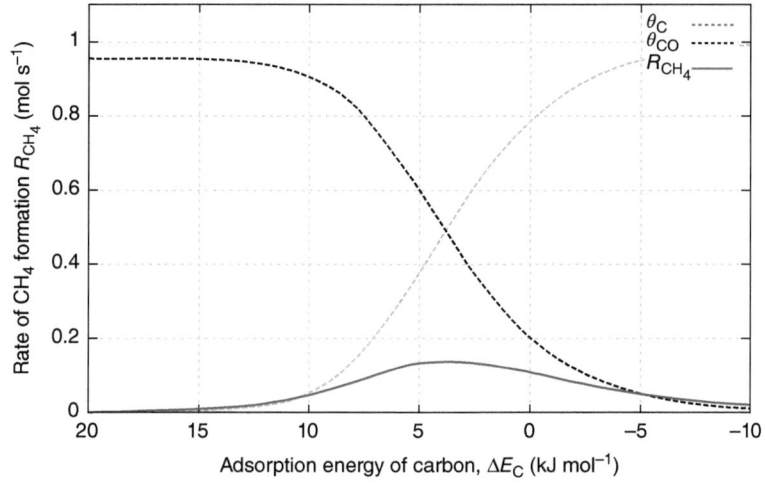

Figure 4i.1 Rate of CH_4 formation R_{CH_4} versus change in adsorption energy ΔE_c $k_H^0 = 1$, $k_{diss}^0 = 0.01$, $k_{des} = 10$, $k_{ads} = 1000$, $\beta_H^C = \beta_d^C = \beta_d^O = 1/5$.

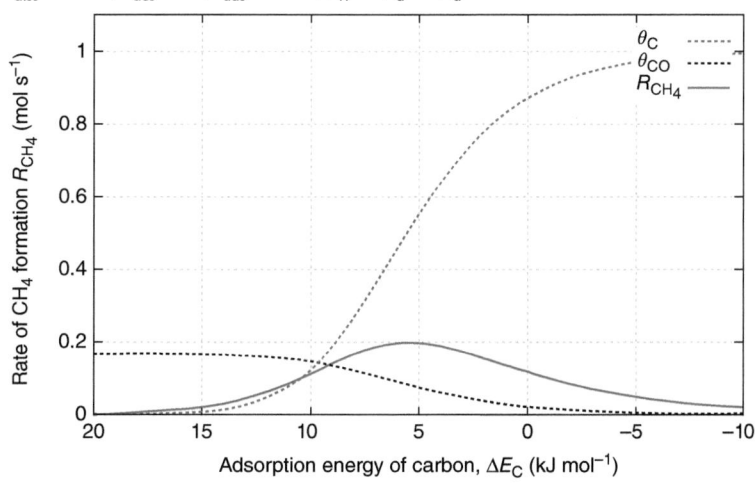

Figure 4i.2 Rate of CH_4 formation R_{CH_4} versus change in adsorption energy ΔE_c $k_H^0 = 1$, $k_{diss}^0 = 0.01$, $k_{des} = 1000$, $k_{ads} = 1000$, $\beta_H^C = \beta_d^C = \beta_d^O = 1/5$.

(Continued)

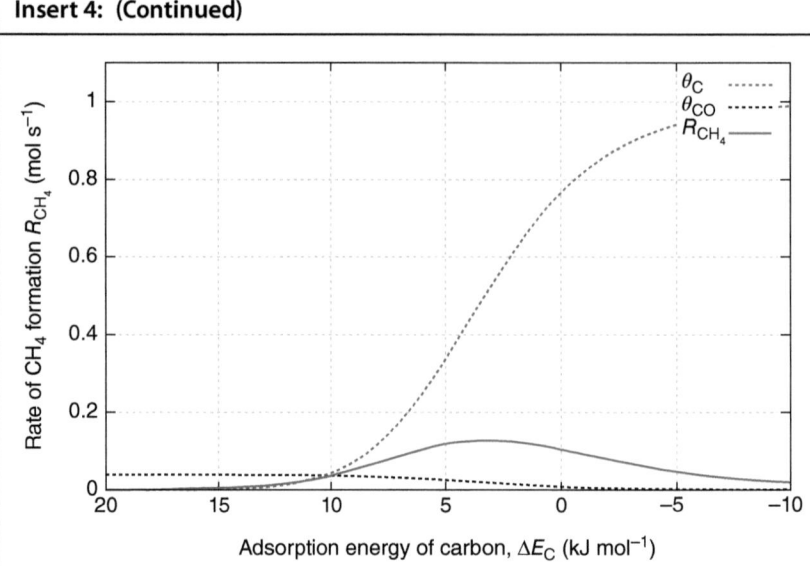

Figure 4i.3 Rate of CH_4 formation R_{CH_4} versus change in adsorption energy ΔE_c $k_H^0 = 1$, $k_{diss}^0 = 0.01$, $k_{des} = 5000$, $k_{ads} = 1000$, $\beta_H^C = \beta_d^C = \beta_d^O = 1/5$.

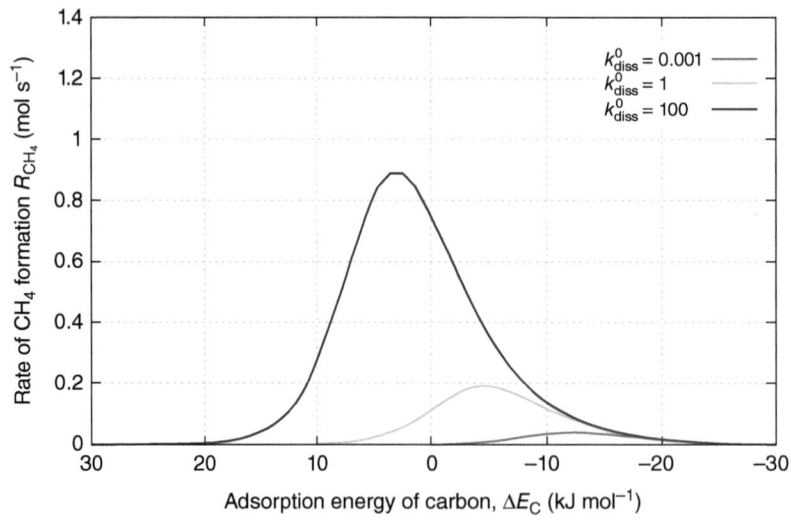

Figure 4i.4 Rate of CH_4 formation R_{CH_4} versus change in adsorption energy ΔE_c $k_H^0 = 1$, $k_{des} = 1000$, $k_{ads} = 1000$, $\beta_H^C = \beta_d^C = \beta_d^O = 1/5$.

expressions for the variation of elementary rate constants as a function of adatom adsorption energy are given in Eqs. (4i.1)–(4i.3) of Insert 4. For convenience of interpretation, the variation in adsorption energies of C and O have been assumed to be similar, and the CO adsorption energies are insensitive to a

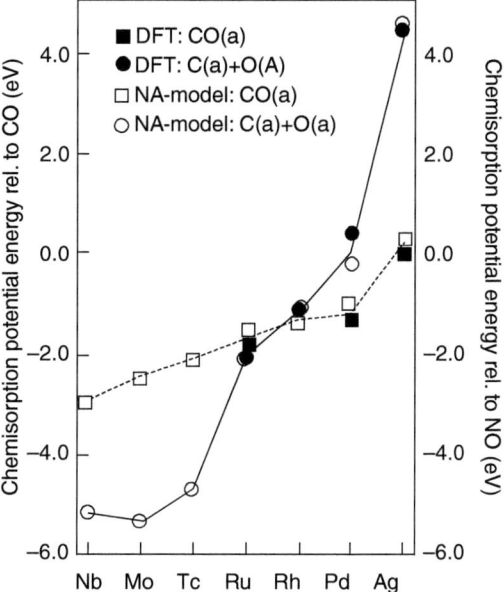

Figure 3.12 Calculated and model estimates of the variation in the adsorption energy of molecular CO compared with atomically adsorbed C and O for the most closely packed surface of the 4d-transition metals. (Hammer and Nørskov 2000 [25]. Reproduced with permission of Elsevier.)

change in catalyst material. Parameter values have been selected for didactic purposes and do not directly refer to experimental systems. Kinetics simulations based on quantum-chemical results of physically representative catalyst models that also include differences in variations of M—C and M—O bond energies are presented in Figure 3.13.

It is reasonable to assume that the adsorption energies of CO do not vary significantly with a change in M—C bond energies. As we will explain later, this variation is minimal because the electron donative and back donative chemical bonding interactions between molecule and surface counteract and nearly cancel. This is well illustrated in Figure 3.12, which compares the variation in the sum of the respective M—C and M—O adatom bond energies as computed by DFT for different metals.

As we will discuss in detail in Chapter 7, the changes in the sum of adatom energies for C and O are dominated by the changes in the oxygen adatom energies; the changes in the adatom C energies are substantially less.

The simulations of Insert 4 assume rapid equilibration of gas phase CO with CO adsorbed to the surface. Figures 4i.1–4i.3 compare simulations with different values of K_{ads}^{CO}, ranging from large in Figure 4i.1 to small in Figure 4i.3. All three figures show a maximum rate of methane production at a comparable value of E_c, the adsorption energy of C. This rate varies slightly with K_{ads}^{CO}. The surface coverages are very different in the three cases. At high CO coverage, the order in CO partial pressure will be negative, while at the low CO coverage it will become positive.

To the right of the volcano maximum, Eq. (3.11a) applies. We observe a strong increase in θ_C with increasing adsorption energy E_c. The rate of C atom removal becomes limiting and the reaction rate becomes insensitive to CO partial pressure.

The major reaction intermediates (MARI) is this case are the surface coverages that are experimentally accessible. The concept of MARI is important. It varies with change in rate controlling step and hence order of reaction.

The position of the volcano maximum is determined by the reference values k_H^0 and k_{diss}^0 as well as the ratio of the respective BEP parameters of the elementary rate constants [26]. These values will vary when surface topology alters. The sensitivity of the rate maximum with respect to the values of k_H^0 and k_{diss}^0 is illustrated by Figure 4i.4. The position of the volcano rate maximum is shifted to a lower value of the adsorption energy E_c and is substantially higher when the value of k_{diss}^0 is increased.

The rates of competing reactions balance with the Sabatier reaction rate maximum. In the reaction shown in Insert 4, the reaction rate constant of CO dissociation becomes comparable to the reaction rate constant of C atom removal. This implies that kinetic methods to determine the volcano cannot be based on an assumption of a rate-controlling step. The LHHW expression Eq. (3.18) that is typical in surface kinetics cannot be applied. Instead, expressions such as Eq. (3.10) that are not based on an assumption of a rate-controlling step must be used. Insert 5 describes a general method to solve the microkinetics ODEs without making an assumption about the rate-controlling step.

Insert 5: The Microkinetics Equations

1. **The Microkinetics Equations** The initial step for constructing the ODEs is to assemble a library of elementary reaction steps and corresponding rate constants. This produces a set of 2R elementary reaction equations (forward and backward) and N compounds. For each unique compound, an ODE is obtained of the form:

$$\frac{d\theta_i}{dt} = \sum_{j}^{2R} \left(v_{j,i} k_j \prod_{q}^{N_i} \theta_{q,j}^{v_{q,j}} \right)$$

where θ_i is the concentration of species i on the surface, $v_{j,i}$ is the stoichiometric coefficient of compound i in reaction j, and k_j is the rate constant of reaction j. The set of 2R ODEs can be solved by integrating over time until a steady state solution is reached, given by:

$$\left\{ \frac{d\theta_i}{dt} = 0 \right\}, \quad \text{for all } i.$$

The algorithm for integrating a set of ODEs over time is called an *ODE solver*.

3.3 Elementary Kinetics

In order to compare microkinetic simulations with experimental observations, the reaction order n_i as a function of partial pressure of reagent i, p_i, is of interest, as is the apparent activation energy E_{act}^{app}. These are given by Eqs. (5i.1) and (5i.2) respectively.

$$n_i = \frac{\partial \ln(r_j^+)}{\partial p_i} \qquad (5i.1)$$

$$E_{act}^{app} = RT^2 \frac{\partial \ln(r_j^+)}{\partial T} \qquad (5i.2)$$

The rate-controlling step can be determined from the rate control of a particular elementary step [27].

$$\chi_i = \left\{ \frac{\partial \ln(r_j^+)}{\partial \ln(k_i^\pm)} \right\}_{k_{i \neq j},\, K_i}, \quad \chi_i \leq 1$$

2. **The Rate Constants of Adsorption and Desorption** A useful expression for the rate constant of adsorption is:

$$k_{ads} = \frac{PA}{\sqrt{2\pi m k_b T}}$$

P is the pressure and A is the surface area of molecules; in the transition state, the molecule rotates freely. The corresponding expression for the rate constant of desorption is:

$$k_{des} = \frac{k_b T}{h} \frac{A(2\pi m k_b T)}{h^2} \frac{8\pi I k_b T}{\sigma h^2} \exp\left[-\frac{\Delta E_{des}}{k_b T}\right]$$

σ is the symmetry number to properly account for the number of distinguishable orientations of the molecule.

For comparison with the lumped kinetics simulations of Insert 4, the result of a first principle microkinetics simulation of the same reaction is shown in Figure 3.13. It displays the simulated CO to methane reaction rate versus M—C and M—O adatom bond energies for a particular surface based on first principle quantum-chemical calculations and microkinetics simulations. These microkinetics simulations are based on reaction models that explicitly contain all intermediates of the reaction. It shows a two-dimensional simulated plot of activity, or turnover frequency (TOF), which is the number of molecules converted per unit surface area and per unit time. In Figure 3.13, the adsorption energy of C decreases to the right along the horizontal axis and that of O decreases moving upward along the vertical axis.

In this figure, the contours of comparable activity are indicated. The optimum is a compromise between elementary reaction rates of CO activation, of CH_x conversion to methane, and of O atom removal by reaction with H_2.

When the adatom adsorption energy of O is too strong, the surface becomes poisoned by adsorbed O, but when the interaction is too weak, CO dissociation

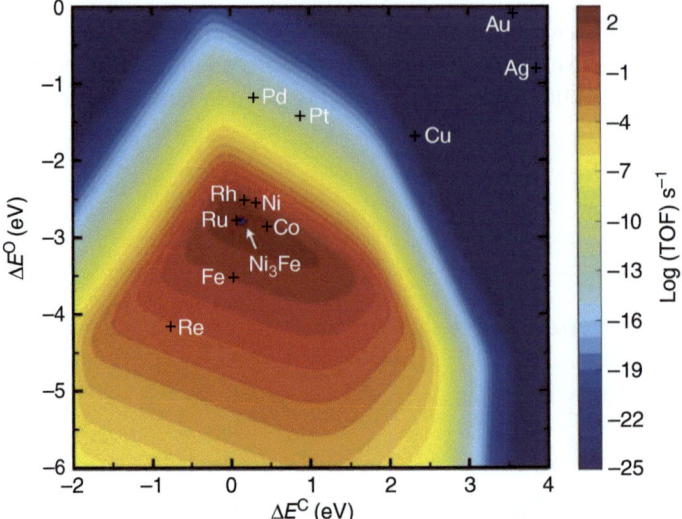

Figure 3.13 Theoretical volcano for the production of methane from syngas, CO, and H_2. The turnover frequency (TOF) is plotted as a function of carbon and oxygen binding energies. The carbon- and oxygen-binding energies for the stepped 211 surfaces of selected transition metals are depicted. Reaction conditions are 573 K, 40 bar H_2, 40 bar CO. (Nørskov et al. 2011 [28]. Reproduced with permission of National Academy of Sciences.)

will be prevented by a high activation energy barrier. We find a very similar dependence on the adsorption energy of the C atoms. A surface that has the composition of Ni_3Fe appears to optimize the TOF of the reaction [28].

The determination of the Sabatier reaction rate maximum as a function of catalyst performance descriptors leads to a rational approach to improve a catalyst. By comparing the experimental results with predicted apparent activation energies and reaction orders one can determine where the experimental system is located on the Sabatier plot. This will predict in which direction surface reactivity parameters must be changed in order to increase the reaction rate. In the above example, optimization of these parameters has been done by determining the metal combination that would give the O adatom and C adatom adsorption energies near the reaction rate maximum.

Chemical changes in surface reactivity are often realized by adding promoting compounds to the system, which alter the reactivity descriptors. Changes in surface topology can be achieved experimentally by altering the particle size of the reactive particles.

3.3.2.3 The Reaction Energy Diagram of the Catalytic Reaction Cycle

The reaction energy diagram displays the succession of energies of reaction intermediates and their respective activation energies when a reaction proceeds through its catalytic cycle. As an example, Figure 3.14 shows the two reaction energy diagrams including transition state energies calculated for the reaction of CO to methane on two Ni surfaces. One of them has been used to construct

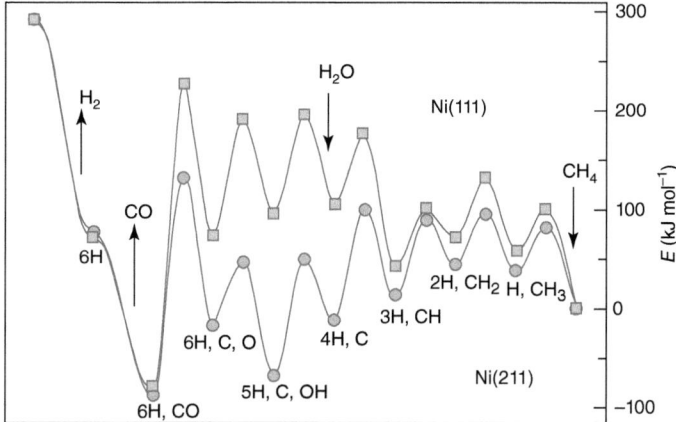

Figure 3.14 Reaction energy diagram of the methanation reaction. DFT-computed energies and activation energies are plotted with progress of reaction. Energies are given for the species adsorbed on Ni(211) and Ni(111) surfaces. The Ni(111) surface is a dense non-stepped surface, the Ni(211) surface is a more reactive stepped surface (see [7]). All energies are relative to CH_4 and H_2O in the gas phase and calculated using the results for the individual species. (Bengaard et al. 2002 [29]. Reproduced with permission of Elsevier.)

Figure 3.13. The reaction energy diagram shows the energy changes of all proposed reaction intermediates.

In order to construct a reaction energy diagram one must assume a particular reaction mechanism. In this case, it is assumed that CO adsorbs and dissociates directly. The adsorbed C and O atoms are hydrogenated in consecutive elementary reaction steps.

An alternative mechanism on Ni surfaces (which we will discuss in Chapter 7, Insert 7) is the initial addition of a H atom to CO and subsequent HC—O bond cleavage. H_2O can also be removed through recombination of two surface hydroxyls instead of by the direct addition of a hydrogen atom to adsorbed OH.

In Figure 3.14, the energy changes are given relative to gas phase CO and H_2. The reaction products are methane and water. We observe that this reaction is slightly endothermic.

In the reaction energy diagram shown above, the energy of the system is initially lowered because the gas phase molecules CO and H_2 adsorb to the surface. H_2 dissociation has no activation barrier. The CO dissociation has a high barrier and is slightly endothermic, which raises the energy. Another local minimum is seen for adsorbed OH. From this surface state, the energy gradually increases until methane desorbs. When the reverse reaction (the steam reforming of methane) is studied, the same reaction energy diagram can be used, but the energy changes must be followed in the opposite direction.

Methane only adsorbs weakly. The initial high barrier for CH activation may indicate that CH_4 activation is a slow step. However, these reaction steps actually compete with the recombination of the adsorbed O (generated by dissociative adsorption of H_2O) and C, to give CO and the desorption energy of CO.

Reaction energy diagrams are constructed from energies calculated at 0 K and do not contain information on activation entropies, but entropy changes are sometimes essential to consider.

With respect to the gas phase, adsorbing molecules lose entropy, but desorbing molecules gain energy.

This can have important consequences for kinetics. At a finite temperature, the free energy of activation of the dissociative adsorption of CH_4 will be higher than its activation energy barrier because of the loss in entropy with respect to the gas phase, and CO desorption has a substantially lower free energy of desorption compared to its desorption energy. On the other hand, there is hardly any difference between the free energy of activation and the activation energy for the surface reaction of CO dissociation because of the low activation entropy of the surface reaction. For steam reforming this makes methane activation the rate-controlling step. Insert 7 (pp.97) provides an example of the simulated kinetics of these reactions as a function of temperature.

The reaction energy diagram provides insights into the condition where the heterogeneous catalytic reactions has its maximum rate. As the reaction progresses, the energy of the system goes through a minimum and the surface reaches a state where it is covered with the MARI. To recover the vacant catalyst sites, the energy must climb from its minimum in order to produce molecules that desorb at the end of the reaction cycle.

The role of the catalytically active surface is to stabilize intermediate reaction products that result from the phase of cleavage of chemical bonds. There is an energy cost to remove reaction fragments from the surface. The optimum surface condition for the maximum rate of a catalytic reaction is reached when the interaction energy with reaction intermediates is low enough to cause desorption of the product at a low activation energy, but still high enough to activate them.

This leads to the conditions for the maximum Sabatier reaction rate based on general considerations of the reaction energy diagram, summarized in Insert 6. The deviations of the minimum of the reaction energy diagram below the equilibrium free energy and the maximum activation energy of the reaction above the equilibrium free energy of reaction are to be minimized with respect to the equilibrium free energy of the reaction.

Insert 6: Minimum Deviation from the Equilibrium Free Energy Requirement

An endothermic reaction can be achieved at a temperature where free energies G of the reactant and product are the same. The reaction-free energy diagrams are sketched below for three endothermic cases. For each of the three cases (a–c), the expression for overall rate R and apparent activation energy E_{app} are given. Case (d) gives the reaction energy diagram for an exothermic reaction that is not at equilibrium (Figure 6i.1).

- *Endothermic reaction*

    Case (a): $\left|G^B_{ads}\right| > E^\#$; $R = k^B_{des}\theta_B$; $E_{app} = \left|E^B_{ads}\right|$

    Case (b): $\left|G^B_{ads}\right| < E^\#$; $\theta_A \approx 1$; $R = k^A_r$; $E_{app} = E^\#$

Case (c): $|G^B_{ads}| < E^{\#}$; $\theta_A \ll 1$; $R = k_r^A \theta_A$; $\theta_A = K^A_{ads} P_A \ll 1$; $E_{app} = E^{\#} - |E^{\#}_{ads}| > 0$
- Exothermic reaction
  - Case (d): $|E^A_{ads}| > E^{\#}$; $\theta_A \ll 1$; $R = k_r^A \theta_A$; $E_{app} = E^{\#} - |E^A_{ads}| < 0$.

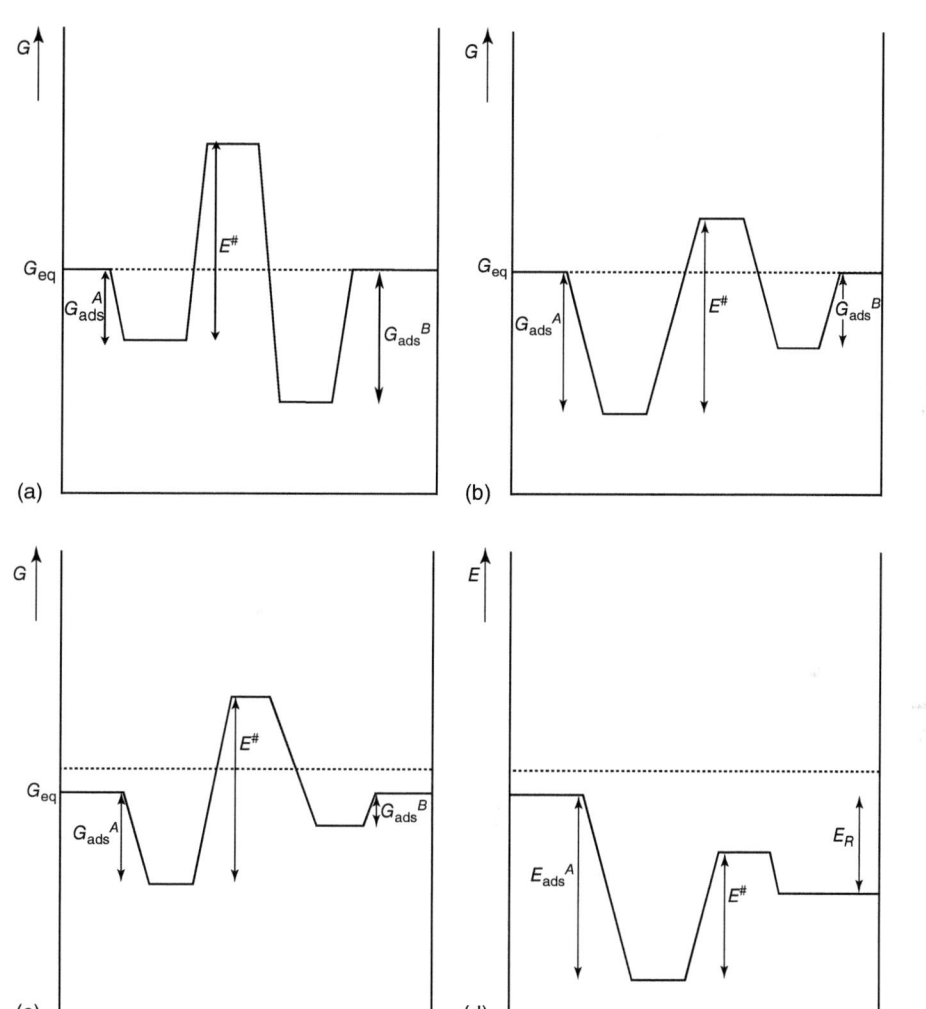

Figure 6i.1 The reaction free energy diagrams for three endothermic cases (a–c) and for an exothermic reaction that is not at equilibrium (d).

In case (a), desorption of product B is rate controlling, so that $E_{app} = E_{ads}^B$. In case (b), the rate constant k_b indicates that product desorption is fast, and the coverage of reactant A is high. The apparent activation energy becomes $E_{app} = E^{\#}$, the activation energy of reaction with respect to the adsorbed state of A.

(Continued)

> **Insert 6: (Continued)**
>
> Case (a) is the high temperature case when $\theta_A \ll 1$, so that $E_{app} = E^{\#} - |E_{ads}{}^A|$.
>
> The reaction rate has its maximum value when $E_{app} \approx 0$. The cases together show that deviations of activation energies in adsorption surface energies should be minimized with respect to the equilibrium free energy values of the reactant and product.
>
> Case (d) gives an example of an exothermic reaction energy diagram. Again, when $\theta_A \approx 1: E_{app} = E^{\#}$ and when $\theta_A \ll 1: E_{app} = E^{\#} - |E_{ads}|$. Since a high level of surface coverage is beneficial to a reaction, in several catalytic systems $|E_{ads}| > E^{\#}$. This implies that at high temperature, the apparent activation energy may become negative.

An insightful way to appreciate the need for minimizing the difference between the free energy of the overall reaction and the free energies of intermediate adsorption or activation is provided by the energetics of the electro-catalytic oxygen evolution reaction (OER) of water by transition metal oxide electrodes, to be discussed in the next section.

3.3.2.4 Electrocatalysis and Sabatier Principle Optimum

The overpotential of the electro-catalytic H_2O decomposition to H_2 and O_2 is mainly caused by the low rate of the O_2 evolution reaction at the anode. The half reaction that corresponds to this reaction is:

$$2H_2O \leftrightarrow O_2 + 4H^+ + 4e^- \tag{3.21}$$

The surface reactivity descriptor in this case is the adsorption energy of the O atom, which through bond order conservation (see Section 3.3.2.1.1) can be correlated with the interaction energies of the other reaction intermediates OH and OOH.

The four reaction steps of the reaction are as follows:

$$H_2O + * \leftrightarrow HO^* + H^+ + e^- \tag{3.22a}$$

$$HO^* \leftrightarrow O^* + H^+ + e^- \tag{3.22b}$$

$$H_2O + O^* \leftrightarrow HOO^* + H^+ + e^- \tag{3.22c}$$

$$HOO^* \leftrightarrow * + O_2(g) + H^+ + e^- \tag{3.22d}$$

An elegant thermodynamic procedure has been devised [30] that enables a quantum-chemical calculation of the reaction steps in Eqs. (3.22), with the replacement of $H^+ + e^-$ by $½H_2$. Since the energy difference is that of the hydrogen electrode, it is the equivalent of zero electrochemical potential. The contribution to the electrode potential of the reaction can be found by the additional use of Faraday's constant to relate free energy to electrode potential:

$$\Delta\mu = Z \cdot F \cdot \Delta U \tag{3.23}$$

$\Delta\mu$ is the change in chemical potential, Z is the number of electrons transferred, F is the Faraday constant, and ΔU is the electrochemical potential contribution. When an external potential is used, the free energies of reaction change by $-Z \cdot U$. Since each of the reactions generate one proton, the differences in free energy are independent of proton concentration.

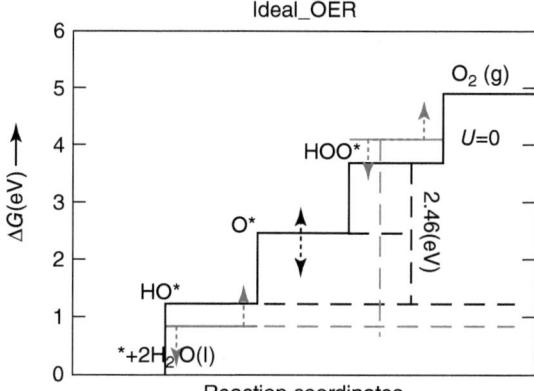

Figure 3.15 The free energy diagram for the OER ideal catalyst. All free energies of the intermediate steps have the same magnitude and are equal to 1.23 V (thick black line). Gray lines indicate the actual HOO* and HO* levels for the catalysts, which, in order to approach the ideal, must be moved down and up, respectively, by about 0.37 eV. Unfortunately, when the metal oxide catalyst is changed, these two levels tend to move in the same direction with the same magnitude (as indicated by the gray and black arrows). For the OER, the position of the O* level between the HOO* and HO* levels governs the potential determining step. (Valdés et al. 2012 [30]. Reproduced with permission of Elsevier.)

A schematic illustration of the thermodynamic reaction energy diagram one obtains for the reaction sequence of Eq. (3.22) is shown in Figure 3.15.

This figure shows an ideal curve according to the Sabatier optimum (the black curve) as well as a curve typical for a practical case (the gray curve).

In the electrocatalytic reaction, the role of the apparent activation energy is superseded by the overpotential η. When the overpotential η is zero, the electro-potential of the reaction is equal to its thermodynamics value. For the OER, the thermodynamic electro-potential is 1.23 V. In Figure 3.15, the free energy of O_2 formation is plotted and corresponds to 4×1.23 eV. The overpotential of reaction is found when the potential U is shifted downward by $Z \times 1.23$ eV for each electron transferred. Z is the number of electrons transferred.

For the ideal curve shown in Figure 3.15, this shifts the free energies of all intermediates on the reaction equilibrium line. In this ideal curve, the free energy changes in each reaction step are the same. As long as one ignores activation energies of the surface elementary steps, one predicts that the electrochemical reaction does not require any additional potential to proceed because the overpotential for the ideal case is predicted to be zero. (A critical discussion on the assumptions involved can be found in Section 11.6.2). The ideal reaction energy plot in Figure 3.15 therefore represents the diagram that corresponds to the Sabatier optimum of the reaction.

This ideal diagram cannot be realized in practice due to bond order conservation relations that constrain the energies of some reaction intermediates. This holds true for the bonds of OH and OOH with the catalyst surface that are chemically equivalent [31]. The energy difference that exists between the two relates to the presence of an O—OH bond in the peroxide versus an O—H bond in the hydroxyl. Therefore, changes in adsorbate M—O bond energies will affect the energies of both intermediates similarly. The difference in the free energies of those intermediates that are part of Figure 3.15 can therefore never be reduced. Calculations indicate that the overpotential of this reaction can never become less than an estimated 0.4 V, which is a substantial energy loss compared to the equilibrium electrode potential of the reaction.

3.3.2.5 Temperature Dependence of Catalytic Reaction Rate

While the rate of a non-catalytic homogeneous reaction will increase with temperature until equilibrium is reached, the temperature dependence of a catalytic reaction is very different. A heterogeneous catalytic reaction will always go through a maximum conversion rate as a function of reaction temperature and then decrease. This is schematically illustrated in Figure 3.16.

At a low temperature the reaction cannot occur because the surface is saturated with adsorbate. The reaction will reach a finite rate only when temperature is increased to cause desorption thereby creating vacant sites to accommodate the molecular fragments generated by dissociative activation of the adsorbed molecules. Alternatively, surface vacancies are needed to adsorb co-reactant molecules.

At higher temperatures, the rate of reaction is again inhibited. Reagents will no longer adsorb to the catalyst surface for reaction, because surface coverage becomes limited by the reduced adsorption equilibrium constant. At the highest reaction temperature, the surface sites may become completely vacant or alternatively become covered by deactivating adsorbed species due to non-selective reactions (for instance, surface carbon generated as a by-product in hydrocarbon conversion reactions). Generally, the reaction order of the reaction will be negative or zero at the low temperature edge of the temperature–reaction rate (T–R) plot, but positive at the high temperature edge of the T–R plot.

In the low-temperature area of the T–R plot the apparent activation energy E_{app} is positive. It decreases gradually until it becomes zero at the T–R plot maximum and then becomes negative to the right of the T–R maximum. Case (d) of Insert 6 shows a reaction energy diagram that explains how the apparent activation energy can become negative. This occurs when surface coverage becomes low and the

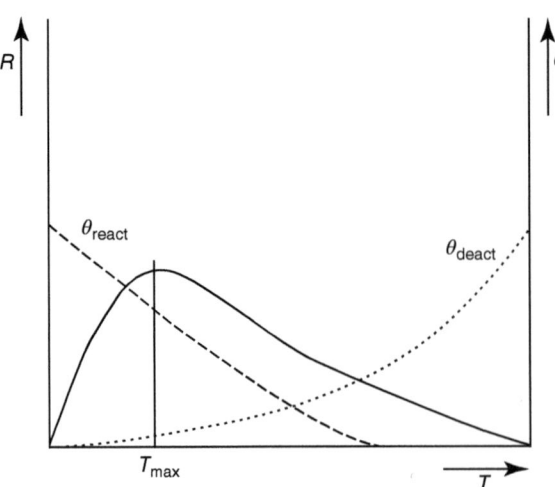

Figure 3.16 Schematic representation of the dependence of reaction rate R of a heterogeneous catalytic reaction on temperature T. The dependence of reactant surface coverage θ_{react} is also shown (dashed line). The possible increase in surface concentration of a deactivating compound θ_{deact} that is formed with higher temperature is indicated (dotted line). When no deactivating compounds are formed, there is an increase in surface vacancy concentration.

energy of adsorption is higher than the activation energy of the rate-controlling reaction.

As we have shown in Eqs. (3.11) and (3.12) and illustrated in Insert 6, the negative value of E_{app} beyond the rate maximum can be directly related to a decrease in surface coverage. For a highly covered surface, the apparent activation energy relates to elementary reaction activation energies with respect to the adsorbed state, but with increasing temperature the surface coverage decreases and the apparent activation energy is related to elementary reaction activation energies with respect to the gas phase.

In Insert 7, this behavior of rate versus temperature is illustrated by microkinetics simulations of the methanation reaction and the steam reforming reaction (Figures 7i.1–7i.4).

> **Insert 7: The Rate of a Heterogeneous Catalytic Reaction as a Function of Temperature**
>
> Microkinetic simulations based on quantum-chemical elementary reaction rate data are presented for the methanation reaction and the steam reforming reaction (see [7] for surface structures) (Figures 7i.5 and 7i.6).
>
> 1. *The methanation reaction:* $CO + 3H_2 \rightarrow CH_4 + H_2O$, catalyzed by the Ru(0001) surface [34]. The Ru(0001) surface is the dense surface of the hexagonal close-packed metals structure of Ru [7]. The adsorption energy of CO has been adapted to 120 kJ mol^{-1}, in order to predict the rate maximum at the proper temperature. DFT calculations tend to predict strong CO adsorption.
>
>   Figures 7i.1 and 7i.2 show simulated temperature dependence of product formation and the corresponding surface coverages. At low temperature, the
>
>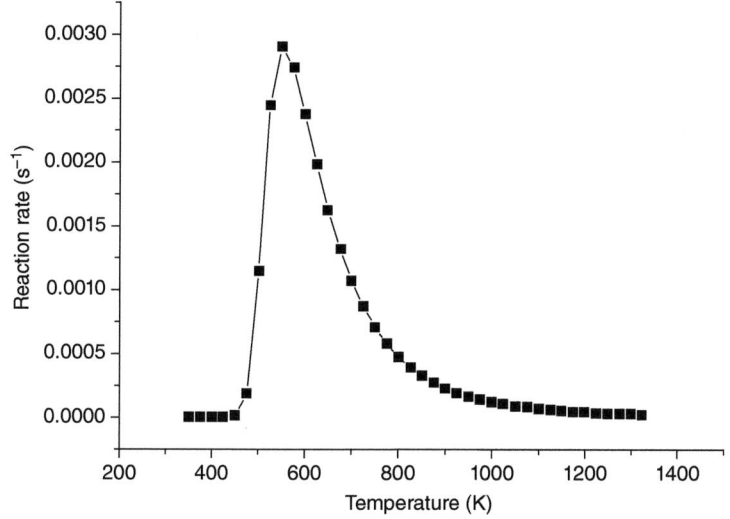
>
> **Figure 7i.1** Simulation of the production rate of methane in the methanation reaction catalyzed by Ru(0001) surface at 20 bar pressure, with a hydrogen:CO ratio of 3. The Ru(0001) surface is the dense surface of the hexagonal close-packed metals structure of Ru. (After [32, 33].)

(Continued)

Insert 7: (Continued)

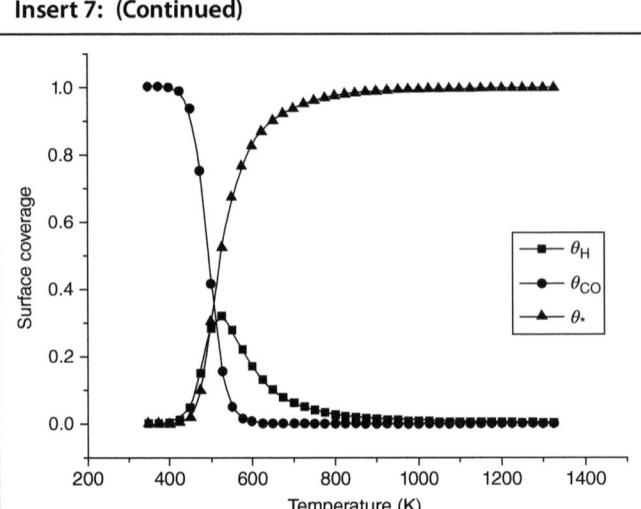

Figure 7i.2 The corresponding surface coverages of Figure 7i.1 at 20 bar pressure, with a hydrogen:CO ratio of 3. (After [32, 33].)

surface is covered with CO that is initially replaced by hydrogen and then by vacancies at higher temperatures.

2. *The steam reforming reaction:* $H_2O + CH_4 \rightarrow CO + 3H_2$, is catalyzed by the open Rh(211) surface of the Rh face-centered cubic structure.

   We compare two scenarios. In one case, the elementary reaction rate of CH_4 activation is relatively fast [33], and in the other case the elementary reaction rate of methane activation is slow.

   a. Figures 7i.3 and 7i.4 show the results of microkinetic simulations with a relatively fast rate of methane activation.
      The steep decline in reaction rate at high temperature is due to replacement of surface vacancies by deactivating C atom coverage.
   b. Figures 7i.5 and 7i.6 show the results of microkinetic simulations with a relatively slow rate of methane activation.

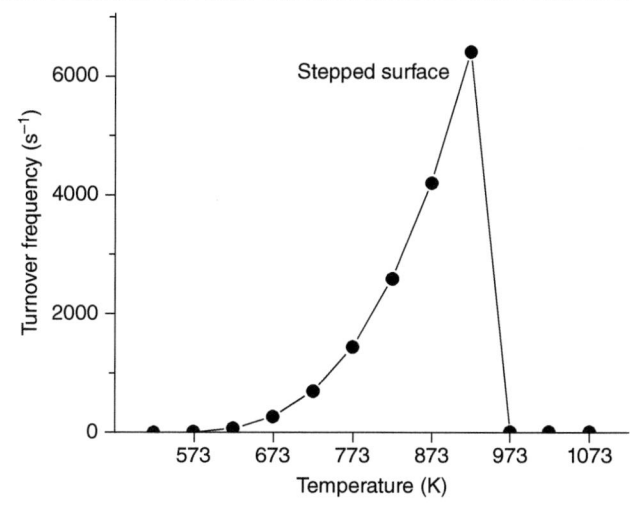

Figure 7i.3 Turnover frequency (in s^{-1}) of methane in steam methane reforming on the Rh(211) surface as a function of temperature (in K). CH$_4$ activation is fast, pressure is 1 bar, with CH$_4$:H$_2$O 1 : 3. (After [32, 33].)

Figure 7i.4 The steady-state surface coverage (in ml) as a function of temperature (in K) corresponding to Figure 7i.3. The surface coverages for the species not indicated in the graph are less than 0.01 ml. Methane activation rate is fast. (Adapted from [33].)

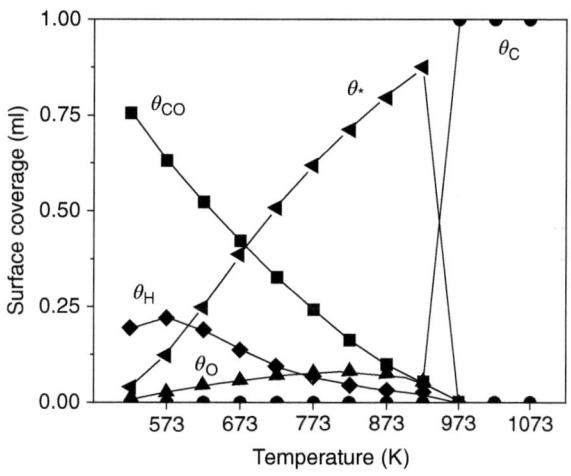

(Continued)

Insert 7: (Continued)

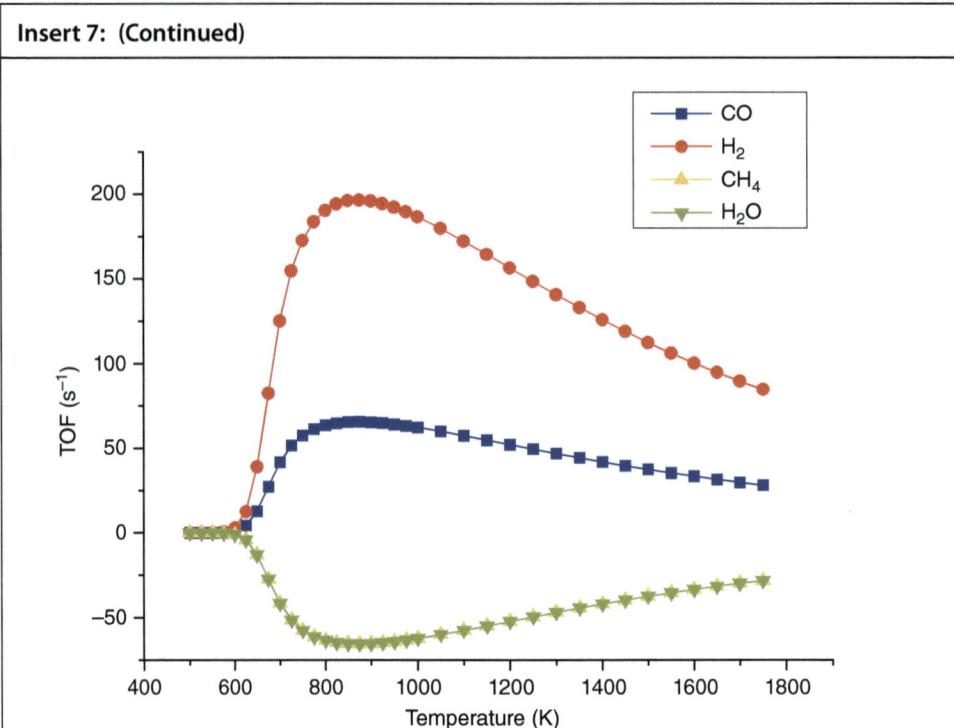

Figure 7i.5 Microkinetics simulations of the steam reforming reaction catalyzed by Ru surface assuming slow methane activation. Methane and H_2O consumption overlap. Reaction conditions are similar to those used in Figures 7i.3 and 7i.4. (After [34].)

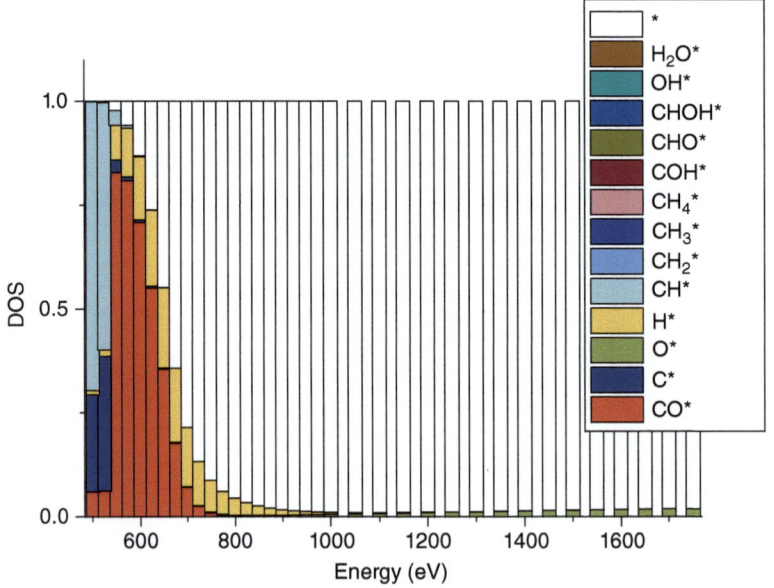

Figure 7i.6 Surface coverages as a function of temperature that correspond to the simulation results of Figure 7i.5. (After [34].)

In the methanation reaction, CO is converted with H_2 to methane. Its lumped sum kinetics scheme has been discussed in Section 3.3.1.2. The data shown in Insert 7 are derived from microkinetic simulations based on all elementary reaction steps.

An example of a reaction energy scheme based on a particular assumed mechanism for this reaction is shown in Figure 3.14. In contrast to that mechanism, the simulations shown in Insert 7 also include additional steps to direct CO dissociation as C—O bond cleavage through intermediate formyl formation and H_2O formation through OH—OH recombination [32, 34].

In the figures of Insert 7 the changes in surface concentration of key reaction intermediates are also shown. For the CO methanation reaction, the dominant surface intermediate is CO. The rate-controlling step of the reaction is the CO dissociation reaction. The lowest barrier for C—O bond cleavage is provided by a reaction path that proceeds through the hydrogen-activated C—O bond cleavage reaction path (see Section 7.3.2). At high temperatures, the rate of reaction will decrease. The apparent activation energy of the overall reaction will become negative, since the activation energy of hydrogen-activated C—O bond activation of 110 kJ mol^{-1} is less than the desorption energy of CO of 120 kJ mol^{-1} (condition (d) of Insert 6).

In Insert 7, two cases are compared for the steam reforming reaction: the case where CH_4 activation is relatively fast as found for Rh, and the case where CH_4 activation is relatively slow, as found for Ru.

In the first case, the recombination reaction of CH_x and O is initially rate controlling. This reaction competes with the rate of O atom removal which has the higher activation energy. Thus, at the higher temperature surface coverage with O decreases with increasing temperature, whereas that of CH_x continues to increase. CH_x can no longer be removed from the surface, so the surface becomes covered with carbon that poisons the reaction. In the second case, the maximum reaction rate relates to the increasing coverage of the surface by CO and its subsequent desorption. Activation of methane competes with that of CO desorption.

3.3.2.6 Summary: The Order of Reaction Rate

The dependence of reaction rate on reaction mixture concentrations is often expressed by a power-law expression of the reaction rate:

$$R_{PL} = k_{app}[A]^x \cdot [B]^y \qquad (3.24a)$$

For the reaction of reagents A and B in Eq. (3.24a), the powers x and y determine the concentration dependence. When x and y are equal to one such an expression would represent the mass-action law rate expression of the chemical reaction.

$$A + B \rightarrow P \qquad (3.24b)$$

For a catalytic reaction the reaction rate is a complex function of surface coverage, so that the powers x and y as well as k_{app} are not true constants of the reaction rate expression. Instead, they are a strong function of reaction conditions and can only be expected to behave ideally (with a constant apparent activation energy and orders of reaction independent of condition) in a very limited temperature and pressure regime.

The change in kinetics in a Sabatier volcano curve plot as a function of surface reactivity is an example of a change in reaction orders, when the interaction of the reaction intermediates with the surface is varied. When the interaction energy is weak, the rate of reaction is positive order in the reagent, but when the interaction energy increases, the reaction order becomes zero or even negative. This change in reaction order is accompanied by an increase in the surface concentration of reaction intermediates.

The reaction rate of the catalytic reaction exhibits a maximum as a function of temperature. At low temperature, the rate of the reaction has a zero or negative order in the concentration of one of the reagents, but beyond the maximum this partial reaction order becomes positive. The surface concentration of this component changes from high to low.

Nearly all catalytic transformation reactions (except truly monomolecular reactions such as isomerization reactions) are reactions between at least two components.

For the monomolecular isomerization reaction, the rate expressions are given by Eqs. (3.14a) and (3.14b). The concentration dependence of the reaction order follows the Langmuir expression of Eq. (3.14b) as shown in Figure 3.5. With increasing concentration, the reaction order decreases from 1 to 0 and the surface concentration saturates. Insert 5, Case (d) provides an analysis of the change in concentration of the reagent.

For non-monomolecular reactions, the reaction rate will always show a maximum as a function of pressure of reagents due to competitive adsorption of reagents. This is illustrated in Figure 3.17 for the case where the adsorption of one of the reaction components is stronger than the adsorption of the other component.

At low pressure, both reaction orders are positive, but while the reaction order of the component with weaker adsorption remains positive at increased pressure, it becomes negative for the component with stronger adsorption. At higher pressure, this is accompanied by dominant surface concentration of the more strongly adsorbing component that suppresses adsorption of the other component.

The relationship between surface coverage and reaction order can be readily understood by comparing Eq. (3.24a) with the Langmuir–Hinshelwood rate expression in Eq. (3.25a).

$$R_{LH} = \frac{k_r K_{ads}^A [A] \cdot K_{ads}^B [B]}{(1 + K_{ads}^A [A] + K_{ads}^B [B])^2} \quad (3.25a)$$

$$R_{LH} = k_r \theta_A \theta_B \quad (3.25b)$$

$$\theta_A + \theta_B + \theta_v = 1 \quad (3.25c)$$

We must remember that the bimolecular LH expression is a relation determined by surface concentrations and the elementary rate constant. The LH expression can be rewritten as a power law rate expression as in Eq. (3.26), but

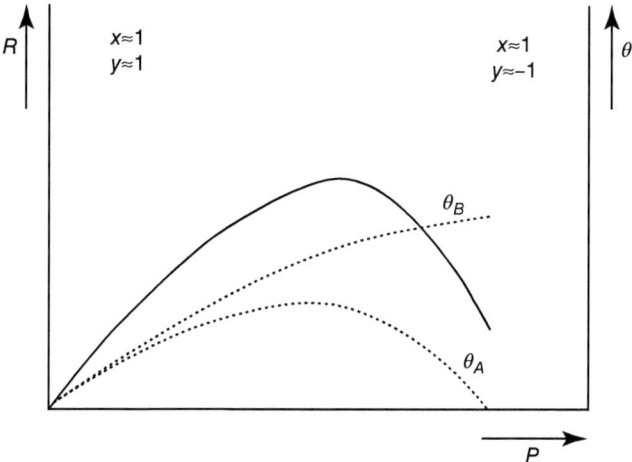

Figure 3.17 Schematic representation of change in rate R (unbroken line) with change in pressure P, with $K_{ads}^B \gg K_{ads}^A$, for a two-component reaction mixture of A and B. x and y are the exponents of [A] and [B] in the power law rate expression.

generally has a complex form:

$$R'_{PL} \approx k_r (K_{ads}^A)^x (K_{ads}^B)^y [A]^x [B]^y$$
$$(x = f(y); y = g(x)) \tag{3.26}$$

The reaction orders x and y depend on surface coverages. Thus, the apparent activation energy and the reaction orders vary with surface coverages and cannot be considered independent.

Equation (3.26) will reduce to the following expressions at various limiting conditions, for example, at low surface concentration or when one adsorbed species dominates. Representative situations are compared in Eqs. (3.27a)–(3.27c):

$$R'_{LH} \approx k_r K_{ads}^A (K_{ads}^B)^{-1} [A][B]^{-1}, \quad K_{ads}^B [B] \gg K_{ads}^A [A]; \ \theta_A \ll 1; \ \theta_B \approx 1 \tag{3.27a}$$

$$R'_{PL} = k_r (K_{ads}^A)^x K_{ads}^B [A]^x [B], \quad x \approx 1 - \theta_A; \ \theta_B \ll 1 \tag{3.27b}$$

$$R'_{PL} = k_r \theta_A (1 - \theta_A) = k_r \theta_B (1 - \theta_B), \quad \theta_v = 0 \tag{3.27c}$$

Equation (3.27a) gives the rate expression that corresponds to the right part of the schematic Figure 3.17. One notes that k_{app} is equal to the elementary reaction rate k_r multiplied by K_{ads}^A / K_{ads}^B. For comparison, in Eq. (3.27b) the rate expression is given for the case where the surface coverage of component B is small and surface coverage of A varies. Now the apparent rate constant is proportional to K_{ads}^B and $(K_{ads}^A)^x$.

Equation (3.27c) gives the rate expression for the case where pressure is high and the surface has no vacancies. This expression shows that the rate will have a maximum related to pressure and that this maximum will occur when $\theta_A = \theta_B = 1/2$.

We will conclude with an example of a competitive adsorption reaction as a function of reactant mixture concentration changes. The ring-opening reaction

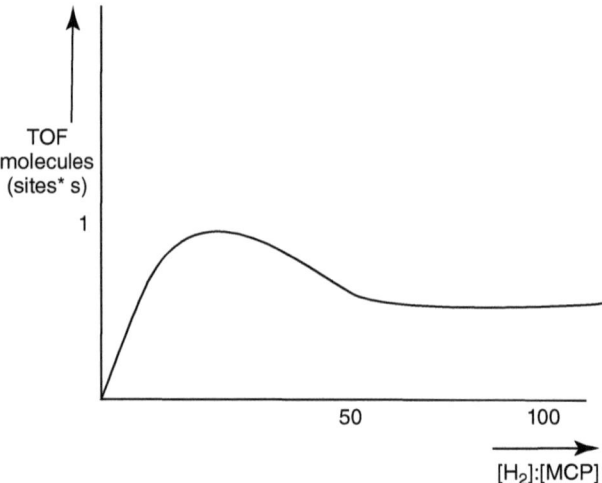

Figure 3.18 Activity of the ring opening of methyl cyclopentane as a function of [H$_2$]/[MCP] for platinum supported on Al$_2$O$_3$ (schematic). (After [35].) (See also [36].)

of methylcyclopentane to give linear and branched alkanes is this type of reaction, catalyzed by Pt. We will study its rate as a function of H$_2$/HC ratio.

As Figure 3.18 shows, the rate of hydrogenation has a maximum as a function of the H$_2$/MCP partial pressure ratio.

At low hydrogen partial pressure, there is an expected positive order in hydrogen because H$_2$ is needed for the reaction. At high partial pressure of hydrogen, the reaction rate becomes negative order in H$_2$, because the adsorption of hydrogen now suppresses olefin adsorption. The positive order of hydrogen at low partial pressure can also be partially due to the need for removing catalyst-deactivating carbonaceous residue formed at low hydrogen pressure.

The dependence of reaction order on condition implies that the order may be very different when partial and total pressures or temperature is varied. For this reason, attempts to deduce mechanistic information from reaction orders have only limited value. Reaction rate expressions based on parameters derived by experiment can only be applied within the reaction condition regime where they were determined and have no *a priori* predictive value outside this regime.

3.4 Transient Kinetics: The Determination of Site Concentration

The kinetic expressions predict a rate of reaction that is proportional to the number of catalytically reactive sites. This prediction is valid only as long as diffusion does not limit the transport from the reagent phase to the catalytically reactive site, which is the condition of intrinsic kinetics. When diffusion is limiting, we must consider extrinsic kinetics that accounts for diffusion. This will be discussed

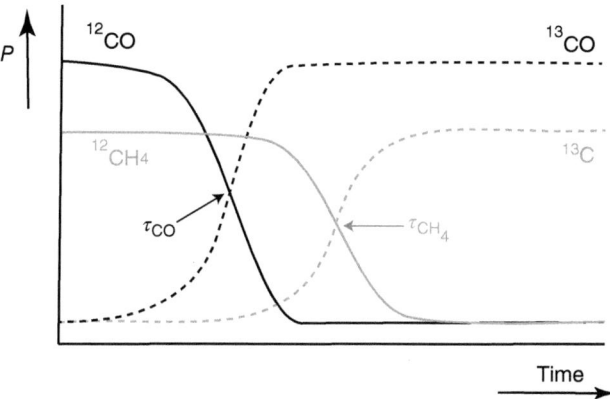

Figure 3.19 Example of transient behavior for CO methanation following an isotope switch, illustrating how residence half-lives and surface coverages change.

in the next section. The rate of reaction normalized to the number of reactive surface atoms and expressed as the number of molecules converted per unit of time is called the *turnover frequency*.

For a practical catalyst, the actual fraction of catalytically active sites on a reactive surface is rarely known. This fraction can be quite small compared to the total number of exposed surface atoms. This is because the catalytic reactivity may also be a strong function of the surface topology of the reaction center. Practical catalysts will have a distribution of different surface structures exposed, and this distribution sometimes relates to the distribution of particle size of the reactive particles.

As long as intrinsic kinetics apply, transient kinetic measurements in combination with steady state kinetics can be used to decouple the kinetics of the reaction from the number of sites.

Since the surface state of the catalyst depends on reaction conditions that are not necessarily maintained after the reaction, the number of reaction centers must be determined during the reaction by a so-called *in situ* experiment. A chemical probe used for this purpose is an isotope switch in the reagent. This has only a small effect on the chemistry of the reaction (when H/D exchange is used, the activation energies are slightly altered), so that the TOF is not essentially affected.

Such an isotope switch event is schematically illustrated in Figure 3.19 for the CO hydrogenation reaction to methane.

For reactions that include an isotope-switch event, a plug flow reactor is used at conditions of minimized back-mixing. When the reaction is at steady state, a reagent molecule is switched to its isotope. The rate of disappearance of the non-labeled reagent and product molecule and the appearance of labeled product is tracked by mass spectroscopy.

In the example of the CO to methane reaction, ^{12}CO can be switched to ^{13}CO and the $^{12}CH_4$ and $^{13}CH_4$ product concentrations are followed as a function of time. The concentrations of labeled and non-labeled products will decrease exponentially with time. In addition, an inert gas is usually injected and the response times of product molecule and inert gas are compared. This helps to determine the time constant of the reactor that must be subtracted from the transient product signal.

The response times τ_x are independent of the number of reactive centers. In the example in Figure 3.19, τ_{CO} measures the replacement of ^{12}CO by ^{13}CO. This is determined by the inverse of the rate constant of CO desorption $k_{des,CO}^{-1}$, as long as the equilibration of CO between the gas phase and the surface is fast compared to the rate of reaction R. The rate of replacement of CH_4 by its isotope-labeled product is different, and given by the inverse of the rate of CH_4 production normalized per site, which is equal to the TOF^{-1}. Once the rate per site is known the number of sites N_s can be calculated since the overall rate of methane production is given by $R_{CH_4} = N_s \cdot TOF$.

The relation between τ and the elementary reaction rates depends on the rate-controlling step. When the rate constant of CO dissociation is rate controlling, τ_{CH_4} is determined by the rate of CO dissociation, but when hydrogenation of adsorbed C atoms is rate-controlling it is determined by the rate of C_{ads} removal.

Steady-state isotopic-transient kinetic analysis (SSITKA) measurements can only be useful as long as diffusion is fast. This prevents their application to microporous systems where pore diffusion becomes a significant factor.

3.5 Diffusion

3.5.1 Concentration Profiles

Heterogeneous catalysts are porous particles with a high internal surface area. The catalytically reactive particles are distributed over this internal surface. Molecules require time to travel the length of the pores to reach the catalytically reactive particles and surfaces. This causes a gradient in the concentration of reacting molecules across the micropores when the rate of reaction is fast compared to the diffusion rate. The time of diffusion τ_D needed to traverse the micropores is compared with the reaction time τ_R, of intrinsic kinetics, equal to R_i^{-1}. As the following equations illustrate, diffusion becomes limiting when the reaction rate is fast, the micropore is long, and the diffusion coefficient is small.

The diffusion time τ_D is given by the Einstein relation:

$$L^2 = 2D\tau_D \tag{3.28}$$

L is the length of the micropore and D is the diffusion constant. Diffusion becomes relevant when:

$$\tau_D = \frac{L^2}{2D} > \tau_R = \frac{1}{R_i} \tag{3.29}$$

R_i is the intrinsic rate constant in the absence of diffusion-limitation.

The time τ_r of catalytic reaction must be small compared to the diffusion time τ_D required to reach the reactive center. Diffusion may be expected to be important at high temperatures when the rate of reaction is also high.

The relation shown in Eq. (3.28) is valid for the case of Knudsen diffusion. The rate of diffusion is determined by collisions with other molecules, and thus the activation energy of diffusion will be low compared to that of a chemical reaction.

Differences in diffusion rates are related to the mass of the particles by $\frac{1}{\sqrt{m}}$ (since $(1/2)mv^2 = (3/2)kT$).

When the size of the molecule equals the size of the micropore, the diffusion rate is dominated by collisions with the pore wall. This is exhibited in the micropores of zeolitic systems, which we will discuss in Section 3.5.3.

3.5.2 Effectiveness Factor

Diffusion will limit the overall rate of a reaction, and this is represented by an inhibiting factor called the *effectiveness factor* η_f. The expression for the overall rate of the reaction R_d when diffusion is limiting becomes:

$$R_d = \eta_f R_i \tag{3.30a}$$

For a spherical particle the effectiveness factor η_f is related to the Thiele modulus ϕ_t [1].

$$\eta_f = \frac{3}{\phi_t}\left[\frac{1}{\tanh \phi_t} - \frac{1}{\phi_t}\right] \tag{3.30b}$$

For a monomolecular reaction whose rate is diffusion-limited ($L^2/D \gg 1/R_i$), the expression for the effectiveness factor η_f reduces to:

$$\eta_f \approx \frac{1}{\phi_t} = \left(\frac{2D}{L^2 R_i}\right)^{1/2} \tag{3.31}$$

Insert 8 gives an insightful derivation of this expression.

Insert 8: Effectiveness Factor Derivation and Diffusion-Related Change of Apparent Activation Energy

Derivation of Effectiveness Factor for Monomolecular Reaction A simple heuristic argument can be used to deduce an expression for the Thiele modulus (after [3]). When pore diffusion is limiting, the rate of product formation is given by

$$\frac{d[P]}{dt} \approx R_d[c_0]L \tag{8i.1}$$

$[c_0]$ is the concentration of reagent at the pore mouth, which by diffusion becomes distributed over the pore of length L.

Since reaction is diffusion-limited, instead of the inequality in Eq. (3.29), the equality in Eq. (8i.2) applies:

$$\frac{L^2}{2D} = \frac{1}{R_i} \tag{8i.2}$$

L can be substituted into Eq. (8i.1) to give

(Continued)

Insert 8: (Continued)

$$\frac{d[P]}{dt} \approx R_i[c_0]L\sqrt{\frac{2D}{R_iL^2}}$$

$$= R_d[c_0]L \qquad (8i.3)$$

The effectiveness factor follows from the comparison of Eq. (8i.3) with Eq. (3.30a).
The Apparent Activation Energy Change by Diffusion Equation (3.34) leads to the conclusion that when a reaction becomes diffusion-limited the apparent activation energy E_{app} becomes half of the apparent activation energy of the intrinsic (not controlled by diffusion) activation energy E_{app}^{intr}. The rate of reaction also becomes proportional $\sqrt{N_c}$ instead of being linearly dependent on the number of active sites N_c. The apparent activation energy as a function of temperature is shown in Figure 8i.1.

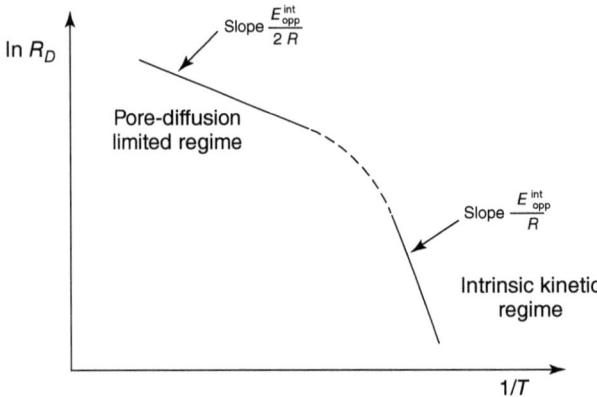

Figure 8i.1 Illustration of the change in activation energy for a reaction that becomes limited by diffusion at high temperature.

From Eqs. (3.31) and (3.30a), the expression for the diffusion-limited rate of the reaction R_d becomes:

$$R_d = \sqrt{\frac{2D \cdot R_i}{L^2}} \qquad (3.32)$$

Interestingly, we see in Eq. (3.33) for the diffusion-limited rate that the apparent rate now depends on the square root of the intrinsic apparent activation energy. Therefore, the apparent activation energy for the diffusion-limited case E_{app}^{d} now becomes half of that of the apparent activation energy that is not diffusion-limited:

$$E_{app}^d = \frac{1}{2}E_{app}^{intr} \qquad (3.33)$$

When the rate of a reaction catalyzed by porous materials is studied as a function of increasing temperature, it may change from non-diffusion-limited to diffusion-limited. Consider the case where, at a low temperature, the rate of reaction is slow and therefore not diffusion-limited. When temperature increases, the rate of reaction also increases and may reach a point where diffusion becomes limiting. The activation energy at high temperature is half that of the activation energy at low temperature, as illustrated in Figure 8i.1.

Experimentally, the decrease in the activation energy of a reaction cannot be considered proof that the reaction is pore diffusion-limited, because even when the catalytic reaction is not diffusion-limited the intrinsic apparent activation energy of the reaction rate may decrease with temperature. This is due to the change in surface concentrations which will affect intrinsic kinetics (see Figure 3.16).

A hallmark of pore diffusion-limitation is the change in rate of reaction when catalyst particle size changes. Because the effectiveness factor depends on pore length, the rate of the diffusion-limited reaction will be affected, but if reaction rate is non-mass transfer limited it will remain unchanged.

Pore diffusion may also change the relative concentration of a gas mixture as a function of pore depth, because of differences in diffusion constant. A heavy molecule has a smaller diffusion constant than a lighter one. This is important in hydrogenation reactions where the hydrogen partial pressure will increase deeper in the pore due to the higher diffusion constant of H_2.

Pore diffusion may affect not only the conversion rate, but also the selectivity of a reaction. If the desired product of a reaction is an intermediate that must not be converted by consecutive reaction steps, this intermediate should be quickly removed from the neighborhood of the reactive center. Since pore diffusion-limitation will limit the rate at which the intermediate product molecule is moved out of the catalyst pore, it will reduce the selectivity of the reaction. The increase of the residence time of the intermediate product in the pore will increase its conversion rate.

An example of this scenario is selective oxidation, where the undesirable consecutive reaction is the total combustion of the initially formed product. For this reason, selective oxidation reactions are catalyzed using catalysts with low pore volume. This must be balanced against the need for enough surface area to achieve an acceptable overall rate of reaction.

In the case of the Fischer–Tropsch reaction (see Section 2.2.3.2) the diffusion constant of H_2 is substantially faster than that of CO. This increases the relative hydrogen concentration deep in the catalysts pore, which will increase the relative rate of undesirable light gas formation by enhancing the rate at which adsorbed hydrocarbon intermediates will become hydrogenated and desorb from the catalyst surface.

When the rate of reaction becomes so fast that a concentration depletion layer of reagent molecules forms around the catalyst particle, the overall rate of reaction is controlled by the diffusion of molecules through this layer. This is called *film diffusion*, and it can occur in reactions at very high temperature or in electro-catalytic reactions at low temperature in the liquid phase. The apparent activation energy of the reaction is then equal to the activation energy of molecular diffusion on the order of only a few kilojoules per mole. One way to verify these interphase concentration or related temperature effects is to vary the space velocity of reactor feed streams using a constant contact time. This can be done by adjusting the amount of catalyst material in the reactor. Film diffusion will cause the observed rate of reaction to change with space velocity. (For a more in-depth discussion refer to Dautzenberg [13].)

3.5.3 Diffusion in Zeolitic Micropores

Unlike Knudsen diffusion, diffusion of molecules in a microporous zeolite does not necessarily depend on the mass of the molecules that diffuse. When the radii of molecules and micropores match closely and the micropore wall is smooth (e.g., in a linear pore) the diffusion constant may become independent of molecular mass. For instance, in the siliceous mordenite zeolite (Figure 3.20) that has linear micropores, diffusion of alkanes becomes independent of alkane chain length and much faster than expected from Knudsen diffusion estimates. Diffusion in this case can be considered to be floating diffusion (Figure 3.20).

The adsorption energy of hydrocarbons with the siliceous wall of the zeolite occurs by physical adsorption. This interaction is based on the dispersive interaction between molecules and zeolite channel atoms, and is proportional to the polarizability of the atoms that are in contact. In the zeolite, this is dominated by the interaction with large oxygen atoms: the polarizability is proportional to

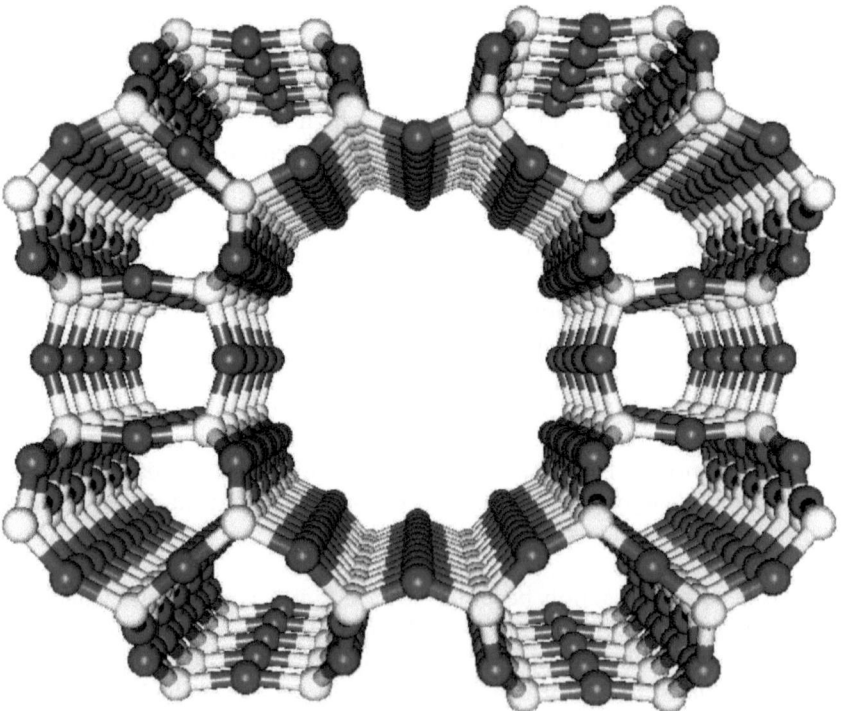

Figure 3.20 Zeolite mordenite structure.

the volume of an atom, and is hence dominated by the larger O anions with little contribution from the smaller cations. The interaction is linear in the number of alkane CH_2 units. It is on the order of 5–10 kJ mol^{-1} per CH_2 unit in contact with a zeolite wall O atom, but decreases when the microchannel chain radius increases. While the adsorption energy increases linearly with alkane length, the diffusion constant remains independent because the area of contact with the zeolite wall does not change.

Figure 3.21a illustrates the independence between the diffusion constant and the length of the hydrocarbons for hydrocarbons of varying length. The importance of the topology of the micropore structure and the shape of the microcavities is illustrated by Figure 3.21b for the faujasite structure. The faujasite structure is three-dimensional and contains large cavities. The average distance between the molecules and the zeolite wall is larger and floating diffusion is absent. The simulated diffusion constants now decrease with the increasing mass of the alkanes. The diffusion rate of molecules in the micropores of a zeolite can vary substantially as a function of micropore filling. Yet, the diffusion time through the inter-crystallite space of the zeolite that consists of the larger meso- and micropores is often longer than the diffusion time through the smaller zeolite micropores. This is because the inter-crystal pores are longer and have a larger relative volume than the micropores of the small zeolite particles.

The interaction of hydrocarbons with the zeolite micropore wall is surprisingly strong, despite the fact that it is based on weak physical dispersive van der Waals interactions. It increases linearly with the number of CH_2 units. For a narrow pore siliceous zeolite with a 10-ring channel diameter such as Ferrierite, the adsorption energy of hexane is 60 kJ mol^{-1}. This is because in the small zeolite micropore the hydrocarbon CH_2 unit makes contact with several O atoms.

This can cause high hydrocarbon occupation of the micropores when high pressure of reaction is used at relatively low temperatures. The diffusion rate decreases because the mobility of molecules is hindered by the presence of co-adsorbed molecules. Equilibration of the reaction intermediates within the micropore becomes fast compared to the equilibration with the external reaction medium outside the zeolite. Selectivity of the reactions is then determined by the adsorption-free energies of reaction intermediates within the micropores. There is a bias for the formation of molecules with a shape that can be adapted to the shape of a particular zeolite micropore (for instance, linear molecules are favored in the case of linear microchannels). The result is shape-selective catalysis. We will discuss this in detail in Sections 9.2 and 9.3.

Another cause of shape-selective catalysis is when molecules are too large to enter the microcavities of the zeolite; in a mixture of molecules with different sizes, only the smaller molecules will be converted. This size restriction also prevents the formation of reaction intermediates that are too large to be accommodated in the micropores. We described this phenomenon earlier for the suppression of coke formation in the catalytic cracking reaction, where formation of large aromatic molecules is suppressed by the size of the zeolite micropores.

When cations are located in the zeolite channels, their high electrostatic field may cause relatively high adsorption energies for polarizable molecules or

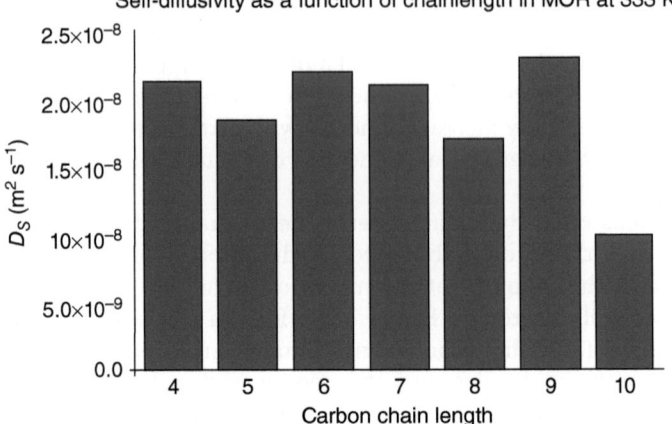

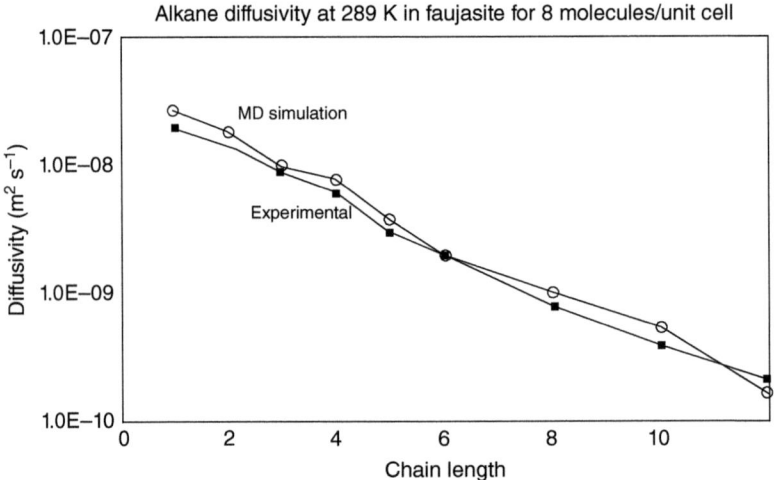

Figure 3.21 (a) Self-diffusivity as a function of chain length in siliceous mordenite at 333 K. (b) Simulated diffusion constants in siliceous faujasite as a function of hydrocarbon chain length. (After [9, 37].)

molecules with a large dipole moment. In this case, diffusional motion depends strongly on temperature. For instance, in zeolites exchanged with earth alkali cations strong sieving is observed for mixtures of benzene and hexane. The polarizable benzene molecule will adsorb strongly and therefore diffuse slowly, but the hexane molecule will have a high diffusion constant.

A unique phenomenon that occurs at high micropore occupation in the narrow one-dimensional micropores of a zeolite is single-file diffusion. The small micropore size restricts the passage of molecules. This is important in hydroisomerization catalysis where the preferred catalyst is mordenite which contains narrow one-dimensional micropores of a diameter comparable to that of linear hydrocarbons (Section 2.2.2.1.3). Mobility is possible only when all molecules that have no vacant position between them displace at the same time. The number of molecules in a file is at its maximum when it fills the zeolite micropore between two catalytically reactive centers. Its dependence on pore occupation θ is [38, 39]:

$$D_s = D_0 \frac{1-\theta}{\theta} \frac{l}{L} \tag{3.34}$$

D_0 is the self-diffusion constant of a single particle and θ the partial occupancy of the pore, L is the length of the pore, and l is the average distance between the catalytic reaction centers in the micropore.

Single-file diffusion leads to a specific relationship between the rate of reaction and the partial pressure of reagent molecules [40].

The kinetics of the hydroisomerization reaction catalyzed by the bifunctional Pt-promoted mordenite zeolite can be approximately described by the lumped kinetics expression of the monomolecular reaction in Eqs. (3.14). As a function of partial pressure of the alkane, the rate will have a Langmuir-type dependence on pressure. For regular micropore diffusion, the intrinsic rate shown by Eqs. (3.14a) and (3.14b) will depend on the effectiveness factor η_f (Eq. (3.31)), as illustrated in Figure 3.22.

Pore diffusion will cause a deviation from the Langmuir-type partial pressure dependence, but the rate will still monotonically increase with partial pressure of alkane. This is not the case for single-file diffusion. Instead, the apparent diffusion constant becomes zero at high coverage, and the rate shows a maximum as a function of partial pressure with a subsequent decrease at high pressure.

In principle, small zeolite crystallites can be used to minimize this strong diffusional resistance, but this is not realized in practice. Instead, catalyst activity is improved by creating mesoporous channels within the zeolite crystallite, so that the effective micropore channel length decreases and the transport of molecules becomes efficient. These channels can be created by partial leaching of Al present in the mordenite crystals that are synthesized with a high Al/Si ratio.

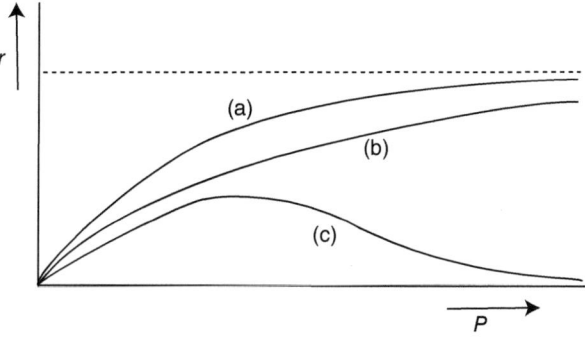

Figure 3.22 (a) The rate of a monomolecular zeolite-catalyzed reaction as a function of pressure (no diffusional limitation). (b) The rate of the monomolecular reaction that is diffusion-limited (pore diffusion, no single-file diffusion). (c) The rate of a monomolecular reaction that is limited by single-file diffusion. (van Santen and Neurock 2006 [9]. Reproduced with permission of John Wiley & Sons.)

In catalytic reactions, the use of such bimodal porous zeolitic systems containing a combination of mesoporous and microporous channels is also effective for reducing the rate of deactivation that is caused by zeolitic micropore blockage because the mesochannels allow a larger fraction of the zeolite particle interior to remain accessible.

References

1. Levenspiel, O. (1999) *Chemical Reaction Engineering*, 3rd edn, John Wiley & Sons, Inc., New York.
2. Carberry, J.J. (2001) *Chemical and Catalytical Reaction Engineering*, Courier Corporation.
3. Boudart, M. and Djéga-Mariadassou, G. (1984) *Kinetics of Heterogeneous Catalytic Reactions*, Princeton University Press.
4. Ferrari, A.M., Neyman, K.M., Mayer, M., Staufer, M., Gates, B.C., and Rösch, N. (1999) Faujasite-supported Ir 4 clusters: a density functional model study of metal–zeolite interactions. *J. Phys. Chem. B*, **103**, 5311–5319. doi: 10.1021/jp990369e
5. Somorjai, G.A. (1994) *Introduction to Surface Chemistry and Catalysis*, John Wiley & Sons, Inc., New York.
6. Ertl, G. (2009) *Reactions at Solid Surfaces*, John Wiley & Sons, Inc., Hoboken, NJ.
7. Somorjai, G.A. and Li, Y. (2010) *Introduction to Surface Chemistry and Catalysis*, 2nd edn, John Wiley & Sons, Inc, Hoboken, NJ.
8. Nørskov, J.K., Studt, F., Abild-Pedersen, F., and Bligaard, T. (2014) *Fundamental Concepts in Heterogeneous Catalysis*, John Wiley & Sons, Inc., Hoboken, NJ.
9. van Santen, R.A. and Neurock, M. (2006) *Molecular Heterogenous Catalysis*, Wiley-VCH Verlag GmbH.
10. Nilsson, A., Pettersson, L.G.M., and Nørskov, J.K. (2008) *Chemical Bonding at Surfaces and Interfaces*, Elsevier, pp. 255–322.
11. Thomas, J.M. and Thomas, W.J. (1997) *Principles and Practice of Heterogenous Catalysis*, Wiley-VCH Verlag GmbH.
12. van Santen, R.A. (2012) in *Catalysis* (eds M. Beller, A. Renken, and R.A. van Santen), Wiley-VCH Verlag GmbH, pp. 143–145.
13. Dautzenberg, F. (2012) in *Catalysis* (eds M. Beller, A. Renken, and R.A. van Santen), Wiley-VCH Verlag GmbH, pp. 536–561.
14. Brunauer, S., Emmett, P.H., and Teller, E. (1938) Adsorption of gases in multimolecular layers. *J. Am. Chem. Soc.*, **60**, 309–319. doi: 10.1021/ja01269a023
15. Niemantsverdriet, J.W. (1993) *Spectroscopy in Catalysis: An Introduction*, 1st edn, Wiley-VCH Verlag GmbH.
16. Chester, A.W. and Derouane, E.G. (2009) in *Zeolite Chemistry and Catalysis* (eds A.W. Chester and E.G. Derouane), Springer, Dordrecht, http://link.springer.com/10.1007/978-1-4020-9678-5 (accessed 30 September 2014).
17. Weckhuysen, B. (2012) *Catalysis*, Wiley-VCH Verlag GmbH, pp. 493–514.

18 Glasstone, S., Laidler, K., and Eyring, H. (1941) *Theory of Rate Processes*, 1st edn, McGraw Hill.
19 van Santen, R.A. and Niemantsverdriet, J.W. (1995) *Chemical Kinetics and Catalysis*, Plenum Press.
20 Rice, O.K. (1967) *Statistical Mechanics, Thermodynamics and Kinetics*, W.H. Freeman and Co..
21 Dahl, S., Logadottir, A., Jacobsen, C.J.H., and Nørskov, J.K. (2001) Electronic factors in catalysis: the volcano curve and the effect of promotion in catalytic ammonia synthesis. *Appl. Catal., A*, **222**, 19–29, http://linkinghub.elsevier.com/retrieve/pii/S0926860X01008262 (accessed 16 October 2014).
22 Moulijn, J.A., Makkee, M., and van Diepen, A.E. (2013) *Chemical Process Technology*, 2nd edn, John Wiley & Sons, Ltd, Chichester.
23 van Santen, R.A., Neurock, M., and Shetty, S. (2010) Reactivity theory of transition-metal surfaces: a Brønsted–Evans–Polanyi linear activation energy–free-energy analysis. *Chem. Rev.*, **110**, 2005–2018.
24 Shustorovich, E. (1986) Chemisorption phenomena: analytic modeling based on perturbation theory and bond-order conservation. *Surf. Sci. Rep.*, **6**, 1–63, http://linkinghub.elsevier.com/retrieve/pii/0167572986900038 (accessed 31 October 2014).
25 Hammer, B. and Nørskov, J.K. (2000) Impact of surface science on catalysis. *Adv. Catal.*, **45**, 71–129.
26 van Santen, R.A., Ciobîcă, I.M., van Steen, E., and Ghouri, M.M. (2011) in *Advances in Catalysis* (eds B.C. Gates and H. Knozinger), Academic Press, pp. 127–187.
27 Stegelmann, C., Andreasen, A., and Campbell, C.T. (2009) Degree of rate control: how much the energies of intermediates and transition states control rates. *J. Am. Chem. Soc.*, **131**, 8077–8082.
28 Nørskov, J.K., Abild-Pedersen, F., Studt, F., and Bligaard, T. (2011) Density functional theory in surface chemistry and catalysis. *Proc. Natl. Acad. Sci. U.S.A.*, **108**, 937–943, http://www.pubmedcentral.nih.gov/articlerender.fcgi?artid=3024687&tool=pmcentrez&rendertype=abstract (accessed 10 July 2014).
29 Bengaard, H., Nørskov, J.K., Sehested, J., Clausen, B.S., Nielsen, L.P., Molenbroek, A.M., and Rostrup-Nielsen, J.R. (2002) Steam reforming and graphite formation on Ni catalysts. *J. Catal.*, **209**, 365–384, http://linkinghub.elsevier.com/retrieve/pii/S0021951702935797 (accessed 8 October 2014).
30 Valdés, Á., Brillet, J., Grätzel, M., Gudmundsdóttir, H., Hansen, H.A., Jónsson, H., Klüpfel, P., Kroes, G.-J., Le Formal, F., Man, I.C. et al. (2012) Solar hydrogen production with semiconductor metal oxides: new directions in experiment and theory. *Phys. Chem. Chem. Phys.*, **14**, 49–70, http://www.ncbi.nlm.nih.gov/pubmed/22083224 (accessed 16 October 2014).
31 Koper, M.T.M. (2011) Thermodynamic theory of multi-electron transfer reactions: implications for electrocatalysis. *J. Electroanal. Chem.*, **660**, 254–260, http://linkinghub.elsevier.com/retrieve/pii/S1572665710003917 (accessed 28 January 2014).
32 van Grootel, P.W. (2012) A theoretical study on the structure dependence of the steam methane reforming reaction by rhodium. PhD thesis. Eindhoven University of Technology.

33 Zhu, T., Van Grootel, P.W., Filot, I.A.W., Sun, S.G., Van Santen, R.A., and Hensen, E.J.M. (2013) Microkinetics of steam methane reforming on platinum and rhodium metal surfaces. *J. Catal.*, **297**, 227–235.

34 Filot, I.A.W. (2015) Quantum chemical and microkinetic modeling of the Fischer–Tropsch reaction. PhD thesis. Eindhoven University of Technology.

35 Vaarkamp, M. (1993) The structure and catalytic properties of supported platinum catalysts. PhD thesis. Eindhoven University of Technology.

36 Paál, Z. (1980) Metal-catalyzed cyclization reactions of hydrocarbons. *Adv. Catal.*, **29**, 273–334.

37 Snurr, R.Q. and Kärger, J. (1997) Molecular simulations and NMR measurements of binary diffusion in zeolites. *J. Phys. Chem. B*, **101**, 6469–6473. doi: 10.1021/jp970242u

38 Fedders, P. (1978) Moment expansions and occupancy (site) correlation functions for a one-dimensional hopping. *Phys. Rev. B*, **17**, 2098–2109. doi: 10.1103/PhysRevB.17.2098

39 Nelson, P.H. and Auerbach, S.M. (1999) Self-diffusion in single-file zeolite membranes is Fickian at long times. *J. Chem. Phys.*, **110**, 9235, http://scitation.aip.org/content/aip/journal/jcp/110/18/10.1063/1.478847 (accessed 25 November 2014).

40 de Gauw, F.J.M., van Grondelle, J., and van Santen, R. (2001) Effects of single-file diffusion on the kinetics of hydroisomerization catalyzed by Pt/H–mordenite. *J. Catal.*, **204**, 53–63, http://linkinghub.elsevier.com/retrieve/pii/S0021951701933755 (accessed 14 November 2014).

4

The State of the Working Catalyst

4.1 Introduction

In this chapter, we will introduce the chemistry of the catalytically reactive phase that develops when a catalyst is exposed to reagents during a catalytic reaction. This is a story of the interplay between the inorganic chemistry of the catalyst and its changing surface state, from its initial state when freshly prepared, through its intermediate stages during catalyst activation, and then to its operational state when it executes the catalytic reaction, and finally to its deactivating state.

The catalytically reactive phase is rarely in its thermodynamic global minimum. The small particles of catalytically reactive particles are thermodynamically quasi-stationary, because there is a driving force for their growth into large stable particles. A high surface area support can be considered thermodynamically unstable as well, because its higher density induces more favorable chemical potential. The diffusion of the catalyst particle's atoms provides a barrier that stabilizes the catalyst in its local minimum, but external conditions resulting from the catalytic reaction may reduce this barrier to change. The state of the catalyst can be affected during catalytic reaction by local increases in temperature caused by the heat of the reaction, as well as by changes in the composition of the catalyst surface due to contact with the reacting gases and the formation of non-selective products. A local temperature increase will increase the mobility of the catalyst's atoms on the particle fragments, which deactivates the catalyst.

Adsorption of reactant and reaction intermediates may lead to substantial reconstruction of the exposed surfaces, since this presents a different driving force to a new state of equilibrium. As we will see, this can lead to either activation or deactivation of the reactive catalyst particles. Some examples of non-selective catalyst deactivating reactions are the carbonaceous residue forming reactions that poison surface reactive sites. We will begin this chapter with an introduction to surface structure reconstruction of the transition metal surfaces.

Reaction may not only change the surface of the catalyst, but may also change its bulk state. An example is the change of the originally metallic iron catalyst to carbide in the Fischer–Tropsch reaction. The rate of such changes will be sensitive to variations in catalyst preparation and activation conditions. We will discuss these factors in the following subsections.

The changes that the catalyst undergoes when it reacts affect the structure and concentration of the reactive centers responsible for the desired catalytic reaction. Therefore, it is essential to understand the relationship between the structure of the catalyst's reactive centers and the catalyst's reactivity, which must also include an understanding of particle size dependence. Promoting compounds and catalyst modifiers are often added to the reaction mixture to inhibit catalyst deactivation or increase the selectivity of a reaction. We will discuss alloy catalysts as an example of systems that can have different catalytic performance than the components. This is not only by composition change, but also by the preferential exposure of particular sites. In this chapter, we will also introduce chemical theories of reactivity–structure relations. We will apply molecular theory to explain particle size dependence of transition metal-based catalysts. The difference in particle size dependence as a function of reaction is of special interest.

There is a wide selection of catalytically important materials based on the oxides. High surface area oxides of low intrinsic reactivity are often used as supports. Despite their low reactivity, the chemistry support surface and catalytically active precursor particles is important to catalyst synthesis as well as to catalyst activation. These factors affect both catalyst performance and catalyst lifespan. Advanced spectroscopies have dramatically increased our understanding of the changes that occur within catalyst materials during activation (see [1]).

Solid acids such as the zeolites are used extensively in a variety of catalytic reactions (see Chapter 2). The micropore structure of the zeolite will affect stability as well as activity of the catalyst, since its pore size in the case of hydrocarbon conversion will affect the deposition of carbonaceous residue. The formation of carbonaceous disordered or graphitic material on the catalyst can lead to catalyst deactivation by covering reactive centers or blocking micropores. When these materials are exposed to steam in order to regenerate them, large-scale structural and compositional changes can occur. Additives to the catalyst can be used to stabilize it and prevent these changes. A final subsection will illustrate structural changes at the surface of reducible mixed oxides that are applied as selective oxidation catalysts.

The catalytic system can be considered to exist in a state of slow transition, stationary only at specific intervals. It is transformed from its initial pre-reaction state to a stationary primary reactive state that is frequently at optimum performance, but over time the catalyst will deactivate. The rate of deactivation will differ, as it is dependent on the particular cause of deactivation. Initially, these changes may increase or decrease the activity and the selectivity of the catalyst, and will ultimately affect its stability.

In the first subsection, we will review structural changes of catalysts in relation to their catalysis. Since these effects are particular to the specific catalytic system, some basic features of catalytic reactivity in relation to catalyst composition will also be introduced. These topics will provide an introductory basis for part II where we provide a more detailed molecular description of catalyst reactivity.

4.2 Surface Reconstruction

In a catalytic system, strong metal atom adsorbate bonding competes for the electrons needed to bind the catalyst surface metal atoms together. The weakening of the surface metal–metal bonds will initially lead to only small displacements of the surface atoms. However, when the surface concentration of adatoms increases, surface layer tension caused by the interaction between the slightly expanded surface sites increases. This tension may push the surface overlayer into a new structure that has a long range order of surface metal atoms very different from the free surface. At the atomic level, the basis for reconstruction is the combination of the effects of weakened surface bonds and increased surface tension due to lateral interaction between the adsorbate and metal atoms.

A well-known example of this phenomenon occurs when the Pt(100) surface is exposed to the CO oxidation reaction. In the bulk metal, the Pt atoms are arranged in a face-centered cubic (fcc) structure (see Figure 4.1).

In the CO oxidation reaction, the catalytic cycle includes surface reconstruction. In a vacuum, the Pt (100) surface is unstable; the more stable surface is the (100) hex reconstructed surface (see Figure 4.1). An overlayer of Pt atoms with a (111) surface structure forms over the (100) terminated surface. The nearest neighbor coordination of the surface metal atoms within the (100) surface is 4, while within the (111) surface it is 6. This provides a driving force for the outer layer surface atoms to obtain (111) surface packing, but this comes at a cost: there is a slight mismatch of the reconstructed overlayer hexagonal (111) surface with the supporting cubic (100) second layer Pt atoms (Figure 4.2).

When CO adsorbs to the (100) hex reconstructed surface, metal–metal atom bonds weaken and the Pt(100) hex surface is transformed to the non-reconstructed (100)1 × 1 surface. The cost of the creation of this less stable surface (in a vacuum) is overcome by the increased bond energy of CO with the more reactive, metal surface atoms of the non-reconstructed (100) surface with lower coordinative saturation.

The Pt(100) hex system is interesting because catalysis will only occur on the non-reconstructed (100)1 × 1 surface that is generated during the reaction. While CO molecules will adsorb to the less reactive (100) hex reconstructed surface, O_2 molecules will not dissociate from this non-reactive surface phase. Transformation to the non-reconstructed (100) surface occurs when at least five

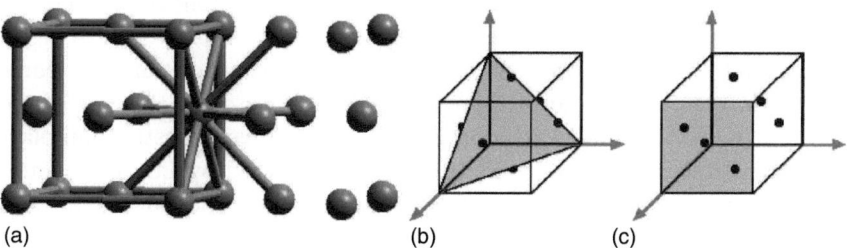

(a) (b) (c)

Figure 4.1 (a) Pt bulk structure indicating the bulk metal atom coordinated to 12 nearest neighbors as well as nearest neighbor atoms over a cube. (b) The (111) cross-section through the fcc cube. (c) The (100) cross-section through the fcc cube.

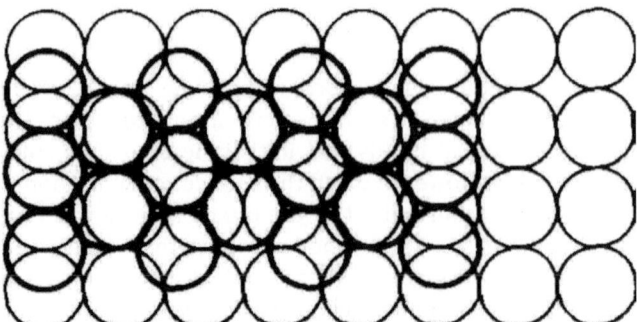

Figure 4.2 Model of the (100) hex reconstructed surface. The (111) surface overlayer is situated on top of the (100) surface of Pt. (Van Hove *et al.* 1981 [2]. Reproduced with permission of Elsevier.)

CO molecules are adsorbed near each other. On this more reactive Pt(100)1 × 1 surface, O_2 will dissociate so that CO oxidation to CO_2 occurs by recombination with the O atoms. The rate of reaction is enhanced by the CO adsorption induced reconstruction! At the end of the catalytic reaction cycle, the surface returns to its initial non-reactive state. This phenomenon of transient site generation by the catalytic reaction itself is called *self-organization*. A detailed discussion of the consequences of self-organization to kinetics will be presented in Chapter 5.

Reconstruction of surfaces, when covered with a high concentration of adsorbates appears to be a general phenomenon. C adatoms and O adatoms adsorbed at high coverage to a dense metal surface induce reconstruction to a more open surface.

An example of the reconstruction of a more stable surface is demonstrated by the change in surface state of the Co(111) surface of a small Co particle that has preferentially the fcc structure. When C adsorbs with increasing surface concentration, this surface changes to the more open Co (100) surface. With increasing concentration of adsorbed C, there is an increasing repulsive interaction between the sites with adsorbed C atoms. The strong interaction with the C atoms weakens the Co–Co metal bonds, which gives a driving force to increase their bond distances. The increased strain within the surface layer is released by transformation to the more open (100) surface in which the Co atom density is also lower than in the (111) surface (see Figure 4.3).

In addition to the reduction of the surface strain, the coordination of C atoms to four surface atoms instead of three also helps to overcome the cost of the energy loss in metal atom–metal atom coordination. This reconstruction plays an important role in Fischer–Tropsch catalysis, catalyzed by Co. It generates the step-edge sites essential for selective reaction (see Sections 4.4.1 and 5.3).

An instructive illustration of a similar surface reconstruction phenomenon is provided by the missing row reconstruction of the Cu(110) surface (see Figure 4.4).

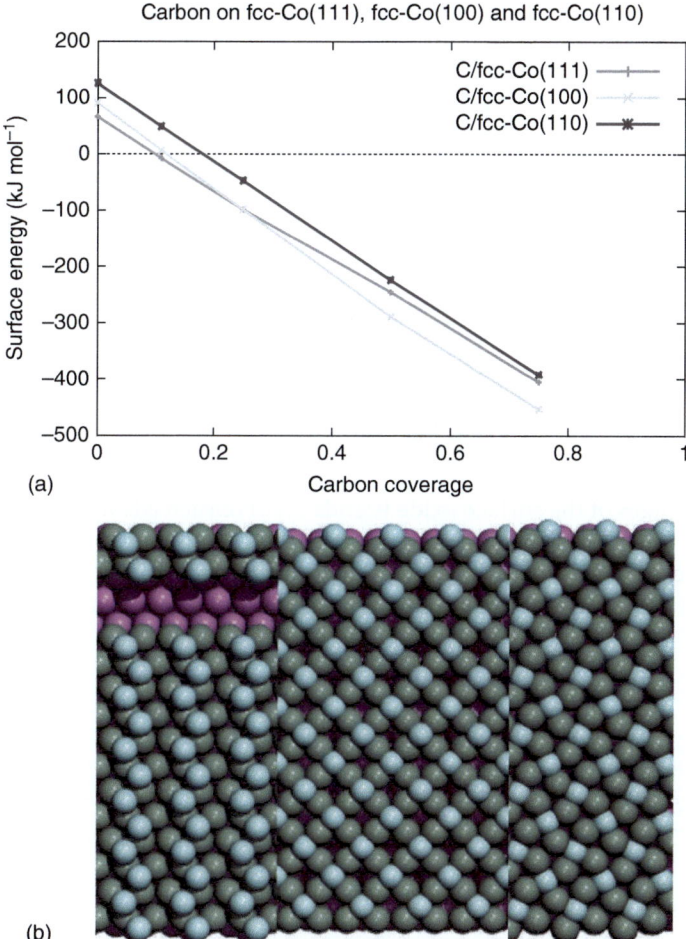

Figure 4.3 (a) The relative stability of the Co(111), Co(100), and Co(110) in the absence and in the presence of adsorbed carbon. The C-covered Co(100) surface becomes more stable than the C-covered Co(111) surface when the C coverage exceeds 25%. (b) Surface structures of carbon adsorption on the reconstructing Co(111) surface (the unit cell) was doubled to put together the missing rows (left side). This shows that the density of surface atoms on the (100) surface is lower than that of the (111) surface, which will lead to surface corrugation and the generation of step-edge sites (see Section 4.4.1); the (100) surface reconstruction with C, without additional displacement of the metal atoms in this plane (middle); the final clock reconstruction of the (100) surface which is the stable structure of the highly C-covered Co(100) surface (right side). (Ciobîcă et al. 2008 [3]. Reproduced with permission of Elsevier.)

In this figure, the different energy cost and energy gain contributions according to DFT calculations are shown for the surface reconstruction that occurs on the highly covered Cu surface. With increasing O coverage the strain in the surface increases, and is then reduced by the removal of rows of Cu atoms. The resulting surface oxide structure is very stable compared to non-ordered O adsorption surface structure. In the Cu-catalyzed oxidation of methanol to formaldehyde, reaction between adsorbed methanol and surface oxygen will

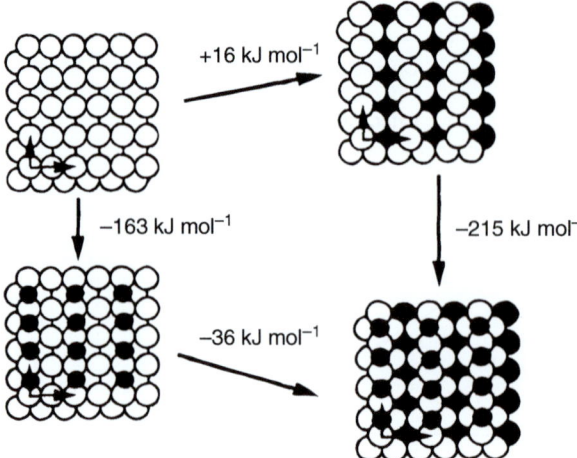

Figure 4.4 Changes in overlayer energies of O adsorbed on a Cu(110) surface compared with the corresponding changes in transition metal surface energies. (Frechard and van Santen 1998 [4]. Reproduced with permission of Elsevier.)

only occur at the edges of the surface oxide islands [5] of oxygen adsorbed to the Cu surface.

An additional example of the changes in stable surface phase with changing surface composition is shown in Figure 4.5. It considers the structure the Pt

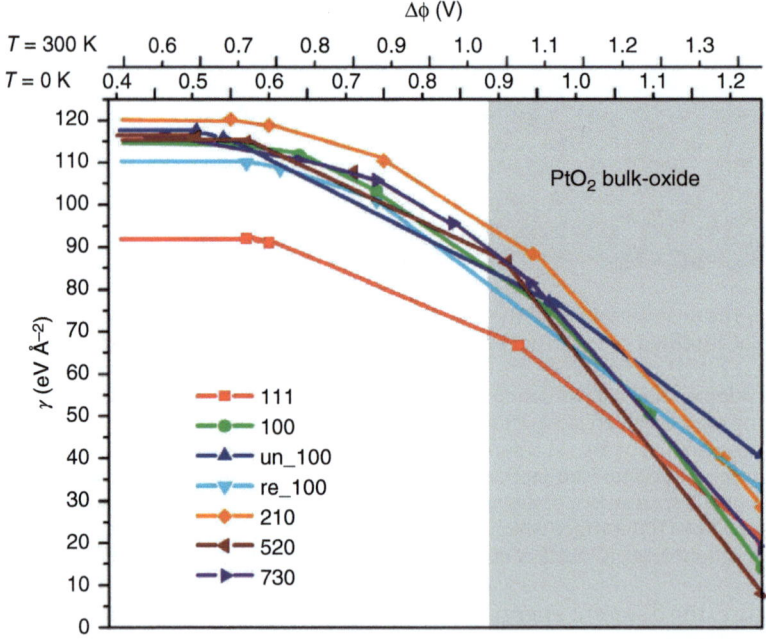

Figure 4.5 Computed electrochemical phase diagram for the electro-oxidation of Pt surfaces showing the surface stability γ as a function of the electrode potential $\Delta\phi$ at zero and room temperature for $P_{H_2} = 1$ atm. The phase diagram shows the conditions under which the different phases become thermodynamically stable. The region at which the bulk-oxide (PtO_2) is stable is in gray. (Zhu et al. 2013 [6]. Reproduced with permission of Royal Society of Chemistry.)

electrode used for water decomposition. The Pt anode will generate an increasing concentration of adsorbed O atoms on its surface, which leads to different surface structures as a function of electrode potential. The loss in the density of Pt surface atoms when the low density open Pt surfaces become stable leads to roughening of the surface to accommodate for the Pt atoms that have to find a new position.

In this case, the Pt(111) surface always remains the most stable but the relative stabilities of the other surfaces are strongly dependent on surface concentration. Interchanges of relative stability occur between the respective (100) and (110) surfaces.

We conclude that different reaction conditions will affect not only reaction rates, but also the surface structure of the catalytic reactive phase during reaction. Both of these changes must be considered in a predictive theory of catalysis, since the state of the surface has a large effect on catalyst performance.

4.3 Compound Formation: Activation and Deactivation

In the previous subsection, we have seen that surfaces will reconstruct when the coverage of strongly interacting adsorbate is high. This may also initiate bulk compound formation such as the state of iron in the Fischer–Tropsch reaction. In this reaction, CO dissociates to give "C_1" intermediates that are inserted into the growing hydrocarbon chains, where the C atoms will react with Fe as a site reaction. In the Fischer–Tropsch reaction, the iron is transformed into a carbide phase as Fe_5C_2 (Hägg iron carbide) and catalysis occurs not by the transition metal, but instead by this carbide phase. This appears to be beneficial to the selectivity of the reaction.

This transformation of a reactive metal by catalytic reaction into a bulk compound is common when reactive gasses such as H_2S or O_2 are used. The transition metals are converted into the corresponding sulfides by desulfurization of sulfur-containing process streams or into the corresponding oxides in oxidation reactions. For this reason MoS_2- and WS_2-based catalysts are used instead of metals such as Ni or Pt which become poisoned by the sulfur. As we will discuss in Chapter 11, the cations of reactive metals such as W and Mo have a hydrogenation activity in the absence of sulfur comparable to metals such as Pt. Similarly, active oxide catalysts are based on the oxides of metals located in the same region of the periodic table.

In addition to bulk compound formation, volatile compounds may also be formed, which can lead to deactivation of the catalyst or to a different catalytic reaction. An example of this change is in the formation of metal carbonyl complexes at high pressures of CO (see Figure 4.6). The formation of volatile metal carbonyls at high pressure prevents the use of metals such as Ni, Co, or Fe in the Fischer–Tropsch reactions. Since Ru has a much lower tendency to form carbonyls, it can be used at high pressures, where it remains very active and can be used to produce long hydrocarbon chain waxes.

Intermediate carbonyl formation may lead to sintering in supported metal catalysts and can thus cause catalyst deactivation.

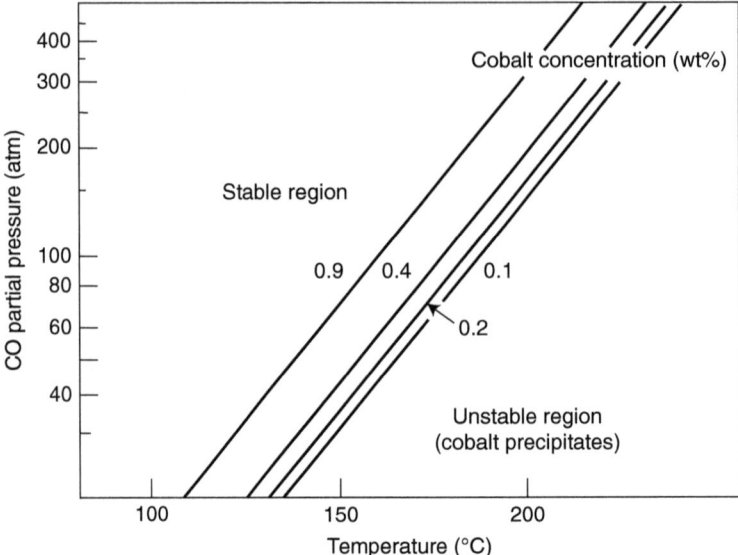

Figure 4.6 Instability of a cobalt carbonyle complexes [$Co_2(CO)_8$ + $HCo(CO)_4$] as a function of temperature and equilibrium carbon monoxide partial pressure with respect to bulk Co. This defines the region where the homogenous complexes can be used in the hydroformylation reaction (see Section 2.2.3.3). The figure can also be used to deduce the stability region of bulk Co as a function of temperature and pressure. (Falbe 1970 [7]. Reproduced with permission of Springer.)

This phase diagram of Figure 4.6 shows that high CO pressure and low temperature stabilizes Co_2CO_8, while the metal Co phase can be stable at the same pressure but at higher temperatures.

Oxidation may also lead to evaporation of a catalyst. For example, in MoO_3 formation, evaporation of the MoO_3 molecule (starting at 700 °C) can lead to deactivation.

4.4 Supported Small Metal Particles

High surface area non-reducible oxides such as alumina or silica are used as supports for small catalytically active transition metals or other catalytically active compounds. Even when catalysis is not pore-diffusion-limited overall, the support structure and its surface may have a significant effect on total catalysis because the active catalyst particles may have a very different size distribution. When an equal amount of transition metal per weight of support material is distributed on supports with differing surface areas, the particle size distribution of catalytically active material can vary markedly. Catalysis will be affected because the distribution of the various catalytically active sites on a particle will differ for large particles and small particles. For instance, the distribution of corner, edge and terrace sites will be dissimilar. Catalytic reactions that are structure-dependent will be especially affected. We will discuss structure–reactivity relations due to differences in particle size in Section 4.5.1.

4.4.1 Nature of the Support Material

The nature of the support will also affect the particle size distribution because its chemistry determines the intermediate complexes that are formed in catalyst synthesis. In solution, alumina will behave as a base material with a surface covered by basic hydroxyls. In contrast, a silica surface will also be terminated by OH groups, but these will be weakly acidic (see Section 9.2.2.1). As a result, an alumina will react with negatively charged complexes such as $PtCl_4^{2-}$ by ion exchange with the negatively charged OH^- surface hydroxyls, and when dried, this catalyst will initially contain Pt that is well distributed over the support surface. If $Pt(NH_3)_4^{2+}$ salt is used instead, no ion exchange can occur. After drying, the Pt particles would form into large Pt containing crystals that transform into large Pt particles upon reduction. When a similar salt solution is used with silica, the positively charged $Pt(NH_3)_4^{2+}$ cations exchange with the silica protons, resulting in a high dispersion of Pt particles. These examples illustrate that the colloid chemical properties of the support must be tuned to the catalyst synthesis procedure in order to control the size distribution of the particles at reaction conditions (see Insert 1).

Insert 1: Catalyst Preparation

The point of zero charge (PZC) is the pH at which the net charge of the surface is zero in water solution (Figure 1i.1).

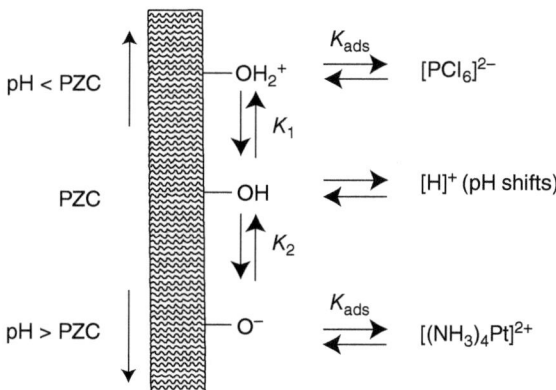

Figure 1i.1 Surface charging of oxides in aqueous systems depending on pH and PZC, and equilibration with Pt cations or anions in solution. (Regalbuto 2009 [8]. Reproduced with permission of John Wiley & Sons.)

In an acidic solution (pH < 7) surface hydroxyls can become protonated and hence adsorption of solvated anions becomes possible to the now bare surface cations or to the positively charged Lewis acid surface cation–water complex. Similarly, when the solution is basic (pH > 7), the surface may deprotonate and solvated cations will adsorb. The latter will become the dominant adsorption mode for predominantly acidic surfaces

(Continued)

> **Insert 1: (Continued)**
>
> (low PZC), whereas the former is the dominant adsorption mode for materials composed predominantly of basic hydroxyls (high PZC) (after [9]). A comparison of the PZC's for different materials in shown in Table 1i.1.
>
> **Table 1i.1** PZC of common oxide and carbon supports.
>
> | Support | PZC |
> | --- | --- |
> | MoO_3 | <1 |
> | Nb_2O_5 | 2–2.5 |
> | SiO_2 | 4 |
> | Oxidized activated carbon | 2–4 |
> | Graphitic carbon | 4–5 |
> | TiO_2 | 4–6 |
> | CeO_2 | 7 |
> | ZrO_2 | 8 |
> | Co_3O_4 | 7–9 |
> | Al_2O_3 | 8.5 |
> | Activated carbon | 8–10 |
>
> Source: Regalbuto 2009 [8]. Reproduced with permission of John Wiley & Sons.

The surface distribution of Brønsted basic OH groups and Brønsted acidic protons will be very different for dried materials compared to the water-phase exposed surfaces of Insert 1. As we will see, aluminas applied in gas phase catalysis may have considerable solid acid reactivity, despite their dominant distribution of surface hydroxyls in the water phase. Surface reactivity of the oxides will be extensively discussed in Section 11.2.

Chemical interaction between catalyst precursor and catalyst support may also be important in consecutive catalyst activation steps after wet synthesis. Compound formation with metal cations will affect reducibility as well as reactivity, for example, in the formation of Co aluminates when Co catalysts are prepared. Reduction of these complexes may require high temperatures that cause the formation of large Co particles on the catalyst. The high temperature increases the mobility of the Co atoms on the support surface, leading to their aggregation. In addition, the formation of these compounds may prevent complete reduction, so that reduced metal will be in close contact with cationic Co ions. This may generate unique catalytic sites between the edge of the metal particle and the catalyst support. The ions themselves may be catalytically active as well, as observed for ions of Au or Rh when reducible supports such as CeO_2 are used [10].

In the preparation of sulfide catalysts, the active catalyst must be fully in the sulfide state when used in desulfurization catalysis. The oxidic precursors of these catalysts that are based on Co and Mo may form intermediate compounds with the alumina supports that are used. These compounds are difficult to reduce, and thus calcination (oxidation) which leads to the formation of these compounds must be prevented in the drying step. This is done by using a H_2S atmosphere during drying so that sulfides are formed directly from the active catalyst precursors.

Insert 2 shows how the structure of this catalyst changes with reaction and the catalyst treatment for commercially used platforming catalysts based on Pt/Sn (see also Section 2.2.2.1.2).

Insert 2: The Pt/Sn Platforming Catalysts

See Figure 2i.1.

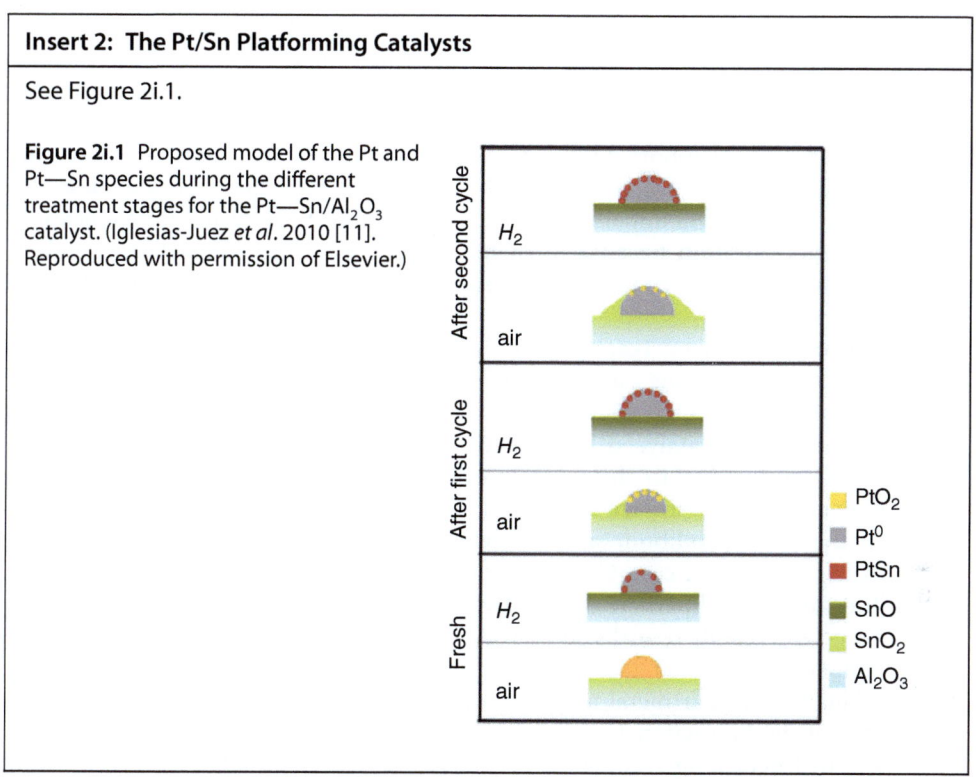

Figure 2i.1 Proposed model of the Pt and Pt—Sn species during the different treatment stages for the Pt—Sn/Al_2O_3 catalyst. (Iglesias-Juez *et al.* 2010 [11]. Reproduced with permission of Elsevier.)

Advanced *in situ* spectroscopies [1] provide details of the structural changes of the Pt/Sn on an Al_2O_3 support as a function of catalyst pretreatment. This is illustrated by Figure 2i.1. After synthesis, Pt is initially present as an oxidic phase, which becomes reduced by H_2 exposure. When this is followed by an oxidation step, SnO_2 starts to attach to the partially oxidized Pt. When the particle is again reduced, Sn starts to attach to the Pt particle.

4.4.2 Reactivity and Stability

The reducibility of a small metal particle will differ from that of a large metal oxide particle even when the reaction is not stabilized by intermediate compound formation with the support. As we will see in Chapter 5, a transition metal particle has high surface energies compared to an oxide, which affects the thermodynamics of oxide particle reduction by requiring higher temperatures for slam metal particle reduction. This also makes small metal particles easier to oxidize, which may lead to rapid deactivation of the initially reduced metal particles when oxygen is present.

An example is the use of small metal particles in the reforming reaction of methane by steam to produce synthesis gas. As we will discuss in Chapter 5, this reaction is structure-dependent and is preferentially catalyzed by small metallic particles. The high fraction of reactive corners and step-edges on these particles leads to the ready activation of methane, which is the rate-controlling step of this reaction. However, the atoms of the reactive small particle transition metal atoms not only activate methane, but also strongly bind C and especially O atoms that are generated by the dissociation of water. Highly active catalysts based on small Rh and Ru metal particles a few nanometers in size and distributed over an oxidic support deactivate rapidly because of their easy oxidation, but this will not occur for catalysts with larger metal particles. Catalysts based on larger particles will initially be less active, but may be preferred for applications of longer duration because of their increased stability. When these catalysts are used for longer times they are relatively more active than the small-particle catalysts.

In other catalytic reactions, the preferred oxidic state of the small metal particles in oxidizing conditions may be beneficial. An example is the use of Rh catalysts for CO oxidation. This reaction requires a high temperature of reaction when catalyzed by reduced Rh particles because the high strength of CO adsorption to the metal prevents low temperature reactivity. At low temperature, the surface is covered with CO, which inhibits O_2 dissociation. At the same condition, small metal particles of Rh rapidly convert to their oxide, and in this case CO oxidation can be performed at low temperatures. The small oxidic particle is an active catalyst because CO adsorption is weak, hence CO readily desorbs so that O_2 dissociation is not hindered at the lower temperatures.

Small molecular metal particles such as Pt or Pd are present in the micropores of the bifunctional zeolite catalysts that are used for hydroisomerization and hydrocracking reactions (see Chapter 2). The high reactivity of the small nanometer-sized particles will initially lead to the coverage of the reactive particles with a carbonaceous overlayer, producing a carbidic particle with reduced reactivity. This may be beneficial, since hydrogenation and dehydrogenation activity is maintained, but non-selective hydrogenolysis reactions are suppressed.

4.4.3 Summary

The high surface area support used to distribute reactive catalyst particles may affect their reactivity for several different reasons. The mode of catalyst preparation and intermediate surface compound formation will affect the particle size distribution and the state of the catalyst particle, and these may have important consequences for activity as well as stability.

4.5 Structure Sensitivity of Transition Metal Particle Catalysts

There are two opposing views on the structure dependence of transition metal catalysts. According to the classical view of Taylor [12], a practical catalyst contains an inhomogeneous distribution of sites that differ in their local arrangement of surface metal atoms. High catalytic reactivity usually requires a very specific arrangement of surface atoms in the catalytically reactive site. According to this view, only a fraction of the exposed surface atoms will be catalytically active. In the alternative view based on the checkerboard surface model of Langmuir [13], sites are uniformly distributed and all of the surface atoms are active.

It is now known that most catalytic systems have site distributions that agree with the Taylor view. Catalyst preparation will lead to particles terminated by relatively stable surfaces with low reactivity. The dominant reactive sites will be found at the terminating edges of the surfaces where the surface atoms display lower coordinative unsaturation leading to higher reactivity.

As shown in Figure 4.7, one can distinguish three kinds of structure dependence when catalyst reactivity is studied as a function of catalyst particle size. The rate of reaction, normalized per exposed surface metal atom (turnover frequency, TOF) can increase, decrease, or show a maximum when reduced transition metal particle size is varied at the nanometer scale.

An important theoretical insight arises from the relationship between site preference and the type of chemical bond that must be activated. We will provide an introductory discussion here. A detailed explanation of the quantum-chemical basis of this surface-site-dependent reactivity can be found later in Chapter 11.

4.5.1 Particle Size and Structure Dependence of Heterogeneous Catalytic Reactions

The three types of particle size dependence are as follows: in the hydrogenolysis reaction, C—H and C—C bonds are activated; in the hydrogenation reaction, C—H bonds are formed; in the methanation reaction the C=O bonds are cleaved.

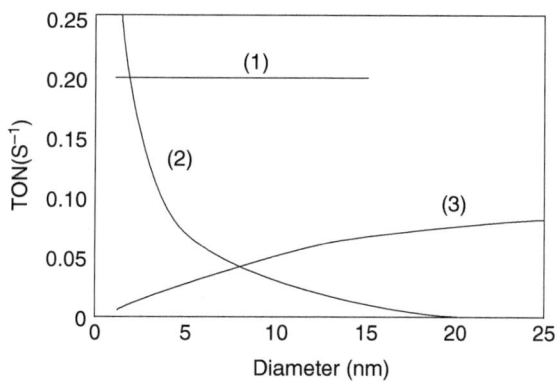

Figure 4.7 Turnover number for selected reactions versus particle size. (1) Benzene hydrogenation on Pt/SiO$_2$, (2) Ethane hydrogenolysis on Pt/SiO$_2$, (3) CO hydrogenation on Ru/Al$_2$O$_3$ [14].

There can be several reasons for the wide variation in the dependence on transition metal particle size of a catalytic reaction, and these factors are discussed in this chapter. When the reaction mechanism and rate controlling step do not change with variation in particle size changes in the rate of a elementary reaction step are responsible for reactivity changes. A different particle size will alter the relative concentration of surface sites. Surface sites vary in topology and coordinative unsaturation of the surface atoms. We will consider here the differences in particle size dependence as a function of the types of elementary reaction steps one needs to distinguish.

Insert 3: Structures of Transition States and Corresponding Transition State Energies

In this insert, we compare the transition state structures of methane, and CO and NO activation on different surfaces respectively.

The transition state and activation barriers of methane are shown in Figures 3i.1. Methane dissociates over a single metal atom. This CO dissociation transition state and dissociation energetics are shown in figures 3i.2. CO dissociation as well as NO dissociation occurs over an ensemble of surface atoms. The transition state of NO dissociation is shown in figure 3i.3 and the reaction energetics is illustrated in figures 3i.4.

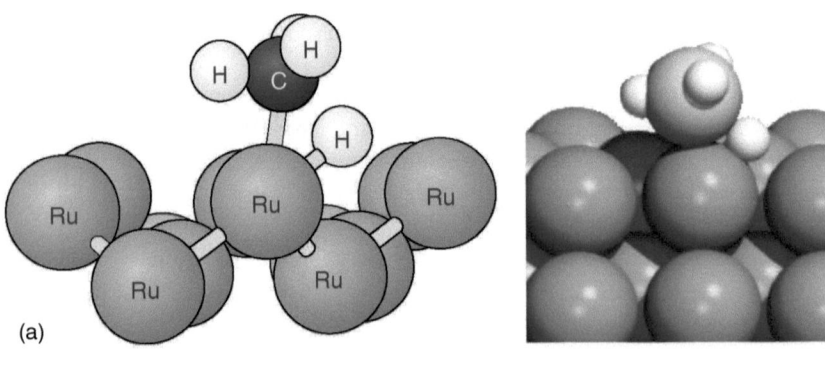

	$E_{diss}^{\#}$ (kJ mol^{-1})
Ru (0001)	76
Ru (1120)	56
Rh (111)	67
Rh step	32
Rh kink	20
Pd (111)	66
Pd step	38
Pd kink	41
Pd atom	5

Figure 3i.1 (a) Transition state structure of methane activated by a Ru atom. (b) Methane activation energies as a function of coordinative unsaturation of metal surface atoms in kJ mol^{-1}. (Ciobîcă and van Santen 2002 [15]. Reproduced with permission of American Chemical Society.)

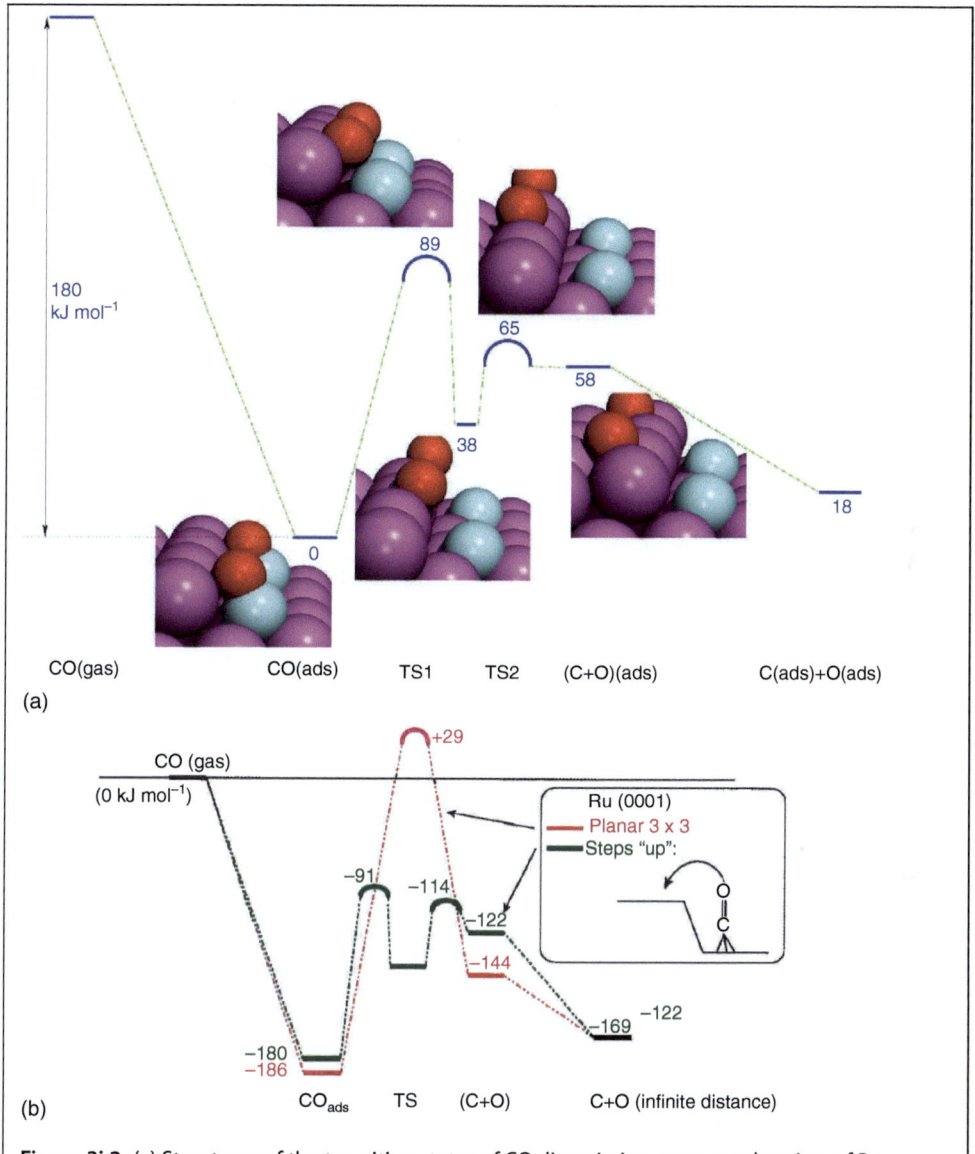

Figure 3i.2 (a) Structures of the transition states of CO dissociation on step-edge sites of Ru. (b) Comparison of reaction energies for the CO dissociation on a stepped and non-stepped surface. (Ciobîcă and van Santen 2003 [16]. Reproduced with permission of American Chemical Society.)

(Continued)

Insert 3: (Continued)

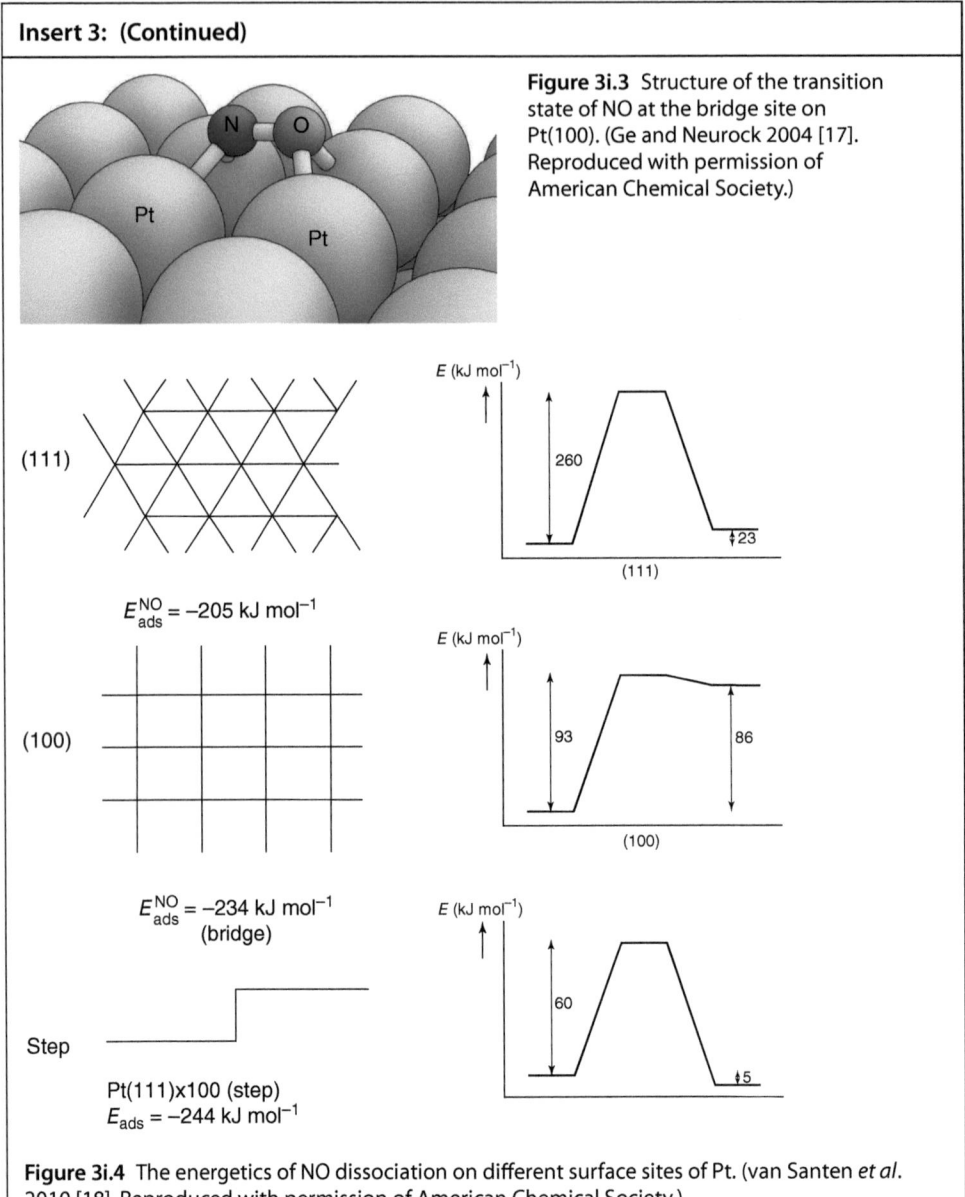

Figure 3i.3 Structure of the transition state of NO at the bridge site on Pt(100). (Ge and Neurock 2004 [17]. Reproduced with permission of American Chemical Society.)

Figure 3i.4 The energetics of NO dissociation on different surface sites of Pt. (van Santen et al. 2010 [18]. Reproduced with permission of American Chemical Society.)

As shown in Insert 3, the σ-type C—H bonds of the alkanes dissociate preferentially over a single surface metal atom. The barrier of activation relates to the intrinsic activity of the activating transition metal atom, with reactivity increasing as the number of metal atom neighbors of the reacting surface atom decreases. In contrast, the activation of π-bonds such as in CO, NO, or N_2 requires stabilization of the atoms that are generated by dissociation, such as C or O. These atoms only bind strongly when attached to several surface metal atoms. Diatomic molecules with π-bonds will only dissociate with a low barrier of activation when a surface ensemble of several metal atoms is available.

A second important requirement for low activation energy activation of molecules with π bonds is that the adatoms generated by dissociation should not be destabilized by strong lateral repulsion at the site of dissociation. Step-edge sites present on a surface step (Figure 3i.3) or on a square (100) surface (Figure 3i.4) satisfy this condition. The dissociated atoms do not interact at these sites because they occur at the bottom or top of the step-edge site or on different sides of the square in the (100)-type site. CO activation or N_2 activation preferentially occurs at step-edge type sites on the group VIII transition metals. NO activation can also occur on a surface terrace with a square (100) structure.

When particle size decreases, the number of coordinatively unsaturated edge or corner sites compared to coordinatively saturated terrace atoms increases. Since C—H activation preferentially occurs over a single metal atom, and corner and step-edge sites are more reactive, small transition metal particles are preferred as C—H bond-activation catalysts.

4.5.2 Site Generation

Step-edge sites are unstable on particles smaller than 1 nm [19, 20] because the surface of exposed terraces is too small to accommodate these sites. On larger surface step-edge sites can be stabilized or created by surface reconstruction processes. For instance, this will occur when C-atoms generated by a catalytic reaction (such as CO dissociation, methane activation or ethylene decomposition) induce surface reconstruction. As we have seen in Section 4.1, a non-reactive surface reconstructs to a more reactive surface. At the same time, surface corrugation increases because the density of surface atoms is lower on the more reactive surface, so that step-edge centers are generated during the reconstruction (see Figure 4.8).

These site generation processes occur only on larger particles, and thus reactions that require step-edge sites will have a low rate when catalysts with a small metal particle size are applied. These reactions will reach their maximum rate when catalyst particles large enough to support the step-edge centers are used.

Reactions that are preferentially catalyzed by the (100)-type surfaces are well suited for the use of cubic particles of the fcc metals that are terminated by (100) surfaces.

4.6 Alloys and Other Promotors

To improve catalyst performance, small additions of particular compounds are often added to the catalyst as promotors or moderators.

Here, we will discuss the use of alloys that are formed by the addition of a second non-reactive metal to transition metal catalysts. These alloys are formed by the combination of a transition metal that can activate H_2 or C—H bonds in an organic molecule, with a metal such as Au or Sn that is non-reactive with respect to these elementary reaction steps. Alloy catalysts are used in reactions

Figure 4.8 (a) The initial state of carbon adsorbed to a Co(0001) surface. (b) Reconstructed Co(0001) surface showing the step-edge sites on the carbon-containing surface that have been generated to release strain induced in the surface by adsorbed carbon. (Zhang et al. 2015 [19]. Reproduced with permission of American Chemical Society.)

to improve the selectivity of hydrocarbon conversion, such as in the platforming process discussed in Section 2.2.2.1.2.

Alloying can affect catalysis in two ways: by changing the organization of the reactive surface atoms, or by changing the chemical reactivity of the reactive surface atoms. In the case of C—H bond activation, the dominant effect of using an alloying non-reactive metal is the change in surface structure.

There are two types of changes to surface structure that effect catalytic reactivity. They can be best understood by comparing the Langmuir checkerboard view [21] (see also Section 3.2.1) of the catalyst surface that assumes sites are uniform with the Taylor view that assumes an arrangement of unique sites. The step-edge sites we discussed earlier in Section 4.2 are an example of these unique sites.

The Langmuirian view proposes an ensemble effect. Alloying with a non-reactive metal will change the selectivity of hydrocarbon conversion

Figure 4.9 (a) Structure of a partially dehydrogenated hexane molecule adsorbed in α–γ coordination to a transition metal surface (schematic). (b) α–β-Adsorbed intermediate (schematic).

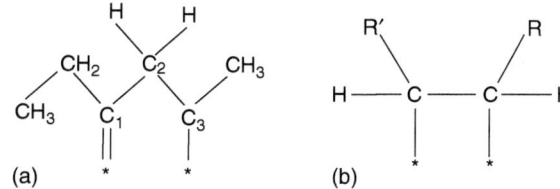

because it will suppress reactions that require a large ensemble of transition metal atoms. This is the case for the hydrogenolysis reaction that may require activation by a large number of reactive transition metal atoms to decompose the longer hydrocarbon chains into shorter ones.

As illustrated in Figure 4.9, in order to activate the C—C bond, at least two carbon atoms must directly attach to the transition metal surface after the cleavage of some of their C—H bonds. Figure 4.9a shows the intermediate of an α–γ carbon adsorbed species. Molecules such as ethane cannot form such an intermediate, but will hydrogenolyze through an α–β-adsorbed intermediate (see Figure 4.9b).

In subsequent elementary reaction steps, the C—C bonds will cleave and hydrocarbon fragments smaller than the original hydrocarbon can become hydrogenated to form reaction products. Alloying of the surface with a non-reactive metal will reduce the surface atom ensemble size of the reactive atoms, and hence suppress the bond cleavage reactions that require a large reactive surface metal atom group. With respect to alkane or alkene activation, this implies increased selectivity of dehydrogenation or isomerization reactions that are thought to be catalyzed by a site of only one or two transition metal atoms. Hydrogenolysis reactions by continued dehydrogenation of adsorbed hydrocarbon intermediates will lead to graphitic- and graphene-type carbon deposition, which will deactivate the catalyst. Suppression of the hydrogenolysis reaction will slow this deactivation, which will increase the overall rate of conversion of hydrocarbons.

An analogous explanation can be used for the beneficial effects of adding sulfur to the catalytic platforming reaction (see Section 4.4.1). This bifunctional reaction is catalyzed by Pt particles on an acidic support. It has been discovered that adding sulfur-containing compounds to the feed at the ppm level benefits the stability of the reaction. The S atoms generated by decomposition of the moderating sulfur-containing compound will decorate the Pt surface [22], and as a metal alloyed with non-noble metals it will decrease the Pt surface metal atom ensemble size. Re is also added as a promoting metal. Since Re is a reactive metal, it will become a sulfide when exposed to the sulfur-containing feed. At high temperature, the "ReS"-species dispersed over the Pt particle are more efficient than S alone at suppressing the hydrogenolysis reaction.

The Taylor view of the hydrogenolysis reaction leads to a different, but related explanation of the alloying effect. According to this view, the hydrogenolysis reaction requires step-edge sites, whereas hydrogenation or dehydrogenation can also be catalyzed by surface metal terrace atoms. Non-reactive atoms such as Au, Cu, or Sn will preferentially locate at the edges or corners of metal particles because their metal bond energies are weaker than those of the group VIII transition metals. The cleavage of these weaker metal–atom bonds causes their surface energy to be smaller, so that these atoms tend to adsorb at positions where coordinative saturation of the surface metal atoms is lowest. As a result, these metals will preferentially adsorb at the step-edge sites that are the supposed sites of the hydrogenolysis reaction. The suppression of C—H bond activation at step-edge sites has been experimentally observed for the activation of methane on stepped Ni surfaces [23, 24]. It has been found that with the addition of Au, the activation energy for methane increases from 82 kJ mol^{-1} (its value when activation takes place at a Ni step-edge) to 101 kJ mol^{-1} (its value for activation by a surface terrace atom). In Section 8.2.2.3, we will describe the single-site mechanism proposed for the hydrogenolysis reaction that is consistent with the step-edge site condition of this reaction.

Alloying can also be beneficial when one of the reactants prevents co-adsorption of a second reactant that is necessary for the reaction. An example is the use of Pd/Au alloys to oxidatively catalyze the formation of vinyl acetate from ethylene and acetic acid. In this case, acetate adsorbs strongly to Pd, thus preventing co-adsorption of ethylene. The presence of Au atoms on the metal surface creates surface sites that do not adsorb acetate, but instead allow ethylene to adsorb ([25], see also section 8.3.1.4.2, p334). Then, in a consecutive reaction step, ethylene can react with the acetate and oxygen already adsorbed to Pd, and the catalytic reaction cycle closes by vinyl acetate desorption. The final surface state will be different than expected based on the state after reduction, due to the strong adsorption of oxygen and acetate to the Pd metal atoms. This changes the balance of surface energies of the Pd-enriched sites versus the energies of the Au-enriched sites at the cost of Au. During reaction, the system will show a relative increase in Pd concentration.

The electronic structure of the surface metal atoms will also be affected by the presence of a second alloying metal. This has been extensively documented in surface science studies, especially those concerning overlayers of transition metals on a substrate of a different metal [26]. This condition leads to altered metal–metal atom distances of the reactive surface layer atoms, as well as an altered electronic distribution. We will discuss the effects of changes in electronic structure on adsorption in the next chapter. Although the adsorption energies and reactivity of the transition metal atoms are influenced by such ligand effects in catalysis, the geometric ensemble effects tend to dominate. In chemical terms, the chemical nature of the individual metal atoms remains largely unchanged.

4.7 The Working Zeolite Catalysts

The nanoporous zeolitic materials that we discussed earlier in Chapters 2 and 3 (see Figures 2.1 and 3.20) are mainly applied as solid acid catalysts in refinery-related processes. The mechanism of solid acid catalyzed reactions will be discussed in detail in Chapter 9. Zeolites activated by reducible cations are also used in selective oxidation reactions, which are treated in Chapter 10.

Deactivation of zeolites can be due to structural changes or the deposition of carbonaceous residue. Upon reaction, the composition of a zeolite may change considerably, leading to a more reactive state. Here, we will discuss some of these changes.

As we have already discussed in Chapter 2, reacting organic molecules activated by solid acidic protons will not only produce desirable reactions, but may also lead to carbonaceous residue due to oligomerization of the reaction intermediates, which leads to catalyst deactivation by site and pore blocking.

In the catalytic cracking operation, this requires the continuous removal of a portion of the catalyst material from the reactor (see Section 2.2.2.1), so that it can be regenerated by oxidative carbon removal using high-temperature steam. In this process, protonic sites are regenerated, but the catalyst's reactivity also changes. High-temperature exposure of the alumino-silicate lattice to steam induces the reaction of water with the Al cations which may become removed from their positions in the framework, thereby decreasing the number of zeolite protons.

One might expect that this treatment would reduce the reactivity of the catalytic material. Indeed, the number of active protons may decrease, but the intrinsic activity of the catalyst per proton may actually increase. This occurs for two reasons when materials with high alumina content are used.

First, when the Al concentration within the zeolite framework decreases, its negative charge density (generated by the substitution of Si^{4+} with Al^{3+}) also decreases. This reduces attraction to the zeolitic protons and hence their intrinsic acidity increases. Second, the alumina oligomers now located in the zeolite micropores may actually increase the reactivity of the protons left on the zeolite framework by a Lewis acid-type activation [27] of the oxygen atoms to which the protons are attached.

The framework of the zeolite can become destabilized by Al leaching. For this reason, framework-stabilizing rare earth cations such as La^{3+} are added to the zeolite Y catalytic cracking catalysts that display the faujasite structure Al/Si = 2/3. The rare earth oxy cations occupy positions within the lattice spaces of the zeolite such as the spacious cavities produced by the double six rings or the cubohedral sodalite cavities, which are in between the open zeolite cavities that are accessible to the reacting organic molecules [28]. Just as with the alumina oligomers located in extra-framework positions, the addition of La to the zeolite will enhance the intrinsic acidity of the zeolitic protons.

In contrast to their role as a potential culprit in catalyst deactivation, organic aromatic intermediates can also be beneficial to catalyst performance. An example is the conversion of methanol to ethylene or propylene (see Section 2.2.3.1). In this process, it is the aromatic intermediates that perform the essential catalysis instead of the zeolitic protons. These aromatics lead to oligomerization of the methanol molecules through the intermediate alkylation of methanol to the aromatic precursor molecules, which in turn act as organocatalysts in the SAPO materials used for the methanol conversion reaction (see Figure 4.10).

The paring reaction is initiated by proton-catalyzed methylation of methanol to benzene. The reaction then proceeds through the paring reaction cycle or branches into the side reaction cycle. Within the paring reaction cycle, C—C bonding occurs by intermediate cyclopentyl formation. Within the side-chain reaction cycle, a second methanol molecule is added to protonated toluene.

Similar cationic organo-catalysis in acidic protonic materials has been proposed for the isomerization of n-butene to isobutene (see Section 9.3.2.1).

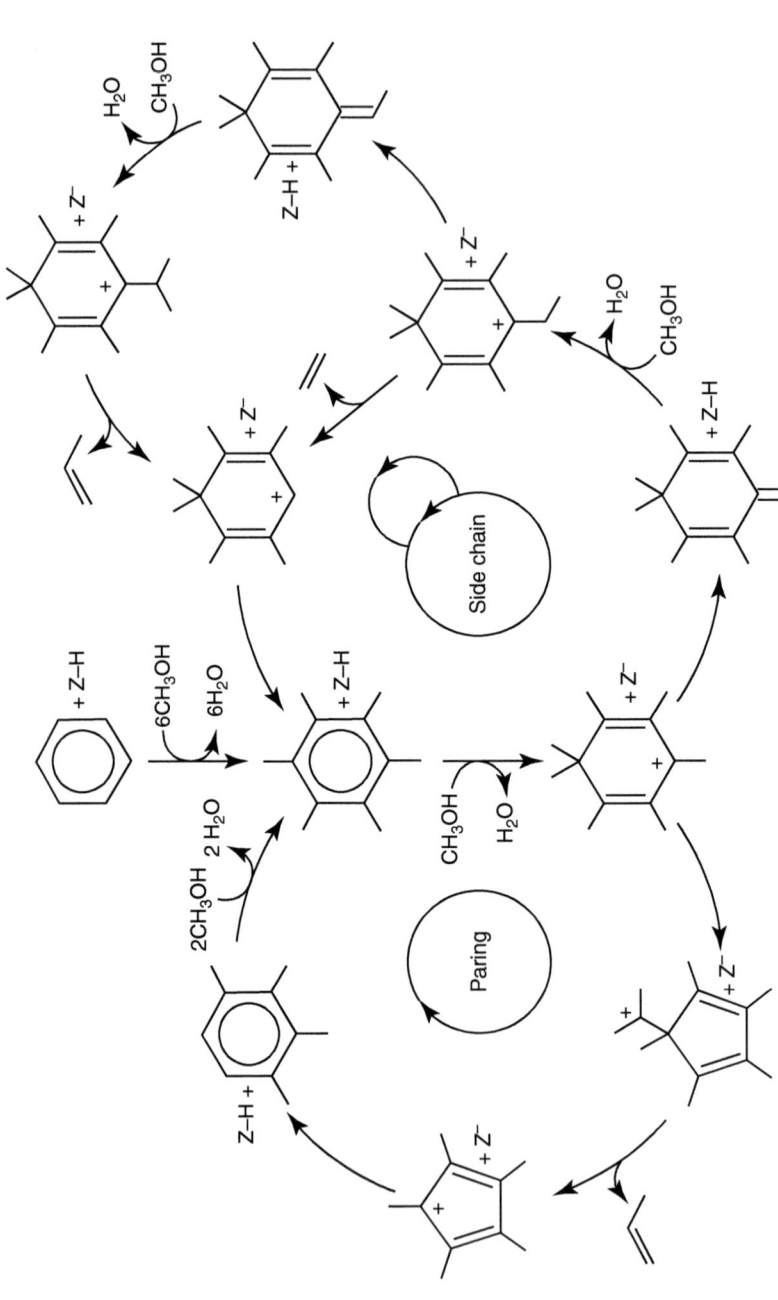

Figure 4.10 Carbon–carbon bond formation through alkylation reactions of an aromatic molecule. The two reaction cycles, paring and side-chain reaction, are shown. (Olsbye et al. 2012 [29]. Reproduced with permission of John Wiley and Sons.)

When single-site framework oxidation catalysts are applied in the liquid phase, deactivation similar to the deactivation of framework alumina can be induced by reaction with water. Zeolitic systems that contain the reactive Co or Mn ions will form soluble complexes of such cations with water and may rapidly lose their reactivity (see Chapter 10). For this reason, it is preferable to use these catalysts in the gas phase.

The zeolitic siliceous Ti catalyst, in which some framework Si is substituted by Ti, is used to epoxidize propylene by H_2O_2 (Section 10.6.4). The catalyst is selective as long as Ti remains located in the framework of the zeolite. When small Ti oligomers are formed by reaction with H_2O or in the synthesis of the material, they can occupy the zeolite micropores, and the remaining accessible Ti sites will act non-selectively to decompose H_2O_2 into O_2 and H_2O.

Selective catalysis can also be achieved by reducible cations located in the zeolite micropore. An example is the oxidation of benzene to phenol by N_2O. This reaction is selectively catalyzed by Fe^{2+} cations that compensate for the negative framework charge generated by the substitution of Si with Al in the ZSM-5 material. The Fe^{2+} cations are unique in that they will not activate phenol for oligomerization, and this deactivates the catalyst (Section 10.4.3). The preparation of a catalyst with single Fe^{2+} centers that could activate phenol has not yet been accomplished. The rate of catalyst deactivation by phenol oligomerization is strongly dependent on the zeolite structure. Improved catalysts have been designed, which are composed of biporous materials that contain a combination of beneficial attributes. Despite having narrow pores due to the small particle size [30] required for catalysis, the pores of these zeolitic materials are adequately sized in all three dimensions to enable the continuous transport of reagent and product molecules. This increases the effective life of the catalyst considerably, because the catalytically reactive centers remain more accessible than in other unimproved catalysts.

4.8 The State of the Mixed Oxide Surface

The process of butane oxidation by $VOPO_4$ illustrates the surface restructuring of a working oxide catalyst. $VOPO_5$ is a catalyst that oxidizes butane to maleic acidic anhydride with high efficiency (see Chapter 2, Section titled "Butane to Maleic Acid Oxidation"). Oxidation of butane reduces V^{5+} to a lower valency, and the reduced cationic vanadium is oxidized by oxygen. Based on *in situ* and model studies, it has been shown [31] that this ongoing change of composition of the surface produces complex disordered structures. These structural changes are illustrated in Figure 4.11. This catalyst satisfies an important requirement for stable catalytic performance – the reversibility between the various phases that are formed on the catalyst during reaction.

4.9 Summary

The examples presented in this chapter illustrate the various aspects of the practical catalytic system which are crucial for its operation, but that often relate only

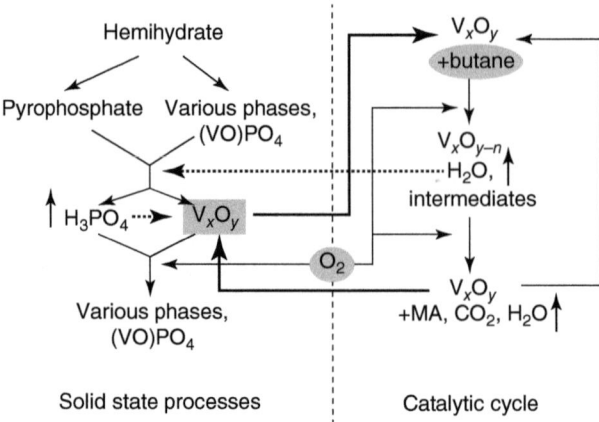

Figure 4.11 Structural changes in the VOPO$_4$ catalyst under reaction conditions. H$_2$O that is produced by oxidation of butane segregates a VO$_2$ phase with vanadium +4 cations. The particles in this face remain small because phosphoric acid inhibits crystalline growth. Reduction of butane prevents over-oxidation of the V^{4+} overlayer. (After [31, 32].)

indirectly to the mechanism of the reaction being catalyzed. We have introduced these aspects as a separate chapter precisely because of this seeming disconnect.

We have seen that synthesis design is critical for preparing a catalyst in the proper state before a reaction. The prepared catalyst undergoes further changes when it is activated by drying, reduction, or oxidation.

In most heterogeneous catalytic systems the most active and selective phase is formed during the reaction. Catalyst performance may initially increase or decrease with time, but then achieves a period of stability.

After a period of stable operation, deactivation of the catalyst will begin, caused by a variety of processes. The most common reasons for catalyst deactivation are poisoning by small amounts of impurities in the feed or poisoning by the products of non-selective reactions that accumulate over time. In addition, changes in the inorganic chemistry of the catalyst can occur, such as sintering of the catalytically reactive phases by particle growth, alteration of the catalyst support, or leaching of the catalytically active phase when reaction occurs in a solvent.

The chemistry of the evolutionary changes of the working catalysts is complex. One of the great remaining challenges in the current science of heterogeneous catalysis is to understand and control the transient state of the heterogeneous catalyst.

References

1. Weckhuysen, B. (2012) *Catalysis*, Wiley-VCH Verlag GmbH, pp. 493–514.
2. Van Hove, M.A., Koestner, R.J., Stair, P.C., Bibérian, J.P., Kesmodel, L.L., Bartoš, I., and Somorjai, G.A. (1981) The surface reconstructions of the (100) crystal faces of iridium, platinum and gold. *Surf. Sci.*, **103**, 189–217. http://linkinghub.elsevier.com/retrieve/pii/0039602881901072 (accessed 14 November 2014).

3 Ciobîcă, I.M., van Santen, R.A., van Berge, P.J., and van de Loosdrecht, J. (2008) Adsorbate induced reconstruction of cobalt surfaces. *Surf. Sci.*, **602**, 17–27. http://linkinghub.elsevier.com/retrieve/pii/S0039602807009260 (accessed 14 November 2014).

4 Frechard, F. and van Santen, R.A. (1998) Theoretical study of the adsorption of the atomic oxygen on the Cu(110) surface. *Surf. Sci.*, **407**, 200–211. http://linkinghub.elsevier.com/retrieve/pii/S0039602898001800 (accessed 14 November 2014).

5 Jones, A.H., Poulston, S., Bennett, R.A., and Bowker, M. (1997) Methanol oxidation to formate on Cu(110) studied by STM. *Surf. Sci.*, **380**, 31–44. http://linkinghub.elsevier.com/retrieve/pii/S0039602896015816 (accessed 25 November 2014).

6 Zhu, T., Hensen, E.J.M., van Santen, R.A., Tian, N., Sun, S.-G., Kaghazchi, P., and Jacob, T. (2013) Roughening of Pt nanoparticles induced by surface-oxide formation. *Phys. Chem. Chem. Phys.*, **15**, 2268–2272. http://www.ncbi.nlm.nih.gov/pubmed/23303314 (accessed 14 November 2014).

7 Falbe, J. (1970) *Carbon Monoxide in Organic Synthesis*, Springer-Verlag, pp. 3–77.

8 Regalbuto, J.R. (2009) in *Synthesis of Solid Catalysts* (ed. K.P. de Jong), Wiley-VCH Verlag GmbH, pp. 33–58.

9 de Jong, K.P. (2012) Characterization of supported catalysts, in *Catalysis* (eds M. Beller, A. Renken, and R.A. van Santen), Wiley-VCH Verlag GmbH.

10 Fu, Q., Saltsburg, H., and Flytzani-Stephanopoulos, M. (2003) Active nonmetallic Au and Pt species on ceria-based water-gas shift catalysts. *Science*, **301**, 935–938. http://www.ncbi.nlm.nih.gov/pubmed/12843399 (accessed 17 November 2014).

11 Iglesias-Juez, A., Beale, A.M., Maaijen, K., Weng, T.C., Glatzel, P., and Weckhuysen, B.M. (2010) A combined in situ time-resolved UV–Vis, Raman and high-energy resolution X-ray absorption spectroscopy study on the deactivation behavior of Pt and PtSn propane dehydrogenation catalysts under industrial reaction conditions. *J. Catal.*, **276**, 268–279. http://linkinghub.elsevier.com/retrieve/pii/S0021951710003295 (accessed 28 November 2014).

12 Lennard-Jones, J.E. and Taylor, P.A. (1925) Some theoretical calculations of the physical properties of certain crystals. *Proc. R. Soc. A, Math. Phys. Eng. Sci.*, **109**, 476–508. doi: 10.1098/rspa.1925.0139

13 Langmuir, I. (1918) The adsorption of gases on plane surfaces of glass, mica and platinum. *J. Am. Chem. Soc.*, **40**, 1361–1403. doi: 10.1021/ja02242a004

14 Che, M. and Bennett, C.O. (1989) The influence of particle size on the catalytic properties of supported metals. *Adv. Catal.*, **36**, 55–272.

15 Ciobîcă, I.M. and van Santen, R.A. (2002) A DFT study of CHx chemisorption and transition states for C–H activation on the Ru(1120) surface. *J. Phys. Chem. B*, **106**, 6200–6205.

16 Ciobîcă, I.M. and van Santen, R.A. (2003) Carbon monoxide dissociation on planar and stepped Ru(0001) surfaces. *J. Phys. Chem. B*, **107**, 3808–3812.

17 Ge, Q. and Neurock, M. (2004) Structure dependence of NO adsorption and dissociation on platinum surfaces. *J. Am. Chem. Soc.*, **126**, 1551–1559. http://www.ncbi.nlm.nih.gov/pubmed/14759214 (accessed 28 November 2014).

18 van Santen, R.A., Neurock, M., and Shetty, S. (2010) Reactivity theory of transition-metal surfaces: a Brønsted–Evans–Polanyi linear activation energy – free-energy analysis. *Chem. Rev.*, **110**, 2005–2018.

19 Zhang, X.-Q., van Santen, R.A., and Hensen, E.J.M. (2015) Carbon-induced surface transformations of cobalt. *ACS Catal.*, **5** (2), 596–601.

20 Honkala, K., Remediakis, I.N., Logadottir, A., Carlsson, A., Dahl, S., Christensen, C.H., Nørskov, J.K., and Hellman, A. (2005) Ammonia synthesis from first-principles calculations. *Science*, **307**, 555–558.

21 Langmuir, I. (1922) Part II: heterogeneous reactions: chemical reactions on surfaces. *Trans. Faraday Soc.*, **17**, 607–620.

22 Shum, V.K., Butt, J.B., and Sachtler, W.M.H. (1985) The effects of rhenium and sulfur on the activity maintenance and selectivity of platinum/alumina hydrocarbon conversion catalysts. *J. Catal.*, **96**, 371–380. http://linkinghub.elsevier.com/retrieve/pii/0021951785903070 (accessed 22 November 2014).

23 Bengaard, H., Nørskov, J.K., Sehested, J., Clausen, B.S., Nielsen, L.P., Molenbroek, A.M., and Rostrup-Nielsen, J.R. (2002) Steam reforming and graphite formation on Ni catalysts. *J. Catal.*, **209**, 365–384. http://linkinghub.elsevier.com/retrieve/pii/S0021951702935797 (accessed 8 October 2014).

24 Besenbacher, F., Chorkendorff, I., Clausen, B.S., Hammer, B., Molenbroek, A.M., Nørskov, J.K., and Stensgaard, I. (1998) Design of a surface alloy catalyst for steam reforming. *Science*, **279**, 1913–1915. doi: 10.1126/science.279.5358.1913

25 Neurock, M. (2003) Perspectives on the first principles elucidation and the design of active sites. *J. Catal.*, **216**, 73–88. http://linkinghub.elsevier.com/retrieve/pii/S002195170200115X (accessed 22 November 2014).

26 Kitchin, J., Nørskov, J., Barteau, M., and Chen, J. (2004) Role of strain and ligand effects in the modification of the electronic and chemical properties of bimetallic surfaces. *Phys. Rev. Lett.*, **93**, 156801. doi: 10.1103/PhysRevLett.93.156801

27 Almutairi, S.M.T., Mezari, B., Filonenko, G.A., Magusin, P.C.M.M., Rigutto, M.S., Pidko, E.A., and Hensen, E.J.M. (2013) Influence of extraframework aluminum on the Brønsted acidity and catalytic reactivity of faujasite zeolite. *ChemCatChem*, **5**, 452–466. doi: 10.1002/cctc.201200612

28 Schüßler, F., Pidko, E.A., Kolvenbach, R., Sievers, C., Hensen, E.J.M., van Santen, R.A., and Lercher, J.A. (2011) Nature and location of cationic lanthanum species in high alumina containing faujasite type zeolites. *J. Phys. Chem. C*, **115**, 21763–21776. doi: 10.1021/jp205771e

29 Olsbye, U., Svelle, S., Bjrgen, M., Beato, P., Janssens, T.V.W., Joensen, F., Bordiga, S., and Lillerud, K.P. (2012) Conversion of methanol to hydrocarbons: how zeolite cavity and pore size controls product selectivity. *Angew. Chem. Int. Ed.*, **51**, 5810–5831.

30 Li, Y., Feng, Z., van Santen, R.A., Hensen, E.J.M., and Li, C. (2008) Surface functionalization of SBA-15-ordered mesoporous silicas: oxidation of benzene to phenol by nitrous oxide. *J. Catal.*, **255**, 190–196. doi: 10.1002/cctc.201200612

31 Bluhm, H., Hävecker, M., Kleimenov, E., Knop-Gericke, A., Liskowski, A., Schlögl, R., and Su, D.S. (2003) In situ surface analysis in selective oxidation catalysis: n-Butane conversion over VPP. *Top. Catal.*, **23**, 99–107. doi: 10.1023/A:1024824404582

32 van Santen, R.A. (2012) in *Catalysis* (eds M. Beller, A. Renken, and R.A. van Santen), Wiley-VCH Verlag GmbH, pp. 143–145.

5

Advanced Kinetics: Breakdown of Mean Field Approximation

5.1 Introduction

This chapter deals with advanced aspects of the kinetics of heterogeneous catalytic reactions. The kinetics methods discussed in Chapters 3 and 4 are based on the mean field assumption. This assumption supposes that the surface concentration of reaction intermediates on the catalytically active surface is independent of position, resulting in a uniform distribution of the surface intermediates. This is followed by the implicit assumption that the rate of surface diffusion is fast compared to the rate constants of the elementary reactions.

Lateral attractive or repulsive interactions between the adsorbed intermediates become especially important when a surface is highly covered with adsorbates. This may lead to the inhomogeneous distribution of reaction intermediates through the formation of surface overlayer islands or even surface compounds. Such phenomena can be considered as the two-dimensional analogs of phase separation and, hence, phases will demix above a specific critical temperature. Therefore, we expect that overlayer island formation may result in substantially different kinetic behavior than predicted by mean field microkinetics, especially at low temperatures.

The next section introduces the kinetic Monte Carlo (KMC) method, which is useful for the prediction of kinetics beyond the mean field approximation, followed by an introduction to single molecule dynamics. Single molecule spectroscopy gives direct access to the stochastic kinetics of a single molecular event. This stochastic behavior is fundamental to the KMC method, which can predict collective deterministic behavior for a multiparticle system. The use of the KMC method will be illustrated in the next section by its application to the kinetics of CO oxidation by a RuO_2 surface.

When surface reconstruction (and hence surface reactivity) depends strongly on surface overlayer concentration, unique self-organized features such as the oscillatory time dependence of reaction rate can arise. This will be discussed in Section 5.4.

Modern Heterogeneous Catalysis: An Introduction, First Edition. Rutger A. van Santen.
© 2017 Wiley-VCH Verlag GmbH & Co. KGaA. Published 2017 by Wiley-VCH Verlag GmbH & Co. KGaA.

5.2 The Kinetic Monte Carlo Method: RuO$_2$ Catalyzed Oxidation

The KMC method is a stochastic method that calculates the rate of change of a system from the probabilities of its inherent individual transitions. It can simulate the rate of surface phase transitions such as island formation of an adsorbate overlayer or the transformation rate of one surface phase to another. Calculating the rate of such a transition is complex because it is induced by movements of surface atoms, which are due to the collective motion of the adsorbates.

Stochastic simulation techniques calculate the change in state of a system from the hopping or reaction probabilities of its individual components (reaction intermediates). In the case of heterogeneous catalysis, these hopping probabilities are the elementary rate constants or diffusion probabilities, and can also include the probability of movement of the surface atoms. These simulations can be performed on a fixed or movable grid that represents the surface sites. Details of this method are given in Insert 1.

Insert 1: Details of the Kinetic Monte Carlo Method

$$\frac{dP_\alpha}{dt} = \sum_\beta [W_{\alpha\beta}P_\beta - W_{\beta\alpha}P_\alpha] \qquad (1i.1)$$

In the KMC method the master equation (Eq. (1i.1)) is solved [1]. In these equations t is time, α and β are configurations of the adlayer, P_α and P_β are their probabilities and $W_{\alpha\beta}$ and $W_{\beta\alpha}$ are the transition probabilities per unit time that specify the rate with which the adlayer changes due to reactions.

There are several possible approaches used to solve these equations by Monte Carlo statistical methods. As an example, we will describe the variable step size method (VSSM). An excellent reference for further reading is A.P.J. Jansen, *An introduction to kinetic Monte Carlo simulations of surface reactions*, Springer 2012 [2].

The time of first reaction can be deduced from the expression:

$$\exp[-R_{\alpha\alpha}t'] = r_1 \qquad (1i.2)$$

The left-hand side of Eq. (1i.2) gives the probability that the system is still in state α at the time. $R_{\alpha\alpha}$ is related to the reaction probabilities by:

$$R_{\alpha\beta} = \begin{cases} 0, & \text{if } \alpha \neq \beta, \\ \sum_\gamma W_{\gamma\beta}, & \text{if } \alpha = \beta. \end{cases} \qquad (1i.3)$$

r_1 is a random variable to be chosen from the unit interval.

Figures 1i.1 and 1i.2 illustrate the difference between kinetics as derived by mean field kinetics (MFK) and the KMC methods according to Hess and Over [3]:

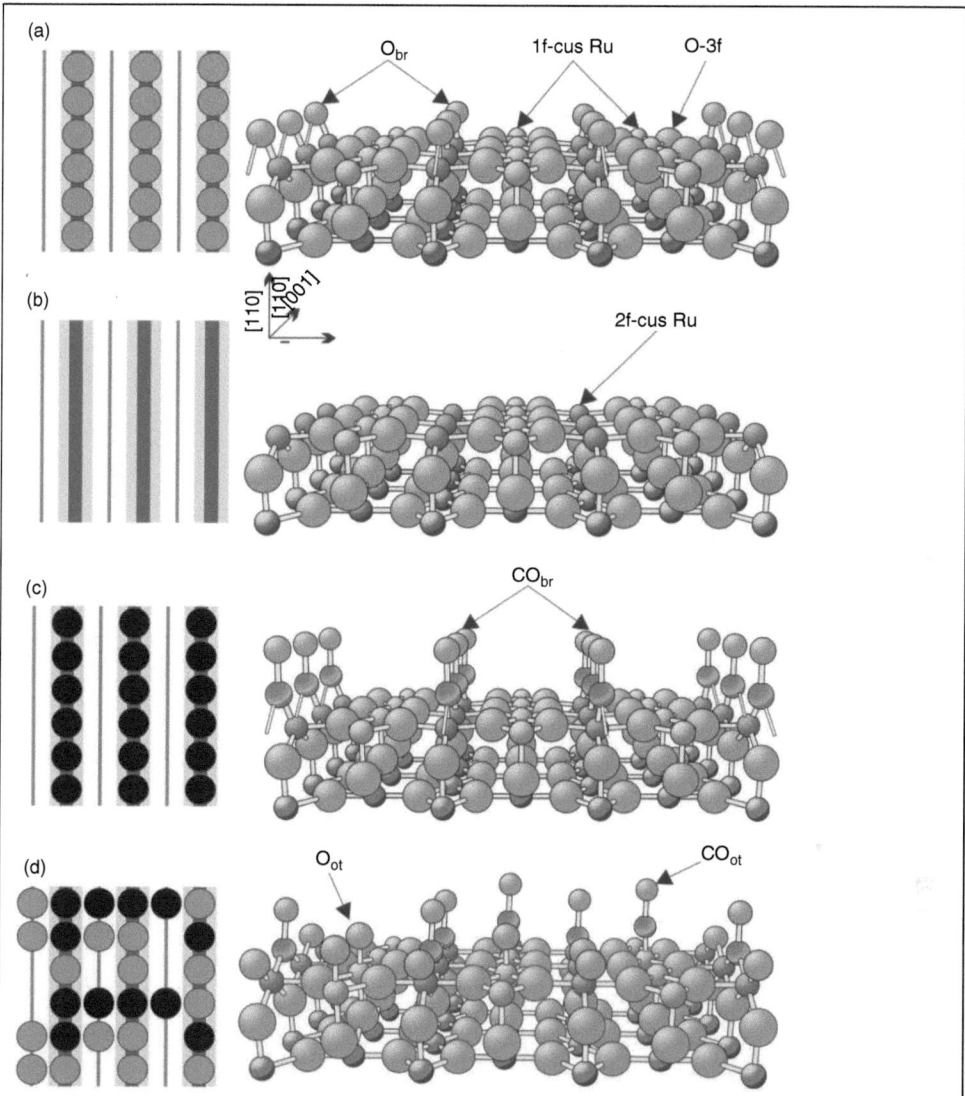

Figure 1i.1 Ball and stick models of the $RuO_2(110)$ surface. At the stoichiometric surface (a) there are two types of chemically active sites: the bridging O atoms (O_{br}) and the onefold coordinatively unsaturated Ru sites (1f-cus Ru). Removal of the O_{br} atoms leads to a mildly reduced $RuO_2(110)$ surface (b) where the twofold unsaturated 2f-cus Ru sites are exposed. For the rest of the discussion, we use a simple representation of the mildly reduced surface by gray and dark stripes for the rows of 1f-cus and 2f-cus sites, respectively (c). (d) Example of a coadsobate structure of bridging O and CO together with on-top CO and on-top O. (Hess and Over 2014 [3]. Reproduced with permission of Royal Society of Chemistry.)

(Continued)

Insert 1: (Continued)

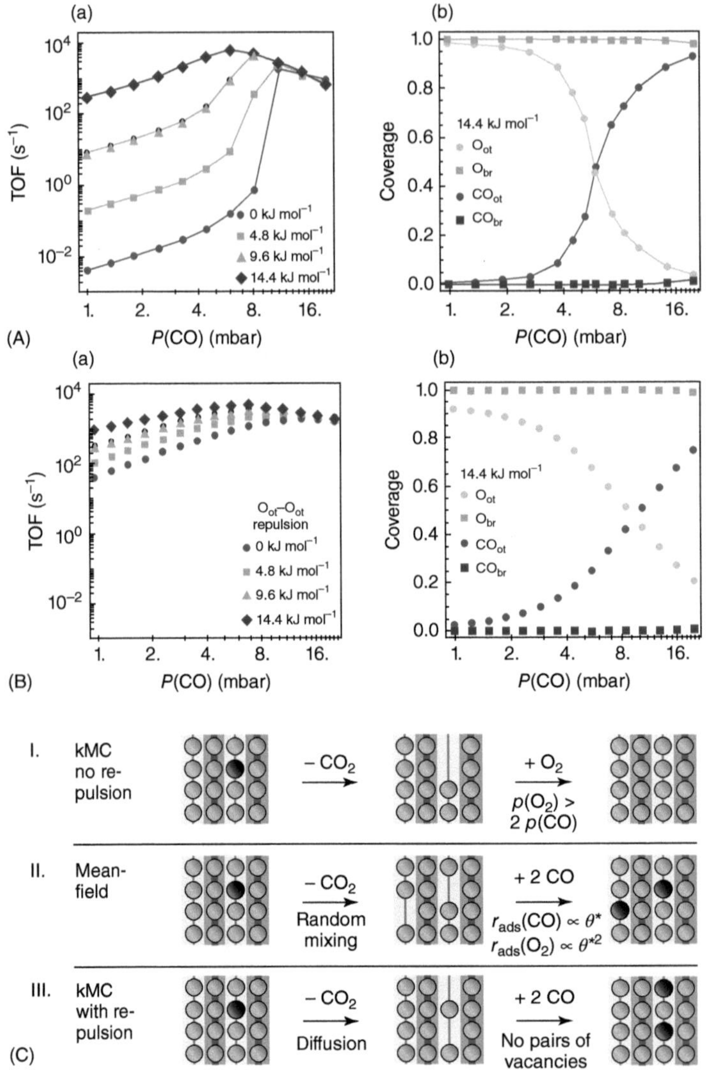

Figure 1i.2 Comparison between mean-field approach (MFA) and kinetic Monte Carlo (KMC) methods for the oxidation of CO over RuO_2 (110) at $T = 510$ K and $p(O_2) = 5.5$ mbar. The $O_{ot}-O_{ot}$ repulsion was varied between 0 and 14.4 kJ mol^{-1}. Coverage values are shown only for the 14.4 kJ mol^{-1} case. (A) TOFs (a) and coverage values (b) for KMC simulations, (B) TOFs (a) and coverage values (b) for MFA calculations (see Figure 1i.1 for notation) for the most likely scenarios under oxidizing conditions for the three modeling approaches. (C) After formation of a CO_2 molecule, the O_{ot} overlayer reorders in different ways, depending on the modeling approach. This leads to I. Complete poisoning of the surface with O_{ot} in KMC without repulsion, II. A sustained activity because CO adsorption happens faster than O_2 adsorption for a random distribution of adparticles in the MFA, and III. A sustained activity because the pair correlation of vacancies is lifted due to a repulsive interaction, preventing poisoning by dissociative O_2 adsorption in KMC simulations. (Hess and Over 2014 [3]. Reproduced with permission of Royal Society of Chemistry.)

KMC simulations will duplicate the results derived using mean field-approach kinetics when the hopping frequencies of a diffusive step of adsorbed reaction intermediates are high compared to those of the elementary rates of reaction. In this case, the concentration of reaction intermediates is uniform over the surface. The probability distribution of surface sites is especially important for elementary reactions that require two adjacent surface vacancies, such as surface dissociation and association reactions.

The KMC results of the oxidation of CO by O_2 on a $RuO_2(110)$ surface shown in Figure 1i.2 [3] provide an illustration of this approach. The Monte Carlo grid in these simulations is provided by the adsorption sites of O atoms to the fixed Ru atom positions of the $RuO_2(110)$ surface (Figure 1i.1). On this type of surface, O atom position vacancies must exist in order for O_2 to dissociate. The recombination reaction of an O atom with CO requires that the two atoms be adsorbed on adjacent sites. When diffusion is absent and O_2 dissociates, the neighboring Ru sites will become occupied. A vacancy for CO adsorption next to adsorbed O will be generated when an O atom moves. A repulsive lateral interaction between surface O atoms will provide a driving force for this diffusion step, and will substantially increase the overall rate of reaction (Figure 1i.2a). The oxygen adatom surface concentration distribution then becomes more uniform and hence the rate converges to that of the mean field solution (Figure 1i.2b). In mean field theory, lateral repulsion of the O atoms also increases the overall rate of reaction because it increases the rate constant for the O—CO recombination reaction. The transition state occurs late in this reaction, so that its structure is close to that of the free CO_2 molecule. The lateral interactions cause a decrease in the activation energy for recombination due to the enhanced destabilization of surface O atoms and adsorbed CO as the surface concentration increases.

Despite the stochastic nature of individual molecular reaction events, the overall rate of reaction will generally be stationary and independent of time. In the next section, we will discuss single molecule events in which the stochastic behavior can actually be followed. After that section, we will return to discussing the kinetics of an ensemble of molecules and oscillatory heterogeneous catalytic reactions where surface diffusion plays an essential role.

5.3 Single Molecule Spectroscopy

Single molecule fluorescence microscopy allows for the observation of individual molecular events. When applied to the study of heterogeneous catalytic reactions, a catalytic reaction is used to produce a fluorescent molecule [4]. Reaction must occur along the edges of catalyst particles so that the fluorescence can be observed by a microscope. By observing the bursts of fluorescence as a function of time, subsequent product formation, and desorption from the surface can be followed.

Insert 2: Single Molecule Fluorescence Microscopy

This insert closely follows the description by Smiley and Hammes [5] on the relationship between the time dependence of fluorescent emission and the rate of reaction.

Rather than using concentration, the probability of a given state is used in equations closely related to conventional kinetics. This will be explained for the one-step mechanism:

$$A \xrightarrow{k} B \tag{2i.1}$$

where k is the ensemble rate constant. The rate equation for a single molecule transitioning from state A to state B is given by:

$$dP_a/dt = -kP_a \tag{2i.2}$$

here P_a is the probability of the molecule being in state A and t is the time. This equation can be solved with the initial condition that $P_a = 1$ when $t = 0$, and $P_a = P(T)_a$ when $t = \tau$, the reaction lifetime. The result is:

$$P_a = \exp(-k\tau) \tag{2i.3}$$

Thus, for this simple mechanism the probability distribution of reaction lifetimes follows an exponential function, and the ensemble rate constant is directly related to this distribution. The probability density, $f(\tau)$, is defined by the relationship $dP_a/dt = -kP_a = f(\tau)$ or:

$$f(\tau) = k\exp(-k\tau) \tag{2i.4}$$

Note that $f(\tau)d\tau$ is the probability of A switching to B during the interval τ and $\tau + d\tau$. The distribution function $f(\tau)$ is determined experimentally in single molecule studies. In Figure 2i.1, a schematic representation of the trajectory of a molecule is shown as the molecule switches between states of high and low fluorescence. The reaction lifetimes can be tabulated from such trajectories, and the probability density of the lifetime falling within a given range can be plotted as a bar graph as shown in Figure 2i.2. The probability density is simply the number of events with a lifetime τ, $N(\tau)$, divided by the total number of events, N_{total}.

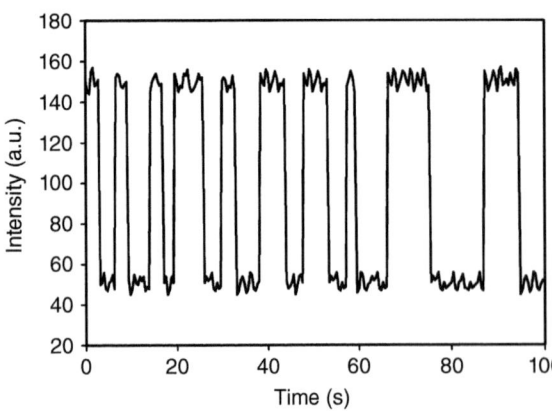

Figure 2i.1 Hypothetical trajectory for a single molecule transitioning between high and low fluorescence states. The fluorescence versus time is plotted. (Smiley and Hammes 2006 [5]. Reproduced with permission of American Chemical Society.)

Figure 2i.2 Histogram of the number of events with a lifetime τ, $N(\tau)$, versus τ for the lifetimes of a single molecule moving from a high fluorescence state, A, to a low fluorescence state, B (Eq. (2i.1)). The line has been calculated with Eq. (2i.4) and $k = 0.2\,s^{-1}$. (Smiley and Hammes 2006 [5]. Reproduced with permission of American Chemical Society.)

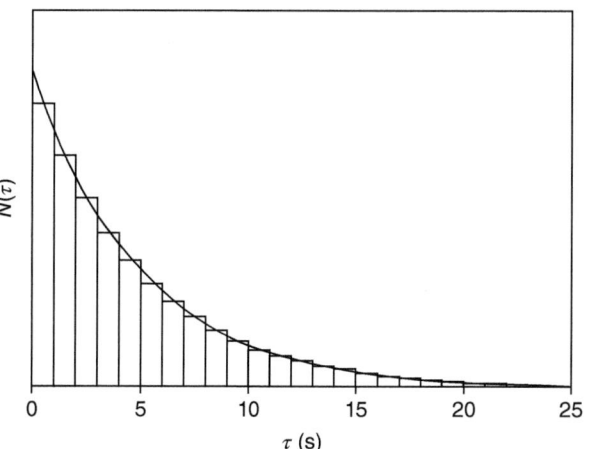

Figures 2i.3 and 2i.4 give experimental results of single molecule spectroscopy experiments catalyzed by Au nanoparticles.

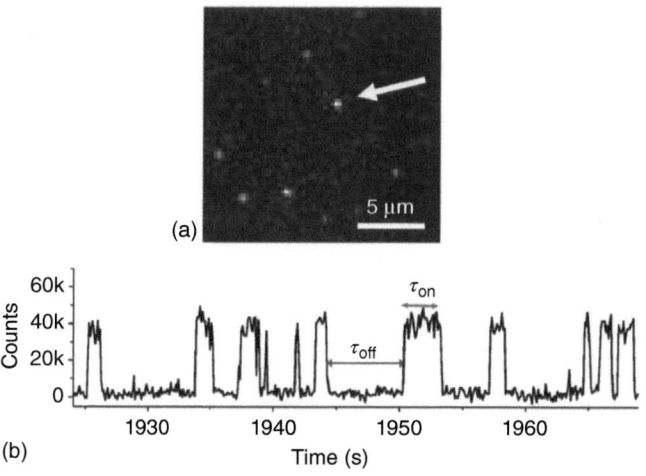

Figure 2i.3 Single-nanoparticle catalysis at single-turnover resolution. (a) Single molecule fluorescence image of product molecules on 6 nm pseudospherical Au nanoparticles while catalyzing the N-deoxygenation reaction. (b) Segment of the fluorescence intensity trajectory from the fluorescence spot marked by the arrow in (a). (c) Experimental scheme of single-molecule microscopy of fluorogenic catalytic reactions on single nanoparticles for two fluorogenic catalytic reactions. (d) Schematic of the kinetic mechanism. Au_m: Au nanoparticle; S: resazurin; and P: resorufin. Au_m-S_n represents a Au nanoparticle with N-adsorbed substrate molecules at adsorption equilibrium. The fluorescence state (on or off) of the nanoparticle is indicated at each reaction stage. $\gamma_{eff} = kn_T$ and represents the combined reactivity of all surface catalytic sites of a nanoparticle. k is a rate constant representing the reactivity per catalytic site for the catalytic conversion. N_T is the total number of surface catalytic sites on one Au nanoparticle. yS is the fraction of catalytic sites that are occupied by substrates and equals $K_1[S]/(1+K_1[S])$, where K_1 is the substrate adsorption equilibrium constant. (Chen *et al.* 2014 [6]. Reproduced with permission of Royal Society of Chemistry.)

(Continued)

Insert 2: (Continued)

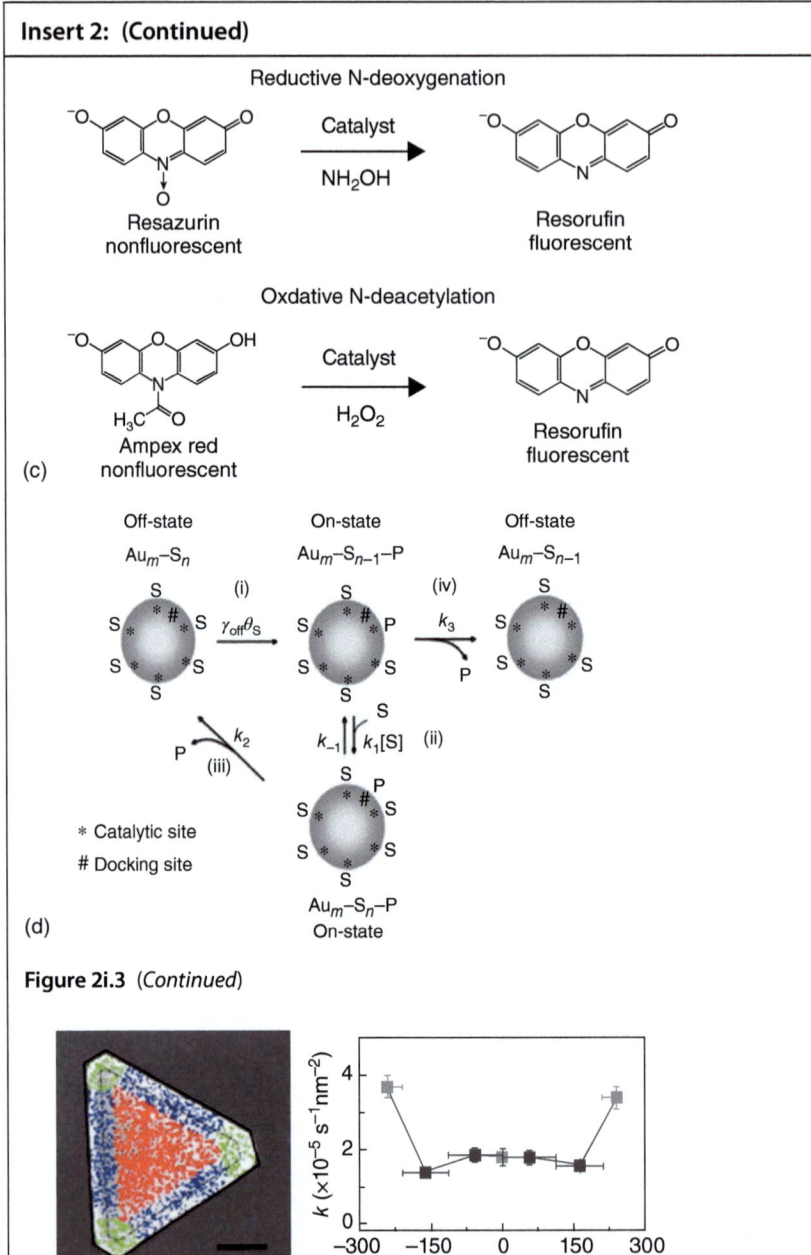

Figure 2i.3 (Continued)

Figure 2i.4 Site-specific catalytic activity on single-shaped nanocrystals. (a) Locations of catalytic products mapped onto the SEM image of a single Au@mSiO$_2$ nanoplate. Each dot is a product molecule. The product locations are color coded based on their regions: corners (green), edges (blue), and top facets (red). (b) Specific catalytic rate constant k of each segment of the nanorod (Chen et al. 2014 [6]. Reproduced with permission of Royal Society of Chemistry.)

The typical time dependence of a single molecule fluorescence signal is schematically illustrated in Figure 2i.1, which shows a stochastic distribution of on and off times. As outlined in Insert 2 a plot of these distributions as a function of time provides the rate constants of the reactions.

In Figure 2i.3 fluorescence spots on a Au nanoparticle are measured by a microscope. The catalyzed reactions that produce fluorescent molecules are given in Figure 2i.3c. The sequence of on and off fluorescence times observed in such experiments is shown in Figure 2i.3b. τ_{on} is the time scale of the generation of the fluorescent particles that remain adsorbed to the surface and τ_{on} is the time scale of desorption. Based on the expectation that the catalytic reaction sequence includes at least two fluorescence-producing reaction steps, the rates of the two steps can be determined.

When measurements are defined as a function of reactant concentration, Michaelis–Menten (see Chapter 3) kinetics is observed. When averaged over many nanoparticles, both $\langle \tau_{off} \rangle^{-1}$ and $\langle \tau_{on} \rangle^{-1}$ show saturation kinetics as a function of reagent concentration S, and the rates are quantitatively described by the following equations [6]:

$$\langle \tau_{off} \rangle^{-1} = \frac{\gamma_{eff} K_1 [S]}{1 + K_1 [S]} \tag{5.1a}$$

$$\langle \tau_{on} \rangle^{-1} = \frac{k_2 K_2 [S] + k_3}{1 + K_2 [S]} \tag{5.1b}$$

The kinetic parameters here are defined in Figure 2i.3d and $K_2 = k_1/(k_{-1} + k_2)$. When $[S] \to \infty$, $\langle \tau_{off} \rangle^{-1}$ equals γ_{eff} and $\langle \tau_{on} \rangle^{-1}$ equals k_2, where γ_{eff} is the single-particle catalytic rate constant that represents the combined reactivity of all surface catalytic sites, and k_2 is the product dissociation rate constant in the substrate-assisted pathway. With single nanoparticle measurements, the temporal resolution of τ_{off} and τ_{on} enables independent determination of the rates of product formation and desorption.

Au nanoparticle experiments have been conducted on metal particles of different sizes, and a variation in rate constants as a function of the ratio of corner, edge, or terrace sites can be observed (Figure 2i.4). As expected, the corner sites are the most reactive.

In a conventional experiment, the molecular events that occur at different locations on the catalyst are averaged and, hence, no fluctuations in product distribution are observed. Thus, in conventional kinetics, transient isotope exchange kinetics methods such as steady-state isotopic transient kinetic analysis (SSITKA) must be used to unravel the rates of the consecutive reaction steps (see Section 3.4).

5.4 Catalytic Self-Organizing Systems

5.4.1 Introduction

Oscillatory phenomena, exploding reactions, or runaway reactions that cause hot spots are familiar occurrences in catalytic reactor technology. Hot spots can occur when heat or mass transfer becomes limiting but the reaction rate increases simultaneously, and this increase is amplified by the exothermicity of the reaction, which increases its temperature. Similar phenomena also occur in catalytic reaction–diffusion coupled systems when autocatalytic reactions at different locations in a reactor are synchronized with respect to heat or mass transfer [7].

In addition, in heterogeneous catalytic systems at isothermal conditions kinetic self-organization phenomena may occur that result in oscillatory time dependence of reaction rate as well as time-dependent surface state patterns on the catalyst. Surface intermediates may become organized in pulsing or spiraling patterns.

These self-organizing systems are well understood [8, 9], and their manifestation in heterogeneous catalysis will be discussed in the following subsection. The section will conclude with KMC simulations of catalytic self-organizations that indicate the conditions necessary for self-organization at the molecular level.

5.4.2 Heterogeneous Catalytic Self-Organizing Systems

In heterogeneous catalysis several systems have been identified that show self-organization at isothermal conditions. These systems must at least satisfy the following two macroscopic kinetics conditions:

1. At least one reaction step must be autocatalytic. An autocatalytic reaction is defined as a reaction in which a molecule multiplies itself. This gives a kinetics expression, formulated as:

$$A + X \rightarrow 2X$$

2. At least one additional reaction step of the system must be rate-inhibiting. In a rate-inhibiting step X is removed from the reaction mixture by another reaction.

These two conditions define an excitable system in which amplification and damping processes occur. Excitable kinetics systems can display classical steady-state kinetic behavior as long as they are close to equilibrium. Once a threshold value is exceeded, they can show oscillatory or even chaotic dependence.

Catalytic systems all display self-organized conditions where different surface states are stable. At these conditions, surface states alternate and, hence, oscillations appear in the reactivity pattern. When self-organization occurs at isothermal conditions, systems can display either of the following behaviors:

1. Alternating between two different surface phases
2. Competitive reaction steps preventing reactive phase formation.

Self-organization is modeled by microkinetics so that it can be studied at the atomic level. In the next section, CO oxidation catalyzed by single crystal Pt(110) will be discussed, which is an example of catalysis using a transition metal catalyst. In the final section, a second example of self-organization is presented for a zeolitic system.

5.4.2.1 Alternation Between Two Different Surface Phases: Spiral Wave Formation

5.4.2.1.1 CO Oxidation by Pt(110) Surface

We will introduce the topic of catalytic self-organization by a detailed discussion of CO oxidation by the Pt(110) single-crystal surface because this system is understood in great detail.

The Pt(110) surface can expose a non-reactive (1×2) phase and a reactive (1×1) phase. Adsorption with CO stabilizes the reactive (1×1) phase with respect to the (1×2) phase, which is the most stable when vacant. The CO oxidation step of a catalytic reaction shows an oscillatory time dependence at specific conditions [10, 11]. This is an example of self-organization that induces dependence of the catalyst on the reactive state at a particular time during the reaction. This relationship disappears when reaction stops.

Figure 3i.1a illustrates the two surfaces. The (1×2) surface phase is the missing-row reconstructed Pt(110) surface. This surface displays decreased reactivity that prevents dissociation of O_2 molecules, and consequently prevents CO oxidation. However, when CO is adsorbed in excess of a particular concentration, reconstruction to the more reactive non-reconstructed Pt(110) (1×1) phase occurs. On this surface, O_2 can dissociate and CO oxidation will occur. Reaction again stops when so much CO has been removed that the (1×1) surface phase reconstructs back to the non-reactive (1×2) phase. At the oscillatory condition, reaction resumes when enough CO has been adsorbed, and the cycle continues. The inhibiting reaction is the transition from the reactive (1×1) phase to the non-reactive (1×2) phase, and the reaction is autocatalytic in the number of surface vacancies. This is illustrated in Figure 3i.1b. Imagine the reactive (1×1) phase completely covered with CO. In this condition, O_2 dissociation is only possible when CO desorbs and two next-neighbor surface vacancies are created. This is possible only when the temperature of reaction is high enough for CO to desorb. When two surface vacancies are present, O_2 can dissociate, the O atoms can react with CO to give CO_2 and four surface vacancies are created. Therefore, this reaction is autocatalytic in Pt (1×1) surface vacancy concentration.

> **Insert 3: Self-Organization of the Ertl System: CO Oxidation**
>
> Figures 3i.1–3i.3 concern the oscillatory time-dependent behavior of the CO oxidation reaction catalyzed by the Pt(110) surface.

(Continued)

Insert 3: (Continued)

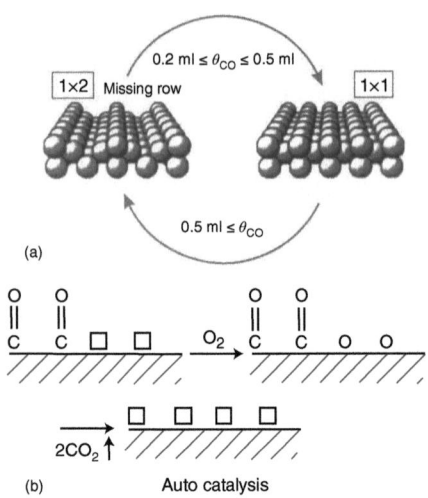

(a)

(b) Auto catalysis

Figure 3i.1 (a) The CO-induced $1\times2 \rightarrow 1\times1$ structural transformation of the Pt(110) surface. (b) Illustration of autocatalytic generation of surface vacancies in heterogeneous catalyzed oxidation of CO. Two surface vacancies next to adsorbed CO are needed to dissociate O_2. Upon reaction with adsorbed O this generates an additional two surface vacancies.(Ertl 2009 [11]. Reproduced with permission of John Wiley & Sons.)

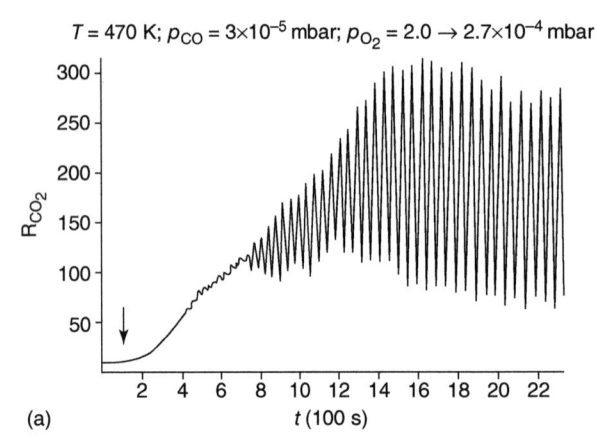

(a)

$2CO + O_2 \rightarrow 2CO_2$ / Pt(110)

Mechanism:

CO + * $\leftrightarrows$ CO_{ad}
O_2 + 2* $\rightarrow$ $2O_{ad}$
O_{ad} + CO_{ad} $\rightarrow$ CO_2 + 2*

Pt (1×2) $\xrightarrow{CO_{ad}}$ (1×1)

(b)

Kinetic modeling:

$\theta_{CO} \equiv u$; $\theta_O \equiv v$; $\theta_{1\times1} \equiv w$ ($\theta_{2\times1} \equiv 1-w$)

$\dfrac{du}{dt} = s(CO)\, p_{CO} - k_2 u - k_3 uv$

$\dfrac{dv}{dt} = s(O_2)\, p_{O_2} - k_3 uv$

$\dfrac{dw}{dt} = k_5[f(u) - w]$

$s(CO) = k_1(1-u^3)$

$s(O_2) = k_4[s_1 w + s_2(1-w)](1-u-v)^2$

Figure 3i.2 (a) Onset of kinetic oscillations in the rate of CO oxidation at a Pt(110) surface. (b) Kinetic modeling of the rate oscillations in CO oxidation on Pt(110). (c) Solution of the set of ordinary differential equations presented in (b) describing the kinetic oscillations in CO oxidation on Pt(110). (Ertl 2009 [11]. Reproduced with permission of John Wiley & Sons.)

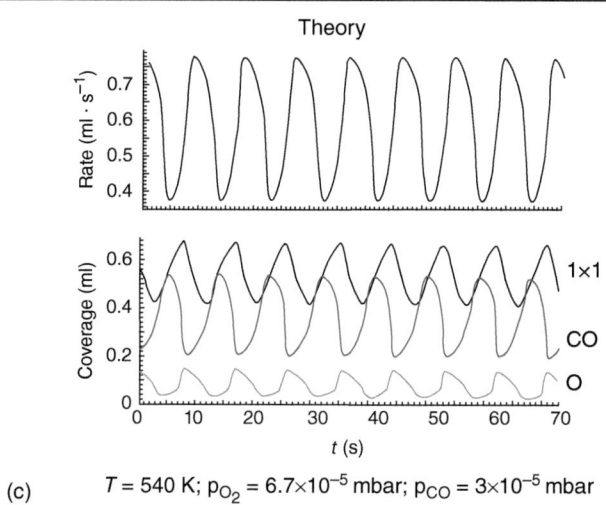

(c) $T = 540$ K; $p_{O_2} = 6.7 \times 10^{-5}$ mbar; $p_{CO} = 3 \times 10^{-5}$ mbar

Figure 3i.2 (*Continued*)

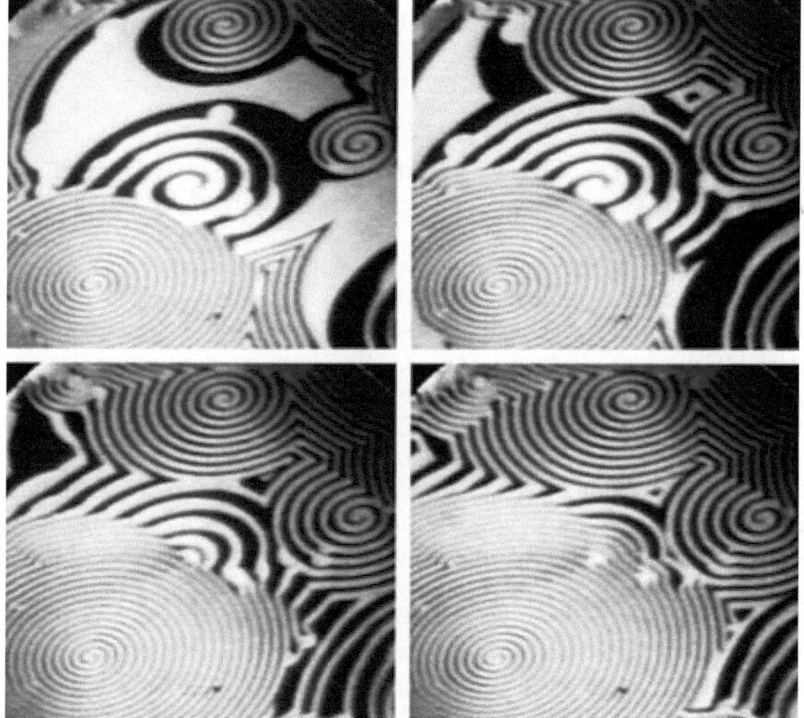

Figure 3i.3 A sequence of photoemission electron microscopy (PEEM) images from a Pt(110) surface during spiral wave formation in CO oxidation taken at intervals of 30 s. $T = 448$ K, $pO_2 = 4 \times 10^{-4}$ mbar, $pCO = 4.3 \times 10^{-5}$ mbar. (Ertl 2009 [11]. Reproduced with permission of John Wiley & Sons.)

An experiment that illustrates the temporal behavior of the CO oxidation reaction is shown in Figure 3i.2a. At the onset of the reaction, the CO pressure is slightly increased (from 2×10^{-3} to 2.7×10^{-3} mbar). The oscillatory dependence that develops can be simulated using coupled mean-field equations as shown in Figure 3i.2b. An example of simulated time-dependent coverage of CO and O and their phase relation with the CO oxidation rate is shown in Figure 3i.2c. The maxima in the rate of reaction are found to correspond with the maxima in O coverage. Measurement of such time-dependent coverage changes has become possible by the application of photoemission electron microscopy [12] at low-pressure reaction conditions.

Self-organization phenomena occur even at stationary conditions. For example, at isothermal conditions when an excitable system such as the CO oxidation reaction catalyzed by the Pt(110) surface is reacting, reaction that occurs at different locations on the surface is communicated through mass transfer. When the distance-related diffusion delay time is greater than the oscillation rate, the reaction becomes stationary, but temporal reaction concentration patterns are generated (see Figure 3i.3).

It is possible to simulate this spiraling phenomenon using mean field rate expressions by considering the surface concentrations of Figure 3i.2b,c as a function of position. The rate expressions for the change in adsorbate concentration must be extended with a diffusion term. Equation (5.2) provides an example of such a reaction-diffusion equation:

$$\frac{\partial \theta_c}{\partial t} = F_i(\theta_j, P_c) + D_i \nabla^2 \theta_i \qquad (5.2)$$

with θ_i the concentration of surface adsorbate I, F_i the rate of θ_i, production, P_c, the partial pressure of gas phase components, and D_i the diffusion coefficient of surface intermediate i.

The parameters in the resulting partial differential equations (PDEs) can then be derived by fitting the simulated solutions with experimental data.

Simulation based on the input of molecular elementary rate data is possible with the KMC method (see Insert 4 [1, 13, 14]). These simulations require a grid where there is a change in both the adsorbed reaction intermediates as well as in the state of surface sites that depend on CO concentration (see Figure 4i.2). We will use the results of these simulations to illustrate the importance of synchronization between reaction events that occur at different surface locations. Since collisions of the reactant molecules with the surface are stochastic, the catalytic cycles that evolve are otherwise out of phase and no global oscillatory behavior is observed.

Oscillatory behavior will only arise when the temperature of the reaction exceeds the temperature of maximum reaction rate. This is illustrated by the simulation shown in Figure 4i.1. This figure shows the expected dependence of the rate of reaction on temperature. At low temperature, the CO oxidation reaction is poisoned because the catalyst surface is covered by adsorbate, and the reaction can only begin at a temperature where molecules start to desorb. In the case of the CO oxidation reaction, the surface is initially covered with CO. The rate maximum occurs at a temperature where a further increase in temperature will decrease the surface coverage with reactive species.

While the rate of reaction is stationary at the lower temperature, beyond its maximum the reaction rate may start to oscillate as indicated in Figure 4i.1. Oscillation also requires a conducive pressure regime. In Figure 4i.1, the rate maxima and minima in this regime are indicated. The surface state now oscillates between a reactive state covered with CO and a vacant non-reactive state as sketched in Figure 4i.2. This figure also shows the dependence of self-organization on an order parameter, which in this case is temperature. When the temperature is low, the reaction is at steady state and the surface is a statistical mix of different coverages and surface phases. When the rate becomes fast and reaction conditions depart significantly from equilibrium, the probability of finding the surface in one of the two reactive phases shows an oscillatory time dependence.

The importance of diffusion in the occurrence of this globally synchronized behavior is illustrated in Figure 4i.3, which compares the case where the diffusion rate of adsorbed CO is zero with the case where its diffusion rate is significant. At the self-organizing condition where oscillatory time dependence is observed, the amplitude of the oscillation is plotted as a function of grid size L. One observes that when diffusion $D = 0$ the amplitude of oscillations rapidly declines with increasing grid size. When grid size is small there will only be a single site of reaction that will oscillate at the condition of self-organization. When grid size increases, the number of sites with an oscillating reaction will increase, but the phases of the oscillations are stochastically determined and will thus be out of phase. When averaged over time, this will lead to a cancelation of intensities and hence decreases in oscillation amplitude. A state of global synchronization with all oscillating reaction cycles in-phase is reached only when the diffusion is fast enough to allow reactive sites at a distance L to communicate with a time delay shorter than the oscillation time.

The results reported in Figures 4i.4 and 4i.5 are based on simulations with the following model [14]: CO can adsorb and desorb from a free surface site on phase α and β. The reactions are independent of the surface phase to which the site belongs. In addition, CO can diffuse via hopping onto a vacant nearest neighbor site. The sticking coefficients of O_2 to the two surface phases is smaller on the α phase than on the β phase. The smaller sticking coefficient for the α phase leads to a displacement of oxygen by CO on the α phase. O_2 adsorbs dissociatively onto two nearest-neighbor sites on the reactive β phase. The $CO + O \rightarrow CO_2$ reaction occurs with a rate constant R, when CO and O are on nearest-neighbor sites. CO_2 desorbs immediately, forming two vacant sites. The α surface to β surface phase transition is modeled as a front propagation with a rate constant V. For two nearest-neighbor surface sites in the state $\alpha\beta$, the transition $\alpha\beta \rightarrow \alpha\alpha$ occurs if neither of these two sites is occupied by CO. Therefore, without CO the substrate reconstructs to the α phase, and with CO this reconstruction is lifted and the substrate converts to the β phase.

Insert 4: Kinetic Monte Carlo Simulations of Catalytic Self-Organization

Figures 4i.1–4i.5.

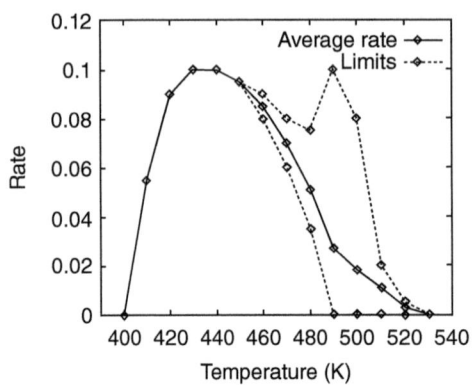

Figure 4i.1 CO_2 production rate (molecules CO_2 per Pt atom per second) versus temperature. The solid line indicates the average rate. In the oscillatory region, two extra dashed curves are drawn: one for the maximum rate (top of oscillation peaks) and one for the minimum rate. The amplitude of the oscillations in this region is then given by the difference between the two dashed curves. The grid size was 256 × 256. Larger grids did not change the plot significantly. (Gelten et al. 1998 [13]. Reproduced with permission of American Institute of Physics.)(Gelten et al. 1998 [13]. Reproduced with permission of American Institute of Physics.)

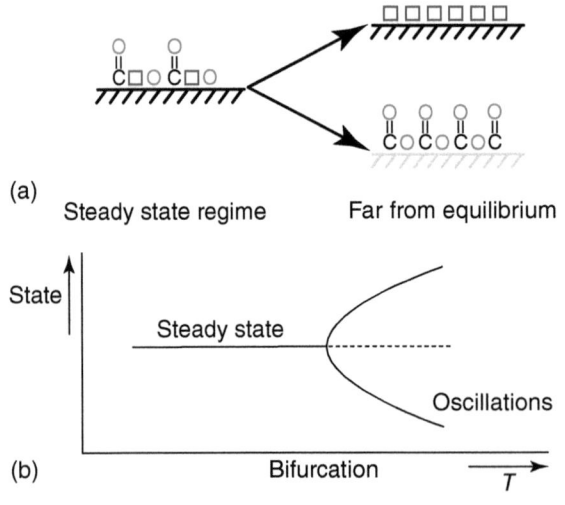

Figure 4i.2 (a) The two state coverages between which the reaction oscillates at the self-organized condition. (b) Illustration of the transition from steady state to the self-organized state when a reaction order parameter increases. In the example the order parameter is the temperature.

5.4 Catalytic Self-Organizing Systems | 161

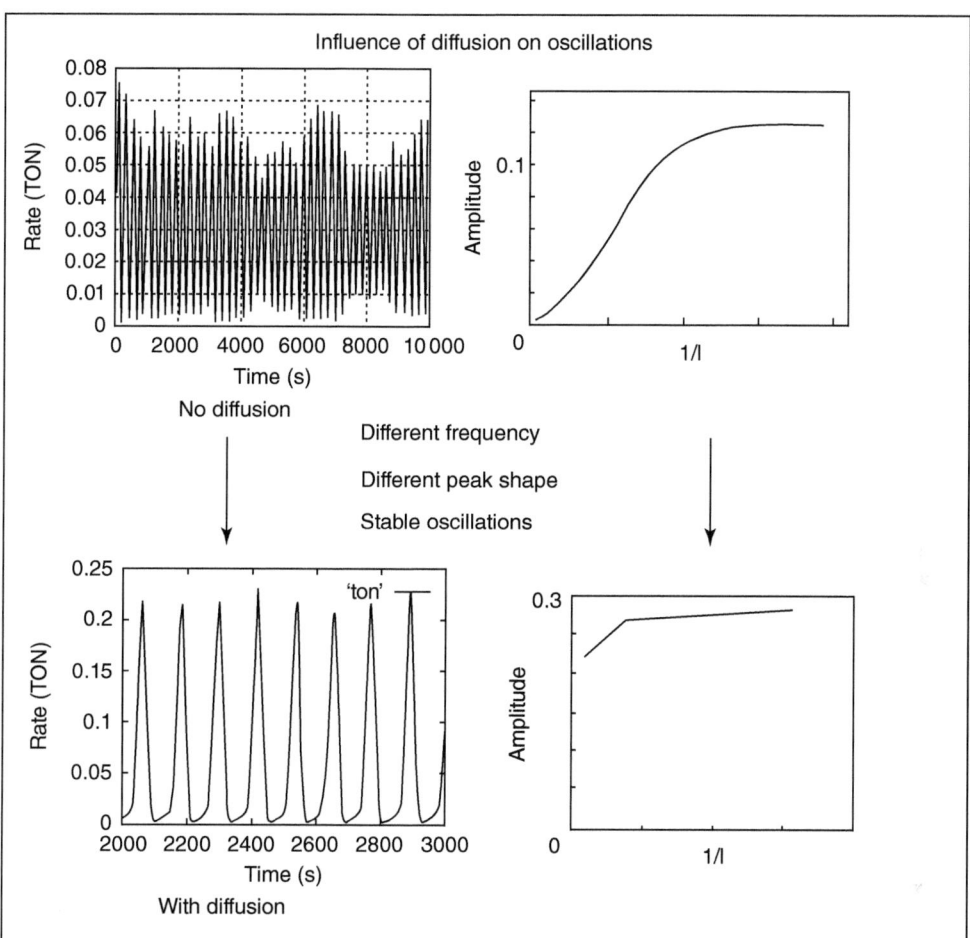

Figure 4i.3 Example of simulated oscillations in the production rate of CO_2 (left) and its power spectrum averaged over 10 simulation runs (right), grid size = 1024 × 1024 unit cells. Turn over numbers (TONs) are molecules produced per Pt atom per second. The cases of no diffusion, upper part, and with diffusion, lower part are compared. (Gelten et al. 1998 [13]. Reproduced with permission of American Institute of Physics.)

(Continued)

Insert 4: (Continued)

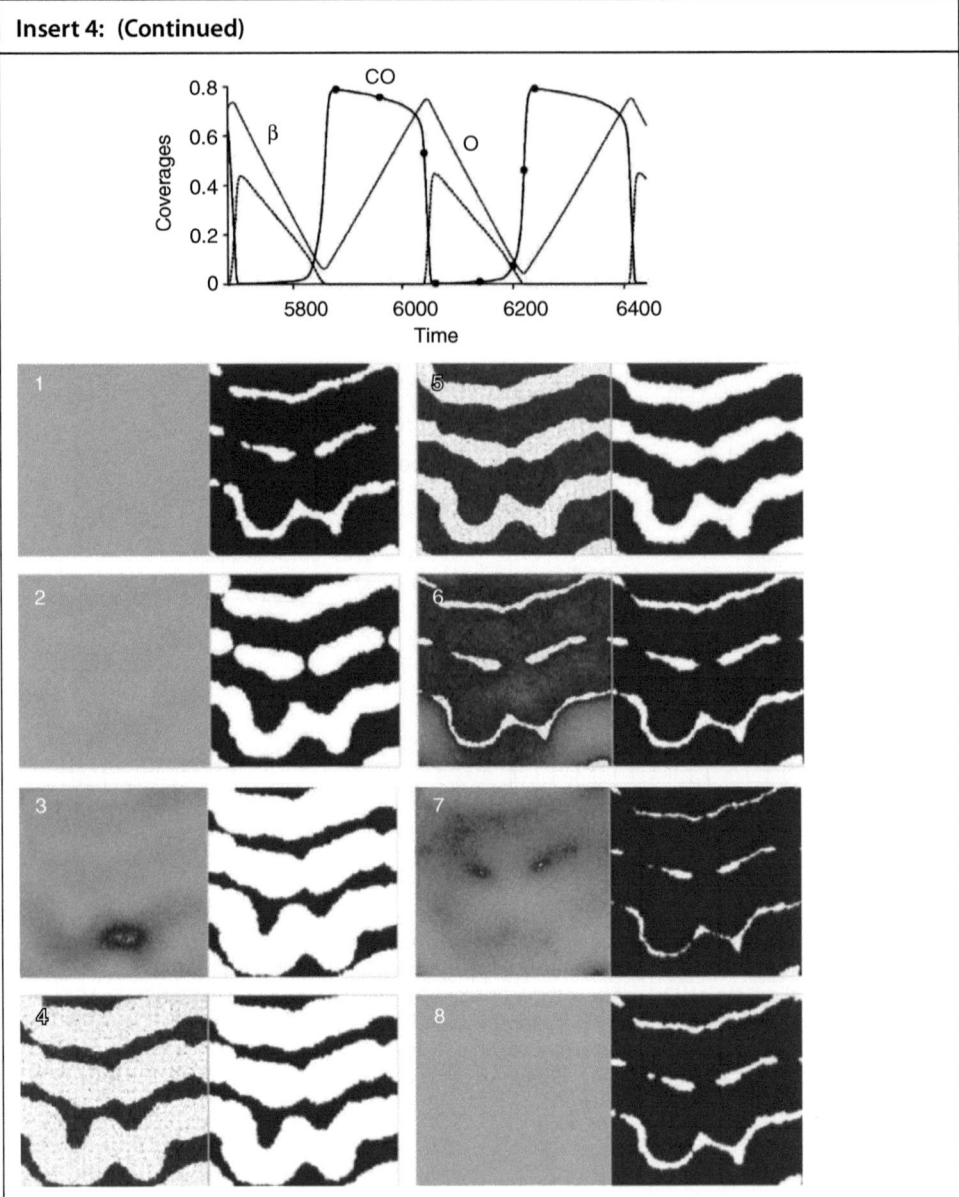

Figure 4i.4 Simulated self-organized oscillations of the coverage CO, oxygen, and surface phases in the CO oxidation reaction. Sequence of temporal snapshots corresponds to the points in the upper figure that give coverage as a function of time. Each snapshot has two parts: In the left part we plot the chemical species; CO particles are gray, O particles are white, and empty sites are black. The right part shows the structure of the surface; α phase sites are black, and β phase sites are white. (Salazar et al. 2004 [14]. Reproduced with permission of American Institute of Physics.)

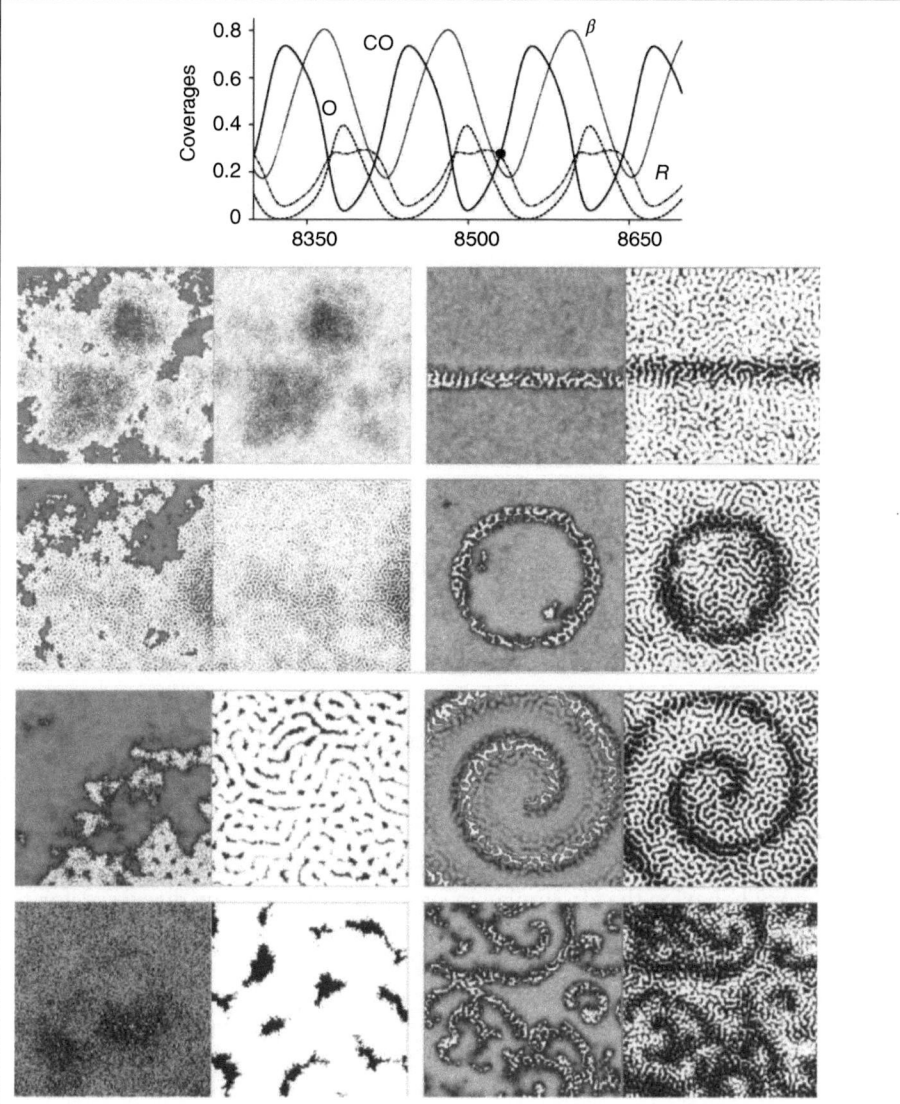

Figure 4i.5 Global oscillations and pattern formation. The whole system and sections of the upper-left corner with different length scales are shown on the left side. The snapshots correspond to the dot in the temporal plot at the top. It shows the same information as Figure 4i.4, but it also includes the rate of CO_2 formation R. On the right side we have a wave front, a target, a spiral, and turbulence (all from simulations with $L = 2048$), which can be obtained with different initial conditions. (Salazar et al. 2004 [14]. Reproduced with permission of American Institute of Physics.)

Figure 4i.5 shows the temporal dependence of the probability of occurrence of one of the two phases, when the diffusion rate of surface intermediates is fast compared to the oscillation time. However, when diffusion rate is slower and only partial synchronization is possible, surface adsorbate and state patterns appear.

In the simulations of Figure 4i.5, the diffusion constant has been defined by the experimental diffusion parameter D_A by $D_A = a^2 D/zt_0$, where a is unit cell size, $t_0 = 10^{-2}$ s and the unit of grid length L is a. In Figure 4i.5, however, a simulation is shown on the left where the diffusion length is much smaller than the lattice size, but larger than the correlation length. In this case the diffusion synchronizes the oscillations on neighboring islands, and the self-organization leads to global oscillations. When the diffusion length is reduced to less than the correlation length the result is pattern formation. In cases where the synchronization is larger than the correlation length, islands tend to group and form parallel rolls.

In summary, a reaction will show stable oscillations only if the following conditions are satisfied:

- The reaction mechanism must include an autocatalytic elementary reaction step.
- The reaction mechanism must also include an inhibiting reaction step.
- Synchronization between reactions at different sites by mass or heat transfer must occur.
- Reaction conditions must depart significantly from equilibrium.

Self-organization is a complex phenomenon that is observed for many non-linear physical systems where both oscillatory time dependence and chaotic behavior can occur. Introductory references about this phenomenon are Schroeder [15] and Prigogine [8].

5.4.2.1.2 Additional Self-Organizing Heterogeneous Catalytic Systems

Oscillatory behavior that is similar in origin to the behavior discussed for the CO oxidation reaction has been found for reactions such as $H_2 + O_2$ recombination and NO reduction using H_2 or NH_3 catalyzed by Pt, Ir, or Rh [16]. In these systems, autocatalytic surface vacancy formation and the presence of different surface phases also play a role. For instance, in NO reduction with H_2, a vacant surface alternates with a surface phase of NO or N islands.

Related self-organization with oscillatory time dependence has been observed for reactions in which the surface state varies between the oxidic state and the metallic state. Catalysis of CO oxidation by RuO_2 is an example of this type of reaction. At low temperature, the oxidic state of the reaction is active because it is not poisoned by CO, but when reduced to the metallic state the reaction becomes poisoned. This reaction is again autocatalytic in surface site vacancies [17].

The oxidation of propane by NiO provides another example reaction. Propane activation is suppressed by NiO, but it is accelerated on the metallic Ni after the reduction of NiO [17, 18].

In addition, the presence of a subsurface oxide layer may induce oscillatory behavior of the CO oxidation reaction, as has been found for Pd [19]. In this reaction, oscillations occur between the metallic surface phase that is poisoned by CO but efficiently dissociates O_2, and the surface that includes a subsurface oxide with weaker CO bonding which dissociates O_2 at a suppressed rate. Again, the reaction of CO and adatom O atoms is autocatalytic in surface vacancy concentration. CO will also react with O from the subsurface oxide layer. The inhibiting reaction step is the slow diffusion of surface oxygen adatoms into the subsurface layer, which decreases the rate of dissociative O_2 adsorption.

5.4.2.2 Competitive Reactive Steps Prevent Reactive Phase Formation

Cu^+ ions located in the microchannels of the HZSM-5 zeolite (see figure 2.7, p. 26) are efficient catalysts for the decomposition of N_2O to N_2 and O_2. (After [20].) There are two mechanistic routes for this decomposition reaction: The decomposition of N_2 by a Cu^+ ion (steps 1 and 2) gives $(CuO)^+$ and N_2. Then, the $(CuO)^+$ ions can recombine to produce O_2 (step 3a) and regenerate free Cu^+ ions. Alternatively, the $(CuO)^+$ cationic compound reacts with a second N_2O to give N_2 and O_2. The inhibiting step in the mechanism is the reaction of $(CuO)^+$ with N_2O. Because this reduces the concentration of $(CuO)^+$ it suppresses the recombination reaction of the two $(CuO)^+$ oxidations (step 3b). The latter step results in the autocatalysis Cu^{+1} formation.

Step 1: $2N_2O + 2Cu^+ \rightarrow 2[N_2O^- - Cu^{2+}]$
Step 2: $2[N_2O^- - Cu^{2+}] \rightarrow 2N_2 + 2[O^- - Cu^{2+}]$
Step 3a: $2[O^- - Cu^{2+}] \rightarrow O_2 + 2Cu^+$
Step 3b: $2N_2O + 2[O^- - Cu^{2+}] \rightarrow 2N_2 + 2O_2 + 2Cu^+$.

The $(CuO)^+$ cationic oxide cluster phase alternates with the Cu^+ cationic oxide phase. When O_2 formation by recombination of O atoms becomes too slow, the inhibiting reaction dominates and the oscillatory temporal behavior disappears. The interesting aspect of this reaction scheme is that self-organization is intimately connected with the inorganic chemistry of the catalytic system.

References

1 Gelten, R.J., van Santen, R.A., and Jansen, A.P.J. (1999) in *Molecular Dynamics* (eds P.B. Balbuena and J.M. Seminario), (Elsevier), pp. 737–784.
2 Jansen, A.P.J. (2012) *An Introduction to Kinetic Monte Carlo Simulations of Surface Reactions*, (Springer).
3 Hess, F. and Over, H. (2014) Kinetic Monte Carlo simulations of heterogeneously catalyzed oxidation reactions. *Catal. Sci. Tech.*, **4**, 583. doi: 10.1039/C3CY00833A
4 De Cremer, G., Sels, B.F., De Vos, D.E., Hofkens, J., and Roeffaers, M.B.J. (2010) Fluorescence micro(spectro)scopy as a tool to study catalytic materials in action. *Chem. Soc. Rev.*, **39**, 4703. doi: 10.1039/C0CS00047G
5 Smiley, R.D. and Hammes, G.G. (2006) Single molecule studies of enzyme mechanisms. *Chem. Rev.*, **106**, 3080–3094. doi: 10.1021/cr0502955
6 Chen, P., Zhou, X., Andoy, N.M., Han, K.-S., Choudhary, E., Zou, N., Chen, G., and Shen, H. (2014) Spatiotemporal catalytic dynamics within single

nanocatalysts revealed by single-molecule microscopy. *Chem. Soc. Rev.*, **43**, 1107–1117. doi: 10.1039/C3CS60215J

7 Schüth, F., Henry, B.E., and Schmidt, L.D. (1993) Oscillatory reactions in heterogeneous catalysis. *Adv. Catal.*, **39**, 51–127. http://linkinghub.elsevier.com/retrieve/pii/S0360056408605775 (accessed 24 September 2016).

8 Prigogine, I. (1980) *From Being to Becoming*, W.H. Freeman and Co.

9 Haken, H. (1983) *Synergetics: An Introduction*, (Springer).

10 Krischer, K., Eiswirth, M., and Ertl, G. (1992) Oscillatory CO oxidation on Pt(110): modeling of temporal self-organization. *J. Chem. Phys.*, **96**, 9161. http://scitation.aip.org/content/aip/journal/jcp/96/12/10.1063/1.462226 (accessed 24 September 2016).

11 Ertl, G. (2009) *Reactions at Solid Surfaces*, John Wiley & Sons, Inc., Hoboken, NJ.

12 Engel, W., Kordesch, M.E., Rotermund, H.H., Kubala, S., and von Oertzen, A. (1991) A UHV-compatible photoelectron emission microscope for applications in surface science. *Ultramicroscopy*, **36**, 148–153. http://linkinghub.elsevier.com/retrieve/pii/030439919190146W (accessed 24 September 2016).

13 Gelten, R.J., Jansen, A.P.J., van Santen, R.A., Lukkien, J.J., Segers, J.P.L., and Hilbers, P.A.J. (1998) Monte Carlo simulations of a surface reaction model showing spatio-temporal pattern formations and oscillations. *J. Chem. Phys.*, **108**, 5921. http://scitation.aip.org/content/aip/journal/jcp/108/14/10.1063/1.476003 (accessed 24 September 2016).

14 Salazar, R., Jansen, A., and Kuzovkov, V. (2004) Synchronization of surface reactions via turing-like structures. *Phys. Rev. E*, **69**, 031604. doi: 10.1103/PhysRevE.69.031604

15 Schroeder, M. (1991) *Fractals, Chaos and Power Laws*, Dover Publications.

16 Gorodetskii, V., Elokhin, V., Bakker, J., and Nieuwenhuys, B. (2005) Field electron and field ion microscopy studies of chemical wave propagation in oscillatory reactions on platinum group metals. *Catal. Today*, **105**, 183–205. http://linkinghub.elsevier.com/retrieve/pii/S0920586105002488 (accessed 24 September 2016).

17 Ota, A., Armbrüster, M., Behrens, M., Rosenthal, D., Friedrich, M., Kasatkin, I., Girgsdies, F., Zhang, W., Wagner, R., and Schlögl, R. (2011) Intermetallic compound Pd2Ga as a selective catalyst for the semi-hydrogenation of acetylene: from model to high performance systems †. *J. Phys. Chem. C*, **115**, 1368–1374. doi: 10.1021/jp109226r

18 Kaichev, V.V., Gladky, A.Y., Prosvirin, I.P., Saraev, A.A., Hävecker, M., Knop-Gericke, A., Schlögl, R., and Bukhtiyarov, V.I. (2013) In situ XPS study of self-sustained oscillations in catalytic oxidation of propane over nickel. *Surf. Sci.*, **609**, 113–118. http://linkinghub.elsevier.com/retrieve/pii/S0039602812004177 (accessed 24 September 2016).

19 Imbihl, R. and Ertl, G. (1995) Oscillatory kinetics in heterogeneous catalysis. *Chem. Rev.*, **95**, 697–733. doi: 10.1021/cr00035a012

20 Ciambelli, P., Garufi, E., Pirone, R., Russo, G., and Santagata, F. (1996) Oscillatory behaviour in nitrous oxide decomposition on over-exchanged Cu-ZSM-5 zeolite. *Appl. Catal., B*, **8**, 333–341. http://linkinghub.elsevier.com/retrieve/pii/0926337395000658 (accessed 24 September 2016).

Part II

Molecular Heterogeneous Catalysis

Introduction

In this second part of the book, we will embark on a detailed discussion of the relationship between surface chemical reactivity of catalysts and the mechanism of catalytic reactions. Due to the rapid increase in computational heterogeneous catalysis studies during the last decade, the molecular basis for the performance of many heterogeneous catalytic reactions has been determined in great detail. In addition, new reactions and catalytic systems have been discovered using the modern tools of catalyst synthesis, characterization, and computation. We will discuss these reactions in successive chapters, grouped by related catalyst systems. The systems we will cover are mainly the transition metals and reducible and non-reducible oxidic systems, but we will also include the sulfides and chlorinated materials in our discussions. We have introduced some of these reactions in previous chapters, but we will now investigate them further by examining the molecular chemistry of their elementary reaction steps, and the mechanistic networks related to the reactivity of their surface sites as a function of composition and structure. In most cases, we will include information about relevant experimental studies of kinetics or selectivity for the reaction.

Because an understanding of surface reactivity is essential, we have included in most chapters introductory sections on spectroscopic and computational characterization, followed by sections dealing with the specific reactions for most systems.

The first two chapters are different in character. These chapters serve to introduce the quantum chemistry of the chemical bond in transition metals, which is basic to understanding the reactivity trends of the transition metal surface.

The quantum chemistry of the chemical bond involves the use of unique tools and concepts. A proper introduction to modern methods for analyzing the output of state-of-the-art quantum chemical calculations is indispensable for understanding this chemistry. For this reason, Chapter 6 provides an introductory chapter on chemical bonding theory, starting with bonds between individual molecules and finishing with the chemical bonds of transition metals. This material is presented in an elementary way, without a detailed discussion of the

Modern Heterogeneous Catalysis: An Introduction, First Edition. Rutger A. van Santen.
© 2017 Wiley-VCH Verlag GmbH & Co. KGaA. Published 2017 by Wiley-VCH Verlag GmbH & Co. KGaA.

quantum-chemical methods on which the electronic structural information is based, but in sufficient depth to provide the basic background required to understand the quantum-chemical aspects of surface reactivity discussed in the following chapters. For an introduction to quantum-chemical density functional theory (DFT) methods, please refer to the references at the end of this chapter [1–6].

The reaction mechanism of a catalytic reaction consists of a network of elementary reaction steps that lead from reagent to product. In cases where there is more than one reaction product, several competing reaction paths are followed. In these catalytic systems we are interested not only in the activity of the catalyst (which measures the rate of reagent conversion), but also in the selectivity of product formation. In the following chapters, we will present the mechanism of reaction for most important heterogeneous catalytic systems. We will also introduce the surface chemistry that relates the reaction mechanism to the catalyst composition and structure.

In practice, the selectivity of a reaction is measured as a fraction, calculated as the weight of desired product divided by the total amount of raw material converted. A useful measure of chemical efficiency is the gram atom fraction of resulting desirable product, normalized with respect to the total number of gram atoms that theoretically could have been converted to this product. This defines the atom efficiency of the reaction (see Insert 1).

Insert 1: Atom Utilization

The concept of atom utilization can be used to calculate the efficiency of converting raw material by a stoichiometric process versus conversion by a catalytic process [7]. We will illustrate this using the ethylene epoxide (EO) formation from ethylene reaction (see Figure Ii.1).

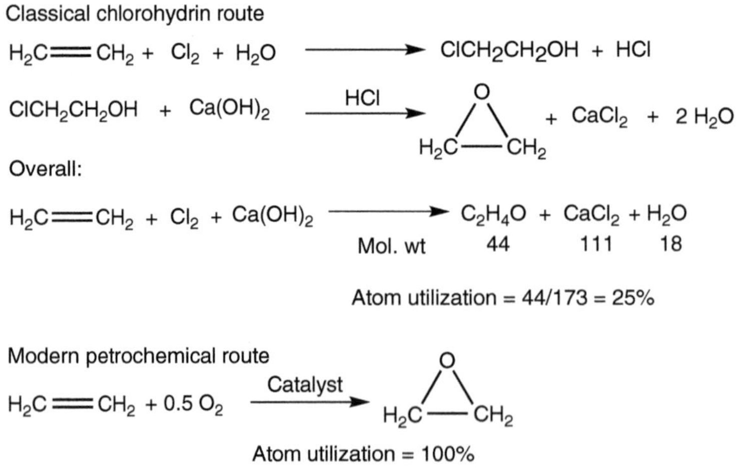

Classical chlorohydrin route

$H_2C{=}CH_2 + Cl_2 + H_2O \longrightarrow ClCH_2CH_2OH + HCl$

$ClCH_2CH_2OH + Ca(OH)_2 \xrightarrow{HCl} \underset{H_2C-CH_2}{\overset{O}{\triangle}} + CaCl_2 + 2 H_2O$

Overall:

$H_2C{=}CH_2 + Cl_2 + Ca(OH)_2 \longrightarrow C_2H_4O + CaCl_2 + H_2O$

Mol. wt   44   111   18

Atom utilization = 44/173 = 25%

Modern petrochemical route

$H_2C{=}CH_2 + 0.5\,O_2 \xrightarrow{Catalyst} \underset{H_2C-CH_2}{\overset{O}{\triangle}}$

Atom utilization = 100%

Figure 1i.1 The two approaches to ethylene epoxide production.

> Classically, EO has been produced using Cl_2 and $Ca(OH)_2$ as reactants with $CaCl_2$ as the waste product. This results in an atom utilization of 25%. In the catalytic process, a silver catalyst is used to promote direct oxidation with oxygen. This reaction is not 100% selective, because non-selective total oxidation also occurs, thus the maximum selectivity is 90% based on ethylene. One can then calculate the "Effective atom utilization":
>
> $$\text{Effective atom utilization} = \frac{EO}{E + 0.5 \times O_2 + 0.1(E + 3O_2)} \times 100\% = 77\% \quad (1\text{i}.1)$$

There can be many reasons for an atom efficiency of less than 100% for the different known catalytic reactions. Improving their efficiency is a considerable incentive for studying catalytic reactions that are currently used on a large scale. To achieve this improvement we have to influence the selectivity controlling rate parameters of the catalytic system. Even small selectivity improvements can result in significant saving of raw materials.

The mechanism of the reaction is intrinsically related to the surface chemical properties of the catalyst, which vary with material type. The type of catalyst materials that we will discuss are:

A. Transition metals.
B. Zeolitic Brønsted and Lewis acidic solid materials.
C. Reducible solid state materials such as the reducible oxides and sulfides.

Computational quantum-chemical studies enable calculation of the relative stability and reactivity of reaction intermediates that are difficult to detect by experiment. The deduction of the mechanism of catalytic reaction starts with a proposal of a set of elementary reactions that form the catalytic cycle. Such a proposal can be based on experimental information on reaction kinetics, catalyst composition and surface state of reaction during reaction. Rarely, direct molecular information on key reaction intermediates formed during reaction is available.

A great aid in defining the reaction mechanism is the use of isotopically labeled molecules (see Section 3.4). This method facilitates tracing the reaction path of atoms through their respective reaction intermediates. We will discuss several example reactions where this has been essential for the identification of elementary reaction steps that are key to the selectivity of a catalytic reaction.

In operando spectroscopic techniques used to identify the state of the catalyst surface conducted simultaneously with product measurement are indispensable. However, the species observed on the surface do not necessarily take part in the reaction cycle, but may instead be spectator species whose presence can play an indirect role by affecting the state of the surface.

Practical systems tend to have an inhomogeneous site distribution, making complementary experiments with catalyst model systems beneficial because the rate constants of elementary reaction steps and relative stability of reaction intermediates may vary markedly for different exposed surfaces. In addition, the use of model experiments with well-defined surfaces is especially important to validate quantum-chemical computations, so that a direct comparison between the molecular and experimental data can be made.

Theory developed through catalysis research is a necessary complement to the often limited experimental information that is available. The connection between theoretical and experimental results is complex, since the exact state of the catalytic reaction center is often unknown. Microkinetics simulations are an important tool to relate reaction kinetics with data on elementary reaction steps as a function of catalytic site structure or composition.

The mechanism of the reaction can be formulated as a reaction scheme that consists of a cycle of reactions involving catalyst and reagents. Examples of reaction schemes have been presented for the CO to methane transformation reaction and the reversed steam reforming reaction in Chapter 3, Insert 7 and for zeolite-catalyzed reactions in section 4.7.

The cyclic reaction scheme is the basis for the formulation of the kinetics equations that can be used as input for kinetic simulations. To develop these simulations we must construct the corresponding reaction energy diagram. Quantum-chemical input is also necessary, since information on the relative stability of all reaction intermediates can rarely be derived by experiment.

Microkinetics tools that allow for the modeling of reaction kinetics without a presumptive choice of the reaction rate controlling step are essential for developing simulations, especially when attempting to simulate the Sabatier principle reaction activity plots. In this case, the rate controlling step changes when overall rates of reaction are plotted as a function of surface reactivity and compared at the left and right of the rate maximum.

The kinetic equations that are formulated based on the reaction energy diagram usually rely on the mean field assumption, as discussed in Chapter 3, Insert 5). These equations must be solved at steady state for selected temperatures as a function of reactant concentrations. This is not a trivial task due to the large number of equations involved and their nonlinear dependence on reaction intermediate surface concentrations. In addition to information on catalyst performance, the solutions of these equations will also provide an approximation of the surface overlayer composition, which in principle can be validated by experiment. For a more detailed discussion we refer to Nørskov et al. [8]. In Chapter 5, we discussed advanced kinetics methods that can be used to describe catalytic systems that are non-stationary or at conditions where ordered overlayers determine kinetics.

The molecular insight into the functioning of the heterogeneous inorganic catalyst makes comparison with other molecular catalytic systems relevant. We will refer occasionally to these related mechanisms of organometallic or enzyme catalyzed reactions.

After the two chapters that introduce the electronic structure of the chemical bond and its relation with surface reactivity, each of the following chapters will discuss a particular catalytic system. The main theme of these chapters is the relationship between catalytic performance and the structure and composition of the catalytic system. Each chapter consists of two parts: the first part deals with the chemical bonding aspects of the chemical system and the second discusses the reaction mechanisms for the various reactions. These sections have been written to be read independently, so that the mechanistic aspects of the reactions can be understood without the deeper insights into the surface chemical bond provided in Chapters 6 and 7.

References

1 Gross, A. (2009) *Theoretical Surface Science*, Springer-Verlag, Berlin.
2 Sholl, D.S. and Steckel, J.A. (2009) *Density Functional Theory: A Practical Introduction*, John Wiley & Sons, Inc., New York.
3 van Santen, R.A. and Neurock, M. (2006) *Molecular Heterogenous Catalysis*, Wiley-VCH Verlag GmbH.
4 Nilsson, A., Pettersson, L.G.M., and Nørskov, J.K. (2008) *Chemical Bonding at Surfaces and Interfaces*, Elsevier, pp. 255–322.
5 Kresse, G. and Furthmüller, J. (1996) Efficiency of ab-initio total energy calculations for metals and semiconductors using a plane-wave basis set. *Comput. Mater. Sci.*, **6**, 15–50, http://linkinghub.elsevier.com/retrieve/pii/0927025696000080 (accessed 14 July 2014).
6 Schleyer, P.v.R. (1998) *Encyclopedia of Computational Chemistry*, John Wiley & Sons, Ltd, Chichester.
7 Sheldon, R.A. (1994) Green chemistry performance metrics: E2 factor. *Chem. Tech.*, **24**, 38–47.
8 Nørskov, J.K., Studt, F., Abild-Pedersen, F., and Bligaard, T. (2014) *Fundamental Concepts in Heterogeneous Catalysis*, 1st edn, John Wiley & Sons, Inc., New York.

6

Basic Quantum-Chemical Concepts, The Chemical Bond Revisited (Jointly Written with I. Tranca)

6.1 Introduction

This chapter provides an introduction to the quantum-chemistry of the chemical bond in molecules and transition metals, along with the necessary tools and concepts to define the relationship between bond energies and their associated structures. This information will serve as a foundation for subsequent chapters concerning trends in surface chemical reactivity.

The two essential electronic structure descriptors we will use are the following:

- The partial density of states (PDOS), which gives the probability of an electron to populate an atomic orbital at a particular electronic energy.
- The bond order overlap population density (BOOPD), which measures the bonding, non-bonding, or antibonding nature of interacting atomic orbital pairs.

These two descriptors and their related variants will be defined in Section 6.2, followed by a section that illustrates their significance and application in the electronic structure of simple molecules. This topic will be familiar to most of our readers and helpful for understanding how the chemical reactivity of a catalyst relates to its electronic structure. This section will conclude with a comparison of diatomic molecules to their corresponding solid state structures. It will illustrate the ability of PDOS and BOOPD to describe the electronic structure in both molecules and solids.

In a final section, we will discuss the electronic structure of the transition metals as a function of their position in the periodic table of elements. This will provide a vehicle for understanding features such as the distribution of valence electrons over bonding and antibonding orbital fragments and the relationship of this distribution to differences in the crystal structure of the transition metals.

In this chapter and in the following chapters, we will use recently computed data made available from state-of-the-art electronic structure calculations based on the Vienna *ab initio* Simulation Package (VASP) code [1–6] and Gaussian references [11]. The computational details are included in an appendix at the end of this chapter for reference.

Modern Heterogeneous Catalysis: An Introduction, First Edition. Rutger A. van Santen.
© 2017 Wiley-VCH Verlag GmbH & Co. KGaA. Published 2017 by Wiley-VCH Verlag GmbH & Co. KGaA.

6.2 The Definitions of Partial Density of States and Bond Order Overlap Population

We can analyze the electronic structure of a chemical bond based on its molecular orbital structure. The molecular orbitals are linear combinations of atomic orbitals. The number of molecular orbitals is determined by the number of atomic orbitals that constitute them, but not all of them will be occupied by electrons. Their electron occupation is determined by the Pauli principle that states that the lower energy orbitals are occupied first, with subsequent occupation of higher energy orbitals as energy increases. Not more than two electrons can occupy the same molecular orbital and these electrons must have opposing spins.

According to quantum mechanics, the electrons in a molecular orbital are delocalized over the atomic orbitals that are combined to produce it. The probability for an electron to occupy a particular atomic orbital is given by the PDOS $\rho_i(E)$ that expresses the probability of finding an electron of energy E in atomic orbital i (see Eqs. (1i.1)). $\rho_i(E)$ is an important property of the electronic structure. Its dependence on electron energy provides a measure of the degree of delocalization of the electron density, which relates to the interaction strength of the atomic orbitals on which it is built. The PDOS is defined by Eqs. (1i.1).

Insert 1: Definition of the Chemical Bond Descriptors Partial Density of States and Bond Order Overlap Population Density

The PDOS measures the probability of finding an electron in a particular atomic orbital as a function of electron energy. The mathematical expression for the PDOS in atomic orbital i is:

$$\text{PDOS}_i(E) \equiv \rho_i(E) \tag{1i.1a}$$

$$= \sum_n \langle \varphi_i | \psi_n \rangle \delta(E - E_n) \langle \psi_n | \varphi_i \rangle \tag{1i.1b}$$

$$= \sum_n |c_i^n|^2 \delta(E - E_n) \tag{1i.1c}$$

The bracket notation $\langle \varphi_i | \psi_n \rangle$ in Eq. (1i.1b) means that the integral over volume must be taken. The coefficients c_i and atomic orbitals φ_i are used in calculating the molecular orbitals ψ_n defined by Eq. (1i.2):

$$\psi_n = \sum_i c_i^n \varphi_i \tag{1i.2}$$

Equation (1i.1c) is strictly valid only when an orthogonal atomic basis set $|\varphi_i\rangle$ is used, so that $S_{ij} = 0$ when $i \neq j$.

The crystal orbital Hamiltonian population (COHP) density [9,10, 17] measures the magnitude and sign of the bond order energy overlap $\rho_{ij}^h(E)$ between atomic orbitals located on different atoms. This enables determination of the bonding versus antibonding nature of the orbital fragments. It is also a measure of the interaction strength between two atomic orbitals. The bond order energy overlap $\rho_{ij}^h(E)$ is equal to $-\text{COHP}_{ij}$.

$$-\text{COHP}_{ij}(E) \equiv \rho_{ij}^h(E) \tag{1i.3a}$$

$$= H_{ij} \cdot \rho_{ij}(E) \tag{1i.3b}$$

$$= H_{ij} \sum_n c_i^n c_j^n \delta(E - E_n) \tag{1i.3c}$$

H_{ij} is the Hamiltonian matrix element between atomic orbitals φ_i and φ_j. As shown in Eq. (1i.3b), COHP is the product of overlap energy H_{ij} and bond order density $\rho_{ij}(E)$.

The integrated crystal orbital Hamiltonian population (ICOHP) value (Eq. (1i.4)) can be considered to be a measure of bond strength. It is not an exact expression for the bond energy, but a good approximation as long as the repulsive energy of the nuclei is canceled by the double-counted electrostatic interactions[16]. It leads to the approximate expression Eq. (1i.5):

$$\text{ICOHP} = \int_{-\infty}^{E_F} dE \sum_{i,j} \text{COHP}_{ij}(E) \tag{1i.4}$$

$$\text{ICOHP} = \frac{1}{2} \sum_{N_i}^{occ} N_i E_i \tag{1i.5}$$

with N_i being the electron occupation of molecular orbital i. In Eq. (1i.5), the orbitals are assumed to be doubly occupied by electrons.

−ICOHP is maximum for a covalent bond when all bonding orbitals are occupied by electrons and all of the antibonding orbitals are vacant. In general, the smaller the value of −ICOHP the more ionic the bond.

The crystal orbital overlap population (COOP) density defined by Hoffmann [7] is related to −COHP$_{ij}(E)$:

$$\text{COOP}_{ij}(E) = \rho_{ij}(E) \tag{1i.6a}$$

$$= S_{ij} \sum_n c_i^n c_j^n \delta(E - E_n) \tag{1i.6b}$$

$$S_{ij} = \langle \varphi_i | \varphi_j \rangle \tag{1i.6c}$$

S_{ij} is the overlap of atomic orbitals φ_i and φ_j.

The values of COOP$_{ij}(E)$ are also a measure for the bonding or antibonding character of an orbital fragment, but the bond order density is now weighted by the atomic orbital overlap S_{ij} instead of the bond energy overlap H_{ij}. As a consequence, it cannot quantitatively analyze the contribution of the bonds to the total energy (as COHP can) but can only determine their character (bonding, antibonding, and non-bonding).

In order to calculate the COHP and the atomic orbital electron occupancies from the PDOS, orthogonalized basis functions must be used; for example, those developed by using the Löwdin orthogonalization method [8].

The orbital overlap density between two atomic orbitals on different atoms is another important function that indicates the bonding or antibonding contribution to the chemical bond strength for a specific molecular orbital fragment at a distinct orbital energy. It can be deduced from the crystal orbital overlap Hamiltonian population density (COHP$_{ij}$), that we denote as $\rho_{ij}^h(E)$ (see Eqs. (1i.3)) or from the (COOP$_{ij}$) density (see Eqs. (1i.6)) that we denote as $\rho_{ij}(E)$.

In $\rho_{ij}^h(E)$, the bond order population density is proportional to the overlap energy contribution of the two atomic orbitals on the different atoms, while, in $\rho_{ij}(E)$, it is only proportional to atomic orbital overlap. Despite this difference, we will see that both expressions lead to the same prediction of the bonding and antibonding electron density regimes of a chemical bond.

It is essential to understand this: at a particular orbital energy, if $\rho_{ij}^h(E)$ or $\rho_{ij}(E)$ is positive, the electronic interaction is bonding; if $\rho_{ij}^h(E)$ or $\rho_{ij}(E)$ is negative then the electronic interaction is antibonding; and if $\rho_{ij}^h(E)$ or $\rho_{ij}(E)$ are zero they describe a non-bonding interaction regime. When the electron occupation of a bonding orbital increases, the strength of the corresponding chemical bond increases. Likewise, when the electron occupation of an antibonding orbital increases, the bond strength is weakened.

In the upcoming sections we will illustrate the use of the PDOS and COHP functions, which can be calculated from the electronic structure of the first principle calculations using the Lobster code [9, 10].

We will begin with a discussion of diatomic molecules that have σ bonds, and follow with a comparison of the non-polar H_2 molecules to the polar molecules such as LiH, NaH, KH, and HCl.

Then, the chemical bond in diatomic molecules with π bonds will be discussed. The diatomic molecules we will consider here are N_2, O_2, and F_2.

Subsequently, we will provide an introduction to the electronic structure of bulk metals and their surfaces, where we will again use the PDOS and COHP functions. These discussions will provide a useful background to the following section on the chemisorptive bond.

6.3 Diatomic Molecules that Have σ Bonds

The valence electron structure of H_2 consists of the $1\sigma_g$ and $1\sigma_u$ molecular orbitals. They form bonds called *σ bonds* because they are symmetrical with respect to the molecular axis. Calculated orbital density plots of the two orbitals are shown in Figure 2i.1. We can deduce the bonding and antibonding nature of the orbitals from the corresponding COHP values.

In H_2, the $1\sigma_g$ orbital is occupied by two electrons of opposite spin (Figure 2i.1). Since this orbital is shown to be bonding by its positive contribution in the −COHP plot, its electron occupation produces an attractive interaction between the two atoms. The $1\sigma_u$ orbital is unoccupied and of antibonding character (negative contribution in the −COHP plot).

The equilibrium distance is determined by the balance between the nuclear repulsion and the attraction from the occupied bonding molecular orbital.

The probability of finding an electron at a particular energy in the H 1s atomic orbital is measured by the PDOS $\rho_i(E)$. Figure 2i.1 illustrates the plot of $\rho_i(E)$ of the H 1s atomic orbital at one atom, as a function of the orbital energies for the H_2 molecule. There is a peak at each of the two respective molecular orbital energies σ_g and σ_u. The values of $\rho_i(E)$ on the two atoms are equal because the molecule is symmetrical.

Insert 2: The Electronic Structures of Diatomic Molecules with σ Bonds

In this Insert, we will provide a compilation of the electronic structure properties, bond energies, and bond distances of diatomic molecules with σ bonds. Calculated results presented here and in the following Inserts of this chapter have been determined by the Gaussian and the VASP software.

Molecules with π bonds are discussed in Insert 3.

A σ bond is produced by electron occupation of a molecular orbital that is symmetrical with respect to the interatomic axis. A π bond is produced by a molecular orbital that is asymmetrical with respect to the molecular axis. We will analyze the electronic structure in terms of the PDOS $\rho_i(E)$ and of the bond order energy overlap densities $\rho_{ij}^h(E)$. These functions were explained earlier in Insert 1. The relaxation of the molecular geometries toward their minimum energy configuration has been calculated with the VASP and Gaussian software [1–6, 11].

The PDOS and COHP values have been computed with the Lobster code [9, 10]. The molecular orbitals and the charges on the atoms were calculated with the Gaussian software.

Figures 2i.1 and 2i.2 show details of the molecular orbitals for the H_2 and LiH molecules. On the left side of these plots the valence electron molecular orbitals are displayed. Next to them, the PDOS and the COHP analysis is shown. We denote the energy difference between corresponding bonding and antibonding molecular orbitals using Δσ.

In Figure 2i.1 the electronic structure of the H_2 molecule is compared at equilibrium distance ($r_{eq} = 0.75$ Å) and at a larger distance ($r' = 0.95$ Å). Here, as well as in Figure 2i.2, the values for the molecular bonding energy (E_b), bonding–antibonding gap (Δσ), and integrated COHP (−ICOHP) are shown below the graphs.

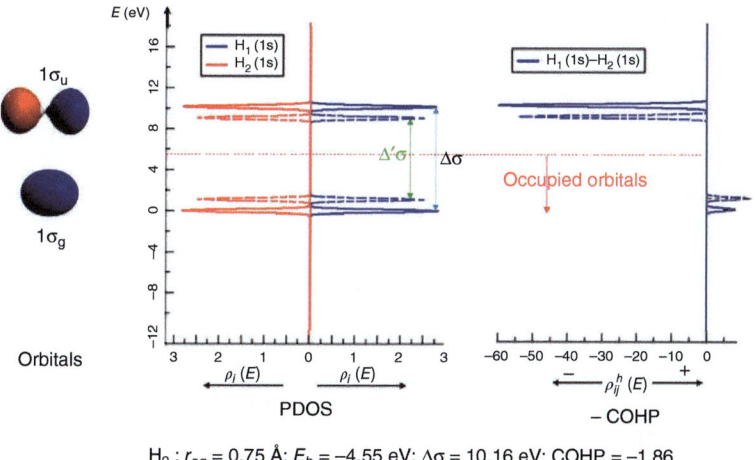

H_2 : $r_{eq} = 0.75$ Å; $E_b = -4.55$ eV; Δσ = 10.16 eV; COHP = −1.86
H'_2 : $r'_{eq} = 0.95$ Å; $E'_b = -4.07$ eV; Δ′σ = 8.94 eV; COHP = −1.72

Figure 2i.1 The structures of the molecular orbitals and the PDOS and the −COHP densities of H_2. Solid lines show values at the equilibrium distance of H_2. Dashed lines show values at a larger distance. The values of the binding energies E_b, Δσ, and −COHP are shown below the plots.

(Continued)

Insert 2: (Continued)

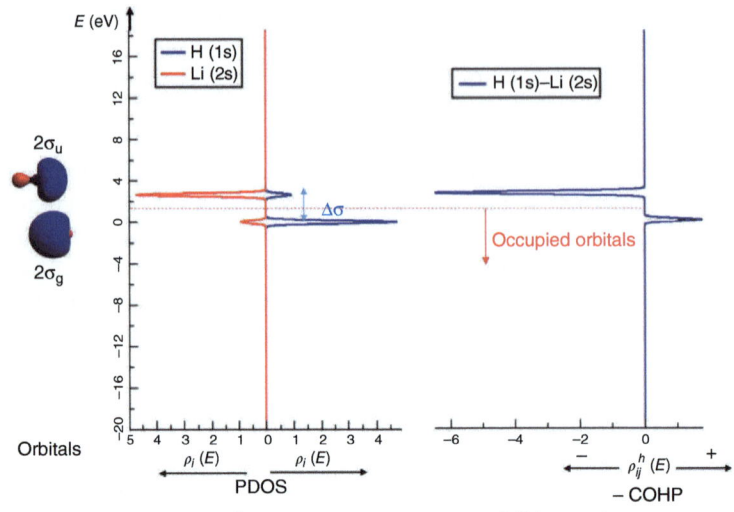

Figure 2i.2 The sigma symmetric electronic orbitals and their energies for the LiH molecule, calculated at its equilibrium distance. The values of equilibrium distance r_{eq}, of the binding energy E_b, bonding–antibonding gap $\Delta\sigma$, and integrated COHP are shown below the plots.

In Table 2i.1, electronic structure data are compared for different alkali halide molecules, as well as for the HCl and NaCl molecules.

Table 2i.1 Comparison of the electronic structure of diatomic polar molecules.

	r_{eq} (Å)	E_b (eV)	Δ_σ	$POL^S_{1,2}$ (E_{HOMO})	ICOHP	Q_H
HLi	1.60	−2.30	2.62	0.67	−0.39	−0.75
HNa	1.91	−1.84	2.26	0.49	−0.45	−0.67
HK	2.27	−1.75	1.74	0.59	−0.30	−0.69
HRb	2.40	−1.69	1.66	0.60	−0.26	−0.68
HCl	1.28	−0.449	10.48	−0.52	−2.91	0.18
NaCl	2.39	−3.91	3.22	−0.90	−0.17	0.77[a]

a) The charge on the Na ion.

The ICOHP is defined in Eq. (1i.4). It is the summation of the COHP values over the occupied molecular orbitals.

The polarity of a molecular orbital between atoms a and b can be deduced from the $POL^S_{ab}(E)$, which is calculated from the difference of the PDOS values of respective atomic orbitals a_i and b_j of an occupied molecular orbital with energy E:

$$POL^S_{a,b}(E) = \frac{\sum_i PDOS_{a_i}(E) - \sum_j PDOS_{b_j}(E)}{\sum_i PDOS_{a_i}(E) + \sum_j PDOS_{b_j}(E)} \qquad (2i.1)$$

$POL^S_{a,b}(E)$ defines the polarity contribution to the chemical bond between atoms a and b of the molecular orbital with energy E.

The atom charges q have been deduced from the calculated dipole moments divided by the equilibrium bond distance.

The electronic structures of HCl and NaCl are shown in Figures 2i.3 and 2i.4.

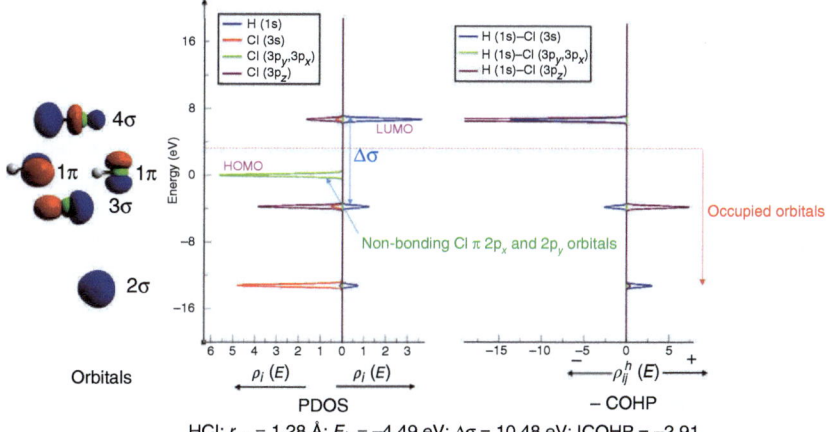

Figure 2i.3 The electronic structure of HCl.

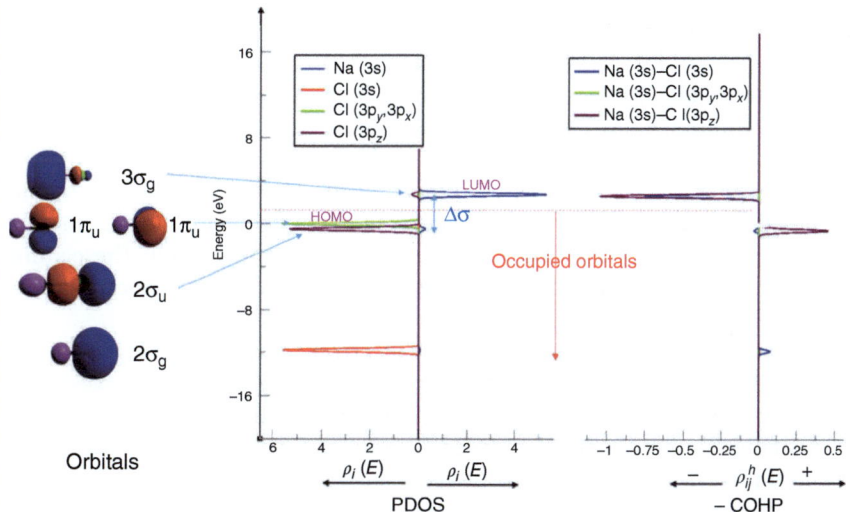

Figure 2i.4 The electronic structure of NaCl.

PDOS illustrates the contribution of the 1s atomic orbital to the two H_2 molecular orbitals. The energy difference $\Delta\sigma$ between the bonding orbital and the corresponding antibonding σ orbital is a measure of the energy overlap or interaction energy between atomic orbitals 1 and 2 of hydrogen, denoted as H_{12}. Figure 2i.1 shows that the orbital energy difference $\Delta\sigma$, becomes smaller when the bond distance is larger.

The larger the difference in energy between complementary bonding and antibonding molecular orbitals, the larger the contribution to the bond energy when the bonding orbital becomes occupied. This can also be observed from a comparison of the respective bond energies (indicated in the legend of Figure 2i.1).

A measure of bond polarity is the charge q on the atoms, which can be deduced for diatomic molecules from the dipole moment and bond distance. (For less symmetrical molecules and adsorbed surface states another method must be used. We will discuss this in a later section.) For the diatomic molecules, we can use $POL^S_{a,b}(E_k)$ defined in Eq. (2i.1) as a measure for the polarity (POL) of a molecular orbital. This is the weighted ratio of the PDOS $\rho_i(E)$ of the atomic orbitals on the two atoms that interact. When the value of $POL^S_{a,b}(E_k)$ is zero, the bond is non-polar and can be considered purely covalent. But when its value is one, the bond is purely ionic.

In heteropolar molecules such as LiH or NaH, that feature different electron affinities of their constituent atoms, electron density will be transferred between the atoms and the chemical bond will be polar. This implies that the chemical bond will have a mixed covalent-ionic bonding nature.

The electronic structures of the different alkali hydride molecules are very similar to the diatomic molecules. Therefore, we have shown only the valence electronic structure of LiH in Insert 2 (see Figure 2i.2). The electronic structure data of the different alkali halide molecules as well as the HCl and NaCl molecules (Figures 2i.3 and 2i.4) are compared in Table 2i.1.

The valence np_x and np_y atomic orbitals, which are asymmetrical with respect to the molecular axis (denoted as π symmetric), do not contribute to the chemical bond energy because they are perpendicular to the molecular bond axis and their differing symmetry prevents overlap with the H 1s atomic orbital. These atomic orbitals must be considered non-bonding, and because they are unoccupied they have not been included in Figure 2i.2.

A major difference between the valence electronic structure of H_2 and LiH, is that for LiH the highest occupied molecular orbital (HOMO) and lowest unoccupied molecular orbital (LUMO) are not its only orbitals. The deeper 1s atomic orbital of Li is doubly occupied. It does not significantly participate in the delocalization of the valence molecular orbitals that consist of the Li 2s and $2p_z$ atomic orbitals in combination with H 1s atomic orbital. The z-axis is chosen along the line of the two atoms of the molecule.

The Li 1s orbital is a core orbital that contains two electrons that screen the 3+ nuclear charge with respect to the valence electrons to a value close to 1+.

In Na, K, and Rb, the respective 3s and $3p_z$, 4s and $4p_z$, and 5s and $5p_z$ orbitals constitute the valence atomic orbitals. Their increasing size is the main reason for the decreasing value of the bond energies (Table 2i.1). In parallel with the bond energies, the value of $\Delta\sigma$ (the difference in energy between bonding and corresponding antibonding molecular orbitals) also decreases, indicative of the decreased interaction between the valence atomic orbitals. The value $\rho^h_{1s,2s}$, which is the bond order density value between the H 1s atomic orbitals and the Li 2s atomic orbitals, respectively, is also plotted in Figure 2i.2. We note the positive value of the contribution of the $2\sigma_g$ orbital and the negative value of the $2\sigma_u$ orbital, indicative of bonding versus antibonding interactions.

The charges $q(H)$ on the hydrogen atom have a substantial negative value. The charge is the highest in LiH, reaches a minimum for NaH, and then shows a small increase for the alkali atoms lower in the column of the periodic table corresponding to a decrease in their respective ionization potentials. The exceptionally high POL of the chemical bond in LiH is caused by the smaller bond distance. Its charge separation is stabilized by increased electrostatic interaction. Reduced electrostatic stabilization results in the very small changes in charge distribution seen for other alkali hydride molecules. Table 2i.1 shows that the polarities of the occupied σ orbitals indicate substantial ionicity and show a trend parallel to the hydrogen charges.

The COHP values are also a measure of the ionicity of a bond. The lower its relative value the larger the ionicity of a bond.

A comparison of the electronic structures of the alkali hydrides with the structures of the HCl and NaCl molecules (Figures 2i.3 and 2i.4) provides interesting information. The main difference in their valence electron regimes is produced by the contribution of the three 3p atomic orbitals of Cl to orbitals that are occupied by electrons. The $3p_x$ and $3p_y$ Cl atomic orbitals that have π symmetry and thus cannot overlap with the H 1s atomic orbital are now doubly occupied. They constitute the HOMOs of the HCl and NaCl diatomic molecules. Their bond strength is mainly determined by the interaction between the σ symmetric molecular orbital comprised of the H 1s or Na 3s and Cl $3p_z$ atomic orbitals. This can be deduced from the $\rho_i(E)$ and $\rho^h_{ij}(E)$ plots of the electronic structures of the two molecules.

On the Cl atoms, the 3s and 3p atomic orbitals have a different energy and their contribution to the chemical bond energy is also different. The lower occupied valence orbitals are molecular orbitals dominated by the contribution of the Cl 3s atomic orbitals, which interact weakly with the H 1s or Na 3s atomic orbitals. As the respective $\rho^h_{ij}(E)$ indicates, the corresponding antibonding contributions to the chemical bond overlap is dominated by the interaction with the Cl $3p_z$ atomic orbitals. The main contribution to the chemical bond is from the occupied Cl $3p_z$-dominated molecular bonding orbital. The related antibonding σ orbital is unoccupied and high in energy.

The ionicity of the chemical bond of the HCl molecule is significantly different from that of the NaCl molecule, as shown by the relatively small positive charge on the H atom in HCl versus the higher positive charge on the Na atom in NaCl. The value of the polarity $POL^S_{a,b}$, leads us to deduce that the ionic bond in NaCl is substantially more polar. In addition, the near-zero value of ICOHP of NaCl supports this assumption. Since COHP is determined by the cross-product of the coefficients c_i on different atoms in the same molecular orbital, when one of the two is nearly zero the value of COHP total becomes very small. In contrast, the ICOHP value of HCl is indicative of a strong covalent bond. In HCl, $\Delta\sigma$ is also large because of the strong overlap energy of the H 1s atomic orbital and the Cl $3p_z$ atomic orbital.

The covalency of the NaCl chemical bond is small, reflected by the low value of $\Delta\sigma$. For an ionic molecule, $\Delta\sigma$ represents the electrostatic energy cost to transfer an electron from Cl^- to Na^+.

The low POL of the HCl chemical bond appears to conflict with the high acidity of water solutions with dissolved HCl, but this acidity reveals itself only in the water phase. In this phase, the HCl bond cleaves into Cl^- and H^+ due to the high dielectric constant of water, and this reduces the electrostatic interaction of the charged dissolved ions. The Cl^- anion is a closed shell ion and the proton is present as a hydronium ion H_3O^+ which acts as the proton donating agent in the water solution.

The POL of the chemical bond of HCl is comparable to that of the OH bond in solid acids such as the zeolites. In the zeolite framework, the protons are part of neutral SiOHAl units. As in HCl, the cleavage of OH causes charge separation. In a zeolite a high temperature for proton activation is required, since no stabilizing electrolyte environment such as water is present: the dielectric constant of a zeolite is around 4, whereas that of water is 80.

6.4 Diatomic Molecules with π Bonds

Trends in bond energies in systems where the electron count per atom varies are determined to a significant extent by the distribution of valence electrons over bonding and antibonding molecular orbitals. To illustrate this, we will compare the electronic structure of chemical bonds in molecules with essentially similar molecular orbitals structure, but with different valence electron count.

Insert 3 illustrates the relationship between changes in a molecule's electronic structure and trends in its chemical bond energies for the diatomic molecules with π bonds (N_2, O_2, and F_2). The valence electronic structure of the three molecules is similar, but their total number of valence electrons increases. The differences in their bond strengths result from the increasing population of electrons in antibonding molecular orbitals.

The atomic orbitals that constitute the valence chemical bond of the N_2, O_2, and F_2 molecules are the 2s and $2p_x$, $2p_y$, and $2p_z$ atomic orbitals respectively.

Insert 3: Diatomic Homoatomic Molecules with π Bonds: N_2, O_2, F_2

The electronic structures of N_2, O_2, and F_2 are shown in Figures 3i.1–3i.3 respectively. Their properties are summarized in Table 3i.1.

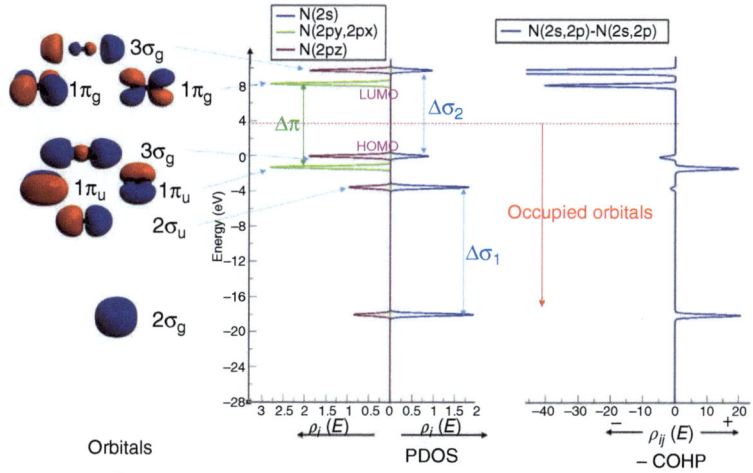

N_2: r_{eq} = 1.10 Å; E_b = −10.04 eV; $\Delta\sigma_1$ = 14.50 eV; $\Delta\sigma_2$ = 9.86 eV; $\Delta\pi$ = 9.46 eV; COHP = −13.05

Figure 3i.1 Electronic structure of N_2.

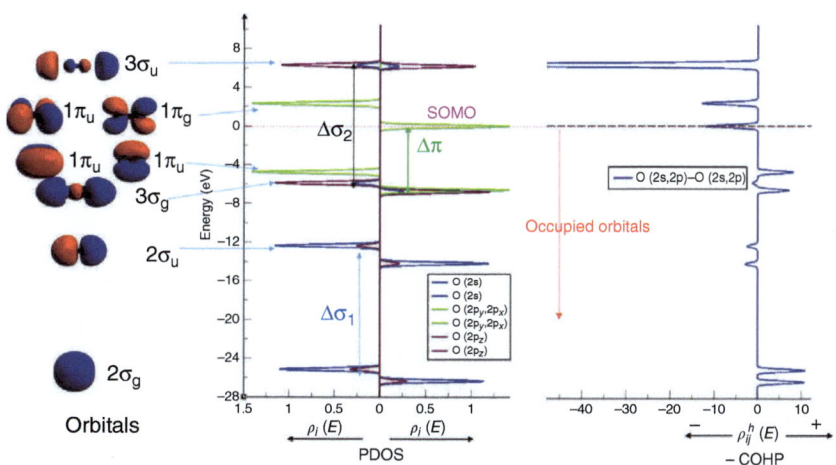

O_2: r_{eq} = 1.21 Å; E_b = −5.72 eV; $\Delta\sigma_1$ = 12.5 eV; $\Delta\sigma_2$ = 12.59 eV; $\Delta\pi$ = 7 eV; ICOHP = −3.12

Figure 3i.2 Electronic structure of O_2.

(Continued)

Insert 3: (Continued)

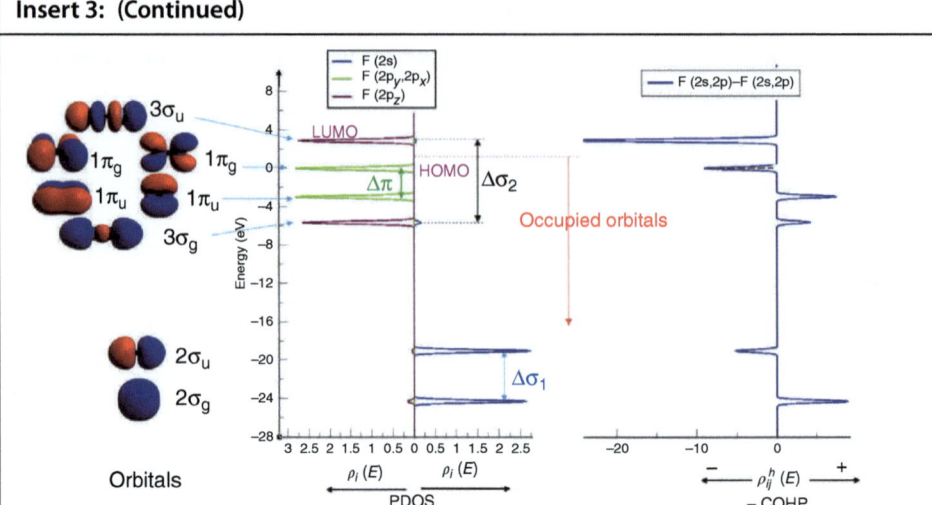

F_2: r_{eq} = 1.46 Å; E_b = −1.88 eV; $\Delta\sigma_1$ = 5.22 eV; $\Delta\sigma_2$ = 8.63 eV; $\Delta\pi$ = 3 eV; ICOHP = −3.35

Figure 3i.3 Electronic structure of F_2.

Table 3i.1 Comparison of bond energies and electronic features for N_2, O_2, F_2, and Cl_2.

	r_{eq} (Å)	E_b (eV)	$\Delta\sigma_1$	$\Delta\sigma_2$	$\Delta\pi$
N_2	1.10	−10.04	14.5	9.86	9.46
O_2	1.21	−5.72	12.5	12.59	7
F_2	1.46	−1.88	5.22	8.63	3
Cl_2	2.00	−2.59	4.3	7.75	2.65

The HOMO orbitals of O_2 are called *SOMOs* because they are occupied by a single electron. The molecule is in a triplet state. Molecular orbitals calculations depend on the spin states of the orbitals. When N_2 is compared to F_2 the number of orbitals double and show small differences in energy. Each orbital in the above diagram has one electron.

Insert 3 shows that three sets of σ symmetric and two sets of π symmetric orbitals determine the valence electronic structure.

The σ symmetric orbitals are linear combinations of 2s and $2p_z$ atomic orbitals. The degenerate π orbitals are linear combinations of the $2p_x$ and $2p_y$ atomic orbitals.

The σ bonds can be considered to be partially hybridized 2s–2p$_z$ combinations of atomic orbitals. The degree of hybridization depends on the energy difference between the 2s and 2p atomic orbitals in the isolated atoms (see also [12], pp. 155–157). Since the energy difference between the 2s and 2p$_z$ atomic orbitals increases with the number of valence electrons, an increase in valence electrons on the atom results in lower hybridization.

This can be readily seen by a comparison of the PDOS plots for F$_2$ and N$_2$ in Insert 3. The 2σ$_g$ and 2σ$_u$ molecular orbitals in F$_2$ are mainly a product of the interaction of the 2s atomic orbitals, whereas in N$_2$ they also receive a substantial contribution from the 2p$_z$ atomic orbitals. As the COHP graph shows, in F$_2$ the 2σ$_g$ and 2σ$_u$ form a separate bonding and antibonding orbital pair, mainly formed by the 2s atomic orbitals.

The 3σ$_g$ and 3σ$_u$ orbitals are the corresponding bonding and antibonding orbital pair, mainly formed by the 2p$_z$ atomic orbitals. In N$_2$, the strong hybridization of the 2s and 2p$_z$ atomic orbitals results in a slightly antibonding character for the 3σ$_g$ orbital. The 3σ$_g$ orbital and 2σ$_u$ orbital in N$_2$ constitute the respective lone pair orbitals on the N$_2$ nitrogen atoms. Because of partial hybridization they are not degenerate in the molecular orbital description of the N$_2$ chemical bond.

In F$_2$, the occupied 3σ$_g$ orbital provides the main attractive contribution to the chemical bond from the σ symmetric orbitals. Its corresponding antibonding 3σ$_u$ orbital is not occupied, and the contribution to the bond energy of the 2σ orbital pair is repulsive.

The increased mixing of 2s and 2p$_z$ atomic orbitals in N$_2$ causes the 3 σ$_g$ to become antibonding, and the attractive contribution due to the electronic occupation of the σ symmetric orbitals comes only from the 2σ$_g$ orbitals. The overall occupation of the σ molecular orbitals will lead to a net attractive interaction because two electrons occupy a bonding orbital and one electron an antibonding orbital.

There are two degenerate 1π$_g$ bonding molecular orbitals and two degenerate 1π$_u$ antibonding orbitals. Each π orbital is a linear combination of the perpendicular 2p$_y$ and 2p$_x$ atomic orbitals, respectively.

The electron distribution over these orbitals is very different for the three molecules N$_2$, O$_2$, and F$_2$. In N$_2$, only the two bonding degenerate 1π$_g$ atomic orbitals are occupied by a total of four electrons, while in O$_2$ each of the antibonding 1π$_u$ orbitals are also occupied by an electron. According to the rule of Hund their spins must be parallel, since the orbitals they occupy are degenerate. This makes O$_2$ magnetic and causes its triplet state. In F$_2$ and in Cl$_2$, the two degenerate antibonding 1π$_u$ orbitals are occupied with a total of four electrons. The electron occupation of the antibonding 1π molecular orbitals causes the O$_2$ and F$_2$ bond strengths to be weaker than the N$_2$ bond strength. In F$_2$ and Cl$_2$, the contribution of the π electron interaction is repulsive. Despite their very similar electronic structures, the bond energy of Cl$_2$ is higher than that of F$_2$, due to its much larger bond distance. This distance is defined mainly by the radii of the atomic orbitals, which are larger for Cl than for F. The smaller bond distance in F$_2$ causes a larger repulsive nuclear interaction, which is responsible for its lower bond energy as compared to Cl$_2$. The comparable electronic structure of

N$_2$ and O$_2$ also clearly demonstrates that the weaker bond of O$_2$ compared to N$_2$ is due to the population of antibonding orbitals in O$_2$ that are unoccupied in N$_2$.

The electron occupation of both the bonding and antibonding orbitals that form a pair, leads to a repulsive interaction, which is called *Pauli repulsion*. The stabilization of the bonding orbitals is always less than the destabilization of the corresponding antibonding orbitals. The name Pauli repulsion is derived from the Pauli principle that states that no electron of the same spin can occupy the same position in space. Pauli repulsion is proportional to the square of the overlap of the two interacting atomic orbitals that build the molecular orbital pair and their energy difference Δ.

6.5 Comparison of the Electronic Structure of Molecules and Solids

As an introduction to the next chapter we will compare in Figures 4i.1 and 4i.2 the electronic structure of the NaH and NaCl solids with that of their corresponding molecules. The crystal structure of the solids is the cubic rock salt structure, with each atom in the center of a hexagon.

Insert 4: The Electronic Structure of Bulk NaH and NaCl

In Figures 4i.1 and 4i.2, the electronic structures of the bulk NaH and NaCl are compared with those of the corresponding molecules. The crystals both have rock salt structures. Each atom is surrounded by six other atoms in its first coordination shell.

In a solid, ionicity is calculated from the following expression:

$$\text{POL}_{ab}^{t} = \frac{\int_{-\infty}^{E_F} dE \left\{ \sum_i \rho_{ai}(E) - \sum_j \rho_{bj}(E) \right\}}{\int_{-\infty}^{E_F} dE \left\{ \sum_i \rho_{ai}(E) + \sum_j \rho_{bj}(E) \right\}} \qquad (4i.1)$$

$$\text{ICOHP} = \int_{-\infty}^{E_F} dE \sum_{i,j} \text{COHP}_{ij}(E)$$

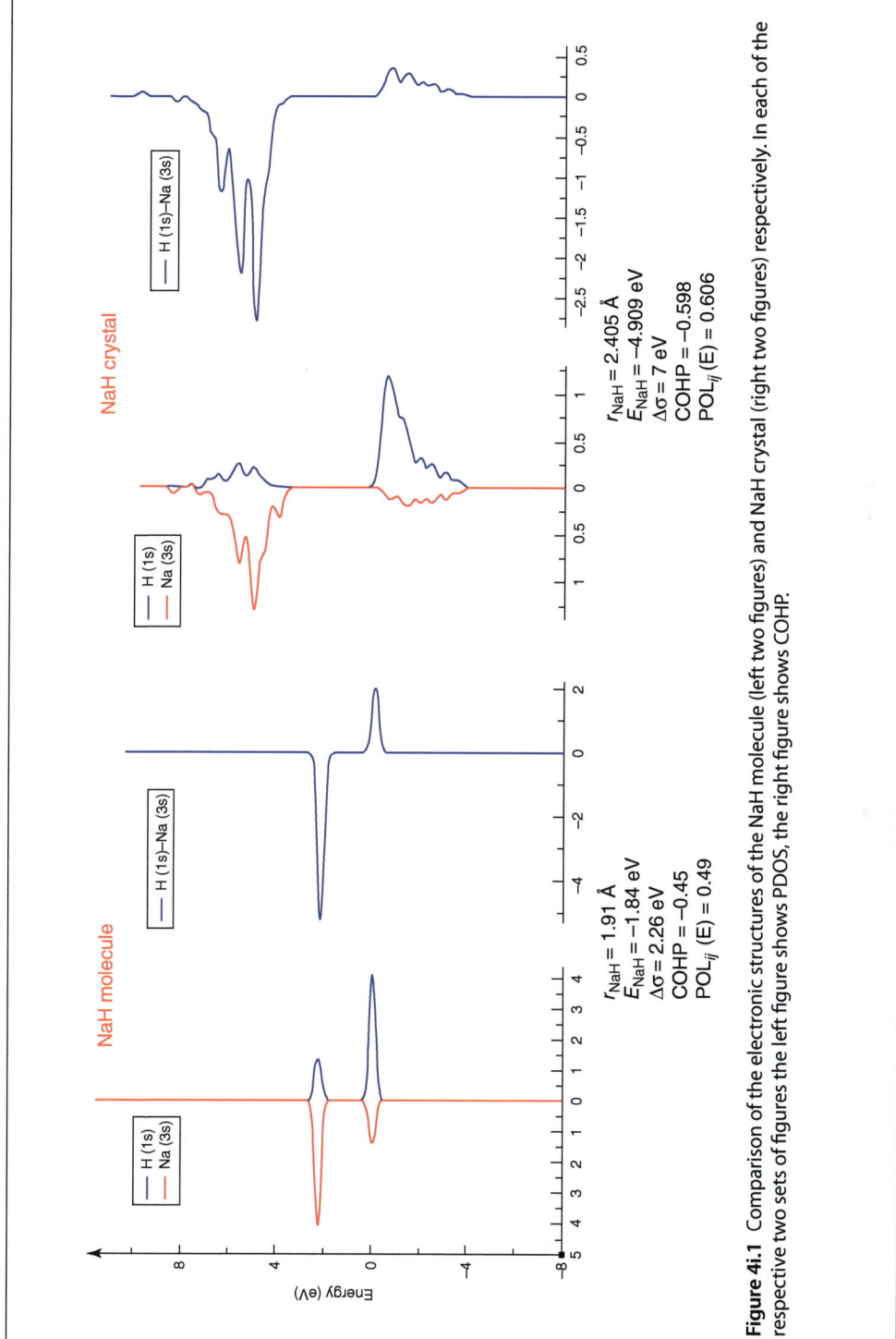

Figure 4i.1 Comparison of the electronic structures of the NaH molecule (left two figures) and NaH crystal (right two figures) respectively. In each of the respective two sets of figures the left figure shows PDOS, the right figure shows COHP.

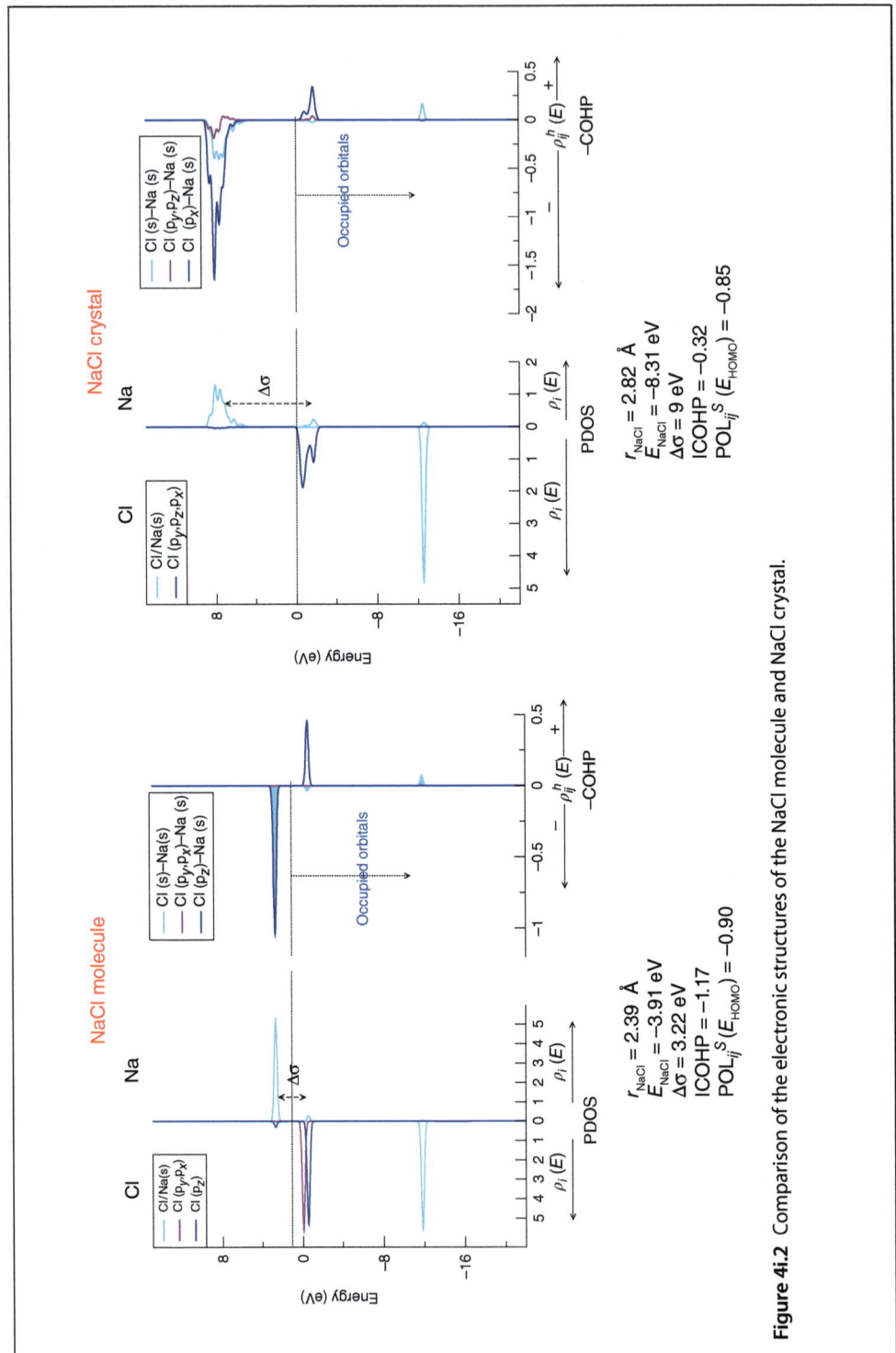

Figure 4i.2 Comparison of the electronic structures of the NaCl molecule and NaCl crystal.

In the solids, the discrete PDOS distributions of the atomic orbitals in a molecule become broadened into an energy band. This is because the molecular orbitals in the crystal are built from an infinite number of atomic orbitals. Instead of one bonding and one antibonding orbital (as in the diatomic molecule), an infinite number of bonding and antibonding orbitals are now formed and located in limited energy regimes. Because the crystal is infinite, the difference in energy between the orbitals in the same set becomes infinitely small. The result is the formation of a continuous density distribution around an average density energy. The width of the PDOS distribution is a measure of the degree of delocalization of the electrons. In NaH, the occupied bonding orbital set is mainly localized on the hydrogen atoms with the antibonding orbital set mainly on the sodium atoms. In NaCl, the occupied bonding orbital set is mainly localized on the Cl atoms and again the antibonding orbitals are mainly localized on the Na atoms. The occupied electron band is called the *valence electron band* and the unoccupied band is called the *conduction band*.

The energy difference between the occupied PDOS distribution and the corresponding unoccupied PDOS orbital density is called the *energy gap between valence and conduction band electrons*. Because of the high POL of the chemical bonds on both compounds, the energy bandgap is essentially the energy cost to transfer an electron from the hydride or chloride anion to the Na cation. The larger bandgap in the crystals compared to the energy difference between bonding and antibonding orbital in the molecule is due to the larger local electrostatic potential in the crystal. The chemical bonds in both solids can be considered to be dominantly ionic. The values of the POL indicate that in the crystal, the NaH bond becomes slightly more polar. Since the bond in the NaCl molecule is already highly ionic there is no change in ionicity from molecule to crystal in this case.

The increased bandgap of the ionic solids compared to that in the molecule can be estimated from the Madelung constant. In ionic solids, the electrostatic contribution to the bulk formation energy is called the *Madelung energy*. It is defined in these equations:

$$U = \frac{1}{2N} \sum_k q_k e \sum_{l=0}^{N} \sum_j \frac{q_j e}{|\bar{r}_l + \bar{r}_j - \bar{r}_k|} \quad (l = 0, \; \bar{r}_l = 0) \tag{6.1a}$$

$$U = \frac{1}{2N} \sum_k e q_k \Phi_k(\bar{r}_k) = \frac{Mq^2 e^2}{r_0} \tag{6.1b}$$

$$M = \frac{r_0}{2qe} [\Phi(\text{cat}) - \Phi(\text{an})] \tag{6.1c}$$

In Eqs. (6.1a) and (6.1b), U is the ionic energy, $q_k e$ is the charge in Coulombic units on the ions, and r_k represents the atom positions. In Eq. (6.1c), the number of anions and cations is assumed to be equal. The Madelung constant is a property that only depends on crystal structure. Using the Madelung constant definition Eq. (6.1b) for crystals of the same structure, the ionic energy only varies due to the difference in charge and nearest neighbor cation–anion distances (see also Chapter 11).

The electrostatic energy correlates weakly with the coordination numbers of cations or anions. But because of the wide range of the electrostatic potentials, it also depends strongly on the interaction with the ions in the higher coordination shells, which partially screens the attractive interaction between nearest neighbors. As Eq. (6.1c) illustrates, the Madelung constant is proportional to the difference of the electrostatic potential at the cation position $\Phi(cat)$ and the electrostatic potential at the position of the anion $\Phi(an)$. Therefore, bandgap is proportional to the Madelung constant. For the rock salt structure of NaH and NaCl, the constant's value is 1.75. This value is close to the ratio of the bulk band gap and to NaH and NaCl σ symmetric HOMO–LUMO energy differences.

The difference in the bandgap between the bonding and antibonding valence electrons in the ionic solid, compared with the same difference in the isolated molecule is proportional to the Madelung constant.

6.6 Chemical Bonding in Transition Metals

In the two following sections we will apply the same analysis tools used in the previous sections to describe the electronic structure of the transition metals, and thus understand trends in cohesive energies and the electronic bases for the various structures of these metals. For additional reading refer to: Levitin [13], Lejaeghere et al. [14], and Harrison [15]. It is important to note that the transition metals differ from ionic bulk systems in that their bonding is covalent.

6.6.1 The Electronic Structure of the Transition Metals

The three valence atomic orbitals of the transition metals are the nd, and the $(n+1)$s and $(n+1)$p atomic orbitals. In the transition metal atom the number of valence electrons varies, as does their distribution over the respective valence s, p, and d atomic orbitals. The metal $(n+1)$p atomic orbitals remain essentially unoccupied.

In a bulk metal there are only two valence electron bands: a relatively narrow d-valence electron band and a broad s–p valence electron band. This is illustrated by the PDOS distribution of Ru in Figure 6.1.

The valence electron structure of bulk Ru consists of the d-valence electron band that contains seven electrons per atom with a bandwidth of 7 eV, which is overlapped by a broad s-valence electron band that contains approximately one electron per atom. In Ru metal, all bonding molecular orbitals are occupied and the Fermi level is located in the antibonding valence electron energy region (see Figure 5i.2b).

We can also distinguish bonding and antibonding molecular orbitals for the transition metals as in the molecules. COHPs and PDOS for all the 3d, 4d, and 5d transition metals are shown in Figure 5i.2.

In contrast with ionic solids, there is no energy gap between the bonding and antibonding orbitals in the transition metals. This is a characteristic feature of

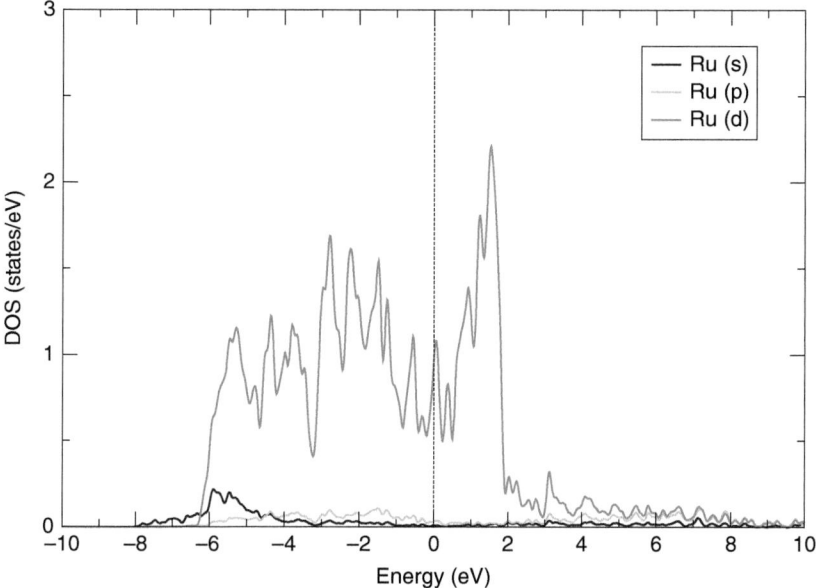

Figure 6.1 The PDOS of bulk hcp Ru.

conducting materials. The energy to excite an electron to an unoccupied orbital where it can contribute to conduction is infinitely small. The overall bandwidth is a measure of the degree of delocalization of the d-valence electrons. It is also a measure of the strength of the covalent interaction analogous to Δ, the energy difference between bonding and antibonding orbitals in the diatomic homonuclear molecules.

The PDOS exhibits no bandgap and as we will discuss so the peak maxima can be associated with bonding, nonbonding, and antibonding energy density regimes.

Differences in chemical and physical properties of the transition metals are mainly determined by the variation of electron occupation in the d-valence electron band. This is the basic property of the electronic structure that determines trends in the cohesive energies of the transition metals as a function of their positions in the periodic table.

Trends in the cohesive energies of the transition metals result from the difference between the energy of the bulk metal and that of the free metal atom. The minima found in experimental cohesive energies for 3d, 4d, and 5d transition metals (see Figure 5i.1a) are due to the magnetic properties of the free metal atoms used as a reference. These dips in the cohesive energies of the transition metals disappear when the cohesive energies are compared to the non-spin polarized free metal atom (compare Figure 5i.1b,c). In this case, all three cohesive energy curves show the maximum in bond energy for a d-valence electron count of five electrons. As we will discuss next, this illustrates that like the molecules, maximum bond energy for the transition metals along a row of the periodic table is also found when all the bonding orbitals are occupied.

Insert 5: Trends in Stability and Electronic Structure of the Transition Metals

A measure for the stability of the transition metals is the cohesive energy. For the transition metals of the third, fourth, and fifth row of the periodic table, experimental and calculated cohesive energies are provided in Figure 5i.1.

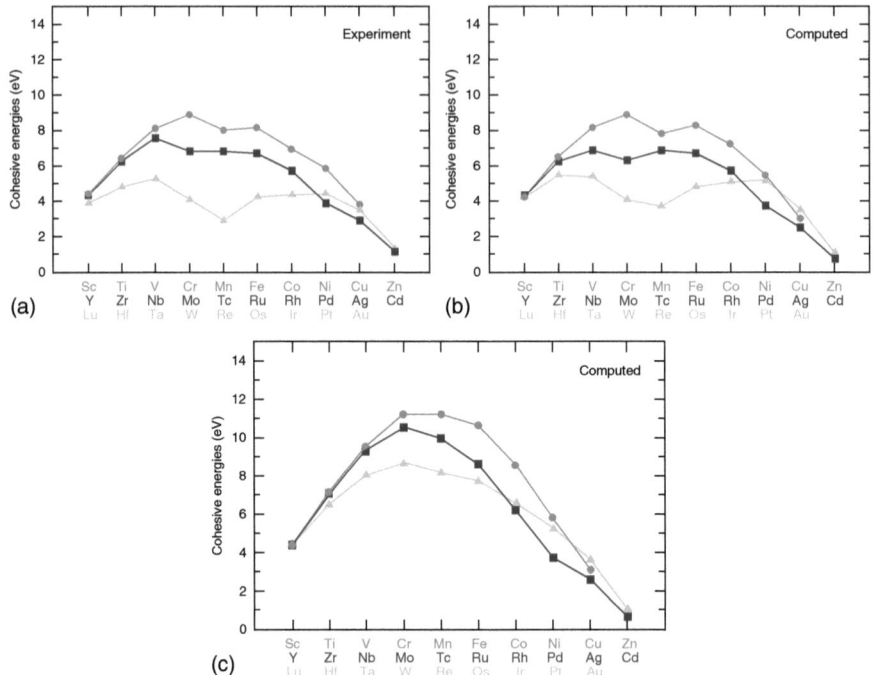

Figure 5i.1 (a) The experimental cohesive energies of the third, fourth, and fifth row transition metals. (Lejaeghere et al. 2014 [14]. Reproduced with permission of Taylor & Francis.) (b,c) DFT-computed cohesive energies of the spin-polarized metal bulk relative to the spin-polarized atom (b) and DFT-computed cohesive energies of the spin-polarized metal bulk relative to the non-spin-polarized atom (c). Triangles indicate transition metals of the third row, the squares indicate transition metals of the fourth row, and the circles indicate the transition metals of the fifth row of the periodic system.

For the transition metals positioned in the periodic table according to their d-valence electrons, Figure 5i.2 shows the calculated PDOS and COHP data for their geometry-optimized minimum energy structures.

Figure 5i.2a–c presents the PDOS and COHP distributions as calculated for zero-spin state for transition metals in three rows of the periodic table. Figure 5i.2d shows a comparison of the respective spin and non-spin polarized systems for the magnetic transition metals in the third row of the periodic table. s-Valence electrons are shown in dark gray and d-valence electrons are shown in light gray. The energy regime of unoccupied orbitals is shown with white shading and the energy regime of the occupied orbitals is shaded gray.

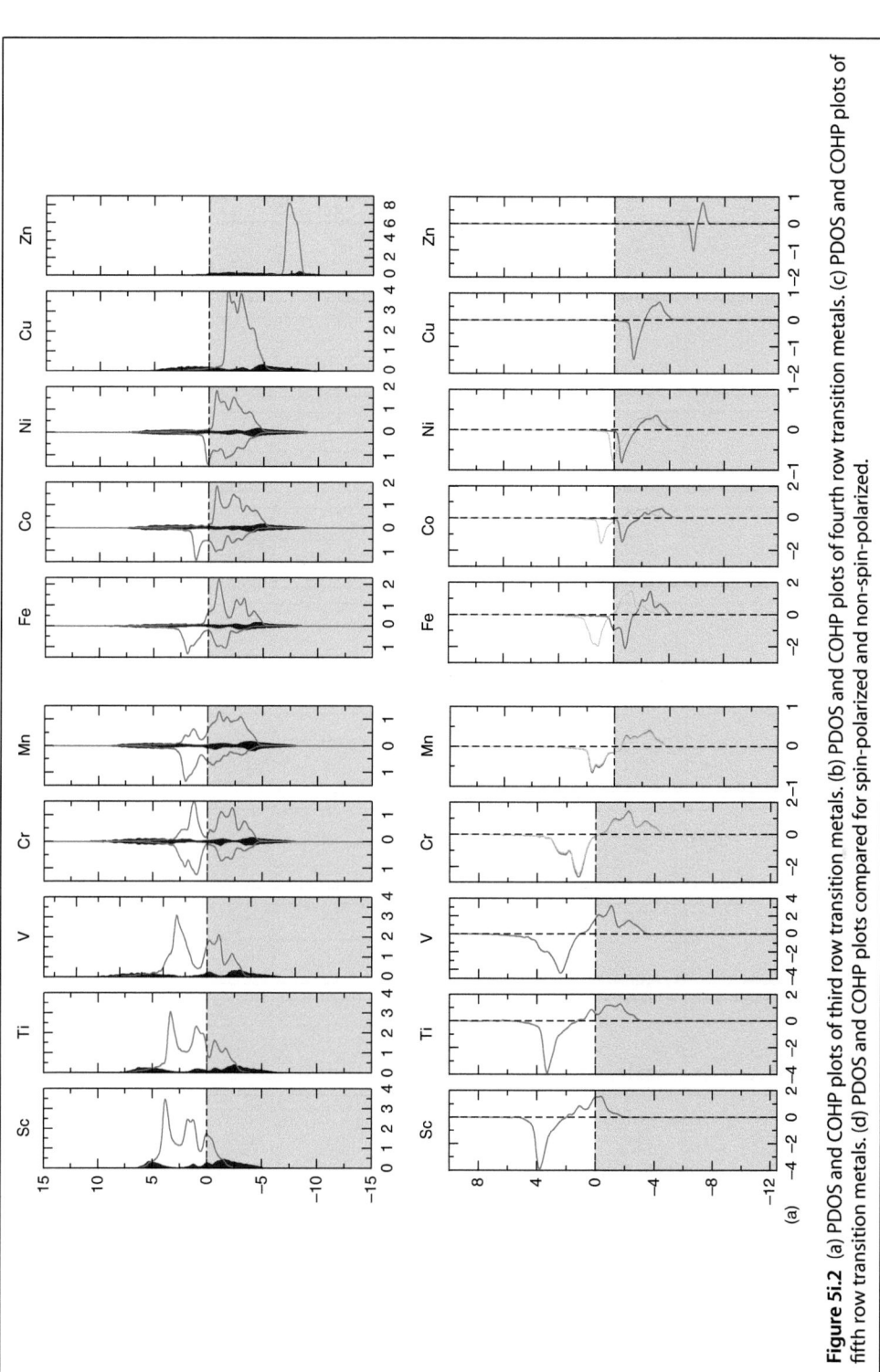

Figure 5i.2 (a) PDOS and COHP plots of third row transition metals. (b) PDOS and COHP plots of fourth row transition metals. (c) PDOS and COHP plots of fifth row transition metals. (d) PDOS and COHP plots compared for spin-polarized and non-spin-polarized.

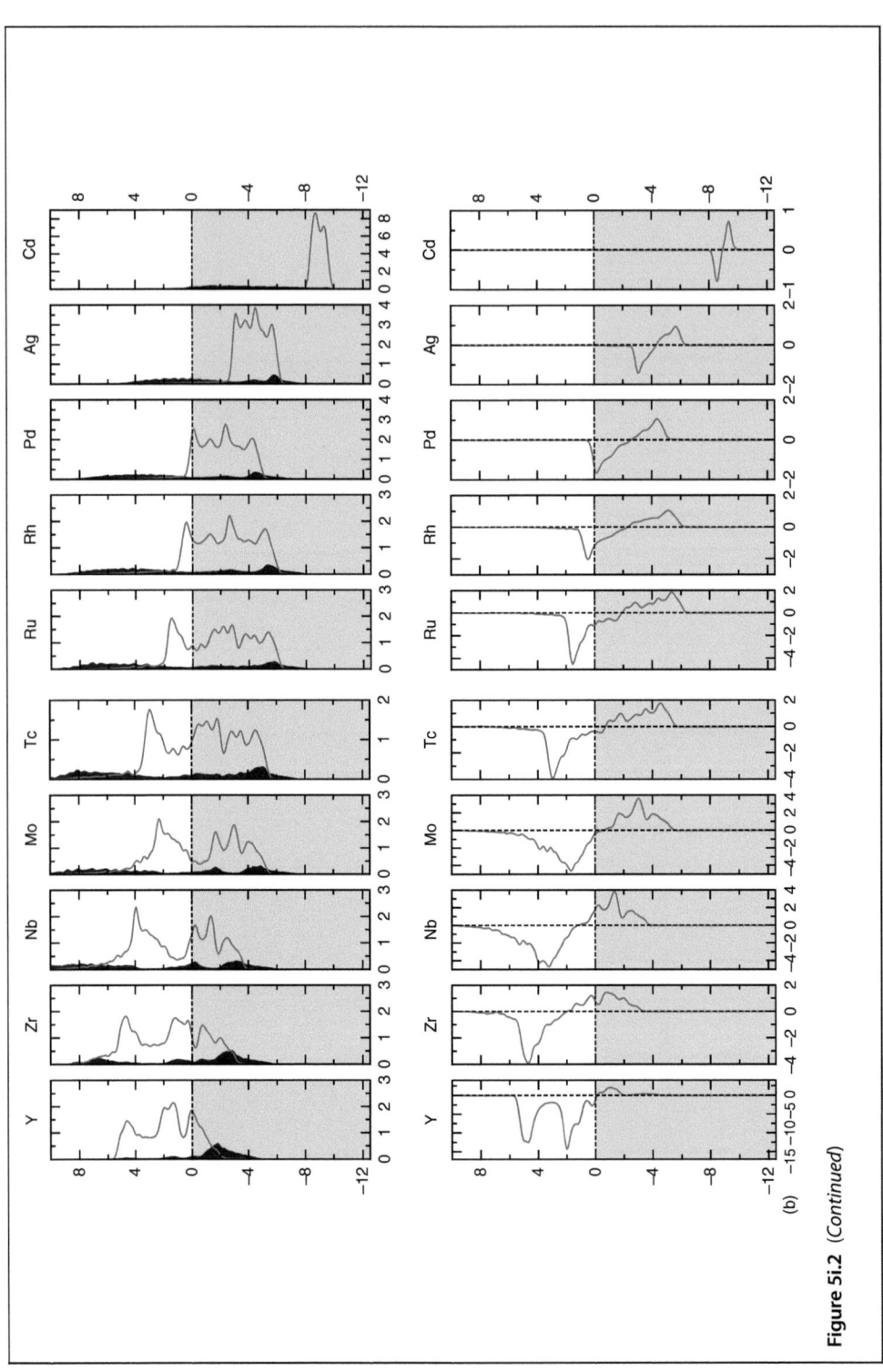

Figure 5i.2 (Continued)

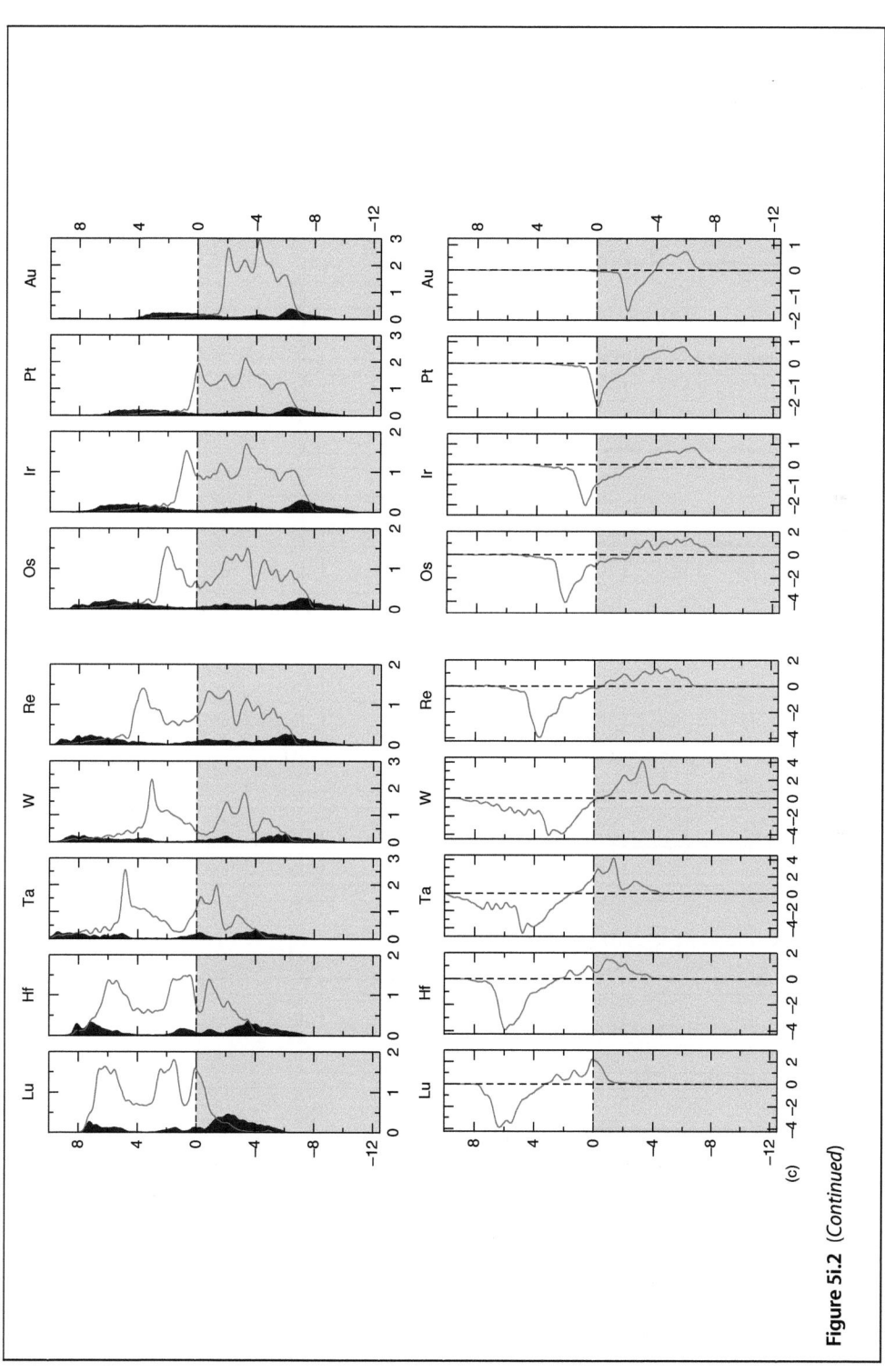

Figure 5i.2 (*Continued*)

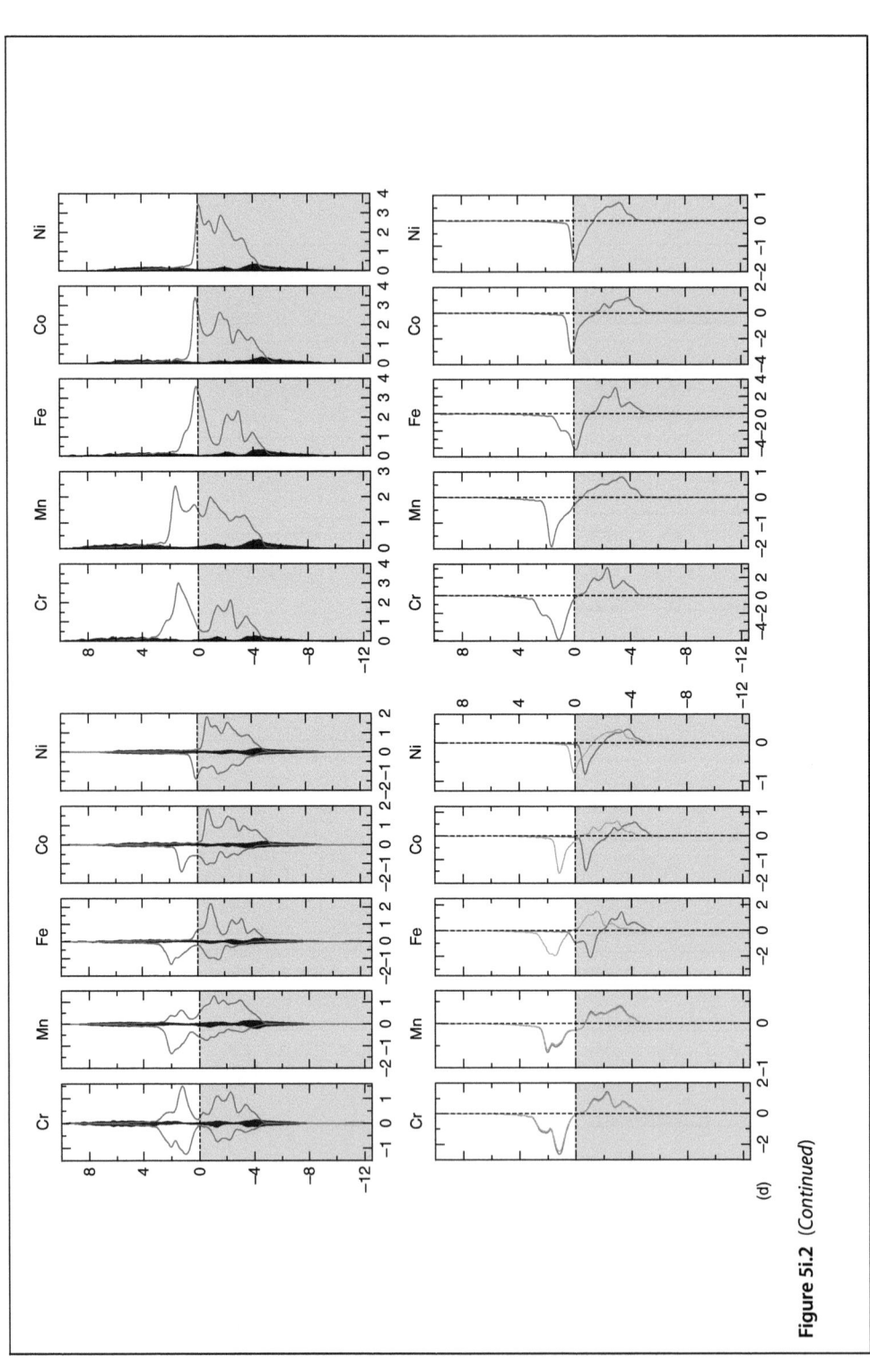

Figure 5i.2 (*Continued*)

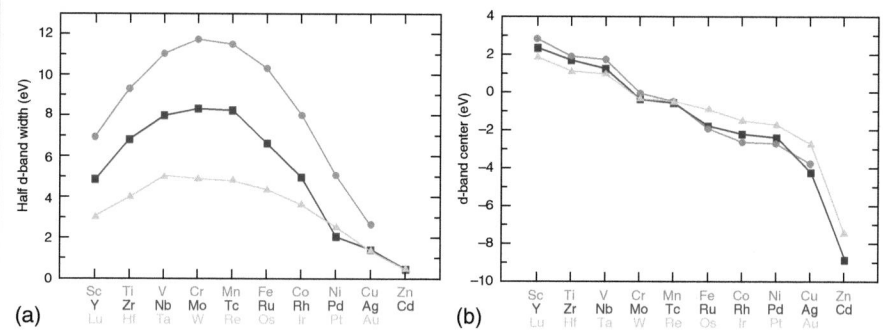

Figure 5i.3 (a) Trends in the d-bandwidth along a row of the periodic table. (b) Average d-band energy with respect to Fermi energy for non-spin-polarized systems. The d-band center has been computed considering an integration over the entire d-valence band spectrum. Triangles indicate transition metals of the third row, the squares indicate transition metals of the fourth row, and the circles indicate the transition metals of the fifth row of the periodic system.

The average d-band energy is calculated as the average energy of the occupied and unoccupied parts of the d-valence electron distribution. The average d-band energy is given by:

$$\overline{E}_d = \int_{-\infty}^{+\infty} dE\, E\rho_d(E) \tag{5i.1}$$

The half d-valence electron band width is calculated from the second moment of the d-valence electron distribution:

$$W_d = \frac{1}{2}\sqrt{\int_{-\infty}^{+\infty} dE(E - E_d)^2 \rho_d(E)} \tag{5i.2}$$

Table 5i.1 shows d-band center energies for the spin-polarized calculations. ICOHP values for spin-polarized and non-spin-polarized cases are compared. Metal atom bond distances in spin-polarized systems are 1–2% larger than in non-spin-polarized systems. When this distance is equal the summed spin-polarized ICOHP decreases in Co to −1.24, but only shows an increase for Ni in the third decimal place.

Table 5i.1 Decomposition per spin component of d-band center and ICOHP.

	Spin 1 d-band center	Spin 2 d-band center	Spin 1 + 2 ICOHP	Non-spin polarized ICOHP
Cr	−0.70	0.17	−5.47	−5.65
Mn	−1.22	0.39	−1.65	−1.90
Fe	−1.86	0.22	−2.63	−2.58
Co	−2.22	−0.57	−1.11	−1.19
Ni	−2.00	−1.34	−0.25	−0.29

Figure 5i.2 provides an overview of the PDOS and −COHPs for the transition metal valence electron structures as a function of transition metal position in the periodic table, ordered as a function of row position.

Comparing the PDOS and COHP distributions from left to right along a row of the periodic table leads to three observations:

1. Initially, only the bonding d-valence electron bands are occupied. When electron occupation increases, antibonding orbitals become occupied. The crossover point is when there are around five d-valence electrons per atom. For the non-spin-polarized case, the maximum in covalent energy occurs when all the bonding orbitals are occupied.
2. When electron count increases, the average position of the d-valence electron band energy, which is initially above the Fermi-level, becomes lower than the Fermi-level (see Figure 5i.3b). This is another sign that the d-valence electron band becomes occupied when the atom moves from left to right along a row of the periodic table.
3. The d-valence electron bandwidth (a measure for delocalization and interaction energy, Figure 5i.3a) has a maximum at the same d-valence electron count (five d-valence electrons) as the theoretical cohesive energy. This variation in bandwidth corresponds with a change in. At the maximum in bond energy the metal atom bond distance reaches a minimum. It does not relate to the variation in free atom size when metal position varies in a row of the periodic system. Different from the band width the radius of free atoms uniformly decreases from left to right along the row of the periodic table due to the increase in effective nuclear charge. The d-valence electron bandwidth increases from the third row to the fifth row within a column. In this case this trend is consistent with an increase in cohesive energy and the larger size of the metal atoms within the column.

Figure 5i.2d illustrates the different PDOS calculations of the spin orbitals for the magnetic metals in the third row of the periodic table. The differing values of average d-electron energies of the respective spin orbitals reflect the varying distribution of bonding and antibondingmolecular orbitals states.

As a consequence of this difference in electron distribution, while in one spin state the contribution to the bond energy is small of repulsive, in the other state the contribution can be substantially more attractive.

It becomes favorable to have a difference in spin polarization, when the repulsive energy of two electrons of oppisite spin in the same orbital is larger than the energy cost to excite one electron to an unoccupied orbital.

This becomes important when d-valence atomic orbitals are relatively small as is the case for group VIII transition metal of the third row of the periodic system.

This is one reason for the magnetic properties of these metal [16, 18]. The overall result then is an average increase in the population of bonding orbitals and higher bond strength in the spin-polarized system.

Bond strength can also be deduced by comparing the ICOHPs for the spin- and non-spin-polarized systems. As Table 5i.1 shows, a comparison of these values at different equilibria positions shows that ICOHP for the spin-polarized case is still greater. For Mn or Cr, the non-magnetic state is more stable. For Co and Ni, the ICOHP values in both states are similar. The preference for the spin-polarized state is obscured by changes due to differences in equilibrium distance. The contribution of the electron occupation of the s-electron valence band (which is bonding) to the bond energy can be estimated from the bond energies of Cu, Ag, and Au. For these metals, bonding as well as antibonding orbitals are occupied by d-valence electrons, and the contribution of the interaction between the d-valence electrons becomes repulsive. The attractive contribution to the chemical bond must be due to the s, p-valence electron band which produces at least 30% of the total cohesive energy.

6.6.2 The Relative Stability of Transition Metal Structures

The face-centered cubic (fcc) and hexagonal close packed (hcp) structures provide the highest density packing for hard spheres. Both the fcc and hcp structures exhibit 12 nearest neighbors, but they differ in the ordering of their hexagonally packed layers. In the fcc structure, three layers are shifted successively and then repeat, whereas in the hcp structure double layers alternate. While these structures are common for the metals, not all metals have these structures. Another common structure for the transition metals is the body centered cubic (bcc) structure (see Figure 6i.1) in which each atom has eight nearest neighbors.

There is a clear trend in the relative stabilities of the different structures, which occurs as a function of transition metal position along a row in the periodic table. Moving left from the noble metals at the right of the table, the structures vary from fcc, to hcp, to bcc, to hcp, and back to fcc. The more complex structure of Mn and the magnetic metals Fe and Co break this order, which is preserved for 4d and 5d. The alkali metals exhibit the bcc structure at room temperature, but at low temperature they shift to the fcc structure [13, 19, 20].

Insert 6: The Electronic Basis of Lattice Stability Differences

Schematic PDOS and COHP distributions on different crystal lattices with one s-atomic orbital per atom are shown in Figure 6i.2. The densities of states calculated based on their first and second moments display elliptic shapes [21].

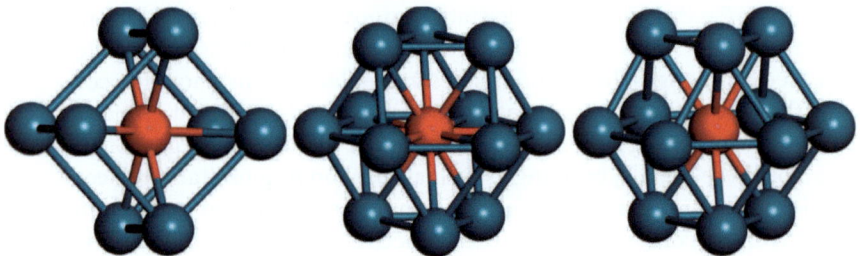

Figure 6i.1 The coordination of the transition metal atoms in bcc, fcc, and hcp structures, respectively. Nearest neighbors atoms positions are shown: for bcc in (110) direction, for fcc in (111) direction and for hcp in (0001) direction.

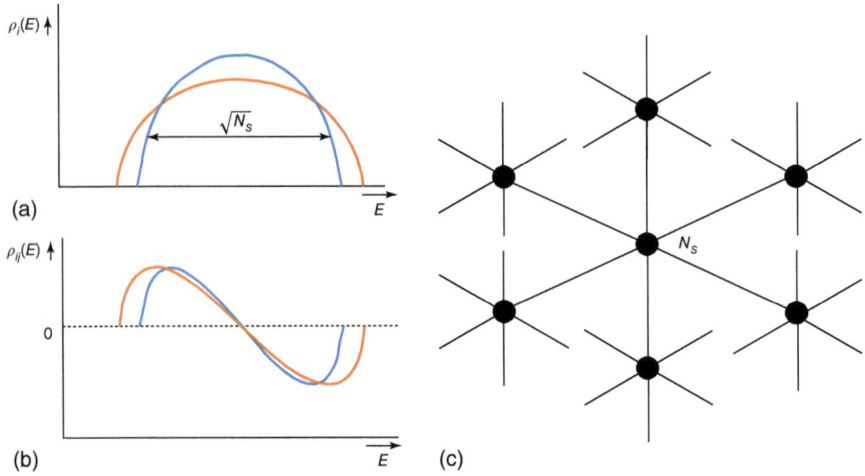

Figure 6i.2 Schematic (a) PDOS $\rho_i(E)$ and (b) COOP $\rho_{ij}(E)$ distributions (see Eq. (1i.6)) for Bethe lattice model (c). In the Bethe lattice, N_s is the number of nearest neighbors. The nearest neighbor atoms do not share bonds. The bandwidth is proportional to $\sqrt{N_s}$ red represents the electron density and bond overlap of atoms with more nearest neighbours than the blue case.

As in Figure 6i.2, only one s-type atomic orbital is assumed per metal atom. For the fcc/hcp structures, the Bethe lattice structure shown in Figure 6i.2c has been adapted to also incorporate the next nearest neighbor atom–atom interactions. These next nearest

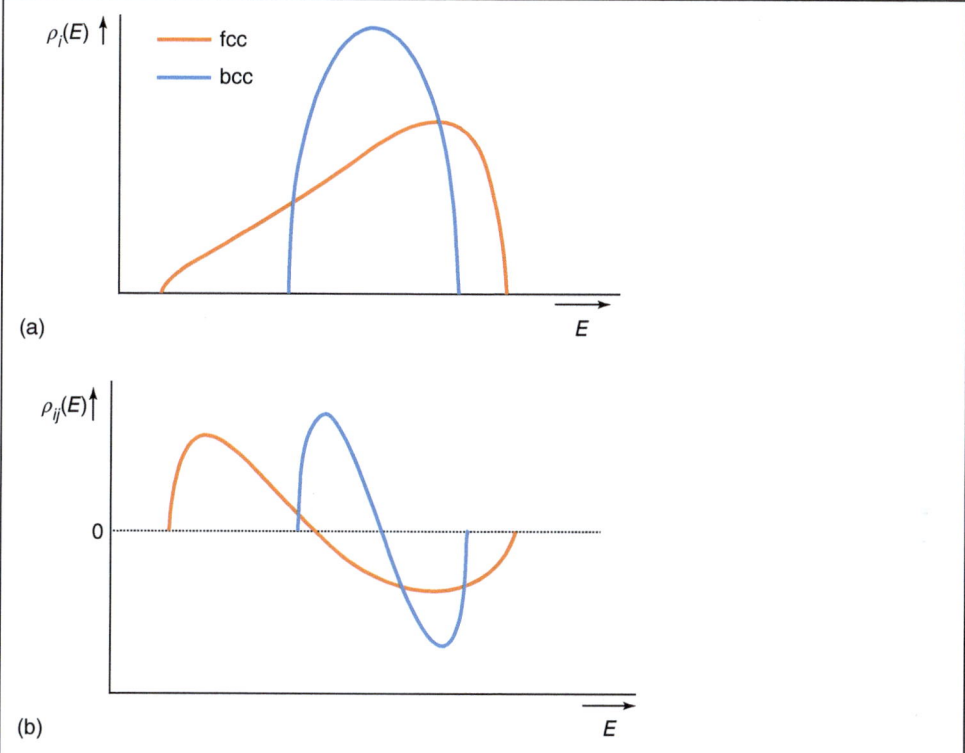

Figure 6i.3 (a,b) Schematic of electronic structure differences between fcc/hcp structures versus those of bcc structures.

neighbor atom–atom interactions cause the density of states to be asymmetrical. The hcp and fcc structure are closely related, so their electronic structures will be nearly similar (not shown here). The larger bandwidth of the fcc/hcp lattice as compared to the bandwidth of the bcc lattice relates to the difference in the number of nearest neighbor atoms (12 vs 8) [21, 22].

The reason for this dependence is the sensitive relationship between relative population of antibonding orbitals and preferred structure. Spin polarization in magnetic metals decreases the relative population of antibonding orbitals.

Figure 6i.5 summarizes the concept that the partial occupation of bonding and antibonding orbitals determines which bulk structure is most stable.

(Continued)

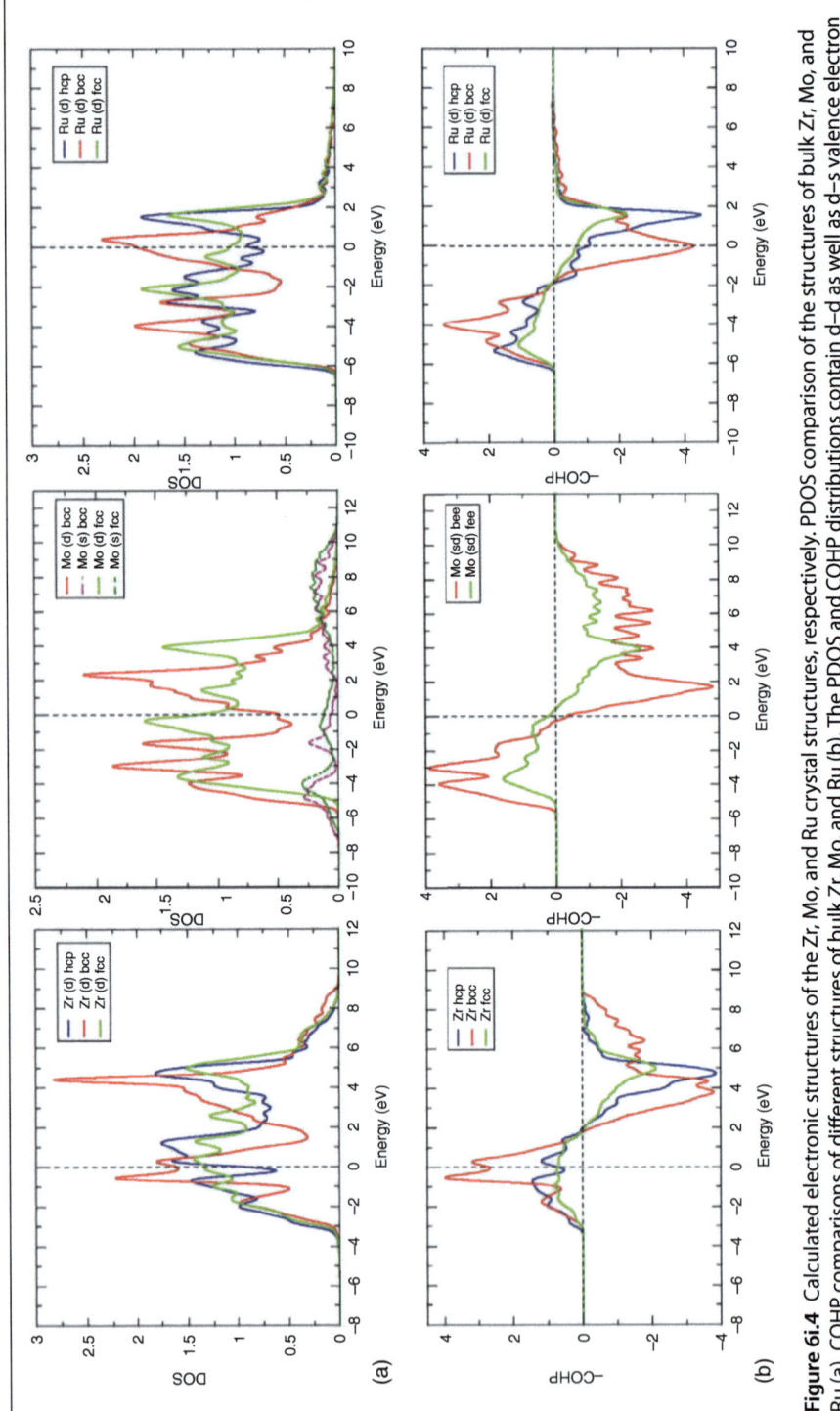

Figure 6i.4 Calculated electronic structures of the Zr, Mo, and Ru crystal structures, respectively. PDOS comparison of the structures of bulk Zr, Mo, and Ru (a). COHP comparisons of different structures of bulk Zr, Mo, and Ru (b). The PDOS and COHP distributions contain d–d as well as d–s valence electron interactions. Because the COHP distributions also contain energetic interactions, the relative values of bonding and antibonding orbital overlap densities must be compared rather than their absolute values.

Table 6i.1 Occupied bonding (r+) and antibonding (r−) fractions of the respective −COHP distributions.

Elements	hcp	bcc	fcc
Zr	r+ 0.70	r+ 0.625	r+ 0.657
Mo	—	r+ 0.002	r− 0.012
Ru	r− 0.182	r− 0.380	r− 0.361

Zr has hcp structure, Mo has bcc, and Ru has fcc.

Table 6i.2 The lowest energy structure as a function of spin state.

	Spin state	
Elements	Spin-polarized	Non-spin-polarized
Fe	bcc	hcp
Co	hcp	fcc

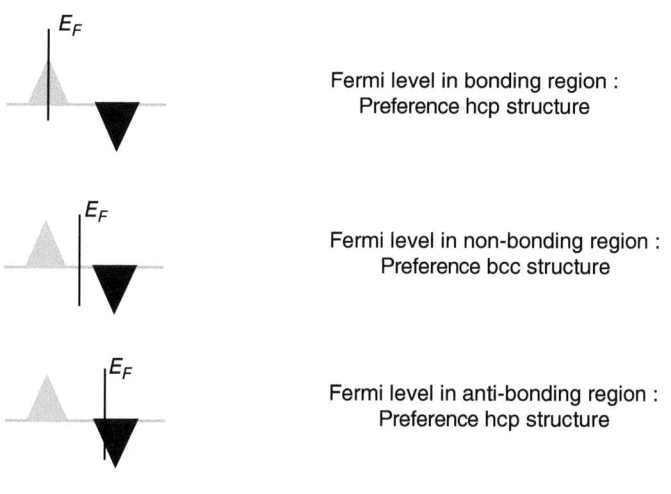

Fermi level in bonding region : Preference hcp structure

Fermi level in non-bonding region : Preference bcc structure

Fermi level in anti-bonding region : Preference hcp structure

Fermi level in the upper anti-bonding region : Preference fcc structure

Figure 6i.5 Relationship between the position of the Fermi level in COHP and the crystal structure for the transition metals. Light gray represents the bonding domain in the transition metals, black the anti-bonding domain with the non-bonding domain between.

The number of nearest neighbors in the bcc structure is eight and these atoms do not share a bond. In the fcc and hcp structures, the number of nearest neighbors is 12, and 5 of these atoms share a bond (see Figure 6i.1). In the fcc structure, the nearest neighbor atom configurations are uniform, while in the hcp structure there are two different nearest neighbor configurations distinguished by their connection type. This difference in nearest-neighbor configuration results in small differences in PDOS for the fcc and hcp structures.

PDOS and COHP distributions as calculated by VASP and Lobster for different structures of the same metal are shown in Figure 6i.4. The PDOS distributions are similarly shaped for different metals with the same structure. The bcc PDOS typically has two major peaks that correspond to the bonding and antibonding energy density regimes, whereas the fcc and hcp PDOS distributions are characterized by three peaks that correspond to bonding, non-bonding, and antibonding energy density regimes. The averaged bandwidth of the fcc and hcp structures tends to be broader than that of the bcc structure, as shown by the larger difference between the upper and lower peak maxima of the COHPs in the fcc and hcp structures. The respective COHP curves clearly show that the fcc and hcp structures are slightly skewed toward the higher energies.

Figure 6i.3 displays a sketch of the broader PDOS bandwidth and skewed density distributions of the fcc/hcp structures compared to the bcc structure. These schematics are based on lattice models that only contain one s-atomic orbital per site as shown in Figure 6i.1. The fcc/hcp structure is wider due its larger number of nearest neighbor atoms. The skewed distribution is due to the sharing of bonds between the next nearest neighbors in the fcc/hcp lattices, which is absent in the bcc lattice.

The consequence of the difference in energy dependence of the bond order density distributions is also shown in Figure 6i.3b. This schematic reproduces the observed experimental trends of the relative stability of the structures as a function of their d-valence electron occupation. At low levels of band filling, fcc/hcp has a higher occupation of bonding orbitals. At a high level of band filling the bcc structures are preferred, since only their bonding orbitals are occupied, while in the fcc/hcp structures antibonding orbitals become occupied. When the antibonding orbitals of the bcc structure also become occupied they are filled earlier. At high electron occupation, the fcc/hcp structure is again most stable (Figures 6i.4 and 6i.5).

The changes in relative structural stabilities correlate with the relative fractions of bonding and antibonding orbitals that are occupied (see Table 6i.1). The most stable structure exhibits a maximum occupation of bonding orbitals or a minimum occupation of antibonding orbitals.

In general, at low or high d-valence electron occupation, structures with high nearest neighbor coordination are more stable than the bcc structure with intermediate valence electron occupation. A clear example is Ru for which the hcp structure is more stable than the bcc structure due to the high density occupation of the antibonding valence electron for the bcc structure. For the hcp structure, the antibonding maximum density peak is at a higher energy.

The striking relationship between spin polarization of the third row metals and their preferred structure has a similar origin (see Table 6i.2). When spin is non-polarized, the crystal structures in the respective columns of the third, fourth, and fifth row of the periodic table are similar. But when spin is polarized the structure tends to shift to the most stable structure of the non-spin-polarized system with lower valence electron occupancy. This associates the spin-polarized systems with a lower electron occupation of the antibonding orbitals.

Appendix

In order to identify the most stable molecular and bulk configurations we have performed first-principles total energy calculations using the gradient corrected (PBE) density functional theory as implemented in the VASP code. The electronic wave functions have been expanded into plane waves up to an energy cutoff of 550 eV for the molecular systems, and up to 400 eV for the bulk systems. A projected-augmented-wave scheme has been used in order to describe the interactions between the valence electrons and the nuclei (ions). The minimum energy configurations for the subject structures have been considered reached when the forces on each atom of the systems were less than 0.02 eV Å^{-1}. In order to span the Brillouin zone, $17 \times 17 \times 17$ k-points have been used for the bulk simulations. Due to the large size of the unit cell considered, only 1 k-point – the Gamma point – was necessary for the molecular systems. High accuracy settings have been enforced throughout.

In order to get a more complete description for the molecular systems (e.g., plot the wave functions, etc.) in addition to the VASP calculations, Gaussian 09 simulations have been also performed. The B3LYP functional has been used with the AUG-cc-pVTZ basis set throughout these calculations. In order to check the impact of the basis set on the results obtained, def2-TZVPP basis set computations have been also performed on each of the molecular systems. No significant variations have been observed.

The COHP and the DOS calculations have been performed for all the systems using the Lobster code.

References

1 Kresse, G. and Furthmüller, J. (1996) Efficiency of ab-initio total energy calculations for metals and semiconductors using a plane-wave basis set. *Comput. Mater. Sci.*, **6**, 15–50, http://linkinghub.elsevier.com/retrieve/pii/0927025696000080 (accessed 14 July 2014).
2 Kresse, G. and Hafner, J. (1993) Ab initio molecular dynamics for liquid metals. *Phys. Rev. B*, **47**, 558–561. doi: 10.1103/PhysRevB.47.558
3 Kresse, G. and Hafner, J. (1994) Ab initio molecular-dynamics simulation of the liquid-metal-amorphous-semiconductor transition in germanium. *Phys. Rev. B*, **49**, 14251–14269. doi: 10.1103/PhysRevB.49.14251

4 Kresse, G. and Furthmüller, J. (1996) Efficient iterative schemes for ab initio total-energy calculations using a plane-wave basis set. *Phys. Rev. B*, **54**, 11169–11186. doi: 10.1103/PhysRevB.54.11169
5 Kresse, G. and Joubert, D. (1999) From ultrasoft pseudopotentials to the projector augmented-wave method. *Phys. Rev. B*, **59**, 1758–1775. doi: 10.1103/PhysRevB.59.1758
6 Perdew, J.P., Burke, K., and Ernzerhof, M. (1996) Generalized gradient approximation made simple. *Phys. Rev. Lett.*, **77**, 3865–3868. doi: 10.1103/PhysRevLett.77.3865
7 Hoffmann, R. (1971) Interaction of orbitals through space and through bonds. *Acc. Chem. Res.*, **4**, 1–9. doi: 10.1021/ar50037a001
8 Löwdin, P.-O. (1950) On the non-orthogonality problem connected with the use of atomic wave functions in the theory of molecules and crystals. *J. Chem. Phys.*, **18**, 365. doi: 10.1063/1.1747632
9 Deringer, V.L., Tchougréeff, A.L., and Dronskowski, R. (2011) Crystal orbital Hamilton population (COHP) analysis as projected from plane-wave basis sets. *J. Phys. Chem. A*, **115**, 5461–5466.
10 Maintz, S., Deringer, V.L., Tchougréeff, A.L., and Dronskowski, R. (2013) Analytic projection from plane-wave and PAW wavefunctions and application to chemical-bonding analysis in solids. *J. Comput. Chem.*, **34**, 2557–2567.
11 Frisch, M. J., Trucks, G. W., Schlegel, H. B., Scuseria, G. E., Robb, M. A., Cheeseman, J. R., Scalmani, G., Barone, V., Mennucci, B., Petersson, G. A., Nakatsuji, H., Caricato, M., Li, X., Hratchian, H. P., Izmaylov, A. F., Bloino, J., Zheng, G., Sonnenberg, J. L., Hada, M., Ehara, M., Toyota, K., Fukuda, R., Hasegawa, J., Ishida, M., Nakajima, T., Honda, Y., Kitao, O., Nakai, H., Vreven, T., Montgomery, Jr., J. A., Peralta, J. E., Ogliaro, F., Bearpark, M., Heyd, J. J., Brothers, E., Kudin, K. N., Staroverov, V. N., Kobayashi, R., Normand, J., Raghavachari, K., Rendell, A., Burant, J. C., Iyengar, S. S., Tomasi, J., Cossi, M., Rega, N., Millam, J. M., Klene, M., Knox, J. E., Cross, J. B., Bakken, V., Adamo, C., Jaramillo, J., Gomperts, R., Stratmann, R. E., Yazyev, O., Austin, A. J., Cammi, R., Pomelli, C., Ochterski, J. W., Martin, R. L., Morokuma, K., Zakrzewski, V. G., Voth, G. A., Salvador, P., Dannenberg, J. J., Dapprich, S., Daniels, A. D., Farkas, Ö., Foresman, J. B., Ortiz, J. V., Cioslowski, J., and Fox, D. J. (2009), *Gaussian 09*, Revision D.01, Gaussian, Inc., Wallingford CT.
12 van Santen, R.A. and Neurock, M. (2006) *Molecular Heterogeneous Catalysis*, Wiley-VCH Verlag GmbH.
13 Levitin, V. (2014) *Interatomic Bonding in Solids: Fundamentals, Simulation and Applications*, Wiley-VCH Verlag GmbH.
14 Lejaeghere, K., Van Speybroeck, V., Van Oost, G., and Cottenier, S. (2014) Error estimates for solid-state density-functional theory predictions: an overview by means of the ground-state elemental crystals. *Crit. Rev. Solid State Mater. Sci.*, **39**, 1–24. doi: 10.1080/10408436.2013.772503
15 Harrison, W.A. (1989) *Electronic Structure and the Properties of Solids*, Dover Publications.
16 Landrum, G.A. and Dronskowski, R. (2000) The orbital origins of magnetism: from atoms to molecules to ferromagnetic alloys. *Angew. Chem. Int. Ed.*,

39, 1560–1585. doi: 10.1002/(SICI)1521-3773(20000502)39:9<1560::AID-ANIE1560>3.0.CO;2-T

17 Dronskowski, R. and Bloechl, P.E. (1993) Crystal orbital Hamilton populations (COHP): energy-resolved visualization of chemical bonding in solids based on density-functional calculations. *J. Phys. Chem.*, **97**, 8617–8624. doi: 10.1021/j100135a014

18 Aldén, M., Skriver, H.L., Mirbt, S., and Johansson, B. (1992) Calculated surface-energy anomaly in the 3d metals. *Phys. Rev. Lett.*, **69**, 2296–2298. doi: 10.1103/PhysRevLett.69.2296

19 Watson, R.E. and Weinert, M. (2001) Transition-metals and their alloys. *Solid State Phys.*, **56**, 1–112, http://linkinghub.elsevier.com/retrieve/pii/S0081194701800187.

20 Skriver, H.L. (1985) Crystal structure from one-electron theory. *Phys. Rev. B*, **31**, 1909–1923. doi: 10.1103/PhysRevB.31.1909

21 van Santen, R.A. (1991) *Theoretical Heterogeneous Catalysis*, 1st edn, World Scientific.

22 Cyrot-Lackmann, F. (1968) Sur le calcul de la cohésion et de la tension superficielle des métaux de transition par une méthode de liaisons fortes. *J. Phys. Chem. Solids*, **29**, 1235–1243, http://linkinghub.elsevier.com/retrieve/pii/0022369768902163.

7

Chemical Bonding and Reactivity of Transition Metal Surfaces

7.1 Introduction

In the first part of this chapter, the chemical bonding aspects of the surface chemical reactivity and its quantum chemistry will be described. We will use the tools of the previous chapter to analyze the electronic reasons for the differences in the relative stability of reaction intermediates adsorbed to the surfaces of different transition metals.

The second half of this chapter discusses chemical bond activation of adsorbed molecules. The chemical nature of the transition states of elementary reaction steps and its relationship with the surface structure of activating reaction centers will be presented.

The adsorption energies of reaction intermediates have an intimate relationship with surface reactivity. This may appear surprising at first, since surface reactivity is a kinetic property but adsorption energy is a thermodynamic property related to the rate of reactions.

Thermodynamics and kinetics have no direct fundamental link, but instead present approximate and very useful relationships such as the Brønsted–Evans–Polanyi (BEP) principle that we discussed earlier in Chapter 3. This principle correlates the activation energy of an elementary reaction step to the adsorption energies of adsorbed reaction intermediates, which is beneficial because trends in adatom bond energies are much easier to study than trends in activation energies.

The electronic structure of the surface chemical bond determines its strength. An important reference for this topic is the book by Hoffmann [1] that we recommend for additional reading.

Chemisorption of adatoms and molecules will be discussed in the first subsections of this chapter. An initial discussion of the electronic structure of the surface chemical bond and trends in the adsorption energies for adatoms will be followed by information about the chemisorption of molecules.

We conclude this topic on the electronic structure of the surface chemical bond with a subsection on scaling laws, which relate the adsorption strength of molecular fragments to the adsorption strength of related surface adatoms. This relationship enables us to use the adsorption energies of surface adatoms as surface reactivity descriptors.

The final sections of this chapter discuss the elementary reaction steps of adsorbed molecules or molecular fragments. Initially, we will discuss the nature of the transition state as a function of the chemical bond that is activated for the C—O bond of CO and C—H or N—H bonds in CH_4 and NH_3. Direct activation of chemical bonds as well as bond activation promoted by interaction with co-adsorbed atoms such as H or O will be introduced. These reactions will also be used to illustrate the changes in the chemical reactivity of a metal surface covered with an overlayer. Most notably, we will illustrate how the reactivity of adsorbed oxygen may change for oxygen insertion into the π bond of an olefin and how a unique reactivity between surface intermediates can evolve.

We recommend the book, *Fundamental Concepts in Heterogeneous Catalysis*, Nørskov et al. [2] for additional reading on related topics.

7.2 The Nature of the Surface Chemical Bond

7.2.1 The Electronic Structure of the Transition Metal Surface

In a transition metal, the surface metal atoms have fewer neighbors than the subsurface atoms. The surface energy is related to the energy cost of breaking the metal atom bonds to form the surface, and thus it relates to the transition metal bulk cohesive energy. The surface energy will also increase with the decreasing coordination number of surface metal atoms as illustrated in Figure 1i.1, which displays this trend for different surfaces of the fcc and bcc bulk crystal structures. The coordination number of atoms in the fcc bulk structure decreases from 12 to 9 on the most dense (111) surface and to 8 on the (100) surface, respectively. On the (110) surface of the fcc structure, the two different surface atoms have 7 and 11 coordination numbers, respectively. In the bcc structure, the atoms on the (110) surface have six nearest neighbors and on the most dense (100) surface they have five nearest neighbors. This represents a lowering from the eight nearest neighbor atoms found for the bulk of the bcc metal atom.

Insert 1: The Surface Energy

Calculated surface energies (Figure 1i.1) and work functions (Figure 1i.4) are compared as a function of the position of a transition metal in the periodic table. Figure 1i.2 compares the partial density of states (PDOS) for surface and bulk atoms. Figure 1i.3 illustrates the change in average d valence band position with d valence electron bandwidth. In Figure 1i.4, the work functions of the transition metals are compared.

Equation (1i.1) gives an approximate relationship between the change of the average d valence band energy $\Delta \varepsilon_d$ and transition metal atom surface coordination number N_{surf}, and that of bulk N_{bulk}, the electron occupation of the surface valence electron band n_d^s and the effective one-center electron–electron repulsion energy U_{dd}.

$$\Delta\varepsilon_d \approx (n_d^s - 5)U_{dd}\left(1 - \sqrt{\frac{N_{surf}}{N_{bulk}}}\right) \quad (1i.1)$$

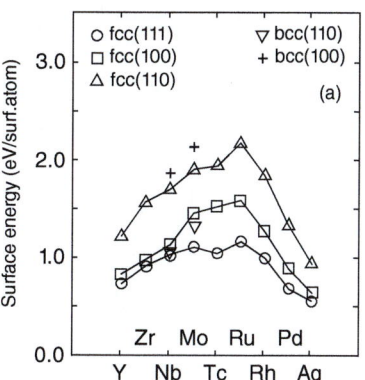

Figure 1i.1 Calculated surface energies in electron volts per surface atom. (Methfessel et al. 1992 [3]. Reproduced with permission of American Physical Society.)

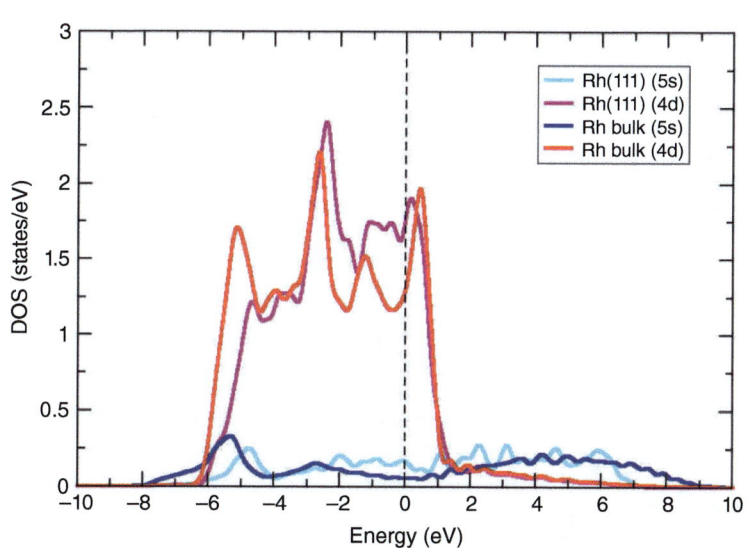

Figure 1i.2 Comparison of the PDOS of the valence atomic orbitals of a Rh atom at the Rh(111) surface and of the Rh bulk. The PDOS of the respective valence atomic orbitals are indicated in the figure.

(Continued)

Insert 1: (Continued)

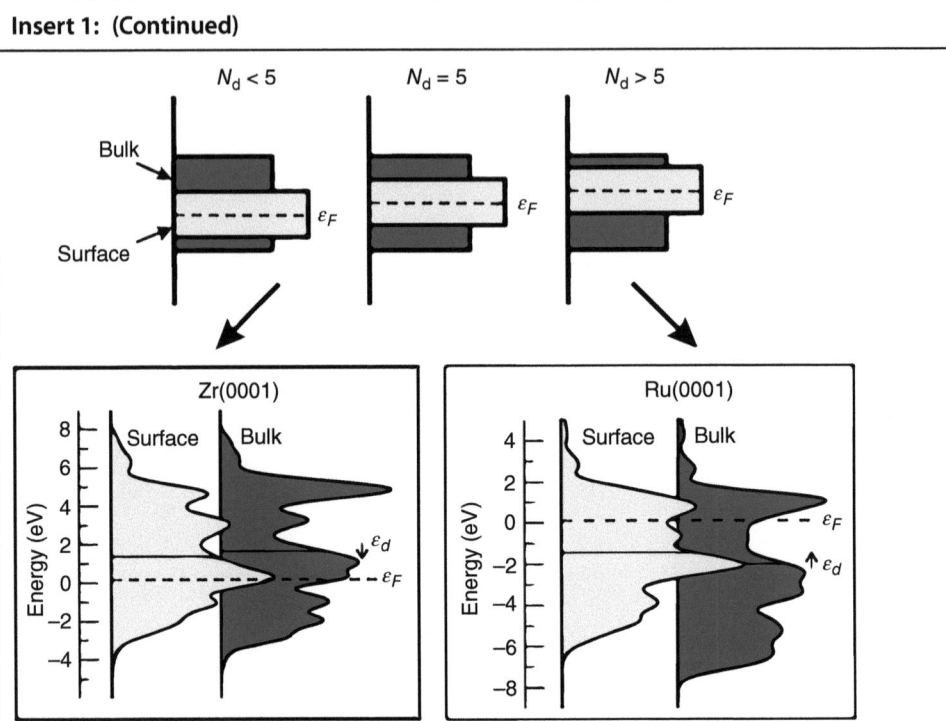

Figure 1i.3 (a) Schematic illustration of d-band narrowing at transition metal surfaces and its consequences for the energy levels of the surface atoms with less than 5d electrons ($N_d < 5$), exactly 5d electrons ($N_d = 5$), and more than 5d electrons ($N_d > 5$). (b) PDOS for the d orbitals in bulk Zr and Ru and at the surface of the Zr(0001) and Ru(0001), as computed from a plane-wave pseudopotential DFT calculation within the LDA. ε_d is the d-band center of the d PDOS and the small arrows indicate that ε_d is lower at the surface than in the bulk for Zr whereas ε_d is higher at the surface than in the bulk for Ru (computed by [4]). (Michaelides and Scheffler 2012 [4]. Reproduced with permission of John Wiley & Sons.)

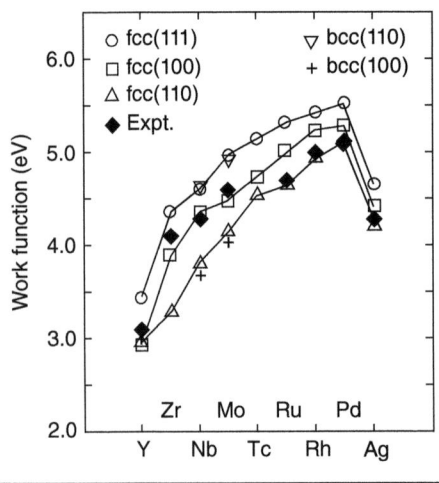

Figure 1i.4 Calculated surface-resolved work functions in electron volts as compared to the experimental polycrystalline values. (Methfessel et al. 1992 [3]. Reproduced with permission of American Physical Society.)

This decrease in the coordination number from bulk to surface is accompanied by a decrease in the d valence electron bandwidth. This decrease in bandwidth can be seen for the d valence electron on the Rh(111) surface when compared with the bandwidth for the bulk Rh atom (see Figure 1i.2). The Rh(111) surface atom has 9 neighbors versus the bulk atom with 12 neighbors.

The consequences of bandwidth narrowing on the electron occupation of the surface d valence electron bands are schematically illustrated in Figure 1i.3. When the number of d-electrons in the bulk is larger than half and the average d-band position would not change, the number of d-electrons at the surface must increase. In the opposite case, that is, when the number of d-electrons in the bulk is less than half, the number of d-electrons will decrease. This is because of equilibration of electron density because chemical potential at the surface and in the bulk must be equal. As a consequence, there will be a change in the electron–electron repulsion energies that is proportional to the square of the electron occupation. An increase in electron occupation will increase electrostatic repulsion, while a decrease in occupation decreases repulsion. This causes the average atomic orbital energies to shift so that surface valence electron band occupation above and below Fermi level will reduce the changes in surface d valence band electron occupation. The direction of the shift in the average d valence electron energy position at the surface depends on electron occupation. It is proportional to the change in bandwidth, which is in turn proportional to the difference of the square root of the number of nearest neighbors between surface and bulk (Eq. (1i.1)).

The work function determines the transition metal ionization potential, or affinity. Calculated values are shown in Figure 1i.4. The work function depends on the surface dipole moment, which depends on the electron distribution of the surface atoms and on the bulk chemical potential. It tends to increase when the transition metal atom position in the periodic table moves from left to right along a row, and decreases again for the group Ib metals. As expected, the trend is based on differences in the ionization potential of the gas phase atoms. The more open surfaces that are coordinatively less saturated and corrugated have a slightly decreased work function due to their altered surface dipole moment.

This section introduced the electronic features of the free surface which we will discuss more extensively in the next section for surfaces containing adsorbed atoms or molecules,

For further information on the electronic structure of the transition metal surfaces we refer to the books, *Theoretical Surface Science,* Gross [5] and *Concepts in Surface Physics,* Desjongueres and Spanjaard [6].

7.2.2 Chemisorption of Atoms and Molecules (This Section has been Jointly Written with I. Tranca.)

Several electronic structure properties determine trends in the adsorption energies of adsorbates [7]. An important factor is *the redistribution of electrons over the bonding and antibonding orbitals* of the surface chemical bond between the adsorbate and metal surface atoms.

The polarity of the chemical bond can also contribute to the bond energy through electron transfer. This relates to the work function of the surface and to the electron affinity of the adsorbate. In addition, *the weakening of metal–metal atom bonds* adjacent to metal atoms that are connected with adsorbed atoms contributes to the adsorbate interaction energy, and this is responsible for the surface reconstruction that is sometimes observed when the surface becomes covered with an overlayer of adsorbates.

As we will see, the interaction between the adsorbate and surface metal atoms is significantly stronger than the bond strengths of the metal–metal atom bonds. Viewing these interactions in the context of a molecular complex of adsorbate and metal atoms embedded in the lattice of the solid is useful.

On the metal atoms, the valence atomic orbital size as well as their energies will vary for the different metals, which will also affect electron affinity and, hence, the chemisorptive bond.

We will discuss these four energetic and electronic contributions to the adatom and admolecular surface chemical bond energy in detail in the next sections.

We will begin with an analysis of the surface chemical bond of the adsorbed H atom, which is a σ type bond. This will be followed by a discussion of trends in the adsorption energies of adsorbed C and O atoms, which in addition to σ bonds also form π bonds with the surface. We will distinguish the features produced when covalent bonding of adsorbate to the metal surface (such as for the C atom) dominates versus features that result from bonding dominated by polar interactions (as is the case for the O atom).

This section on the quantum chemistry of the surface chemical bond will also contain a discussion of the adsorption of molecular fragments such as CH_x ($x = 1$, 2, or 3) and the molecules NH_3 and CO. It will be concluded with the scaling laws that relate the adsorption energies of different adsorbates.

7.2.2.1 The H Adatom: The Covalent Surface Bond

7.2.2.1.1 The Electronic Structure of Atop and Threefold Coordinated H

In this subsection, we will analyze the adatom chemical bond of the H atom adsorbed to a transition metal surface by examining the PDOS and the crystal orbital Hamiltonian population (COHP) densities of the electronic structure of a H atom adsorbed atop or threefold to the Rh(111) surface [8]. The PDOS and COHP densities have been introduced in Chapter 6 (Insert 1).

Figure 7.1 illustrates the different adsorption sites on the (111) surface.

The PDOS of the H 1s atomic orbitals and the valence electron atomic orbitals of the Rh surface atom attached to the H atom are shown in Figure 2i.1. At around

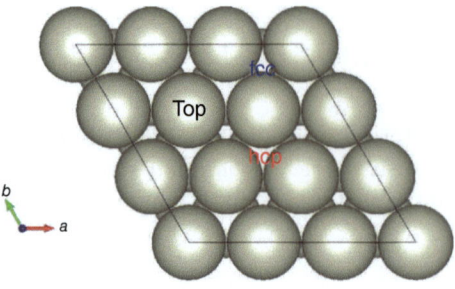

Figure 7.1 The different adsorption sites on the (111) surface of the fcc structure or on the (0001) surface of the hcp structure. The FCC site is defined as the threefold site with a subsurface atom below its center, while the HCP site has no atom in the second layer below its center.

−4.5 eV below the Fermi level, a broad PDOS of the H 1s atomic orbital is present that is indicative of the delocalization of the H 1s atomic electron density caused mainly by interaction with the d valence electrons of the Rh surface atom.

We expect the surface chemical bond to exhibit both bonding and antibonding surface orbitals. The corresponding electron density changes can be observed by comparing the d valence electron distribution on the Rh surface atom attached to H both in the presence and absence of the H atom (Figure 2i.1c). The increased densities at the lower energy (−6 to −4 eV energy interval, see Figure 2i.1c) correspond to bonding electron density. The antibonding electron density is the increased electron density near and above the Fermi level. These changes appear in the $4d_{z^2}$ valence electron density of the Rh surface atom.

Symmetry dictates that the H 1s atomic orbital atop of a Rh atom will only interact with the Rh, $4d_{z^2}$, 5s, and $5p_z$ atomic orbitals. The sign of the bond order overlap population density indicates the respective bonding, non-bonding, and antibonding electron density regimes. Figure 2i.1b gives the COHP plots that indicate the signs of the bond order overlap populations as a function of energy.

Near and above the Fermi level the surface electron density becomes antibonding. This implies that the bond energy will decrease when the number of electrons on the metal atom increases. Indeed, the interaction energy of H to a Pd atom that has one more valence electron than Rh decreases (see Figure 2i.3a).

Insert 2: The Chemisorption of Hydrogen

Figures 2i.1 and 2i.2 show the PDOS and COHP plots for the electronic structure of an H atom adsorbed atop (Figure 2i.1) or threefold (Figure 2i.2) on the (111) surface of Rh. In Figures 2i.1c and 2i.2c, a PDOS difference plot constructed from the PDOS of the Rh atom after and before the interaction with the H atom is shown. Figure 2i.3 shows the trend in the H adsorption energies as a function of transition metal position along a row of the periodic table. Figure 2i.4 schematically illustrates the difference in PDOS for a H atom adsorbed atop or threefold to a transition metal surface.

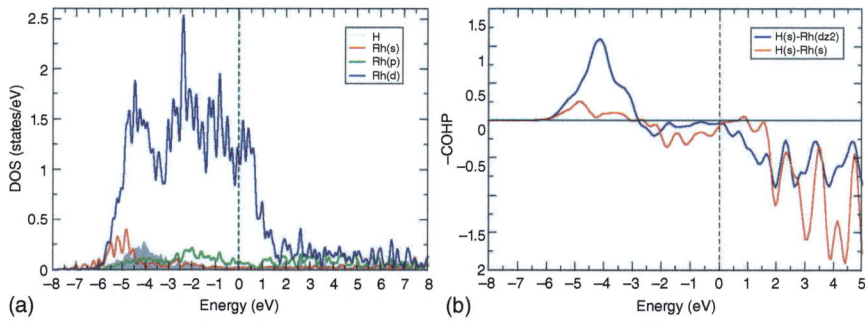

Figure 2i.1 The PDOS and COHP of H adsorbed atop of the Rh(111) surface. (a) The PDOS on the H atom and attached Rh atom. (b) The COHP between the H atom and Rh atom of (a). (c) The difference plot of PDOS of an Rh atom at the free Rh surface and with a H atom attached.

(Continued)

Insert 2: (Continued)

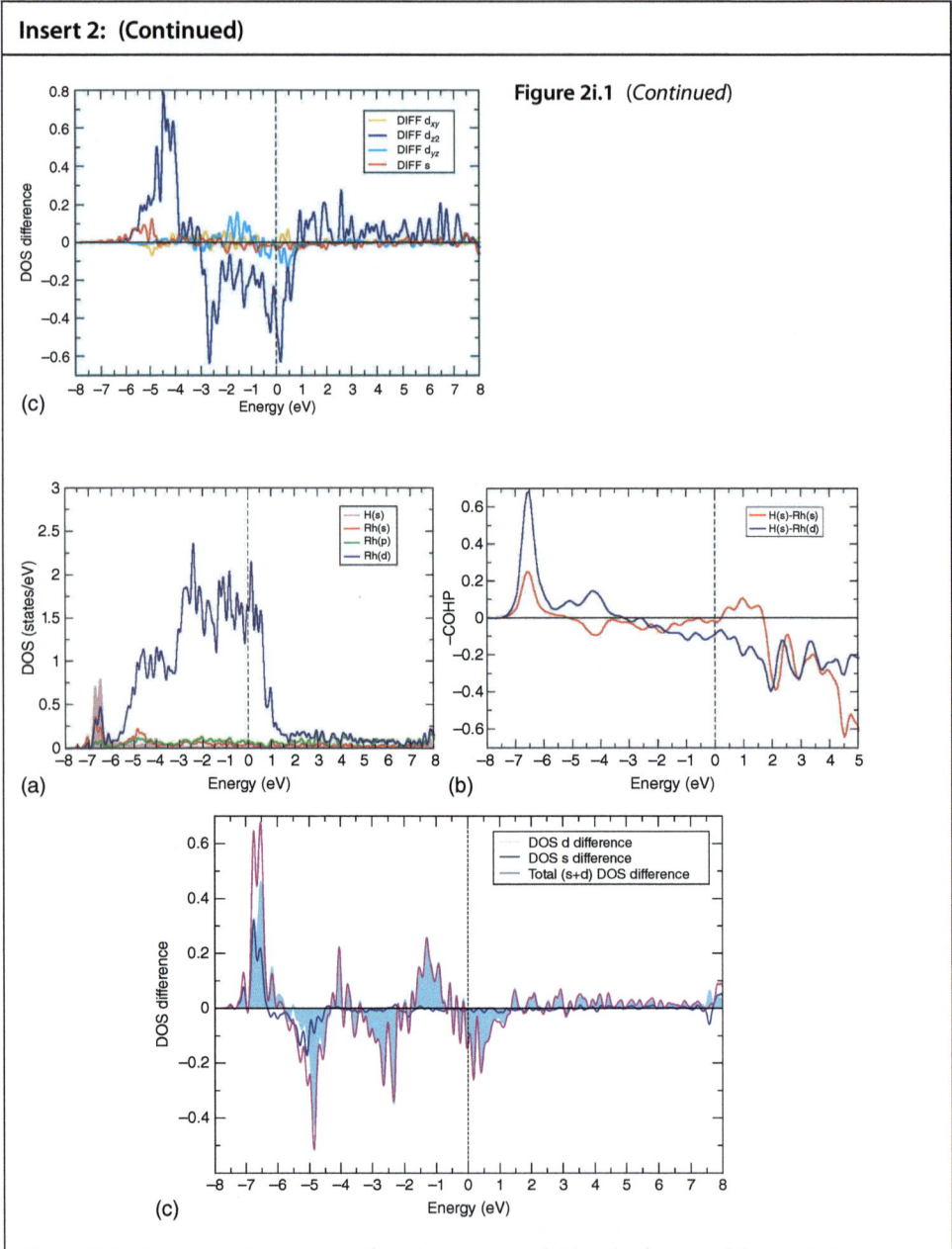

Figure 2i.1 (*Continued*)

Figure 2i.2 The electronic structure of H adsorbed threefold to the fcc site of the Rh(111) surface. (a) The PDOS on the H atom and an attached Rh atom. (b) The COHP between the H atom and Rh atom of (a). (c) The difference plot of PDOS of an Rh atom at free Rh surface and an Rh atom to which a H atom is attached.

Figure 2i.3 Trends in adsorption energies of H attached to the surfaces of a transition metal as a function of the number of their valence electrons. (a) and (b) The most dense surface of the respective transition metal has been chosen. As indicated on the bcc surfaces the twofold site becomes favored over threefold. (c) Comparison of the energy trends calculated for adsorption to the (111) surface when each transition metal is in the optimized fcc structure.

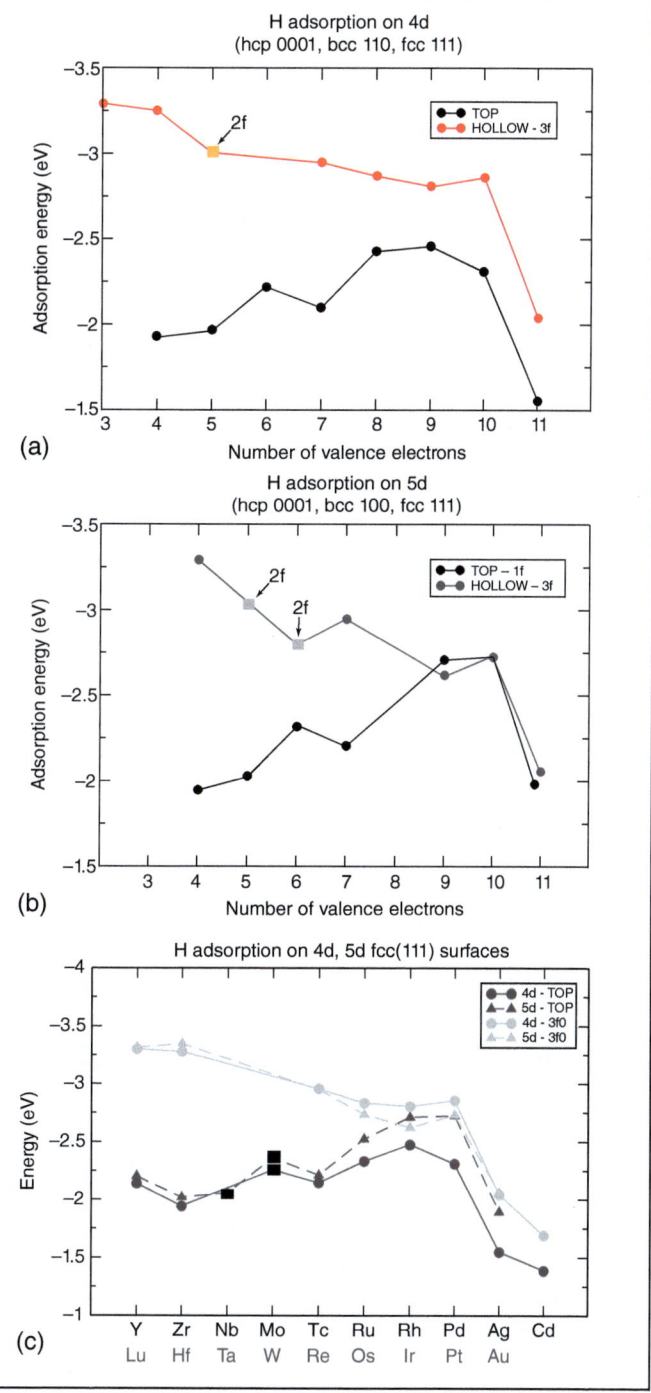

(Continued)

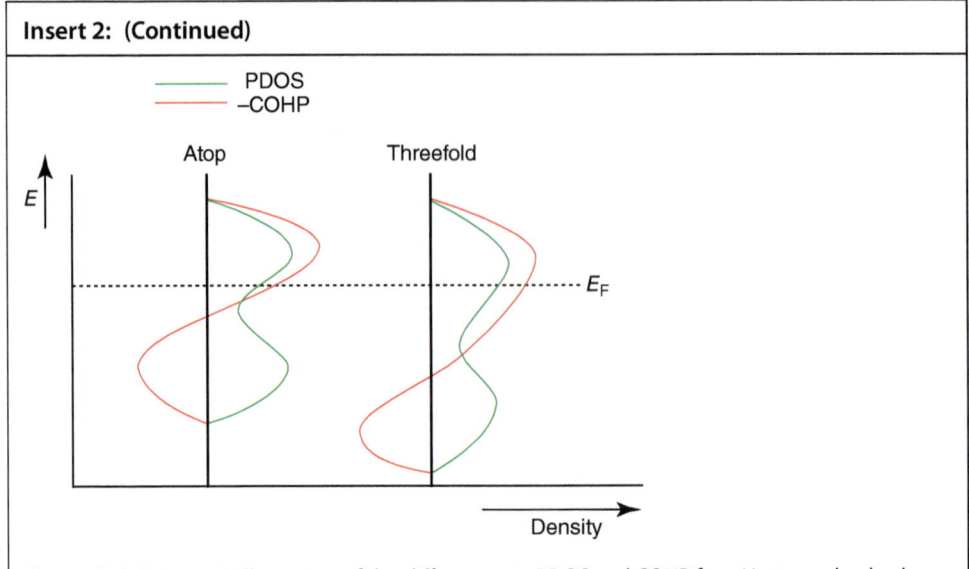

Figure 2i.4 Schematic illustration of the differences in PDOS and COHP for a H atom adsorbed atop or threefold to a transition metal surface.

The comparison in Figure 2i.2 with the threefold-adsorbed H atom indicates a stronger interaction for the threefold site with the metal surface due to the interaction with three surface metal atoms. The hydrogen atom adsorbed to three metal atoms is symmetrically oriented, and it will not only interact with s atomic orbitals of several transition metal atoms, but also with several d orbitals of the transition metal surface atoms. This results in a much more complex d valence electron density difference plot (Figure 2i.2c) than the same plot for the hydrogen atom adsorbed atop to a single metal surface atom.

The stronger interaction for threefold-adsorbed H can also be deduced from the increased contribution of the H 1s atomic orbital density to the PDOS at −6.8 eV below the Fermi level. Compared with atop-adsorbed H the position of the bonding PDOS of the H 1s atomic orbital is lower for threefold-adsorbed H and it is now located below the d valence electron band. In addition, its contribution to the PDOS in the bonding electron density peak has increased.

Another indication of the stronger interactions is the average difference between bonding and antibonding d valence electron energies Δ. This is also larger for threefold-adsorbed H as can be seen in the corresponding COHP curves in Figures 2i.1b and 2i.2b. The ratio of the Δ values (between atop and threefold-adsorbed H) is approximately proportional the square root of 3, the number of coordinating atoms. For the H atom adsorbed atop to the Rh atom we see that Fermi level is located in the non-bonding energy density regime with the d valence electrons. For threefold-adsorbed H, the Fermi level is located in the antibonding electron density regime. The electronic interaction remains antibonding deep below the Fermi level.

The different nature of the respective electron densities at the Fermi level indicates a different change in adsorption energy when d valence electron count decreases. For threefold-adsorbed H when less antibonding orbitals become occupied the interaction energy will increase, but for adatom adsorbed H now bonding d valence electrons will deplete and, hence, the interaction energy will increase. This is the essential reason for the differences in adsorption energy trends observed in Figure 2i.3, that we will discuss in more detail in the next subsection.

The contribution of the interaction with the s valence to the chemical bond energy can be estimated from the adsorption energies to the Ib group metals. For these metals, in addition to the bonding electron density, the antibonding surface orbitals between adsorbate and metal d valence electrons are nearly completely occupied. The interaction with the d valence electron will then be weakly attractive or repulsive. One can deduce from the bond energies of H to Ag and Au shown in Figure 2i.3 that the contribution of the s valence electron interaction to the respective bond energies is at least 30%. The Pauli repulsive interaction, which arises when the d valence electron band has a high electron occupation, will favor atop coordination since this minimizes contact with electrons in the surface d valence electron band.

When one compares adsorption energies for different transition metals along a row of the periodic table, the interaction with the s valence electrons can be considered to be approximately constant. The trends in adsorption energies of the H atoms to be discussed in the next subsection are instead mainly determined by differences in the d valence structure and electron occupation of the respective transition metals.

7.2.2.1.2 Trends in H Atom Adsorption Energies as a Function of Metal Position in the Periodic Table

Calculated trends in the adsorption energies of H adsorbed atop or threefold on the transition metals of the fourth and fifth rows of the periodic table are shown in Figures 2i.3. A comparison is made between the case showing adsorption to the dense surfaces of the transition metals in their state (Figures 2i.3a,b) and the case where all transition metals are optimized with the fcc structure (Figure 2C, 3C). It can be seen that the energies of adsorption change slightly, but the trends in adsorption energies are very similar. Therefore, we will base our discussion in the following on the relationship between electronic structure of the adatom chemical bond and trends in adsorption energies on systems of the bulk fcc structure [8].

Since the interaction with the s valence electrons will not vary substantially, an important electronic factor in the adsorption energy change will be the variation in the distribution of electrons over the bonding and antibonding d-surface valence electron densities. A trend in adsorption energy can now be predicted if one also assumes that the d-PDOS structure does not vary significantly when adsorption to different metals is compared. This is called the rigid band assumption and is approximately correct as long as the structure of the adsorption site does not change and transitionmetal elements are compared along the same row of the periodic table.

When the transition metal is varied, electrons are redistributed over the bonding and antibonding electron densities of the respective chemical bonds. This electron ditribution can be deduced from the PDOS and COHP curves calculated for the metal atom–metal atom or metal atom–adsorbate chemical bonds.

When one compares adsorption to a transition metal with a filled d valence electron band with adsorption to a transition metal with less d valence electrons, the adsorption energy is expected to increase because antibonding orbitals become depleted. The maximum in adatom binding energy will be reached when the further depletion of electrons of the d valence electron band induces depopulation of electrons from the bonding surface–adatom orbitals fragments. The position of the bond strength maximum will depend on the nature of the atomic orbitals and their energy distribution.

Figures 2i.3 show that for atop-adsorbed H, the adsorption energy has the expected maximum when the d valence electron count decreases from its maximum of 10 to lower values. The trend for threefold-adsorbed H is very different, showing an ongoing increase in energy with a decrease in electron occupation of the d valence electron band.

A consequence of these distinct adsorption energy trends is that there is a minimum in adsorption energy differences between atop and threefold-adsorbed hydrogen systems for the metals with nearly filled d valence electron bands. The nearly similar energies for atop and threefold-adsorbed hydrogen imply that their activation energies for diffusion will be very low and that the H atoms will be highly mobile.

When the d valence electron band occupation decreases, threefold-adsorbed hydrogen becomes increasingly more stabilized than atop-adsorbed hydrogen. The increasing difference in adsorption energy of atop and threefold-adsorbed H with decreasing d valence electron count is due to the difference in the occupation of bonding and antibonding surface fragment orbitals in the two adsorption modes. This is schematically illustrated in Figure 2i.4. Because the bonding surface's part for threefold coordination is deeper, the complementary antibonding part is also located lower with respect to the Fermi level than for atop-adsorbed hydrogen. This causes the relative electron occupation of the antibonding electron density part of threefold-adsorbed H to be higher than that of onefold-adsorbed H. Hence, when electron occupation decreases, the adsorption energy of the threefold hydrogen atom will continue to increase, whereas the adsorption energy of atop-adsorbed hydrogen may decrease because of the depopulation of its bonding surface fragment orbital electron density. With high valence electron occupation, atop-adsorbed hydrogen can have the lower antibonding electron density occupation and hence may even adsorb stronger than threefold adsorbed hydrogen.

7.2.2.2 Adsorption of the Carbon Atom: The Surface Molecular Complex

The essential difference between the adsorption of a H atom and atoms such as C and O is that the C and O atoms have more valence atomic orbitals and, hence, more possibilities for binding.

We will also consider in this subsection the difference between the surface chemical bond and that of the corresponding molecule of the isolated surface complex molecule. We will use the Born–Haber cycle of the chemisorptive bond that provides an estimate of the relative contribution of bond weakening of the metal atom bonds next to the surface adsorption complex to the chemisorptive bond.

We will consider the adsorption of C-atom here first (see figures 3i.1, 3i.2 and 3i.3). In the chemisorptive bond of adsorbed C, the C 2s atomic orbitals contribute little to the surface chemical bond; the main contributions are from the C $2p_z$ and $2p_x$ and $2p_y$ atomic orbitals. In the PDOS plots in Figures 3i.1 and 3i.2 of atop and threefold-adsorbed C, one recognizes the difference in the interactions of the 2p atomic orbitals, which are respectively parallel (π-symmetric with respect to the surface normal) and perpendicular (σ symmetric with respect to the surface normal). Again, we note a large separation between the bonding and antibonding surface fragment orbitals, and this is significantly larger than for the top adsorption case. The substantial C 2p atomic orbital contribution in the antibonding electron interaction region indicates that bonding between C atomic orbitals and surface d valence electrons is mainly covalent.

We can observe that there is a substantial interaction between C atom 2p atomic orbitals and metal surface s valence electrons for threefold-coordinated C, especially in the bonding electron density regime. In threefold coordination, the π-symmetric $2p_x$ and $2p_y$ atomic orbitals of the C atom strongly interact with the metal valence s atomic orbitals. This is mainly responsible for the large difference in energies of threefold and atop-adsorbed C and the systematically stronger bonding of the C atom to the high coordination site. The interaction of the π-symmetric 2p atomic orbitals with the metal s valence atomic orbitals is not possible in the atop configuration due to the limitations imposed by symmetry. (This is explicitly shown for the case of oxygen adsorption in Figure 4i.1c.)

Compared to threefold coordination, fourfold coordination with C located in the same plane of the metal atoms optimizes the interaction of the C $2p_x$ and $2p_y$ atomic orbitals with transition metal s valence electrons and is, therefore, a very stable adsorption mode. This mode is found for the fcc (100) surfaces as well as for the bcc structures, where surface atoms are positioned in a square.

Insert 3: The Adatom Bond of Chemisorbed C

The PDOS and COHP plots of the electronic structure of C-adsorbed atop and threefold to the Rh(111) surface are shown in Figure 3i.1. Figure 3i.2 gives the COHP plots for C atop and threefold adsorbed to Tc. In Figure 3i.3, the trends of the adsorption energies of C are shown as a function of transition metal position in the periodic table. Figures 3i.4 and 4i.5 compare the electronic structure of C adsorbed to the four transition metals Pd, Ag, Pt, and Au. In Table 3i.1, for the values of molecular complex energy of formation and for the metal–metal bond weakening are given for the adsorption of C to Rh and to Tc according to the Born–Haber analysis.

(Continued)

Insert 3: (Continued)

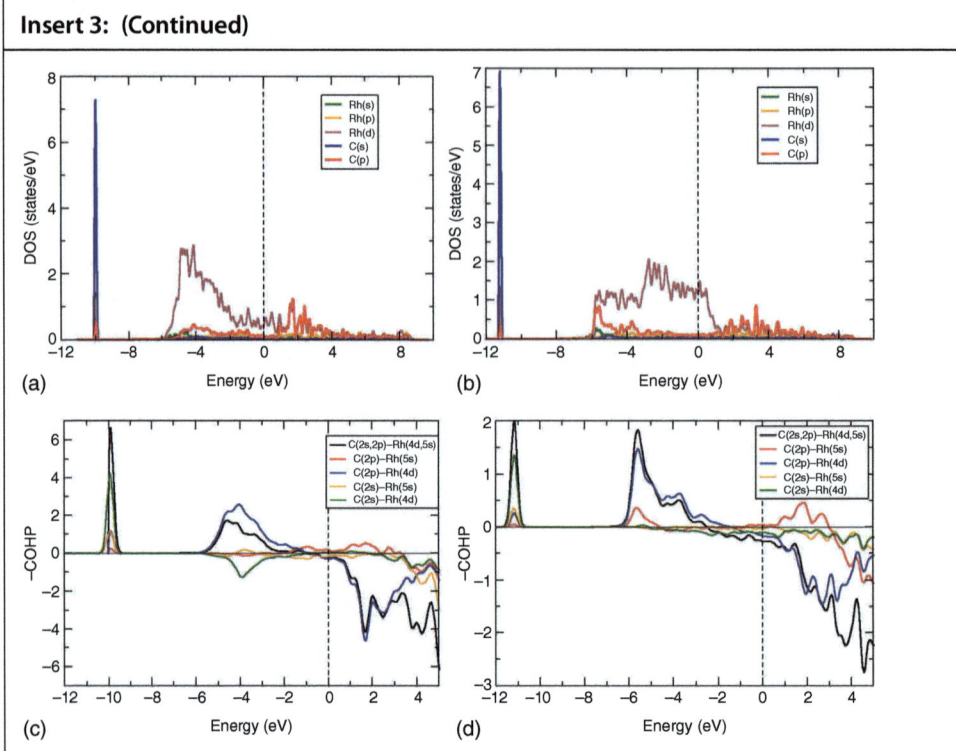

Figure 3i.1 The electronic structure of C adsorbed to the Rh(111) surface. (a) PDOS of C adsorbed atop Rh. (b) PDOS of C adsorbed threefold to Rh. (c) COHP for atop C on Rh(111). (d) COHP of C adsorbed threefold to Rh.

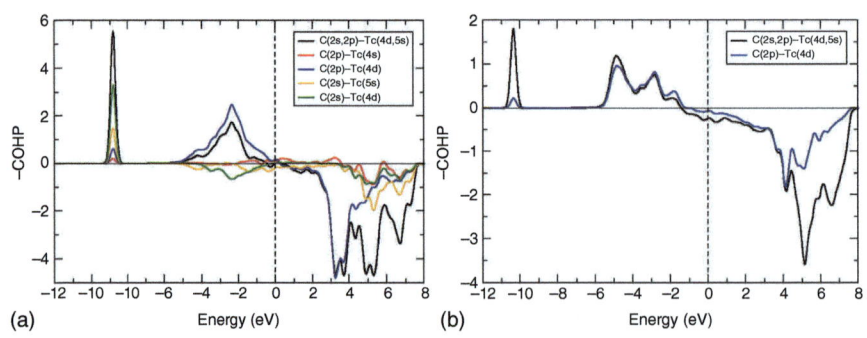

Figure 3i.2 (a) COHP for C adsorbed atop Tc(111). (b) COHP for C adsorbed threefold on Tc(111).

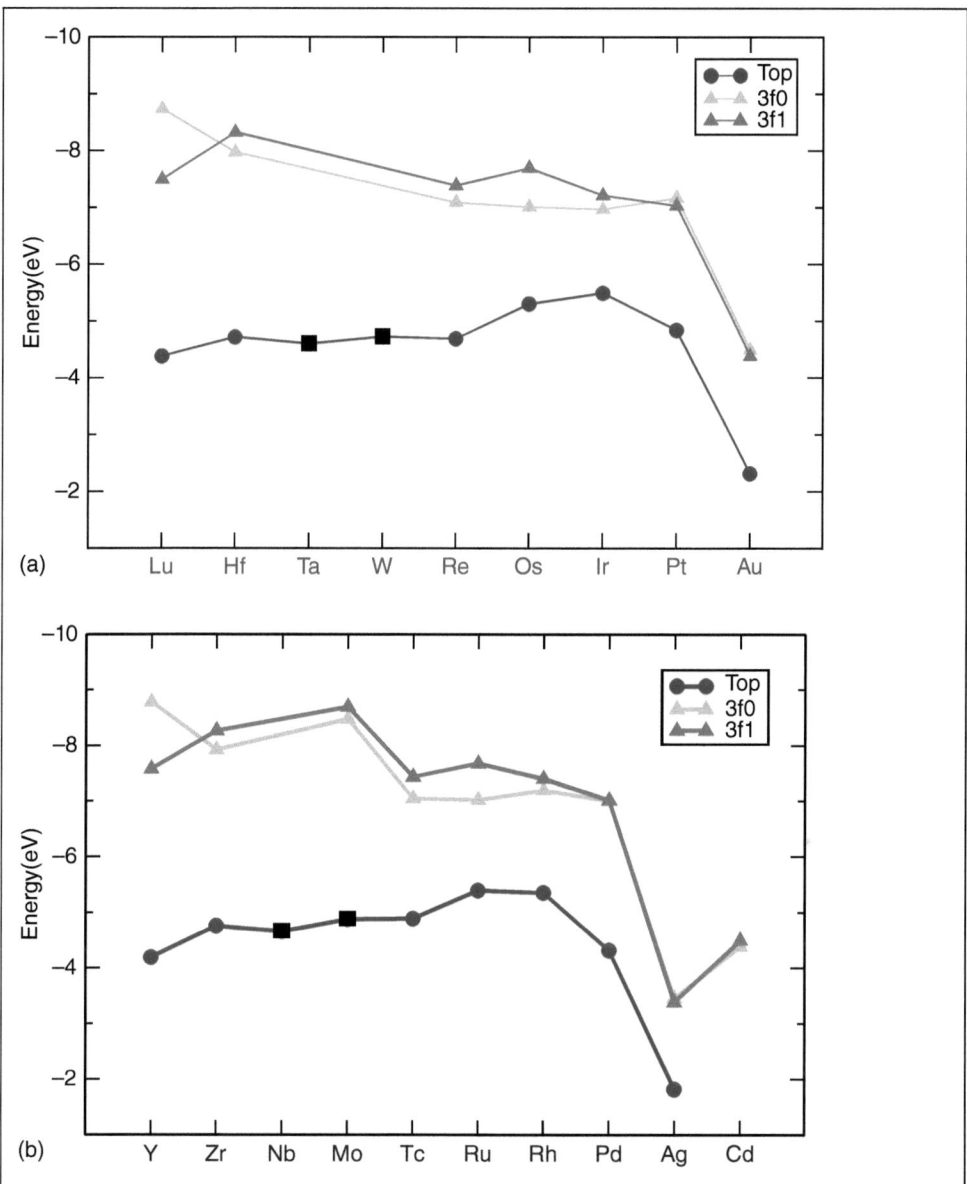

Figure 3i.3 Trends of chemisorption energy of C when adsorbed atop and threefold to the (111) surfaces of the fcc transition metal structures of Tc and Rh. Black squares designate adsorption on the bcc (110) surfaces. (a) Transition metals of the fifth row of the periodic system. (b) Transition metals of the fourth row of the periodic system.

(Continued)

Insert 3: (Continued)

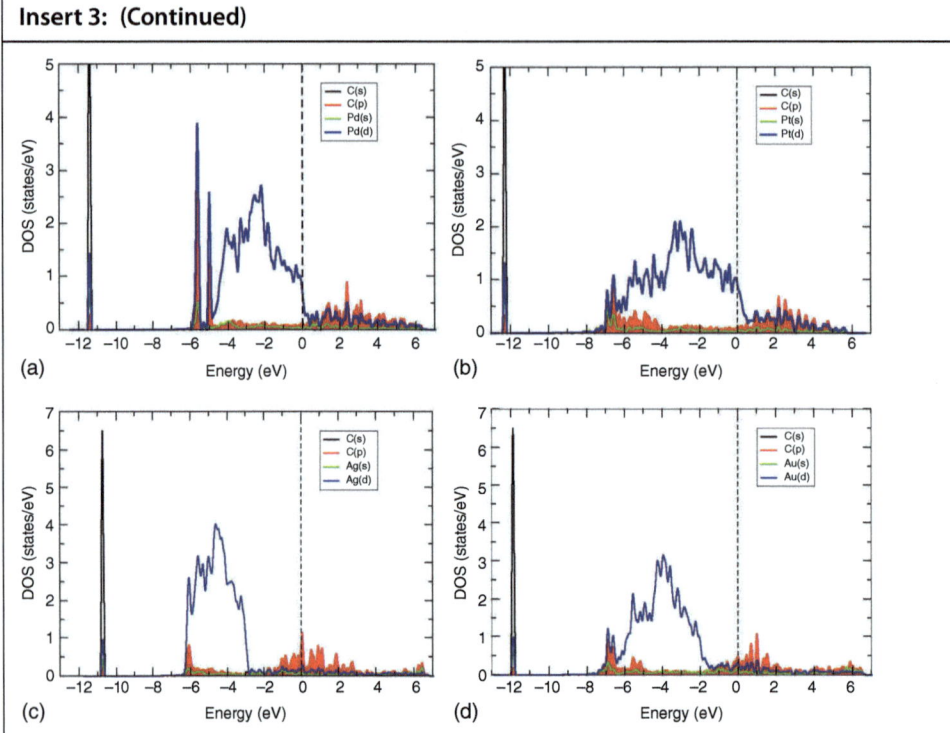

Figure 3i.4 PDOS of carbon atoms absorbed to a threefold site of the (111) surfaces on (a) Pd, (b) Pt, (c) Ag, and (d) Au. The respective energy differences Δ_x between average bonding and antibonding electronic energy density regimes are: $\Delta(\text{Pd}) = 7.5$ eV, $\Delta(\text{Pt}) = 9.2$ eV, $\Delta(\text{Ag}) = 6.0$ eV, $\Delta(\text{Au}) = 8.0$ eV.

Table 3i.1 Born–Haber cycle analysis of C atoms adsorbed atop to Rh and Tc(111) surfaces, respectively.

Atop on C	Reinsert	E_{vac}	E_{reinst}	E^s_{rel}	E_{emb}	E^w_{M-M}	E_{ads}
Rh		−7.010	7.092	−5.901	0.411	−6.312	0.780	−6.23
Tc		−7.098	9.232	−8.285	0.383	−7.902	1.186	−5.911

In Figures 3i.4 and 3i.5, the PDOS and COHP's are compared for the C atom adsorbed threefold to the (111) surface of the transition metals Pd, Pt, Ag, and Au. In the following subsection, these are compared with the electronic structures of O adsorbed to similar sites of these metals (see Insert 4). The adsorption energies of adsorbed O and C will also be compared in Insert 4. This comparison illustrates nicely the role of the different orbital interactions that cause the differences in relative bond energies. Figures 3i.4 and 3i.5 similarly illustrate the increased occupation of antibonding energy levels when adsorption to Pd or Pt is compared with adsorption to Ag or Au. This is consistent with the decrease in respective adsorption energies. As opposed to O, C binds more strongly to Au and Ag because the stronger interaction with the fifth row metal increases the energy difference between bonding and antibonding electronic energies. The antibonding transition metal d- carbon p atomic interaction regime has a lower electron occupation for Au than for Ag. Interestingly, the repulsive interaction due to the occupation of this antibonding electron density is partially compensated by increased electron population of the bonding s valence electron density.

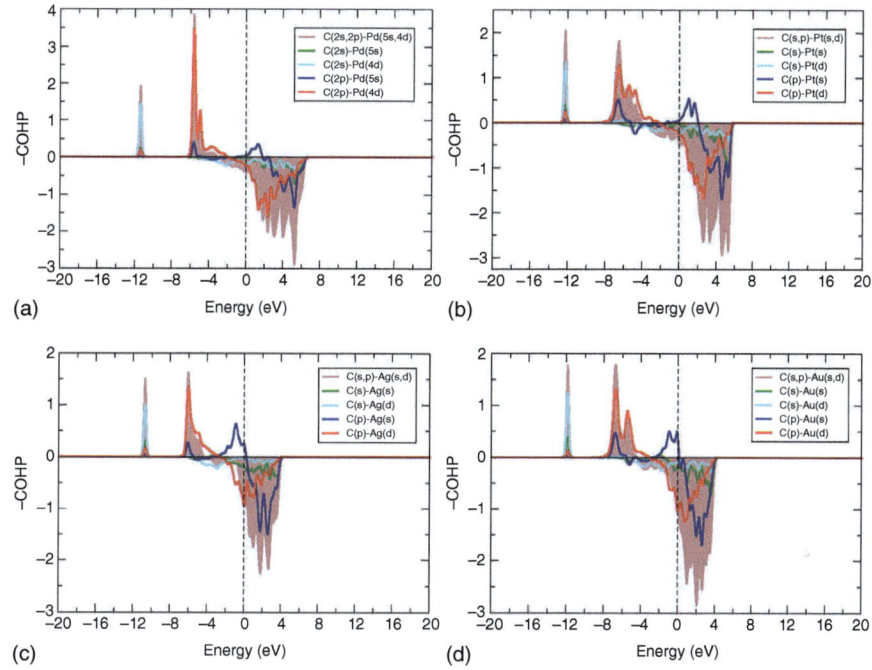

Figure 3i.5 The COHP plots of C adsorbed threefold to the (111) surfaces of (a) Pd, (b) Pt, (c) Ag, (d) Au, respectively.

Figure 3i.3 shows the adsorption energy trends of C adsorbed atop and threefold to the (111) of fcc transition metals. At high d valence electron occupation, the trends are comparable to those found for adsorbed H, except that threefold-adsorbed C always remains more stable than atop-adsorbed H. At low d valence electron occupation, the adsorption energy of threefold-adsorbed C becomes very strong compared to the metal atom–metal atom interaction energies of the corresponding transition metals. This induces surface reconstruction. The energy gain of the transformation of threefold coordinated C to fourfold coordination is so large that it overcomes the cost of the surface to reconstruct to the (100) structure. The cost of reconstruction in this case is low because the metal–metal bonds for these elements are relatively weak. Fourfold coordination of C has also been found for compounds such as $C(Al)_4^{2-}$ [9].

In Section 4.1, the clock reconstruction that occurs on the dense Co(111) surface at higher C coverage was discussed. In this surface reconstruction, threefold-adsorbed C is also converted into fourfold-adsorbed C.

Similar as for H adsorption the d-electron density of states at the Fermi level is substantially antibonding when C is adsorbed atop, whereas for atop-adsorbed C to Rh the Fermi level is located in the non-bonding density regime. The interaction with the adatom π atomic orbitals of C with transition metal s valence electrons when adsorbed in high coordination is different. This Fermi level is located in the bonding electron density regime. This is the reason for the general rule that interaction with π-symmetric orbitals directs an adsorbate atom or molecule to a high coordination site. This contrasts with the interaction with σ symmetric adsorbate orbitals, which directs the adsorbate to low coordination sites. When interaction with the d valence electrons is relatively strong the d valence electron band is nearly completely occupied by electrons. It directs the adsorbate to low coordination sites as we have seen for H. In this low coordination site the antibonding electron density regime is least occupied and, hence, Pauli repulsion minimized.

For the group VIII metals along a row of the periodic table (such as Pd, Rh, or Ru) the changes in bond energies are relatively small. The trends in energy are similar to those found for atop-adsorbed H. From Rh to Ru or Pt to Ir, there is initially a small increase in bond energy when d valence electron count decreases, but there is a decrease in bond energy when d valence electron count reduces further, as in the case of atop-adsorbed H.

The COHP distributions on the Rh(111) or Tc(111) surface (Figures 3i.1 and 3i.2) show that the interaction of the C atom with d valence electrons for the group VIII transition metals are nearly non-bonding at the Fermi level for the C atom adsorbed atop, but significantly antibonding for C adsorbed in threefold coordination. Since the adsorption energy increases when fewer valence electrons occupy the antibonding orbitals one predicts that the interaction energy of the adsorbate that is threefold-adsorbed will increase faster than the stability of the adsorbate that is atop-adsorbed. Indeed the difference in adsorption energies of threefold-adsorbed C over atop-adsorbed C increases when the number of transition metal valence electrons decreases.

The decrease in adsorption energy from Rh to Tc for atop-adsorbed C or the nearly constant value of the C adsorption energy for threefold-adsorbed C cannot be explained within the rigid band model.

The deviation from the expected behavior based on surface orbital bonding and antibonding electron density occupation is mainly due to the larger embedding energy of the surface complex molecules of with Tc compared to Rh. This follows from a comparison of the interaction energies in the surface complex molecule to the energies of the adsorbed state that we will discuss now.

The adsorbed state can be considered to be a surface complex molecule embedded in the surface of the transition metal by using the Born–Haber cycle of adsorption illustrated in Figure 7.2.

Chemisorption is a process that occurs in three steps as we will illustrate these for atop adsorption. In the first step of the Born–Haber cycle, a metal atom is taken from the surface (the surface vacancy formation energy E_{vac}). This free metal atom can then react with the adsorbing atom (the molecule formation

Figure 7.2 Born–Haber cycle decomposition of the different contributions to the adsorption energy. E_{vac} is the energy cost to desorb a metal atom or a metal atom complex from the surface, E_{mol} is the energy of formation of the free molecular complex between adsorbate and desorbed metal atoms, E_{emb} is the energy gained by the resubstitution of the molecular complex into the surface.

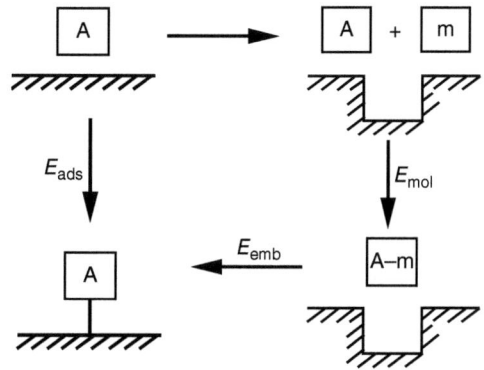

energy E_{mol}). This is the second step. The third step is the reinsertion of the M—A molecule into the vacant site at the surface (the embedding energy contribution E_{emb}). For an adsorbate in a higher coordination site, the vacancy formation energy must be computed for the removal of the metal atoms from the surface in direct contact with the adsorbate, and molecule formation is studied with these metal atoms in the configuration at the surface. Table 3i.1 shows a comparison of the calculated values for the different energy components for adsorption to Rh and Tc(111) surfaces. In the calculations [8], the structural coordinates have been taken as calculated for the adsorbed surface state. This implies that E_{emb} is the sum of two terms:

$$E_{emb} = E_{reinst} + E_{rel}^s \qquad (7.1)$$

In Eq. (7.1), E_{reinst} is the energy gain of the reinsertion of the surface complex molecule M—A into the surface vacancy generated in step one. The second term in Eq. (7.1), E_{rel}^s is the surface energy difference between the surface configuration generated by the adsorbate (but without the adsorbate) and the free surface before adsorption. Adsorption will weaken the metal–metal atom bonds of the metal atoms next to the adsorption site, which will lead to small displacements of the surface atoms. While this will increase the adsorption energy overall, it will also have a small energy cost.

Table 3i.1 gives the different energy contributions of the Born–Haber cycle analysis for C adsorbed atop to the (111) surfaces of Tc and Rh, respectively. The energy of adsorption is equal to the energy of formation of the free surface molecular complex minus the metal–metal bond weakening energy of the surface. This metal–metal bond weakening decreases the adsorption energy by approximately 15% compared to the energy the surface molecule. The decrease in energy is smaller for C adsorbed atop to Rh than Tc because the bond weakening cost is higher for the metal with the stronger metal atom–metal atom bonds.

The interpretation of E_{M-M}^w as a weakening of the surface atom bond energies is only strictly valid when one assumes that the bonding within the A—M$_x$ surface complex molecule does not change upon embedding in the surface complex.

The Born–Haber analysis illustrates that the adsorbed state can be considered a surface molecule embedded in a surface. A significant part of bonding between the adsorbate and the surface is already present in the free surface complex molecule. The proviso is that the structure of the free surface complex molecule is that of the embedded complex. The structure of the surface complex is constrained by embedding in the surface. The structure or energy of a free surface complex molecule would be very different it were allowed to transform to its true local energy minimum.

7.2.2.3 The Oxygen Adatom: The Polar Surface Bond

Just as for C, the interaction of the O atom with the metal surface is dominated by the O 2p atomic orbitals. To the transition metals located to the right in their row of periodic system. The binding of the C atom is substantially stronger than that of the O atom, because the latter experiences a significantly larger Pauli repulsion with the highly electron occupied d-valence electrons. For adsorbed O, the Fermi level is located in the antibonding surface fragment orbital electron density regime over a substantially wider energy range than for adsorbed C

The PDOS and COHP of O adsorbed atop and threefold to the Rh(111) surface are shown in Figure 4i.1. There is a striking difference when these are compared to the electronic structures of adsorbed C. For adsorbed C, the surface chemical bond electron density is non-bonding or weakly antibonding at the Fermi level and the main antibonding peak is far above the Fermi level, while for adsorbed O the Fermi level is near the center of the antibonding electron density. This will weaken the M—O adsorbate bond more than the M—C bond. The bond strength of the M—O bond remains weaker than the M—C bond until the orbital's electron occupation has been reduced so far that only bonding surface adsorbate fragment orbitals become occupied.

When bonding to the transitionmetals from the right to towards the left is compared along a row of the periodic system a steep increase in the bond energy of the M—O bond is observed as shown in Figure 4i.3. Because the location of the Fermi level is close to the maximum of the antibonding electron surface fragment density for a transition metal located at the right in a row of the periodic system, antibonding orbitals become depleted when an O atom attaches to a transition metal with decreasing electron occupation.

Similar to the adsorbed C atoms, for O atoms threefold coordination is always preferred over atop adsorption. Overlap of the O $2p_x$ and $2p_y$ atomic orbitals with the asymmetric combination of the s atomic orbitals is then possible, which results in the strong preference for adsorption to the high coordination site (see Figure 4i.1c). The difference in energy between atop and threefold-coordinated O is nearly constant when the transition metal varies along a row of the periodic table. The energy changes are dominated by differences in the distribution of electrons over the d valence orbitals.

The polarity of the M—O bond can be deduced from the changes in work function, as shown in Figure 4i.2 as well as PDOS ratios of the respective atomic orbitals of the occupied electron densities of the adatom chemisorptive bond. The work function differences of the different adsorbates in Figure 4i.2 demonstrate that the charge on the O atoms is higher than on the C or H atoms. These differences decrease for the low work function metals, because of their increased electron donation capability.

The PDOS plots of the respective M—O surface bonds shown in Figures 4i.1 and 4i.4 can also be used as indicators of the polarity of the M—O bond.

In Figure 4i.1, the PDOS on the O atom is compared for the atop and threefold-adsorbed atom. In the atop-adsorbed configuration, the interaction with the O $2p_z$ shows strong polarity (its intensity in the bonding regime is higher than in the antibonding regime), but the interaction with the O $2p_x$ and O $2p_y$ atomic orbitals is still dominantly covalent. This changes dramatically for O that is threefold-adsorbed. Then, due to the increased interaction with the metal s atomic orbitals, the high polarity of the M—O chemical bond is indicated by not only the O $2p_z$ orbital, but also by the O $2p_x$ and $2p_y$ orbitals. This follows by comparing the peak intensities in the respective bonding and antibonding density regimes.

The transition metal work functions increase when one compares transition metals along a column of the periodic table from top to bottom, such as for Ni, Pd, or Pt. The polarity of the bond will decrease in this order, as expected.

The high electron occupation of the antibonding surface adatom orbitals implies a substantial contribution to the M—O bond strength from Pauli repulsion. This is counteracted by the bonding interaction with the metal valence s electrons. This Pauli repulsion is stronger for the interaction with Pt than with Pd or Ni because of the wider extent of the d valence atomic orbitals. For adsorbed O this causes the adsorption energy to decrease along a column of the periodic table. This trend differs from the case of covalent C adsorption (Table 4i.1). As we will see in the next subsection, it also biases O to become twofold-adsorbed on the Pt metal surface.

Insert 4: The Electronic Structure of the Adsorbed O Atom

Figure 4i.1 shows the PDOS and COHP plots for O adsorbed to the Rh(111) surface. Figure 4i.2 shows the computed change in work function when adsorption of H, C, and O is compared. Trends in adsorption energies of O on various surface sites are compared in Figure 4i.3, and Figures 4i.4 and 4i.5 give PDOS and COHP values of O threefold-adsorbed to Pd, Pt, Ag, and Au. Table 4i.1 presents a comparison of the adsorption energies of C and O adsorbed to the (111) surfaces of the same metals.

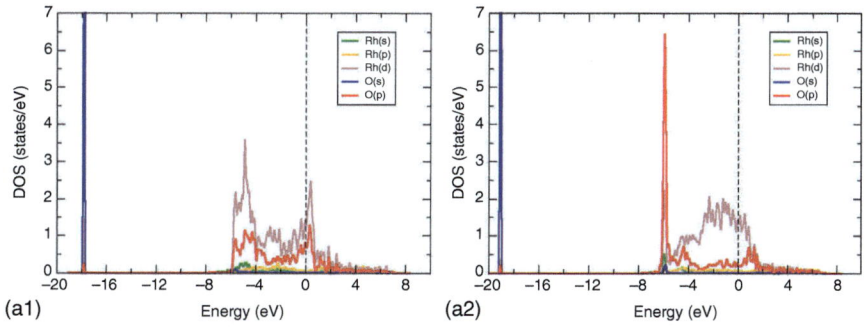

Figure 4i.1 The electronic structure of O adsorbed to the Rh(111) surface. (a1) PDOS of O adsorbed atop. (a2) PDOS of O adsorbed threefold. (b1) COHP of O adsorbed atop. (b2) COHP of O adsorbed threefold. (c) COHP comparison of the interaction of O with metal 5s valence electrons; atop and threefold adsorption comparison.

(Continued)

230 | *7 Chemical Bonding and Reactivity of Transition Metal Surfaces*

Insert 4: (Continued)

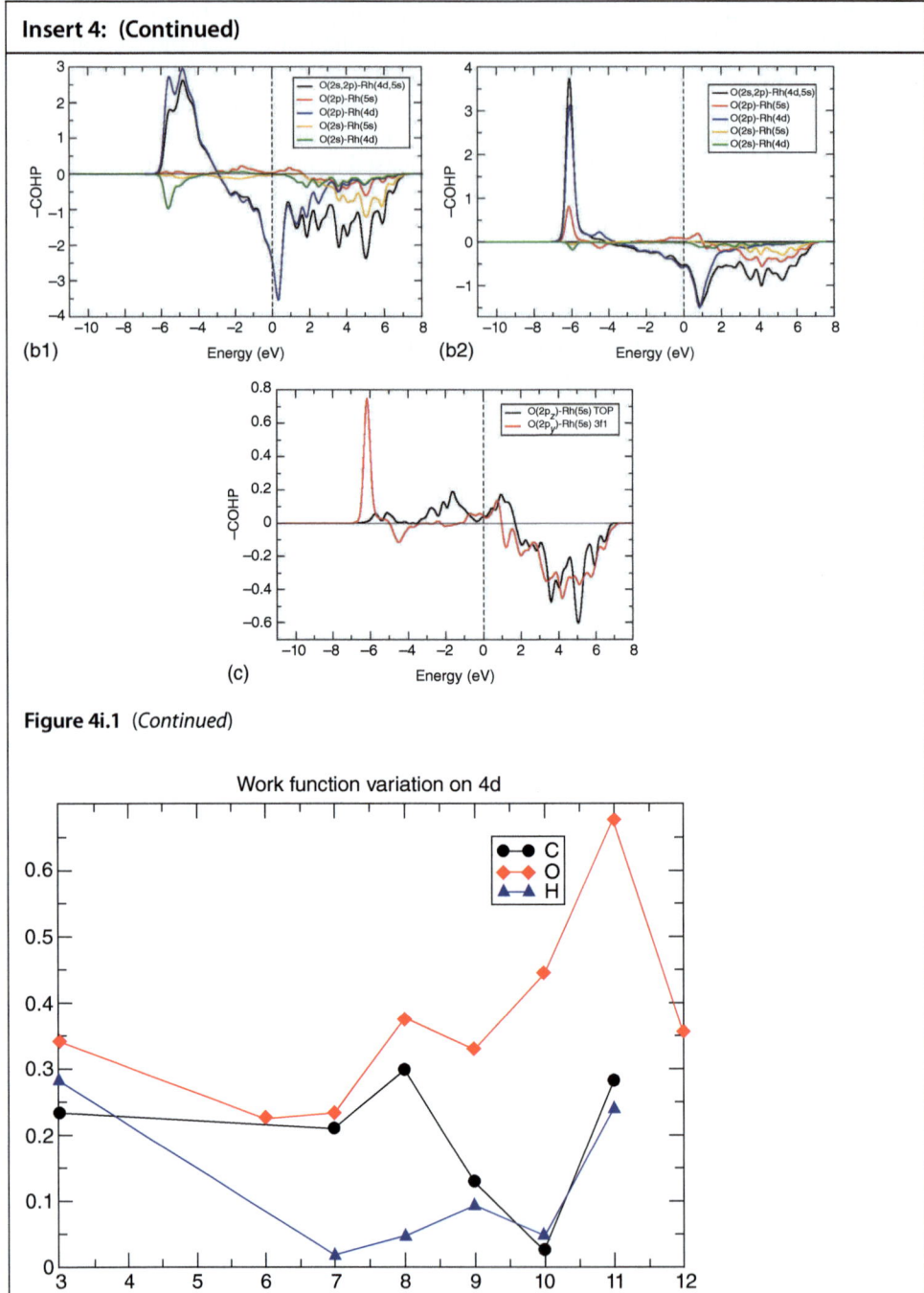

Figure 4i.1 (Continued)

Figure 4i.2 Change in work function for atop-adsorbed adatoms to the (111) surface of different metals of the fourth row of periodic table, plotted as a function of the number of transition metal valence electrons. An overlayer of 3×3 density is used in the DFT calculations.

7.2 The Nature of the Surface Chemical Bond

Table 4i.1 DFT adsorption energies (in electron volts) of carbon and oxygen adsorbed atop and threefold on the (111) surfaces of Pd, Pt, Ag, and Au.

	Carbon		Oxygen	
	Atop (eV)	Threefold (eV)	Atop (eV)	Threefold (eV)
Pd	−4.32	−7.01	−2.7	−4.41
Pt	−4.84	−7.03	−2.82	−4.18
Ag	−1.82	−3.38	−2.00	−3.54
Au	−2.10	−4.50	−1.76	−3.17

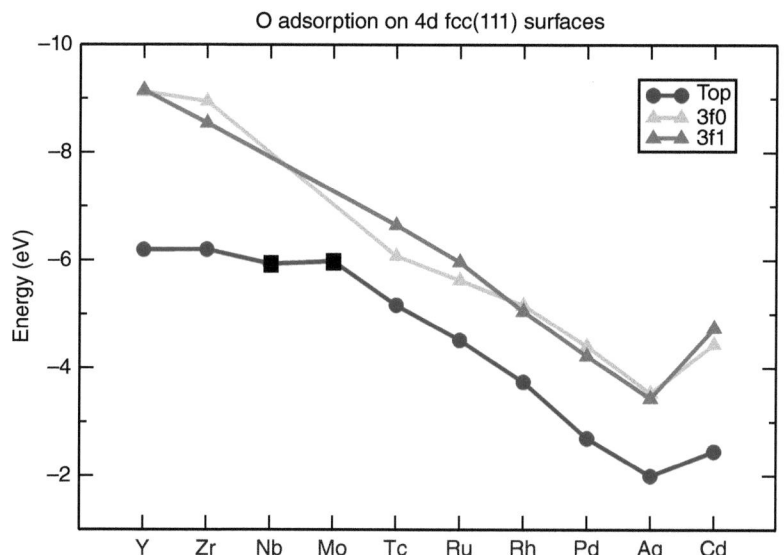

Figure 4i.3 Adsorption energy trends for O adsorbed to the three coordination sites on the (111) surfaces of the transition metal fcc structures of the fourth row of the periodic systems.

(Continued)

Insert 4: (Continued)

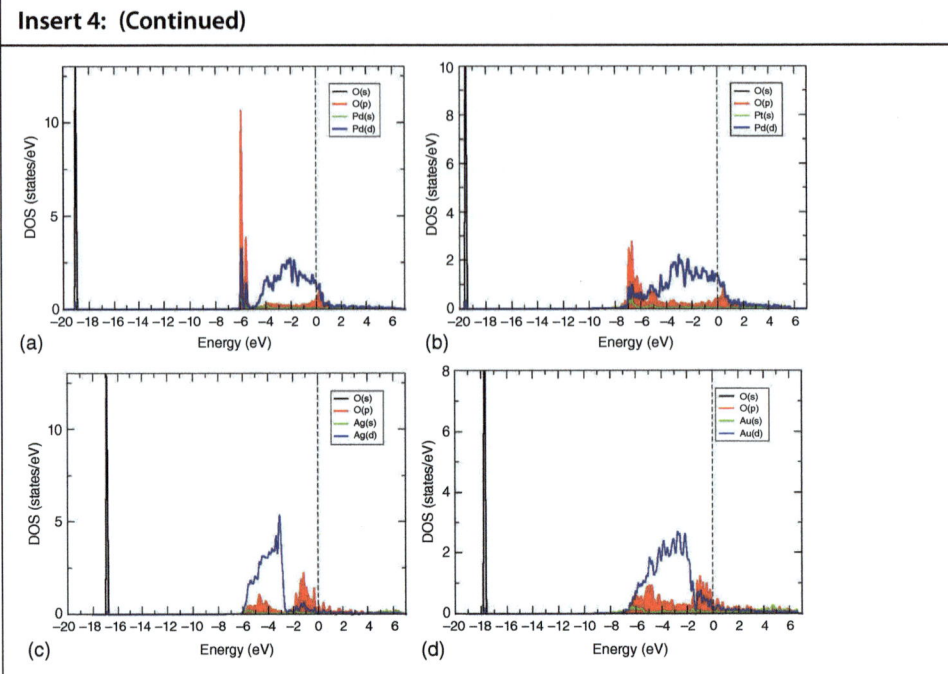

Figure 4i.4 PDOS of oxygen atoms absorbed to a threefold site of the (111) surfaces on (a) Pd, (b) Pt, (c) Ag, and (d) Au. The respective energy differences Δ_x between average bonding and antibonding electronic energy regimes are: $\Delta(Pd) = 6.3$ eV, $\Delta(Pt) = 7.2$ eV, $\Delta(Ag) = 3.4$ eV, $\Delta(Au) = 4.1$ eV.

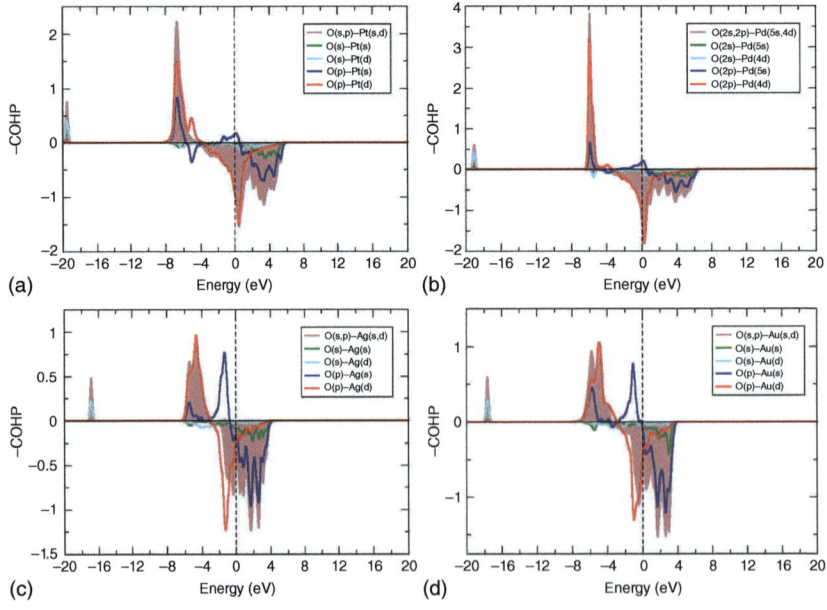

Figure 4i.5 The COHP plots of O adsorbed to the (111) surfaces of (a) Pt, (b) Pd, (c) Ag, and (d) Au.

The electronic structures of C adsorbed to Pd, Pt, Ag, and Au can be compared with the structures for O adsorbed to these same metals in Figures 3i.4, 3i.5, 4i.4, and 4i.5. The respective adsorption energies are compared in Table 4i.1. For adsorbed O on Ag and Au the antibonding adsorbate electron density is nearly completely occupied with the d valence electrons, and this results in a Pauli repulsion interaction proportional to Δ, the energy difference of corresponding bonding and antibonding energy levels. This repulsion will be stronger for Au than for Ag because of the larger Δ value for Au. The overall attractive interaction is now mainly due to the occupation of the bonding orbital interaction with the s valence electrons.

7.2.3 Adsorption Site Preference as a Function of Accessible Free Valence

In this section, we will consider the coordination of molecular fragments such as CH_x, NH_x, or OH_x. With a change in x, the balance of σ-symmetric versus π-symmetric interaction of the respective chemisorption bonds differs, which will affect the preference for a particular adsorption site.

It is important to distinguish between σ-symmetric and π-symmetric orbital interactions, as they may result in opposing forces that determine either low- or high-coordination site adsorption. One must also distinguish the interaction of orbitals with s valence electrons from interaction with d valence electrons.

The interaction of adsorbate orbitals with the metal s valence electrons leads to electron-occupied bonding orbital fragments, which always direct adsorbate to a high coordination site. This preference for high coordination is stronger for the interaction with adsorbate π-symmetric orbitals than for σ-symmetric orbital fragments because no π-symmetric interaction is possible in the atop sites. In these sites, the π-symmetric $2p_x$ and $2p_y$ atomic orbitals cannot interact with the s valence electrons, since they exhibit different symmetry.

The interaction of σ-symmetric adsorbate orbitals with the metal d valence electrons of high electron occupation will favor low-coordination site adsorption, since the antibonding orbital fragments in the surface adsorbate have become occupied. When adsorbed atop, the Pauli repulsive interactions that evolve are reduced compared to threefold coordination by the minimal contact with the surface lattice.

For CH_3, in addition to the formation of a metal atom–carbon atom bond, C—H bond stabilization occurs by interaction with the metal surface through so-called agostic interactions. We will compare their relative importance for CH_3 adsorption and NH_3 adsorption. When the agostic interaction occurs, there is a contribution to the chemical bond energy of the π-symmetric orbitals that are part of the C—H or N—H bonds in addition to the interaction through their lone pair σ HOMO.

7.2.3.1 Chemisorption of Molecular Fragments CH_x, NH_x, and OH_x Species: Coordination Preference as a Function of Accessible Free Valence

For transition metals such as Pt, a clear difference in preferred adsorption site is seen for adsorption of CH_x, NH_x, or OH_x fragments. This is due to the varying balance of σ-symmetric versus π-symmetric interactions. Variation in

Table 7.1 Chemisorption energies (eV) for CH_x, NH_x, and OH_x fragments adsorbed to Pt(111) surface.

Site	CH	CH_2	CH_3	NH	NH_2	NH_3	OH	H_2O
hcp	6.76	–	1.25	3.76	1.52	–	1.82	–
fcc	6.84	–	1.38	4.05	1.72	–	2.00	–
Bridge	6.10	4.06	1.39	3.15	**2.58**	–	2.27	0.21
Top	3.92	3.19	**2.05**	1.26	1.93	**1.05**	2.23	**0.34**

Source: Michaelides and Hu, 2001 [10]. Reproduced with permission by Kluwer.

the atop-directing interaction of σ-symmetry is mainly responsible for these differences, which are significant when the interaction with the d valence electrons is repulsive, and may dominate when the electron occupation of the transition metal d valence is high and the spatial extent of the d-atomic orbitals is large.

When the adsorption of CH, CH_2, or CH_3 or the analog intermediates NH, NH_2, NH_3 or OH, and H_2O are compared, a shift in the preferred adsorption site is observed with increasing H atom coordination of the central atom (Table 7.1). On the Pt (111) surface, coordination occurs according to the free valence of the different intermediates: CH prefers threefold coordination, CH_2 prefers a twofold coordination and CH_3 a onefold coordination. The NH_x and OH_x species follow a similar sequence when x increases. This tendency can be understood by considering the balance between the σ-symmetric and the π-symmetric orbital interactions. The σ-symmetric orbitals have a high antibonding electron density at the Fermi level, which directs the surface fragments to low coordination sites, while the π-symmetric orbital interactions have mainly bonding electron density at the Fermi level, which directs fragments to high coordination sites. Interaction with π-symmetric atomic orbitals on the adsorbate increases when the coordination number of H atoms on C, N, or O decreases. While CH prefers threefold coordination due to the strong interaction between its π-symmetric orbitals and the s valence electrons, this interaction is weaker for CH_2 and NH_2 that have only one π-symmetric orbital available. Therefore, the force directing fragments to high coordination sites is also weaker. The X—H_2 fragment is instead adsorbed between two surface metal atoms with the hydrogen atoms perpendicular to the bond that connects the two surface atoms, because the resulting interaction with the metal s valence electrons is optimum in this configuration.

The preference of sites as a function of adsorbate free valence that is exhibited by Pt disappears gradually when transition d valence electron occupation decreases because the associated decrease in the fraction of antibonding electron density of the surface adsorbate chemical bond above the Fermi level results in a weaker atop-directing interaction.

There are subtle differences in the relative stability for different adatoms, even when these atoms are adsorbed threefold to a (111) surface. Table 7.2 illustrates this. On the Pt(111) surface, the C atom with the higher coordinative unsaturation prefers the hcp site where it is coordinated with three metal atoms in

Table 7.2 Chemisorption energies in electron volts for C, N, O, and H adsorption the Pt(111) surface slab.

Site	C	N	O	H
hcp	**6.86**	4.27	3.59	2.79
fcc	6.85	**4.43**	**3.97**	**2.85**
Bridge	6.08	3.56	3.41	2.83
Top	4.49	1.92	2.51	2.83

The most stable sites are indicated in bold type.
Source: Michaelides and Hu, 2001 [10].
Reproduced with permission of Kluwer.

the outer surface layer and a fourth in the subsurface layer. N, O, and H prefer adsorption to the fcc site where they only coordinate to the three metal atoms in the outer surface layer. The energy difference between twofold coordination and threefold coordination decreases from nitrogen to oxygen, and is negligible for adsorbed H.

7.2.3.2 CH$_3$ and NH$_3$ Chemisorption: The Agostic Interaction

The interaction energy of CH$_3$ adsorbed at the threefold site of the transition metal surface depends on the orientation of the C—H bonds with respect to the transition metal atom bonds (see Figure 5i.2). When the two sets of bonds are parallel, an additional attractive interaction occurs called the agostic interaction [11, 12] which overrules the atop preference of CH$_3$. A similar agostic stabilization has been seen for CH$_2$ adsorbed at a high coordination site on a metal such as Ru [13].

We see in Figure 5i.4a that non-agostic CH$_3$ adsorbed threefold has a weaker interaction with the surface than atop-adsorbed CH$_3$, and also that threefold-adsorbed non-agostic CH$_3$ becomes more stable than atop-adsorbed CH$_3$ only for Tc. The COHP curves of adsorbed CH$_3$ (Figure 5i.3) show that the relative increase in adsorption energy of threefold-adsorbed CH$_3$ with decreasing transition metal valence electron occupation is due to electron depletion of the antibonding surface molecule electron density. For atop-adsorbed CH$_3$, the Fermi level is located near the non-bonding electron density regime, so that the change will be substantially less when electron occupation decreases.

The trend in chemisorption energy of atop-adsorbed CH$_3$ is remarkably similar to that of the atop-adsorbed C or H atom. This is an example of the scaling laws between adsorption energies of related adsorbates that we will discuss in Section 7.2.7.

The electronic structures of CH$_3$ adsorbed atop and threefold (non-agostic and agostic) are shown in Figure 5i.3a–c. The main bonding interaction of CH$_3$ is through its lone pair 2p$_z$ orbital, which can be seen to be slightly hybridized with the C 2s atomic orbital. This generates the out of plane deformation of the hydrogen atoms, compared to the planar structure of the isolated CH$_3$ radical. At the Fermi level, for atop-adsorbed CH$_3$ the interaction with the C 2p$_z$ atomic orbital is essentially non-bonding. In addition to this interaction, there is also the interaction with the C $2p_x$ and $2p_y$ atomic orbitals, which participate in the bonding and antibonding CH orbitals. CH_3 decreases in energy when adsorbed atop to Tc, as do H and C due to the lower electron affinity of Tc.

CH_3 adsorbed at the threefold non-agostic site (Figure 5i.3b) exhibits two important chemical bonding differences from atop-adsorbed CH_3.

First, the interaction with the lone pair CH_3 $2p_z$ orbital is greater, which is reflected in its downward shift with respect to the bonding CH peak. Second, compared to the case of atop-adsorbed CH_3 the corresponding antibonding electron density shifts downward with respect to the Fermi level, which destabilizes threefold-adsorbed CH_3. The C—H bond interacts with the Rh atoms mainly in their outward-oriented d_{xz} orbitals. On the C atom the π-symmetric C $2p_x$ and C $2p_y$ atomic orbitals that participate in the C—H bond interact with the transition metal atomic orbitals). As indicated by the COHP plots, the corresponding antibonding features are present below the Fermi level. This indicates Pauli repulsion between the bonding CH orbital and the metal atom. As we will see, this changes for the agostic interaction.

For agostic CH_3, there is a stronger interaction with the d_{z^2} atomic orbitals in Rh metal. This pushes the antibonding electron density between the C $2p_y$ and C $2p_x$ atomic orbitals and the metal atomic orbitals above the Fermi level, and causes a reduction of the Pauli repulsion. Electrons from between the C—H bond and the metal atoms are donated to the transition metal. The altered C—H bond distances show that these bonds are weakened, which gives an overall increase in M—C bond interaction that can be deduced from the ICOHP values in Table 5i.1.

The preference of CH_3 to adsorb atop of Rh is due to the position of the Fermi level in the non-bonding electron density regime. For threefold CH_3, the Fermi level is located in the antibonding electron density regime. This agrees with the schematic scheme shown in Figure 2i.4, which illustrates the higher location of antibonding electron density for the atop-adsorbed case compared to the threefold case with respect to Fermi level.

In the atop site, the CH_3 structure is closely related to the structure of gas phase NH_3. While in CH_3 the C atom $2p_z$ as well as $2p_x$ and $2p_y$ atomic orbitals interact with the transition metal surface atoms, this differs for ammonia which interacts mainly with the surface atom through its lone pair HOMO orbital that is dominated by the $2p_z$ atomic orbital electron density. The occupied NH orbitals of NH_3 are deeper in energy than those of CH_3, which causes them to interact more weakly with the transition metal valence orbitals. The Fermi level is located deep in the antibonding electron density regime of the interacting ammonia lone pair orbital (see Figure 5i.6).

While the ammonia molecule binds strongly atop, it binds very weakly in the threefold site. Its adsorption strength does not change when the three hydrogen atoms atop rotate, unlike threefold-adsorbed CH_3.

Threefold-adsorbed ammonia adsorbs only weakly, and its electronic structure differs from the expected structure based on threefold-adsorbed CH_3 (see Figure 5i.6). The interaction with the lone pair orbitals is shifted up in energy for threefold adsorption instead of down in energy as for atop-adsorbed NH_3. The main electronic interaction is with the Rh s valence atomic orbitals. The antibonding orbital density that corresponds to the interaction with the N $2p_z$ atomic orbitals is nearly completely occupied.

This difference in electronic structure is due to large changes in the physical structure of threefold-adsorbed ammonia versus atop-adsorbed ammonia. For threefold adsorption, the ammonia molecule (see Figure 5i.5) inverts its three hydrogen atom umbrella. The Rh—N bond distance for atop-adsorbed ammonia is 2.4 Å, but for threefold-adsorbed ammonia it is 3.5 Å. A large repulsive interaction would evolve for ammonia adsorbed threefold in the same configuration as atop-adsorbed ammonia, due to the occupation of antibonding M—N electron density. This is replaced by the weak attractive interaction of the nitrogen atom $2p_z$ atomic orbital with the Rh atom.

Insert 5: Electronic Structure of Adsorbed CH_3 and NH_3

Figure 5i.1 shows the electronic structure of ammonia. The electronic structure of CH_3 is related to that of ammonia, but in the gas phase its structure is planar and it is a radical. Figure 5i.2 gives the different adsorption modes of CH_3 or NH_3 to the (111) surface of an fcc transition metal. Figure 5i.3 gives the electronic structure of adsorbed CH_3 and Figure 5i.4 shows the trends in adsorption energies of CH_3 and NH_3 for the group VIII transition metals of the fourth row of the periodic table. Figure 5i.5 compares the calculated structures of ammonia atop and threefold-adsorbed and Figure 5i.6 gives the electronic structure of ammonia in these two adsorption configurations. Table 5i.1 compares calculated ICOHP values of adsorbed CH_3 and NH_3 and Table 5i.1 gives the Born–Haber analysis data of CH_3 adsorbed to Rh and Tc.

Table 5i.1 ICOHP (eV) values of atom–atom bonds of adsorbed CH_3 (a) and NH_3 (b).

(a)

	C—Rh	C—H	H—Rh
CH_3 (atop)	−1.65 ($d = 2.08$ Å)	−3.13 ($d = 1.09$ Å)	0.00
CH_3 (threefold non-agostic)	−0.50 ($d = 2.38$ Å)	−3.44 ($d = 1.10$ Å)	0.01
CH_3 (threefold agostic)	−0.65 ($d = 2.25$ Å)	−3.21 ($d = 1.12$ Å)	0.08

(b)

	N—Rh	N—H	H—Rh
NH_3 (atop)	−0.87 ($d = 2.14$ Å)	−4.24 ($d = 1.02$ Å)	−0.00
NH_3 (threefold)	−0.17 ($d = 3.51$ Å)	−3.72 ($d = 1.02$ Å)	−0.00

Bond distances are indicated as well.

(Continued)

Insert 5: (Continued)

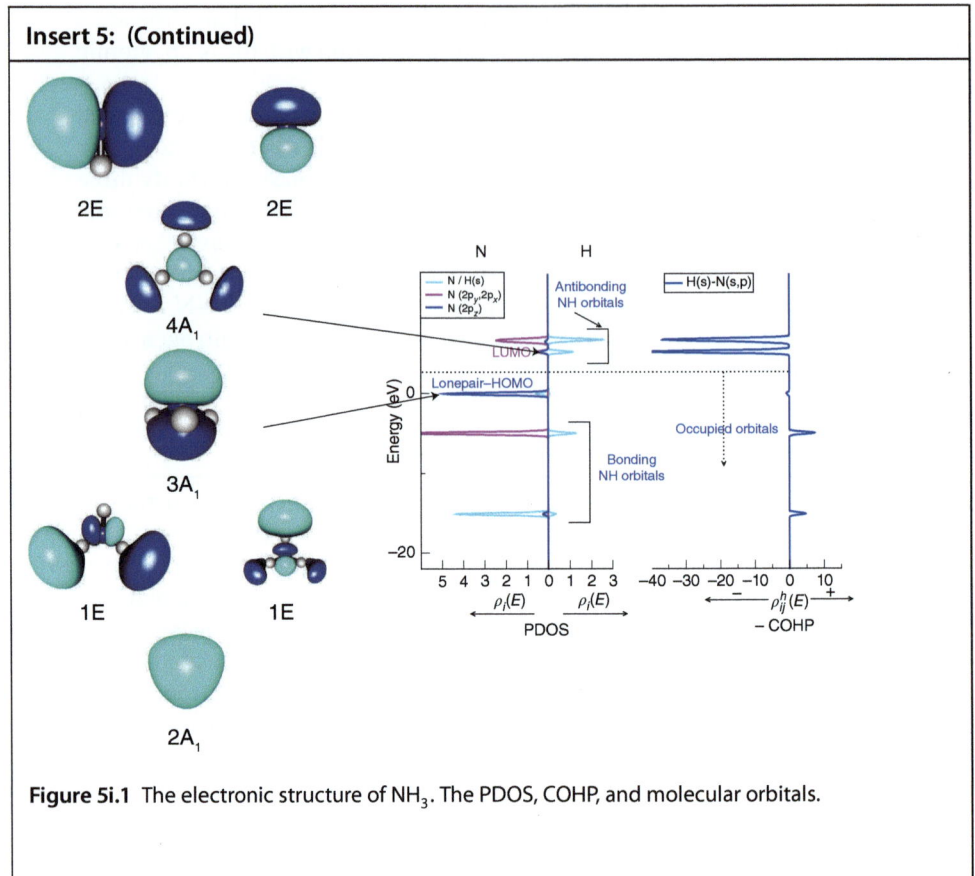

Figure 5i.1 The electronic structure of NH_3. The PDOS, COHP, and molecular orbitals.

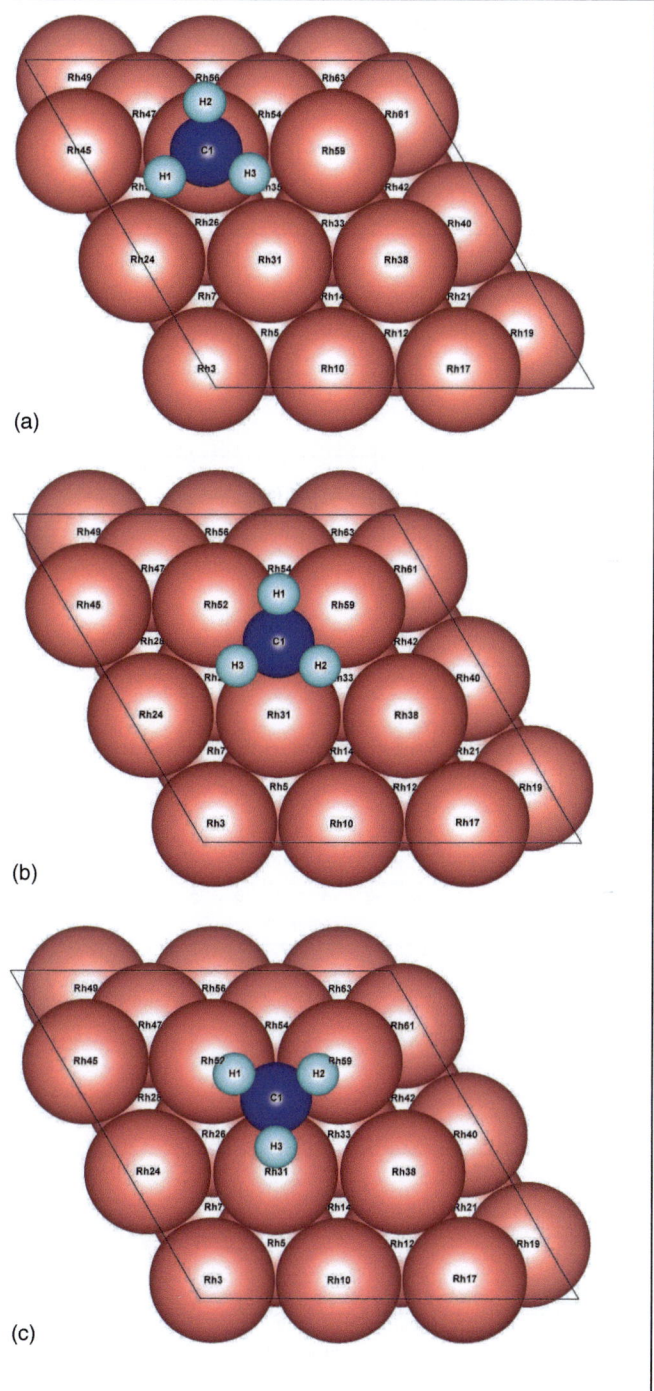

Figure 5i.2 Adsorption modes of CH_3 adsorbed to the (111) surface of an fcc transition metal. (a) Atop-adsorbed CH_3. (b) Threefold-adsorbed CH_3, non-agnostic; (c) agnostic threefold-adsorbed CH_3.

(Continued)

240 | *7 Chemical Bonding and Reactivity of Transition Metal Surfaces*

Insert 5: (Continued)

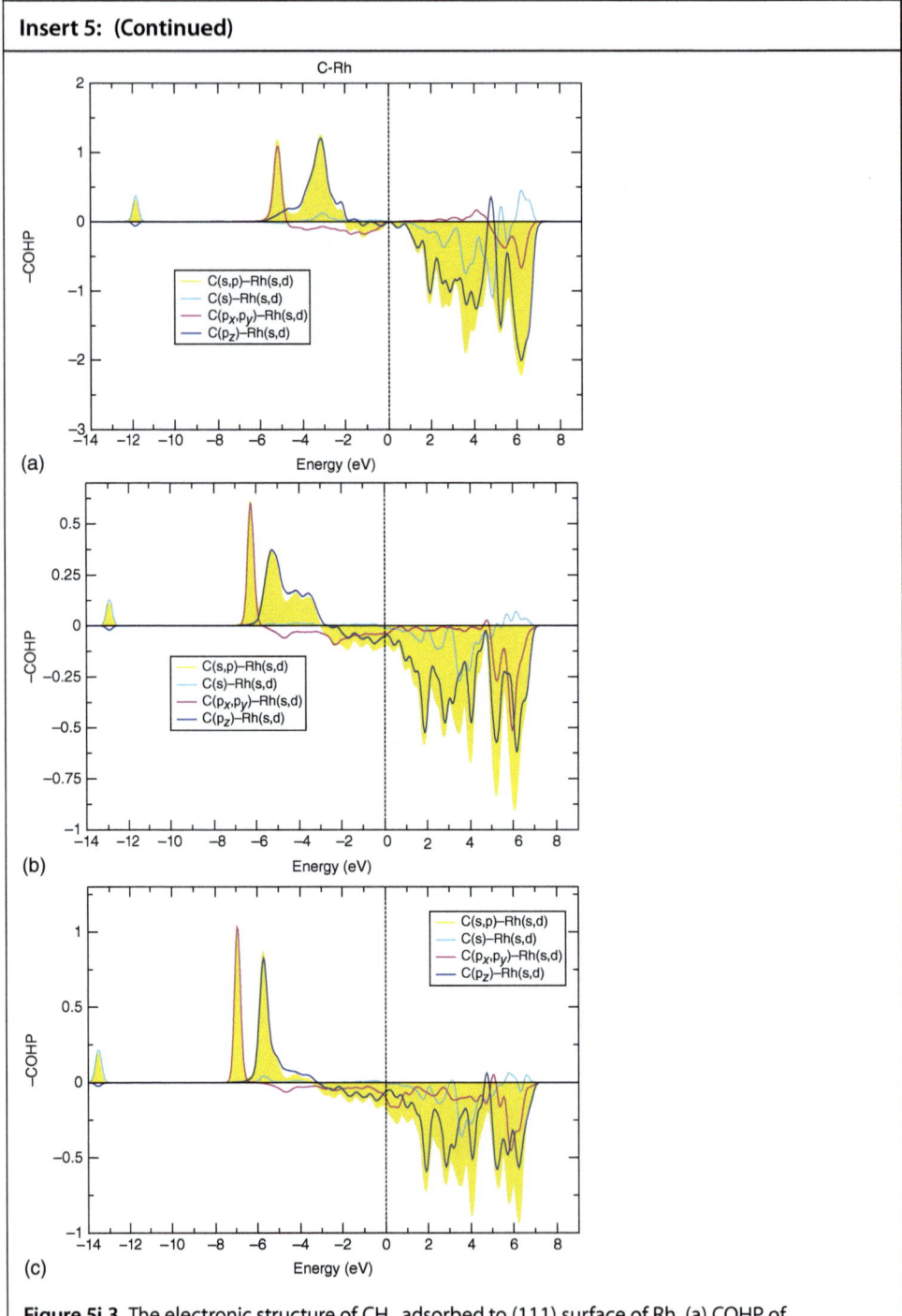

Figure 5i.3 The electronic structure of CH_3 adsorbed to (111) surface of Rh. (a) COHP of atop-adsorbed CH_3. (b) COHP of CH_3 on non-agostic threefold coordination site. (c) COHP of CH_3 adsorbed in agostic conformation.

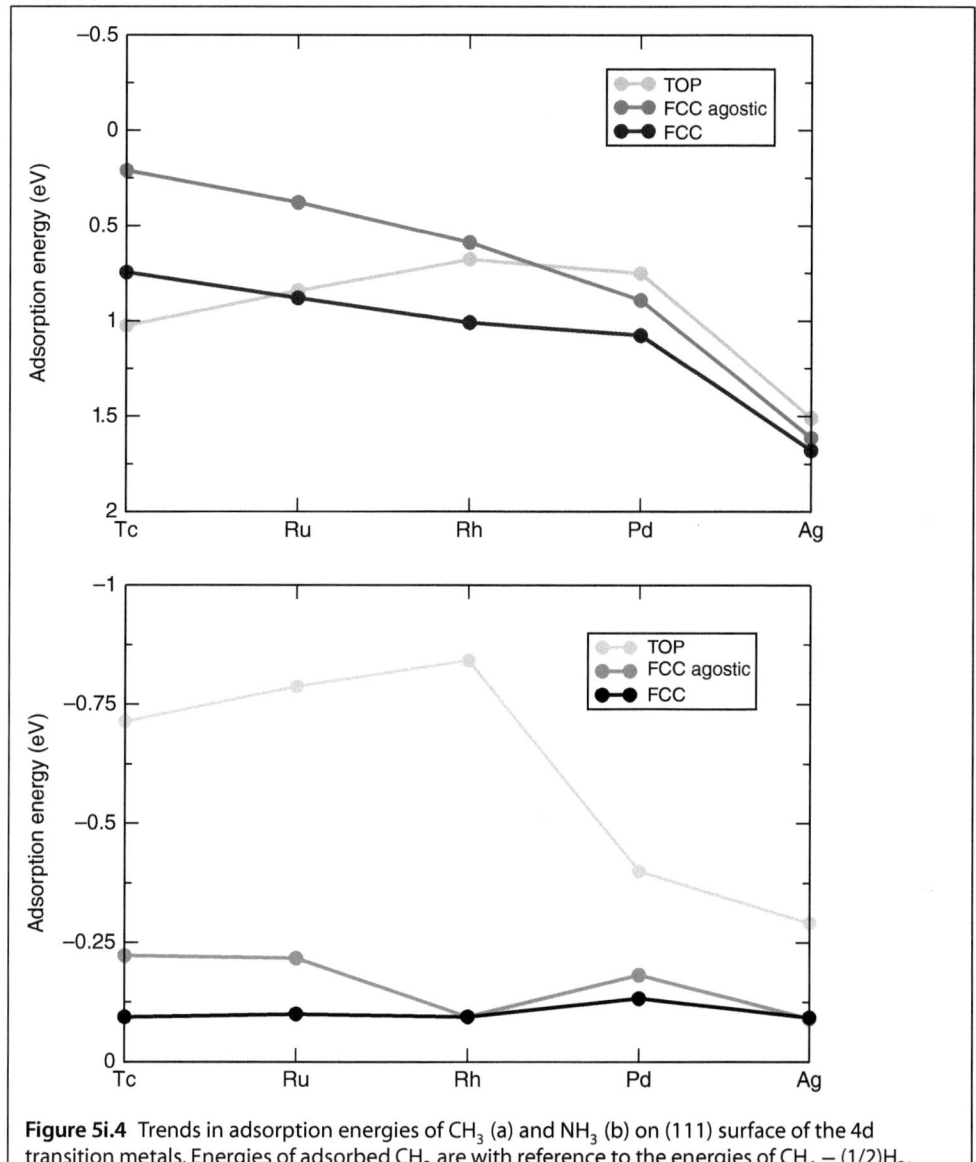

Figure 5i.4 Trends in adsorption energies of CH_3 (a) and NH_3 (b) on (111) surface of the 4d transition metals. Energies of adsorbed CH_3 are with reference to the energies of $CH_4 - (1/2)H_2$. Note that when ammonia is initially agostically adsorbed it will convert to the atop position.

(Continued)

Insert 5: (Continued)

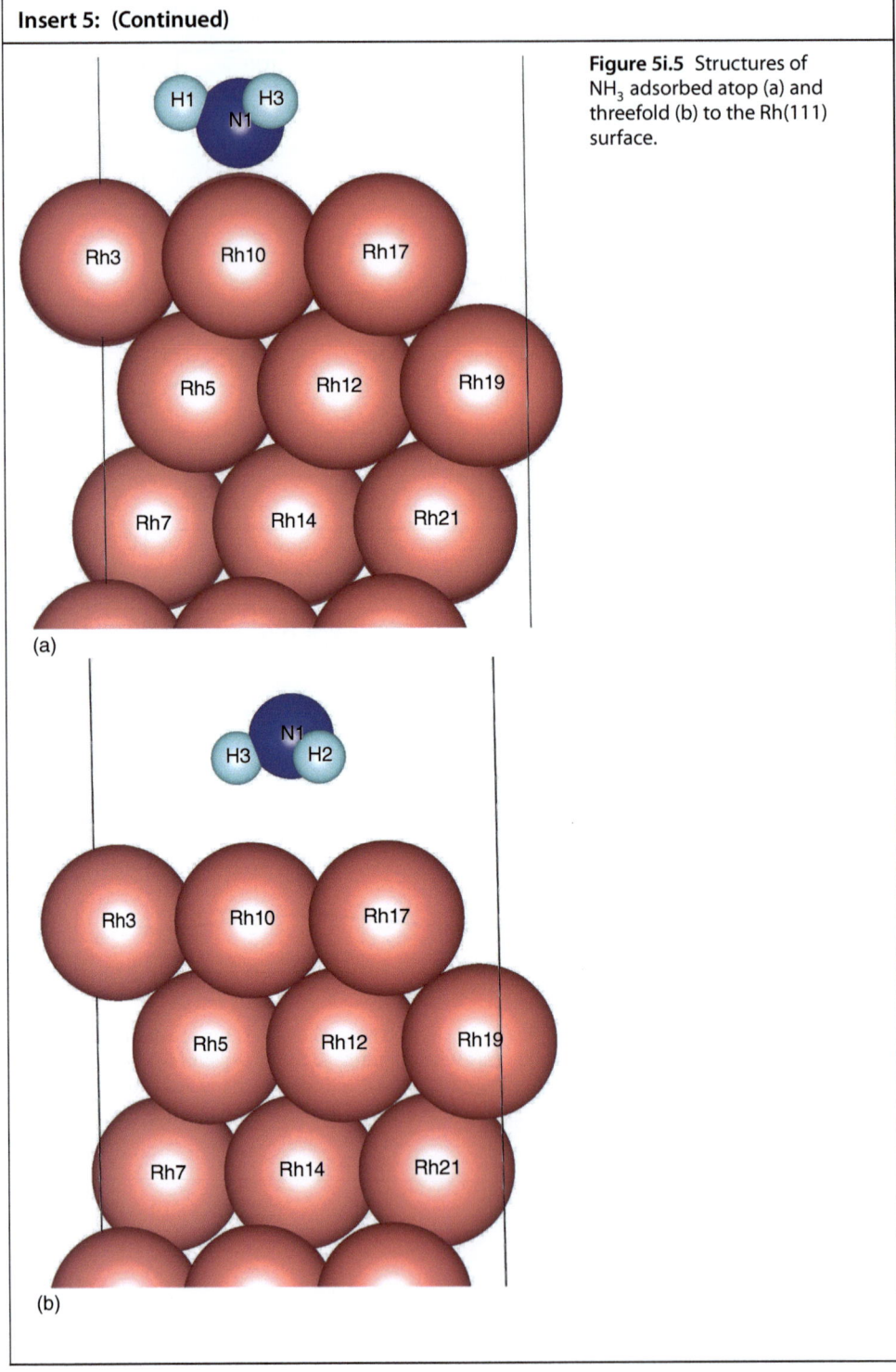

Figure 5i.5 Structures of NH_3 adsorbed atop (a) and threefold (b) to the Rh(111) surface.

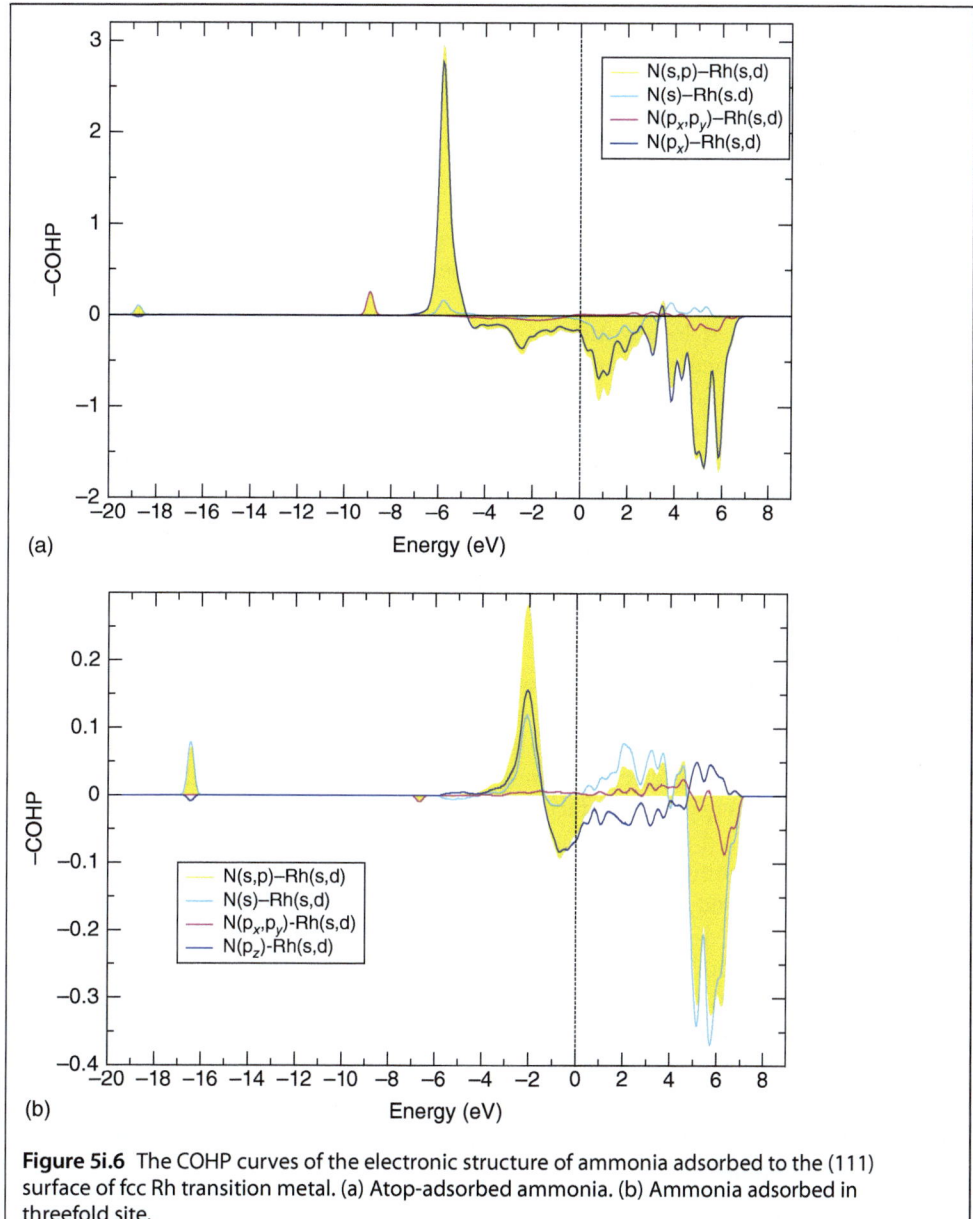

Figure 5i.6 The COHP curves of the electronic structure of ammonia adsorbed to the (111) surface of fcc Rh transition metal. (a) Atop-adsorbed ammonia. (b) Ammonia adsorbed in threefold site.

7.2.4 Adsorption as a Function of Coordinative Unsaturation of Surface Atoms: Relation with d Valence Band Energy Shift

The Born–Haber cycle analysis (Section 7.2.2.2) indicates that surface metal–metal atom bonds weaken when adsorbates interact. The energy cost is proportional to the interaction energy between metal atoms in contact with the adsorbate and their surrounding metals atoms and, therefore, relates to the number of bonds between the coordinating surface atom and neighboring surface metal atoms. As coordinative unsaturation of the surface metal atoms increases, the number of metal atom–metal atom bonds that weaken decreases, and adsorption strength increases.

This can also be understood based on differences in the electronic structure: when metal surface atom coordination number decreases its electron density becomes less delocalized, resulting in decreased electron localization energy cost. Table 3.1 shows the resulting increase of adatom bond energy when the transition metal surface atom coordination number decreases for C adsorbed to different Ru surfaces.

This increased localization of valence electrons can be deduced from the decreased valence electron bandwidth in the PDOS of the more coordinatively unsaturated surface atom. In Figure 6i.1, the PDOS on a Ru(1000) surface with nine neighbors is compared with the PDOS of a Ru atom of the Ru(1120) surface with seven neighbors. Table 6i.1 compares the d valence electron bandwidths of the Ru atoms with varying coordination.

In Insert 1 in this chapter, we discussed the consequences to the electronic structure of a metal atom when its coordination number decreases. The d valence electron bandwidth will decrease and there is an increase in average d valence electron band position. Just as for the group VIII transition metals, when the d valence electron occupation is greater than half of the valence electron band capacity, band narrowing increases the repulsive electron–electron interactions. Therefore, the average d valence electron energy increases slightly. This shift can be considered as a measure of the electron delocalization, because of its relation to the d valence electron bandwidth. Figure 6i.3 illustrates that the change in adsorption energy with surface atom coordination can indeed be related to this average shift in d valence electron band position.

Figure 6i.2 compares the differences in d valence electron densities when a C atom is adsorbed atop a Ru surface atom. The energy difference Δ between the average of the M—C bonding and antibonding energies for the metal atom that has nine neighbors (such as Ru(0001)) is 5.2 eV, while it is 6 eV for an Ru atom with seven neighbors (such as Ru(1120)). The larger Δ is, the stronger the interaction between adatom and metal surface atom. The weaker interaction on the Ru(0001) surface is reflected in the smaller value of Δ.

Insert 6: Electronic Structure of Chemisorption as a Function of Surface Atom Coordination Number

Figure 6i.1 shows the decrease in d valence bandwidth in the PDOS of a surface atom at the Ru surface when the surface coordination number decreases. Figure 6i.2 shows the increase in Δ, the energy difference between adatom bonding and antibonding electron density when the adatom adsorbs to a surface atom with lower coordination. Figure 6i.3 illustrates the relation between increased adsorption energy and the average position of the d valence electron band energy. In Table 6i.1, this is also compared with the differences in d bandwidth. Figure 6i.4 provides a comparison between experimentally measured PDOS and computed PDOS of a N atom adsorbed to a Ni and Cu surface.

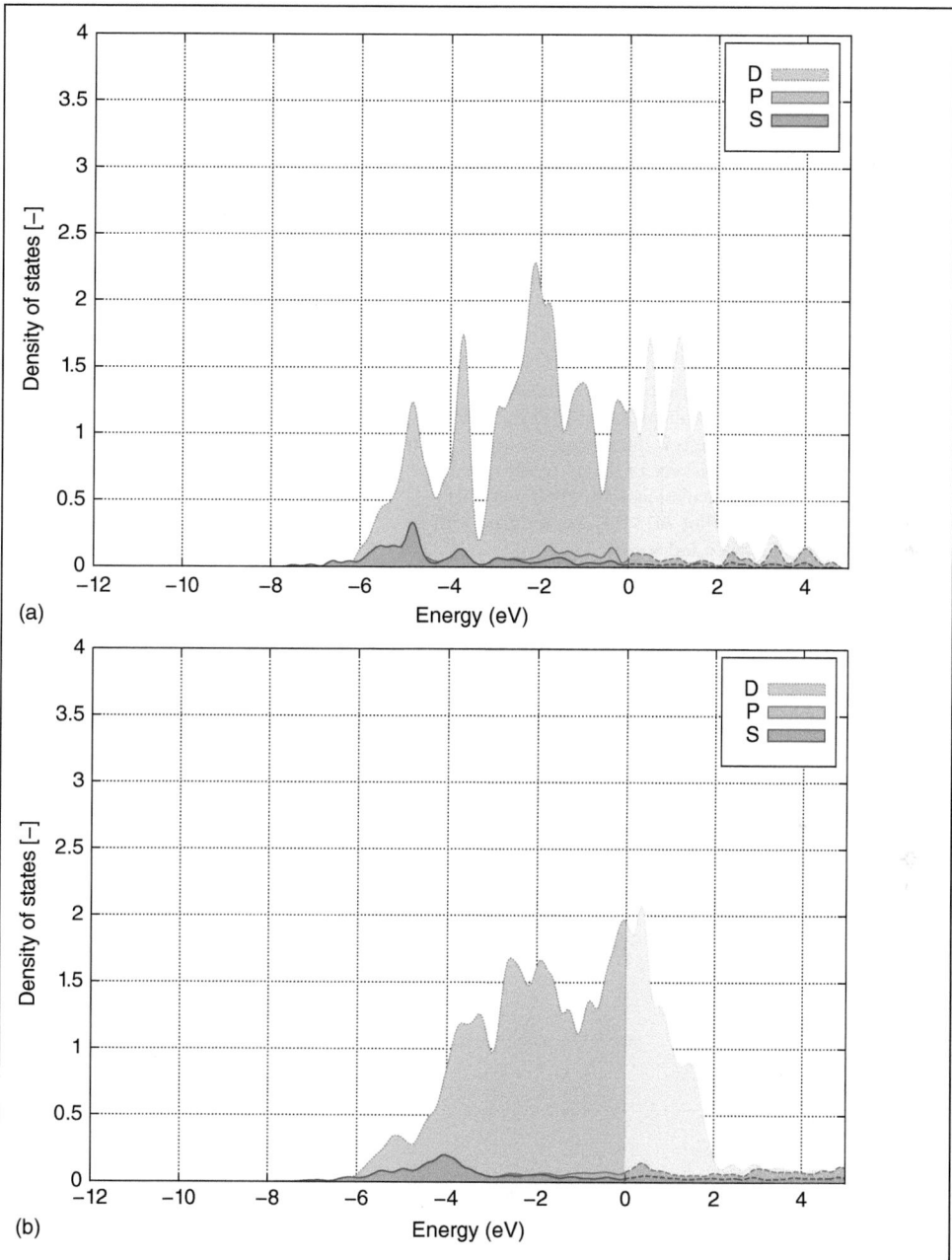

Figure 6i.1 (a) PDOS of the valence electrons of the Ru(0001) surface atom, dark areas denote electron occupied density. Light areas indicate non-electron occupied valence bands.
(van Santen and Filot 2013 [14]. Reproduced with permission of John Wiley and Sons.)
(b) PDOS of Ru atom of Ru(1120) surface, dark areas indicate electron occupied valence bands, light areas indicate non-electron occupied valence bands.

(Continued)

Insert 6: (Continued)

Table 6i.1 Comparison of average d valence electron energies ($\bar{\epsilon}_d$), d valence electron bandwidth (W), and adsorption energies of C(E^c_{ads}). $\bar{\epsilon}_d$ and W are compared for the bulk Ru atom, an Ru atom of (0001) surface and (1120) surface, respectively.

N_s	$\bar{\epsilon}_d$ (eV)	W (eV)	E^C_{ads} (kJ mol^{-1})
12	−2.94 (−1.83)	4.58	
9	−2.55 (−1.62)	4.06	−657
7	−2.19 (−1.39)	3.60	−705

C is adsorbed atop to the Ru(0001) and Ru(1120) surfaces, respectively. $\bar{\epsilon}_d$ with brackets is the average energy for the full valence bond, without brackets for the occupied valence electrons. DFT/VASP calculations with PBE functional.
Source: van Santen and Filot 2013 [14]. Reproduced with permission of John Wiley and Sons.

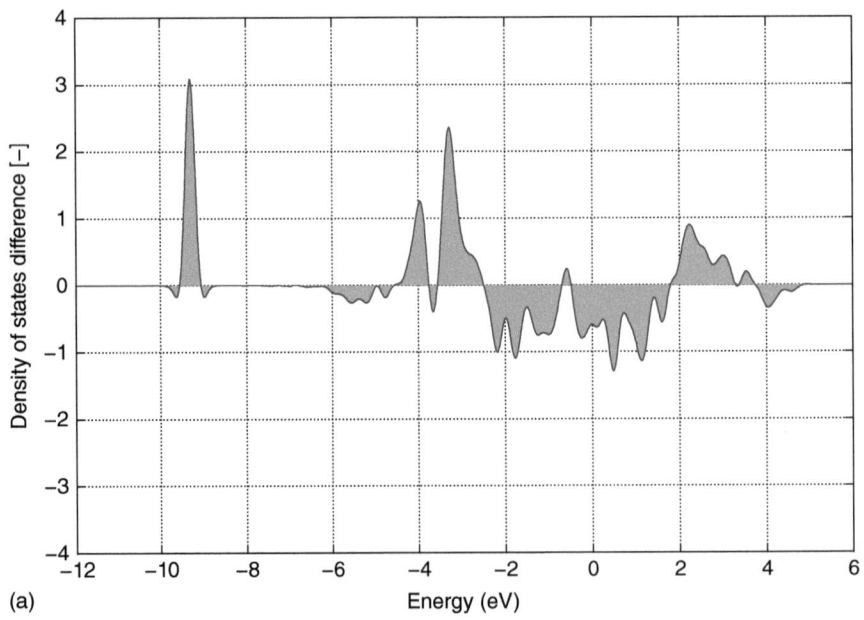

(a)

Figure 6i.2 (a) The difference in partial densities of states (PDOS) of a Ru atom on the Ru(0001) surface in the presence and absence of an atop-adsorbed C atom according to DFT (VASP) computations with PBE functional. (van Santen and Filot 2013 [14]. Reproduced with permission of John Wiley and Sons.) (b) PDOS difference of an Ru atom on Ru(1120) surface in the presence and absence of an atop-adsorbed C atom.

7.2 The Nature of the Surface Chemical Bond | 247

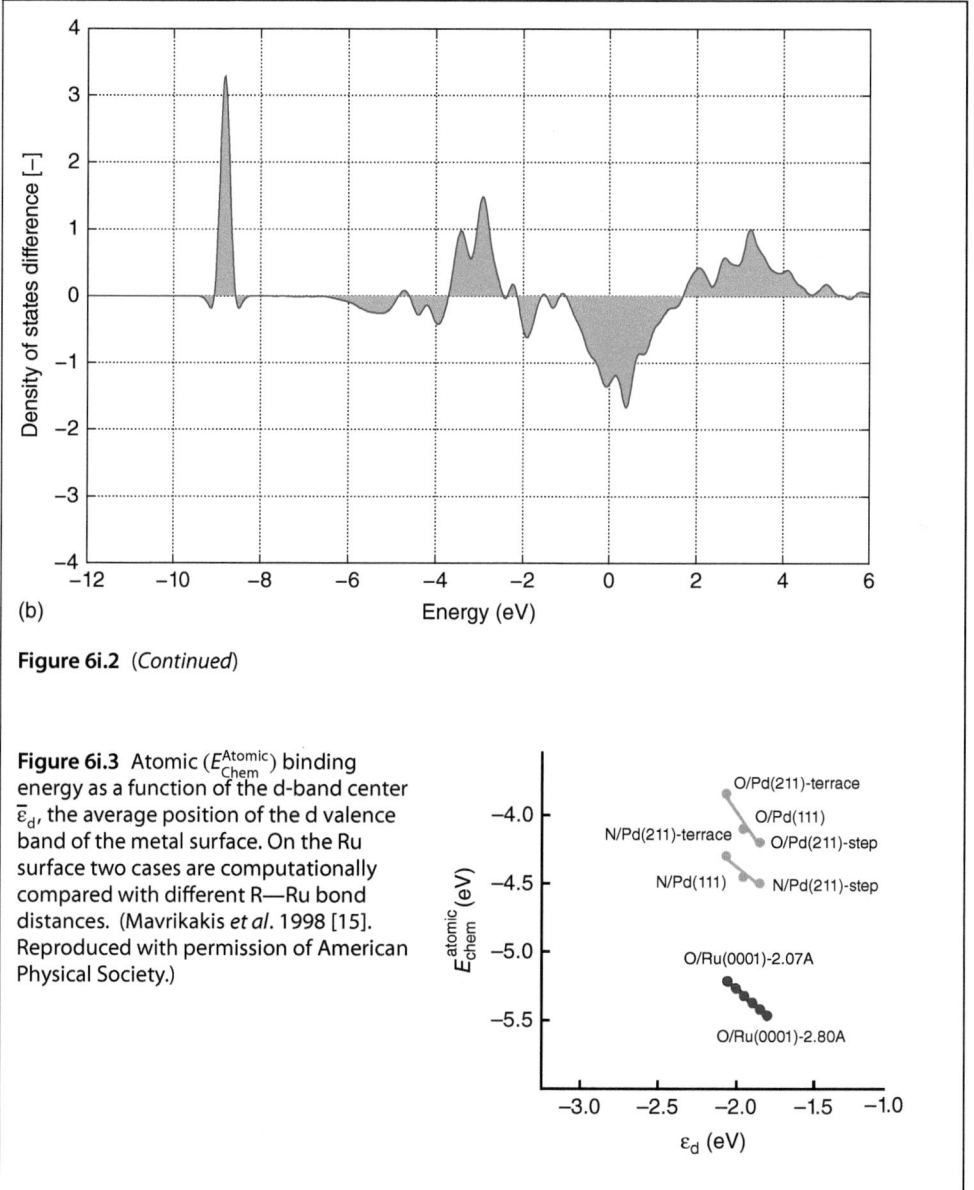

(b)

Figure 6i.2 (Continued)

Figure 6i.3 Atomic (E_{Chem}^{Atomic}) binding energy as a function of the d-band center $\bar{\varepsilon}_d$, the average position of the d valence band of the metal surface. On the Ru surface two cases are computationally compared with different R—Ru bond distances. (Mavrikakis et al. 1998 [15]. Reproduced with permission of American Physical Society.)

(Continued)

Insert 6: (Continued)

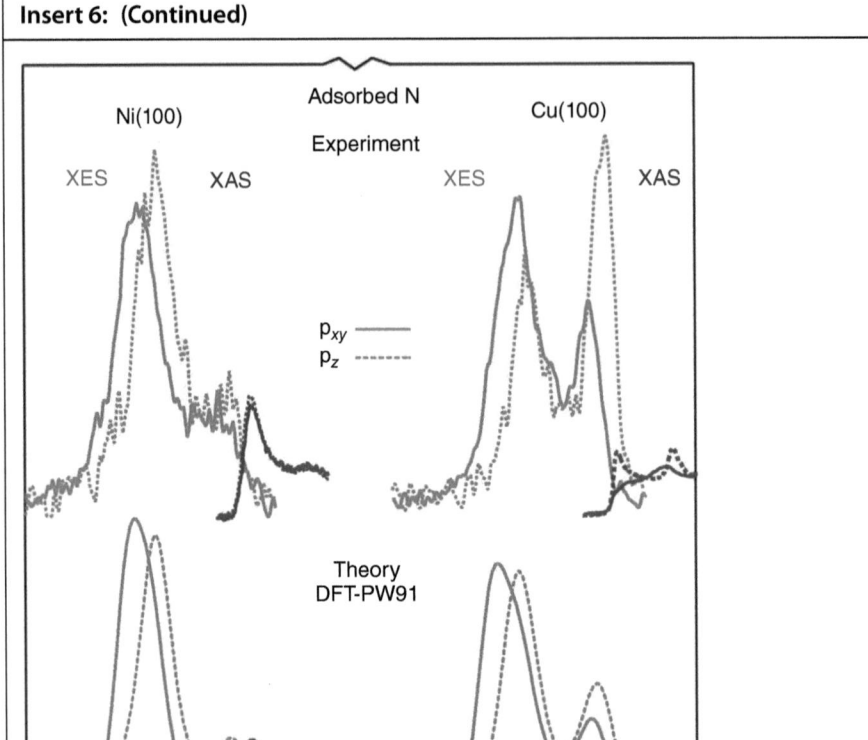

Figure 6i.4 Experimental X-ray emission (XES) and X-ray adsorption (XAS) spectra for N adsorbed fourfold to a Ni(100) and Cu(100) surface respectively. The peak maxima can be interpreted as bonding and antibonding states originating from the adsorbate $2p_x$, $2p_y$, and $2p_z$ states are clearly seen. (Nilsson et al. 2005 [16]. Reproduced with permission of Springer.)

Experimental confirmation of the presence of electron-occupied bonding and non-occupied antibonding adatom-surface electron densities is provided by X-ray emission and adsorption data. The PDOS of the nitrogen atom contribution to the valence of electron band structure is measured, as shown in Figure 6i.4. For an adsorbed N atom, a comparison is made of the experimentally measured PDOS on the N atom, below and above the Fermi level. The emission data refer to the energy density below the Fermi level, the adsorption data refer to the energy density above the Fermi level. The respective bonding and antibonding nature of the electron densities derive from the comparison with calculations.

7.2.5 Chemisorption of CO: Donative and Backdonative Interactions

In order to create a strong chemical bond between the adsorbate and surface, the adsorbing molecule must be able to donate or accept electrons. Thus, the HOMO and LUMO orbitals of the molecule must be at energies close to the Fermi level energy of the surface.

Here, we will discuss the surface chemical bond of the CO molecule, which is representative of a molecule that interacts through both strong σ and π bond interactions. There are several different but related ways to describe the nature of this bond and, therefore, several different representations of the CO surface bond are shown in Insert 7.

Figure 7i.1 shows the electronic structure of the CO molecule. The HOMO is an antibonding 5σ lone pair orbital, mainly localized on the carbon atom. The 4σ orbital, with a lower energy, is the counterpart to the lone pair orbital localized on the oxygen atom. The 3σ orbital is the occupied orbital responsible for the strong interaction between C and O, and features a combination of 2s and $2p_z$ atomic orbitals of low energy.

The two 1π molecular orbitals, built from the $2p_x$ and $2p_y$ atomic orbitals, respectively, are slightly deeper in energy than the HOMO 5σ orbital, with their energy located slightly above that of the 4σ orbital. The corresponding unoccupied antibonding orbitals are the $2\pi^*$ LUMO orbitals.

An elementary view of the interaction of CO with the transition metal surface is the donation–backdonation model of chemical bonding (Figure 7i.2), or the so-called Chatt–Dewar model [17, 18]. According to this model, the surface chemical bond of CO consists of electron donation from its 5σ HOMO orbital into the empty d valence electron orbitals of the transition metal surface and a backdonation of electrons from the transition metal into the empty LUMO $2\pi^*$ orbitals of CO.

On most of the transition metals, the CO molecule binds perpendicularly through its carbon atom to the surface because the energy of the lone pair 5σ orbital on the carbon atom is closer to the Fermi level than the energy of the oxygen 4σ orbital. Therefore, it can more readily interact with the transition metal surface.

Orientation parallel to the surface will have the advantage of greater overlap with π type orbitals. It is usually less stable than the perpendicular orientation, because the repulsive interaction between doubly occupied 1π electrons and doubly occupied d valence electrons (Pauli repulsion) counteracts the interaction with the unoccupied $2\pi^*$ CO orbitals.

The parallel adsorption mode of CO is favored by reactive metals such as Fe that have a low work function and can easily transfer electrons to the LUMO of a molecule. For other molecules such as O_2, their higher electron affinity also makes backdonation more favorable, and side-on adsorption is more common.

Insert 7: The Electronic Structure of Chemisorbed CO

Figure 7i.1 shows the electronic structure of the free CO molecule. The elementary donation–backdonation chemisorption model is shown in Figure 7i.2. Figure 7i.3 displays the calculated trends in adsorption energies of CO. Figures 7i.4–7i.9 present different views of the electronic structure of CO. Figure 7i.4 is a schematic illustration of the Blyholder model. Figure 7i.5 provides a quantitative translation of the Blyholder model in terms of DFT theoretical results. Figure 7i.6 presents the CO adsorption hybridization model. Figure 7i.7 shows the corresponding COHP plots of the CO chemisorptive bonds. Figure 7i.8 shows that excellent agreement is possible between experimental and DFT-calculated PDOS values for CO. Figure 7i.9 displays an orbital representation of CO adsorbed atop of a Ru surface. Figure 7i.10 shows the calculated CO frequency downward shifts of CO adsorbed in different coordination sites. In Table 7i.1, the calculated $2\pi^*$ occupation as a function of adsorption site is given. The ICOHP values of Table 7i.1 show the weaker M—C as well C—O interaction of CO adsorbed in threefold coordination to Rh compared to atop-adsorbed CO. Table 7i.2 gives calculated changes in $2\pi^*$ orbital occupation as a function of CO coordination number.

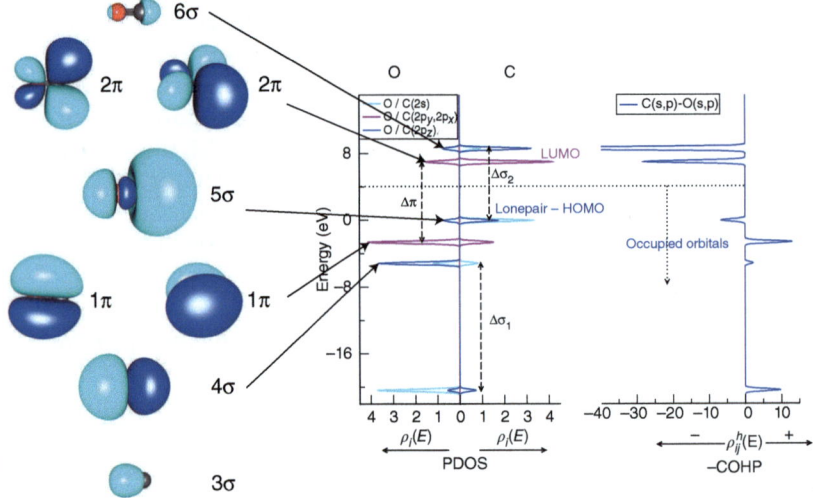

Figure 7i.1 The molecular orbital scheme of CO. Note the antibonding nature of the CO 5σ lone pair orbital.

7.2 The Nature of the Surface Chemical Bond

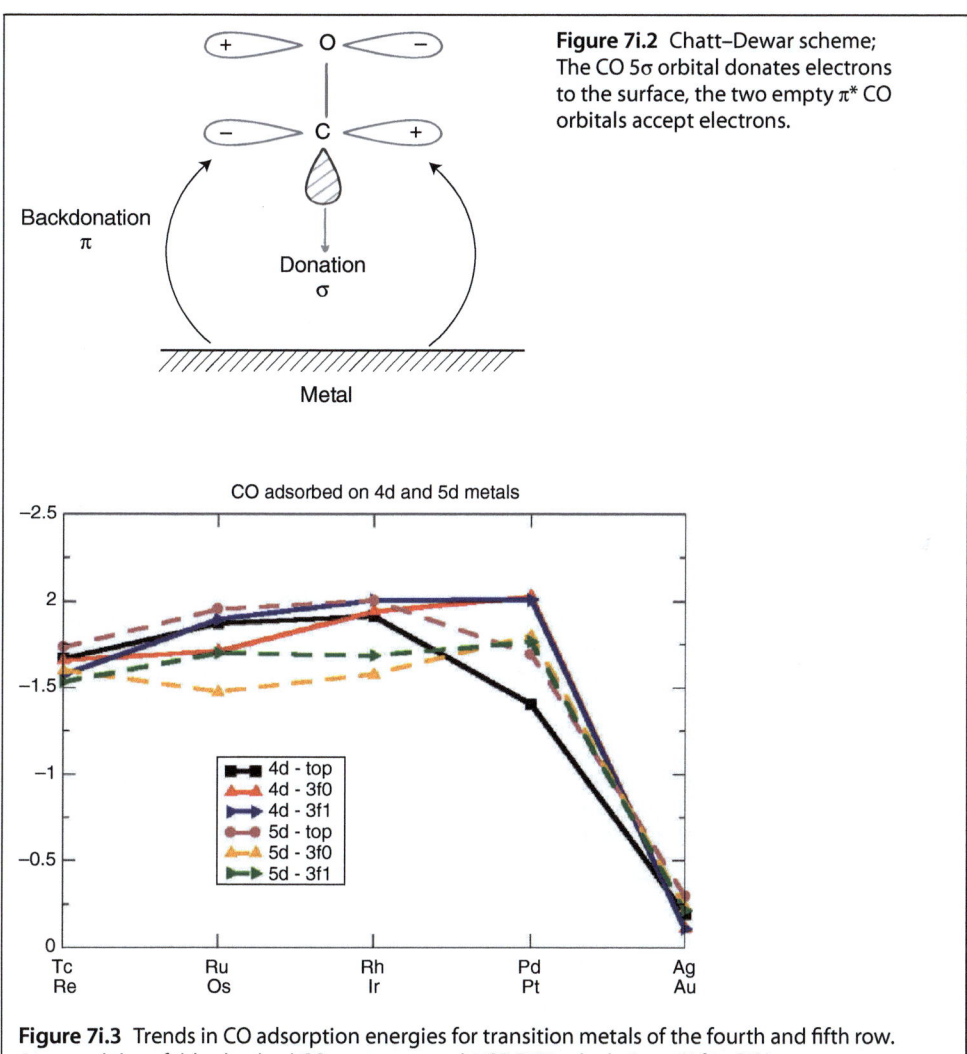

Figure 7i.2 Chatt–Dewar scheme; The CO 5σ orbital donates electrons to the surface, the two empty π* CO orbitals accept electrons.

Figure 7i.3 Trends in CO adsorption energies for transition metals of the fourth and fifth row. Atop and threefold-adsorbed CO are compared. VSP DFT calculations. (After [8].)

(Continued)

Insert 7: (Continued)

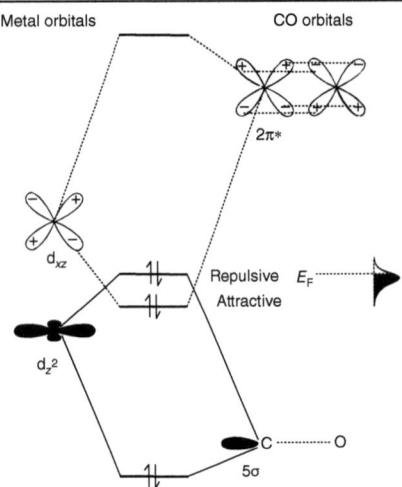

Figure 7i.4 Blyholder's schematic view of the chemisorption of CO to a transition metal surface [19]. Schematic for interaction of CO with Cu atom. In the antibonding regime of the σ symmetric interaction, the partial occupation of the antibonding σ symmetric electron density in the metal is indicated. The donating interaction consists of a complementary bonding and antibonding pair of surface orbitals with σ symmetry. The bonding orbitals are occupied by electrons, the antibonding electron density is partially occupied. The backdonating interaction consists of a complementary orbitals scheme of π symmetry. In this case, the antibonding electron density is unoccupied, but the electron occupation of the bonding electron density varies. (Blyholder 1964 [19]. Reproduced with permission of American Chemical Society.)(Blyholder 1964 [19]. Reproduced with permission of American Chemical Society.)

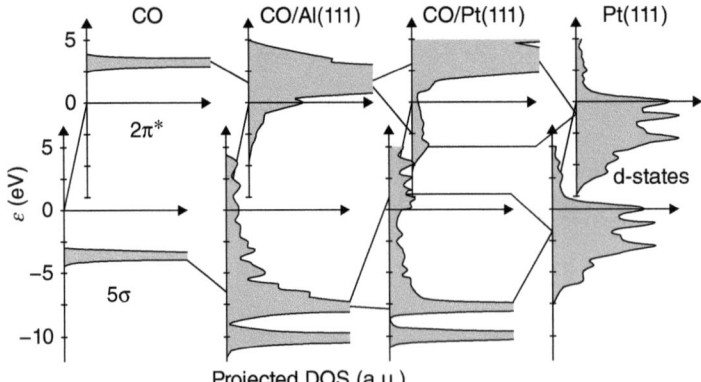

Figure 7i.5 The self-consistent electronic density of states (DOS) projected onto the 5σ and 2π orbitals of CO: in vacuum and over Al(111) and Pt(111) surfaces. Also shown is the DOS from the d bands in the Pt(111) surface. The sharp peaks in the states of CO in a vacuum are seen to broaden into resonances and shift down in energy over the simple metal surface (mixing with the 4σ state causes additional structure in the 5σ resonance). Over the transition metal surfaces the CO resonances further hybridize with the metal d states. This leads to shifts in the 5σ and 2π* levels. At the top of the d bands antibonding 5σ states and bonding 2π* states appear. These states have low weight in the 5σ and 2π* projections shown. (Hammer et al. 1996 [20]. Reproduced with permission of American Physical Society.)

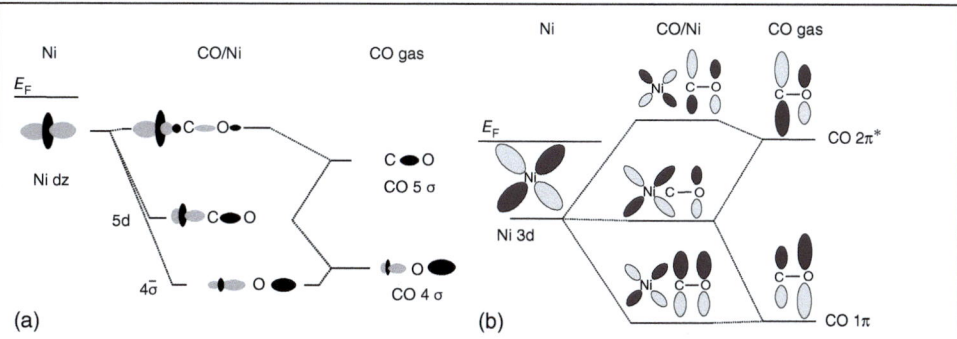

Figure 7i.6 Schematic representation of the changes in hybridization of CO adsorbed to a surface Ni atom. An allylic electron structure is formed in this model of a single CO molecule interacting with a Ni atom. (a) Schematic orbital diagram of the σ-interactions. (b) Schematic orbital diagram of the π-symmetric electronic structure. (Föhlisch *et al.* 2000 [21]. Reproduced with permission of American Institute of Physics.)

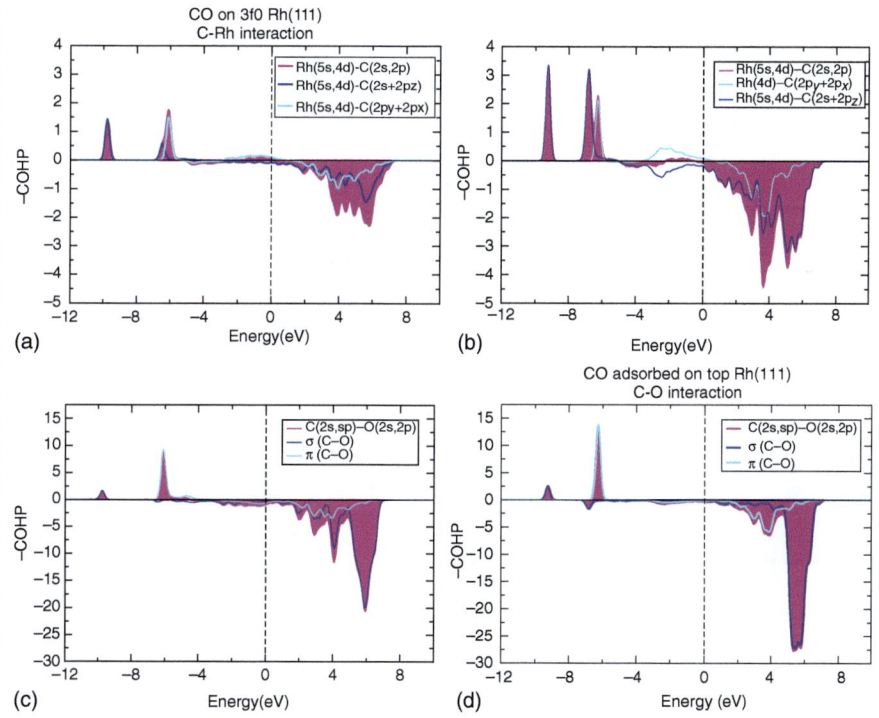

Figure 7i.7 The COHP of CO M—C bond atop- and threefold-adsorbed on the Rh(111) surface. (a,b) Also COHP values of the respective C-O are given (c,d) (a,c) COHP of atop-adsorbed CO, (b,d) COHP of threefold-adsorbed CO.

(Continued)

Insert 7: (Continued)

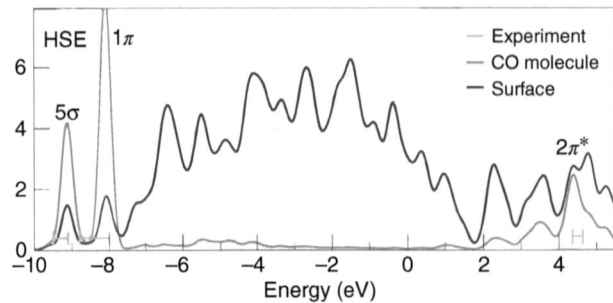

Figure 7i.8 DFT calculation compared with experimental measurements. PDOS of CO adsorbed to a Pt(111) surface according the DFT hybrid functional (HSE) calculation. Light gray is PDOS on the CO molecule, dark partial density of states on a transition metal, the bars mark the corresponding signals from photoemission data. (Schimka et al. 2010 [22]. Reproduced with permission of Nature Publishing Group.)

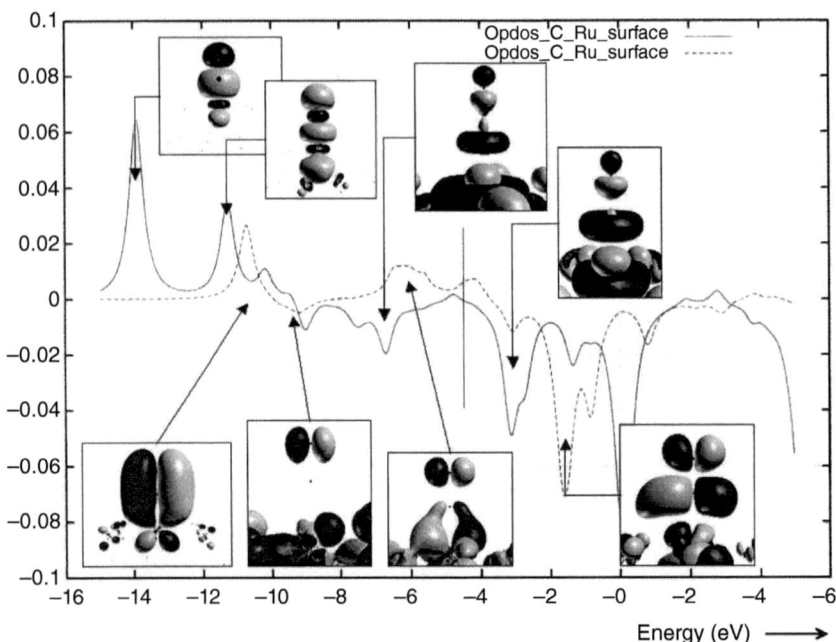

Figure 7i.9 Electronic structures of CO interacting with a Ru19 cluster. Overlap population densities of states are represented for the carbon atomic orbitals of CO interacting with Ru d valence electrons as a function of electron orbital energies. Orbital densities are shown for energies of maximum electron density. Solid lines, σ-Symmetric orbital interactions; dashed lines, π-symmetric orbital interactions. The Fermi level is at −4.8 eV indicated in red. Ru19-CO with CO adsorbed threefold. (van Santen and Filot 2013 [14]. Reproduced with permission of John Wiley and Sons.)

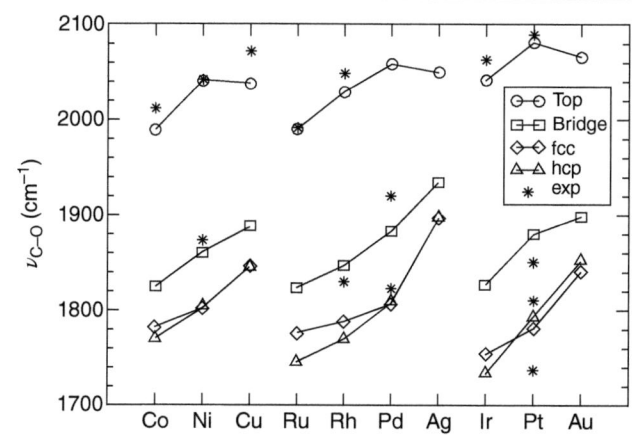

Figure 7i.10 Calculated C—O stretching frequencies (ν_{C-O}) on close-packed transition metal surfaces. The calculated value of ν_{C-O} for the free CO molecule is 2136 cm^{-1} compared to the experimental value of 2145 cm^{-1}. (Gajdo et al. 2004 [23]. Reproduced with permission of IOP Publishing.)

Table 7i.1 Comparison of ICOHP (eV) values for CO adsorbed Rh(111) For threefold-adsorbed CO the ICOHP has to be multiplied by three.

	C—O	Rh-C	Rh-O
Atop	−6.635	−3.00	−0.402
Threefold	−4.543	−1.307	−0.190
Free molecule	−7.34		

Table 7i.2 Fractional occupation of the $2\pi^*$ orbital for the CO molecule in top, bridge, fcc, and hcp sites on the close-packed transition metal surfaces, in %.

Occupation of $2\tilde{\pi}^*$

Site	Co	Ni	Cu	Ru	Rh	Pd	Ag	Ir	Pt	Au
Top	16.0	11.5	8.1	11.4	10.8	8.8	5.3	11.4	10.0	6.0
Bridge	19.5	14.5	11.1	12.5	11.4	10.5	7.0	11.1	10.1	7.4
fcc	29.9	22.1	16.3	18.5	18.5	18.1	11.7	20.5	19.8	15.0
hcp	29.7	21.6	16.1	19.9	19.5	18.2	11.5	21.4	19.7	14.4

Source: Gajdo et al. 2004 [23]. Reproduced with permission of IOP Publishing.

Trends in chemisorption energies of CO adsorbed atop and threefold to the (111) surfaces of the group VIII transition metals are shown in Figure 7i.3. There are two important observations. First, the differences in energy between atop- and threefold-adsorbed CO tend to be small. For several 5d transition metals atop adsorption is even preferred. Second, the trends in adsorption energy approximately follow the trends of atop-adsorbed C.

The Blyholder model [24] schematically illustrated in Figure 7i.4 gives a representation of the CO adsorptive bond in terms of the formation of bonding and corresponding antibonding orbitals. Interaction with a single metal atom is plotted. The 5σ CO orbitals produce bonding and antibonding combinations with d valence electrons where the bonding orbital combination is fully occupied with electrons and in the metal the antibonding combination is only partially occupied. This reflects electron donation from the occupied 5σ CO orbitals to the surface orbitals. When d valence electron occupation increases, the overall contribution to the bond energy will decrease.

The $2\pi^*$ CO orbitals that are unoccupied in the free molecule also form bonding and antibonding orbital combinations. In this case, only the π-symmetric bonding electron density between metal and C atom is occupied. The contribution to the chemical bond strength will increase when the d valence electron count increases. In this process, the $2\pi^*$ CO orbitals become partially occupied.

A useful comparable view of the electronic structure of CO based on early DFT calculations is given by Hammer *et al.* [25] for the chemisorption of CO to Pt in Figure 7i.5. This shows the extension of the donation–backdonation chemisorption model and confirms the early Blyholder model based on semi-empirical calculations [24].

The occupied 5σ orbital forms a bonding orbital density of states at the bottom of the d valence electron band antibonding electron density that is located near the Fermi level. Its electron population will vary when the d valence electron occupation varies. When highly occupied, it will cause a Pauli repulsive interaction due to the σ-symmetric orbital interaction. This provides a force for the molecule to be adsorbed atop a surface atom.

The interaction with the $2\pi^*$ molecular orbitals also gives rise to bonding and antibonding electron densities. The bonding density is located near the Fermi level, and produces an attractive contribution. This attractive interaction will favor high coordination.

In addition to the interaction with 5σ and $2\pi^*$ orbitals, the interaction of surface atomic orbitals with the CO 4σ and 1π orbitals is also important. This interaction causes significant electronic structure rearrangement in the adsorbed molecule. The hybridization model that illustrates this change is shown in Figure 7i.6. The most important physical consequences of this rehybridization are the weakening of the C—O bond the increasing dipole moment of CO, which polarizes in the direction of the O atom due to the low electron desity on the C-atom of adsorbed CO.

Figure 7i.7 shows the COHP plots of CO adsorbed atop and threefold to the Rh(111) surface. The σ-symmetric interaction with the C 2s and $2p_z$ atomic orbitals and the π-symmetric interactions with the C $2p_x$ and $2p_y$ atomic orbitals are easily distinguished. The COHP for atop-adsorbed CO in Figure 7i.7a,b shows that its 5σ molecular orbitals shift to a lower energy than the 1π CO molecular orbital.

The COHP plot of the C—O bond for the σ-symmetric as well as the π-symmetric interactions shows the appearance of a small antibonding density in the d valence band (Figures 7i.7). This weakening of the C—O bond is dominated by the π-symmetric CO orbital interaction with the metal surface. For threefold-adsorbed CO we see a small increase of these antibonding densities.

The COHP plot of the σ-symmetric M—C interaction shows a bonding and antibonding interaction below the Fermi level. These antibonding interactions are stronger for threefold-coordinated CO. They indicate a strong repulsive interaction of the σ orbital interaction with CO, which directs the molecule to atop coordination.

The π orbitals of atop-adsorbed CO can only interact with the Rh d atomic orbitals. Below the Fermi level they create a bonding electron density. This corresponds with the bonding interaction due to the backdonative interaction of the Blyholder model. This bonding characteristic is relatively stronger for threefold-adsorbed CO, but it is not obvious because it is dispersed over a larger energy range.

In the threefold case, the C $2p_x$ and $2p_y$ atomic orbitals will interact not only with the d atomic orbitals, but also with the s valence electrons. For this reason, the shift downward of the 1π CO orbital becomes comparable to that of the 5σ CO orbital, and the 1π and 5σ orbitals nearly overlap. Since the π orbital interaction is bonding, it directs the CO molecule to a high coordination site.

The calculated ICOHP values given in Table 7i.1 show the consequences of the different distributions of electrons in the bonding and antibonding surface orbitals. The C—O interaction of adsorbed CO is weakened compared to the interaction of free CO. This bond weakening is greater for threefold-adsorbed CO than atop-adsorbed CO. On the other hand, the M—C bond between a surface metal atom and the adsorbed CO atom is slightly stronger for threefold-adsorbed CO.

The PDOS spectrum of CO adsorbed atop Pt (Figure 7i.8) shows electron distributions on the C and attached Rh atom. The theoretical prediction is compared with experimental results. The 1π and 5σ CO orbital contributions are readily distinguished, which represent bonding electron density distributions. The PDOS distributions are narrow because they are located outside the metal valence band energy interval. The 5σ orbital density, that is the HOMO on the CO molecules, is now shifted downward compared to the 1π orbital. It has the stronger interaction.

The unoccupied 2π* orbitals are broadened, indicating a strong interaction with the d valence electrons. The complete agreement between the experimental results and high-quality DFT calculations confirms the accuracy of the interpretation of the electronic structure of the CO chemisorptive bond.

A detailed interpretation of the electronic structure of the surface chemical bond between adsorbed CO and metal surface can be derived from the orbital projections shown for atop-adsorbed CO on a large Ru transition metal atom cluster model that simulates the Ru(0001) surface [26], as shown in Figure 7i.9. The σ-symmetric and π-symmetric interactions can be distinguished. σ-Symmetric bonding interactions with the electron occupied 4σ and 5σ CO orbitals can be seen at −14 and −12 eV. At −9, −7, and −2.5 eV the corresponding σ-symmetric antibonding electron density can be distinguished. Similarly to CO adsorbed to Rh, a substantial fraction of antibonding σ-symmetric electron density is occupied for Ru.

The 1π orbitals as well as $2\pi^*$ orbitals of CO form bonding and antibonding orbitals with the surface. This leads to rehybridization of the π electron density. In Figure 7i.9, one notes a bonding orbital π combination at -11 eV, a broad slightly bonding and antibonding electronic energy regime between -10 and -4 eV, and antibonding π-symmetric M—C interactions at higher energies. We note in this case the dominant occupation of bonding orbital fragments between M—C, which can be interpreted as the bonding electron backdonating interaction. This weakens the C—O bond since it effectively increases the occupation of the antibonding CO $2\pi^*$ orbitals.

Electron backdonation increases with increased coordination of CO and hence weakens the C—O bond energy. In Figure 7i.10, this weakening of the CO bond is shown as measured from the lowered vibrational frequencies of the adsorbed molecule. Table 7i.2 illustrates its correlation with the electron occupation of the $2\pi^*$ molecular orbitals.

This increased weakening of the CO bond by π-symmetric orbital interactions does not uniformly relate to the CO adsorption energy, because the σ type interactions also change.

But, as we discussed earlier in Chapter 4 and as we also discuss in Section 7.3.1 increasing electron backdonation into the unoccupied $2\pi^*$ orbitals lowers the activation energy of C—O bond dissociation.

7.2.6 Lateral Interactions

CO adsorbed at higher coverage to a metal surface will have substantially reduced adsorption energies on the order of 30 kJ mol^{-1}. The electron density radius around perpendicularly adsorbed CO is larger than the surface transition metal–metal atom distances. This causes a repulsive interaction of the CO molecules when adsorbed at close distance. An additional repulsive interaction results from the negative charge on CO. The maximum obtainable surface coverage of CO on a dense surface is about 60%.

Lateral interactions can be due to direct steric contact between adsorbates, but can also result from the altered electronic structure with the transition metal surface. When adsorbate atoms such as C or O share binding with a common surface metal atom, a repulsive interaction is produced that can be as large as 80 kJ mol^{-1}. Therefore, adsorbates favor adsorption on alternating sites (see Figure 7.3).

As a result of the interaction with adsorbate metal atom–metal atom bonds are weakened next to the adsorbates. This will increase the reactivity of the metal atoms and may even lead to attractive interactions (see Figure 7.3b).

Electrostatic interactions can also produce attractive interactions when their polarity is different. For instance, this is the case when CO and ammonia co-adsorb [27] since these molecules adsorb with differently directed dipole moments. When adsorbates are adsorbed on neighboring sites, these interactions are on the order of 10–30 kJ mol^{-1} and rapidly decline with distance between the adsorbates.

We have seen that lateral interactions between adsorbed species can be due to direct steric or electrostatic interactions and by indirect interaction mediated by changes in the electronic structure of the surface metal atoms. These interactions can be repulsive or attractive. These can be understood by

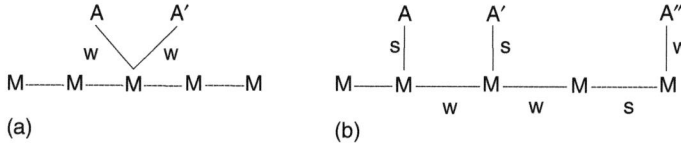

Figure 7.3 Schematic illustration of the use of bond order conservation to predict attractive or repulsive interactions between adsorbates. (a) Adsorbates which bond to the same surface metal atom are weakened by the presence of one another as the result of competition for electron density from the same metal atom. These interactions are repulsive. (b) Adsorbates that are bound to metal atoms which are neighbors have an effective attractive interaction, because of the weakening of the metal–metal bond due to their co-adsorption. Bond order conservation indicates that attractive and repulsive interactions alternate through bonds. Attractive and repulsive interactions may alternate. When the adsorbates are adsorbed with one metal atom in between, the overall interaction will be repulsive. This is indicated in (b) for the interaction between A′ and A″.

using a simple argument based on the bond order conservation (BOC) principle discussed in Section 3.3.2.1.1.

A strong repulsive interaction arises when the adsorbate shares bonding with the same metal atom (Figure 7.3a). According to BOC, the valence electrons of the surface metal atom have to be distributed over one extra metal adsorbate bond. This will weaken the metal–adsorbate bond in the shared metal atom case compared to the case when only one adsorbate interacts with the surface metal atom.

When two adsorbates do not bind to the same surface metal atom but instead bind to neighboring metal atoms, their interaction energies with the surface will be enhanced (the situation A, A′ shown in Figure 7.3b). BOC now predicts that the metal atom–metal atom bond between the two atoms with attached adsorbate weakens compared to atom bonds at the free surface. The surface metal atom that shares a bond with the adsorbate has to redistribute its valence over an additional bond and hence the metal atom–metal atom bond weakens.

When two adsorbates bond to a surface metal atom with one free surface metal between them, the same argument predicts that the interaction between the adsorbates will now be repulsive (the situation A′, A″ shown in Figure 7.3b).

7.2.7 Scaling Laws

Scaling laws are rules that relate the adsorption energies of adsorbed molecules fragments to the energies of the respective surface adatoms that attach them to the surface. They are useful for predicting relative energies without conducting a quantum-chemical calculation for each system. A molecular theory of heterogeneous catalysis can be developed to identify the key surface reactivity performance parameters that either determine the differences in reactivity for various materials, or help discriminate between different reaction mechanisms. They provide a computational vehicle to microkinetics simulations, which aims to determine trends in catalyst performance as a function of catalyst material properties.

For example, scaling laws can predict for an adsorbate X—H_x how its adsorption energy varies with metal surface composition based on the adsorption energy of adatom X. This is useful as long as surface topology does not change, and it is preferable to compare intermediates that are adsorbed in the same site configuration. Scaling laws were discovered by Nørskov et al. [28] and apply for chemisorption to group VIII and IB transition metals. They validate the qualitative results based on the Bond Order Conservation (BOC) rules as proposed earlier by Shustorovich [29], but their application is limited to those systems where the electronic structure of the chemisorption bond does not vary strongly with transition metal type.

In Figure 7.4, DFT-computed interaction energies of X—H$_x$ intermediates are compared with the adsorption energies of the X$_{ads}$ atom as a function of metal substrate for particular surface topologies.

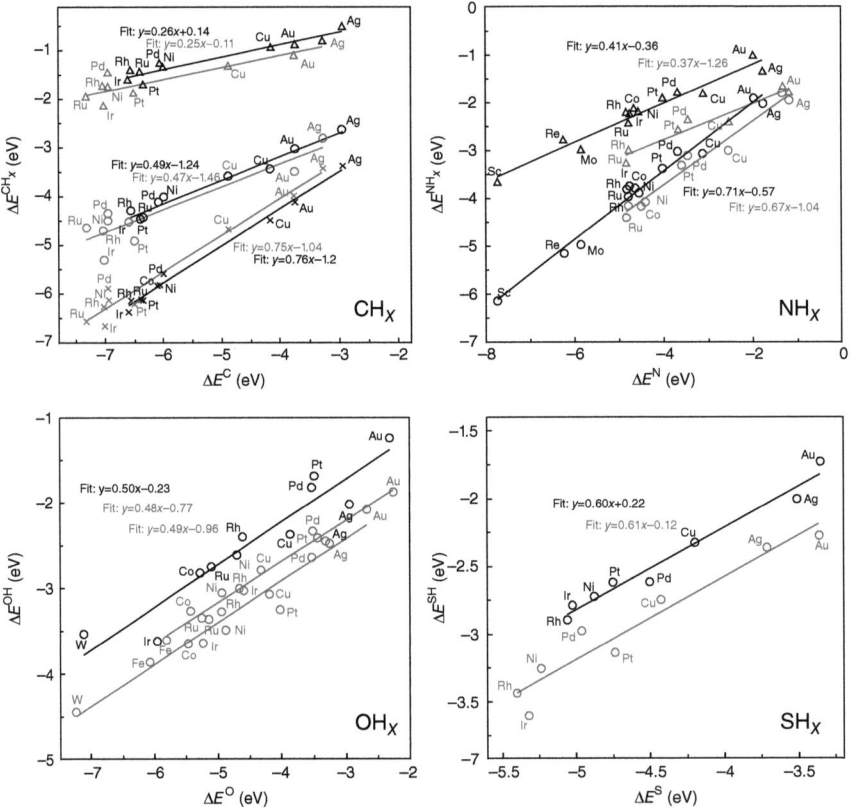

Figure 7.4 Adsorption energies of (a) CH$_x$ intermediates (crosses: $x = 1$; circles: $x = 2$; triangles: $x = 3$), (b) NH$_x$ intermediates (circles: $x = 1$; triangles: $x = 2$), (c) OH, and (d) SH intermediates plotted against adsorption energies of C, N, O, and S, respectively. The adsorption energy of molecule A is defined as the total energy of A adsorbed in the lowest energy position outside the surface minus the sum of the total energies of A in a vacuum and on a clean surface. The data points represent results for close-packed (black) and stepped (open) surfaces on various transition metal surfaces. In addition, data points for metals in the fcc(100) structure (middle line) have been included for OH$_x$. (Abild-Pedersen et al. 2007 [28]. Reproduced with permission of American Physical Society.)

The relative adsorption energies increase more steeply when the number of substituents attached to the adsorbate atom decreases.

According to BOC rules, for a X—H_x fragment an increase in the strength of a bond with a surface atom is accompanied by a decrease in x. The BOC principle is based on the redistribution of the valency of an atom as a function of its coordination. We discussed earlier in Chapter 3 (Insert 3) that this also rationalizes the increase in bond energy of an adsorbate with decreasing surface metal atom coordination. Here, we apply this concept to the valency of atom X. The X atom valence capacity is considered a constant and is distributed over the bonds attached to the X atom when X—H_x bonds are formed. For instance, when one X—H bond is formed, the total valence of X is available and the X—H bond is strong. When two bonds are formed, the total valence must be distributed over both bonds, and thus the bonds are weakened. As more X—H bonds form with the X atom, their individual strengths weaken further. Using Morse potentials as a basis, Shustorovich developed elegant analytical expressions that provide a way to estimate this successive weakening of the X—H bonds. (See Insert 3, Chapter 3 [26, 29].)

Thus, the available free valence of the X—H_x intermediate to bind to the surface will decrease with increasing x. The slopes of the curves in Figure 7.4 show an interesting direct relationship to the available free valence. The slope of the CH_3/C adsorption energies is 1/4, the slope of CH_2/C adsorption energies is 1/2 and for CH/C it is 3/4. The slopes are proportional to the available free valence.

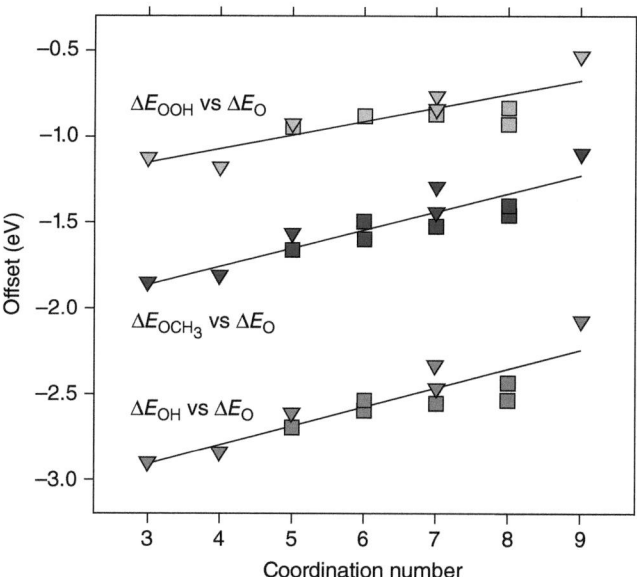

Figure 7.5 Structure–energy relations for atomic oxygen and oxygenates. The slopes of the offsets of the scaling relations between the adsorption energies of *OH versus *O, *OOH versus *O and *OCH_3 versus *O, scale in an approximately linear fashion with the coordination number of the adsorption sites. (Calle-Vallejo *et al.* 2015 [30]. Reproduced with permission of Nature Publishing Group.)

This is also seen for the analog molecules NH_3 and H_2O. The ratios of the adsorption energies of OH versus adsorbed O have a slope of 1/2 and for NH_2 versus N adsorbed on the (111) surfaces the slope is 1/3 [31].

The offset of the different slopes varies with surface structure. The more open surfaces bind the adsorbates more strongly, so that for these surfaces the offset is at the stronger bond energy (lower negative energy value). Adsorption strengths decrease with increasing coordination number of the surface metal atoms. This gives the approximate linear decrease of the offset (higher negative energy value) with increasing coordination number of the surface metal atom illustrated in Figure 7.5.

BOC can also be used to explain changes in adsorbate bond strengths as a function of particle size. Since the coordination number for edge or corner atoms is smaller than the number for surface atoms in a dense surface, the edge or corner atoms will form stronger bonds with their nearest neighbors than with the surface metal atoms. We have used this argument earlier in Chapter 4 to explain differences in catalytic reactivity that are seen with varying transition metal particle size.

However, there is an additional change in the surface electronic structure when transition metal particle size decreases, which explains the changes in bond energy of adsorbates even when the surface coordination number does not change.

A surface edge atom has fewer neighbors, and will therefore according to BOC bind more strongly to its neighbors. As a consequence, the neighboring terrace atoms will experience a stronger metal-mediated metal atom–metal atom interaction energy y. This decreases the relative bonding capacity of the metal atoms next to the surface edge atoms, and the adsorption strength to the terrace atom is weakened.

When we apply these rules to predict changes in adsorption strength when transition metal nanoparticle size decreases, we deduce changes in chemical reactivity based only on changes in the surface coordination number of the atoms at the particle surface. The relative number of coordinatively unsaturated atoms increases, so that the average bond strength will increase per surface atom. The terrace atoms that are nearest neighbors of the surface edge atoms have reduced reactivity because they now have to commit more of their valence to bonding with the edge atoms [30].

Scaling laws used in combination with the Bronsted-Evans-Polanyi (BEP) reaction rate relation are a powerful tool for microkinetics. The overall rate of a catalytic reaction can be modeled as a function of surface reactivity descriptors, which are usually the adsorption energies of adatoms such as C, N, or O. An elementary example of a reaction rate volcano curve as a function of surface reactivity that was constructed in this way can be found in Chapter 3, Insert 4.

7.2.8 In Summary: The Adsorbate Chemical Bond

Here, we will summarize the essential features of the electronic structure of the chemical bond between adsorbate and transition metal surface. The interaction of the delocalized transition metal s and d valence electrons with the localized adsorbate atomic orbitals leads to the chemical bond of the chemisorbed state. While the interaction with the transition metal s valence electrons contributes significantly to the bond energy, trends in surface reactivity are mainly determined by variation in the interaction with the d valence electrons. This is because for the transition metals the number of d valence electrons varies, but the s valence electron band occupation remains approximately constant.

The electronic structure of the interaction of adsorbate orbital and transition metal d valence electron band can be considered to consist of three parts, as is schematically illustrated in Figure 7.6:

- A bonding electron density regime at the bottom of the d valence electron band in which electrons are localized in orbitals between the metal atom and adsorbate.
- A non-bonding electron density regime localized mainly on the transition metal atoms.
- An antibonding regime at the top of the valence band. This is the localized bonding regime between metal atom and adsorbate that is complementary to the bonding regime.

This orbital description of the surface chemical bond implies that the adsorbate–surface interaction can be considered to be a molecular complex of adsorbate and binding surface metal atoms embedded in the surface. An important effect on the surface is that the interaction between the surface metal atoms is weakened upon adsorption. This energy cost to chemisorption

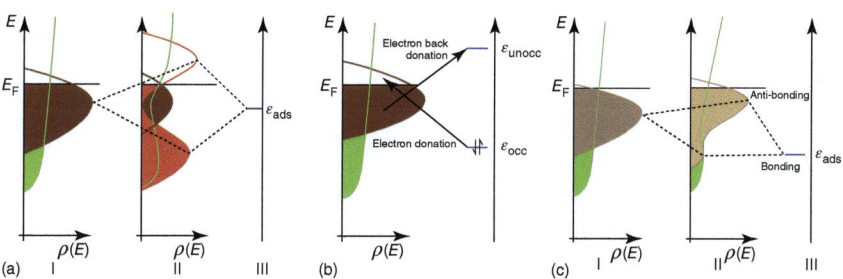

Figure 7.6 Schematic representation of the electronic structure of surface adsorbate chemical bonds. In figures a and c the left side of each figure (I) shows the schematic surface electron density of free metal surface. The PDOS, $\rho(E)$, is plotted as a function of electron energy. The green line denotes the s atomic orbital free electron band of the metal. The brown line denotes the d valence electron density on the surface metal atom. Orbitals below the Fermi level are occupied. In figures a and c the right side of each figure (III) shows the energy and density of a free atomic orbital. In the middle (II), a representation is shown of the electron distribution that results when the two systems interact. (a) The part II shows the PDOS of a chemisorbed atom. In red, the splitting into a bonding and antibonding PDOS of the interaction of with a 2p atomic orbital of an adsorbed atom with metal surface atoms is indicated. (b) Representation of the HOMO–LUMO scheme of chemical bonding of a molecule to a surface. The donative, occupied molecule orbital $\varepsilon_{occupied}$, and empty surface electron density interaction and the backdonative interaction from occupied surface density to unoccupied molecular orbital, $\varepsilon_{unoccupied}$, are indicated. (c) Schematic representation of Pauli repulsion due to the interaction of an occupied adatom or admolecule orbital with the nearly completely occupied d valence electron band. Pauli repulsion is reduced in case a when more antibonding electron density is pushed above the Fermi level, so that its electron occupation decreases.

energy is due to the localization of electrons in the metal atoms involved in the chemisorption bond is formed.

For the group VIII transition metals, a decrease in d valence electron band occupation generally decreases electron occupation of the antibonding surface metal atom–adatom bond, and hence the interaction strength will increase.

However, as has been extensively discussed in the previous sections, this rule has to be applied with care because additional electronic factors may also play a role and sometimes modify this trend.

In Figure 7.6a formation of a strong adatom bond is illustrated. The chemisorptive bond consists of an occupied bonding orbital and an antibonding orbital with low electron occupation. In an adsorbed molecule, chemisorption will weaken the molecular bond. In Figure 7.6b, we schematically indicate the two chemical bonding interactions that are responsible. They are best described as donative and backdonative. The donative interaction is due to the interaction of an occupied molecular orbital with the empty part of the d valence band, the backdonative interaction leads to electron occupation of a previously unoccupied antibonding molecular orbital by interaction with the part of the d valence electron band occupied by electrons. This situation occurs when the interaction between the molecule and surface is rather weak, but before dissociation of the molecular bond. The molecular bond is weakened both because electrons occupy the molecular antibonding orbital, and because electron occupation of the bonding orbital decreases. Usually, the increased electron occupation of the antibonding molecular orbital is the dominant cause for intramolecular chemical bond weakening. The overall charge on the molecule will remain relatively low because the electron occupation changes in the bonding and antibonding molecular orbitals tend to counteract each other.

Figure 7.6c schematically illustrates the interaction of an electron-occupied molecular lone pair orbital with the surface valence band. This is typical for the lone pair σ orbital of ammonia or CO.

The interaction of a doubly occupied molecular orbital and a metal atom with a nearly completely occupied d valence electron band is shown. Bonding and antibonding surface complex orbitals will be formed. On a surface with low reactivity and highly delocalized orbitals, the bonding as well as antibonding surface complex orbitals become occupied and, hence, the interaction with the d valence electrons will be repulsive. This Pauli repulsion results from electron occupation of bonding as well as antibonding orbitals.

When Pauli repulsion dominates the chemical bond, it will favor binding to a single metal atom, since this will minimize the repulsive interaction. This is in contrast to the case where binding is attractive, when bonding orbitals are mainly occupied. Then, binding to a high coordination site is favored.

7.3 The Transition States of Elementary Surface Reactions

The chemical bonding interactions that create low activation barriers for elementary reactions differ for the activation of molecular σ-bonds (such as in methane) and for activation of molecular π bonds (such as in CO). The dependence of the activation energy on surface topology also differs significantly for the two types of bonds. The relationship between reactivity and structure and its implications for the activation energy of bond cleavage reactions will be the topic of this section.

Activation will also depend on whether a molecule or molecular fragment has a lone pair orbital available to bind to the surface. A saturated molecule as an alkane will initially interact with the surface only through dispersive van der Waals interactions. The molecular bond angles must deform significantly to become activated. This is different for activation of an ammonia or water, where direct coordination to the N atom or O atom is possible through a lone pair orbital. We will show that in methane and ammonia activation this leads to important differences of C—H bond activation versus N—H bond activation.

We will first discuss the case of σ adsorbate bond activation and then discuss π adsorbate bond activation.

7.3.1 Adsorbate σ-Bond Activation

7.3.1.1 Activation of Methane

Activation of the C—H bonds of alkanes or alkenes preferentially occurs through interaction with a single surface metal atom.

Insert 8: The Transition State of Methane Activated by a Transition Metal Surface Atom

The structure of the transition state of methane can be found in Chapter 4, Insert 3.
  Table 8i.1 summarizes DFT-calculated activation energies of methane catalyzed by the dense surfaces of transition metals.

Table 8i.1 DFT-computed activation energies in kJ mol^{-1} for CH_4 decomposition to CH_3 computed in a (2 × 2) unit cell.

	100 Co(0001) [32, 33]	118–127 Ni(111) [32, 34]
77–82 Ru(0001) [35, 36]	67–70 Rh(111) [37, 38]	66 Pd(111) [35]
	40$^{a)}$–91 Ir(111) [39, 40]	74–80 Pt(111) [40, 41]

a) Activation energy of methane on reconstructed surface.
Source: Henkelman and Jónsson 2001 [39]. Reproduced with permission of American Physical Society.

Figure 8i.1 compares the activation energies of CH_4 dissociation by a surface terrace metal atom with high surface coordination and by a more reactive step edge metal atom of low surface coordination atom. It illustrates the BEP relation between change in forward activation energies and change in reaction energies.

(Continued)

Insert 8: (Continued)

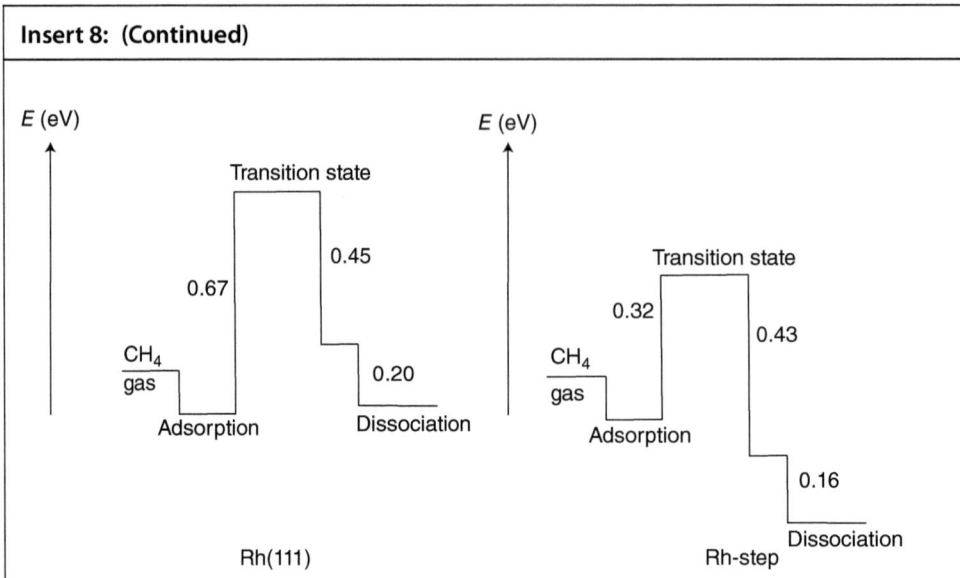

Figure 8i.1 Activation of methane. Changes in energy of a dissociating methane molecule activated by a surface atom on a Rh(111) surface and on a step-edge atom of lower coordination. (Liu and Hu 2001 [42]. Reproduced with permission of American Institute of Physics.)

In Table 8i.2, the bondorders of the chemical bonds of the transition, reactant, and product states are compared for a methane molecule that dissociates atop a Rh(111) surface atom. The bond order has a maximum value of 1 for a strong σ bond.

Table 8i.2 Bond orders of the bonds involved in the methane activation C—Rh, C—H, and H—Rh at the reactant, transition, and product states.

Bond	Reactant state	Transition state	Product state
C—H	0.99	0.29	< 0.2
C—Rh1		0.43	0.54
H—Rh1		0.38	0.42
H—Rh2		< 0.2	0.49

Source: van Santen *et al.* 2010 [31]. Reproduced with permission of American Chemical Society.

Alkane adsorption to the transition metal surface has a low adsorption energy. There are no free valences in the molecule, hence the adsorbed state is physisorbed and the interaction energy is determined by dispersive van der Waals type interactions. This implies that the adsorption strength of alkanes to transition metals is rather insensitive to structure and type of metal. The adsorption energy will increase about 5 kJ mol^{-1} per CH_2 unit.

We will discuss in detail the bond cleavage of CH_4, which can be considered a prototype molecule that has σ bonds and no free valence. A substantial chemical interaction with the surface is necessary to widen the HCH angle and facilitate dissociation. Table 8i.1 shows that the least reactive metals are the transition metals in the third row, followed by the fourth row metals. In the fifth row, Pt again has a higher activation barrier, but a very low value is reported for Ir.

One expects that the activation barrier of C—H bond cleavage will relate to the bond energies of reaction fragments H and CH_3. In the fourth row of the periodic table for atop-adsorbed CH_3, the M—CH_3 interaction is very similar for Pd and Rh but decreases for Ru (Figure 5i.4a). The interaction with the third row metals is weaker [11]. The higher barriers on Ru as well as on Ni and Co relate to this weaker M—CH_3 interaction. For Pt compared to Pd, there is no such correlation between the activation energy and the M—CH_3 bond energy of atop-adsorbed CH_3.

For transition metals such as Pt or Pd that have a highly occupied d valence electron band the energy cost to approach the surface makes a substantial contribution to the activation energy. This is due to the Pauli repulsion between the occupied transition metal d valence orbitals and occupied methane σ orbitals.

For efficient electron backdonation into the unoccupied methane molecular orbitals that will weaken the C—H bonds, the methane molecule must approach the transition metal surface closely so that the respective valence atomic orbitals can overlap. This is counteracted by the Pauli repulsion energies, which are stronger on Pt than on Pd because of the larger spatial extent of the Pt atomic orbitals.

Larger Pauli repulsion is also the reason why adsorption of H prefers atop adsorption on the Pt surface but high coordination on the Pd surface (see Figure 2i.3). The higher coordination possible for H on Pd these lowers the activation energy of methane activation compared to that on Pt. The more difficult approach of methane to the Pt surface and the less favorable energy of multiple coordination on Pt are complementary and both relate to the increased Pauli repulsion.

The relatively low barrier of Ir reported in the table is due to reconstruction of the Ir surface. Upon contact with the dissociating methane molecule, the Ir atom becomes slightly tilted from the surface thus increasing its reactivity [39].

In the transition state, the chemical bond between C and H is nearly broken. This can be concluded from the bond orders for the C—H bond of adsorbed CH_4 in its transition state and gas phase CH_4 of the same geometry (Table 8i.2). The bond orders of the respective M—C and M—H bonds in the transition state are close to those of the dissociated fragments when they are still adsorbed in the states where they still share bonding with the same metal atom. This is not the most stable final state, because sharing the bonding with the same surface metal atom causes a destabilizing lateral interaction.

For methane dissociation on the Rh terrace versus the Rh step-edge one finds that increased coordinative unsaturation of the surface atoms decreases the activation energy (Figure 8i.1). This is to be expected from the increased reactivity of the transition metal surface atom that has lower coordination with surface metal atoms. The activation energy barrier decreases by an amount similar to the final state. This indicates a linear relation between change in transition state energy and change in reaction energy. This is represented by the BEP relation (see Eq. (3.20a)) with a proportionality constant α (akoga BEP) value close to one. In the transition state of Rh, any similarity to the molecularly adsorbed initial state is absent, thus it can be considered a late transition state.

7.3.1.2 The Oxidative Addition and Reductive Elimination Model

In organometallic chemistry, the analogous process of dissociative adsorption is called oxidative addition. Dissociative adsorption of methane creates adsorbed H and CH_3 radicals, which on the surface can formally be considered H^- and CH_3^- species. When adsorbed to single metal atom, this implies that the metal atom gets a 2+ charge. Hence, the elementary step can be considered an oxidation. The reverse process of H and CH_3 adsorption fragment recombination is called reductive elimination.

We can schematically illustrate the quantum chemistry of how oxidative addition and reductive elimination electronically change the cleavage of a σ bond by using the example of the H_2 molecule in Figure 7.7.

At the right of this figure, the energies of the bonding and antibonding orbitals of the free H_2 molecule are sketched. Upon contact with the surface, the electron densities of these two orbitals will broaden. When the H_2 molecule is adsorbed parallel to the surface and dissociates atop the surface atom (as in the Blyholder model) the σ orbital will be broadened by interaction with the transition metal s

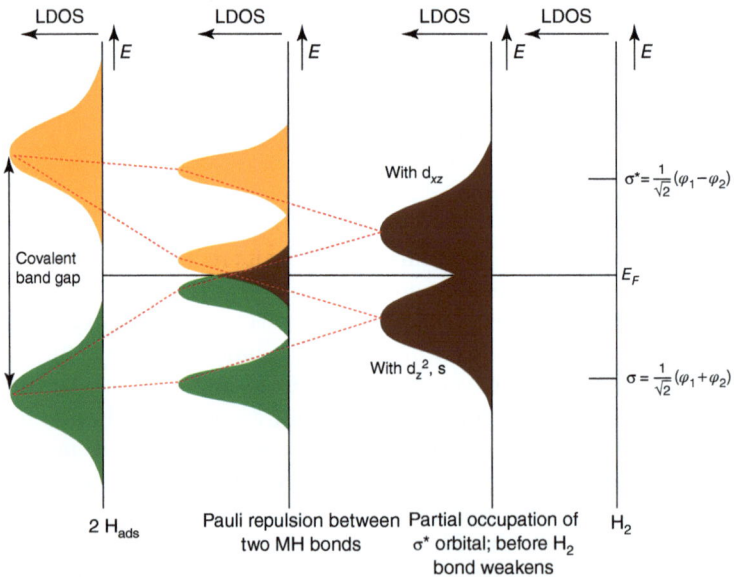

Figure 7.7 Schematic illustration of the changes in electronic structure of the hydrogen molecule that reacts with the metal surface. Oxidative addition of H_2 and reductive elimination of two hydrogen atoms are illustrated.

and d_{z^2} valence atomic orbitals and the σ^* orbital will broaden by interaction with the d_{xz} or d_{xy} transition metal valence atomic orbitals.

When the interaction becomes so strong that the PDOS of the σ orbitals is pushed above the Fermi level and the PDOS of the σ^* orbital is pushed partially below the Fermi level, electron occupation in these orbitals will change. The bonding σ H_2 orbital is depleted of electrons and the antibonding orbitals will become occupied. Both interactions will weaken the H_2 bond.

The PDOS of the two M—H bonds formed upon dissociation are shown in the middle part of the figure. Initially, they share bonding with the same metal atom. Their interaction leads to two bonding and antibonding combinations of the respective M—H bonding and antibonding orbitals. The symmetric combination of the bonding M—H orbitals has the lowest energy and the antibonding combination the next higher energy. This antibonding orbital will be partially filled. The result is the Pauli repulsive interaction between the two atoms that are attached to the same atom.

As we discussed in the previous section on methane activation, this Pauli repulsion in the first local minimum found after the transition state contributes substantially to the activation energy.

The next PDOS at the higher energy are the symmetric and anti-symmetric combination of the antibonding M—H orbitals. The symmetric combination may still be partially filled, which will weaken the M—H bond. Finally, as shown in the left part of Figure 7.7, when the H atoms move apart and do not interact the bonding and antibonding M—H orbitals have the same energy and only the bonding M—H orbital is occupied.

The broken lines in Figure 7.7 show how the orbitals of the non-dissociated H_2 molecule transform into those of the adsorbed hydrogen atoms. The symmetric occupied σ orbital of H_2 is transformed into bonding and antibonding M—H orbitals.

According to the Woodward–Hoffmann rules of quantum-chemical theory [43] non-catalyzed dissociation of the H_2 molecule by reaction with an olefinic bond leads to an activation energy of H_2 dissociation, because a high energy antibonding orbital of the product state becomes occupied.

On the transition metal surface this barrier is lowered by orbital interaction with its d valence atomic orbitals that hybridize the initially unoccupied σ^* H_2 molecular orbital with the occupied σ H_2 orbital.

This enables the donative interaction with the transition metal that depletes the bonding H_2 σ orbitals and the backdonation of electrons from the transition metal into the bond. This weakening of σ^* H_2 orbitals in turn weakens the bond strength so that the activation energy of bond dissociation is lowered.

7.3.1.3 The Umbrella Effect

Dissociation of saturated hydrocarbons does not follow the rule that the least stable bond cleaves most easily (bond energy $H_3C-C_3H = 346$ kJ mol^{-1}; energy $H_3C-CH_2-H = 411$ kJ mol^{-1}). Instead, the C—H bond must be activated first since the first contact with the metal surface is through the H atom.

The reverse reaction of C—C bond formation between two CH_3 fragments is also inhibited. The initial contact for two CH_3 fragments on the surface is through the "umbrella" of the hydrogen atoms around the M—C bond. They have to bend away from the C atoms in order to interact. This causes a substantial repulsive energy that must be overcome and thus effectively prevents this recombination. C—C bond formation will occur with a chemically acceptable low barrier of activation energies only when the C atom is less saturated with H atoms. Hence, the only allowed recombination reactions are $CH_{2,ads}$ with $CH_{3,ads}$ (but not $CH_{3,ads}$ with $CH_{3,ads}$) or the analogous alkyl recombination reactions.

7.3.1.4 Activation Entropy

As we have seen, the transition state of CH_4 dissociation can be considered late with respect to the initial state, or early with respect to the dissociated fragments. In the transition state the bond strengths as well as the geometry of the reaction intermediate are close to that of the dissociated state.

When the dissociated fragments are strongly adsorbed they will have a low mobility. This will also be the case for the transition state, which implies that the entropy of the transition state is nearly zero. This is consistent with the strong interaction with the surface required to overcome the bond dissociation energy of the adsorbed molecules

The transition state is also called *tight* because of its low entropy for dissociation. This is different from the transition state of a desorbing molecule, which is in a "loose" state of high mobility. In the transition state of desorption there is free rotation and movement parallel to the surface. These increases in mobility in the transition state compared to the adsorbed state will cause an increase of the pre-exponent of the rate constant of desorption by 10^4–10^6. This increase in rate constant causes the rates of desorption of molecules to be faster than bond activation of the adsorbate. When they have comparable activation energies.

Conversely, the entropy of the transition state of dissociating molecule is highly reduced with respect to the gas phase. The increases of the free energy of activation are due to this mobility loss (at 800 K the decrease in free energy is in the order 80 kJ mol^{-1}). For the dissociating methane molecule, the activation free energy is nearly double of that of the apparent activation energy.

Surface science experiments in a vacuum usually cannot be used to directly measure the dissociation barrier of preadsorbed molecules, since adsorbed molecules will desorb before they will dissociate. For catalysis, a finite reagent pressure is necessary to maintain an acceptable surface coverage for the dissociation of molecules.

The material discussed in this subsection supports the general rule that for a surface reaction the entropy of the transition state can be assumed to be very low. For adsorption or desorption changes in entropy are considerable and must always be considered.

7.3.1.5 σ-Bond Activation of Molecules that Bind Through their Lone Pair Orbital

We will discuss the activation of ammonia as the prototype example of the cleavage of a σ bond attached to a molecule or reaction intermediate that can adsorb through a free valence. Figure 7.8 shows the computed reaction energy diagram of ammonia decomposition when adsorbed on three different Pt surfaces.

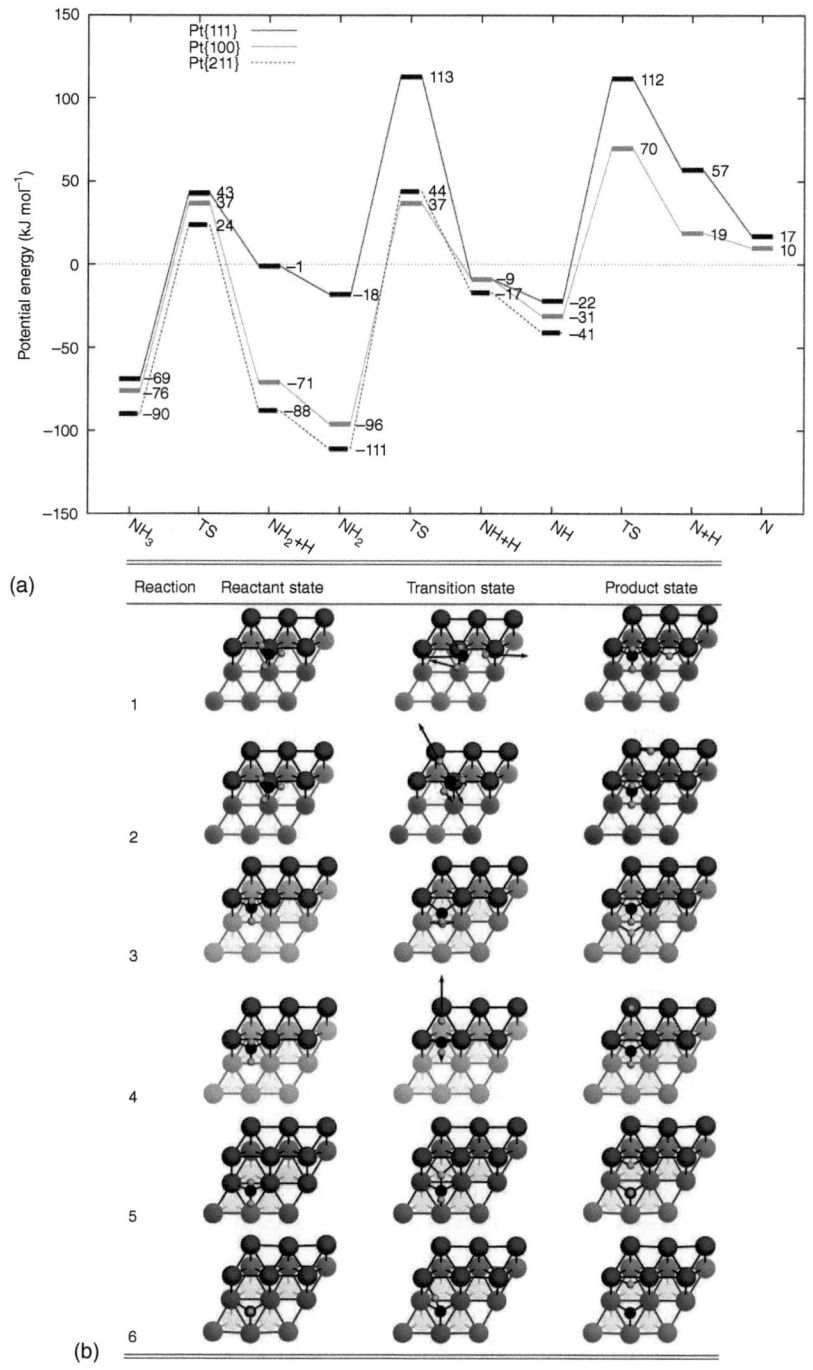

Figure 7.8 (a) DFT-calculated reaction energy diagram for the dissociation of NH_x 1, 2, 3 over Pt(100), Pt(111), and stepped Pt(211) [44]. (All calculations were carried out using a 2×2 unit cell with 25% coverage.). The energies are with respect to NH_3 in the gas phase. (b) NH_3 and NH_2 reaction intermediates for the Pt(111) shown in panels 1–3 and the stepped Pt(111) surfaces shown in panels 4–8. Black, white, and gray spheres indicate N, H, and Pt atoms, respectively. The second layer of the Pt surface has been shaded for clarity. Changes in N and H positions in transition states are indicated [44]. (c) Transition states for NH activation on the Pt(100) surface. (Offermans *et al.* 2006 [45]. Reproduced with permission of Elsevier.)

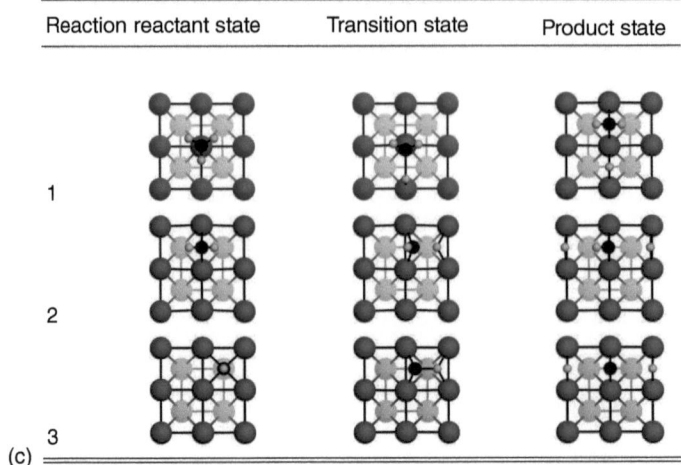

| Reaction reactant state | Transition state | Product state |

1

2

3

(c)

Figure 7.8 (Continued)

While methane activation shows considerable energy differences when activated by different surfaces (see Chapter 4, Insert 3), the activation energy of N—H bond cleavage in ammonia remains essentially unchanged.

The ammonia molecule adsorbs atop the Pt atom through its lone pair orbital. The product fragment NH_2 prefers adsorption in two fold coordination. To convert NH_3 to $NH_{2,ads}$ two bond coordinates change substantially: the N—H as well as the M—N bonds. As illustrated in Figure 7.8b,c the N—H bond cleavage reaction of NH_3 is initiated by the initial stretching of the N—H bond, without a change in M—N bond coordination. In a second reaction step the nitrogen atom moves to twofold coordination, which requires a very low barrier.

In the left part of Figure 7.8a, it can be seen that, when coordinative unsaturation of a surface atom changes, the M—N bond interaction will change similarly in the initial and transition states. The transition state can be considered tight. This causes the activation energy of N—H bond cleavage in ammonia to be independent of the coordinative unsaturation of the surface metal atom. Figure 7.8a also shows the activation energies of consecutive reaction steps. The corresponding structural reaction intermediates are displayed in Figure 7.8b,c.

The activation energies of the N—H bond cleavage of NH_2 are also independent of surface state. In the transition state only the N—H bond is stretched. In a second step, the N—H fragment moves to its preferred threefold coordination site.

Activation of NH at the step edge has a substantially lower energy than of NH adsorbed to a surface terrace. As illustrated in Figure 7.8b6 the NH molecule adsorbed in the "valley" of the edge can donate the hydrogen atom to a nearby edge atom, whereas on the terrace the N—H bond has to stretch and bend before the hydrogen atom is released. The step-edge has the topological advantage in that it can reduce the deformation cost by offering a reaction site with multiple contacts positioned favorably with respect to the reacting adsorption fragment.

7.3.2 Dissociation of Diatomic Molecules with π-Bonds

A molecule such as CO interacts with the transition metal surface through both σ- and π-symmetric interactions (see Section 6.3). On a terrace it will usually adsorb with its C atom perpendicular to the surface, unless the transition metal has a low work function which allows it to adsorb parallel to the surface.

When adsorbed perpendicular to the surface terrace, the CO molecule must bend into the side-on-adsorbed state to cleave the C—O bond in order to dissociate. On a planar surface, the corresponding energy cost contributes considerably to the activation energy of C—O bond cleavage.

In the side-on state it can also interact through the O atom with the transition metal surface, and bond dissociation will occur when the interaction between O atom and surface becomes strong enough. Since the carbon atom and oxygen atom prefer high coordination adsorption sites, several surface atoms (a surface atom ensemble) must be involved for bond activation with low activation energy.

This is a general observation: when a molecular π bond is cleaved, the surface site must be a surface atom ensemble of at least five surface atoms.

A low activation barrier for π bond cleavage not only requires a surface atom ensemble site, but also a site of a particular topology. The activation energy of CO dissociation decreases by more than 100 kJ mol^{-1} when activation by a step-edge site is compared with activation on planar surfaces (see Chapter 4, Figure 3i.2).

The same chemical bonding principle that controls the reduction in activation energy of CO at a step-edge site also applies to other molecules with π bonds such as N_2, NO, or O_2 [46].

There are three reasons for the lower activation energy at a step-edge site:

- At the step-edge site, the molecule is already close to a side-on adsorption configuration. The proximity of the CO oxygen atom to the edge metal atom means that less bending and stretching is needed for the C—O bond cleavage process.
- Electron backdonation from the metal surface orbitals into the 2π* molecular orbitals, which weakens the C—O bond, is increased at the step-edge site. In addition to metal atom orbital overlap via the C atom, increased transition metal atom orbital overlap through the O atom with the surface metal atom orbitals is also now possible.
- In the dissociated state, the O atom will attach to a reactive step-edge metal atom which does not share a chemical bond with the C-atom. The absence of a repulsive interaction that otherwise increases the activation energy helps to lower the activation energy.

Figure 7.9 illustrates the large variation in the activation energies of CO found for the surfaces of different metals. The BEP relation linearly relates the activation energy for C—O bond cleavage with the bond energies of the C and O atoms. Since the variation in bond energies of O is much larger than that of adsorbed C atoms, the activation energies of CO mainly follow the trend of adsorbed O. In contrast to the activation energies, the adsorption energies of CO vary much less and follow the trend of the adsorbed carbon atoms.

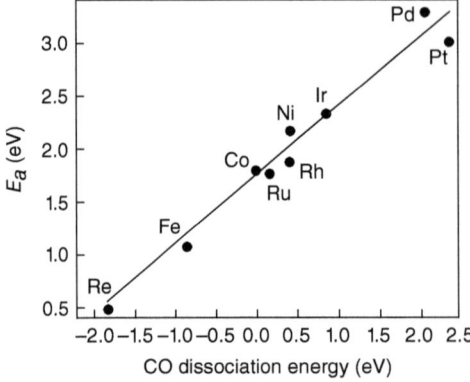

Figure 7.9 Brønsted–Evans–Polanyi relation for CO dissociation over transition metal surfaces. The transition state potential energy E_a, is linearly related to the CO dissociation energy. (Nørskov et al. 2009 [47]. Reproduced with permission of Nature Publishing Group.)

Dissociation barriers are lowest for transition metals such as Fe, Co, and Ru in the upper-left corner of the group VIII transition metal positions in the periodic table.

Due to its strong dependence on the O_{ads} energy, the trend in the activation energy of CO with variation of transition metal is remarkably different from that of methane activation. For instance, although methane has one of its lowest barriers of dissociation on Pd, this metal has one of the highest barriers for CO activation.

7.3.2.1 Principle of Non-Shared Bonding with the Same Surface Metal Atom

Adatoms such as N, O, and C have a strong lateral interaction when they share bonding with the same surface metal atom. Therefore, transition states of low energy of activation will be found for surface topologies in which the atoms generated by dissociation do not share bonding with the same surface metal atom. The step-edge sites discussed in the previous sections are an example.

As seen in Figure 3i.4 of Chapter 4, the dissociation of NO on the (100) activitation of Pt has a lower activation energy than on the Pt(111) surface, but a higher energy than on a step-edge site of the Pt(111) surface. This is because the N atom is adsorbed more weakly to the (100) surface than the (111) terrace (the N atom prefers threefold coordination). For this same reason, the reverse reaction of N and O recombination on the (100) surface is very favorable and has a very low barrier.

The unique reactivity feature of the (100) surface of the fcc transition metal makes it exceptionally suitable for the electrochemical oxidation of ammonia and the reduction of nitrite on Pt(100) [48, 49]. It is also preferred for the electrocatalytic reduction of CO to ethylene catalyzed by the Cu(100) surface [49, 50]. Cubic particles terminated by a surface of the fcc transition metals are found to be uniquely suited to some other reactions as well [51].

7.4 Reactivity of Surfaces at High Coverage

The reactivity of transition metal surface atoms will change when additional adsorbates are present next to the reacting molecule or surface fragment. The interaction energy will generally decrease due to the presence of coadsorbates. Coadsorption of non-reactive adsorbates may also lead to altered site topology, because it will block access to surface atoms. Reactive co-adsorbed atoms can also act as surface co-reagents in bond formation reactions.

The altered surface reactivity of metal atoms by a surface overlayer may by itself lead to new chemistry. For instance, co-adsorbed atoms can assist in the bond cleavage of reactive intermediates or act as acceptors of reaction fragments that would otherwise be less stabilized.

These topics will be covered in the following subsections:

- Decreased surface reactivity and site blocking
- Adatom co-assisted bond activation
- Altered surface chemistry

7.4.1 Decreased Surface Reactivity and Site Blocking

The BEP relation (Eq. (3.20a)) provides an easy way to quantify the effect of decreased reactivity of metal surface atoms since it relates the reaction energy with activation energy. An elementary reaction such as a surface dissociation reaction will have an altered activation energy, because the adsorption energies of reactant or reaction fragments change. This is formulated in Eq. (7.2):

$$E'_{ads,i} = E^0_{ads,i} + \sum_{j \neq i} E^{int}_{i,j} \tag{7.2}$$

$E'_{ads,i}$ is the adsorption energy of adsorbate i at high surface coverage, $E^0_{ads,i}$ is the corresponding adsorption energy of adsorbate i at low coverage and $E^{int}_{i,j}$ is the lateral interaction energy between two adsorbates i and j. $E^{int}_{i,j}$ can be due to direct repulsive or attractive interactions between adsorbed molecules or due to their effect on each other's adsorption energy, altered through surface metal atom interactions. The summation in Eq. (7.2) is over the neighboring adsorbate species i.

Because they experience stronger lateral interactions, the adsorption energies of the dissociation fragments vary more strongly than that of the corresponding molecule. Based on the BEP relation, when the interaction between adsorbed species is repulsive one expects either the dissociation energy of an adsorbed molecule to increase or the association energy of adsorbed molecular fragments to decrease.

Site blocking due to increased overlayer concentration may dramatically affect overall reaction kinetics. As we discussed in Figure 3.18 for the hydrogenolysis of alkanes on transition metals, the rate of reaction has a positive reaction order in H_2 at low pressure, but at higher pressures it has a negative order. At low H_2 partial pressure hydrogen is needed to hydrogenate the reaction intermediates, but at higher H_2 partial pressure H atoms will inhibit the adsorption of the hydrocarbon.

Similarly, at low O_2 pressure oxidation reactions that require surface vacancies for adsorbate bond activation have a reaction rate with a positive order in partial pressure of O_2, which becomes negative order at high O_2 pressures.

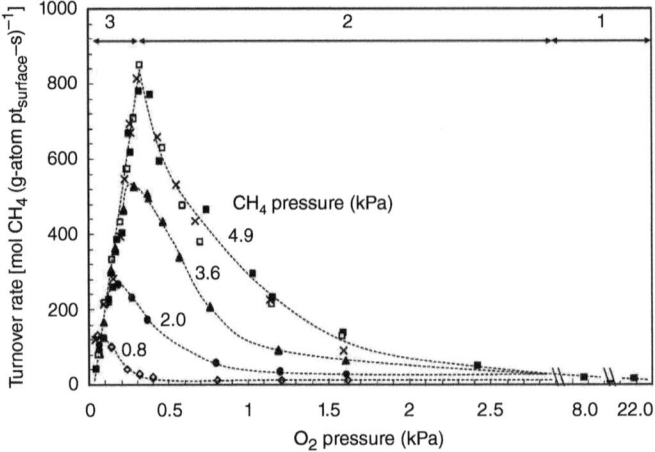

Figure 7.10 CH_4 turnover rates (873 K) during CH_4—O_2 reactions with 0.8 kPa (◊), 2.0 kPa (●), 3.6 kPa (▲), or 4.9 kPa (♦) CH_4 and during [52] CH_4—O_2—CO_2 (×) and CH_4—O_2—H_2O (□) reactions with either 5 kPa CO_2 or 5 kPa H_2O on Pt clusters (8.5 nm average cluster size) as a function of O_2 pressure. (Chin et al. 2011 [53]. Reproduced with permission of American Chemical Society.)

The oxidation of methane to synthesis gas catalyzed by Pt is an example, which is illustrated in Figure 7.10.

Three reaction regimes can be distinguished. An initial regime (labeled 3 at the top of Figure 7.10) has a positive rate of reaction at O_2 partial pressure, and the concentration of O adatoms limits the reaction rate with methane. In regime 2, the reaction becomes poisoned by co-adsorbed O that blocks sites at the higher surface concentration, and reaction rate slows. In the third regime (labeled 1 in Figure 7.10) the surface is oxidized, and reaction proceeds by radical intermediate formation of methane with adsorbed O atoms to create adsorbed OH and the CH_3 radicals. The activation energy of this reaction is higher than direct activation by the Pt metal surface atoms.

7.4.2 Adatom Co-Assisted Activation

Two coadsorbate-activated reactions will be considered in this section. The first concerns the lowered activation energy for bond dissociation through the formation of an intermediate complex with an adsorbed H atom, while the second concerns a decrease of the activation energy of chemical bond dissociation by co-reaction with a surface oxygen atom instead of a transition metal atom.

7.4.2.1 Hydrogen Activated Dissociation

On surfaces of low reactivity the activation energy of the C—O bond cleavage can be reduced by intermediate formyl or carbo-hydroxyl formation. For instance, this occurs in the methanation reaction of synthesis gas.

On the dense terraces of transition metals such as Co, Ru, or Ni the activation energy for direct C—O bond cleavage is higher than 200 kJ mol^{-1}. On these surfaces intermediate formation of a CHO species by the addition of an H atom

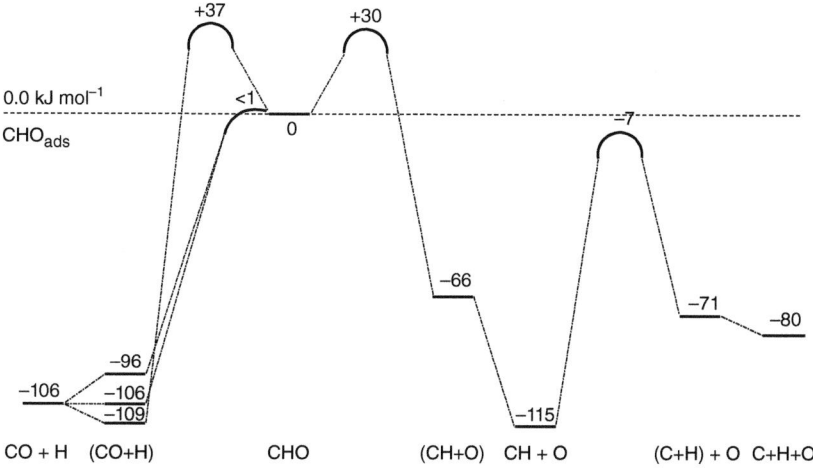

Figure 7.11 The reaction energy diagram of the C—O bond-cleavage reaction via intermediate formyl formation as computed by DFT of the Ru(0001) surface. (Ciobîcă et al. 2003 [54]. Reproduced with permission of American Chemical Society.)

to CO may provide a low activation energy path for C—O bond dissociation. The products will be adsorbed CH and an adsorbed O atom. The energetics of such a reaction is illustrated in Figure 7.11.

As Figure 7.11 illustrates for C—O bond activation on a Ru(0001) surface, the addition of an H atom to adsorbed CO requires 110 kJ mol^{-1}. This weakens the C—O bond, which then only requires an additional activation energy of 30 kJ mol^{-1} to dissociate. On the Ru terrace the overall barrier for C—O bond cleavage has been reduced from 210 to 140 kJ mol^{-1}. In contrast with the dense surface, when dissociation is catalyzed on a reactive step-edge site then direct CO dissociation becomes more favorable than the reaction path via hydrogen addition. This is illustrated in Figure 7.12.

7.4.2.2 Oxygen Assisted X—H Bond Cleavage

When the strength of the metal surface hydrogen bond competes with that of the hydrogen–oxygen bond of adsorbed oxygen, co-adsorbed O may act as a Lewis base with respect to hydrogen and will participate in the X—H bond cleavage reaction. A classic example of this is the activation of alcohols by the oxygen atoms adsorbed to the Ag surface [56–58]. On Pt, which also has a M—H bond that is weaker than the MO—H bond, activation of H_2O will preferentially occur with the adsorbed O atom (which creates two hydroxyls) instead of direct activation by the surface metal atoms.

The barriers for activation are also low due to the intermediate stabilization of the adsorption complex through hydrogen bonding. This is also the reason that, in general, H_2O formation preferentially occurs by the recombination of two adsorbed OH species instead of by the recombination of an adsorbed H atom with an adsorbed hydroxyl. However, on Pt the H—OH recombination reaction competes with OH—OH recombination, due to the weak M—H bond [59].

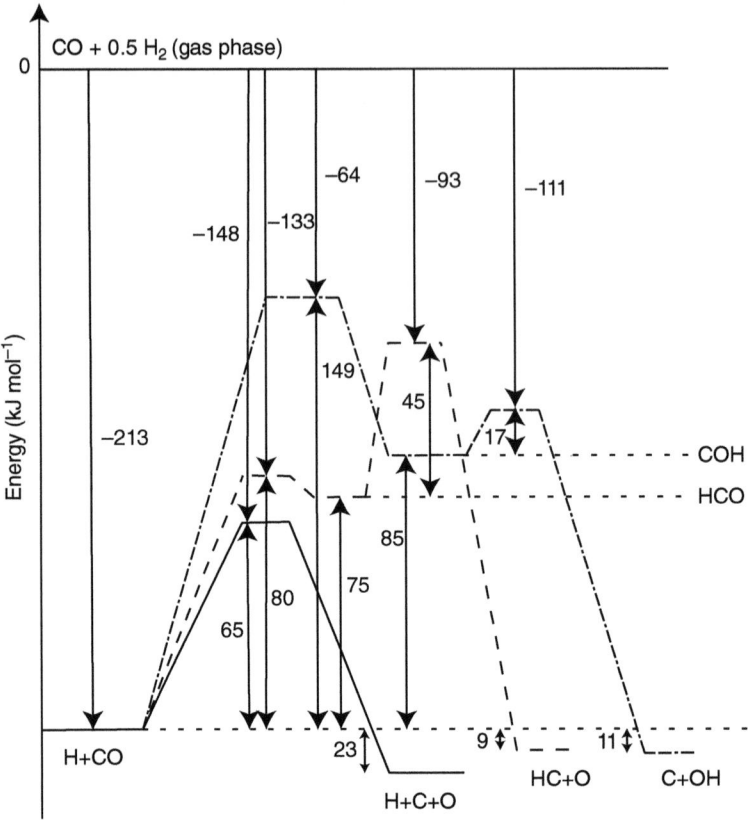

Figure 7.12 Energetics of CO dissociation in the presence of co-adsorbed H on the Ru(1121) surface. (closed line) Direct CO dissociation. (broken lines) CO dissociation via HCO and COH intermediates, respectively. (Shetty et al. 2009 [55]. Reproduced with permission of American Chemical Society.)

Insert 9: NH₃ Activation by Co-Adsorbed O Atoms

In this insert, O activation of NH_X ($X = 1–3$) is studied for ammonia and its reaction intermediates on the Pt(111) surface and Pt(100) surfaces (Figures 9i.1 and 9i.2). Reaction energy diagrams and reaction intermediate structures are compared. Details of activation on oxygen-free surfaces are found in Figure 7.10.

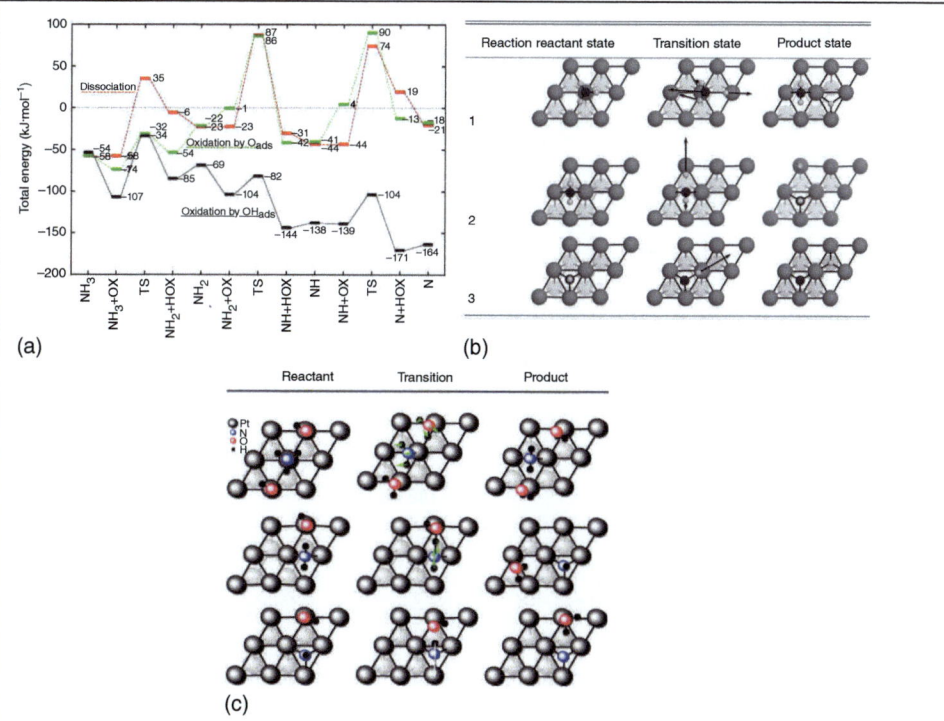

Figure 9i.1 (a) Reaction energy diagram of the dehydrogenation reactions of NH_x on Pt(111). The gray lines connect the reaction energies of respective intermediates NH_x without co-adsorbed oxygen. The dark lines compare reaction energies with co-adsorbed O or OH. All total energies are with respect to NH_3 (g), Pt(111) (s), 3 O_{ads}, and 3 H_{ads} and are zero-point energy corrected. O_x or HO_x on the abscissa mean oxidant or hydrogenated oxidant and can be OH, O, or an empty site or their hydrogenated forms. All NH_x + $(H)O_x$ coadsorbate states are with lateral interactions. In all "single" adsorbate states, we assume no lateral interactions. (b) The structures of reaction intermediates of the oxidation of NH_x by O_{ads} on Pt(111). The vertical projections of the unit cells show the reactant, transition, and product states. The arrows in the transition states are the vertical projections of the imaginary vibration. Hence, they point out the direction of the reaction path at the transition state. Dark, white, light gray, and gray spheres indicate N, H, O, and Pt atoms, respectively. The second layer of the Pt surface has been shaded for clarity. (c) Structures of reaction intermediates of oxidation of NH_x by OH_{ads} on Pt(111). The vertical projections of the unit cells show the reactant, transition, and product states. The arrows in the transition states are the vertical projections of the imaginary vibration. Hence, they point out the direction of the reaction path at the transition state. Since we notice interactions between species of different unit cells in the NH_3 + OH reaction, we also show the mirror images of the OH species in this particular reaction. Black, white, red, and gray spheres indicate N, H, O, and Pt atoms, respectively. The second layer of the Pt surface has been shaded for clarity. (Offermans *et al.* 2007 [60]. Reproduced with permission of American Chemical Society.)

(Continued)

Insert 9: (Continued)

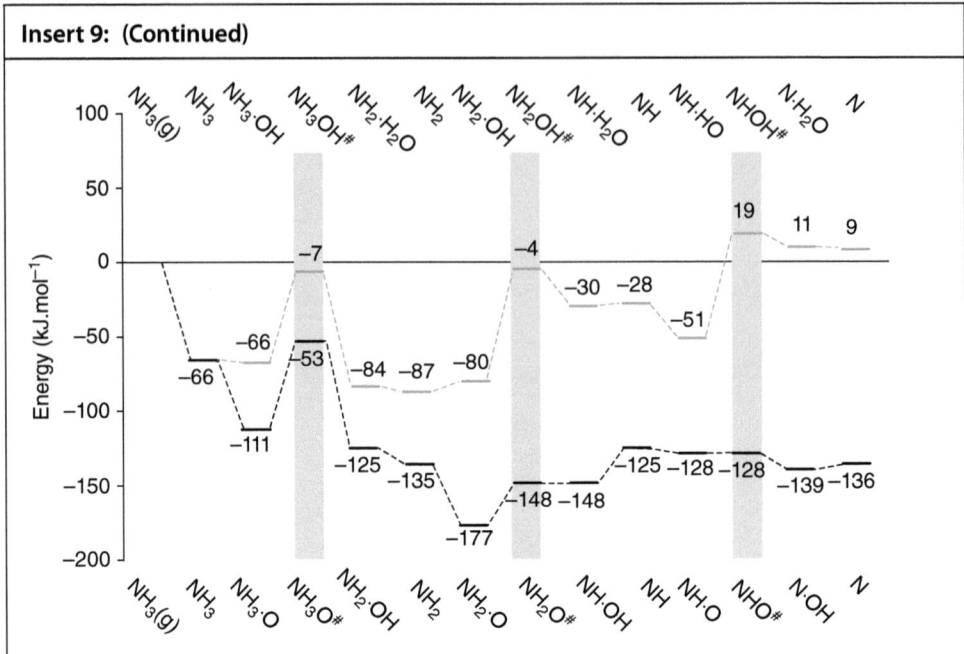

Figure 9i.2 Reaction pathway for the oxidative dehydrogenation of ammonia over Pt(100) assisted by O (black line, bottom legend) and OH (gray line, top legend). The energies are with respect to NH_3 (g) + 3 O_{ads} + 3 OH_{ads}, ZPE corrected. (Novell-Leruth et al. 2007 [61]). Reproduced with permission of American Chemical Society.)

Insert 9 shows a comparison of the very different reactivities of the Pt(111) and Pt(100) surfaces with respect to the activation of NH_3 and adsorbed O or OH. These differences illustrate the sensitivity of surface reactivity with respect to surface topology. On the Pt(111) terrace co-adsorbed O will promote N—H bond cleavage, but it will not promote dissociation of the N—H bond in NH_2 or NH. This is a consequence of the increased activation barrier when a bond with a surface metal atom is shared in the transition state. This is not the case on the Pt(100) surface, where O is twofold-coordinated to two metal atoms. Due to this square arrangement of the surface atoms they do not share boding with the same metal surface atom in the transition state. On the Pt(111) surface when x of NH_x decreases from 3 to 2 to 1 the NH_x fragment likewise moves consecutively from atop coordination to twofold then threefold coordination. Since the O atom adsorbs twofold, shared metal atom bonding must occur when it reacts with NH_2 and NH. In contrast, the (100) surface or a step-edge site will provide low activation energy paths for the reaction of these intermediates with adsorbed O, again because no sharing of bonds with the same surface metal atom occurs. Interestingly, the OH intermediate that is adsorbed atop a Pt atom will activate the NH and NH fragments on the (111) and (100) surfaces for the same reason (non-sharing of the same surface metal atom). Especially on the Pt(111) surface, the reaction with co-adsorbed OH leads only to the final formation of adsorbed N atoms.

When the reactivity of the metal surface increases, the strength of the M—H bond increases, and due to BOC the strength of the MO—H decreases. In this case, the promoting reaction with co-adsorbed O may disappear. For example, this occurs for the oxidation of ammonia on the Rh(111) surface [62]. On this surface the metal–oxygen bond is sufficiently strong to cause a dependence of NH_3 activation with O that is similar to methane activation on Pt. This is shown in Figure 7.10.

A similar observation has been made for the activation of the alcohol or H_2O and the formation of surface methoxy species or hydroxyl. In the case of methanol adsorption, O_{ads} will promote methanol OH bond cleavage and methoxy adsorption on Pt, but this reaction has a higher activation energy on the Rh(111) surface compared to direct OH bond cleavage by the surface metal atom [63].

7.4.2.3 Reactivity of the Oxide Overlayers

In Chapter 11, the chemical bonding aspects of the surface chemistry of reducible oxides will be discussed in detail. Here, we will focus on the reactivity of oxygen atoms that make up the oxide surface compared to their reactivity in an oxide overlayer of a bulk metal. We will distinguish substrate activation by a cation versus activation by an oxygen atom. We will discuss the differences in the reactivity of the cation compared to the metal atom for the activation of methane by the PdO(110) surface.

Another case we will study is a comparison of the surface reactivity of an O atom in an oxide overlayer versus an oxygen atom adsorbed to the metal surface. As an example, we will consider an oxygen atom in the oxide overlayer of an Ag particle. The oxygen atom reactivity in the overlayer is electrophilic, altered from its nucleophilic reactivity when it is adsorbed to the metal surface.

As we will see, the changes in reactivity in both cases are due to a reduction of Pauli repulsion between reagent and reacting surface ion.

7.4.2.3.1 Methane Activation by PdO

Methane activation by the PdO(110) surface has a lower barrier than activation on the transition metal surface. The chemical bonding features that lower the barrier are discussed in this subsection.

The increased reactivity of this transition metal oxide surface is unique and this feature is exploited in the use of Pd as the preferred catalyst for the combustion of methane in turbines. This process enables the combustion of methane in air at a temperature where NO_x formation can be prevented. The unique activity of PdO relates to its rather weak M—O bond energy and the sufficiently strong interaction of the Pd—CH_3 bond.

> **Insert 10: The Activation of Methane on PdO**
>
> Figure 10i.1 shows PdO bulk and PdO(101) surface structures. In the bulk, the Pd^{2+} ion has a square coordination with the oxygen atoms and each O atom is tetrahedrally surrounded by Pd ions. At the surface, two kinds of Pd ions are present. Half of them have

(Continued)

Insert 10: (Continued)

the same coordination as in the bulk, while the remaining Pd cations have three instead of four coordination.

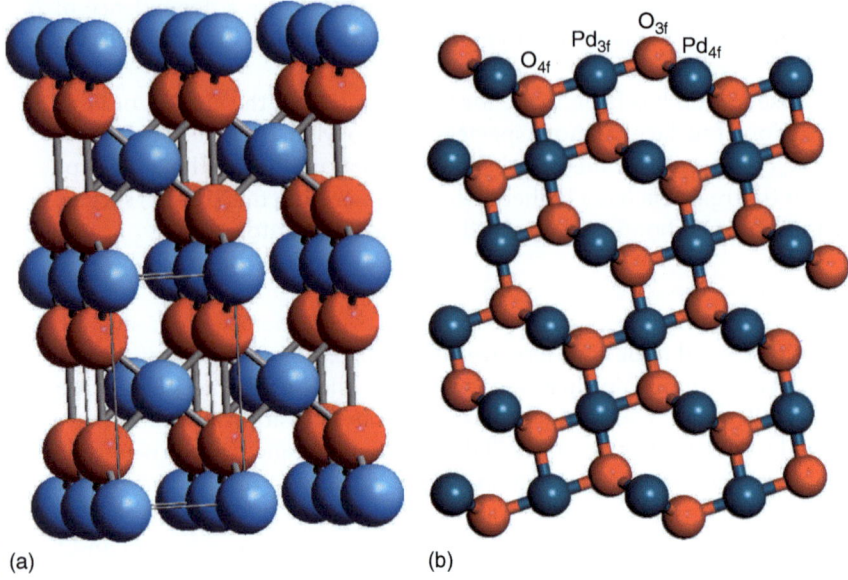

(a)   (b)

Figure 10i.1 (a) Bulk structure of PdO. (b) Structure of PdO(101) surface. Blue is palladium atom, red is oxygen atom.

In Figure 10i.2, the transition state structures of methane are compared for three systems.

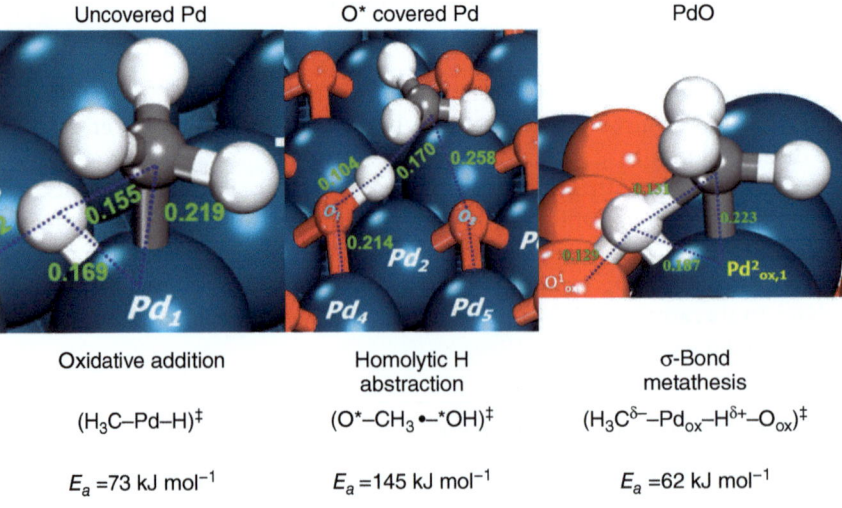

Uncovered Pd   O* covered Pd   PdO

Oxidative addition   Homolytic H abstraction   σ-Bond metathesis

$(H_3C-Pd-H)^{\ddagger}$ $(O^*-CH_3 \bullet -^*OH)^{\ddagger}$ $(H_3C^{\delta-}-Pd_{ox}-H^{\delta+}-O_{ox})^{\ddagger}$

$E_a = 73$ kJ mol^{-1} $E_a = 145$ kJ mol^{-1} $E_a = 62$ kJ mol^{-1}

Figure 10i.2 From left to right: comparison of the transition states for methane activation on Pd(111) surface in the absence (a) and presence (b) of adsorbed O and activation of methane on PdO(101) surface. (Chin *et al*. 2013 [64]. Reproduced with permission of American Chemical Society.)

Figure 10i.3 shows the PDOS and COHP of PdO. For the comparison with the Pd metal electronic structure we refer to Chapter 6, Figure 5i.2b. In PdO, the PDOS of the Pd d valence electrons becomes distributed over three electronic energy regimes. Below the reduced d valence electron band density in the energy interval similar to that of the metal d-electrons, new electronic density features appear due to bonding Pd—O molecular orbitals. At the top and above the metal d valence electron band density, the corresponding antibonding Pd—O molecular orbital density is found [65].

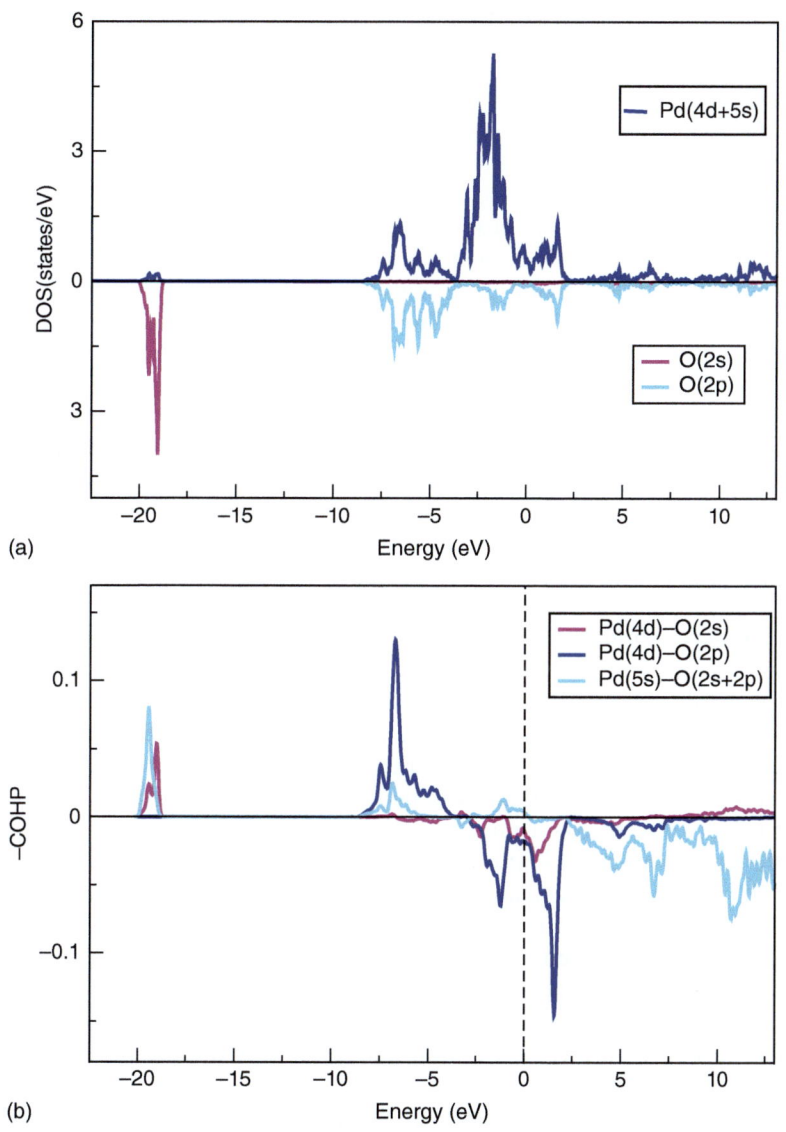

Figure 10i.3 (a) Partial density of state distributions of the valence electrons in bulk PdO. (b) Bond order densities (−COHP) distributions of the valence electrons of bulk PdO.

As illustrated in Insert 10 for palladium oxide, the oxygen atoms form strong bonding and antibonding molecular orbital fragments with the Pd atoms. This reduces the electron density on the Pd atoms and they become cations with a formal 2+ charge. This can be deduced from the appearance of an unoccupied antibonding PdO valence electron density above the Fermi level and a corresponding occupied bonding PdO density below the d valence electron band in the oxide.

The transition metal 5s and 4d valence electron intensity near the Fermi level is substantially reduced compared to the Pd metal.

The weak interaction between methane and the Pd metal surface is caused by the Pauli repulsive interaction between the doubly electron occupied CH orbitals of methane and the electron occupied d and s valence orbitals near the Fermi level at the metal surface. The s valence electrons have a wider extent, which chiefly determines the large Pd metal atom–CH methane distance. When the methane molecule interacts with the Pd cation, the Pauli repulsion with the s and d valence electrons is substantially lessened. Consequently, the CH_4 molecule can occupy a space closer to the Pd ion at the PdO surface than to the Pd atom on the metal surface. This enables the d valence atomic orbitals of Pd^{2+} to overlap more strongly with the methane H atomic orbitals compared with the overlap of the d valence atomic orbitals of the Pd metal. The methane C—H bonds in contact with the Pd^{2+} ion are weakened more than they would be on the metal surface. This increased interaction is reflected in the large (114°) HCH bond angle of methane adsorbed to the Pd^{2+} cation, which contrasts with the 109° bond angle for gas phase methane [69]. Since the contact with the Pd^{2+} ion already substantially activates the methane C—H bond in the adsorbed state, the activation energy required for subsequent C—H bond cleavage by reaction with a surface oxygen atom to produce Pd—CH_3 and adsorbed OH is substantially lower than cleavage by an oxygen atom adsorbed to the Pd metal surface.

7.4.2.3.2 Ethylene Epoxidation: Reactive Electrophilic and Nucleophilic Surface Oxygen

The selectivity of epoxide formation by the oxidation of ethylene catalyzed by Ag is a direct function of the state of the metal (oxide) surface. The process has been introduced in Chapter 2, Section titled "Ethylene Epoxidation: Ag catalysis,", and mechanistic details of this reaction will be presented later Chapter 8, Insert 9. Here, we will discuss the electronic basis for the strong dependence of the reactivity of the surface oxygen atoms on the effective charge of the Ag surface atoms.

The main difference between an oxygen atom adsorbed at low coverage to the Ag metal and an oxygen atom within the Ag_2O surface is that the adsorbed O atom is in an electron-rich environment while the atom within the surface must compete for electrons with other O atoms.

On the silver metal surface, the oxygen atom exhibits nucleophilic reactivity and prefers reaction with the CH proton instead of insertion into the electron rich π-bond.

In contrast, on the silver oxide surface, the Ag atom is in a cationic state, and hence electron donation to the oxygen atom is reduced. The oxygen atom will now exhibit electrophilic reactivity by inserting preferentially into the ethylene π bond, leading to the epoxide molecule as the preferred product.

As illustrated in Insert 11, on the metallic cluster the adatom oxygen atom orbitals are nearly fully occupied, which produces a strong Pauli repulsion with the electron occupied π C=C orbital. On the other hand, when the Ag atom is part of the oxide surface, the Ag—O bond is highly polarizable and electrons can be readily donated into the empty Ag cation d valence orbitals. In this way, Pauli repulsion is reduced since electron density can be removed from the space between the oxygen atom and ethylene when they interact. This reduced Pauli repulsive interaction is then converted into an attractive interaction [70].

Insert 11: Pauli Repulsion Decrease of Oxygen Atoms at Silver-Oxide Surface

In the figures of this Insert, we show the differences in the adatom Ag—O and corresponding O/C=C chemical bonds when the environment of the M—O bond is metallic or oxidic. The figures give DFT-calculated crystal bond order overlap populations (COOP) of the respective chemical bonds. The COOP is related to the COHP's that we used in previous sections (see Chapter 6, Insert 1). A positive sign for the COOP function means that the interaction is bonding, a negative sign means that it is antibonding. The interaction is weakened when a higher fraction of antibonding orbital fragments becomes occupied.

Figure 11i.1a and b display the interaction of ethylene with an oxygen atom adsorbed to the electron-rich Ag surface. Figures 11i.1c and d show the corresponding changes in COOP for an oxygen atom adsorbed to the oxidized surface.

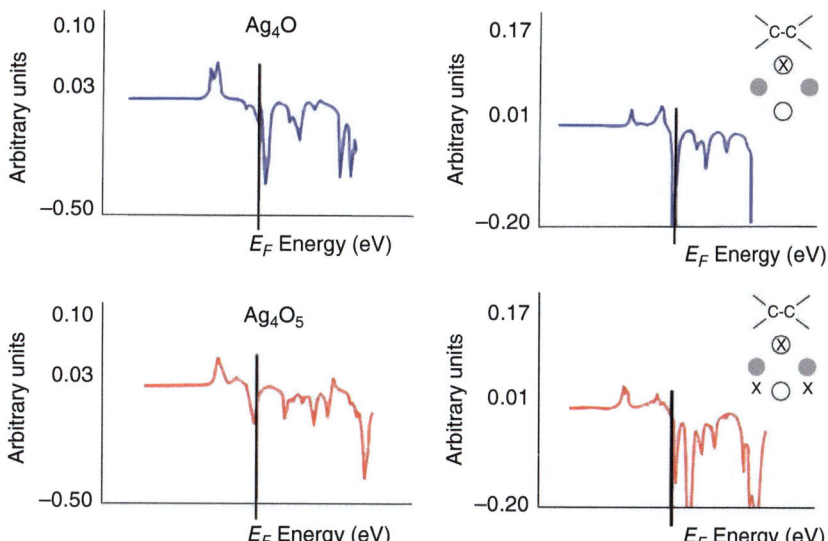

Figure 11i.1 COOP curves of the interaction between chemisorbed oxygen and ethylene. (a, b) The COOPs between O 2p atomic orbitals and the interacting Ag 5d atomic orbitals for the metallic cluster (a) and oxidized cluster (c), respectively. (c, d) The COOP curves of the interaction between adsorbed oxygen and the carbon–carbon 2p atomic orbitals of ethylene for the metallic (c) and oxidized cluster (d), respectively. (Van den Hoek *et al.* 1989 [71]. Reproduced with permission of American Chemical Society.)

Figure 7.13 Schematic representation of the two spin states of the oxygen atom adsorbed to the Ag surface. Of the two proposed states of atomic oxygen adsorbed at a high coordination site on the Ag surface, the oxyradical anion is thought to be the active species for ethylene epoxidation. (Carter and Goddard 1989 [72]. Reproduced with permission of Elsevier.)

di-σ   OXYRADICAL

A comparison of Figure 11i.1a and c shows that a large unoccupied antibonding electron density exists above the Fermi level of the Ag—O bond on the metallic particle. It shifts below the Fermi level and hence becomes occupied in the oxidized system. This is an indication of the weakening of the Ag—O bond on the oxidized particle compared to the bond on the metallic particle. The O–ethylene interaction shows the opposite, as illustrated by Figures 11i.b and d. An unoccupied antibonding density appears above the Fermi level in the oxidized system, but this remains occupied for the metallic case. Hence, the O–ethylene π bond interaction is the stronger interaction on the oxidized particle, which will lead to insertion of the oxygen atom into the ethylene π bond.

When the oxygen atom interacts weakly with a transition metal surface atom, its spin state can become a triplet as in the free atom. In contrast, a strongly adsorbed oxygen atom that is twofold or threefold coordinated to the transition metal surface is in a singlet state, which is less reactive.

It has been suggested that an oxygen atom that is in the triplet state [72] and weakly adsorbed to Ag is selective for ethylene epoxidation, whereas an oxygen atom adsorbed in the singlet state is non-selective. In Figure 7.13 these O adsorbed states are schematically shown and described as di-σ and oxyradical respectively. The di-σ state is representative for the state of O adsorbed threefold to the metallic Ag surface. It is nucleophilic because the three 2p atomic orbitals are doubly occupied. In contrast, the oxygen atom in the triplet state has partially occupied atomic orbitals and hence is electrophilic and will insert into the ethylene π-bond. According to this view, the difference in reactivity of weakly and strongly adsorbed oxygen atoms on the silver surface (as determined by the degree of surface oxidation) relates to their spin states.

References

1 Hoffmann, R. (1988) *Solids and Surfaces*, VCH Verlag, Weinheim.
2 Nørskov, J.K., Studt, F., Abild-Pedersen, F., and Bligaard, T. (2014) *Fundamental Concepts in Heterogeneous Catalysis*, John Wiley & Sons, Inc.
3 Methfessel, M., Hennig, D., and Scheffler, M. (1992) Trends of the surface relaxations, surface energies, and work functions of the 4 d transition metals. *Phys. Rev. B*, **46**, 4816–4829. doi: 10.1103/PhysRevB.46.4816
4 Michaelides, A. and Scheffler, M. (2012) An introduction the theory of metal surfaces, in *Textbook of Surface and Interface Science*, vol. **1** (ed. K. Wandelt), Wiley-VCH Verlag GmbH.
5 Gross, A. (2009) *Theoretical Surface Science*, Springer-Verlag, Berlin.

6 Desjongueres, M.C. and Spanjaard, D. (1996) *Concepts in Surface Physics*, Springer-Verlag.
7 van Santen, R.A. (1991) *Theoretical Heterogeneous Catalysis*, 1st edn, World Scientific.
8 Tranca, I. and van Santen, R.A. (2016) How molecular is the chemisorption bond, *Phys. Chem. Chem. Phys.*, **18**, 20868.
9 Alexandrova, A.N., Nayhouse, M.J., Huynh, M.T., Kuo, J.L., Melkonian, A.V., Chavez, G., Hernando, N.M., Kowal, M.D., and Liu, C.-P. (2012) Selected AB42 (A = C, Si, Ge; B = Al, Ga, In) ions: a battle between covalency and aromaticity, and prediction of square planar Si in SiIn42. *Phys. Chem. Chem. Phys.*, **14**, 14815, http://xlink.rsc.org/?DOI=c2cp41821e (accessed 21 September 2016).
10 Michaelides, A. and Hu, P. (2001) in *Theoretical Aspects of Heterogeneous Catalysis* (eds M.A. Chaer and M.A.C. Nascimento), Kluwer, pp. 199–215.
11 Dunnington, B.D. and Schmidt, J.R. (2015) Molecular bonding-based descriptors for surface adsorption and reactivity. *J. Catal.*, **324**, 50–58, http://linkinghub.elsevier.com/retrieve/pii/S0021951715000275 (accessed 21 September 2016).
12 Brookhart, M., Green, M.L.H., and Parkin, G. (2007) Agostic interactions in transition metal compounds. *Proc. Natl. Acad. Sci. U.S.A.*, **104**, 6908–6914. doi: 10.1073/pnas.0610747104
13 Ciobîcă, I.M. and van Santen, R.A. (2002) A DFT study of CH_x chemisorption and transition states for C–H activation on the Ru(1120) surface. *J. Phys. Chem. B*, **106**, 6200–6205.
14 van Santen, R.A. and Filot, I.A.W. (2013) in *The Chemical Bond* (eds G. Frenking and S. Shaik), Wiley-VCH Verlag GmbH, pp. 269–335.
15 Mavrikakis, M., Hammer, B., and Nørskov, J.K. (1998) Effect of strain on the reactivity of metal surfaces. *Phys. Rev. Lett.*, **81**, 2819–2822. doi: 10.1103/PhysRevLett.81.2819
16 Nilsson, A., Pettersson, L.G.M., Hammer, B., Bligaard, T., Christensen, C.H., and Norskov, J.K. (2005) The electronic structure effect in heterogeneous catalysis. *Catal. Lett.*, **100**, 111–114. doi: 10.1007/s10562-004-3434-9
17 Dewar, M. (1951) A review of π complex theory. *Bull. Soc. Chim. Fr.*, **18**, C79.
18 Chatt, J. and Duncanson, L.A. (1953) 586. Olefin co-ordination compounds. Part III. Infra-red spectra and structure: attempted preparation of acetylene complexes. *J. Chem. Soc.*, 2939. doi: 10.1039/jr9530002939
19 Blyholder, G. (1964) Molecular orbital view of chemisorbed carbon monoxide. *J. Phys. Chem.*, **68**, 2772–2777. doi: 10.1021/j100792a006
20 Hammer, B., Morikawa, Y., and Nørskov, J.K. (1996) CO chemisorption at metal surfaces and overlayers. *Phys. Rev. Lett.*, **76**, 2141–2144. doi: 10.1103/PhysRevLett.76.2141
21 Föhlisch, A., Nyberg, M., Bennich, P., Triguero, L., Hasselström, J., Karis, O., Pettersson, L.G.M., and Nilsson, A. (2000) The bonding of CO to metal surfaces. *J. Chem. Phys.*, **112**, 1946. doi: 10.1063/1.480773
22 Schimka, L., Harl, J., Stroppa, A., Grüneis, A., Marsman, M., Mittendorfer, F., and Kresse, G. (2010) Accurate surface and adsorption energies

from many-body perturbation theory. *Nat. Mater.*, **9**, 741–744. doi: 10.1038/nmat2806

23 Gajdo, M., Eichler, A., and Hafner, J. (2004) CO adsorption on close-packed transition and noble metal surfaces: trends from ab initio calculations. *J. Phys. Condens. Matter*, **16**, 1141–1164, http://stacks.iop.org/0953-8984/16/i=8/a=001?key=crossref.daee232491c288ed683660d82e87347e (accessed 21 September 2016)..

24 Cimino, A., Kemball, C., Burwell, R.L., Bond, G.C., Blyholder, G., Schwab, G.-M., van Reijen, L.L., Ertl, G., Rozendaal, I.A., Cossee, P. et al. (1966) General discussion. *Discuss. Faraday Soc.*, **41**, 249. doi: 10.1039/df9664100249

25 Besenbacher, F., Chorkendorff, I., Clausen, B.S., Hammer, B., Molenbroek, A.M., Nørskov, J.K., and Stensgaard, I. (1998) Design of a surface alloy catalyst for steam reforming. *Science*, **279**, 1913–1915, doi: 10.1126/science.279.5358.1913.

26 van Santen, R.A. and Neurock, M. (2006) *Molecular Heterogenous Catalysis*, Wiley-VCH Verlag GmbH.

27 Sueyoshi, T., Sasaki, T., and Iwasawa, Y. (1996) Coadsorption of NO and NH 3 on Cu(111): the formation of the stabilized (2 × 2) coadlayer. *J. Phys. Chem.*, **100**, 13646–13654. doi: 10.1021/jp9606265

28 Abild-Pedersen, F., Studt, F., Rossmeisl, J., Munter, T.R., Moses, P.G., Skúlason, E., Bligaard, T., Nørskov, J.K., and Greely, J. (2007) Scaling properties of adsorption energies for hydrogen-containing molecules on transition-metal surfaces. *Phys. Rev. Lett.*, **99**, 16105–16109.

29 Shustorovich, E. (1986) Chemisorption phenomena: analytic modeling based on perturbation theory and bond-order conservation. *Surf. Sci. Rep.*, **6**, 1–63, http://linkinghub.elsevier.com/retrieve/pii/0167572986900038 (accessed 31 October 2014).

30 Calle-Vallejo, F., Loffreda, D., Koper, M.T.M., and Sautet, P. (2015) Introducing structural sensitivity into adsorption – energy scaling relations by means of coordination numbers. *Nat. Chem.* doi: 10.1038/nchem.2226

31 van Santen, R.A., Neurock, M., and Shetty, S. (2010) Reactivity theory of transition-metal surfaces: a Brønsted–Evans–Polanyi linear activation energy – free-energy analysis. *Chem. Rev.*, **110**, 2005–2018.

32 Kratzer, P., Hammer, B., and Nørskov, J.K. (1996) A theoretical study of CH_4 dissociation on pure and gold-alloyed Ni(111) surfaces. *J. Chem. Phys.*, **105**, 5595. doi: 10.1063/1.472399

33 Huang, W., Sun, L., Han, P., and Zhao, J. (2012) CH4 dissociation on Co(0001): a density functional theory study. *J. Nat. Gas Chem.*, **21**, 98–103, http://linkinghub.elsevier.com/retrieve/pii/S1003995311603393 (accessed 21 September 2016).

34 Watwe, R.M., Bengaard, H.S., Rostrup-Nielsen, J.R., Dumesic, J.A., and Nørskov, J.K. (2000) Theoretical studies of stability and reactivity of CH_x species on Ni(111). *J. Catal.*, **189**, 16–30, http://linkinghub.elsevier.com/retrieve/pii/S0021951799926994 (accessed 21 September 2016).

35 Liu, Z.-P. and Hu, P. (2003) General rules for predicting where a catalytic reaction should occur on metal surfaces: a density functional theory study of C–H and C–O bond breaking/making on flat, stepped, and kinked metal

surfaces. *J. Am. Chem. Soc.*, **125**, 1958–1967, http://www.ncbi.nlm.nih.gov/pubmed/12580623 (accessed 17 December 2014)..

36 Ciobîcă, I.M., Frechard, F., van Santen, R.A., Kleyn, A.W., and Hafner, J. (2000) A DFT study of transition states for C–H activation on the Ru(0001) surface †. *J. Phys. Chem. B*, **104**, 3364–3369. doi: 10.1021/jp993314l

37 Wang, B., Song, L., and Zhang, R. (2012) The dehydrogenation of CH4 on Rh(111), Rh(110) and Rh(100) surfaces: a density functional theory study. *Appl. Surf. Sci.*, **258**, 3714–3722, http://linkinghub.elsevier.com/retrieve/pii/S0169433211019039 (accessed 21 September 2016).

38 Bunnink, B. and Kramer, G. (2006) Energetics of methane dissociative adsorption on Rh{111} from DFT calculations. *J. Catal.*, **242**, 309–318, http://linkinghub.elsevier.com/retrieve/pii/S0021951706002132 (accessed 21 September 2016).

39 Henkelman, G. and Jónsson, H. (2001) Theoretical calculations of dissociative adsorption of CH4 on an Ir(111) surface. *Phys. Rev. Lett.*, **86**, 664–667. doi: 10.1103/PhysRevLett.86.664

40 Qi, Q., Wang, X., Chen, L., and Li, B. (2013) Methane dissociation on Pt(111), Ir(111) and PtIr(111) surface: a density functional theory study. *Appl. Surf. Sci.*, **284**, 784–791, http://linkinghub.elsevier.com/retrieve/pii/S0169433213014864 (accessed 21 September 2016).

41 Zhang, R., Song, L., and Wang, Y. (2012) Insight into the adsorption and dissociation of CH4 on Pt(hkl) surfaces: a theoretical study. *Appl. Surf. Sci.*, **258**, 7154–7160, http://linkinghub.elsevier.com/retrieve/pii/S0169433212006629 (accessed 21 September 2016).

42 Liu, Z.-P. and Hu, P. (2001) General trends in the barriers of catalytic reactions on transition metal surfaces. *J. Chem. Phys.*, **115**, 4977. doi: 10.1063/1.1403006

43 Hoffmann, R. and Woodward, R.B. (1967) *The Conservation of Orbital Symmetry*.

44 Frechard, F., van Santen, R.A., Siokou, A., Niemantsverdriet, J.W., and Hafner, J. (1999) Adsorption of ammonia on the rhodium (111), (100), and stepped (100) surfaces: an ab initio and experimental study. *J. Chem. Phys.*, **111**, 8124. doi: 10.1063/1.480146

45 Offermans, W.K., Jansen, A.P.J., and van Santen, R.A. (2006) Ammonia activation on platinum {111}: a density functional theory study. *Surf. Sci.*, **600**, 1714–1734, http://linkinghub.elsevier.com/retrieve/pii/S0039602806000665 (accessed 22 December 2014).

46 Van Santen, R.A. (2009) Complementary structure sensitive and insensitive catalytic relationships. *Acc. Chem. Res.*, **42**, 57–66, http://www.ncbi.nlm.nih.gov/pubmed/18986176 (accessed 27 November 2014).

47 Nørskov, J.K., Bligaard, T., Rossmeisl, J., and Christensen, C.H. (2009) Towards the computational design of solid catalysts. *Nat. Chem.*, **1**, 37–46. doi: 10.1038/nchem.121

48 Vidal-Iglesias, F., García-Aráez, N., Montiel, V., Feliu, J., and Aldaz, A. (2003) Selective electrocatalysis of ammonia oxidation on Pt(100) sites in alkaline medium. *Electrochem. Commun.*, **5**, 22–26, http://linkinghub.elsevier.com/retrieve/pii/S1388248102005210 (accessed 19 December 2014).

49 Hori, Y., Takahashi, I., Koga, O., and Hoshi, N. (2002) Selective formation of C2 compounds from electrochemical reduction of CO_2 at a series of copper single crystal electrodes. *J. Phys. Chem. B*, **106**, 15–17. doi: 10.1021/jp013478d

50 Gattrell, M., Gupta, N., and Co, A. (2006) A review of the aqueous electrochemical reduction of CO_2 to hydrocarbons at copper. *J. Electroanal. Chem.*, **594**, 1–19, http://linkinghub.elsevier.com/retrieve/pii/S0022072806002853 (accessed 9 July 2014).

51 Somorjai, G.A. and Li, Y. (2010) *Introduction to Surface Chemistry and Catalysis*, 2nd edn, Wiley-VCH Verlag GmbH.

52 Prettre, M., Eichner, C., and Perrin, M. (1946) The catalytic oxidation of methane to carbon monoxide and hydrogen. *Trans. Faraday Soc.*, **42**, 335b. doi: 10.1039/tf946420335b

53 Chin, Y.-H., Buda, C., Neurock, M., and Iglesia, E. (2011) Reactivity of chemisorbed oxygen atoms and their catalytic consequences during CH_4–O_2 catalysis on supported Pt clusters. *J. Am. Chem. Soc.*, **133**, 15958–15978.

54 Ciobîcă, I.M. and van Santen, R.A. (2003) Carbon monoxide dissociation on planar and stepped Ru(0001) surfaces. *J. Phys. Chem. B*, **107**, 3808–3812.

55 Shetty, S., Jansen, A.P.J., and van Santen, R.A. (2009) Direct versus hydrogen-assisted CO dissociation. *J. Am. Chem. Soc.*, **131**, 12874–12875.

56 Madix, R.J. (1980) Reaction kinetics and mechanism on metal single crystal surfaces. *Adv. Catal.*, **29**, 1–53.

57 Brown, N.F. and Barteau, M.A. (1992) Reactions of unsaturated oxygenates on rhodium(111) as probes of multiple coordination of adsorbates. *J. Am. Chem. Soc.*, **114**, 4258–4265. doi: 10.1021/ja00037a032

58 Xu, X. and Friend, C.M. (1991) Partial oxidation without allylic carbon–hydrogen bond activation: the conversion of propene to acetone on rhodium(111)-p(2.times.1)-O. *J. Am. Chem. Soc.*, **113**, 6779–6785. doi: 10.1021/ja00018a010

59 Hickman, D.A. and Schmidt, L.D. (1993) Production of syngas by direct catalytic oxidation of methane. *Science*, **259**, 343–346, http://www.ncbi.nlm.nih.gov/pubmed/17832347 (accessed 20 January 2015).

60 Offermans, W.K., Jansen, A.P.J., VanSanten, R.A., Novell-Leruth, G., Ricart, J.M., and Perez-Ramirez, J. (2007) Ammonia dissociation on Pt{100}, Pt{111}, and Pt{211}: a comparative density functional theory study. *J. Phys. Chem. C*, **111**, 17551–17557. doi: 10.1021/jp073083f

61 Novell-Leruth, G., Valcárcel, A., Pérez-Ramírez, J., and Ricart, J.M. (2007) Ammonia dehydrogenation over platinum-group metal surfaces. Structure, stability, and reactivity of adsorbed NH_x species. *J. Phys. Chem. C*, **111**, 860–868. doi: 10.1021/jp064742b

62 Popa, C., VanSanten, R.A., and Jansen, A.P.J. (2007) Density-functional theory study of NH_x oxidation and reverse reactions on the Rh(111) surface. *J. Phys. Chem. C*, **111**, 9839–9852. doi: 10.1021/jp071072g

63 Chen, M., Bates, S.P., van Santen, R.A., and Friend, C.M. (1997) The chemical nature of atomic oxygen adsorbed on Rh(111) and Pt(111): a density functional study. *J. Phys. Chem. B*, **101**, 10051–10057. doi: 10.1021/jp971499v

64 Chin, Y.-H.C., Buda, C., Neurock, M., and Iglesia, E. (2013) Consequences of metal-oxide interconversion for C-H bond activation during CH4 reactions on Pd catalysts. *J. Am. Chem. Soc.*, **135**, 15425–15442.

65 van Santen, R.A., Tranca, I., and Hensen, E.J.M. (2015) Theory of surface chemistry and reactivity of reducible oxides. *Catal. Today*, **244**, 63–84, http://linkinghub.elsevier.com/retrieve/pii/S092058611400488X (accessed 22 September 2016).

66 Soler, J.M., Artacho, E., Gale, J.D., García, A., Junquera, J., Ordejón, P., and Sánchez-Portal, D. (2002) The SIESTA method for ab initio order-N materials simulation. *J. Phys. Condens. Matter*, **14**, 2745–2779, http://stacks.iop.org/0953-8984/14/i=11/a=302?key=crossref.8ed2406c09184bcd143191af26e9f492 (accessed 17 July 2014).

67 Izquierdo, J., Vega, A., Balbás, L., Sánchez-Portal, D., Junquera, J., Artacho, E., Soler, J., and Ordejón, P. (2000) Systematic ab initio study of the electronic and magnetic properties of different pure and mixed iron systems. *Phys. Rev. B*, **61**, 13639–13646. doi: 10.1103/PhysRevB.61.13639

68 Robles, R., Izquierdo, J., Vega, A., and Balbás, L. (2001) All-electron and pseudopotential study of the spin-polarization of the V(001) surface: LDA versus GGA. *Phys. Rev. B*, **63**, 172406. doi: 10.1103/PhysRevB.63.172406

69 Hellman, A., Resta, A., Martin, N.M., Gustafson, J., Trinchero, A., Carlsson, P.-A., Balmes, O., Felici, R., van Rijn, R., Frenken, J.W.M. et al. (2012) The active phase of palladium during methane oxidation. *J. Phys. Chem. Lett.*, **3**, 678–682.

70 Jacobsen, C.J.H., Dahl, S., Boisen, A., Clausen, B.S., Topsøe, H., Logadottir, A., and Nørskov, J.K. (2002) Optimal catalyst curves: connecting density functional theory calculations with industrial reactor design and catalyst selection. *J. Catal.*, **205**, 382–387, http://linkinghub.elsevier.com/retrieve/pii/S0021951701934426 (accessed 20 January 2015).

71 Van den Hoek, P.J., Baerends, E.J., and Van Santen, R.A. (1989) Ethylene epoxidation on silver(110): the role of subsurface oxygen. *J. Phys. Chem.*, **93**, 6469–6475. doi: 10.1021/j100354a038

72 Carter, E.A. and Goddard, W.A. (1989) Chemisorption of oxygen, chlorine, hydrogen, hydroxide, and ethylene on silver clusters: a model for the olefin epoxidation reaction. *Surf. Sci. Lett.*, **209**, A41–A42.

8

Mechanisms of Transition Metal Catalyzed Reactions

8.1 Introduction

This chapter discusses the mechanisms of the various reactions catalyzed by transition metals, with a focus on the molecular and atomistic aspects. The catalytic systems chosen introduce and illustrate relevant concepts that are basic to the physical organic chemistry of transition metal catalysis. We will address the question how the state of the metal surface changes at reaction conditions. Several of these reactions have been introduced previously in Chapter 2 on catalytic processes. When possible the mechanism of reactions catalyzed by transition metal surfaces is compared with that of related enzyme or coordination complex catalyzed reactions.

8.2 Hydrogenation Reactions

8.2.1 Ammonia Synthesis

8.2.1.1 Heterogeneous Catalytic Reaction

The reaction using the Haber–Bosch process that converts N_2 with H_2 to give ammonia (see Chapter 1, Insert 1) is at present one of the catalytic heterogeneous processes that is best understood at the molecular level. Surface science model studies and computational studies have provided a nearly complete understanding of the reaction. Essential molecular information can be found in Insert 1.

> **Insert 1: Molecular Basis of Ammonia Synthesis Reaction**
>
> In this Insert simulated volcano curves of the ammonia synthesis reactions are presented in Figure 1i.1. In Figure 1i.2, an experimentally determined reaction energy diagram of the elementary reactions of this reaction is shown for a Fe catalyst. Figure 1i.3 gives
>
> *(Continued)*

Modern Heterogeneous Catalysis: An Introduction, First Edition. Rutger A. van Santen.
© 2017 Wiley-VCH Verlag GmbH & Co. KGaA. Published 2017 by Wiley-VCH Verlag GmbH & Co. KGaA.

Insert 1: (Continued)

experimental reactivity data from single crystal experiments on Fe. Figure 1i.4 shows a calculated reaction energy diagram of N_2 activation for the stepped and nonstepped Ru surface.

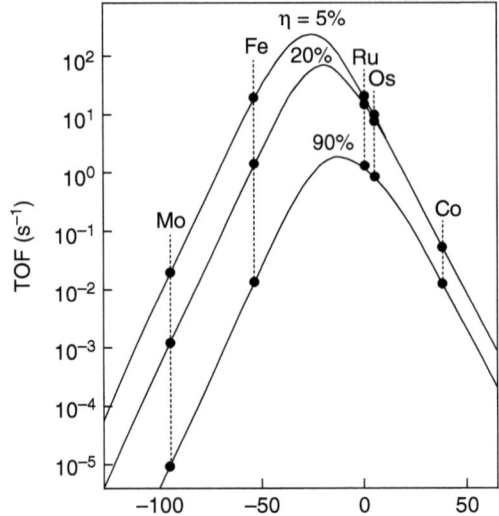

Figure 1i.1 Calculated volcano curves at 420 °C, 80 bar, 2 : 1 H_2:N_2, equilibrium 17.4% NH_3 as a function of the N adsorption energy. (Jacobsen et al. 2002 [1]. Reproduced with permission of Elsevier.)

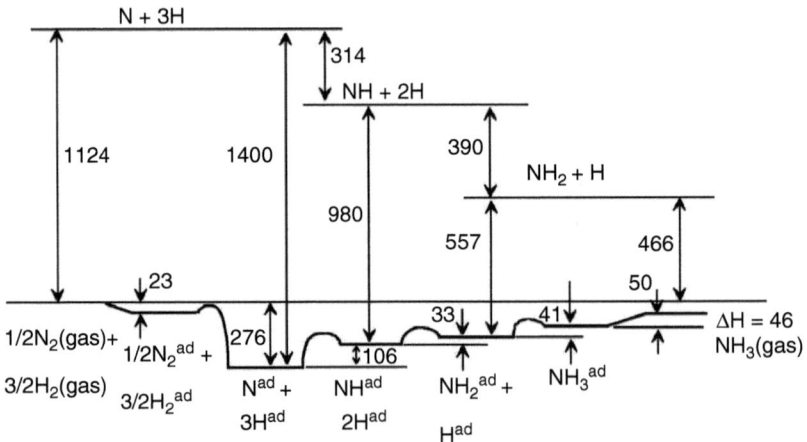

Figure 1i.2 Schematic energy profiles for ammonia synthesis on a promoted iron catalyst with energies in kJ mol^{-1}. (Ertl 1983 [2]. Reproduced with permission of Springer.)

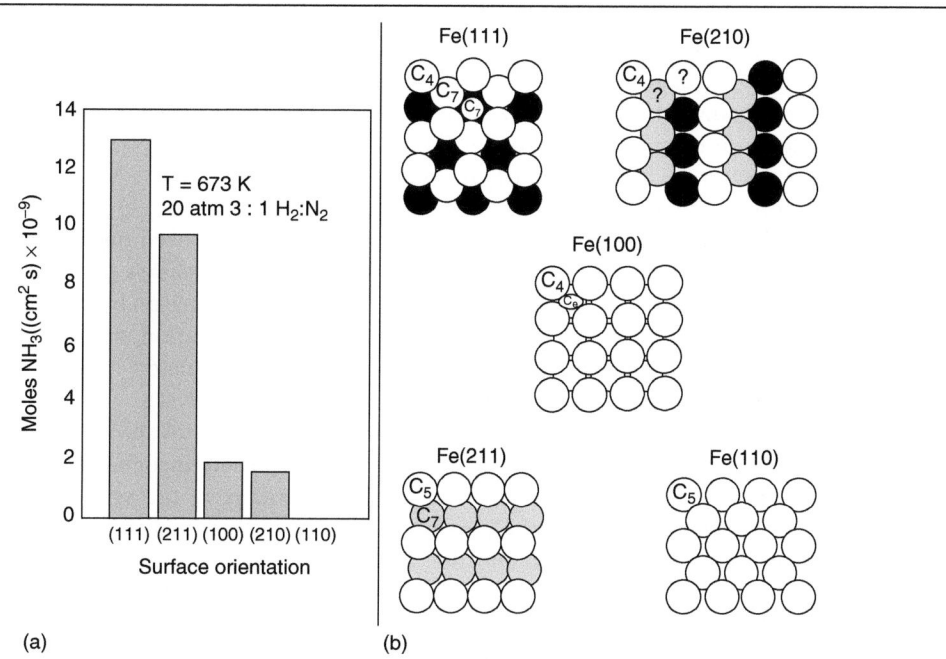

(a) (b)

Figure 1i.3 (a) Rates of ammonia synthesis over five iron single-crystal surfaces with different orientations ((111), (211), (100), (210), and (110)). (b) Schematic representation of the idealized surface structures of an iron single crystal. The coordination number of each atom is indicated. (Spencer 1982 [3]. Reproduced with permission of Elsevier.)

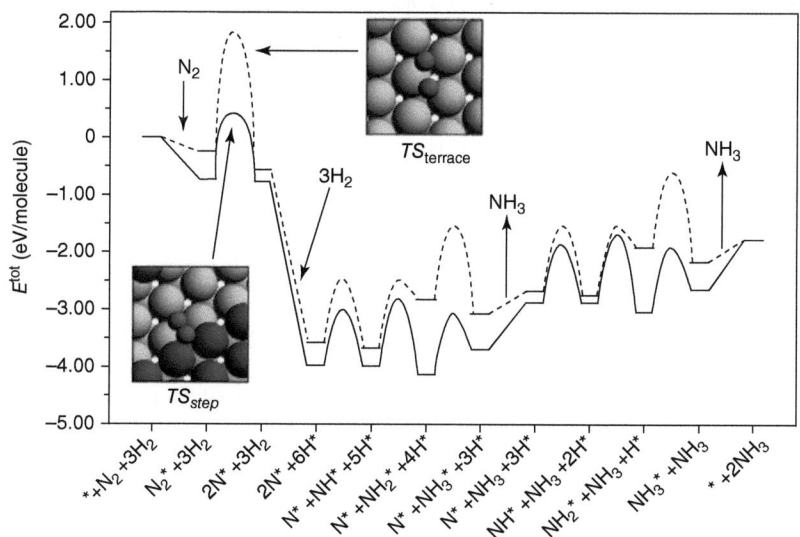

Figure 1i.4 The calculated potential (or reaction) energy (E_{tot}) diagram for NH_3 synthesis from N_2 and H_2 over close-packed (001) and stepped Ru surfaces. A * denotes an empty site and X* denotes an adsorbed species. The configuration of the transition states (TS) for N_2 dissociation over the terrace and step sites is shown in the insets [4].

The microkinetics simulations shown in fig. 1i.1 based on the use of Brønsted–Evans–Polanyi (BEP) relations and quantum-chemical results as shown in Figure 1i.4 allow us to relate the rate of the ammonia synthesis reaction with the N adatom bond energy. The adsorption energy of the N atom is a reactivity performance descriptor for the reaction. From known N_2 adsorption energies this can be used to identify the metal that has the optimum M—N atom interaction, which will produce the Sabatier rate maximum.

While the most common catalysts in commercial use are based on Fe or Co, the computationally constructed volcano curve (Figure 1i.1) shows that the Ru catalyst is predicted to be more active. Note that direct comparison between the experimental and practical systems should be done with caution, since a specific surface topology is used for the simulations and the simulated catalysts are nonpromoted. This is in contrast to the original Fe based catalyst.

The calculated reaction energy diagram shown in Figure 1i.4 shows that the N_2 and H_2 molecules adsorb dissociatively and the adsorbed nitrogen atoms are hydrogenated in consecutive steps. The minimum energy of the system is for adsorbed nitrogen, so that at the maximum of the volcano curve the competing reactions are N_2 activation and the ammonia formation from adsorbed N atoms. Note that the barrier for N_2 activation is high on the Ru(0001) terrace but is reduced for a stepped surface.

The rate of reaction is found to decrease sharply when the size of the Ru particles is less than 3 nm. This is because step-edge sites are not stable on particles that are too small (see Section 4.4.1).

The surface science data of Somorjai et al. (Figure 1i.3) also illustrate the structure sensitivity of the reaction. The rate of reaction increases significantly with coordinative unsaturation of the surface atoms. The data indicate that N_2 dissociation is rate-controlling at the conditions used in the Somorjai experiment.

The kinetics of the reaction is described by the Temkin kinetic expression [5, 6]

$$R_t = r_f[N_2]\left\{\frac{[H_2]^3}{[NH_3]}\right\}^m - k_b\left\{\frac{[H_2]^3}{[NH_3]}\right\}^{1-m} \tag{8.1}$$

As expected, the rate of reaction is first order in nitrogen partial pressure and has a positive order in hydrogen. The reaction becomes poisoned by ammonia adsorption at high pressures of ammonia.

The experimentally determined reaction energy diagram for the commonly used Fe/K system is also shown (see Figure 1i.2). On this catalyst, the barrier of N_2 dissociation is comparable to that of the stepped Ru surface, but ammonia formation from adsorbed N has an overall higher barrier.

The next subsection presents a comparison of the mechanism of the heterogeneous catalytic reaction with the mechanism of the analog enzyme catalyst, which is based on catalysis by the cofactor Fe_7MoS_6 cluster.

8.2.1.2 Enzyme Catalysis

In the biological realm, ammonia formation by plants occurs at room temperature. It is catalyzed by the nitrogenase enzyme which reacts N_2 with protons

Figure 8.1 The structure of the FeMo cofactor of the nitrogen-fixing enzyme nitrogenase. (Branden and Tooze 1991 [7]. Reproduced with permission of John Wiley and Sons.)

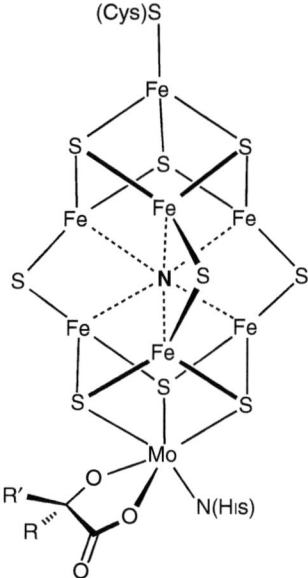

and electrons. Rather than temperature, the driving force for this reaction is the electrochemical potential of the system that reduces the protonated reaction intermediates. The catalytically active cofactor of nitrogenase is a Fe_7MoNS_8 cluster (see Figure 8.1) that is embedded in the enzyme molecule peptide shell. The actual reaction that occurs is:

$$N_2 + 8e^- + H^+ \rightleftharpoons 2NH_3 + H_2$$

An extra H_2 molecule is formed in this reaction. The mechanism of the enzyme reaction is different from the mechanism of the heterogeneous catalytic reaction because, at the enzyme reaction's low temperature, N_2 cannot be activated to directly dissociate. NH_3 formation occurs by consecutive addition of a proton and an electron to the N_2 molecule. N—N bond cleavage occurs only once hydrazine has been formed.

As is illustrated in Figure 8.1, the N_2 binding part in the cofactor is a coordination complex that contains seven Fe atoms and one Mo atom. N_2 is activated through adsorption into the cavity provided by six Fe atoms. Interestingly, the same metal is used in the heterogeneous catalytic process as ammonia synthesis catalyst.

8.2.2 Synthesis Gas Conversion to Methane and Liquid Hydrocarbons

A short historic introduction to Fischer–Tropsch (FT) reaction can be found in Section 2.2.3.2. The Fischer–Tropsch reaction converts synthesis gas that consists of CO and H_2 into a mixture of product molecules dominated by linear long chain hydrocarbons. An important question is how this reaction competes with the production of undesirable light gasses such as methane or ethane, and how the production of undesirable products can be suppressed.

Historically, several reaction mechanisms have been proposed based on different theories about the formation and growth of the hydrocarbon chain.

In essence, two distinct mechanistic proposls can be distinguished. The issue is whether the C—O bond of CO must first cleave with subsequent insertion of a CH_x species into the growing hydrocarbon chain or whether chain growth proceeds by initial CO insertion into the growing hydrocarbon chain with subsequent cleavage of the C—O bond. The two mechanisms are called the carbide mechanism and CO insertion mechanism, respectively (see Insert 2). Identifying the proper mechanism has important consequences for understanding the reactivity–structure relationship and the kinetics of the reaction.

Insert 2: The Two Competing Catalytic Mechanisms of Fischer–Tropsch

Figures 2i.1 and 2i.2 schematically present the catalytic cycles according to the carbide mechanism and the CO insertion mechanism, respectively. Key surface intermediates and relative rates of elementary reaction steps are indicated for high selectivity of long chain hydrocarbons and low selectivity of methane.

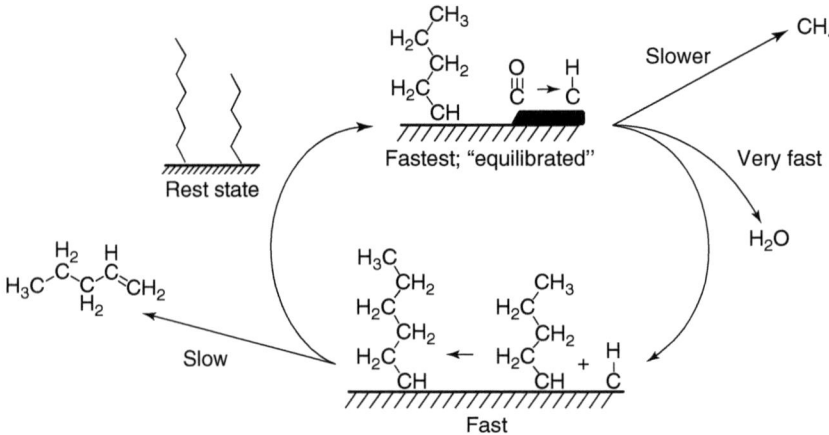

Figure 2i.1 Schematic illustration of the carbide mechanism. Relative rates of elementary rate constants for high chain growth are indicated.

The carbide mechanism of Figure 2i.1 can be described by one catalytic cycle. The role of H_2 is not explicitly indicated. The catalytic cycle is initiated by the dissociation of CO. CH intermediates are formed that compete for insertion into the growing hydrocarbon chain or for methane formation. The hydrocarbons desorb from the surface as olefins, but oxygenates can also be formed when termination occurs after the CO insertion step (not shown).

The CO insertion chain growth mechanism (Figure 2i.2) consists of an initiation step, which is illustrated at the right of the figure, followed by the chain growth reactions illustrated at the left. In the initiation step, CO dissociates to form the C_1 species on which the next chain growth steps occur. In this step, the CH_x species also reacts with CO to produce a CH_xCO fragment. Once the C—O bond cleaves, a C_2 intermediate is generated that can initiate the chain growth reaction by consecutive CO insertion, hydrogen addition, and C—O bond cleavage. The chain growth step can also be terminated by a competitive desorption step that generates an oxygenated hydrocarbon or alkane or alkene.

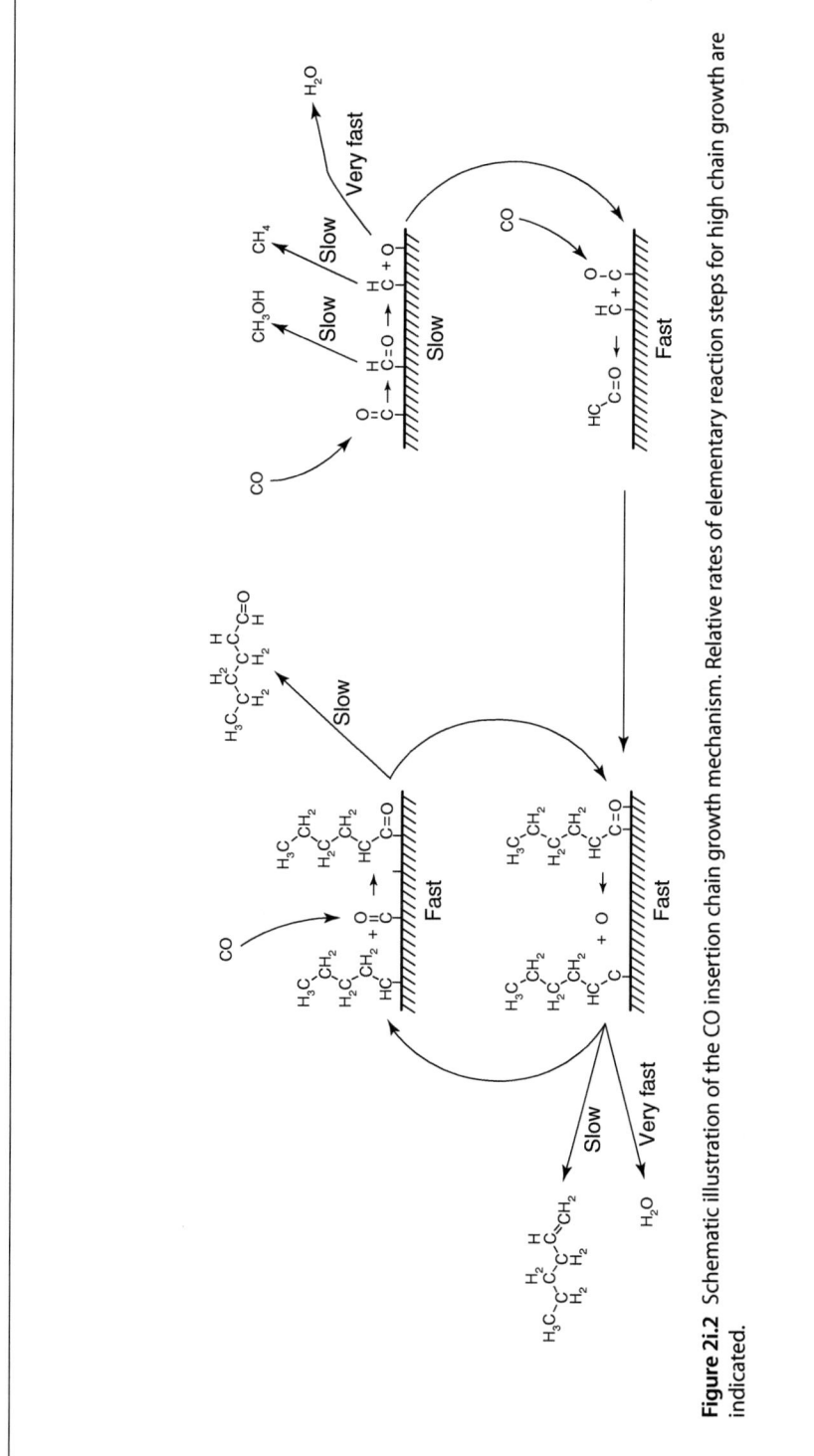

Figure 2i.2 Schematic illustration of the CO insertion chain growth mechanism. Relative rates of elementary reaction steps for high chain growth are indicated.

In the carbide mechanism, the Fischer–Tropsch reaction is initiated by CO decomposition into a C and O surface-adsorbed species. This activation can be either direct or assisted by hydrogen (see Section 7.4.2.1). In the hydrogen-assisted case, a CH intermediate is initially formed, but in the direct case the C atom is hydrogenated to a CH intermediate after the CO dissociation reaction step. Other hydrogen atom addition reactions will occur, allowing the CH_x intermediates to recombine and insert into the growing hydrocarbon chain. O_{ads} can be removed by reaction with CO to give CO_2 (a reaction that occurs on Fe), or by reaction with adsorbed hydrogen atoms (the dominant reaction on Co, Ni, or Ru). The growing hydrocarbon chain is reacted away from the surface as an alkene an alkane, or by a terminating CO insertion reaction that will produce an oxygenated hydrocarbon. This terminating reaction has a mechanistic similarity with the hydroformylation reaction (see Chapter 2, Insert 5).

In contrast, the CO insertion mechanism assumes that chain growth occurs through the insertion of CO into the growing hydrocarbon chain. C—O bond cleavage to produce a CH_x intermediate occurs only in the first reaction step. Then, $CH_{x,ads}$ and CO recombine to give an intermediate that will produce the corresponding alcohol or aldehyde upon hydrogenation. Alternatively, in the precursor molecular fragment the C—O bond cleaves, and a partially hydrogenated C_2H_x species is formed. This intermediate is ready for subsequent chain growth by a consecutive CO insertion step, followed by hydrogen atom addition and C—O bond cleavage. Chain growth can continue with a repetition of these steps until the chain terminates as an oxygenate or a hydrocarbon. There are many variations of this mechanism of chain growth through the insertion of CO; for example, HCO or H_2COH insertion into the growing hydrocarbon chain has also been proposed.

Currently, the carbide mechanism is predominantly accepted as the mechanism for the chain growth reaction [8]. In this mechanism, the rate constant of C—O bond cleavage reaction step must be relatively fast and the rate of hydrogenation of CH_x intermediates to produce methane must be slow. Surfaces of Co and Ru that contain step-edge sites satisfy these conditions, as does the surface of more reactive Fe, which is however in a carbidic phase during the reaction.

Ni (which is next to Co in the same row in the periodic table) is not a Fischer–Tropsch catalyst since its barrier for C—O bond cleavage is too high and the rate of hydrogenation of adsorbed CH_x species is too fast. Instead, it is a good methanation catalyst. Rh (next to Ru in the table) has a higher barrier for CO activation, which may explain why it also produces a substantial fraction of oxygenates in addition to alkanes and alkenes.

An elegant (and now classical) isotope exchange experiment that supports the carbide mechanism proceeds as follows: ^{13}CO is dissociated on a reduced Ru catalyst and the ^{13}C species left on the surface is hydrogenated in a mixture of gas phase ^{12}CO and H_2. In this experiment, the resulting hydrocarbons are found to contain at least two ^{13}C atoms per hydrocarbon chain. This proves that a C_1 carbon atom generated on the catalyst surface before the reaction begins can insert into the growing hydrocarbon chain [9].

Microkinetics simulations based on quantum-chemical reactivity data on a reactive surface of Ru provide insight into the dependence of this mechanism on surface chemical reactivity [10], see Insert 3.

Insert 3: Microkinetics Simulation of the Fischer–Tropsch Reaction

Simulated kinetic data are presented for a reactive stepped Ru(1121) surface. In Figure 3i.1, density functional theory (DFT) calculated reaction energies and activation energies of reaction intermediates are presented. Figures 3i.2 and 3i.3 show simulated reaction performance data based on the DFT-calculated results. In Figure 3i.4, the data of Figure 3i.1 have been used as input for microkinetics simulations based on BEP extrapolation. The final Figure 3i.5 schematically illustrates the selectivity window of the reaction, which is provided by an optimum M—C bond energy as long as the rate of oxygen removal is fast.

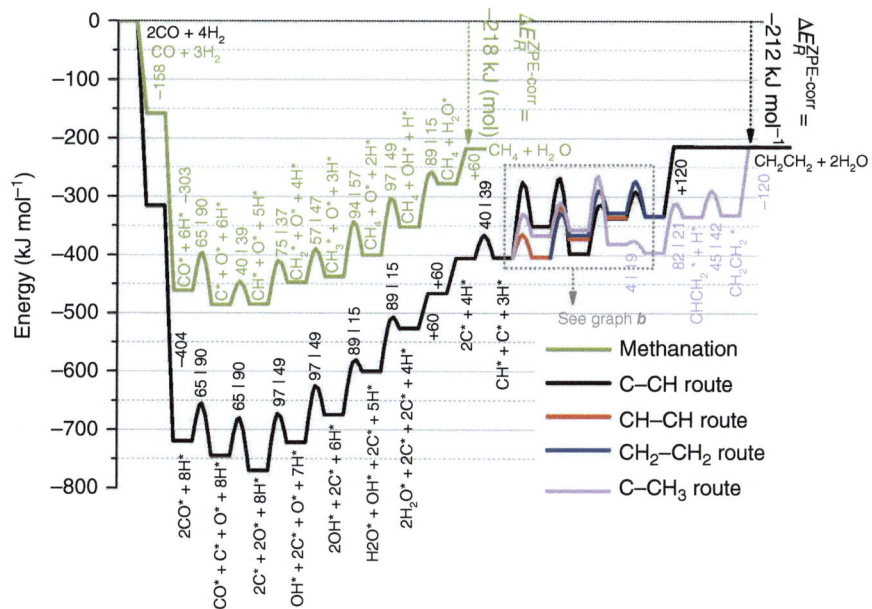

Figure 3i.1 Reaction energy diagram. DFT-calculated reaction energies and activation energies of adsorbed reaction intermediates are shown for methane formation and several chain growth routes involving different reaction intermediates.

(Continued)

Insert 3: (Continued)

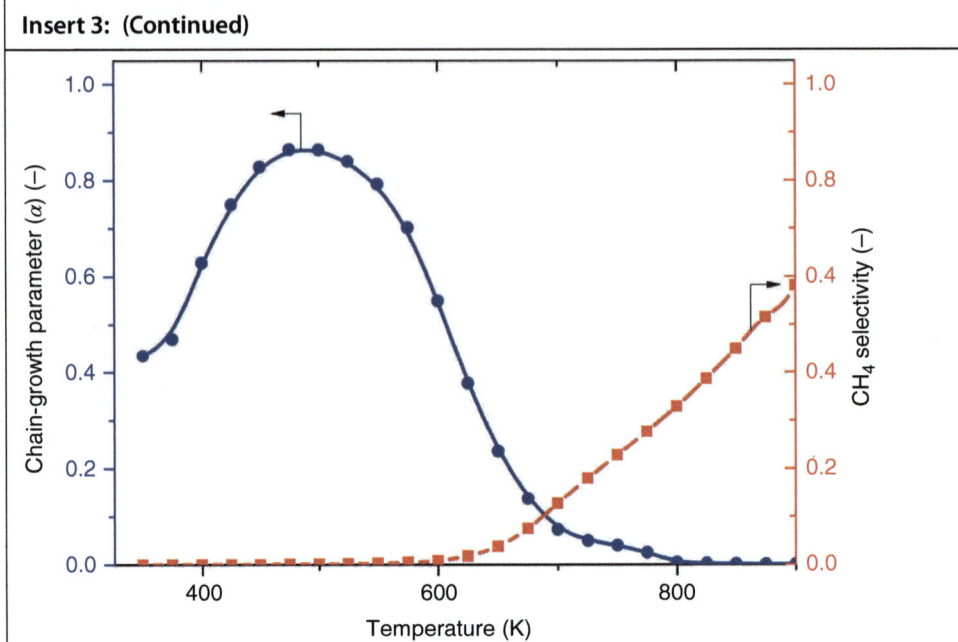

Figure 3i.2 Microkinetics-simulated chain growth parameter and methane selectivity and Anderson–Schulz–Flory product distributions.

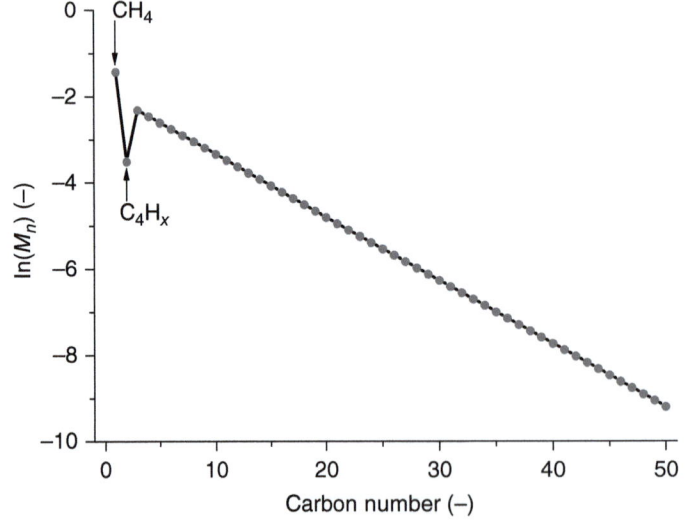

Figure 3i.3 The detailed simulated product distribution at reaction condition of Figure 3i.1.

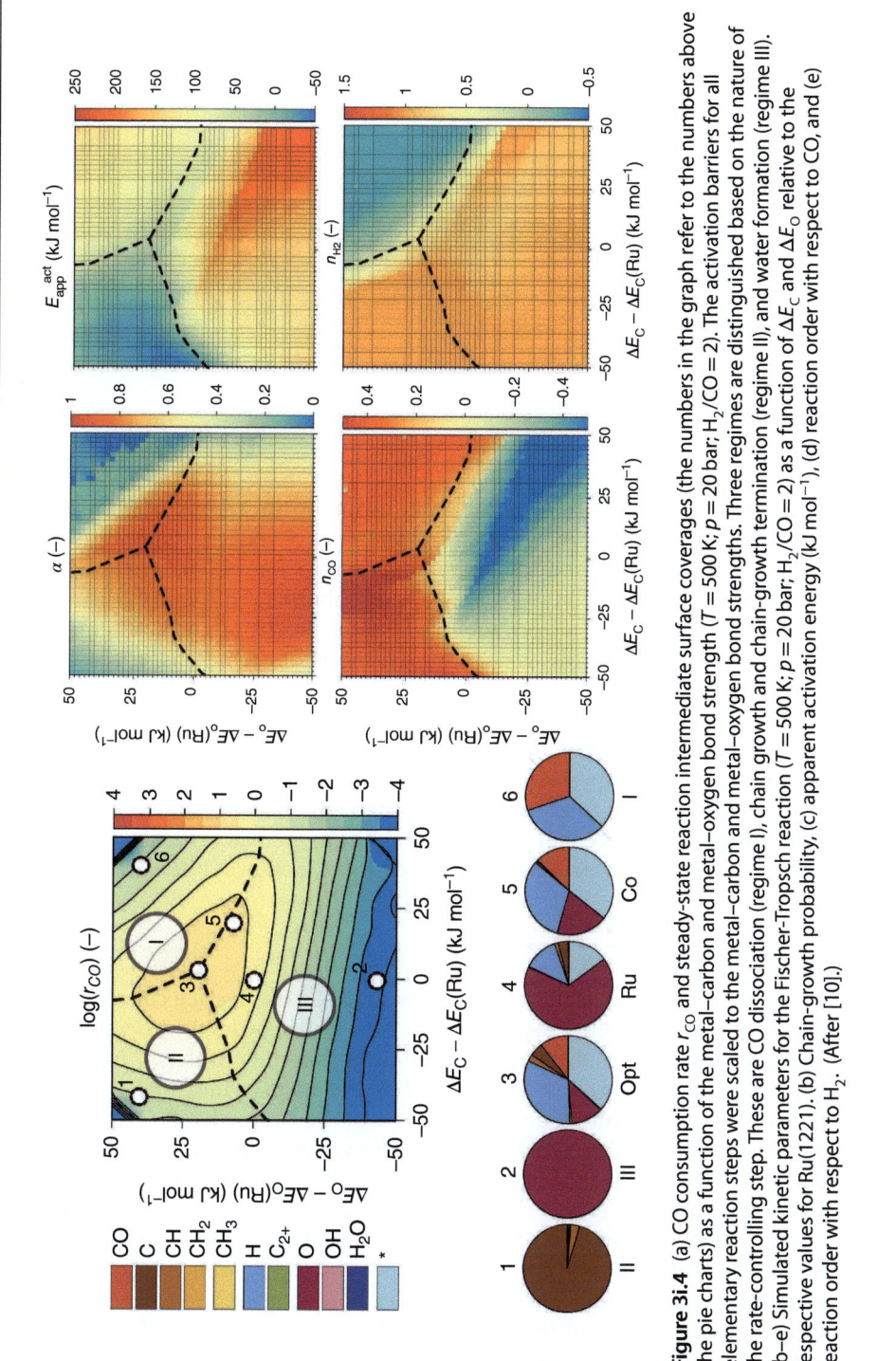

Figure 3i.4 (a) CO consumption rate r_{CO} and steady-state reaction intermediate surface coverages (the numbers in the graph refer to the numbers above the pie charts) as a function of the metal–carbon and metal–oxygen bond strength ($T = 500$ K; $p = 20$ bar; $H_2/CO = 2$). The activation barriers for all elementary reaction steps were scaled to the metal–carbon and metal–oxygen bond strengths. Three regimes are distinguished based on the nature of the rate-controlling step. These are CO dissociation (regime I), chain growth and chain-growth termination (regime II), and water formation (regime III). (b–e) Simulated kinetic parameters for the Fischer–Tropsch reaction ($T = 500$ K; $p = 20$ bar; $H_2/CO = 2$) as a function of ΔE_C and ΔE_O relative to the respective values for Ru(1221). (b) Chain-growth probability, (c) apparent activation energy (kJ mol^{-1}), (d) reaction order with respect to CO, and (e) reaction order with respect to H_2. (After [10].)

(Continued)

Insert 3: (Continued)

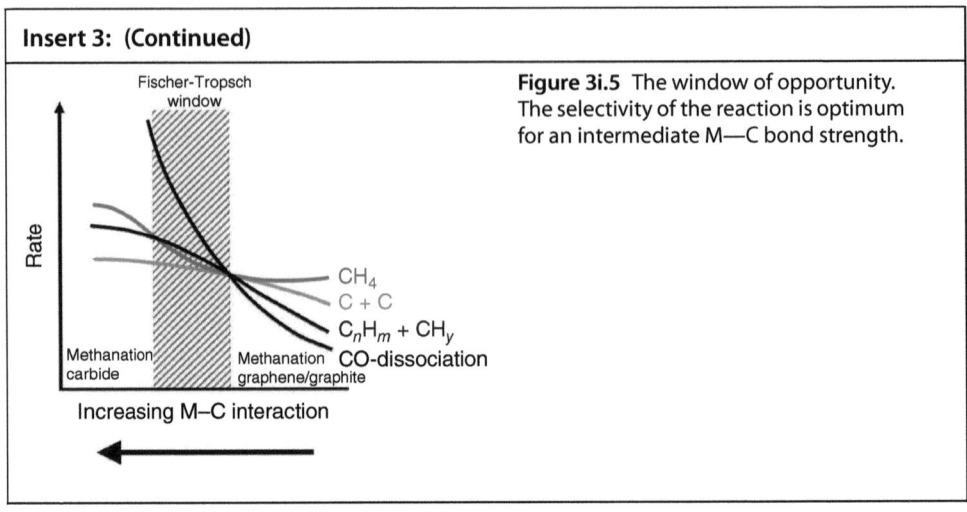

Figure 3i.5 The window of opportunity. The selectivity of the reaction is optimum for an intermediate M—C bond strength.

The product distribution of the long chain hydrocarbons of the Fischer–Tropsch reaction exhibits a linear dependence when the logarithm of the molecular fraction of product is plotted as a function of chain length. This is called the Anderson–Schulz–Flory (ASF) distribution. This type of distribution is found when the rate of production is independent of chain length. In the example in Figure 3i.3, this is seen for hydrocarbon chain lengths longer than three carbon atoms.

Although the microkinetics simulations of Insert 3 include all elementary reactions steps, it is also useful to analyze the results in terms of lumped elementary rate constants. In lumped kinetics, several elementary reaction steps are mathematically combined into a single reaction step.

The lumped kinetics chain growth probability α_c is equal to the slope of the plot of the logarithm of the product distribution as a function of its chain length. One deduces:

$$\alpha_c = \frac{r_c(n-1, n)\theta_c(1)}{r_c(n-1, n)\theta_c(1) + r_t} \tag{8.2}$$

The chain growth probability α_c depends on $\theta_c(1)$, the surface concentration of the C_1 intermediates that insert into the adsorbed growing hydrocarbon chain. The rate constant r_c is the rate constant of the insertion of the C_1 intermediate into the growing chain and r_t is the rate of chain growth termination. A high probability for chain growth occurs when the rate of chain growth $r_c \theta_c(1)$ is large compared to k_t. Within the carbide mechanism the concentration $\theta_c(1)$ depends on the ratio of the rate of dissociative dissociation of CO, the rate of methane formation $r_m \theta_c(1)$ and the rate of chain growth. A strong M—C interaction decreases r_t and r_m, $\theta_c(1)$ and favors a high rate constant of CO activation. For the step-edge Ru surface studied in Insert 3, the rate constant of CO activation is fast compared to r_t and r_m reaction rate. These, in turn, are slow compared to r_c the lumped elementary rate constant of the chain growth. The rate constant of chain growth differs from these other constants because it depends only weakly on the M—C bond energy.

The ASF plot shows a minimum in the production of C_2 hydrocarbons. The primary C_2 product that desorbs is ethylene, and its formation by C_1 recombination is not homologous with the insertion reaction of the longer hydrocarbon chains. It also desorbs more slowly from the surface than the longer alkenes, which suppresses its rate of formation and favors chain growth. Methane formation is also not located on the ASF curve because its selectivity depends on k_m instead of k_t.

The temperature dependence of the Fischer–Tropsch reaction (Figure 3i.2) shows that long chain hydrocarbon formation is favored at low temperature, but at higher temperatures methane formation is dominant r_t and r_m increase quickly compared to the rate constants of C—O bond cleavage r_{diss}^{CO} and r_c.

In contrast, the Ru(0001) terrace only produces methane due to the very high activation energy of C—O bond cleavage on this surface.

The contour plots of the Fischer–Tropsch reaction (see Figure 3i.4) have been constructed based on BEP plots and the scaling laws described in Section 7.2.7. It identify the rate of conversion optimum with a high α_c value.

There are three reactivity regions:

1. The reactivity region with weak M—O bonds and weak M—C bonds (region I).
2. The region (upper left corner) where the M—C bond is strong (region II),
3. The surface reactivity region where the M—O bonds are strong (region III),

In region I, the dominant adsorbate is CO. In region II, the surface becomes covered with C atoms. In region III, water removal is rate-controlling and, hence, the surface is covered with O. The maximum rate is found at the intersection of the three regions, where several elementary rate constants compete. The reactivity regime that corresponds with the experiment is at the interphase of the O poisoned regime and the regime between regions I and III where the rate of CO dissociation is rate-controlling.

We discussed earlier the strong dependence of CO activation on structure (see Section 4.4). The demand for a relatively low barrier of CO activation implies a need for step-edge sites for long chain hydrocarbon formation. This has implications for the particle size dependence of the reaction. When the size of the metal particles becomes too small, the rate of CO consumption as well as the selectivity of chain growth decreases because small metal particles will not support step-edge sites.

The selectivity for producing long chain hydrocarbons in the Fischer–Tropsch reaction is optimum in the reactivity window of transition metal based catalysts (Figure 3i.5) where the rate constant of C—O bond cleavage is relatively fast, the metal–carbon bond is strong enough to prevent methane formation, and the rate constant of chain growth termination is slow. On the other hand, the metal–carbon bond must not be so strong that C—C bond formation becomes endothermic, causing the rate constant of chain growth to decrease and favor methane production.

At relatively low temperature, there is an increase in the selectivity of oxygenate production that yields long chain hydrocarbon alcohols or aldehydes. This implies that the reaction rate constants of CO insertion into the growing hydrocarbon chain and C—O bond cleavage of inserted CO are slow compared to the chain growth reaction, which is another argument in favor of the carbide mechanism.

8.2.3 Hydroconversion of Hydrocarbons

8.2.3.1 Ethylene

According to the Horiuti–Polanyi mechanism [11], ethylene hydrogenation by a transition metal surface occurs by the consecutive addition of H atoms from the surface to the adsorbed ethylene molecule. The sequence of elementary reaction steps is:

$$H_2 + 2* \underset{}{\overset{k_1,k_2}{\rightleftharpoons}} 2H*$$

$$R + 2* \rightleftharpoons *R*$$

$$*R* + H* \rightleftharpoons RH* + 2*$$

$$RH* + H* \overset{k_3}{\rightarrow} RH_2 + 2*$$

An isotope exchange experiment involving D_2 to ethylene hydrogenation catalyzed by Pd shows a rapid exchange with all of the ethylene hydrogen atoms. This implies that the hydrogenation of the ethyl intermediate is rate-controlling and it is the most abundant reaction intermediate (MARI).

At a low H_2 pressure, rapid deactivation of the hydrogenation reaction occurs due to self-hydrogenation of ethylene to yield ethane and acetylene. The acetylene molecule adsorbs strongly and will induce deactivating oligomerization. In general [12, 13], when higher alkenes are hydrogenated a carbonaceous overlayer forms during the reaction (see Insert 4).

Insert 4: Carbonaceous Overlayer Formation

Radiochemical experiments with ^{11}C-labeled hexane molecules demonstrate the formation of carbonaceous residues on freshly reduced small particles. The ^{11}C isotope emits two γ rays of the same energy at the same time, but in different directions. By measuring these emissions simultaneously, the readings can be used to follow the progress of the ^{11}C-labeled hexane pulse through a reactor bed at *in situ* high pressure conditions [12].

In Figure 4i.1, three experiments are compared in which hexane dehydrogenation and hydrogenation is catalyzed by small Pt particles located in the micropores of a protonic mordenite zeolite. As described in Section 3.5.3, this is a bifunctional catalyst that isomerizes hexane to isohexane through intermediate hexane formation by the hydrogen and C—H bonds of the Pt particles. Figure 4i.1a shows the progress of a ^{11}C-labeled hexane pulse as a function of position in the reactor bed, when a small pulse is injected into the reactor while reaction with ^{12}C hexane is ongoing and at steady state. The slope of this curve determines its residence time, which is mainly determined by the adsorption energy of hexane in the zeolite micropore and the number of available free surface sites for adsorption in the mordenite micropores (Fig. 3.20).

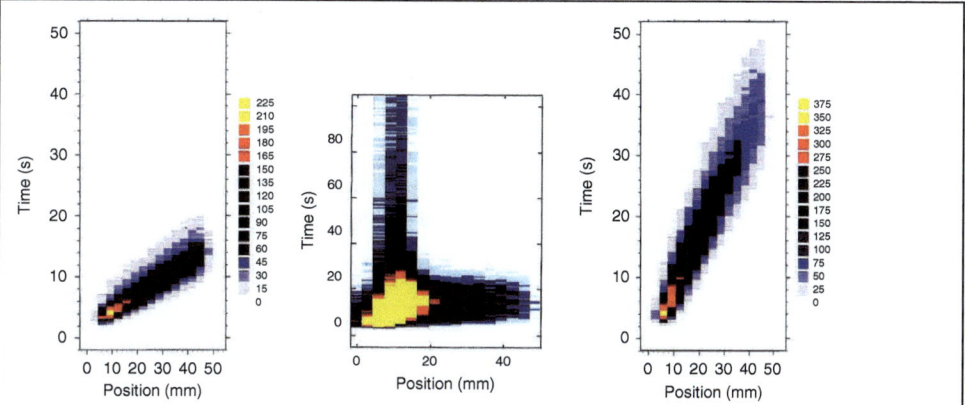

Figure 4i.1 Positron emission profiling (PEP) images of n-^{11}CH$_3$C$_5$H$_{11}$ pulse experiments on a Pt/H-mordenite catalyst bed (l = position along the catalyst bed, t = residence time). The color scale indicates the concentration of the ^{11}C label: the brighter the color, the higher the concentration. Labeled samples were injected into the feed streams (1 atm; 150 Nml min^{-1} total flow rate). (a) The reactor was operating under steady-state conditions at 230 °C with a feed mixture of n-hexane/H$_2$ (1/28 mol). (b) [^{11}C]-n-hexane pulse in hydrogen over the catalyst with freshly reduced Pt at 230 °C. (c) [^{11}C]-n-hexane pulse in hydrogen over the catalyst, which was previously used in the hydroisomerization reaction at 230 °C. (van Santen 1997 [14]. Reproduced with permission of John Wiley & Sons.)

Figure 4i.1b shows the very different behavior when the hydrocarbon pulse is injected into flowing H$_2$ and the metal particle is freshly reduced. The long residence of the ^{11}C signal at the beginning of the reactor bed confirms the formation of a nonremovable species. This is the carbonaceous layer formed on the Pt particle by exposure to the hydrocarbon. When compared to a similar experiment (a reaction based on a catalyst previously used in hydroisomerization) normal steady-state reaction on the catalyst is observed, as shown in Figure 4i.1c.

When the carbonaceous overlayer inhibits reaction, the H$_2$ must dissolve in the overlayer [15] and the reaction is first order in H$_2$. Instead, when no inhibiting overlayer is present, H$_2$ equilibrates with the surface and the rate is half order in H$_2$ pressure as is found for Pd. The latter is consistent with the addition of a H atom to the alkyl intermediate as the rate-controlling step. Reactions are typically first or half order in H$_2$ partial pressure and zero order in olefin.

The reaction energy diagrams for ethylene hydrogenation adsorbed at high or low coverage to the transition metal surface are compared in Figure 5i.2. The high coverage case can be compared with the surface at reaction conditions, when the surface is partially carbided. At low coverage, the first and second hydrogen addition steps compete, but at high coverage the hydrogenation of the alkyl intermediate is the slow reaction step, consistent with the Horiuti–Polanyi mechanism.

Insert 5: Adsorption and Hydrogenation of Ethylene

Figure 5i.1 shows schematic representations of π- and σ-adsorbed ethylene, respectively. Figure 5i.2 compares the respective reaction energy diagrams for hydrogenation on a Pd(111) surface at low and high coverage.

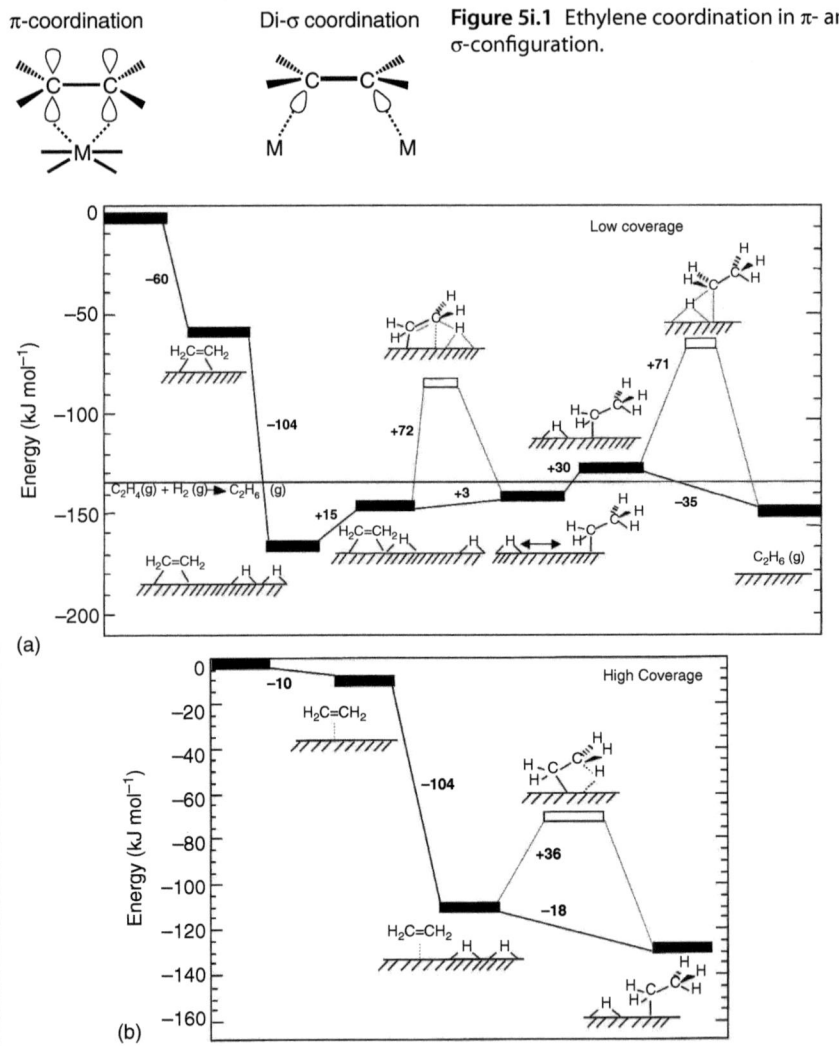

Figure 5i.1 Ethylene coordination in π- and σ-configuration.

Figure 5i.2 (a) The overall potential energy diagram for ethylene hydrogenation over Pd(111) at low surface coverages. The solid boxes refer to reactant, intermediate, and product states, while the two unfilled boxes denote the transition state structures and energies for the surface reactions of ethylene to ethyl and ethyl to ethane. (Neurock and van Santen 2000 [16]. Reproduced with permission of American Chemical Society.) (b) The overall potential energy diagram for ethylene hydrogenation over Pd(111) at high surface coverages. The solid boxes refer to reactant, intermediate, and product states, while the two unfilled boxes denote the transition state structures and energies for the surface reactions of ethylene to ethyl and ethyl to ethane.

The difference between the two reactions is the stability of the adsorbed ethylene. It can adsorb in two ways to the transition metal surface: in the di-σ mode, which features a strong interaction where two of the ethylene carbon atoms attach to two metal surface atoms (the atomic orbitals around the C atom hybridize as in ethane), and in a second, weakly adsorbed mode in which the ethylene is adsorbed through the π-electrons of the C=C bond atom of a metal atom. A high coverage with adsorbate (this can be ethylene as well as C_{ads}, H_{ads}, or other co-adsorbed species) will push the ethylene toward the weakly adsorbed π mode where it interacts with a single metal atom.

The weaker interaction of π-bonded ethylene compared with the interaction of di-σ-adsorbed ethylene implies that the π-bonded case has a smaller energy cost (less M—C interaction weakening) for the addition of H to one of the ethylene molecule's C atoms. The intermediate interaction of alkyl attached to the surface with a strong M—C σ bond is not affected by co-adsorbed intermediates because it favors adsorption to a configuration that covers the surface metal atoms.

Similarly, the selectivity of H addition to a molecule with several unsaturated bonds is controlled by differences in the interaction strength of the respective unsaturated bonds with the surface. Those atoms that have the weaker interaction with the transition metal surface will become more easily hydrogenated. This is especially relevant for competitive hydrogenation of an aldehyde versus olefinic bond [17]. On Au, which has a high activation energy for H_2 dissociation, direct addition of an H from the H_2 molecule can occur. The H addition to the unsaturated molecule will generate highly reactive adsorbed H atoms on the Au surface [18].

Hydrogenation and dehydrogenation reactions show no particle size dependence. This is ascribed to carbonaceous overlayer formation, which will obscure structure-sensitive behavior. Alloying of a reactive transition metal with a nonreactive metal such as Au or Sn increases the rate of hydrogenation as well as dehydrogenation. Dehydrogenation requires higher temperatures because it is endothermic. Alloying isolates the transition metal atoms in smaller ensembles of atoms, which will suppress nonselective hydrogenolysis of hydrocarbons and excessive carbon deposition. The small ensemble size will also favor π-adsorbed ethylene.

8.2.3.2 Acetylene

The selective hydrogenation of acetylene to ethylene is an important technical process that uses Pd as the preferred catalyst. While metallic Pd is a very efficient catalyst for pure ethylene hydrogenation, it is less effective in the competitive adsorption process with acetylene [19, 20]. For this reason, the Pd catalyst is modified by the addition of protonating metals such as Ag, Au, or Ga.

Ethylene adsorbs with an adsorption energy of approximately 60 kJ mol^{-1}, but the acetylene molecule has a higher adsorption energy of 150 kJ mol^{-1}, which will suppress the adsorption of ethylene. This is the essential reason for the high selectivity of acetylene hydrogenation to ethylene instead of ethane. The activation-free energy of ethyl hydrogenation is rate-controlling at high coverage, and thus must be higher than the desorption energy of ethylene.

The Pd surface layer is altered during reaction by the diffusion of carbon to subsurface positions. The presence of this subsurface carbon on the catalyst encourages high selectivity to ethylene. Pd metal forms a hydride at high H_2 pressures, but subsurface carbon reduces contact between the surface and the bulk hydride resulting in reduced hydrogenation. In addition, the presence of both subsurface and surface carbon reduces the reactivity of the Pd surface atoms in an essential way, as shown in Figure 8.2a for the chemisorption of CH_3. Figure 8.2b shows that a scaling law exists between ethylene and acetylene adsorption and the heat of adsorption of CH_3. It indicates a significant weakening of the adsorption energies of ethylene and acetylene. In the calculations ethylene is assumed to be di-σ-adsorbed.

Figure 8.2 shows that the variation in the bonding of acetylene is larger than that of ethylene. However, yet because the surface coverage remains dominated by acetylene, the decrease in surface concentration of acetylene will be much less than the decrease for ethylene.

The second cause of the improved selectivity induced by co-adsorbed carbon or other promoting atoms is that the ethylene admolecule bond is further weakened when it moves from the di-σ adsorbed state to the more weakly interacting π-bonded state.

Promotors such as Ag, Au, or Ga are added to the catalyst because they do not activate H_2 and bind only weakly to ethylene or acetylene. As can be observed from Figure 8.2, the role of these promotors is similar to that of subsurface C: they weaken the interaction with acetylene and ethylene.

8.2.3.3 Hydrogenolysis and Isomerization

For a short introduction to the mechanism of skeletal transition metal hydrocarbon transformation, we refer to Section 4.5. The only transition metal that can selectively isomerize alkanes is Pt, and only if the transition metal particle is at least 5 nm. The other transition metals are too reactive or require higher temperatures for C—H-bond activation, so they mainly produce methane (see Figure 8.3). The ensemble effect of alloying Pt with nonreactive metals such as Sn or Au suppresses hydrogenolysis and will enhance its selectivity for dehydrogenation.

The isomerization reaction is thought to proceed through the formation of a metallocyclobutane with a carbene intermediate, analogous to the metathesis reaction illustrated in Figure 8.4.

Figure 8.4 summarize the proposed mechanisms of the hydrogenolysis of an alkane molecule to short-chain hydrocarbon intermediates [23]. The essential difference between these mechanisms is the size of the surface metal atom ensemble that is required for the hydrogenolysis reaction. In Figure 8.4, schemes a and c refer to hydrocarbon activation by a surface atom ensemble of several metal atoms and scheme d refers to a reaction site of a single metal atom.

When a C—C bond is activated by a surface atom ensemble, there is either cleavage of the C—C bond between atoms α and β, or cleavage of the C—C bond through metal–carbon contact of atoms α and γ. It has also been proposed that the initial C—H activation can give adsorbed alkyl and H_2 by direct interaction

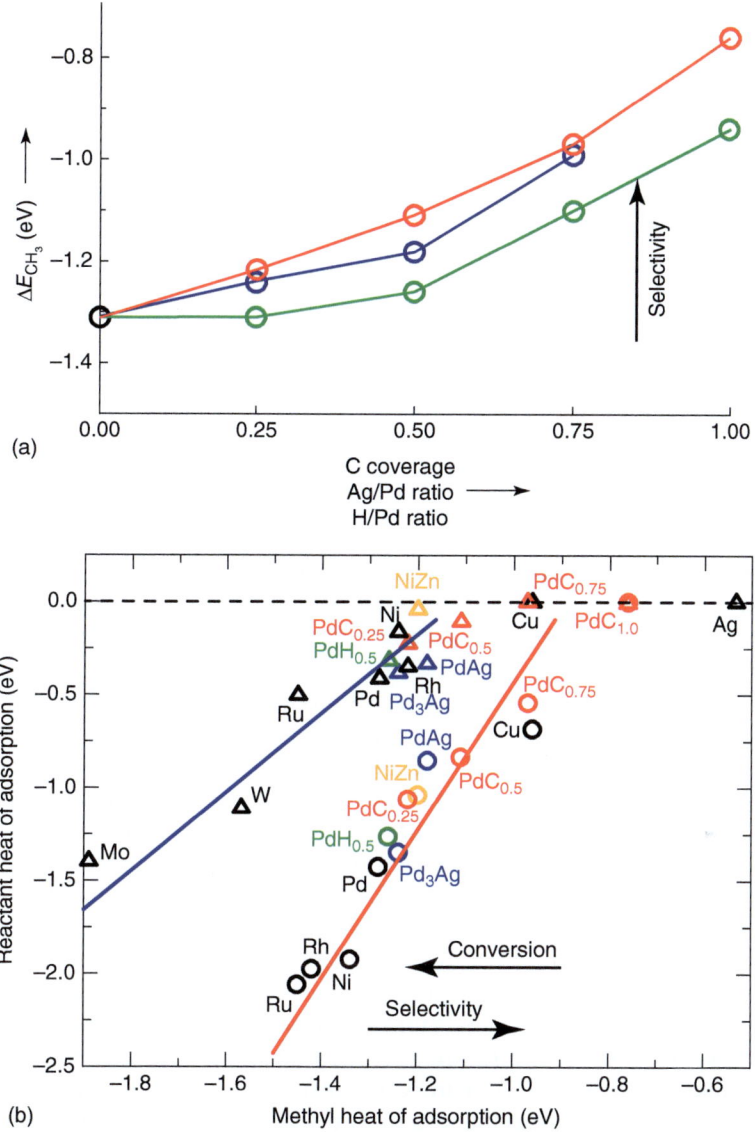

Figure 8.2 (a) Differential adsorption energy of a methyl group as a function of the carbon coverage as well as the Ag/Pd and H/Pd ratios. Red: Palladium with subsurface carbon, blue: Pd–Ag alloys, and green: palladium hydrides. Weakening of the adsorption energy of methyl groups should increase the selectivity toward ethylene. (Studt et al. 2008 [20]. Reproduced with permission of John Wiley and Sons). (b) Adsorption energies for acetylene (circles) and ethylene (triangles) plotted against the adsorption energy of a methyl group. The adsorption energies of acetylene and ethylene represent the most stable adsorption sites and are defined as the total energy of the surface with the adsorbed species minus the sum of the total energy of the clean surface and the adsorbate in a vacuum. The solid lines show the predicted acetylene (red line) and ethylene (blue line) adsorption energies from scaling. The adsorption geometry of ethylene on Ni and Rh surfaces has been accounted for by specifically choosing the methyl binding site corresponding to the site of adsorption of the carbon atoms in ethylene. The dotted lines indicate the energy window. (Studt et al. 2008 [20]. Reproduced with permission of John Wiley and Sons.)

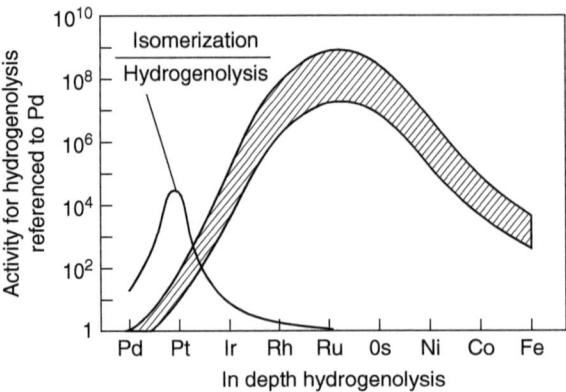

Figure 8.3 Schematic evolution of the activity in hydrogenolysis and the ratio isomerization/hydrogenolysis. (After [21] with kind permission.)

with vacant surface metal atom sites or in an associative reaction with adsorbed hydrogen to (Figure 8.4, a). The second reaction will be favored when the M—H bond is weak, such as on the dense surfaces of Ni or Pt.

As indicated in schemes of Figure 8.4, C—H bond cleavage is followed by steps where α–β adsorbed hydrocarbon intermediates dehydrogenate by β C—H bond cleavage (the reverse of the Horiuti–Polanyi route of ethylene hydrogenation) or undergo hydrogen shuffling reaction steps that also involve α C—H bond cleavage. In the final reaction steps C—C bonds may also cleave.

Watwe et al. [24] showed that on Pt these reactions favor step-edges as opposed to terrace sites. The β C—H bond cleavage step has the highest barrier to activation (E_{act} = 193 kJ mol^{-1}), whereas α C—H bond activation has the lowest barrier. The C—C bond activation reaction at the step-edge has a calculated barrier of 100 kJ mol^{-1}, which is 73 kJ mol^{-1} lower than the barrier for the (111) surface.

The α–γ C—C bond activation can proceed on a terrace as indicated in scheme c in Figure 8.4 or on single surface atoms as shown in scheme d. C—C bond activation proceeds either through C—C bond cleavage with formation of a carbene intermediate and olefin, or via a hydrogen atom addition to give adsorbed alkyl and the olefin. The carbene species results from the intermediate formation of a metallocycle attached to a single metal atom and its subsequent activation.

After the intermediate carbene-forming step, isomerization and hydrogenolysis reaction steps compete [21]. Alloying the transition metal with nonreactive metals such as Sn or Au enhances the selectivity toward isomerization and dehydrogenation. This may indicate that the surface atom ensemble size needed for the isomerization reaction is smaller than the surface ensemble size needed for the hydrogenolysis reaction. Alternatively, step-edge sites are the preferred reaction sites for the hydrogenolysis reaction.

The hydrogenolysis reaction is particle size dependent, as has been shown for the ring-opening reaction of methylcyclopentane catalyzed by Pt [25, 26]. While on small particles this reaction is nonselective, on larger particles only the branched isomers are produced through reactions that proceed through α–β adsorbed intermediates such as ethane.

DFT quantum chemical calculations indicate that the differences in selectivity as a function of site structure relate to differences in the stereochemistry of

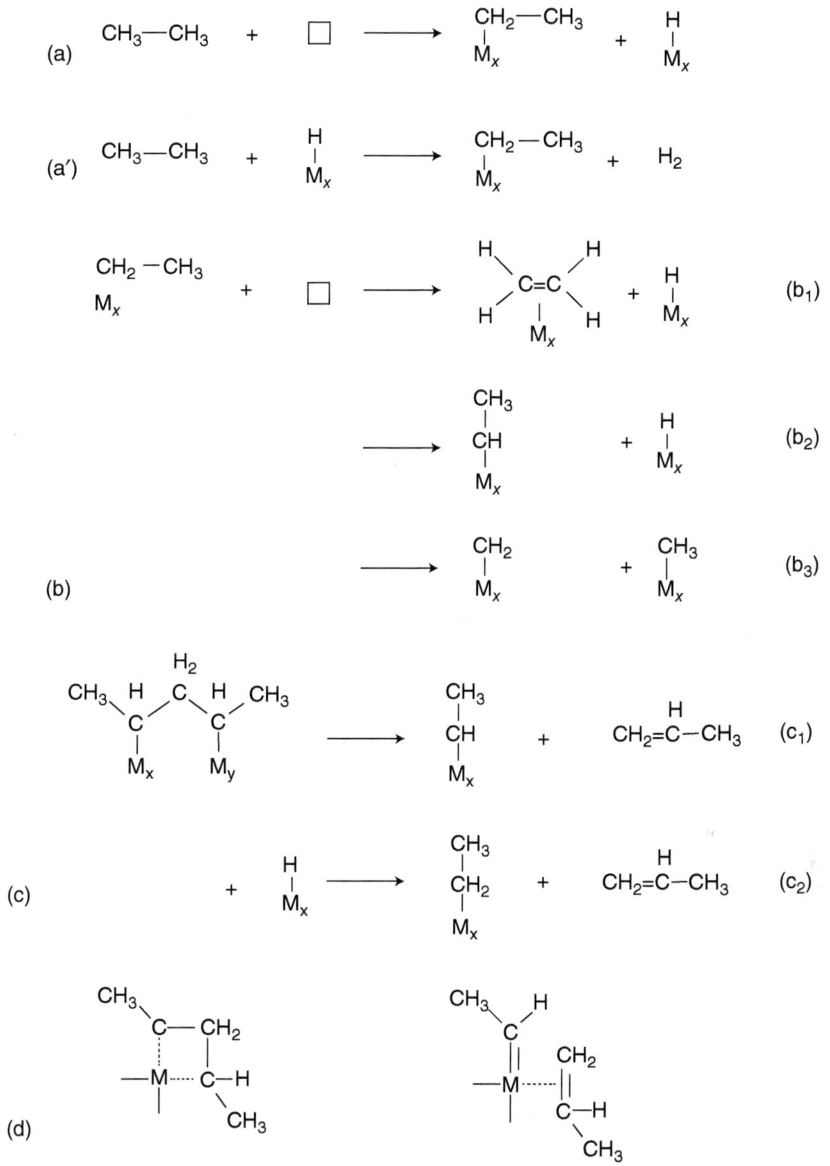

Figure 8.4 Summary of some of the proposed elementary reaction steps involved in the hydrogenolysis of an alkane. (I) α–β bond cleavage (van Santen et al. 2010 [22]. Reproduced with permission of American Chemical Society.) (II) α–γ bond cleavage.

the adsorbed reaction intermediates on surface sites of different topologies. Figure 8.5 shows reaction intermediates for 3-methylpentane (1) and hexane (2) formation. In both cases, the reaction proceeds through an α–β adsorbed state.

For hexane formation, the activation energy of C—C bond cleavage on the stepped surface Pt(211) is reduced to 94 kJ mol^{-1}, compared to its value of 116 kJ mol^{-1} on the nonstepped Pt(111) surface. In comparison, the C—C

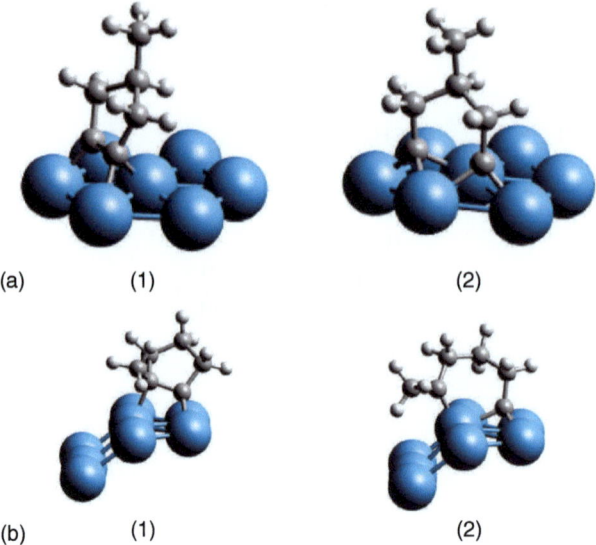

Figure 8.5 (a1) α–β-Adsorbed methylcyclopentyl intermediate. (a2) Reaction intermediate after ring-opening intermediate. (b) Reaction intermediates of the methylcyclopentane ring-opening reaction to hexane on Pt(211) surface. (b1) Structure of adsorbed methylcyclopentane. (b2) Ring-opened methylcyclopentane intermediate.

bond cleavage activation energies for 2-methylpentane and 3-methylpentane formation are both 100 kJ mol^{-1} on the stepped surface, which is slightly higher than the values found for the Pt(111) surface.

The barrier of activation of n-hexane formation is lower due to the reduced steric repulsion of the methyl group of methylpentane in the corresponding transition state complex at the step-edge compared to the transition state complexes at terrace sites. No steric repulsion is present in the transition states (TS) that lead to the branched products, because the methyl group is far from the surface.

Small nanoparticles have an increased ratio of edge over terrace atoms. The observed increased selectivity of hexane formation for such small particles is consistent with the DFT-calculated reduction in activation energy at the step-edge sites of the (211) surface.

8.2.4 NH$_3$ and CH$_4$ to HCN

The Andrussow process converts a mixture of methane and ammonia to HCN at a temperature of 850 K using a Pt/RH catalyst. HCN is essential for the production of nylon-6,6 as well as other commodities. A small amount of oxygen is present for H$_2$ combustion, which provides the heat for this endothermic reaction.

At the high temperature of the reaction the Pt/Rh gauze becomes corrugated, so that a high fraction of step-edge sites can be expected.

Formation of N$_2$ competes with HCN formation. Therefore, the coupled reaction between the CH$_x$ and NH$_x$ intermediates must be fast compared to the exothermic formation of N$_2$ from adsorbed N atoms. The use of Pt as catalyst

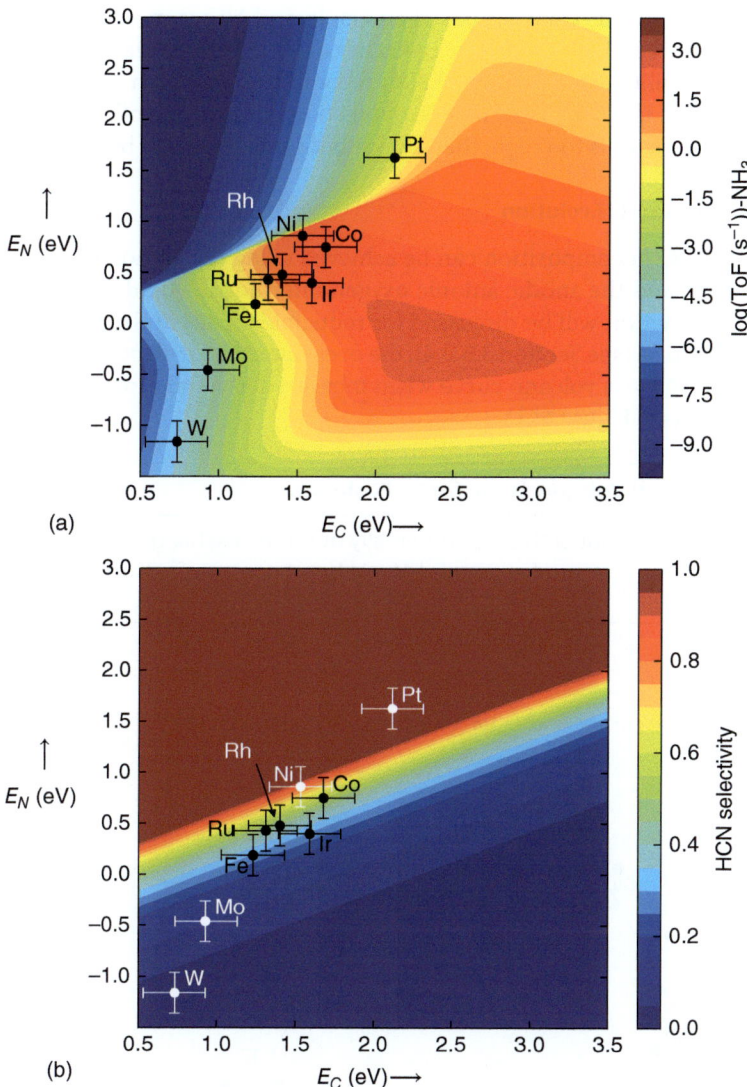

Figure 8.6 (a) Calculated logarithmic turn over frequency (TOF) for NH_3 consumption and (b) selectivity toward HCN production as a function of the carbon (E_C) and nitrogen (E_N) binding energies. The error bars indicate an uncertainty of 0.2 eV for E_C and E_N. (Grabow et al. 2011 [27]. Reproduced with permission of John Wiley and Sons.)

is appropriate because it has a low activation energy for C—H bond activation;, however, a low activation barrier of ammonia requires a more electrophilic catalyst.

Microkinetics simulations using BEP and scaling relations based on quantum-chemical data obtained for the Pt(211) surface indicate a conflict between maximum conversion rate of reagent and maximum selectivity (see Figure 8.6). The adsorption energies of the C and N atoms are used as catalyst reactivity indicators. While catalysts based on Co, Ir, and Ni are most active, Pt is the most selective catalyst. The lower reactivity of Pt for ammonia activation increases the probability of atomic C and NH_2 recombination, which suppresses the formation of N_2. On Pt, ammonia decomposition is rate-controlling: when methane partial pressure is too high, the catalyst becomes poisoned by C.

8.2.5 Electrocatalysis; H_2 Evolution

Electrocatalytic H_2O decomposition can be achieved using the cathodic proton reduction reaction and the anodic anionic oxygen intermediate oxidation reaction. The second reaction will be discussed for reducible oxides in Section 11.6. Similar to O_2 evolution (see Section 3.3.2.4), the overpotential for electrocatalytic H_2 evolution also shows a volcano curve when measured as a function of M—H adsorption energy (see Figure 6i.1).

Insert 6: The Electrochemical Volcano Curve for H_2 Evolution

Figure 6i.1 shows a logarithmic plot of the experimentally observed exchange current for H_2 evolution measured as a function of estimated M—H bond energy of adsorbed H. It

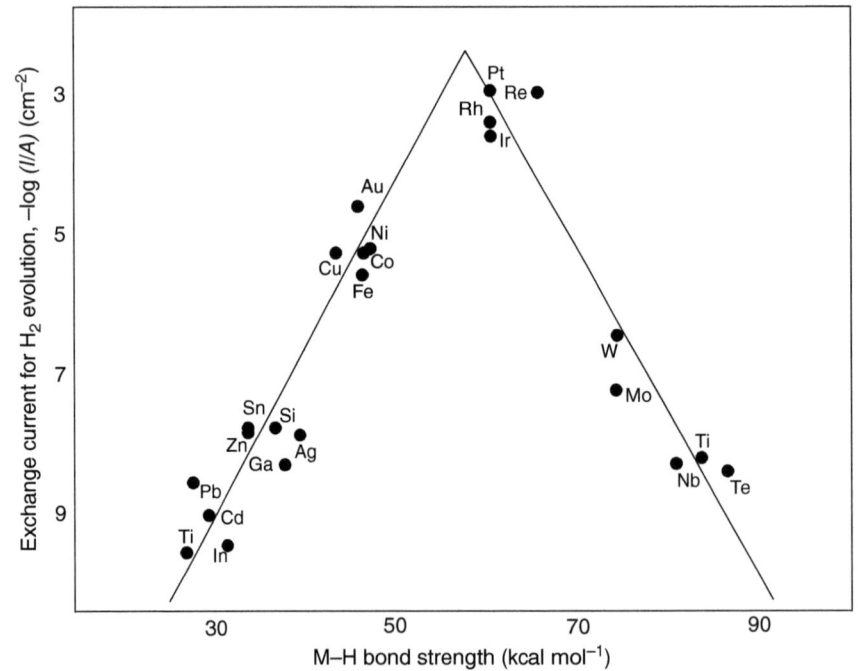

Figure 6i.1 A volcano curve is shown in for the current used for H_2 evolution at a comparable overpotential plotted as a function of the H_2 adsorption energy. (Trasatti 1990 [28]. Reproduced with permission of Elsevier.)

is notable that the extrapolated maximum is near the position of the bond dissociation energy of H_2 of 408 kJ mol^{-1} found for Pt. At this value, the free energy of H_2 adsorption is close to zero.

Experimental curves for H_2 evolution of this type have been criticized by Schmickler and coworkers [29], since the experimental systems are only partially oxidized and hence a strict Sabatier interpretation of the volcano curve is not valid.

The elementary reaction rate theory expressions for hydrogen evolution are shown below.

If one assumes the Volmer reaction to be rate-controlling the expression for the current J is:

$$J = -er$$
$$r = r_f - r_b \tag{6i.1}$$

To the left and right of the volcano curve the following expressions can then be used respectively. The exchange current expressions for the exothermic case for H_2 adsorption at pH = 0 becomes:

$$J_0 = -ek_0 \frac{1}{1 + \exp(-\Delta G_{H^*}/kT)} \tag{6i.2}$$

and for the corresponding endothermic case:

$$J_0 = -ek'_0 \frac{1}{1 + \exp(-\Delta G_{H*}/kT)} \exp(-\Delta G_{H^*}/kT) \tag{6i.3}$$

In these expressions the dependence of k_0 and k'_0 on overpotential is not explicitly considered and these are assumed to be equal.

The results plotted in Figure 6i.2 have been deduced based on Eqs. 6i.2 and 6i.3.

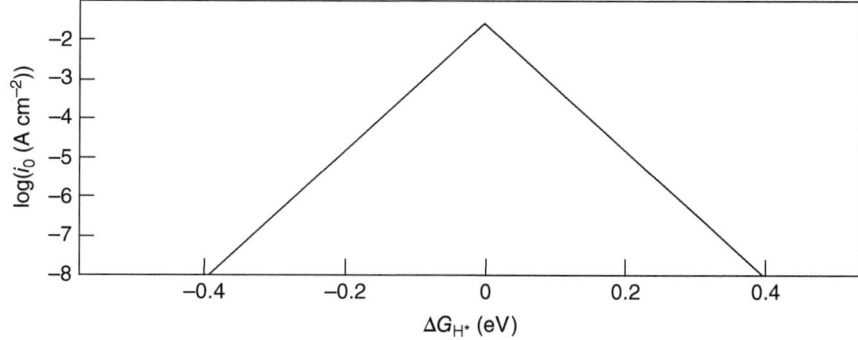

Figure 6i.2 The current in the HER calculated from a simple kinetic model (see below) plotted as a function of the free energy for hydrogen adsorption, ΔG_{H^*} 5ΔE_H = 0.24 eV. (The correction of 0.24 eV is due to entropy and zero point energy correction.) (After [30]. Reproduced with permission of ECS — The Electrochemical Society.)

(Continued)

Insert 6: (Continued)

The alternative to the volcano curve shown in Figure 6i.1 is shown by Figure 6i.3 based on careful single-crystal measurements.

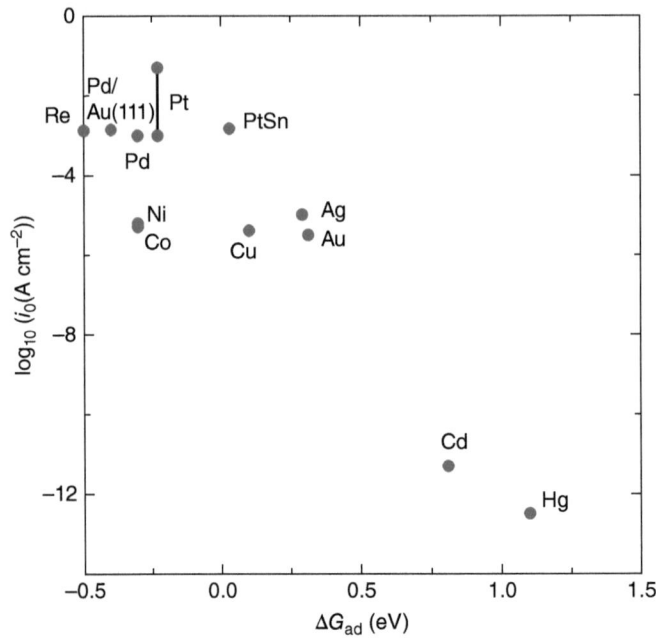

Figure 6i.3 Experimental values for the rate of hydrogen evolution against the free energy of adsorption of hydrogen atom on the densest surface plane at the standard equilibrium potential. For surfaces on which more than one adsorbed species exists, the most strongly bound species has been chosen. (Santos et al. 2012 [29]. Reproduced with permission of Royal Society of Chemistry.)

As is discussed in the main text, this nonvolcano curve dependence is interpreted as evidence that near the maximum current the surface is saturated with H atoms.

The next figure shows a comparison of the free energies of H_2 evolution for a few materials. It also indicates the relatively high activity of MoS_2 (Figure 6i.4).

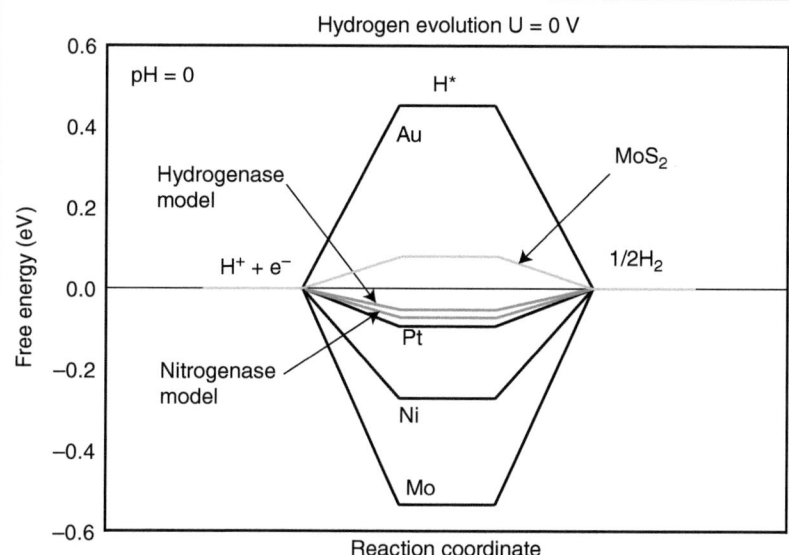

Figure 6i.4 Calculated free energy diagram for hydrogen evolution at a potential $U = 0$ relative to the standard hydrogen electrode at pH = 0. The free energy of $H^+ + e^-$ is by definition the same as that of $1/2 H_2$ at standard conditions. The free energy of H atoms bound to different catalysts is then found by calculating the free energy with respect to molecular hydrogen including zero-point energies and entropy terms. (Hinnemann *et al.* 2005 [31]. Reproduced with permission of American Chemical Society.)

The initial step in the electrochemical generation of H_2 from H^+ is the neutralization of the proton to adsorbed H. This is the Volmer reaction:

$$H^+ + e^- \rightarrow H_{ad} \tag{8.3}$$

Then, one of two reactions can follow, either the Tafel reaction:

$$2H_{ad} \rightarrow H_2 \tag{8.4}$$

or the Heyrovsky reaction:

$$H_{ad} + H^+ + e^- \rightarrow H_2 \tag{8.5}$$

When the hydrogen adsorption varies, the nature of the rate-controlling step may vary. The Tafel slope, which is the overpotential derivative of the current J (Eqs. 6i.1 and 6i.2) can identify the rate-determining steps in electrochemical systems [32]. When the electron transfer step in Eq. (8.3) is rate limiting, the expression for the rate becomes:

$$r_1 = k_1 c_{H^+}(1 - \theta) \tag{8.6}$$

where r_1 is the rate of the forward reaction. This reaction involves the transfer of one electron. When Eq. (8.3) is rate-controlling the hydrogen coverage will be low and r_1 becomes equal to:

$$r_2 = k_1 c_{H^+} \qquad (8.7)$$

The rate constant k_1 is generally assumed to depend exponentially on the overpotential:

$$k_1 = k_{11} \exp\left(-\frac{\alpha_e F}{RT} E\right) \qquad (8.8)$$

F is the Faraday constant. This is analogous to the BEP relation that defines a linear relation between activation energy and reaction energy. The value of α_e is typically chosen 0.5.

The Tafel slope is defined as:

$$-\frac{\delta(\log(-J))}{\delta E} = \alpha_e \frac{F}{2.3RT} = \alpha_e \cdot (60\,\text{mV})^{-1} = (120\,\text{mV})^{-1} \qquad (8.9)$$

The slope of 120 mV^{-1} indicates that the rate-limiting step in the HER reaction is the Volmer reaction.

If on the other hand Eq. (8.2) is rate-controlling, the expression for the current J is given by:

$$-J = 2F k_2 \theta^2 \qquad (8.10)$$

The recombination rate constant k_2 is assumed to be independent of potential and is given by the equilibrium constant K_H of hydrogen adsorption:

$$k_1 = K_H \cdot k_2 \qquad (8.11)$$

and the Tafel slope becomes:

$$-\frac{\delta(\log(-J))}{\delta E} = 2\alpha_e \frac{F}{2.3RT} = (60\,\text{mV})^{-1} \qquad (8.12)$$

When Eq. (8.5) is assumed to be rate-limiting, the equation for the rate becomes:

$$r_3 = k_3 c_{H^+} \theta \qquad (8.13)$$

In this case, the Tafel slope becomes $(40\,\text{mV})^{-1}$.

These three examples show how Tafel slope variations in relation to the kinetics of the electrocatalytic reaction can be used to determine the rate-controlling step of the HER reaction.

To the left of the volcano curve maximum (Figure 6i.2), the Volmer equation (Eq. (8.3)) applies and to the right the Heyrovsky equation (Eq. (8.5)) applies.

However, the curve shown in Figure 6i.2 is not accurate since k_0 and k_0' have been assumed to be equal in the kinetic expressions (Eqs. (6i.2) and (8.3)). Near the current maximum the Heyrovsky reaction occurs. In this reaction, the activation energy for proton transfer depends not only on the hydrogen atom adsorption strength, but also on the changed partial density of state (PDOS) around the Fermi level. More importantly, additional H atoms with substantially altered reactivity will adsorb more weakly adjacent to the strongly adsorbed H [29].

To support this, an experimental curve based on single-crystal experiments is shown in Figure 6i.3 [29]. For the metals investigated, a maximum is no longer found in the reaction curve as a function of the free energy of H adsorption.

Although Ni and Co have strong adsorption energies with a single H atom, they are not efficient electrocatalysts because additional H adsorbs too weakly for efficient electron transfer on the surface with high hydrogen coverage. The spatial extension of their d-atomic orbitals is too small. Ag, Cu, and Au have the same activity because the differences in their adsorption energy and spatial orbital extent compensate.

Interestingly, the fact that the optimum interaction for efficient electro-reduction is zero adsorption free energy of H_2 is consistent with the high activity of non-noble metal catalysts such as MoS_2 [31] (see Figure 6i.4). The thermodynamics of these systems is very similar to that of the Fe_7MoS_8 cluster-based enzyme Nitrogenase used in N_2 reduction, which we discussed in Section 8.2.1.2.

8.3 Oxidation Reactions

8.3.1 Synthesis Gas from Methane

8.3.1.1 Steam Reforming

The steam reforming reaction of methane to synthesis gas is highly dependent on particle size and also on the choice of transition metal (see Figure 8.7).

The strong increase in turn over frequency (TOF) with particle size is caused by an increase in methane activation reaction rate when coordinative unsaturation of surface atoms decreases. Activation of methane occurs over a single metal atom. When particle size decreases, the relative number of coordinatively unsaturated edge and corner sites increases compared to the more highly coordinated metal surface terrace atoms. The lower the coordination number of the metal atom with its surface atom neighbors, the lower its oxygen-surface bond energy.

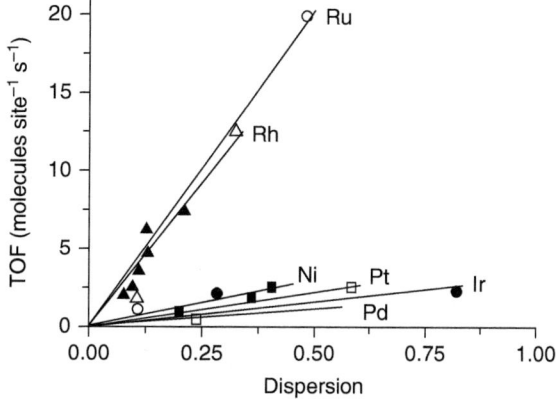

Figure 8.7 Reaction rate as a function of dispersion for CH_4—H_2O reforming (773 K, 0.19 bar CH_4, 0.74 bar H_2O, 0.07 bar H_2). Comparison of turn over frequencies (TOF) of the transition metals Ru, Rh, Ni, Pt, Ir and Pd. (Jones et al. 2008 [33]. Reproduced with permission of Elsevier.)

Table 8.1 Surface metal–oxygen bond strengths in (kJ mol^{-1}) for the densest surfaces of the various transition metals.

Fe (hcp)	Co	Ni	Cu
−714	−550	−496	−429
Ru	Rh	Pd	Ag
−557	−469	−382	−321
Os	Ir	Pt	Au
−533	−428	−354	−270

Metals indicated in green have a high enough surface strength for use in steam reforming, as opposed to metals indicated in red. Yellow indicates a borderline case.

Since the reactivity of a metal particle is determined by the average of the reactivities of the different metal atoms, the reactivity of the metal particle will increase with a decrease in particle size.

Interestingly, Ru and Rh are both more reactive than Pt or Pd but their activation energies of methane dissociation are comparable (Table 7i.1). This is because H_2O must also decompose to provide the O_{ads} atoms for CO formation. A thermodynamic analysis provides insight into the differences in the reactivity of the various metals. In Table 8.1, metals that have a sufficient surface metal–oxygen bond strength for the generation of O by reaction with H_2O are indicated.

Carbondioxide is used instead of H_2O in dry steam reforming. The M—O bond energy necessary to cleave the H_2O bond is 480 kJ mol^{-1}, and the energy to cleave the CO_2 bond is 520 kJ mol^{-1}. One notes from Table 8.1 that this reduces the number of candidate metals for dry reforming compared to those suitable for steam reforming. The DFT-calculated values of Table 8.1 are based on the adsorption energies of the O atom on the (111) or (0001) surfaces of each metal. The adatom bond strengths may become substantially stronger for more open surfaces, and this may make borderline transition metals reactive.

We discussed the kinetics and mechanism of the steam reforming reaction earlier (Section 2.2.5.1 and Chapter 3, Insert 7). The reaction proceeds through the intermediate formation of CH_{ads} that recombines with atomic oxygen. Water formation occurs through recombination of OH hydroxyl species. At high temperature, the reaction becomes poisoned by carbon overlayer decomposition.

When the metal particle size of the active metal catalyst is too small, the catalyst deactivates rapidly because the particles convert easily into stable oxide particles. This has been observed for Rh. [34].

8.3.1.2 Methane Oxidation

Catalytic oxidation of methane by transition metals is used to generate the hot gases necessary for gas turbine operation [35].

Similarly to the reforming reactions, this reaction proceeds through intermediate C_{ads} and CH_{ads} formation. The adsorbed oxygen atoms needed for the consecutive formation of CO, CO_2, and H_2O play different roles. In addition to the recombination with C_{ads} and CH_{ads} species, as we discussed in Section 7.4, adsorbed oxygen atoms can activate CH_4 or block the surface for methane activation. In the latter case, when the O_2 partial pressure increases, the metal surface becomes covered with oxygen and the reaction will become negative order in O_2 partial pressure.

A unique chemistry develops for Pd, which results in the formation of PdO. As long as the reaction temperature remains below 650 °C, the PdO(101) surface remains stable and appears exceptionally suitable to initiate methane activation, which is followed by high-temperature homogenous combustion.

The surface chemistry, which is basic to this chemical process, has been discussed in Chapter 7. In Section 11.3.3, we will more fully discuss the activation of methane by reducible oxides.

8.3.1.3 NH_3 Oxidation to NO and N_2

The Ostwald process invented during the same time period as the Born–Haber process oxidizes NH_3 to NO (see Section 2.2.5.2). The catalyst used in this process is very similar to the one used later in the Andrussow process, which produces HCN from CH_4 and NH_3 (see Section 8.2.2.4). Temperatures of operation for these processes are comparable. As in the Andrussow process, the nonselective reaction that must be prevented in the Ostwald process is N_2 formation.

On the other hand, the reaction of NH_3 to give N_2 is a desirable reaction when NH_3 is decomposed in waste streams. The selectivity of the Ostwald reaction depends strongly on temperature: at low temperature N_2 is formed; at intermediate temperatures N_2O is formed; and at high temperatures NO is formed (see Figure 8.8).

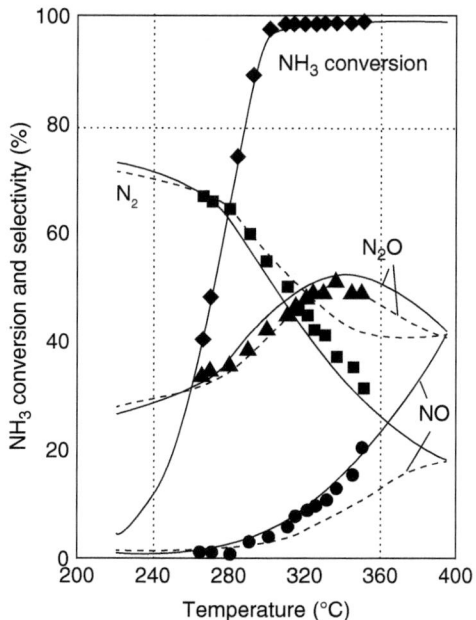

Figure 8.8 Experimental ammonia conversion and selectivities versus simulation results (dashed lines) obtained from a combination of lumped kinetics with heat and mass flow simulations using the hydrodynamic Navier–Stokes model using constant wall temperature as a function of temperature. Reaction conditions: 20 vol%. (Rebrov et al. 2003 [36]. Reproduced with permission of Elsevier.)

There are two essentially different routes for N_2 formation:

1. On the transition metal surface, NH_3 decomposes to N atoms that recombine to N_2 in a consecutive reaction.
2. NH_3 recombines with NO, to give N_2 and H_2O. This is called the Fogel reaction.

Upon decomposition of NH_3 on the Pt surface, two N atoms can recombine to produce N_2. Since adsorbed O atoms are present, this reaction competes with NO formation. Both reactions are structure-sensitive, and efficient on cubic particles. The (100) surfaces that terminate such particles provide both proper thermodynamics and low activation energies Chapter 4, insert 3.

When the temperature of reaction remains low enough so that NO desorbs slowly, NO recombination will give N_2O. N_2O will also be formed by the reaction of NO with an adsorbed N adatom. This reaction then competes with N_2 formation. The undesirable formation of N_2O (a greenhouse gas) can be suppressed by the introduction of excess ammonia, which reduces O surface coverage, thus hindering NO formation. A second beneficial effect of the introduction of excess ammonia is that it will increase the rate of N_2 formation through the Fogel reaction (mechanism b of N_2 formation).

At higher temperature, NO can desorb and it will be the main product.

Metals such as Ag or Cu will become oxidized, and ammonia will be oxidized on them to yield the nitrite or nitrate intermediates [37, 38]. In a consecutive reaction with NH_3 the adsorbed NO_x intermediates will be reduced to N_2 (via intermediate formation of NH_4NO_2) or N_2O (via intermediate formation of NH_4NO_3). This is another example of the Fogel reduction of NH_3 with NO.

A similar reaction is catalyzed by Pt^{2+} ions in the Faujasite zeolite [39, 40] that is analogous to the homogenous reaction [41–43] catalyzed by amine complexes.

In the zeolite, ion-exchanged Pt^{2-}, cations will combine with NH_3 to form a catalytically reactive intermediate, the $Pt^{2+}(NH_3)_4$ amine complex. This complex can activate O_2, which then reacts with an ammonia ligand to give water. A proton is donated to the zeolite in this reaction:

$$Pt(NH_3)_4^{2+} + O_2 \rightarrow [Pt(NH_3)_3NO]^+ + H^+ + H_2O$$

NO^- replaces one of the four ammonia ligands of Pt^{2+}. NO^- can be oxidized with a second O_2 molecule to give NO_3^-, which can react with another ammonia ligand to give N_2O:

$$[Pt(NH_3)_3(NO_3)]^+ + H^+ + 2NH_3 \rightarrow Pt^{2+}(NH_3)_4 + N_2O$$

However, when NO^- reacts with H_2O instead of O_2 the NO_2^- ligand will decompose with ammonia to give N_2:

$$[Pt(NH_3)_3(NO)]^+ + H_2O \rightarrow [Pt(NH_3)(NO_2)]^- + 2H^+$$

In this reaction, the addition of H_2O inhibits N_2O formation. These reaction events are illustrated by the temperature-programmed mass spectrometric measurements shown in Figure 8.9.

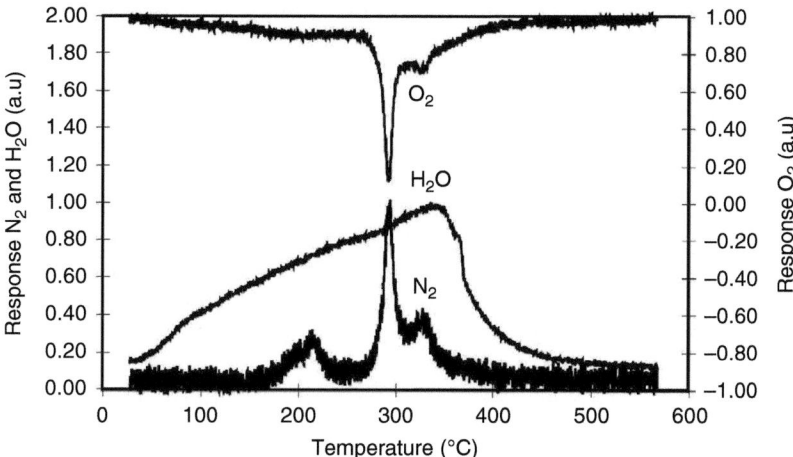

Figure 8.9 Temperature-programmed mass spectrometry of the decomposition of $Pt^{2+}(NH_3)_4$ ion exchanged in zeolite HZSM-5 [39]. The corresponding reaction steps are indicated. Three peaks can be distinguished in the TPD spectrum. Oxygen is consumed in only two of the N_2 formation peaks. The mechanism that explains the occurrence of these three peaks is consistent with proposals made earlier on the homogeneous oxidation of Ru–amine complexes in basic solution [43, 44] and the reaction of NO with Ru–amine or Os–amine complexes [41, 42, 45].

The example in this subsection of catalysis by a transition metal surface and an immobilized metal organic complex immobilized in a zeolite again illustrates the close correspondence between molecular events in the two systems.

8.3.1.4 Selective Oxidation of Ethylene

Three important selective oxidation reactions of ethylene are the formation of acetaldehyde, vinyl acetate formation by reaction with acetic acid, and the epoxidation of ethylene. The discussion of these reactions provides an opportunity to explore differences in their reaction mechanisms at the gas–solid surface interphase and in the liquid phase. This is of special interest when discussing the mechanism of the Wacker reaction.

In a liquid phase medium of water, the Wacker reaction oxidizes ethylene using $PdCl_2$ and $CuCl_2$ salts. Vinyl acetate is selectively catalyzed by metallic Pd and ethylene epoxide is produced in the gas phase by a Ag catalyst activated by Cl. While the Wacker reaction is a homogeneous reaction (which can be catalyzed by an immobilized system), the catalysis of vinyl acetate is believed to be a heterogeneous catalytic reaction. The ethylene epoxidation reaction occurs in the gas phase. Its unique feature is that the selectivity of the catalyst strongly depends on the state of the catalyst surface.

As we will discuss in the next subsection, reactions in the liquid phase catalyzed by organometallic or coordination complexes can proceed through inner or outer coordination shell reactions. In the Wacker reaction, this is the *syn-* versus *anti-*mechanism. Water plays an important role in that it mediates proton shuttling by the hopping of protons through the water coordination shell. Since this is a general feature in many water-catalyzed reactions we will discuss this in some detail.

8.3.1.4.1 Ethylene to Acetaldehyde

The Wacker reaction is an organometallic oxidation reaction that is catalyzed by a water solution containing a mixture of $PdCl_2$ and $CuCl_2$ salts. The Pd^{2+} cation catalyzes formation of a hydroxylated ethyl intermediate by reaction of ethylene with H_2O, which converts to the aldehyde through a β-CH cleavage reaction. This forms intermediate vinyl alcohol, which rapidly isomerizes to acetaldehyde. In this reaction step, Pd^{2+} is reduced to the Pd metal. The Pd^{2+} cation is regenerated by oxidation with O_2 catalyzed by Cu complexes. The overall catalytic reaction scheme is shown in Figure 7i.1.

The Pd cation is part of a planar $PdCl_2OH_2C_2H_4$ complex. The two mechanisms proposed for how this complex catalyzes acetaldehyde formation are shown in Figure 7i.2. In the inner sphere reaction mechanism, the ethylene molecule is initially π-bonded, but shifts to a single M—C bond interacting σ-adsorbed state. Pd-coordinated OH adds to the carbon atom of ethylene that is not attached to Pd [46, 47].

In the alternative outer sphere *anti*-mechanism, OH addition occurs by a water molecule that is not activated by Pd and is not adsorbed in the first coordination shell of Pd [48]. Recently, advanced quantum-chemical dynamic studies have provided strong support for the *anti*-mechanism [48, 49] and β CH cleavage is considered to be rate-controlling. The main evidence for the *anti*-mechanism is the substantial difference in the stability of the *cis*- and *trans*-position of H_2O in the reaction complex. The *trans* complex stability exceeds the *cis* complex stability by 60 kJ mol^{-1} [48].

The outer sphere reaction mechanism can only be properly studied when H_2O molecules are explicitly considered in the simulation model. The water network is essential for providing low-energy paths to generate the OH^- species that react with olefin and for the isomerization of the vinyl alcohol intermediate [48] as schematically illustrated in Figure 7i.3. The proton is shuttled through the water network by consecutive formation of hydronium ions. This proton shuttling, which is common in the water phase, is called the *Grotthuss mechanism of proton transfer*.

In the Wacker reaction, the water molecules concerned are contained in bulk water, but small water clusters near a hydrophobic reaction center (such as in a zeolite or an enzyme) often play the same role in that they enable proton transfer by forming proton-conducting bridges.

8.3 Oxidation Reactions

Insert 7: The Mechanism of the Wacker Reaction

The general scheme of the Wacker reaction is shown in Figure 7i.1. There are two mechanistic options: the syn addition and anti-addition mechanisms respectively sketched in Figure 7i.2. Details of the proton transfer model within the *anti*-mechanism, are given in Figure 7i.3. Figure 7i.4 shows the orbital relation for the different reaction intermediates according to the syn-addition mechanism.

Figure 7i.1 General scheme of the Wacker process. (After [48].)

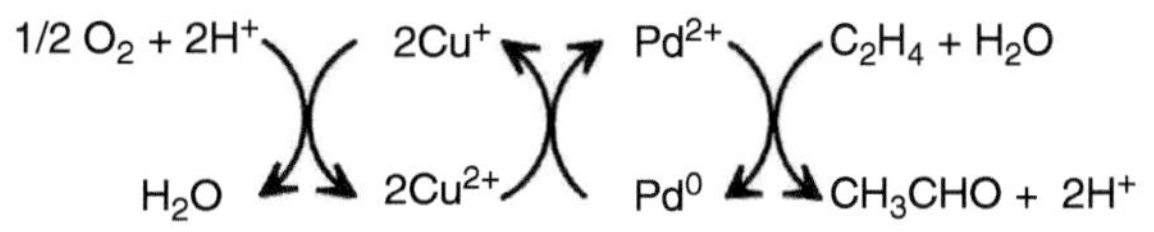

$$1/2\,O_2 + 2H^+ \;\diagup\; 2Cu^+ \;\diagdown\; Pd^{2+} \;\diagup\; C_2H_4 + H_2O$$
$$H_2O \;\diagdown\; 2Cu^{2+} \;\diagup\; Pd^0 \;\diagdown\; CH_3CHO + 2H^+$$

Figure 7i.2 The two postulated reaction pathways for the Wacker process. (After [50].)

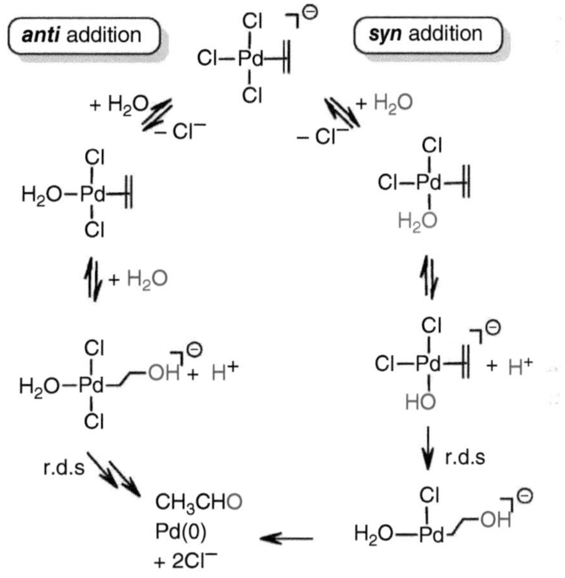

anti addition / *syn* addition

(Continued)

Insert 7: (Continued)

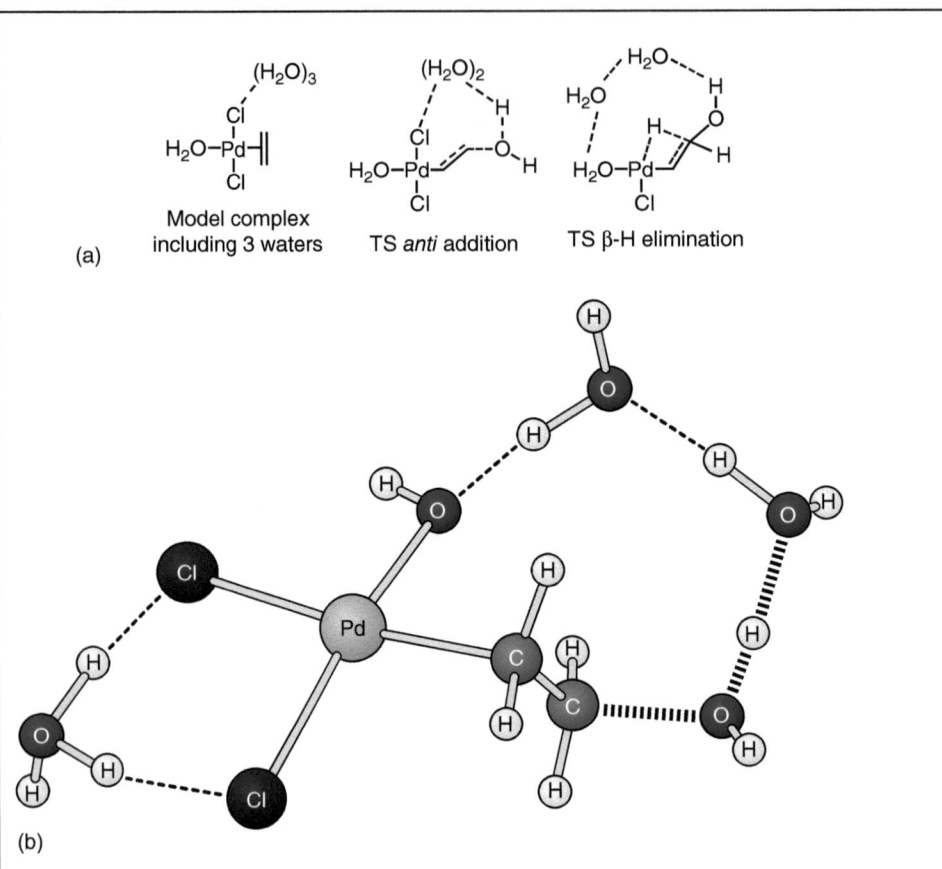

Figure 7i.3 Schematic representation of proton transfer model reaction steps calculated including explicit water molecules. (a) Reaction intermediates; (b) the proton network. (After [51].)

Figure 7i.4 shows the electronic structures for the *syn* insertion reaction intermediates. It illustrates the stabilization of the transition state structure by interaction with the Pd d-valence atomic orbitals located in the plane of the reaction complex. The transition state orbitals of the allyl fragment formed by the C—C—O fragment (second column of Figure 7i.4) have the following energies: -1.99, -1.40, and 0.00 eV which are the bonding orbitals, the nonbonding and the antibonding orbitals, respectively. The -1.99 eV orbital is stabilized by the interaction with the Pd $d_{x^2-y^2}$ orbital.

The -1.40 eV orbital already shows a C—O bonding interaction and the 0.0 eV orbital similarly shows bonding interaction with the Pd—C bond.

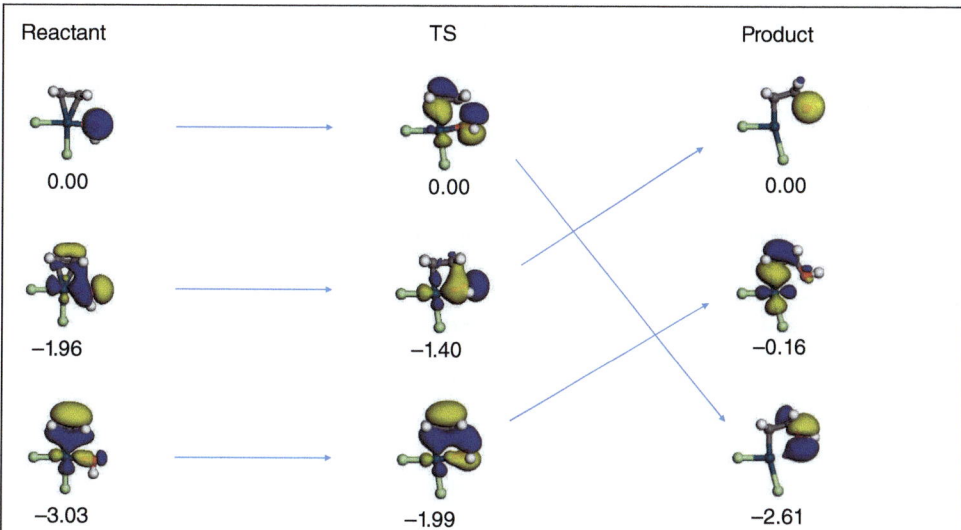

Figure 7i.4 Changes of the occupied molecular orbitals in the syn OH reactionstep [52]. It shows natural bond-orbitals projected in a limited phase space of atomic orbitals localized on one of the four atoms: C, C, Pd, and O. Natural bond-orbitals are found by diagonalizing a density matrix constructed from occupied Kohn–Sham orbitals. The orbitals are projected on the atomic orbitals of the four atoms: C , C , Pd, and O. This is diagonalized to give natural-bond orbitals. The orbital energies are calculated from the Kohn–Sham Hamiltonian equation and are given in electronvolts with respect to the highest occupied molecular orbital. This illustrates how, in the transition state, this allylic intermediate of the C—C—O skeleton is initially stabilized by the $d_{x^2-y^2}$ atomic orbital of the Pd^{2+} cation. (After [52].)

A salient feature of the transition state is the enduring π character of the C—C bond. The transition state can be characterized as early, and the correlation of orbitals of transition state and final state confirms this, since the π bond in ethylene is broken only in the final state.

8.3.1.4.2 Vinyl Acetate

Vinyl acetate is an important intermediate for polymer synthesis. The mechanism of the vinyl acetate synthesis reaction, also catalyzed by Pd, is different from the Wacker reaction since explicit cleavage of the C—H bond is necessary. The reaction shows a large kinetic isotope effect when C_2H_4 is replaced by C_2D_4. In contrast with the Wacker reaction, this reaction proceeds on reduced Pd, present as large clusters in solution or as metal particles on a support. No Cu promotion is needed for O_2 activation because O_2 will dissociate on the large Pd clusters [53].

Insert 8: The Mechanism of Vinyl Acetate Formation

The two mechanistic paths that have been proposed are summarized in Table 8i.1. The essential difference between the two is that they discriminate between the dissociation of the C—H bond in ethylene either before or after the insertion of acetate into the M—C bond.

Table 8i.1 The different surface reaction paths proposed for the synthesis of vinyl acetate monomer (VAM) [54–56].

Nakamura/Moiseev	Samanos
AcOH + * → AcOH*	AcOH + * → AcOH*
C_2H_4 (g) + * → C_2H_4*	½[O_2 (g) + 2* → 2O*]
½[O_2 (g) + 2* → 2O*]	AcOH* + O* → AcO* + OH*
AcOH* + * → AcO* + H*	C_2H_4 (g) + * → C_2H_4*
C_2H_4 + * → C_2H_3* + H*	C_2H_4* + AcO* → C_2H_4OAc* + *
C_2H_3* + AcO* → VAM* + *	C_2H_4OAc* + * → VAM* H*
VAM* → VAM + *	VAM* → VAM + *
O* + H* → OH* + *	OH* + H* → H_2O* + *
OH* + H* → H_2O* + *	H_2O* → H_2O (g) + *
H_2O* → H_2O (g) + *	

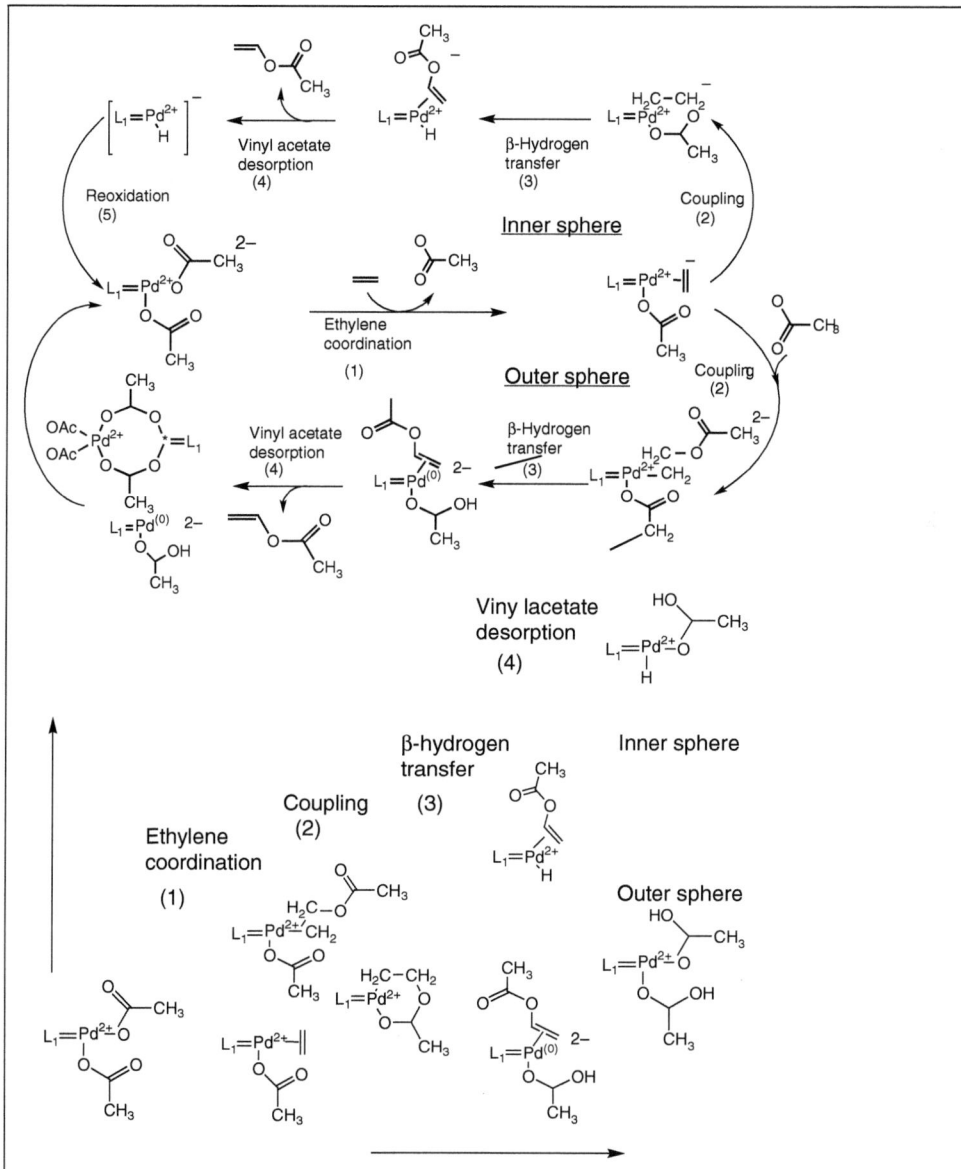

Figure 8i.1 Vinyl acetate formation on a Pd^{2+} ion that is part of $Pd_2(OAc)$ and $Pd_3(OAc)_3$ clusters [57, 58]. This reaction sequence is the analog of the homogenous Wacker reaction and is carried out in glacial acetic acid [55, 56]. (a) Proposed catalytic cycles for the inner- and outer-sphere Wacker-like mechanisms for the acetoxylation of ethylene to vinyl acetate. (b) Energy diagram of the inner- and outer-sphere mechanisms, including solvent effects. Activation barriers are indicated by the small arcs; a barrier as low as the reaction energy is designated by "no act." The structure number refers to models shown in (a).

(Continued)

Insert 8: (Continued)

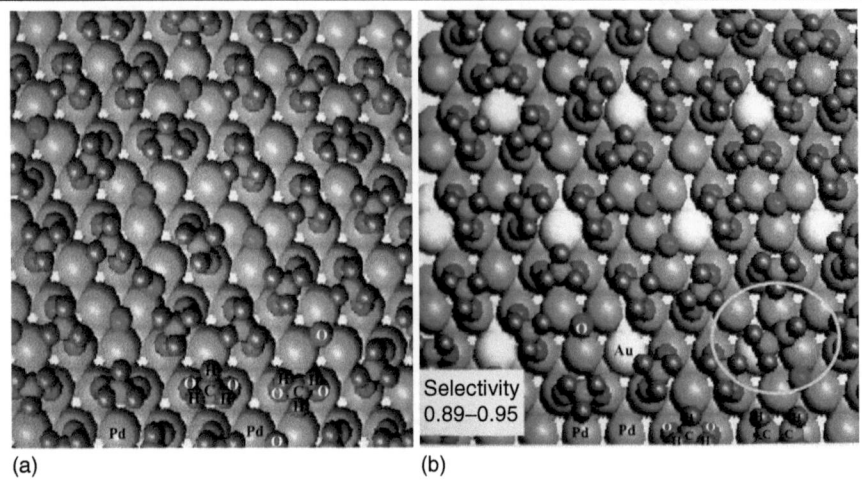

(a) (b)

Figure 8i.2 (a) Snapshot from the steady-state simulation of VAM (vinyl acetate monomer) synthesis over pure Pd(111). (b) Simulation of surface coverage for vinyl acetate synthesis in the presence of PdAu alloys. (After [61].)

The supported Pd catalyst must be promoted by alkali and Au. The high pH shifts the acetic acid–acetate equilibrium in the liquid reaction medium present in the support micropores. Acetate then adsorbs to the surface, without the need to activate the O—H bond.

There is no consensus as to which of the two mechanisms included in Table 8i.1 dominates. It is also debated whether the reaction actually occurs as a surface reaction on large dissolved Pd clusters or on small oxycationic PdO acetate clusters. On small clusters, ethylene reacts with the acetate anion ligand by an outer sphere reaction to give an ethyl acetate ligand, and this in turn gives the vinyl acetate by β-CH cleavage (see Figure 8i.1). Spectroscopic experiments support the Semenov mechanism [59, 60].

Computational studies on the Pd(111) metal surface also support the Semenov mechanism. The activation of ethylene to vinyl and subsequent insertion of acetate into the M—C bond requires $120\,\text{kJ}\,\text{mol}^{-1}$, whereas the route via acetate insertion into ethylene and subsequent β-C—H bond cleavage requires $90\,\text{kJ}\,\text{mol}^{-1}$.

In the Semenov mechanism, a C—O bond is formed between acetate and ethylene, which is analogous to the first step in the hydrogen addition Horiuti–Polanyi reaction. This forms a σ-bonded ethylacetate intermediate that is negatively charged. Then, ethylene acetate is formed by β-C—H bond cleavage, a step similar to the ethyl to ethylene decomposition reaction in the Horiuti–Polanyi mechanism. In this mechanism, oxygen mainly serves to react with adsorbed hydrogen atoms to give H_2O, so that the reaction becomes exothermic.

The main nonselective product is the total combustion of the acetate instead of ethylene. This has been demonstrated by ^{13}C Isotope-labeled experiments.

The main competitive reaction that removes ethylene proceeds through the Wacker reaction: acetaldehyde is oxidized by Pd to acetic acid that reacts with another ethylene molecule to give vinyl acetate.

Kinetic Monte Carlo studies (Figure 8i.2) have demonstrated that during reaction the Pd atoms saturate with acetate and hence on the surface the rate of reaction is limited by the number of vacancy positions available for adsorption of the ethylene molecule. The role of Au is to provide vacant sites for the adsorption of ethylene to make it available for reaction with the acetate.

8.3.1.4.3 Ethylene Epoxidation

The ethylene epoxidation reaction and its kinetics have been introduced in Chapter 2. In section 7.4.2.3 quantum-chemical aspects of the oxygen insertion reaction catalyzed by oxidized surfaces has been discussed. In this section, we will discuss the molecular aspects of the reaction mechanism that determine the selectivity of the reaction.

An important aspect of the catalysis of ethylene epoxidation is the positive effect of Cl coadsorption on selectivity.

Two conflicting mechanistic schemes have been proposed to explain this [62]. According to the molecular epoxidation hypothesis, selective epoxidation occurs by oxygen transfer from molecularly adsorbed O_2 [63]. The O atoms that are left on the surface are consumed in the total combustion reaction. This predicts a maximum selectivity of 6/7, since when an O_2 molecule is used to create the epoxide, the oxygen atom is used for total combustion. To oxidize one ethylene molecule to CO and H_2O six oxygen atoms are needed. So to totally combust one ethylene molecule six epoxide molecules can be formed. This gives a maximum ethylene selectivity to epoxide of 6/7. Within this view the initial selectivity of the reaction is reduced by competitive O_2 dissociation due the presence of unoccupied Ag surface atom sites. When one assumes that Cl inhibits Ag surface vacant site formation. This is consistent with an increase in selectivity at increased co-adsorbate coverage by Cl.

The competitive proposal claimed that atomic oxygen could also selectively epoxidize ethylene, but no oxygen vacancies should be present that would activate the C—H bond of ethylene by direct contact with Ag [64]. According to this proposition, Cl also was thought to reduce vacancy concentration and hence enhance selectivity. According to this proposal 100% selectivity of ethylene conversion to epoxide should be theoretically possible.

An experiment analogous to the ^{13}CO isotope-based proof of the carbide mechanism for the Fischer–Tropsch reaction demonstrates that atomic O can indeed produce the epoxide [65]. When $^{16}O_2$ is dissociatively adsorbed to give a surface oxide on a Ag powder, subsequent reaction of a mixture of ethylene and $^{18}O_2$ demonstrates initial incorporation of ^{16}O into the epoxide. It also proves that the oxidation reaction is a Mars–van Krevelen surface reaction [65]. According to the Mars-van Krevelen mechanism the sites of O_2 dissociation does not relate to the site of the selective oxidation reaction. The oxygen consumed in this reaction is part of the bulk oxide.

Insert 9: The Mechanism of Ethylene Epoxide Formation

The reactivity of O adsorbed to a metallic surface is very different from its reactivity on the Ag_2O surface. Figure 9i.1 illustrates oxametallocycle intermediate formation on the nucleophilic O atoms adsorbed to the Ag surface. The adsorbed species is the intermediate for acetaldehyde formation. Figure 9i.2 illustrates the structure of silver oxide surface and adsorbed ethylene. Figures 9i.3 and 9i.4 show the reaction energy diagrams of ethylene reacting selectively to the epoxide on the closed Ag_2O surface and non-selectively when a surface vacancy is present. The oxygen atoms of the oxide surface are electrophilic and insert into the ethylene π bond only on the nonvacant surface (Figure 9i.3).

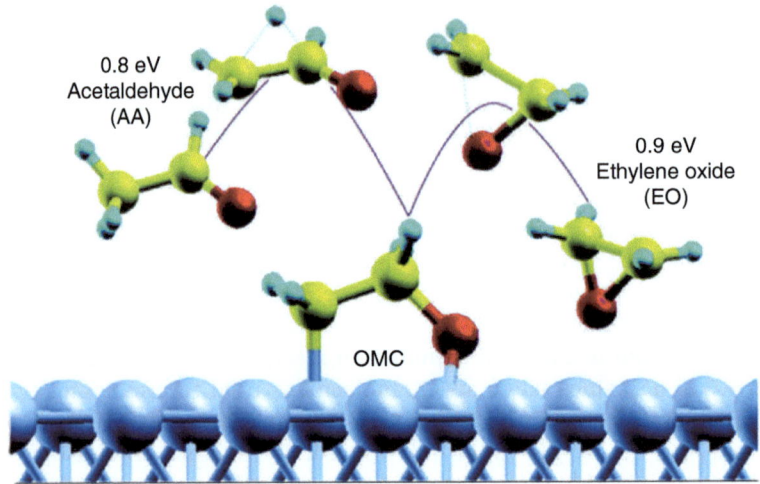

Figure 9i.1 The oxametallocycle intermediate that is formed between adsorbed O and ethylene in a metal Ag surface. Respective activation energies for the decomposition of this complex to give ethylene epoxide (EO) versus acetaldehyde (AA) are indicated [66, 67].

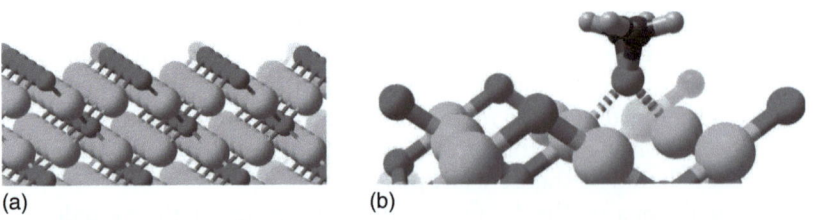

Figure 9i.2 (a) Structure of $Ag_2O(001)$ surface. (b) The adsorbed state of ethylene. The intermediate has a structure close to that of adsorbed epoxide. (Ozbek et al. 2011 [68]. Reproduced with permission of Elsevier.)

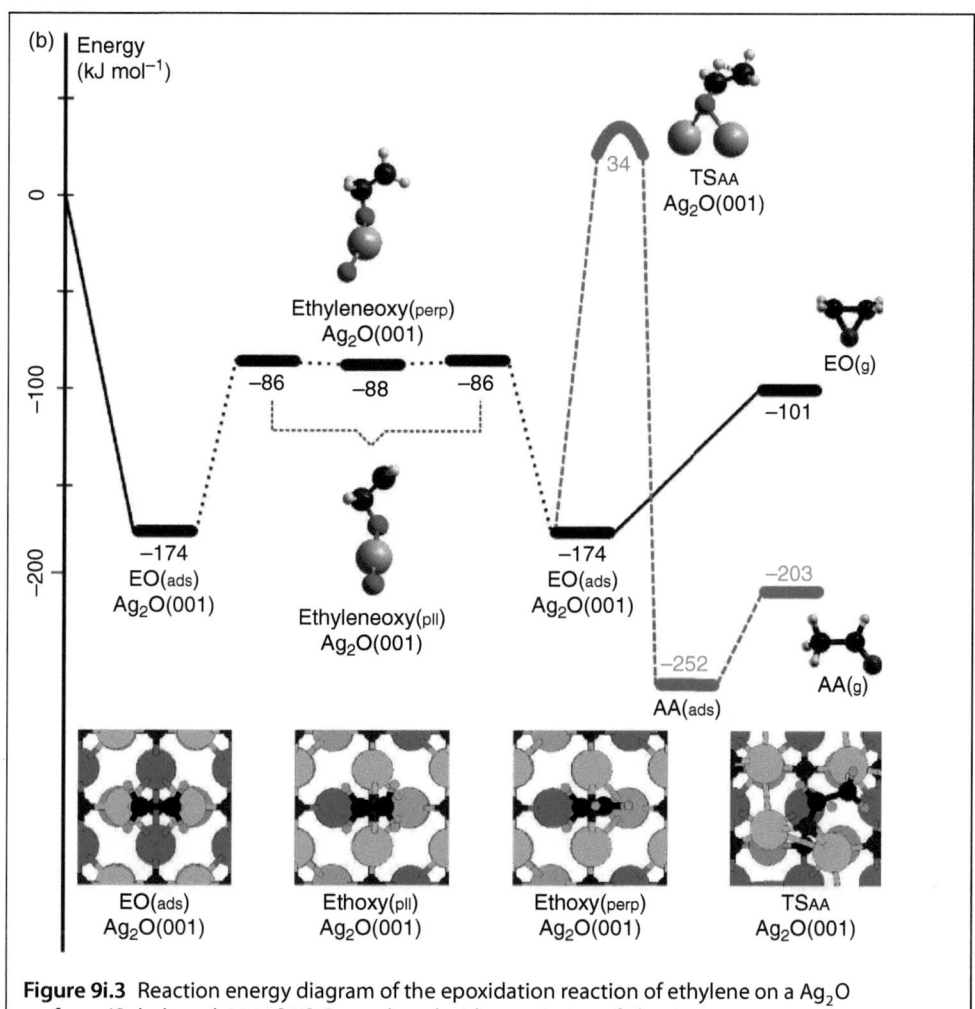

Figure 9i.3 Reaction energy diagram of the epoxidation reaction of ethylene on a Ag$_2$O surface. (Ozbek *et al.* 2011 [68]. Reproduced with permission of Elsevier.)

(Continued)

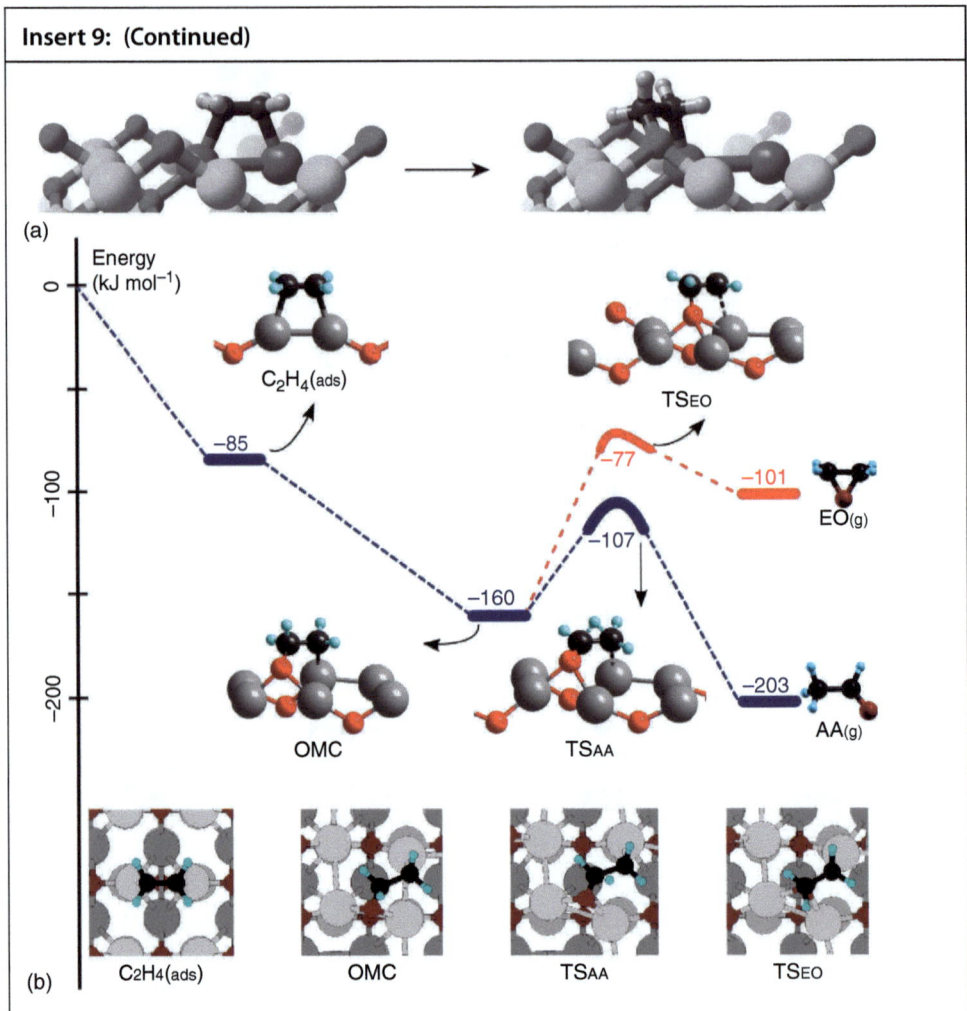

Figure 9i.4 (a) A vacant site of the Ag$_2$O surface with the ethylene oxametallocycle. (b) The corresponding reaction energy diagram. (Ozbek et al. 2011 [68]. Reproduced with permission of Elsevier.)

The kinetics scheme of the ethylene epoxidation reaction of Figure 2.16 (page 42) indicates two nonselective routes for the total combustion of ethylene. One is a parallel route that are discussed above and the other is a consecutive route. In both cases, acetaldehyde formation is responsible for total combustion. Aldehyde combustion is rapid and catalyzed by the Ag catalyst. In the parallel reaction, an oxametallocycle intermediate as shown in Figures 9i.1 and 9i.4 preferentially leads to the acetaldehyde. This reaction requires a surface oxygen atom vacancy in the oxide layer. In the consecutive reaction, the aldehyde is formed by proton-catalyzed isomerization of the epoxide by the α-Al$_2$O$_3$ support surface.

Selective formation of the epoxide occurs by the reaction of ethylene with oxygen within an oxide overlayer on the Ag particle. Results of quantum-chemical calculations shown in Insert 9 demonstrate a significant difference in the reactivity of atomic oxygen with ethylene when the silver oxide is closed (Figure 9i.1) versus its reactivity when a vacancy is present. The barrier for intermediate aldehyde formation is lower than that of epoxide formation when an oxide atom vacancy is present in the oxide overlayer, but higher than the barrier for epoxide formation when no vacancy is present. Reactivity of the oxide overlayer is illustrated by Figures 9i.3 and 9i.4.

In the ethylene epoxidation reaction, Ag reacts in a unique way. The presence of a closed oxide overlayer is a necessity and the oxygen atoms have to react electrophylic. This makes the Ag catalyst unique, since a similar oxide layer is not formed on Au and the oxygen atoms on copper oxide are too strongly bound. Group VIII transition metals cannot be used because they activate the CH in ethylene.

To stabilize the oxidic state of the silver overlayer, Cl co-adsorption is essential since at reaction conditions the oxide overlayer is unstable. Cl adsorbs more strongly than O and will not desorb as Cl_2. However, it can react with olefin and will be slowly removed. For this reason, a Cl-containing molecule is continuously co-fed into the reactor to maintain the Cl on the catalyst surface.

Alkali is also added to the catalyst as a promotor. The main role of alkali is to suppress the presence of acidic protons on the alumina support. During the catalyst preparation process, Brønsted acidic sites develop on the alumina support, which will isomerize the epoxide to acetaldehyde. This is the nonselective consecutive reaction r_3 in (Figure 2.16). Alkali will suppress this acidity and thus prevent the consecutive combustion of ethylene epoxide.

Currently used catalysts operate with a selectivity of over 6/7, which is consistent with the idea that the Mars–van Krevelen mechanism is operational and atomic oxygen is responsible for the oxygen insertion reaction.

8.4 Uniqueness of a Metal for a Particular Selective Reaction

Selective oxidation of a hydrocarbon by transition metals must overcome a variety of challenges:

- In the case of the epoxidation reaction, no C—H or C—C bonds should be cleaved. This requires the use of a transition metal catalyst that has low reactivity with respect to these activation processes. Since the group VIII transition metals readily activate C—H and C—C bonds, it is no surprise that Cu, Ag, or Au are relevant candidates. Au is excluded because the O_2 required for the epoxidation reaction has to dissociate, which is not thermodynamically feasible for metallic Au.
  The only other alternative metal catalyst is Cu. However, its strong Cu—O bond destabilizes O insertion into the π bond and hence competitive C—H bond activation occurs.
- In contrast with ethylene epoxidation, CH activation is desirable in the aldehyde or acetoxylation reaction. This is why a transition metal that is more

reactive than Ag is needed. On the other hand, the metal should not be so reactive that it also cleaves the C—C bond, which would lead to total combustion. Additionally, the acetate should not bind too strongly and poison the catalyst. The high reactivity of the metal will activate the acetaldehyde bonds so that it will be readily oxidized to CO_2 and H_2O. A metal such as Pt cannot be used because it also becomes poisoned by oxygen. Palladium metal is unique, since the Pd ions that are part of an oxidized overlayer are active and selective catalysts (see Section 7.4.4.3) Ni also easily becomes overoxidized, which makes it only active for total oxidation.

References

1 Jacobsen, C.J.H., Dahl, S., Boisen, A., Clausen, B.S., Topsøe, H., Logadottir, A., and Nørskov, J.K. (2002) Optimal catalyst curves: connecting density functional theory calculations with industrial reactor design and catalyst selection. *J. Catal.*, **205**, 382–387, http://linkinghub.elsevier.com/retrieve/pii/S0021951701934426 (accessed 20 January 2015).
2 Ertl, G. (1983) in *Catalysis, Science Technology* (eds J.R. Anderson and M. Boudart), Springer-Verlag, Berlin, Vol **4**, p 273.
3 Spencer, N. (1982) Iron single crystals as ammonia synthesis catalysts: effect of surface structure on catalyst activity. *J. Catal.*, **74**, 129–135, http://linkinghub.elsevier.com/retrieve/pii/0021951782900161 (accessed 20 January 2015).
4 Honkala, K., Hellman, A., Remediakis, I.N., Logadottir, A., Carlsson, A., Dahl, S., Christensen, C.H., and Nørskov, J.K. (2005) Ammonia synthesis from first-principles calculations. *Science*, **307**, 555–558, http://www.ncbi.nlm.nih.gov/pubmed/15681379 (accessed 5 November 2014).
5 Temkin, M. and Pyzhev, V. (1940) Kinetics of the ammonia synthesis on promoted iron catalysts. *Acta Physicochim. URSS*, **12**, 217–222.
6 Temkin, M. and Pyzhev, V. (1939) Kinetics of the synthesis of ammonia on promoted iron catalysts. *J. Phys. Chem. (USSR)*, **13**, 851.
7 Branden, C. and Tooze, J. (1991) *Introduction to Protein Structure*, Garland.
8 van Santen, R.A., Markvoort, A.J., Filot, I.A.W., Ghouri, M.M., and Hensen, E.J.M. (2013) Mechanism and microkinetics of the Fischer–Tropsch reaction. *Phys. Chem. Chem. Phys.*, **15**, 17038–17063, http://www.ncbi.nlm.nih.gov/pubmed/24030478 (accessed 29 January 2015).
9 Biloen, P. and Sachtler, W.M.H. (1981) Mechanism of hydrocarbon synthesis over Fischer–Tropsch catalysts. *Adv. Catal.*, **30**, 165–216.
10 Filot, I.A.W., van Santen, R.A., and Hensen, E.J.M. (2014) The optimally performing Fischer–Tropsch catalyst. *Angew. Chem. Int. Ed.* **53** (47), 12645–12970, doi: 10.1002/anie.201406521.
11 Horiuti, I. and Polanyi, M. (1934) Exchange reactions of hydrogen on metallic catalysts. *Trans. Faraday Soc.*, **30**, 1164. doi: 10.1039/TF9343001164.
12 Davis, S. and Somorjai, G.A. (1980) Correlation of cyclohexene reactions on platinum crystal surfaces over a ten-order-of-magnitude pressure range

*1Variations of structure sensitivity, rates, and reaction probabilities. *J. Catal.*, **65**, 78–83, http://linkinghub.elsevier.com/retrieve/pii/0021951780902791 (accessed 29 January 2015).

13 Thomson, S.J. and Webb, G. (1976) Catalytic hydrogenation of olefins on metals: a new interpretation. *J. Chem. Soc., Chem. Commun.*, 526–527. doi: 10.1039/C39760000526.

14 van Santen, R.A., Anderson, B.G., Cunningham, R.H., Mangnus, A.V.G., Van Ijzendoorn, L.J., and De Voigt, M.J.A. (1997) In situ observation of transient reaction phenomena occurring on zeolite catalysts with the aid of positron emission profiling. *Angew. Chem. Int. Ed. Engl.*, **35**, 2785–2787, http://www.scopus.com/inward/record.url?eid=2-s2.0-0030484751&partnerID=40&md5=1a4ccdc9a60541b63f39f0dd84c759dd (accessed 24 September 2016).

15 Boudart, M. and Djéga-Mariadassou, G. (1984) *Kinetics of Heterogeneous Catalytic Reactions*, Princeton University Press.

16 Neurock, M. and van Santen, R.A. (2000) A first principles analysis of C–H bond formation in ethylene hydrogenation. *J. Phys. Chem. B*, **104**, 11127–11145. doi: 10.1021/jp994082t.

17 Loffreda, D., Delbecq, F., Vigné, F., and Sautet, P. (2009) Fast prediction of selectivity in heterogeneous catalysis from extended Brønsted-Evans-Polanyi relations: a theoretical insight. *Angew. Chem. Int. Ed.*, **48**, 8978–8980, http://www.ncbi.nlm.nih.gov/pubmed/19768819 (accessed 23 November 2014).

18 Yang, B., Gong, X.-Q., Wang, H.-F., Cao, X.-M., Rooney, J.J., and Hu, P. (2013) Evidence to challenge the universality of the Horiuti–Polanyi mechanism for hydrogenation in heterogeneous catalysis: origin and trend of the preference of a non-Horiuti-Polanyi mechanism. *J. Am. Chem. Soc.*, **135**, 15244–15250, http://www.ncbi.nlm.nih.gov/pubmed/24032528 (accessed 29 January 2015).

19 Teschner, D., Borsodi, J., Wootsch, A., Révay, Z., Hävecker, M., Knop-Gericke, A., Jackson, S.D., and Schlögl, R. (2008) The roles of subsurface carbon and hydrogen in palladium-catalyzed alkyne hydrogenation. *Science*, **320**, 86–89, http://www.ncbi.nlm.nih.gov/pubmed/18388290 (accessed 29 January 2015).

20 Studt, F., Abild-Pedersen, F., Bligaard, T., Sørensen, R.Z., Christensen, C.H., and Nørskov, J.K. (2008) On the role of surface modifications of palladium catalysts in the selective hydrogenation of acetylene. *Angew. Chem. Int. Ed.*, **47**, 9299–9302, http://www.ncbi.nlm.nih.gov/pubmed/18833559 (accessed 29 January 2015).

21 Maire, G. and Garin, F. (1984) Catalysis, in *Science and Technology* (eds J.R. Anderson and M. Boudart), Springer.

22 van Santen, R.A., Neurock, M., and Shetty, S.G. (2010) Reactivity theory of transition-metal surfaces: a Brønsted-Evans-Polanyi linear activation energy-free-energy analysis. *Chem. Rev.*, **110**, 2005–2048, http://www.ncbi.nlm.nih.gov/pubmed/20041655 (accessed 22 February 2014).

23 Garin, F. (2006) Site requirements in the kinetics of alkane transformations catalysed by metals. *Top. Catal.*, **39**, 11–27, http://link.springer.com/10.1007/s11244-006-0033-6 (accessed 7 January 2015).

24 Watwe, R.M., Cortright, R.D., Nørskov, J.K., and Dumesic, J.A. (2000) Theoretical studies of stability and reactivity of C2 hydrocarbon species on Pt clusters, Pt(111), and Pt(211). *J. Phys. Chem. B*, **104**, 2299–2310. doi: 10.1021/jp993202u.

25 McVicker, G. (2002) Selective ring opening of naphthenic molecules. *J. Catal.*, **210**, 137–148, http://linkinghub.elsevier.com/retrieve/pii/S0021951702936857 (accessed 17 December 2014).

26 Maire, G., Plouidy, G., Prudhomme, J., and Gault, F. (1965) The mechanisms of hydrogenolysis and isomerization of hydrocarbons on metals I. Hydrogenolysis of cyclic hydrocarbons. *J. Catal.*, **4**, 556–569, http://linkinghub.elsevier.com/retrieve/pii/0021951765901600 (accessed 29 January 2015).

27 Grabow, L.C., Studt, F., Abild-Pedersen, F., Petzold, V., Kleis, J., Bligaard, T., and Nørskov, J.K. (2011) Descriptor-based analysis applied to HCN synthesis from NH_3 and $CH4$. *Angew. Chem. Int. Ed.*, **50**, 4601–4605, http://www.ncbi.nlm.nih.gov/pubmed/21500324 (accessed 29 January 2015).

28 Trasatti, S. (1990) in *Electrochemical Hydrogen Technologies* (ed. H. Wendt), Elsevier, Amsterdam, p 104.

29 Santos, E., Quaino, P., and Schmickler, W. (2012) Theory of electrocatalysis: hydrogen evolution and more. *Phys. Chem. Chem. Phys.*, **14**, 11224–11233, http://www.ncbi.nlm.nih.gov/pubmed/22797577 (accessed 6 February 2015).

30 Nørskov, J.K., Bligaard, T., Logadottir, A., Kitchin, J.R., Chen, J.G., Pandelov, S., and Stimming, U. (2005) Trends in the exchange current for hydrogen evolution. *J. Electrochem. Soc.*, **152**, J23. doi: 10.1149/1.1856988.

31 Hinnemann, B., Moses, P.G., Bonde, J., Jørgensen, K.P., Nielsen, J.H., Horch, S., Chorkendorff, I., and Nørskov, J.K. (2005) Biomimetic hydrogen evolution: MoS2 nanoparticles as catalyst for hydrogen evolution. *J. Am. Chem. Soc.*, **127**, 5308–5309, http://www.ncbi.nlm.nih.gov/pubmed/15826154 (accessed 6 February 2015).

32 Pletcher, D. and Walsh, F.C. (1982) *Industrial Electrochemistry*, Springer.

33 Jones, G., Jakobsen, J., Shim, S., Kleis, J., Anderson, M., Rossmeisl, J., Abild-Pedersen, F., Ligaard, T., Helveg, S., and Hinnemann, B. (2008) First principles calculations and experimental insight into methane steam reforming over transition metal catalysts. *J. Catal.*, **259**, 147–160, http://linkinghub.elsevier.com/retrieve/pii/S0021951708003096.

34 Popa, C., VanSanten, R.A., and Jansen, A.P.J. (2007) Density-functional theory study of NHx oxidation and reverse reactions on the Rh(111) surface. *J. Phys. Chem. C*, **111**, 9839–9852. doi: 10.1021/jp071072g.

35 Geus, J.W. and van Dillen, A.J. (1999) Catalytic combustion, in *Environmental Catalysis* (eds F.F.J.G. Jansen and R.A. van Santen), Imperial College Press.

36 Rebrov, E.V., de Croon, M.H.J.M., and Schouten, J.C. (2003) A kinetic study of ammonia oxidation on a Pt catalyst in the explosive region in a microstructured reactor/heat-exchanger. *Chem. Eng. Res. Des.*, **81**, 744–752, http://linkinghub.elsevier.com/retrieve/pii/S0263876203723636 (accessed 6 February 2015).

37 Gang, L., van Grondelle, J., Anderson, B.G., and van Santen, R.A. (1999) Selective low temperature NH_3 oxidation to N_2 on copper-based catalysts. *J. Catal.*, **186**, 100–109, http://linkinghub.elsevier.com/retrieve/pii/S0021951799925241.

38 Gang, L., Anderson, B., van Grondelle, J., and van Santen, R. (2001) Intermediate species and reaction pathways for the oxidation of ammonia on powdered catalysts. *J. Catal.*, **199**, 107–114, http://linkinghub.elsevier.com/retrieve/pii/S0021951700931543.

39 van den Broek, A.C.M., van Grondelle, J., and van Santen, R.A. (1997) Preparation of highly dispersed platinum particles in HZSM-5 zeolite: a study of the pretreatment process of [Pt(NH)]. *J. Catal.*, **167**, 417–424, http://linkinghub.elsevier.com/retrieve/pii/S0021951797916006 (accessed 7 February 2015).

40 van den Broek, A.C.M., van Grondelle, J., and van Santen, R.A. (1998) Water-promoted ammonia oxidation by a platinum amine complex in zeolite HZSM-5 catalyst. *Catal. Lett.*, **55**, 79–82, http://link.springer.com/10.1023/A:1019066425124.

41 Pell, S. and Armor, J.N. (1972) Production of a dinitrogen complex via the attack of nitric oxide upon a metal-ammine complex. *J. Am. Chem. Soc.*, **94**, 686–687. doi: 10.1021/ja00757a090.

42 Pell, S.D. and Armor, J.N. (1973) Kinetics and mechanism for the production of a dinitrogen complex. *J. Am. Chem. Soc.*, **95**, 7625–7633. doi: 10.1021/ja00804a015.

43 Pell, S.D. and Armor, J.N. (1975) Facile, aerial oxidation of coordinated ammonia. *J. Am. Chem. Soc.*, **97**, 5012–5013. doi: 10.1021/ja00850a044

44 Assefa, Z. and Stanbury, D.M. (1997) Oxidation of coordinated ammonia to nitrosyl in the reaction of aqueous chlorine with cis-[Ru(bpy) 2 (NH 3) 2] 2+. *J. Am. Chem. Soc.*, **119**, 521–530. doi: 10.1021/ja9633629.

45 Buhr, J.D. and Taube, H. (1980) Dinitrogen complexes of osmium(III) haloammines. *Inorg. Chem.*, **19**, 2425–2434. doi: 10.1021/ic50210a049

46 Eisenstein, O. and Hoffmann, R. (1981) Transition-metal complexed olefins: how their reactivity toward a nucleophile relates to their electronic structure. *Science*, **80**, 4308–4320.

47 Keith, J.A., Nielsen, R.J., Oxgaard, J., and Goddard, W.A. (2007) Unraveling the Wacker oxidation mechanisms. *J. Am. Chem. Soc.*, **129**, 12342–12343.

48 Stirling, A., Nair, N.N., Lledós, A., and Ujaque, G. (2014) Challenges in modelling homogeneous catalysis: new answers from ab initio molecular dynamics to the controversy over the Wacker process. *Chem. Soc. Rev.*, **43**, 4940–4952, http://www.ncbi.nlm.nih.gov/pubmed/24654007 (accessed 6 February 2015).

49 Beyramabadi, S.A., Eshtiagh-Hosseini, H., Housaindokht, M.R., and Morsali, A. (2008) Mechanism and kinetics of the Wacker process: a quantum mechanical approach. *Organometallics*, **27**, 72–79. doi: 10.1021/om700445j.

50 van Leeuwen, P.W.N.M. (2004) *Homogenous Catalysis: Understanding the Art*, Kluwer Academic Publishers, Dordrecht, p. 125.

51 Siegbahn, P.E.M. (1995) New perspectives on the nucleophilic addition step in the Wacker process. *J. Am. Chem. Soc.*, **117**, 5409–5410. doi: 10.1021/ja00124a044.

52 Plaisance, C.P. and van Santen, R.A. (2015) Private Communication.

53 Moiseev, I.I. (1995) in *Catalytic Oxidation* (eds R.A. Sheldon and R.A. van Santen), World Scientific, pp. 203–239.

54 Kötz, E.R., Neff, H., and Müller, K. (1986) A UPS, XPS and work function study of emersed silver, platinum and gold electrodes. *J. Electroanal. Chem. Interfacial Electrochem.*, **215**, 331–344, http://linkinghub.elsevier.com/retrieve/pii/0022072886870267 (accessed 7 February 2015).

55 Vargaftik, M.N., Moiseev, I.I., and Syrkin, Y.K. (1963) The Wacker reaction catalyzed by palladium acetate. *Dokl. Akad. Nauk. S.S.S.R.*, **152**, 773.

56 Moiseev, I.I. and Vargaftik, M.N. (1992) in *Perspectives in Catalysis* (eds J.M. Thomas and K.I. Zamaraev), Blackwell Scientific, Oxford, p. 91.

57 Kragten, D.D., van Santen, R.A., Crawford, M.K., Provine, W.D., and Lerou, J.J. (1999) A spectroscopic study of the homogeneous catalytic conversion of ethylene to vinyl acetate by palladium acetate. *Inorg. Chem.*, **38**, 331–339. doi: 10.1021/ic980399g.

58 Kragten, D.D., van Santen, R.A., Neurock, M., and Lerou, J.J. (1999) A density functional study of the acetoxylation of ethylene to vinyl acetate catalyzed by palladium acetate. *J. Phys. Chem. A*, **103**, 2756–2765. doi: 10.1021/jp982956q.

59 Stacchiola, D., Calaza, F., Burkholder, L., and Tysoe, W.T. (2004) Vinyl acetate formation by the reaction of ethylene with acetate species on oxygen-covered Pd(111). *J. Am. Chem. Soc.*, **126**, 15384–15385, http://www.ncbi.nlm.nih.gov/pubmed/15563157 (accessed 7 February 2015).

60 Stacchiola, D., Calaza, F., Burkholder, L., Schwabacher, A.W., Neurock, M., and Tysoe, W.T. (2005) Elucidation of the reaction mechanism for the palladium-catalyzed synthesis of vinyl acetate. *Angew. Chem. Int. Ed.*, **117**, 4648–4650. doi: 10.1002/ange.200500782.

61 Neurock, M. (2003) Perspectives on the first principles elucidation and the design of active sites. *J. Catal.*, **216**, 73–88, http://linkinghub.elsevier.com/retrieve/pii/S002195170200115X (accessed 22 November 2014).

62 van Santen, R.A. and Kuipers, H.P.C.E. (1987) The mechanism of ethylene epoxidation. *Adv. Catal.*, **35**, 265.

63 Kilty, P.A. and Sachtler, W.M.H. (1974) The mechanism of the selective oxidation of ethylene to ethylene oxide. *Catal. Rev.*, **10**, 1–16.

64 Force, E.L. and Bell, A.T. (1976) The effect of dichloroethane moderation on the adsorbed species present during the oxidation of ethylene over silver. *J. Catal.*, **44**, 175–182, http://linkinghub.elsevier.com/retrieve/pii/0021951776903882.

65 van Santen, R.A. and de Groot, C.P.M. (1986) The mechanism of ethylene epoxidation. *J. Catal.*, **98**, 530–539, http://linkinghub.elsevier.com/retrieve/pii/0021951786903416 (accessed 6 February 2015).

66 Linic, S. and Barteau, M.A. (2003) Construction of a reaction coordinate and a microkinetic model for ethylene epoxidation on silver from DFT calculations and surface science experiments. *J. Catal.*, **214**, 200–212, http://

linkinghub.elsevier.com/retrieve/pii/S0021951702001562 (accessed 14 March 2014).

67 Torres, D., Lopez, N., Illas, F., and Lambert, R.M. (2005) Why copper is intrinsically more selective than silver in alkene epoxidation: ethylene oxidation on Cu(111) versus Ag(111). *J. Am. Chem. Soc.*, **127**, 10774–10775.

68 Ozbek, M.O., Onal, I., and van Santen, R.A. (2011) Why silver is the unique catalyst for ethylene epoxidation. *J. Catal.*, **284**, 230–235, http://www.sciencedirect.com/science/article/pii/S0021951711002466 (accessed 23 March 2015).

9

Solid Acid Catalysis, Theory and Reaction Mechanisms

9.1 Introduction

The main part of this chapter describes reactions that are catalyzed by protonic zeolites. This will be preceded by Section 9.2 that deals with the characterization of zeolite Brønsted acidity. Spectroscopic results and chemical bonding aspects will be introduced. This section will also contain an elementary introduction to the Brønsted acidic and basic properties of hydroxylated surfaces.

The main discussion in the following section 9.3 will focus on solid acid catalysis of hydrocarbon conversion reactions by zeolites. We will present the mechanism of zeolite-catalyzed reactions and corresponding catalytic cycles, based on the molecular chemistry and physical organic chemistry of proton-activated elementary reaction steps. We will see that the structure of zeolite microchannels often plays an essential role in the performance of the catalyst. Differences in adsorption and diffusion are important factors that determine the activity as well as the selectivity of these reactions. Reaction kinetics will be explicitly discussed for several reactions.

9.2 Elementary Theory of Surface Acidity and Basicity

9.2.1 The Pauling Charge Excess

According to Pauling, the stability of a complex ionic lattice relates to the electrostatic charge excesses on its cations and anions [1], which indicate electrostatic balances or imbalances in the solid materials. The charge excess is determined by the Pauling valency definition of the chemical bonds in an ionic solid, and it can also be used as an approximate descriptor of the intrinsic Brønsted acidity or basicity of surface hydroxyls.

Modern Heterogeneous Catalysis: An Introduction, First Edition. Rutger A. van Santen.
© 2017 Wiley-VCH Verlag GmbH & Co. KGaA. Published 2017 by Wiley-VCH Verlag GmbH & Co. KGaA.

The Pauling valency is based on the concept that a solid consists of cations and anions that can be considered point charges that only interact through the electrostatic Coulomb interaction between their charges.

The Pauling charge excess is a useful property, because it can be used as an easy discriminator of differences in proton reactivity, but as we will see later, its value as a reactivity descriptor is limited because it does not account for covalent bonding.

The Pauling strength or valency of an ionic bond between neighboring cations or anions is defined as:

$$S^{\pm} = \left| \frac{\text{formal ion charge}}{\text{number of nearest neighbor ions}} \right| \tag{9.1}$$

The charge excesses e^{+} on the respective cations or anions are defined as:

$$e^{+} = Q^{+} - \sum_{i} S_{i}^{-} \tag{9.2a}$$

$$e^{-} = Q^{-} + \sum_{i} S_{i}^{+} \tag{9.2b}$$

$Q^{\pm}$ are the formal charges on cation and anion, respectively. The charge excesses $e^{\pm}$ are a measure of the electrostatic imbalance at a particular ion.

According to Pauling, bulk oxide structures can be considered stable when the charge excesses $e^{\pm}$ on the respective cations or anions are less than $\pm 1/6$. We will see that at an ionic surface, the charge excesses on the ions can be larger than $\pm 1/6$, which indicates that such surfaces must be reactive. For non-hydroxylated surfaces, the charge excesses define Lewis acidity and basicity, while for hydroxylated surfaces they define Brønsted acidity and basicity.

We will illustrate the use of charge excess estimates to characterize the reactivity of the MgO surface in Insert 1.

Insert 1: Surface Hydroxyls of the Stepped MgO(100) Surface

Brønsted Acidity and Basicity

The structure of MgO is the rock salt structure. In this cubic structure the O anions and Mg cations each have six-coordination. At the surface, this six-coordination site is replaced by sites with lower coordination. The resulting charge excesses and Pauling bond strengths on the partially hydroxylated surface are illustrated in Figure 1i.1.

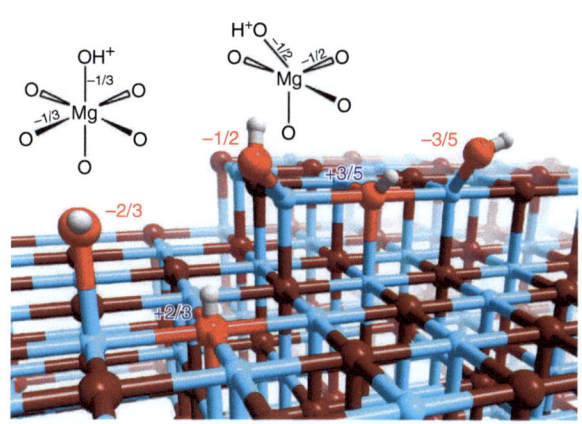

Figure 1i.1 The stepped surface of MgO. Terraces are in (001) orientation. Charge excesses $e^{\pm}$ are indicated on protonated bridging O atoms and hydroxylated Mg ions. The Brønsted basicity is highest on the terraces because electrostatic repulsive interactions by negatively charged oxygen ions around the Mg_{2+} ion that binds the hydroxyl is at a maximum there. Brønsted acidity of the bridging oxygen atom is also the highest at the terrace because the number of cations around the oxygen to which the proton is attached that gives repulsive interaction with the protons is also at a maximum. On the dehydroxylated surface the Lewis acidity and basicity is maximum at step-edge sites because the cation and anion charges are least shielded because of the low coordination with ions of oppostie charge.

Figure 1i.2 and Table 1i.1 compare the calculated and experimental OH frequencies on MgO and Al_2O_3 surfaces. In practice, different surface hydroxyls are stabilized by hydrogen bonding as shown in Figure 1i.3.

Figure 1i.2 An experimental IR spectrum combined with DFT-calculated OH frequencies. The Mg coordination number of the O atoms is indicated as nc. The spectrum is from a partially hydroxylated MgO powder [2].

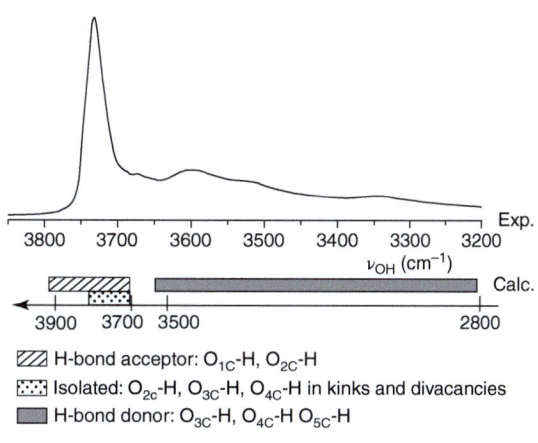

(Continued)

Insert 1: (Continued)

Table 1i.1 Calculated and experimental vibrational stretching frequencies for hydroxyl groups on γ-Al_2O_3 surfaces.

Site	Surface	ω_{calc} (cm^{-1})	ω_{exp} (cm^{-1})
HO-μ_1-Al_{IV}	(110)	3842	3785–3800
HO-μ_1-Al_{VI}	(100)	3777	3760–3780
HO-μ_1-Al_V	(110)	3736	3730–3735
HO-μ_2-Al_V	(110)	3707	3690–3710
HO-μ_3-Al_{VI}	(100)	3589	3590–3650

The coordination numbers of the oxygen atoms on Al ions are indicated as μ_n. The coordination number of Al indicated by the Roman numeral. ω is the vibrational frequency. Note the decrease in vibrational frequency when oxygen atom coordination number increases.
Source: After [3, 4].

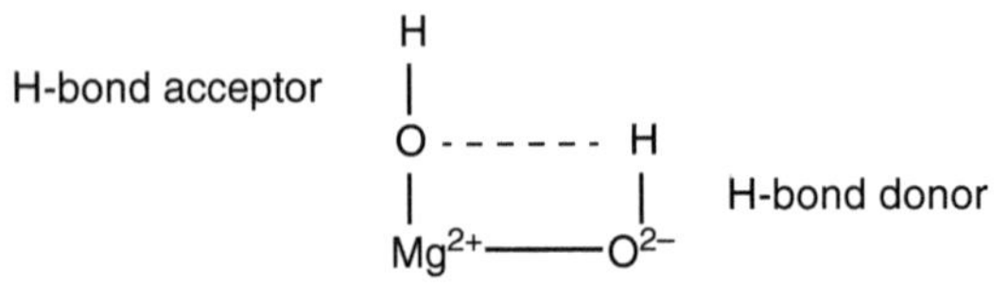

Figure 1i.3 Schematic representation of hydrogen bonding between Brønsted acidic proton and hydroxyl.

Reconstruction of TiO_2 Surfaces by H_2O Dissociation

Figure 1i.4a,b compare the state of dissociated H_2O on two TiO_2 surfaces. The calculated infrared spectra show considerable hydrogen bonding between the two hydroxyls on the TiO_2(001) surface, that reconstruct by Ti-O bond cleavage but little hydrogen bonding between the acidic proton and Brønsted basic OH on the TiO_2(101) surface.

9.2 Elementary Theory of Surface Acidity and Basicity

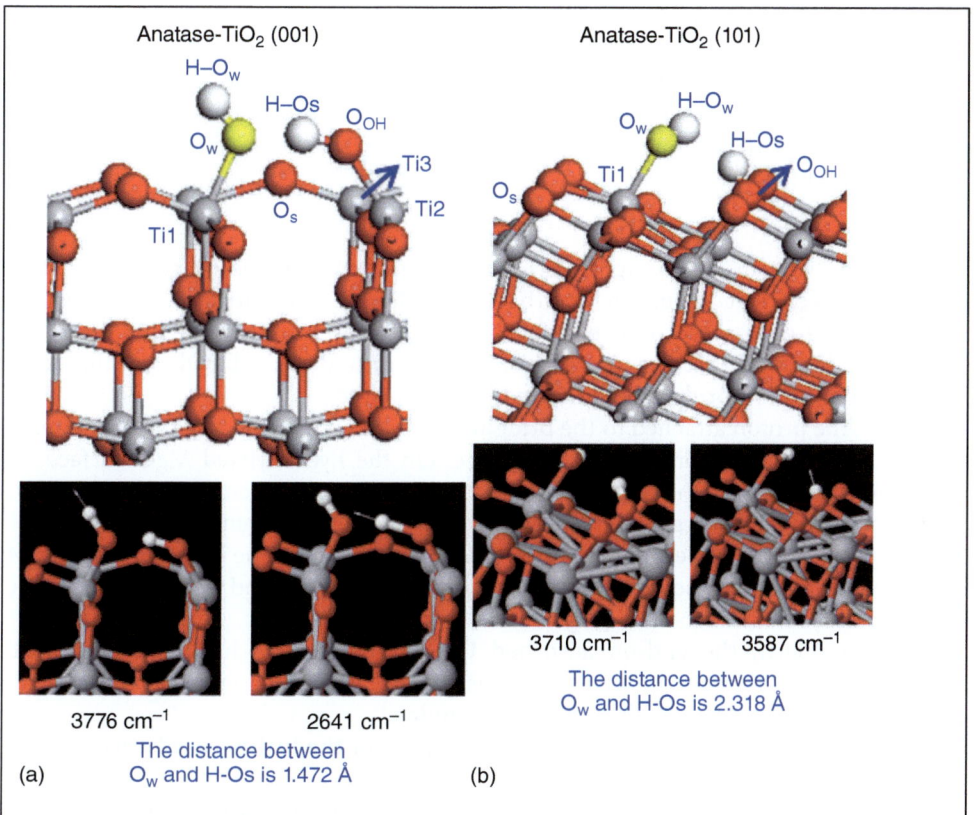

Figure 1i.4 Dissociative adsorption of H_2O on TiO_2(001) (a) and TiO_2(101) (b) surfaces, respectively. The surface structures and computed vibrational stretch frequencies of the OH groups are presented. (After [5].)

Figure 1i.1 compares the charge excesses of the oxygen atoms of hydroxyls present on stepped and non-stepped hydroxylated magnesium oxide surface. The hydroxyl adsorbed atop of a Mg cation is considered, as well as the protons adsorbed to bridging O atoms.

In the bulk of MgO that has a rocksalt structure, each ion is coordinated to six nearest neighbor ions of opposite charge. The Pauling valency of the ionic bonds of the cation or anion are respectively $\pm 1/3$. On the non-hydroxylated MgO(100) surface the ion coordination number is five, which results in charge excesses of $\pm 1/3$ at the MgO (100) surface. At the step-edge sites of the MgO surface the ions have lower coordination. On the step-edges of the MgO surface shown in Figure 1i.1 the Pauling charge excesses on the Mg cation and O anion are respectively $\pm 2/3$.

The charge excesses of surface cations and anions can be regarded as a measure of the respective Lewis acidities (charge excess is positive) and basicities (charge excess is negative) of the corresponding ions. The Lewis acidity and basicity are the highest at the step-edges.

In the ionic bonding scheme, water dissociation generates H^+ and OH^- species on the oxide surface. The proton will attach to a bridging surface oxygen anion and the hydroxyl will attach atop of a Mg cation. Insert 1 compares the respective charge excesses on the different hydroxyls that are formed in this way. The charge excess on the oxygen atom that is attached to a proton and the Mg cations is given by:

$$e_0^- = Q^- + e_{Mg}^+ + 1 \tag{9.2c}$$

As we can see in Figure 1i.1, the charge excess at the oxygen atom of the hydroxyl that is end-on adsorbed to a Mg ion is negative, but it is positive on the protonated oxygen atom that bridges several Mg ions. This shows that the hydroxyl that is end-on adsorbed is Brønsted basic, but the proton that is attached to a bridging O anion is Brønsted acidic.

The proton attached to the bridging O anion on the step-edge is less Brønsted acidic than the proton on the terrace. On the hydroxylated MgO surface, the charge excess at the step-edge of a bridging hydroxyl oxygen atom becomes $+1/3$, where as the charge excess of the bridging hydroxyl oxygen on the (100) surface is $+2/3$. The higher total positive charge excess due to the five coordinating Mg cations around the surface O anion makes the proton bond weaker at the surface terrace than on the step-edge.

Similarly, the end-on adsorbed OH groups have a lower basicity at the step-edge than on the terrace, because the Mg charge is compensated less by the charge from the neighboring oxygen anions.

A comparison of the charge excess values of the hydroxyls adsorbed end-on to the (100) surface ($-2/3$) and at the step-edge ($-1/3$) confirms that the OH bond of the hydroxyls adsorbed atop to the terrace is mostly Brønsted basic. In general, the more reactive surface edge anions or cations have the higher Lewis basicity or acidity, but hydroxylation with water creates weaker Brønsted basic and acidic groups.

The predicted difference in reactivity by charge excess estimates for a hydroxylated surface is confirmed experimentally by a measurement of the infrared spectra of surface hydroxyls. The difference in the vibrational frequencies of the hydroxylated surface is a measure of the strength of its OH bonds. Measured and calculated vibrational spectra are compared in Figures 1i.2 and 1i.3.

Figure 1i.2 shows the infrared spectrum of a hydroxylated MgO powder, where a broad band is observed. The peak at the high end of the spectrum is due to an end-on basic OH. Bands to the right of the maximum are due to acidic protons adsorbed to bridging O atoms, which have lower frequencies. The actual interpretation of the spectrum is complex because the basic OH groups and acidic protons tend to form hydrogen bridges, as schematically illustrated in Figure 1i.3.

The relationship between the coordination of the hydroxyl to the surface of $\gamma\text{-}Al_2O_3$ and its vibrational stretch frequency is similar to that found for MgO.

Table 1i.1 shows the experimentally measured and DFT-calculated OH frequencies of OH groups adsorbed to various sites on the $\gamma\text{-}Al_2O_3$ surface. The hydroxyl adsorbed atop of a Mg cation has higher vibrational frequency. The strong bond indicates Brønsted basicity. Protons attached to bridging oxygen atoms have lower frequencies that become still lower with increased O atom coordination. This implies an increase of Brønsted acidity with increased O coordination, since the H—O bond is increasingly weakened.

It is a general phenomenon that most solid oxide surfaces will exhibit basic as well as acidic hydroxyl when they become hydroxylated. This has important consequences for catalysis by these surfaces. Sometimes the Brønsted acidic protons are desirable, as in the application of alumina in solid acid catalysis, but they can also be detrimental, as we discussed for catalytic ethylene epoxidation (see Chapter 8, page 339).

Hydroxylation of the oxide surface may also disrupt M—O bonds. Figure 1i.4 shows as an example that the hydroxylated TiO_2 surface may reconstruct when disruption by dissociative adsorption of H_2O becomes significant. In this process, the bridging acidic proton is converted into a hydrogen-bonded basic hydroxyl. At high H_2O coverage the MgO surface will also be converted by similar reconstruction into a surface mainly covered with basic OH groups as expected for $Mg(OH)_2$.

The reactivity of these oxide surfaces at the solid/gas interphase is very different than their reactivity in wet chemistry. For instance, in wet chemistry, $Mg(OH)_2$ is considered a basic material because it dissociates in water into hydrated hydroxyl and Mg ions (see also Section 4.3.1), whereas the hydroxylated MgO surface contains basic hydroxyls and acidic protons.

9.2.2 The Chemistry of the Zeolitic Proton

The Pauling charge excess at the hydroxyl oxygen atom is an approximate descriptor of its acidity or basicity. It provides an elementary understanding of the differences in the reactivity of surface hydroxyl groups.

However, it cannot be used to describe differences in proton donation affinity when the charge excesses calculated according to the Pauling valency rule are the same on different systems. In this case, reactivity differences are determined by additional changes in covalent interactions. This is the case for the zeolitic proton, which is the subject of this section.

The structure of zeolites and the basic bonding character of the zeolitic proton are described in Section 2.1. The zeolitic proton is coordinated to a framework oxygen atom that bridges a tetrahedrally coordinated Si^{4+} cation with a threefold charged cation such as Al^{3+} in the neighboring tetrahedron. Alternatively, three-valent framework cations such as Fe^{3+} can be used to compensate for the proton charge instead of Al^{3+}. The Pauling charge excess in both cases is +3/4. Experiments that compare the protonation affinity zeolites with varying framework composition are discussed later in Insert 9i.4. Zeolites with Fe instead of Al in their framework have a lower proton affinity, although their Pauling excess charge estimates are the same.

The polarizability of the OH bond is a more accurate reactivity descriptor of the OH group than the Pauling charge excess. As we noted earlier for hydrogen bonding between adsorbed hydroxyls, additional interaction with the proton will lower its vibrational stretch frequency. The downward shift of the OH bond frequency that results from contact with a probe molecule can be used to measure the OH bond polarizability. This is an indicator of its proton donation affinity. A larger measured downward shift of the OH frequency implies more bond weakening and an increased proton donation affinity.

We will see that when differences in proton donation affinity are relatively small, as is the case for protons in the zeolite, differences cannot be readily determined from the ground state properties of the unperturbed proton. They only become evident when the proton is disturbed by interaction with a probe molecule.

This section will describe the nature of the acidic zeolite OH bond. In Section 9.2.2.1, the vibrational spectroscopy of the OH bond will be presented, followed by a short discussion of its quantum chemistry in Section 9.2.2.2. The section concludes with an introduction to the proton transfer reaction.

9.2.2.1 Vibrational Spectroscopy of the OH Bond

In principle, differences in the polarizability of OH bonds can be measured from their respective IR adsorption intensities. This is illustrated by the experimental results in Figure 2i.1, that can be used to compare the polarizability of the weakly acidic silanol OH bond with that of the more acidic zeolitic OH bond.

Insert 2: Spectroscopy of Acidity Differences of Zeolitic OH

This Insert presents the spectroscopic data for free-surface OH groups and the shift in their frequencies measured by infrared spectroscopy when OH comes in contact with a probe molecule. The differences in response to the probe molecule can be used to characterize the intrinsic proton donation affinity of the proton.

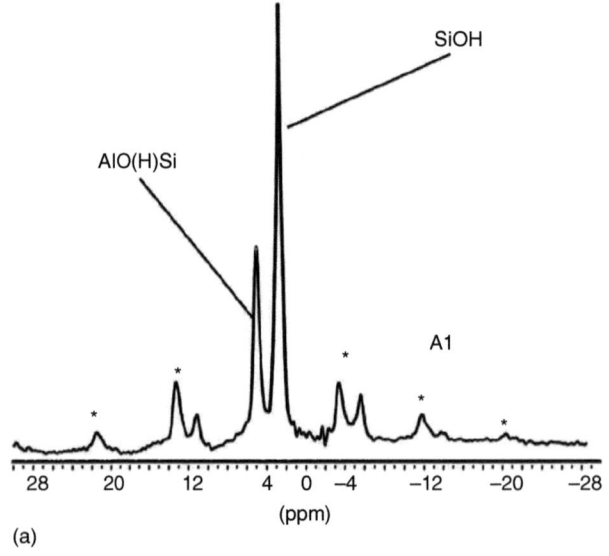

(a)

Figure 2i.1 Spectra of H-ZSM-5 zeolite with Si/Al = 45; (a) ^{1}H solid-state MAS NMR spectrum of OH groups. (b) Diffuse reflection infrared adsorption spectrum. (Kazansky et al. 2003 [7]. Reproduced with permission of Royal Society of Chemistry.)

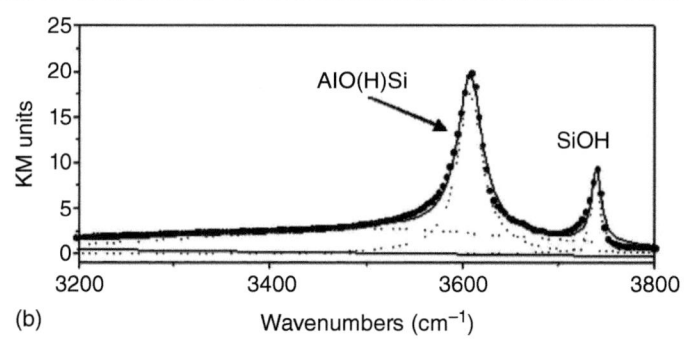

(b)

Figure 2i.1 (Continued)

Figures 2i.1 compares the spectroscopy of a surface silanol with that of the zeolitic proton.

The Pauling excess charge e_0 of the oxygen atom of the silica OH group is 0. The excess charge of the zeolitic O atom attached to the proton +3/4. The structures of the hydroxyls are compared in Figure 2i.2.

Figure 2i.2 Schematic illustrations of the (a) SiOH structure and (b) SiOHAl structure.

In Figure 2i.3, the experimentally measured decrease of the vibrational stretch frequency of a proton perturbed by interaction with the weakly interacting CO molecule is shown. A small shift of the silanol group (initially at 3750 cm^{-1}) but a large downward shift of the zeolitic protonic group (initially at 3615 cm^{-1}) is observed. Table 2i.1 illustrates the differences in the response of hydroxyls to interaction with the probe molecules CO and benzene in different zeolites. Shifts of the hydroxyl by the adsorption of benzene are also included.

(Continued)

Insert 2: (Continued)

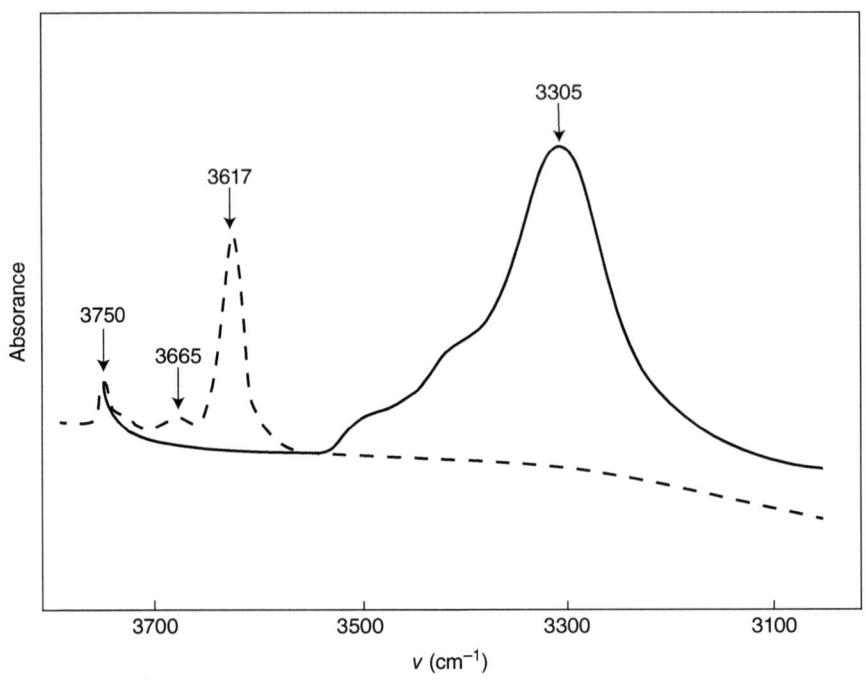

Figure 2i.3 IR spectra of OH groups in the H-ZSM-5 zeolite before (broken line) and after (solid curve) the adsorption of CO. Temperature of measurement was 77 K. (Kustov et al. 1987 [8]. Reproduced with permission of American Chemical Society.)

Table 2i.1 Shifts of bands of OH vibrations caused by the formation of hydrogen-bonded interaction complexes with CO (at 77 K) and C_6H_6 (at 300 K).

Zeolite	v_{OH} (cm^{-1})		Δv_{OH} (cm^{-1})	
	300 K	77 K	CO (77 K)	C_6H_6 (300 K)
ZSM-5	3612–3615	3615–3620	310–320	350
SiO$_2$	3747	3750	90	120
H$_{70}$Na$_{30}$-Y	3647	3652	275	326[a]

Source: Zecchina et al. 1997 [9]. Reproduced with permission of Elsevier.

Figures 2i.4 and 2i.5 show the shifts in infrared spectra when the zeolitic proton interacts with a very basic molecule such as acetonitrile and the zeolites with different framework composition are compared. The ABC spectrum that evolves is shown in Figure 2i.4 for acetonitrile adsorption to different protonic zeolites of varying Al/Si ratios. In Figure 2i.5, the adsorption of acetonitrile is compared to on Al- versus Fe-substituted material.

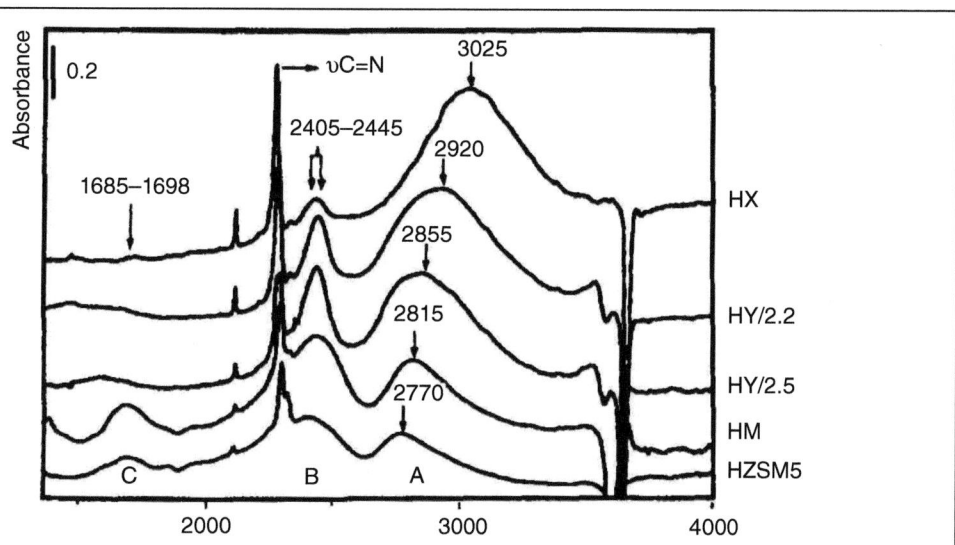

Figure 2i.4 Differences between the IR spectra of the zeolitic proton after adsorption of CD_3CN zeolites with different framework composition are compared. Al/Si ratio decreases from top to bottom. The minima in the ABC spectra (Evans window) result from intensity stealing from the broadened OH stretching band by resonance with low intensity in-plane and out-plane vibrational modes. (After [10, 11].)

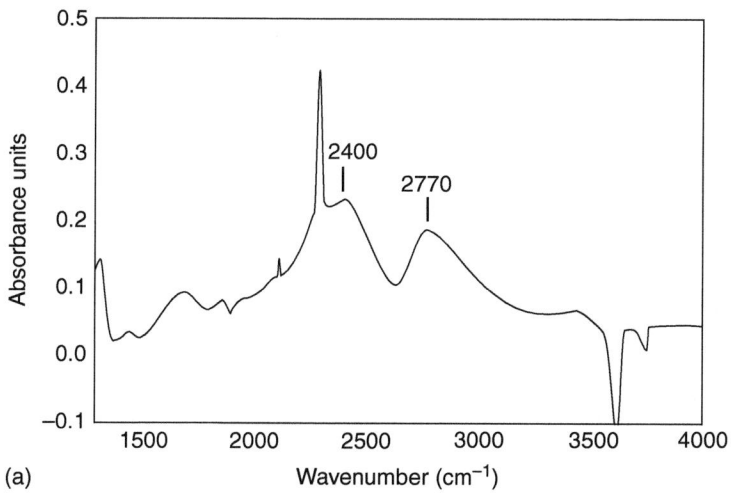

Figure 2i.5 Changes in the 1500–4000 cm^{-1} range of the infrared spectrum H-ZSM-5 (a) induced by CD_3CN adsorption at 295 K and 0.05 mbar and H-FeSil (b) induced by CD_3CN adsorption at 295 K and 0.05 mbar. H—Fe silicalite has the same structure as ZSM-5, but contains Al instead of Fe. (Pelmenschikov et al. 1995 [12]. Reproduced with permission of American Chemical Society.)

(Continued)

Insert 2: (Continued)

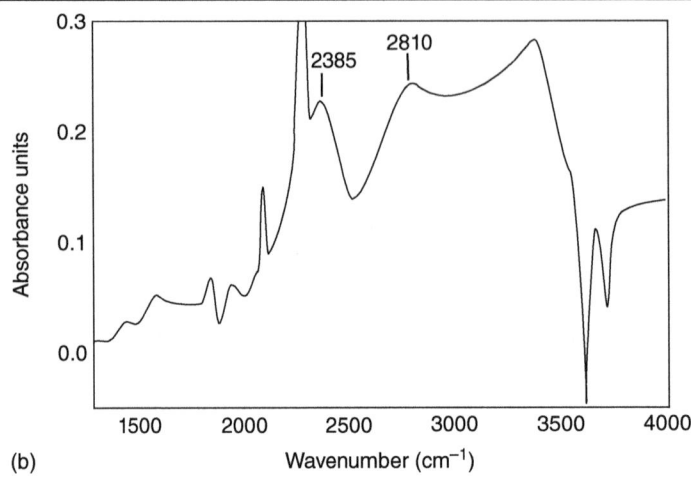

(b)

Figure 2i.5 (Continued)

Figure 2i.6 illustrates how the single-peak spectrum of weakly interacting adsorbates is replaced by a double-peak spectrum within the spectral range of the 3600–1700 cm^{-1} interval.

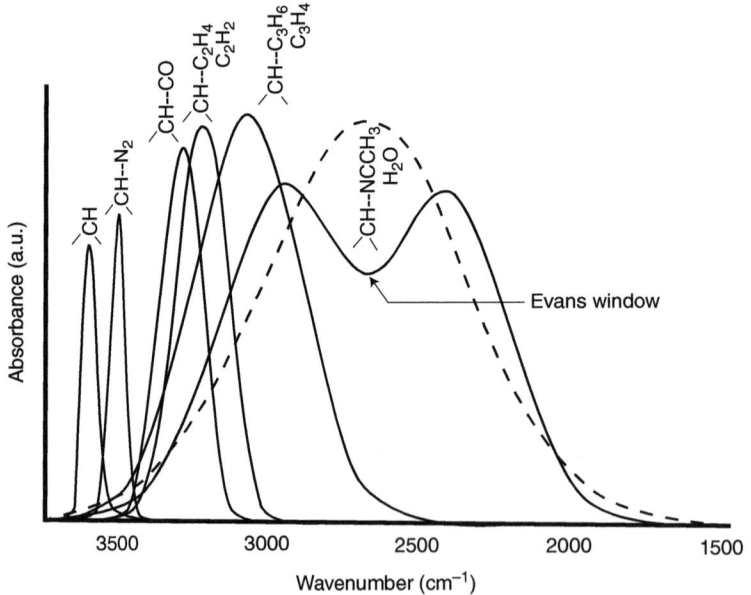

Figure 2i.6 Evolution of the $v(Z-H\cdots B)$ band caused by hydrogen bonds with bases (B) of increasing proton affinities. (Zecchina et al. 1995 [9]. Reproduced with permission of Elsevier.)

Upon hydroxylation part of the framework Si—O bonds at the surface are cleaved. The external surface of the zeolite becomes covered with silanol groups that are attached to Si^{4+} cations (Figure 2i.2a). In Figure 2i.2b the zeolitic proton is sketched attached to the bridging O atom between Si and Al. The experimental results shown in Figure 2.1 compare the measured vibrational stretch intensities of the external surface silanol groups to those of the protons located in the zeolite microchannels. Solid nuclear magnetic resonance (NMR) data can be used to calibrate this to the number of the respective protons. They are distinguished by their different chemical shift.

When normalized per proton, the absorption intensity difference shown in Figure 2i.1b is nearly four times higher for the zeolite proton. This is an indication of the substantially larger polarizability of the zeolitic proton bond compared to the proton bond of the surface silanol. This is consistent with the expected differences in acidity according to Pauling's charge excess estimates. The respective charge excess differences are large: on the silanol proton it is 0, but on the zeolitic proton it is +3/4, so that the zeolitic proton bond is expected to be the weaker.

The Pauling charge excess on the internal zeolite proton remains the same when Al/Si ratio or zeolite structure changes, whereas the proton donation affinity may vary widely with variation in framework structure and composition. Therefore, there is a need for a different bond acidity descriptor than the Pauling charge excess.

Differences in proton donation affinity can be most easily determined experimentally by measuring the weakening of the OH bond using vibrational infrared spectroscopy of the OH bond frequency change when it comes into contact with a probe molecule such as CO, N_2, or methane. The more polarizable the OH bond, the more it is weakened and, hence, the more its proton donation affinity increases. This is illustrated in Figures 2i.3 for adsorption by the CO molecule. The weakening of the OH bond and increased polarizability due to its interaction with CO as indicated by lowered IR frequency and increased absorption intensity. The increase in absorption intensity indicates that charge on the proton becomes increasingly more positive. Table 2i.1 shows the difference in the OH frequency shift of CO and benzene in contact with the zeolitic proton or the silanol group. The interaction with benzene is slightly stronger than with CO and a frequency shift is observed for silanol group that agrees with its weaker proton donation affinity.

The OH bond shifts for zeolites with differing proton donation affinities that are mainly due to a change in the framework Si/Al ratio as illustrated in Figures 2i.4 and 2i.5.

These spectra can be best understood by first examining the trends in vibrational frequency decrease as a function of the Lewis basicity of the probe molecule shown in Figure 2i.6. When the interaction with the probe molecule is initially weak, there is a single adsorption peak at high vibrational proton frequency. As interaction with the probe molecule increases the peak shifts downward in frequency, broadens and increases in intensity. When strongly Lewis basic molecules interact the broadened and downward-shifted peak splits into a broad band with two or three peaks. This is most clearly shown in Figure 2i.4 as seen for acetonitrile adsorbed to H-ZSM-5.

The new peaks that appear are due to partial frequency overlap between the broadened, downward-shifted adsorption peak of the weakened stretching OH vibration and the lower frequencies of the in-plane and out-plane bending modes of OH. The overlapping frequencies are the optically forbidden overtones of the in-plane bending OH mode at 2300 cm^{-1} (normal mode at approximately 1150 cm^{-1}) and the out-plane OH bending frequency at 1600 cm^{-1} (normal mode at approximately 800 cm^{-1}). Overtone excitations are not excited by photons, since they are not related to change in dipole. They borrow intensity from the vibrational bond stretch excitations that have finite transition moments with dipole change [13]. This results in dips on the broadened adsorption peak that appear as new peaks which form the so-called ABC band of strongly disturbed and weakened OH groups [14].

The appearance of an ABC spectrum as shown in Figure 2i.4 is an indication that the proton has not transferred to the probe molecule, but instead remains attached to the zeolite framework. This interpretation of similar infrared spectra of adsorbed molecules such as H_2O or methanol has led to the conclusion that these cannot be considered as adsorbed H_3O^+ or $HOHCH_3^+$ species, but are instead molecules that interact with the zeolitic proton through a hydrogen bond [15, 16].

Figure 2i.4 provides experimental proof that the zeolite OH proton donation affinity increases with decreasing Al/Si ratio in the zeolite lattice framework. The proton donation affinity reaches its maximum when the Al/Si ratio is so low that no tetrahedra containing Al are found in the second coordination shell with respect to the OH group. This typically occurs at an Al/Si ratio of ≤0.1. The decrease in proton donation affinity with increasing Al concentration is due to the increasing average negative charge on the zeolite framework by substitution of Si_{4+} by Al_{3+} ions.

In Figure 2i.5, the effect of substitution of framework Al by Fe can be compared. In Figure 2i.5b, the downward shift of the partial ABC infrared adsorption peak is less dramatic than that shown in Figure 2i.5a, providing experimental proof of the increased OH bond strength of the proton adjacent to a framework Fe atom instead of an Al atom. The Fe cation has a slightly weaker M—O bond than the Al cation, so that the OH bond is stronger as one deduces from the bond order conservation rule (see Section 3.3.2.1.1). This principle applies because of the high covalency of the SiO and AlO chemical bonds in the zeolite framework.

9.2.2.2 Quantum Chemistry of the Zeolite Acidic OH Chemical Bond

In addition to the infrared spectra we discussed above, quantum-chemical calculations also predict that the zeolitic OH bond is highly covalent. These calculations show that for free OH the charge of hydrogen does not exceed +0.2. The chemical bond has an intermediate polarity of 0.6, close to that of gas phase HCl (see Chapter 6, Insert 2), which is consistent with a significantly covalent character.

Insert 3: The Quantum Chemistry of the Zeolitic OH Bond

In Figures 3i.1–3i.3, the PDOS and COHP features (for their definition refer to Chapter 6, Insert 1) for a free zeolitic proton and the zeolitic proton in contact with CO and acetonitrile are compared. Figure 3i.4 gives a schematic representation of the electronic interactions between the probe molecule and OH (E.A. Pidko and C. Liu (2014) private communication) The PDOS plots in Figure 3i.1 show the two electron-occupied bonding regions with contributions from the O 2s atomic orbital and the O $2p_z$ atomic orbitals, respectively. They also show the unoccupied antibonding OH chemical bonding region, which strongly overlaps with the SiO and AlO unoccupied antibonding electron conduction band regimes (not shown). Figures 3i.2 and 3i.3 show the electronic structure changes when CO and acetonitrile interact with the zeolite OH group.

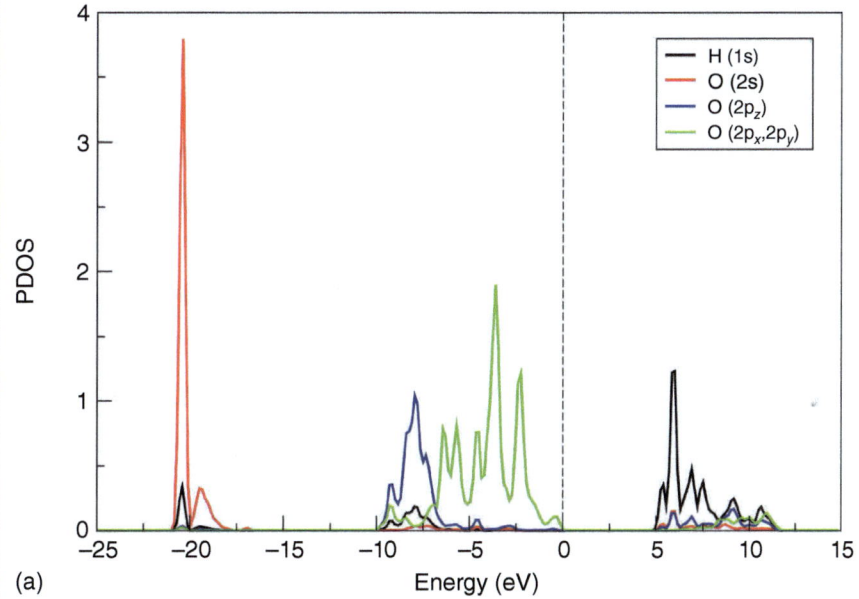

Figure 3i.1 The PDOS (a) and COHP (b) plots of the chemical bond of a zeolitic OH.

(Continued)

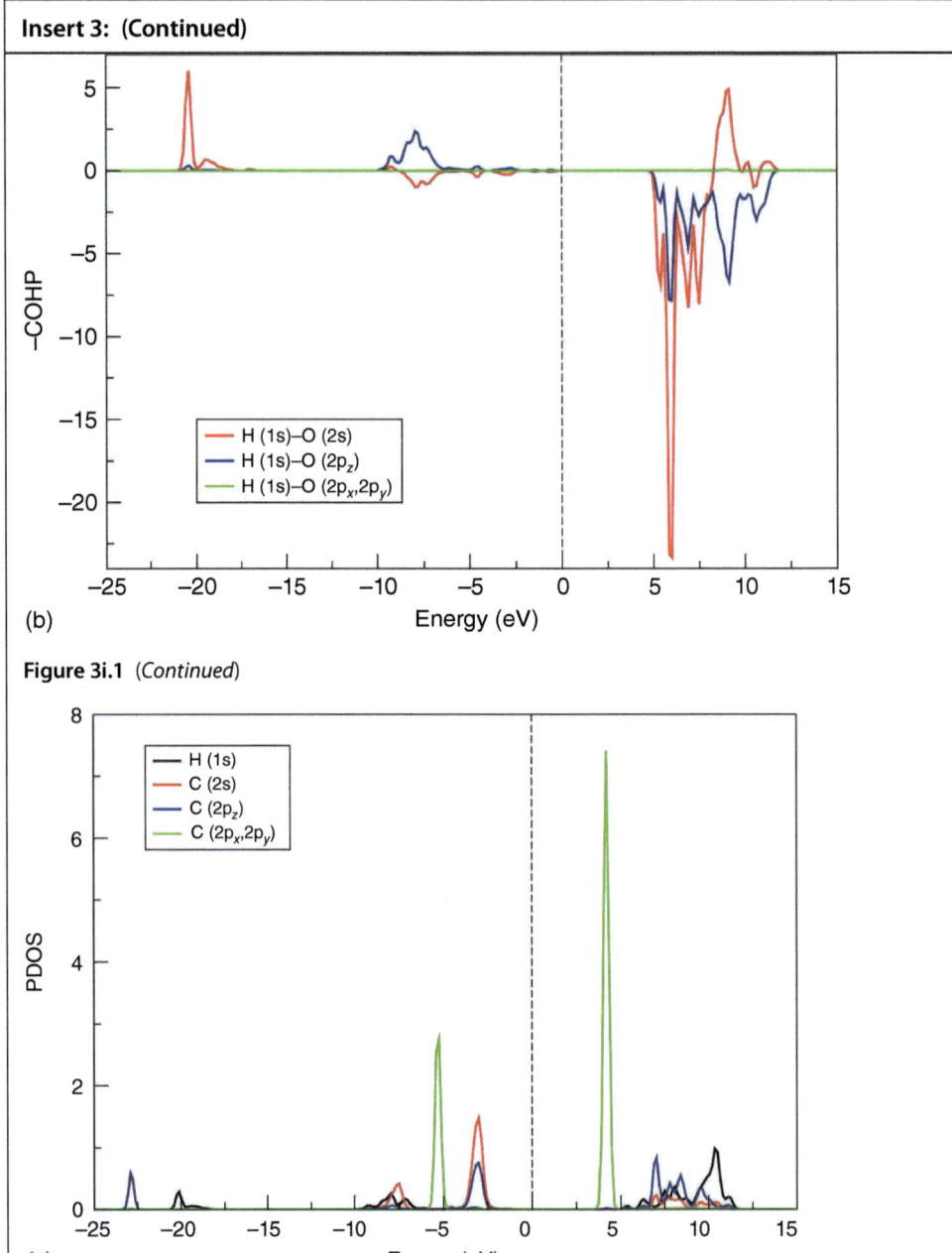

Figure 3i.2 The PDOS and −COHP features of zeolitic OH in contact with CO. Figures (a) and (b) show the electronic features of the C—H interaction. The calculated interaction energy with the CO molecules is $E_{ads} = -38\,\text{kJ}\,\text{mol}^{-1}$ (see Chapter 7, Insert 7 for the electronic structure of CO).

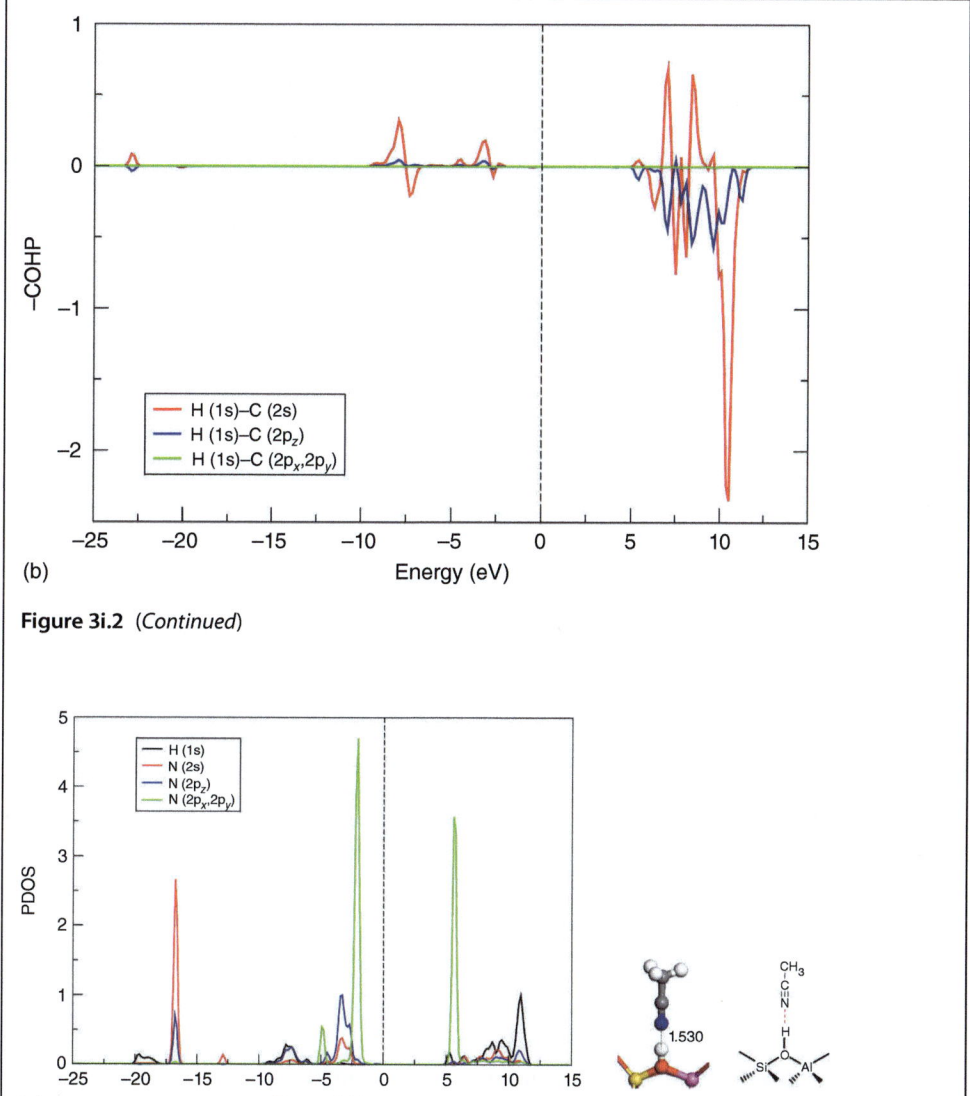

Figure 3i.2 (Continued)

Figure 3i.3 The electronic structure of the system interacting with acetonitrile ($E_{ads} = -94$ kJ mol^{-1}). Figures (a) and (b) show the electronic structures of the N—H interaction. The electronic structures of the two highest occupied orbitals of acetonitrile are also shown in (c).

(Continued)

Insert 3: *(Continued)*

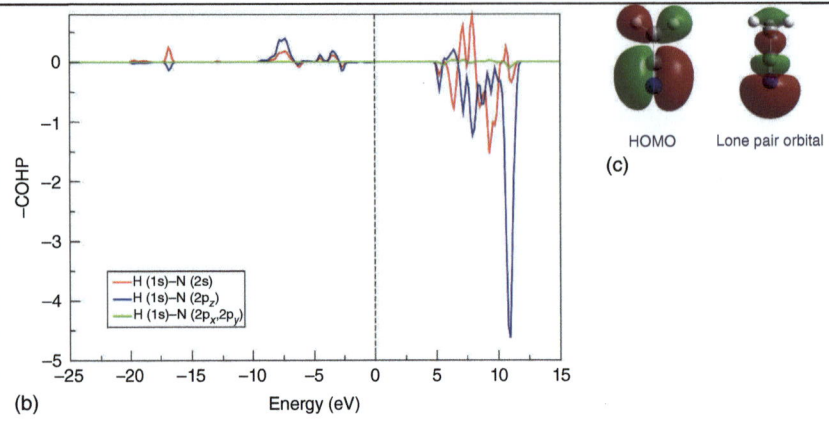

(b)

(c) HOMO, Lone pair orbital

Figure 3i.3 *(Continued)*

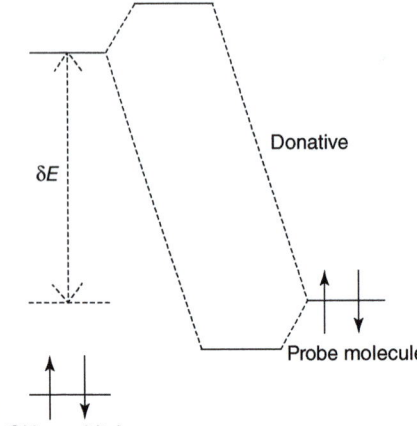

Figure 3i.4 Schematic of the donative electronic interactions between the probe molecules' σ orbital and unoccupied OH σ* orbital. The interaction strength increases when the energy difference δE between the unoccupied OH σ* orbital and the occupied σ probe orbital decreases. The COHP plots in Figures 3i.2 and 3i.3 show that this results in a bonding contribution of the electron densities between the probe molecule and the zeolite proton.

Table 3i.1 provides data on energy-integrated bond order densities (ICOHP) that are a measure of the interaction between the probe contact atom and the zeolite proton.

Table 3i.1 The strength of interaction ICOHP between the probe molecule atom and the zeolite proton.

Si/Al ratio	CO adsorption C—H	CH_3CN adsorption N—H
47	−0.318	−0.847
7	−0.268	−0.665
2.4	−0.247	−0.583

The strong covalent interaction between the proton and the zeolite O atom leads to a large energy difference between the bonding and antibonding orbitals of the OH bond. With the z-axis considered to be oriented along the OH axis we observe in Figure 3i.1 interaction of the proton only with the O 2s and O $2p_z$ atomic orbitals. The strong contributions of O 2s and O $2p_z$ interaction is especially visible in the antibonding OH bond regime, and is indicative of significant hybridization.

The interaction of CO with the zeolitic proton (see Figure 3i.2) is dominated by the interaction of the CO 5σ orbital with the antibonding OH electronic density regime. The chemical bonding interaction can best be described as a weak electron donation from the CO σ electron density into the anti-bonding unoccupied OH bond (see Figure 3i.4).

The increased interaction energy of acetonitrile (see Figure 3i.3) is reflected by the electronic structure features which are larger between hydrogen and nitrogen than between hydrogen and carbon, as shown in Figure 3i.2. Table 3i.1 summarizes the calculated energy-integrated bond orders (ICOHP's) between the probe molecule atom and the zeolite proton, which are a measure of the respective strength in interaction (see Chapter 6, Insert 1). The larger values of ICOHP indicate the stronger interaction of CH_3CN compared with the interaction of CO. There is also a decrease in interaction strength with decreasing Si/Al ratio, indicating a smaller proton donation affinity when Si/Al ratio decreases. This we discussed earlier using data of Figure 2i.4 and is due to the increase in average negative charge on the zeolite framework with decreasing Si/Al ratio. The calculated ICOHP between adsorbate and OH can be considered a reactivity descriptor for proton donation affinity.

Similarly to the OH bonds, the chemical bonds of the zeolite framework also have substantial covalent character.

Differences in the energies of formation are small for siliceous zeolites. Their lattice frameworks are built from four-, five-, or six-ring structures that connect the framework oxygen atom tetrahedra around the framework cations.

Differences in the energy of formation mainly relate to the bond bending energy of the Si—O—Si inter-tetrahedral angle. The calculated variation in the energy of the Si—O—Si angle is only 10 kJ mol^{-1} when the bond angle varies between 120° and 150°. Since the Si—O bond distances vary only slightly, this variation explains the small differences in the energies of formation for siliceous zeolites of varying framework structure. Differences in structure can be accommodated by small changes in the Si—O—Si bond angle without the costly deformation of the angles within the tetrahedral (SiO_4) units or due to Si—O bond stretching.

The preferred angle of the S—OH—Al bonds is close to 120°, as expected for sp^2 hybridization of the O electrons. When the proton transfers to a reactant, the O—Al and Si—O bonds contract and the bond angle increases. The zeolite framework accommodates these changes locally by a variation in the Si—O—Si bond angles of neighboring tetrahedra. This lattice relaxation can contribute small energy gains that aid proton transfer [17].

9.2.2.3 The Proton Transfer Reaction

Temperature-programmed desorption of chemisorbed ammonia is a convenient and extensively used experimental tool for determining the number and strength of acidic sites. The temperature of desorption, typically around 450 K, provides a measure of the interaction strength of the ammonia molecule with the zeolitic proton and of the corresponding ammonium ion. However, direct information on the intrinsic donative affinity of the proton is not readily obtained because local differences in the electrostatic stabilization of the ammonium ion tend to obscure the desorption temperature differences.

Computational studies of proton transfer to ammonia in varying configurations illustrate this [18, 19]. Figure 9.1 shows zeolitic cluster models for three adsorption modes of ammonia. Such clusters can be studied with high quantum-chemical accuracy since they are small [20] and can be embedded in an electrostatic lattice [18, 19] to induce long range electrostatic effects.

When the geometry of the interaction of ammonia with the zeolitic proton is limited only to the interaction of its nitrogen atom with the zeolitic proton (Figure 9.1a), the proton will not transfer. Proton transfer only occurs when ammonium becomes stabilized by additional direct contact with the zeolite site through hydrogen bonding of its hydrogen atoms with several of the negatively charged Lewis basic oxygen atoms around the Al site. Depending on the geometry of the zeolite framework, either twofold or threefold adsorption of ammonium is possible.

The calculated formation energies of twofold- and threefold-coordinated ammonium are 110 kJ mol^{-1}. The interaction energy of ammonia adsorbed to a single bridging O atom is only 60 kJ mol^{-1} [15, 21].

Two conditions must be satisfied to enable proton transfer:

- The geometry of the protonated molecule and the interaction with the negatively charged zeolitic lattice must provide sufficient electrostatic stabilization to overcome the cost of deprotonation of the zeolite framework.
  This cost is partially compensated by the electrostatic interaction between the negative charge left on the zeolite lattice framework with the positively charged protonated cation and the gain in energy due to the protonation of a molecule.
- The proton affinity of the molecule to be protonated must exceed a specific value that depends on charge separation cost and electrostatic stabilization at the zeolite protonation site.

The deprotonation energy of the zeolite framework proton is on the order of 1250 kJ mol^{-1}, while proton affinities of molecules such as pyridine (932 kJ mol^{-1}) or ammonia (854 kJ mol^{-1}) are substantially lower. The energy cost of proton transfer is compensated by the attractive electrostatic stabilization of the protonated molecule and the negative charge of the zeolite lattice. It appears that the minimum proton affinity for protonation is between that of ammonia and acetonitrile (788 kJ mol^{-1}).

This is consistent with the infrared observations that neither H_2O (proton affinity 697 kJ mol^{-1}) nor methanol (proton affinity 761 kJ mol^{-1}) are adsorbed as hydronium or methoxonium ions, but instead are hydrogen bonded by two bonds between the molecule's O atom and zeolite's oxygen atoms and the two OH bonds remain non-equivalent. For the case of methanol Figure 9.2 shows

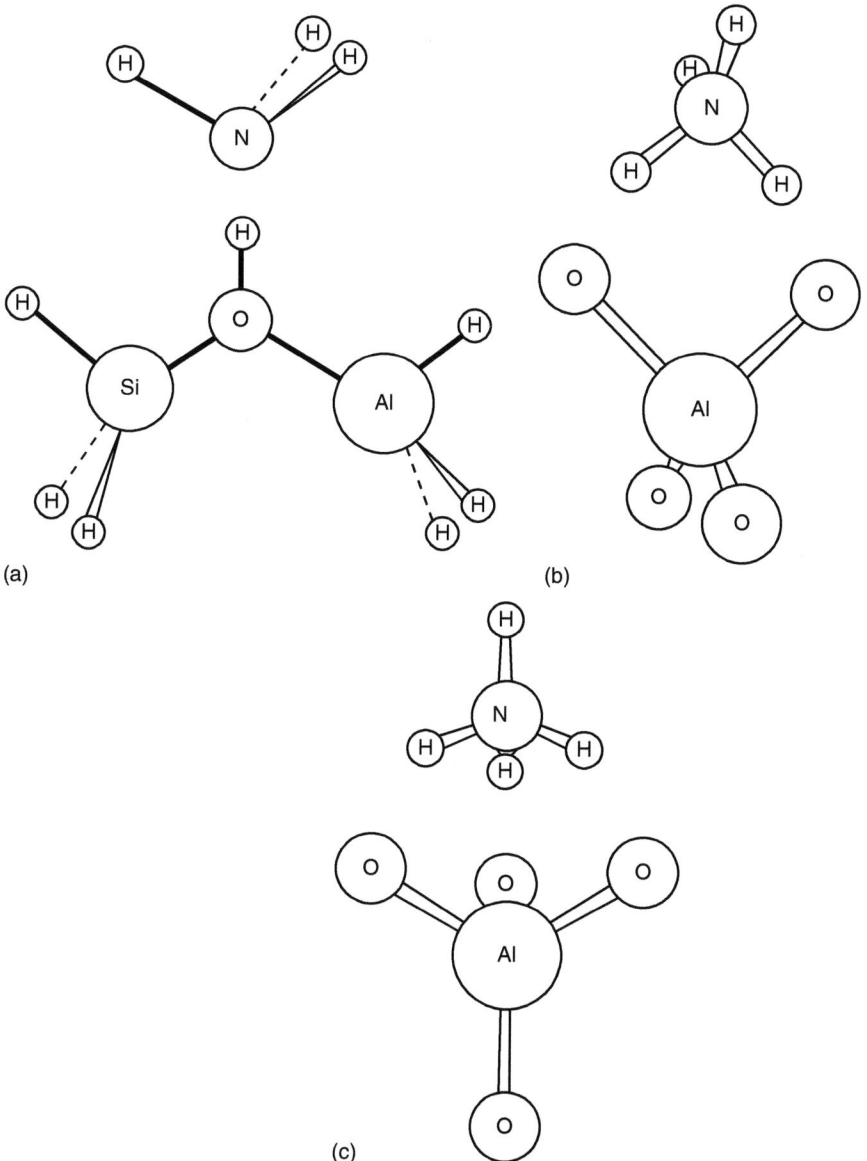

Figure 9.1 Cluster models of ammonia that interacts with zeolitic proton. Case (a) does not show proton transfer. Ammonium is formed in geometries (b) and (c). (a) Ammonia adsorbed through a single H atom within a zeolite framework. (b) Ammonium coordinated by two of its protons with two oxygen atoms of the tetrahedron containing aluminum. (c) Ammonium threefold-coordinated to the tetrahedron containing aluminum. SCF-MP2 calculations on these clusters. (Teunissen *et al.* 1992 [21]. Reproduced with permission of American Chemical Society.)

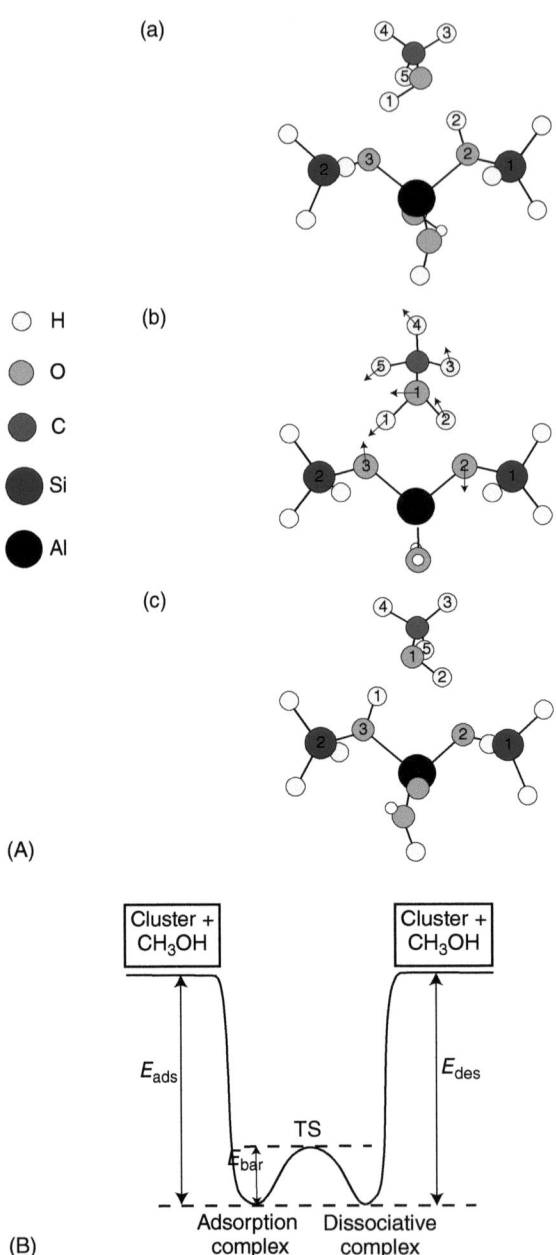

Figure 9.2 Reaction intermediates of the exchange of a zeolitic proton with methanol. (A) The respective adsorbed states of methanol (a, c) and transition state (c). (B) Schematic of the relative energies of adsorbed state and transition state. (Blaszkowski et al. 1995 [22]. Reproduced with permission of American Chemical Society.)

that in the equilibrium-adsorbed state, one proton has a short bond with the molecule while the other remains attached to the zeolite. The activation energy of the proton exchange reaction is only 10–15 kJ mol^{-1}.

The protonated state is a transition state instead of a ground state. The low activation energy arises because protonation is assisted by the backbonding of the methanol proton to a Lewis basic oxygen. Synergy with Lewis base-assisted hydrogen bonding to framework O provides a low barrier to proton transfer.

The situation for an adsorbed water molecule is similar. The hydronium ion only exists as a transition state.

Proton transfer from the zeolite hydroxyl to these molecules will however occur when they are part of a cluster of methanol or water molecules. Then, the protonated molecule is stabilized by hydrogen bonds between the other methanol or water cluster molecules. The energy cost of charge separation is also reduced by the increased dielectric constant that results from the additional presence of the polar molecules in the otherwise low dielectric constant zeolite cavity [23].

9.2.2.4 Chemical Reactivity Probes of Proton Donation Affinity: H/D Exchange Reactions

The hydrogen–deuterium (H/D) exchange reaction is useful for determining the chemical reactivity of acidic protons. Here, we will discuss the H/D exchange reaction of methane [24, 25] and benzene [26]. The reaction with benzene has found practical application in detecting differences in proton reactivity in cases where spectroscopic identification cannot achieved [27].

We will again see that, in the transition state, at least two framework oxygen atoms of the Al tetrahedron interact with a hydrogen atom of the reacting molecule. As in the case of methanol, the transition state energy is lowered by synergy of the proton donative interaction of the zeolite hydroxyl with the proton acceptance capability of an initially free Lewis basic oxygen atom. The oxygen atom accepts a proton from the molecule in the exchange process (see Figures 4i.1a and 9.3b).

In the H/D exchange process, a zeolitic proton is replaced by a hydrogen atom when the reacting molecule is deuterated. After exchange, deuterium moves from the molecule to the other slightly less binding lattice oxygen atom.

Methane is useful in this test because its adsorption energy with the zeolite is small, so that its contribution to differences in reaction rates will be insignificant. On the other hand, benzene is more convenient because it has higher reactivity, so that lower reaction temperatures can be used.

Insert 4: Hydrogen–Deuterium Exchange of Methane and Benzene

Figure 4i.1a shows the calculated transition state for hydrogen–deuterium exchange of methane and a comparison of experimental and calculated activation energies in Figure 4i.1b [28]. The activation energies are close to experimental results. The deviation of TOF from the experimental results is due to the approximate way the activation entropy is calculated using vibration frequencies calculated in the harmonic approximation. Differences in rate by the Faujasite zeolite versus the ZSM-5(MFI) zeolite are due to a slightly weaker adsorption energy of methane in the wider pore Faujasite structure zeolite.

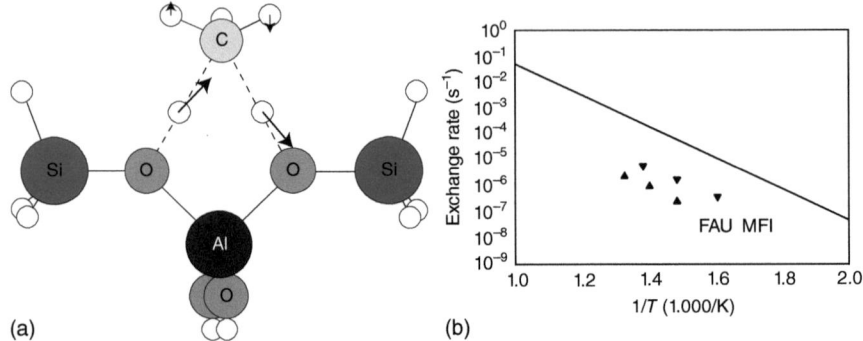

Figure 4i.1 (a) Calculated transition state geometry for the isotope exchange process between methane and an acid zeolite cluster. All small circles represent hydrogen or deuterium atoms. The arrows indicate the main components of the displacement vectors along the reaction coordinate. (b) D/H exchange rate of deuterated methane with the H-forms of zeolites FAU and MFI. Experimental values are given by the filled symbols (triangle for FAU and inverted triangle for MFI). The solid line shows the prediction by transition state theory of the exchange reaction with a zeolite cluster, which predicts a reaction barrier of 122 kJ mol^{-1}. (Kramer et al. 1993 [24]. Reproduced with permission of Nature Publishing Group)

Figure 4i.2 relates to proton/deuterium exchange with benzene. Figure 4i.2a shows the structures of the benzene molecule adsorbed to a zeolite proton in the ground state and Figure 4i.2b the corresponding transition state.

Figure 4i.2 Optimized BLYP/DNP structures and selected bond distances: (a) van der Waals complex (VDW); (b) transition state (TS) for H/D exchange. (Beck *et al*. 1995 [26]. Reproduced with permission of American Chemical Society.)

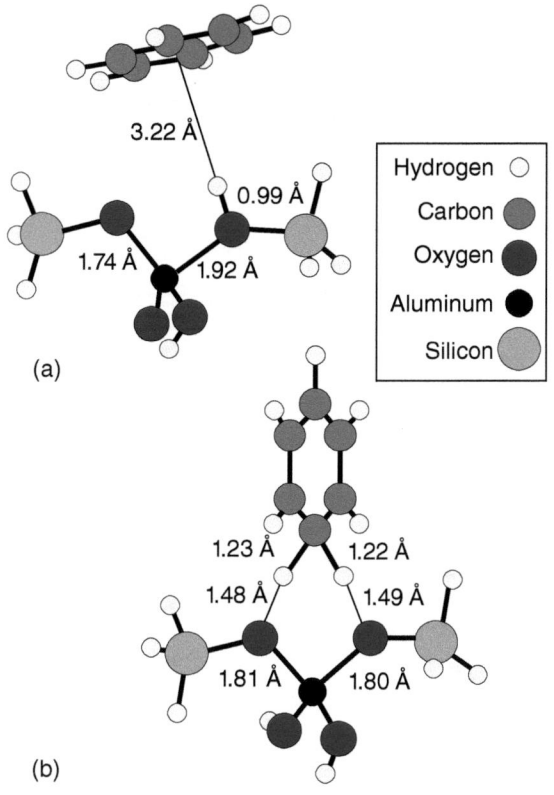

Figure 4i.3 presents experimental data on the hydrogen–deuterium exchange reaction of benzene. In Figure 4i.3a, the progress of the exchange of zeolite protons with deuterated benzene is followed by NMR. Figure 4i.3b shows the measured activation energy plots for different zeolites.

(Continued)

> **Insert 4: (Continued)**
>
>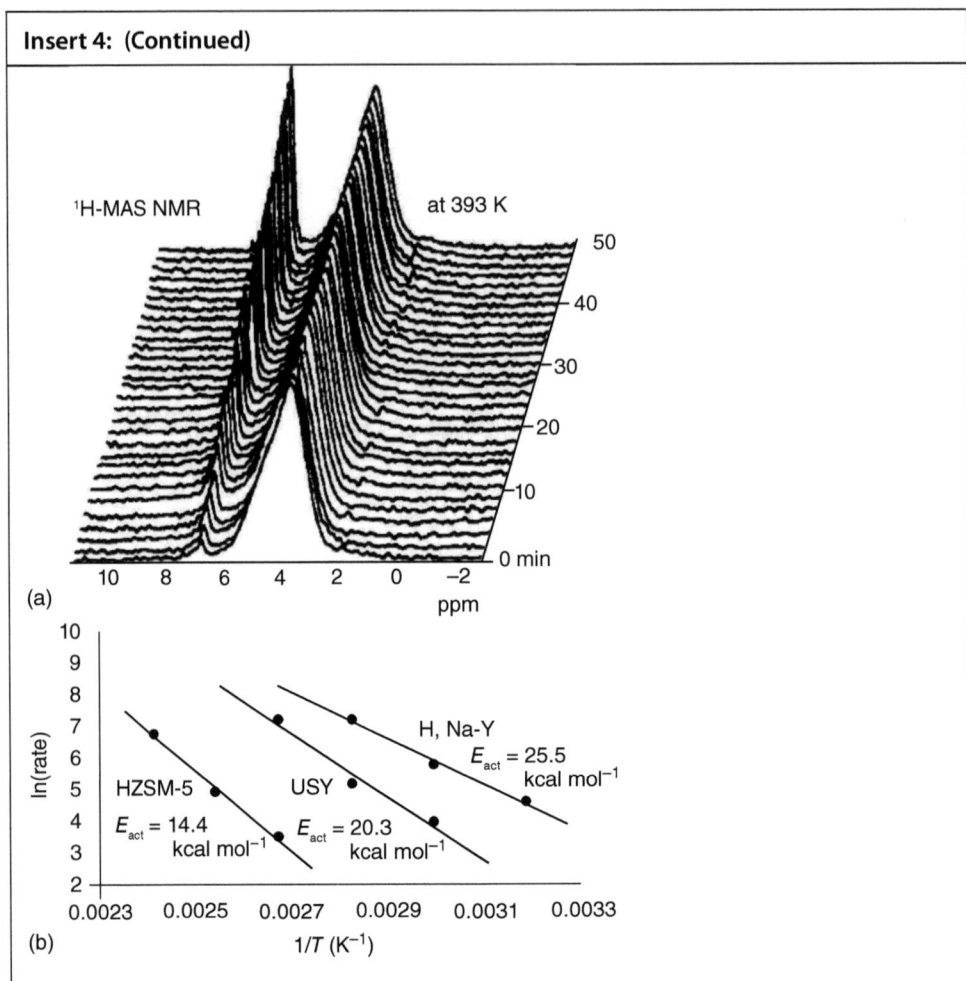
>
> **Figure 4i.3** (a) In situ kinetic study showing the exchange between deuterated benzene and zeolite NaHY at 393 K; 360 MHz. Proton magic angle spinning spectra acquired every 50 s over total measurement times of up to 1 h clearly show the growth of the benzene signal at 7.0 ppm at the expense of the signal at 4 ppm for the Brønsted site. (b) Arrhenius plots constructed from *in situ* kinetic studies at various temperatures for zeolites H-ZSM-5, USY, and NaHY. These zeolites have an increasing framework Al/Si ratio. (Beck *et al.* 1995 [26]. Reproduced with permission of American Chemical Society.)

For the methane H/D exchange reaction, the activation energies are compared for two zeolites of low Al/Si ratio in Figure 4i.1, H/D exchange of methane has an intrinsic activation energy of 122 kJ mol^{-1}. Although their cavity sizes are very different, the activation barrier of the faujasite structure is close to that of MFI (H-ZSM-5).

The structure of the transition state is exemplified by the non-classical carbonium cation CH_5^+. This highly unstable ion consists of a C atom bonded to five hydrogen atoms. The proton affinity of methane is 70 kJ mol^{-1} and its high barrier for the H/D exchange reaction is due to the strong zeolite OH bond of 1250 kJ mol^{-1}. The difference must be overcome by electrostatic interaction and hydrogen bonding with the negatively charged zeolite lattice.

The activation energy will vary with changing deprotonation energy when the lattice composition changes. This is nicely shown in the experiments with benzene for zeolites with different framework Al/Si ratios outlined in Figure 4i.3b.

As long as the steric effects can be excluded, the variation in activation energy ΔE_{act} for a proton-activated reaction relates primarily to variation of the strength of the OH bond. The transition state is mainly stabilized by electrostatic interactions and weak hydrogen-binding interactions that do not vary strongly with zeolite site, hence:

$$\Delta E_{act} \approx \Delta E_{OH} \tag{9.2}$$

The substantial increase in the activation energy of the H/D exchange reaction rate with an increase in Al/Si ratio implies reduced protonation affinities. The increased Al^{3+} substitution increases the average charge on the zeolite lattice, which in turn increases the proton–oxygen interaction energy, and thus makes its framework more basic.

9.3 Mechanism of Reactions Catalyzed by Zeolite Protons

9.3.1 Introduction to Acid-Catalyzed Reactions and Their Mechanism

Thermodynamics determines the reaction conditions for a particular reaction. As illustrated in Table 9.1, exothermic association reactions are favored at low temperatures, while reactions that are mildly exothermic but activated, such as hydrocarbon isomerization of linear molecules to branched products, require intermediate temperatures. High temperatures are necessary for endothermic cracking and dehydration reactions.

The low temperature alkylation of propylene with isobutane requires a liquid superacid catalyst such as HF or H_2SO_4. At an intermediate temperature, butane isomerization to isobutane catalyzed by chlorinated alumina is possible, and at a

Table 9.1 A comparison of reactions, with the required reaction temperatures and preferred acid catalysts.

Temperature (°C)	Catalyst	Reactions
−10	HF	Alkylation Hydride transfer
130	AlCl$_3$	Butane isomerization
250	Pd/H$^+$-zeolite	Alkane hydroisomerization
300	Pd/H$^+$-zeolite	Alkane hydrocracking
550	H$^+$-zeolite	Catalytic cracking

higher temperature of 250 °C bifunctional solid acidic zeolites can be used to catalyze alkane hydroisomerization. At higher temperatures an endothermic reaction is formation of aromatics by alkanes. At higher temperatures aromatics can also be produced from linear hydrocarbons by the catalytic reforming reaction using activated alumina promoted by a noble metal. Since the proton donation affinty of the alimuna is weaker than that of the zeolite, the noble metal is needed to catalyse the cleavage of the alkane CH bonds. At the highest temperature, the catalytic cracking of alkanes using zeolites can be conducted, which produces a mixture of products with a significant fraction of aromatics. Some of the processes based on these reactions we described before in Chapter 2. Here we will discuss the molecular basis of the activity and the selectivity of the corresponding catalytic cycles.

In acid-catalyzed reactions the key reaction intermediates are carbocations such as the carbonium and carbenium ions. These intermediates have been extensively characterized by NMR in low temperature super acid experiments [29]. In the next section, we will discuss how these intermediates can also be formed at high temperatures, but that they will be short lived or present only in transient transition states [14, 29].

The formation of intermediate carbenium and carbonium ions is illustrated schematically for the proton-catalyzed cracking reaction of an alkane in Figure 9.3.

The reaction scheme shown in Figure 9.3a is called the *Haag–Dessau cracking mechanism* [16, 28, 30]. Proton addition to an alkane generates non-classical carbonium cations, so-called because the valency of some of the carbon atoms has become greater than four. Therefore, the carbonium cation is unstable, and it can decompose to a carbenium ion in two ways:

– A carbenium is formed with the same carbon skeleton as the reacting hydrocarbon when two of the carbonium ion hydrogen atoms recombine to produce H_2.
– C–C bond cleavage can occur and a shorter alkane and carbenium ion is formed.

In contrast to the carbonium ions, carbenium ions are stable organic cations with a positively charged carbon atom, which is planar and has three coordination.

In a zeolite, the formation of these cations is endothermic. The zeolite OH bond must be broken and its charge is separated. The cationic intermediate is stabilized by close contact with the negative charge on the framework.

Carbenium ions can be considered the working horse of acid catalysed reactions. The carbenium ions are essential intermediates to many catalytic reactions, such as:

– Isomerization reactions, such as branching and bond-shift reactions that maintain the carbon skeleton;
– C—C bond cleavage reactions, that break the skeleton into parts;
– C—C bond formation reactions with an alkene such as oligomerization and alkylation.

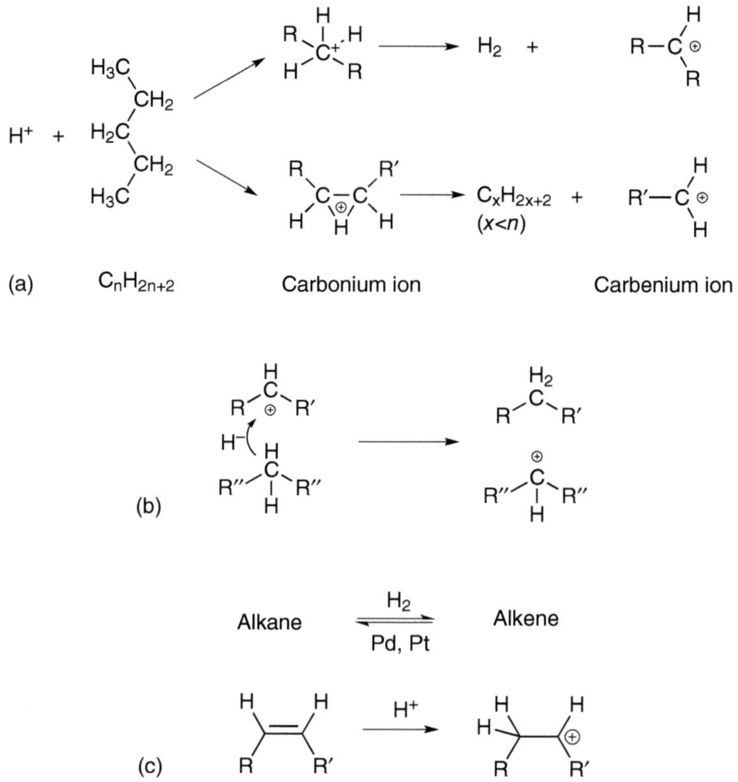

Figure 9.3 (a) The formation of carbonium cation intermediates illustrated by the attachment of a proton to pentane. The two decomposition paths are indicated. One reaction path leads to H_2 formation and a carbenium cation with the same skeleton carbon number. The other reaction path leads to a shorter alkane and a shorter carbenium ion. (b) The carbonium ion mediated processes to carbenium ions can be circumvented by the hydride transfer reaction (b) or by employing a bifunctional catalysis (c). In the latter case, the alkane–alkene equilibrium is maintained by a transition metal such as Pt or Pd or a reactive sulfide and the protons activate the alkene.

Many of these reactions will be discussed in detail in the following sections. The catalytic reaction cycle can close in two ways:

- An alkene is formed by backdonation of the proton to the solid framework of the catalyst and reaction cycle continues through intermediate carbonium ion formation.
- The hydride transfer reaction occurs, where an H anion is transferred from reactant alkane to a carbenium ion, which is converted to a product alkane molecule with formation of a new carbenium ion (Figure 9.3b).

Although an olefin is readily protonated to produce a carbenium ion, proton activation of alkanes to carbonium ions requires a high energy (~200 kJ/mol), but direct alkane transformation to a carbenium ion is also possible through an alternative process. A low activation energy route is provided when the alkane molecule reacts with another carbenium ion and a hydride ion is transferred (see Figure 9.3). This important reaction has a relatively low activation energy and occurs in many different catalytic reactions. In the catalytic cracking process it closes the catalytic reaction cycle direct carbenium ion formation from an alkane without intermediate backdonation of the proton to the solid.

An alternative to the hydride transfer reaction to produce a carbenium ion through a low activitation energy reaction path is formation of intermediate alkene by transition metal catalysed activation of the alkane (Fig 9.3c).

The overall reaction as for instance applied in the hydroisemerisation reaction (see Chapter 2, page 26) occurs in two stages: alkane-alkene equilibration catalysed by a transition metal in the presence of excess H_2 and subsequent proton activation of the alkene that gives a carbenium ion.

In the next subsection we will introduce the molecular chemistry and quantum chemistry of key elementary reactions catalysed by zeolitic protons. This will provide a basis to the presentation in the later part of this chapter of the mechanism of the catalytic reaction cycle of most important zeolitic catalysed reactions.

9.3.2 Elementary Reactions in Acid Catalysis

9.3.2.1 Alkene Protonation

The formation of a carbenium ion by protonation of an alkene is fundamental to acid catalysis, and we will introduce the chemistry of this reaction here.

The reactivity of the alkene π bond depends strongly on its carbon atom coordination. Primary, secondary, or tertiary carbenium ions can be formed upon protonation. Carbenium ions are planar with their σ bonds oriented in a plane. In the classical organic description of the ions, the σ-C—C and C—H bonds are sp^2 hybridized, with an empty $2p_z$ atomic orbital perpendicular to the carbenium ion plane.

A primary carbenium ion has a positive charge at a terminal carbon atom, a secondary carbenium ion has a positive charge with two carbon atom neighbors, and a tertiary carbenium ion has a positive charge with three carbon atom neighbors. In gas phase, the secondary and tertiary carbenium ions become subsequently more stable than the primary ion by 30 and 60 kJ mol^{-1} [31, 32] respectively. In a zeolite the primary and secondary carbenium ions are not stable intermediates, but instead are part of transition states or activated carbocation intermediates (see Figure 5i.1). Only the more stable tertiary carbenium ions can be considered to be true reaction intermediates (see Figure 5i.2). The activation energies of the respective protonation reactions may strongly depend on the zeolite micropore structure. Primary and secondary carbenium ions react with the negatively charged zeolite oxygen atoms to form alcoxy species. This is illustrated in Figure 5i.1.

This figure shows energy changes of the protonation of propene in the zeolite ferrierite (TON) [35] protonation of i-butene is compared in both ferrierite and in mordenite [36] in Figure 5i.2. Mordenite and ferrierite have one-dimensional nano pores formed by rings of 12 and 10 tetrahedra, respectively.

Insert 5: Protonation of Olefins: Primary, Secondary, and Tertiary Carbenium Ion Transition States

In this insert, the activation energies of protonation are compared propylene in Figure 5i.1 and butene in Figure 5i.2. In the case of propylene, the intermediate transition states are primary or secondary carbenium ions. In butene activation, the intermediates are primary and tertiary carbenium ions. In Figure 5i.2b, calculated activation energies are compared for the two zeolites that vary in nanopore dimension.

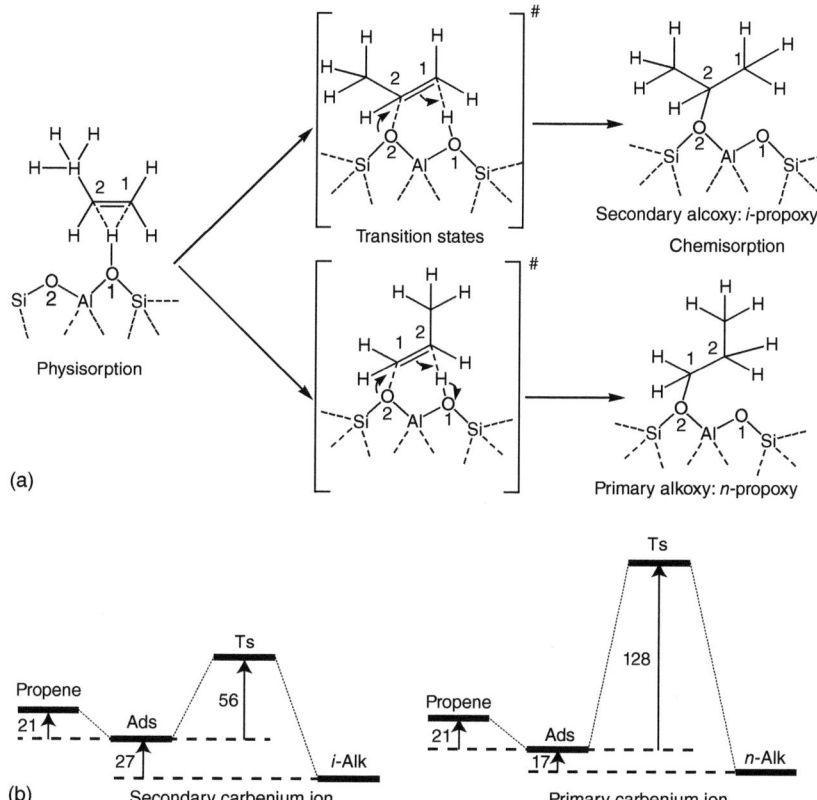

Figure 5i.1 Comparison of primary and secondary carbenium intermediate formation by protonation of propylene. (a) Schematic presentation of initial state, transition state and final alcoxy state. (b) The corresponding calculated activation energies of propylene protonation in the zeolite ferrierite microchannel. The nature of the carbenium ion transition states is indicated. (After [33].)

(Continued)

Insert 5: (Continued)

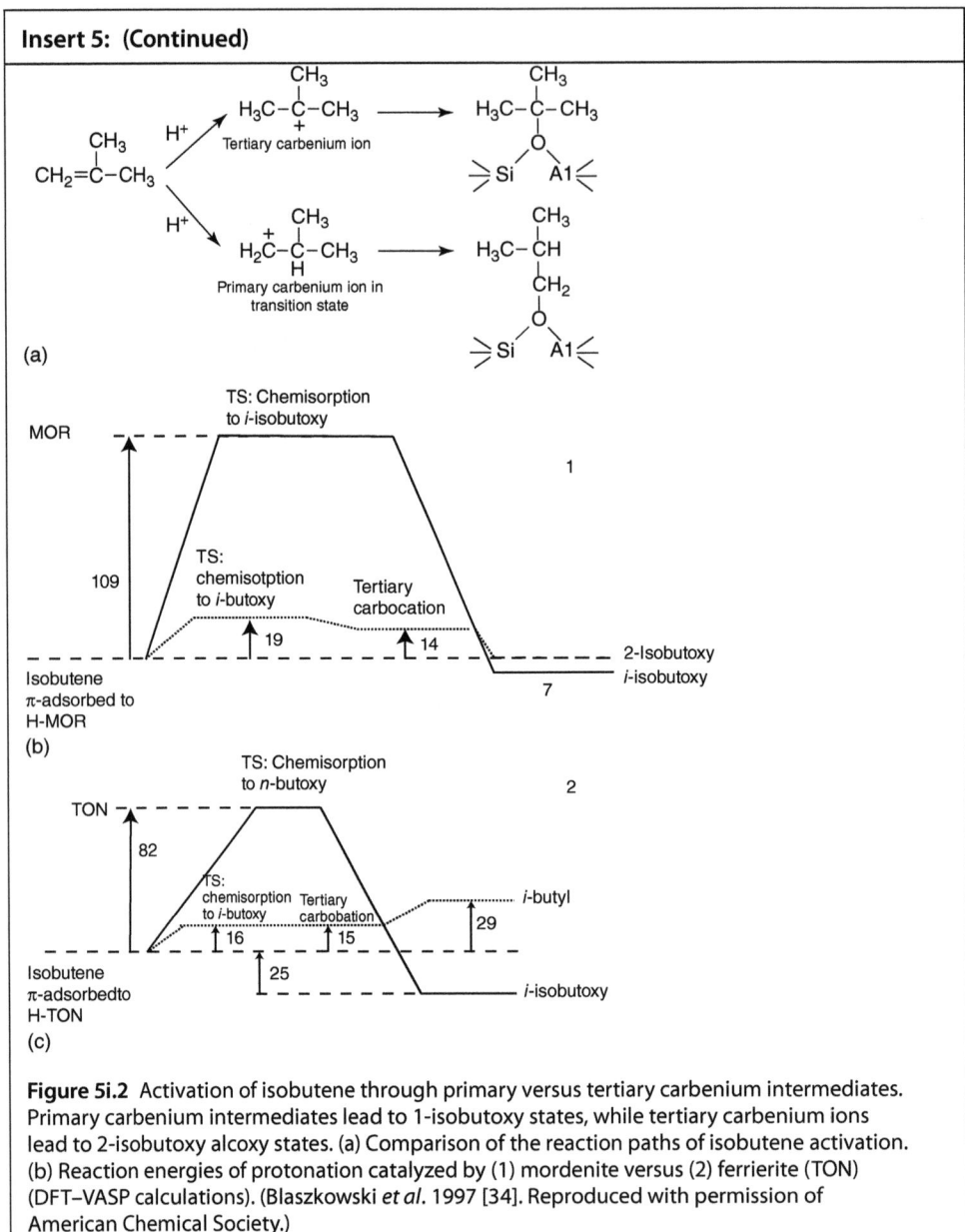

Figure 5i.2 Activation of isobutene through primary versus tertiary carbenium intermediates. Primary carbenium intermediates lead to 1-isobutoxy states, while tertiary carbenium ions lead to 2-isobutoxy alcoxy states. (a) Comparison of the reaction paths of isobutene activation. (b) Reaction energies of protonation catalyzed by (1) mordenite versus (2) ferrierite (TON) (DFT–VASP calculations). (Blaszkowski et al. 1997 [34]. Reproduced with permission of American Chemical Society.)

Figure 5i.2 compares the proton activation paths of isobutene.

Three stages can be recognized in the reaction of the alkene molecule with a zeolite proton. The alkene will initially adsorb to the proton in a weakly interacting π complex. The zeolitic proton is hydrogen bonded to the alkene π electrons. The main contribution to the adsorption energy is from the van der

Waals interaction between the hydrocarbon's carbon and hydrogen atoms and the polarizable zeolite lattice oxygen atoms.

In the next step, the proton is transferred, which results in a substantial energy cost. For the zeolite ferrierite, this step results in activation energies for propene protonation of 56 and 128 kJ mol^{-1} through secondary versus primary carbenium ion transition states, respectively (see Figure 5i.1). The difference in transition state energies reflects the relative stability of the primary carbenium ion versus the secondary ion. Since the primary carbenium ion transition state has the higher energy, formation of the secondary cation is preferred. Both carbocations are unstable and convert into a σ-complex, the analog of an alcoxy intermediate. In this state, their energy difference has essentially disappeared.

This reaction sequence illustrates an important reactivity principle: differences in the chemical reactivity of solid-acid-catalyzed reactions are determined primarily by the difference in the relative stability of their transition states.

The structure of the transition states indicates the importance of the interaction with the Lewis basic oxygen atom next to the oxygen atom to which the proton is initially attached. The positive charge distribution in the transition state cation is directed toward one of the negatively charged oxygen atoms that surround Al. The corresponding transition states of the carbenium ions exhibit the sp^2 hybridization of the threefold-coordinated C atom, which is characteristic of the carbenium ion. In the local minimum of the alcoxy-state, the O—C bond becomes sp^3-hybridized and compensates for the different stabilities of the two cations. This causes the overall reaction energy to be independent of the product state.

As is seen in Figure 4i.2b for isobutene activation, a relatively high activation energy for the formation of the primary carbenium ion also occurs along with the expected substantially lower barrier for the formation of the tertiary carbenium ion. Although protonated isobutoxy is formed as a stable intermediate in the mordenite channel, the curvature of the wall of the narrow pore zeolite (TON) prevents the formation of the alcoxy complex because the tertiary carbenium ion is too bulky to become accommodated. Now, the free isobutyl carbenium ion can also be considered as a stable intermediate. The low energy barriers imply rapid proton exchange at typical temperatures of reaction (550 K).

We observe a substantial decrease in activation energy within the narrow pore system, especially for the high activation barrier of primary carbenium ion formation. In the more narrow pores, the distance between carbenium ion atoms and polarizable zeolite oxygen atoms is less, which gives an extra attractive interaction.

The differences in the activation energies for the formation of primary, secondary, or tertiary carbenium ion intermediates are substantial. The activation energies are 110–130, 70–90, and 10–20 kJ mol^{-1}, respectively. The extremely low activation energy to form tertiary carbenium ions is noteworthy. Although the activation energies for secondary or tertiary carbenium ion formation from olefins are substantially lower, transformation through intermediate primary carbenium formation can also occur at higher temperatures.

The transition states can be considered loose and mobile states without any memory of the initial or final state. With respect to the adsorbed state the activation entropy contribution can be significant. The activation energies depend strongly on the proton donation affinity of the zeolitic proton, the type of carbenium formed, and the cavity size. In the absence of repulsive steric interactions, a smaller cavity may lead to lower activation energies but the activation-free energies may be inversely affected by the loss of mobility of transition state or reaction intermediates in the smaller nanopore.

So far, we have only considered protonation of the alkene, that initiates consecutive reactions as isomerization or C—C bond cleavage that are to be discussed in a later subsections. The activation energies of the overall reaction can be estimated by first considering the formation of the carbenium ion intermediate or transition state from its π adsorbed or σ alcoxy state, and then by adding the activation barrier for the transformation of the carbenium ion to the final product cation. This additional energy can be deduced from carbenium ion transformation reactions calculated in the gas phase. The reactivity of solid acid catalysts is determined by the state of molecules on top of the energy barriers of the corresponding reaction-transition state energy diagram instead of conventional thermodynamics of ground state intermediates.

9.3.2.2 Alkane Activation

Proton-catalyzed alkane activation proceeds through an intermediate carbonium cation. The carbonium cation intermediate can undergo three types of reactions: proton exchange, dehydrogenation, or C—C bond cleavage. Calculations show that each reaction forms a specific type of carbonium intermediate, which is geometrically different from the others.

The three different transition state geometries are shown in Figure 6i.1. The transition state for proton exchange is very similar to the state shown in Figure 4i.1a for the methane H/D exchange reaction. However, the transition states for dehydrogenation and the C—C bond cracking of ethane are very different. In these elementary reactions, a carbenium ion is generated next to H_2 or methane. The characteristic planar structure of the carbenium ion due to its sp^2 hybridization is present in the respective transition states.

Insert 6: Carbonium Cation Type Transition States

Figure 6i.1 compares calculated transition state structures of ethane as obtained from small cluster calculations. Figure 6i.2 illustrates calculated transition states for propane and butane activation in a micropore of chabasite. Figure 6i.3 shows the various intermediates on the reaction path from butane to methane and propyl H-ZSM-5. Table 6i.1 shows experimentally measured data on the cracking and dehydrogenation of propane using various zeolites.

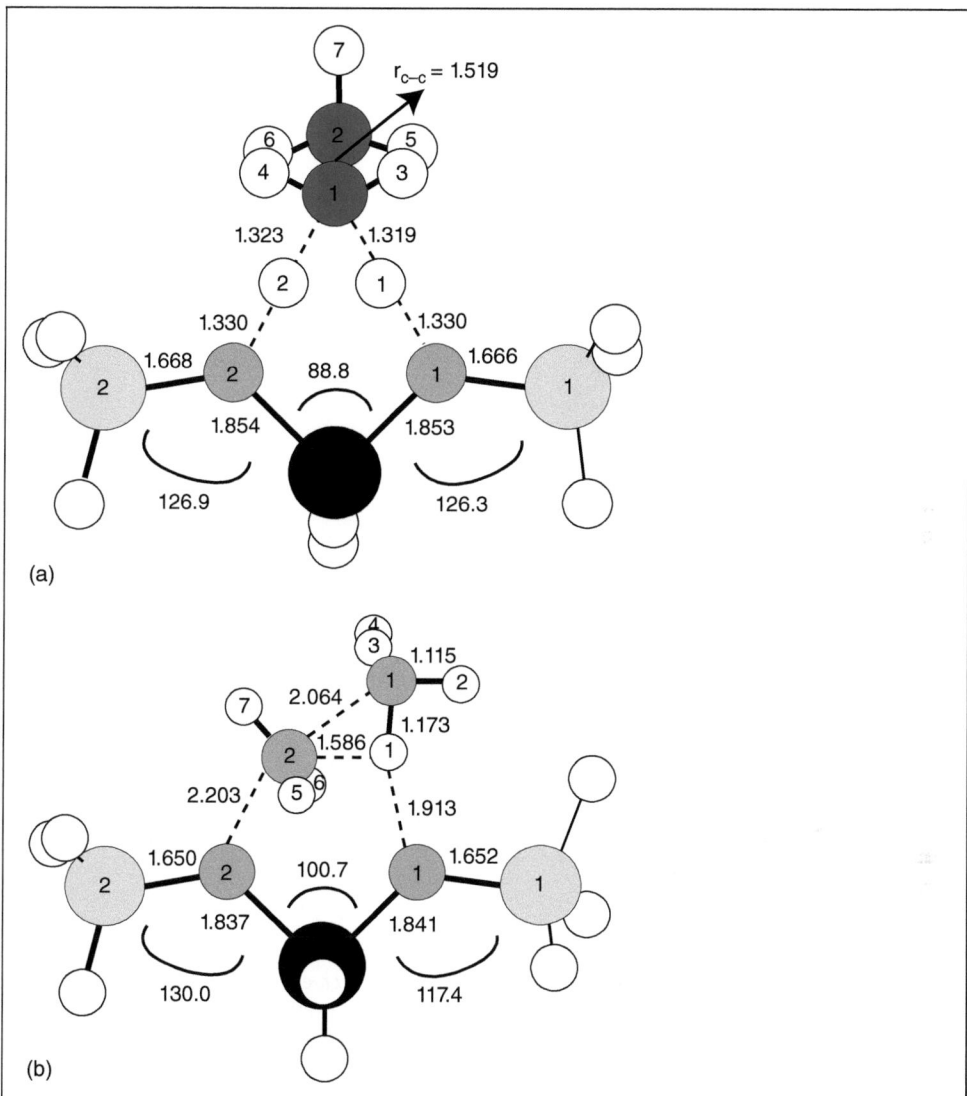

Figure 6i.1 (a) Geometry of the transition state (TS) for the acid-catalyzed reaction H/D of ethane. (b) Geometry of the TS for the acid-catalyzed C—C bond cracking reaction of ethane. (c) Geometry of the TS for the acid-catalyzed dehydrogenation of ethane. Distances in angstroms and angles in degrees. DFT-calculated cluster barriers: H/D exchange 126 kJ mol^{-1}; cracking 300 kJ mol^{-1}; dehydrogenation 305 kJ mol^{-1}. (Blaszkowski et al. 1996 [37]. Reproduced with permission of American Chemical Society.)

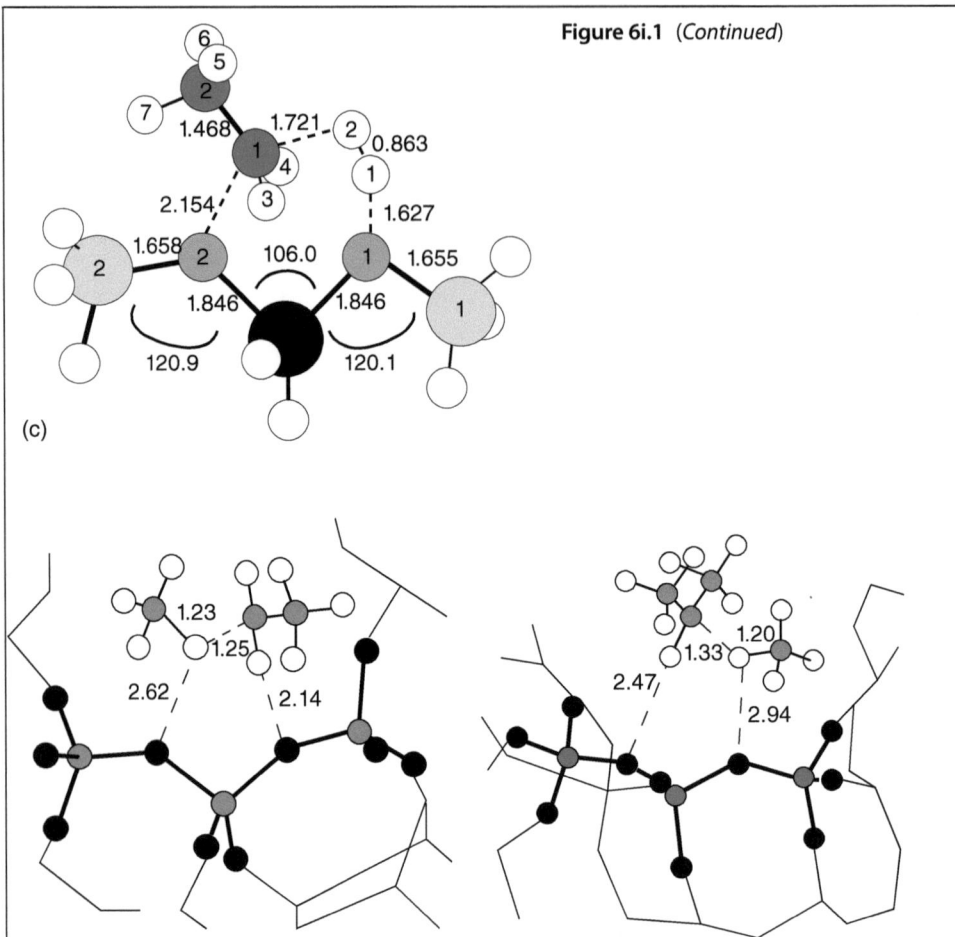

Figure 6i.1 (Continued)

(c)

Figure 6i.2 Transition-state structures of propane and n-butane cracking by a cluster that is part of the chabasite framework. Only the lattice atoms in contact with the substrate molecules are clearly visible. Calculated activation energy for embedded clusters or periodical structure calculations are on the order of 200 kJ mol^{-1}. (Angyan et al. 2001 [38]. Reproduced with permission of Springer.)

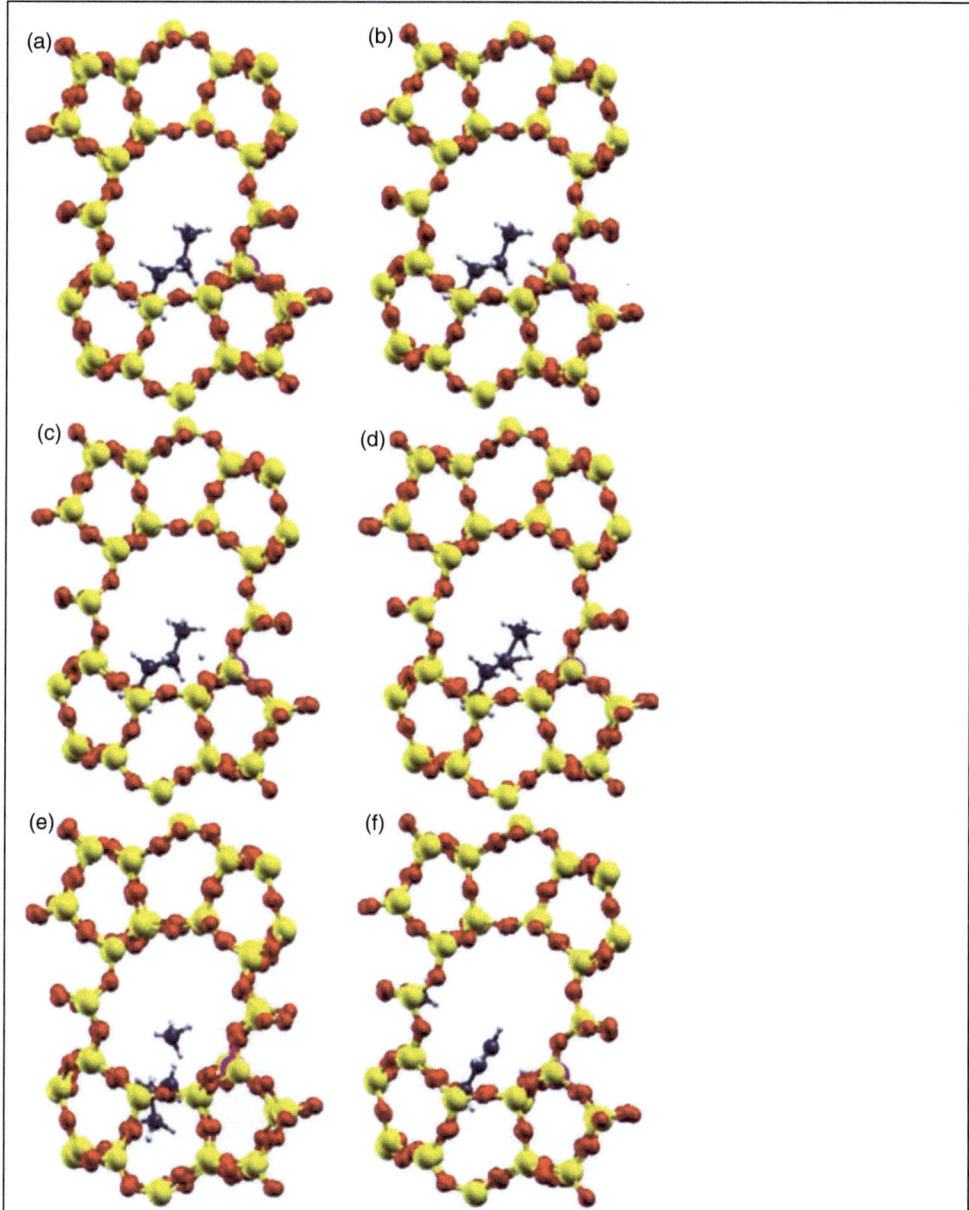

Figure 6i.3 Transition path sampling calculations for butane at $T = 800$ K. (a) Butane interacts with the Brønsted acid site via the methyl group. (b) The distance between the butane molecule and the surface decreases. (c) A proton transfers to the α bond. (d) The five-coordinated alkanium ion is created (TS configuration). (e) The alkanium ion decomposes to the product state. (f) The product state is generated ($C_3H_6 + CH_4$). Colors: carbon, dark gray; hydrogen, white; oxygen, light red; silicon, light orange; aluminum, purple. (Tranca et al. 2012 [39]. Reproduced with permission of American Chemical Society.)

Table 6i.1 Experimental monomolecular propane cracking and dehydrogenation rate constants (k_{meas}), cracking-to-dehydrogenation (C/D) rate ratios at 748 K and measured apparent activation energies (E_{meas}) and entropies (ΔS_{meas}) on acidic zeolites. The symbols of mordenite (Mor-X) refer to different proton sites.

Zeolite	k_{meas} (×10³ mol(molH⁺)⁻¹ s⁻¹ bar⁻¹)		C/D ratio	E_{meas} (kJ mol⁻¹)		ΔS_{meas} (J mol⁻¹ K⁻¹)	
	Cracking	Dehydrogenation		Cracking[a]	Dehydrogenation[b]	Cracking[c]	Dehydrogenation[d]
H-MFI	2.0	2.1	0.9	158	200	−99	−54
H-FER	6.2	3.2	2.0	157	195	−91	−57
H-MOR-T	2.0	3.0	0.7	160	189	−97	−71
H-MOR-S	1.3	1.9	0.7	167	192	−93	−66
H-MOR-Z	1.4	2.2	0.7	160	198	−99	−56

a) Errors are ±5 kJ mol⁻¹.
b) Errors are ±7 kJ mol⁻¹.
c) Errors are ±8 J mol⁻¹ K⁻¹.
d) Errors are ±10 J mol⁻¹ K⁻¹.

Source: Gounder and Iglesia 2009 [40]. Reproduced with permission of American Chemical Society.

The structures shown in Figure 6i.1b are derived from early DFT cluster calculations 20 years ago. Calculated activation energies are on the order of 300 kJ mol^{-1}. The activation energies for H/D exchange tend to be substantially lower due to the intimate contact of the two hydrogen atoms with the zeolite lattice (see Figure 6i.1b). Calculations with larger clusters [41], cluster-embedded calculations (see Figure 6i.2), and full periodical calculations [39] indicate that with respect to the adsorbed state for molecules such as propane and other higher hydrocarbons, the activation energies of C—C bond cleavage are much less than found for the clusters and are of the order of 200 kJ mol^{-1} for cracking reactions. Dehydrogenation reactions tend to have activation energies that can be (20 or 30) kJ mol^{-1} higher. This difference is consistent with the weaker bond energies of the C—C bonds as compared to the C—H bonds. The activation energies shown in Table 6i.1 are apparent activation energies, which also depend on the adsorption energy of the molecule. The intrinsic activation energies can be obtained by adding the adsorption energy (on the order of 40 kJ mol^{-1}) to the observed apparent activation energies. As we will discuss later, differences in the activation energy between homologous molecules tend to be dominated by differences in adsorption energy. The data in Table 6i.1 also show that reaction rates do not always follow the order of activation energies. Especially in the wider pores differences in activation-entropies are significant and bias the dehydrogenation reaction.

Figure 6i.2 showing propane and butane activation in chabasite illustrates that for proton activation of larger molecules, the contact between the positive charge and the zeolite oxygen atom is replaced by the interaction between the zeolite oxygen atom and the hydrocarbon cation hydrogen atoms. This indicates a significant energy contribution due to van der Waals interactions.

The intermediates along a complete computed reaction path for butane activation is shown for a protonic site of H-ZSM-5 in Figure 6i.3. The final state after dissociation of the butane molecule is an adsorbed propyl alcoxy intermediate and methane. The reaction path way is found to be complex. The proton that is initially accepted by a CH bond of the molecule shifts internally to a C—C bond that leads to C—C bond cleavage.

9.3.2.3 Alkene Isomerization

There are several alkene isomerization reactions:

- The branching reaction; the number of branches in a molecule changes.
- The methyl shift reaction; no change in the number of branches.
- The C=C double bond-shift reaction that occurs by subsequent protonation and deprotonation of the alkene.

Proton-activated isomerization reactions that maintain the number of branches and only change branch position are called type A rearrangements. They occur through a sequence of classical alkyl branching and hydride shift reactions. Compared to type B rearrangements that change the number of branches, the rate of type A arrangements tends to be fast.

Selective C=C double bond shift is catalyzed at low temperature by basic catalysts prepared by depositing alkali on alumina. The C=C bond shift reaction proceeds along the hydrocarbon chain by intermediate formation of an allyl cation through hydride transfer to the basic catalyst. C=C bond shift occurs by subsequent hydride backdonation to the other allylic carbon atom [42].

The bond shift reaction can also be catalyzed by protons through the intermediate formation of a carbenium ion, followed by deprotonation of the alkyl cation. However, at reaction temperature this will compete with other acid-catalyzed reactions such as oligomerization and hence will be non-selective.

The type B arrangement branching reaction from linear to branched olefin proceeds through the intermediate formation of the protonated cyclopropyl intermediate (see Figures 7i.1 and 7i.2c).

Insert 7: Isomerization through Protonated Cyclopropyle Intermediate

The protonated cyclopropyle (PCP) intermediate and its reactions according to Brouwer [14] are shown in Figure 7i.1. The figure illustrates isomerisation of butene that is $_{13}C$ labelled at one of its endcarbon atoms.

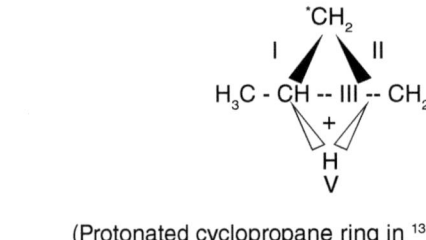

(Protonated cyclopropane ring in ^{13}C isomerization)

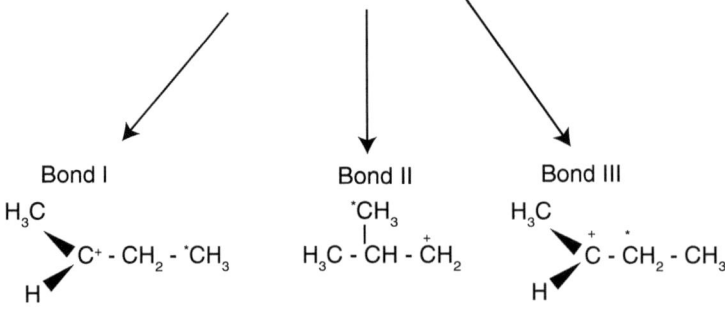

Bond I	Bond II	Bond III
(The original state)	(Skeletal isomerization, forbidden because of primary carbonium-ion formation)	(^{13}C isomerization, experimentally observed)

Figure 7i.1 Structure of a protonated cyclopropyle ring formed from a classical – butyl carbenium ion and the carbenium ions formed by rupture of the indicated C—C bonds. The dot labels the ^{13}C label of butene. (After [43].)

A comparison of calculated reaction intermediates and reaction energies of isomerization of n-pentene through a PCP intermediate versus ethyl bond shift [44] on a protonic site in ferrierite is shown in Figure 7i.2.

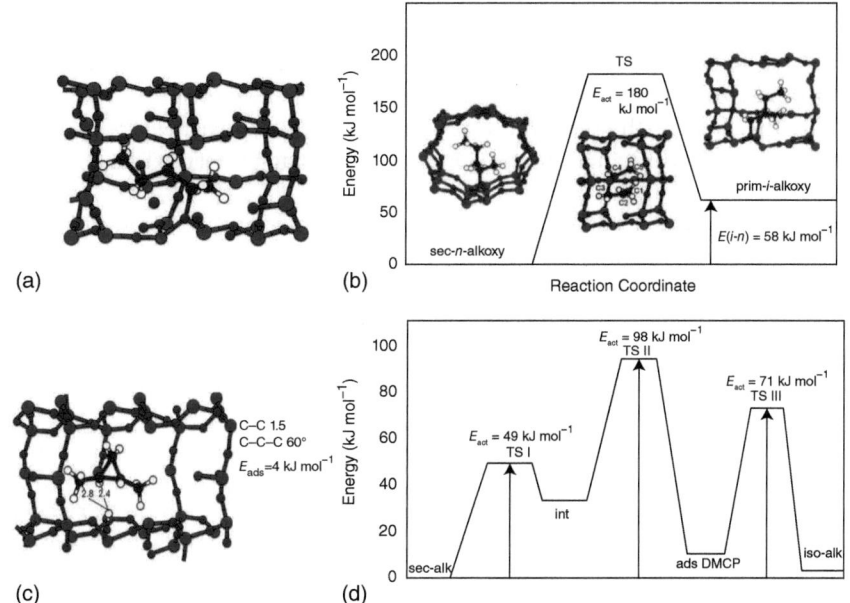

Figure 7i.2 Reaction intermediates and reaction energy diagram of pentene isomerization by a ZSM22 (ferrierite) zeolite proton according to VASP-DFT calculations. (a) Adsorbed pentyl alcoxy intermediate. (b) The reaction energy diagram of the ethyl shift reaction. (c) Structure of the adsorbed protonated cyclopentane intermediate. (d) The activation energies and respective energies of the protonated cyclopropane reaction.(Demuth et al. 2003 [44]. Reproduced with permission of Elsevier.)

Experimental proof of isomerization through the protonated cyclopropyl intermediate is provided by what is now a classical isotope labeling experiment of the isomerization of n-butane-1-^{13}C and n-pentane-1-^{13}C in the superacidic solution of HF-SbF$_5$ [14]. In the superacid, the initial carbenium ions are formed through the protonation and cracking of carbonium ions. Isomerization of the carbenium ions occurs and the reaction chain is propagated through hydride transfer from reagent to product carbenium ion. Hence, after a slow initiation through intermediate carbonium ion formation, carbenium ion formation is rapid and propagation of the reaction is fast.

Catalysis of n-pentane to isopentane conversion by superacid is fast, whereas n-butane isomerization to isobutane is slow. However, the scrambling of the ^{13}C label within the n-butane molecule along the n-butane chain has a rate comparable to the branching isomerization of n-pentane to i-pentane. This suggests the same mechanism for both reactions. The postulate that a PCP intermediate is formed provides a solution. As illustrated in Figure 7i.1 for the butyl cation, the cleavage of its three bonds yields *n*-butyl in two cases, each with the ^{13}C label at different positions along the chain. The carbenium ions then formed are secondary cations. Cleavage of the other bond gives an *i*-butyl cation, which is a primary cation with the positive charge at an end atom. Therefore, the activation energy for this reaction is high, which explains its slow rate. In contrast, when branching isomerization occurs with a longer hydrocarbon than butane (such as pentane), then the branching isomerization would also generate a secondary cation. This enables the reaction to proceed at a rate comparable to the scrambling reaction in *n*-butane, which also proceeds through intermediate secondary cation formation.

Figures 7i.2 show DFT-computed reaction energies and reaction intermediates for branching isomerization of an adsorbed pentyl cation in the micropore of ferrierite. The PCP isomerization pathway (Figures 7i.2c,d) is compared with the competitive alkyl shift reaction (Figures 7i.2a,b). The latter reaction proceeds through an intermediate primary hydrocarbon cation. In the zeolite, the PCP path also appears preferred for the solid acid, as evidenced by its lower overall activation barrier.

9.3.2.4 *n*-Butene Isomerization

In zeolites, protonation reactions through transition states with primary carbenium ions have kinetically reasonable barriers, but they compete unfavorably with transition states that proceed through secondary or tertiary carbenium ions.

n-Butene to isobutene transformation is a desirable reaction, because isobutene is an important feedstock that, when converted to an ether, can be added to fuel to upgrade its octane number. Monomolecular *n*-butene isomerization must proceed through primary carbocation formation. The alternative is a mechanism where the butene dimerizes to an octene. This reaction has a relatively low activation energy. The octene molecule isomerizes in second reaction steps to branched isomers, that can decompose to give *i*-butene through a β C—C cleavage reaction to be discussed in the next subsection. This will circumvent the reaction path through primary cation formation.

This complex reaction network that corresponds to this reaction mechanism is shown in Figure 9.4.

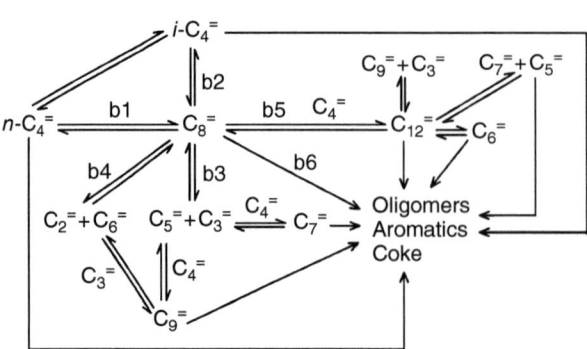

Figure 9.4 Reaction network of *n*-butene on the modified ZSM-5 zeolite catalyst. (Meng *et al.* 2010 [45]. Reproduced with permission of American Chemical Society.)

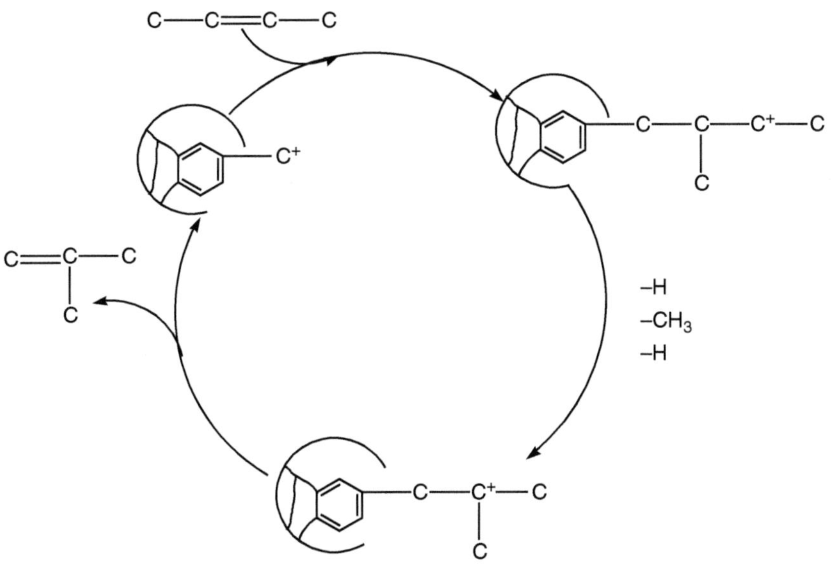

Figure 9.5 Mechanism of 2-butene isomerization over spent FER; the proposed active site is a benzylic carbocation. (After [46].)

When one uses the zeolite ZSM-5 for the conversion of n-butene at the reaction temperature of 700 K many byproducts are formed [45]. Zeolites with wide pores tend to form aromatics at the high temperature of reaction. Only the narrow-pore zeolite ferrierite is selective for the transformation of butene to isobutene [36]. Its narrow pores prevent oligomerization, which requires overly bulky transition states.

Experiments with ^{13}C labeled n-butene can be used to distinguish between the monomolecular and bimolecular pathway. Isobutene formed by dimerization of initially single-labeled n-butene should have a substantial fraction of double-labeled isobutene, and this is indeed found for most systems [47]. This double-labeled fraction decreases strongly with time, which is thought to be due to narrowing of the micropores by the deposition of carbonaceous residue. This pore-narrowing suppresses bimolecular catalysis in favor of monomolecular catalysis.

An alternative mechanism may also occur. In this mechanism, it is proposed that monomolecular selective isobutene formation is catalyzed by an aromatic benzylic organocatalyst intermediate. These aromatic intermediates may be formed as a non-selective coproduct in the zeolite micropore while the reaction is ongoing [48]. The corresponding reaction cycle is shown in Figure 9.5. It is suggested that toluyl cation groups whose primary cations are stabilized by resonance with the aromatic ring react with n-butene.

After isomerization, the isobutene is recovered by β-C—C bond cleavage.

In situ formation of an organocatalyst located in the micropore of a zeolite is proposed for the isomerization of n-butene. In addition, there is extensive evidence that methanol to ethylene or propylene conversion is also catalyzed by organocatalysts formed *in situ* (see Sections 4.5.1 and 9.3.3.4).

9.3.2.5 β-C–C Bond Cleavage

Carbenium ions may undergo C—C bond cleavage at substantially milder conditions than the direct C—C bond cleavage reaction of alkanes via carbonium ion intermediates. This cracking reaction will become especially competitive with direct alkane cracking when initial carbenium ions present. Catalytic cracking through β C—C bond cleavage requires hydride transfer from alkane reagent molecules to cracked carbenium ions, as we will discuss in Section 9.3.3.2.

Carbenium ions can cleave their C—C bonds by a reaction that is the reverse of olefin dimerization. Such C—C bond cleavage reactions take place through the β C—C bond cleavage reaction.

Figure 8i.1 schematically shows how the β-C—C cleavage reaction breaks the C—C bond next to the positively charged C atom in the reactant molecule. A larger carbenium ion is cleaved into a smaller carbenium and an alkene. Two shorter olefins are produced as a final product when the proton is backdonated from the carbenium ion to the zeolite framework.

The carbenium ions are initially adsorbed as alcoxy species to the zeolite framework as can be seen in Figure 8i.2. The β-C—C cleavage reaction occurs in stages in the transition state as shown in the same figure. The two fragments in the transition state both show the characteristic sp^2 hybridized features of a carbenium ion. After reaction, one of the carbenium ions adsorbs as an alcoxy species.

The activation energies of the various possible β-C—C bond cleavage reactions depend on the types of carbenium ions involved in the reaction. The lowest barriers are found for tertiary to tertiary bond cleavage reactions (100 kJ mol^{-1}), followed by secondary to tertiary or secondary to secondary bond cleavage reactions (180 kJ mol^{-1}) (see Figure 8i.3).

Insert 8: β-C—C Bond Cleavage

Figure 8i.1 schematically shows the β-C—C bond cleavage reaction of the *iso*-hexyl cation. Figure 8i.2 shows adsorbed alcoxy intermediates of the carbenium ions in their ground state, the transition state and their relative energies. Figure 8i.3 summarizes the differences in activation energies of the various β-C—C cleavage reactions.

An important conformational requirement is the overlap of the σ bond orbitals of the bond to be broken with the empty electron-accepting C 2p atomic orbital. This is illustrated in Figure 8i.4.

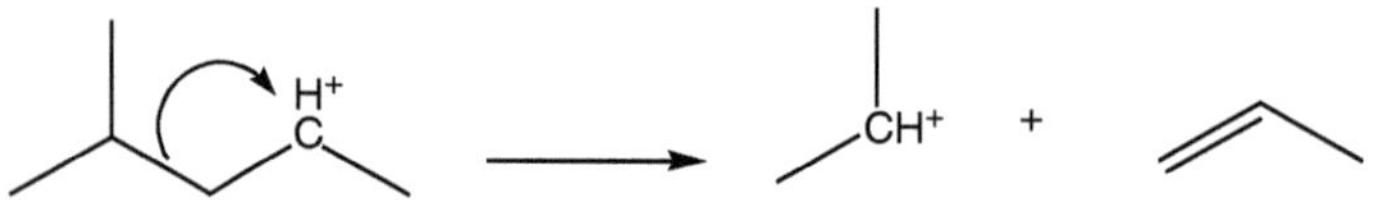

Figure 8i.1 Schematic representation of the β-C—C cleavage reaction. From a secondary carbenium ion, an olefin and another secondary cation is formed. This reaction is only possible for a molecule equal to or larger than hexene. (After [35].)

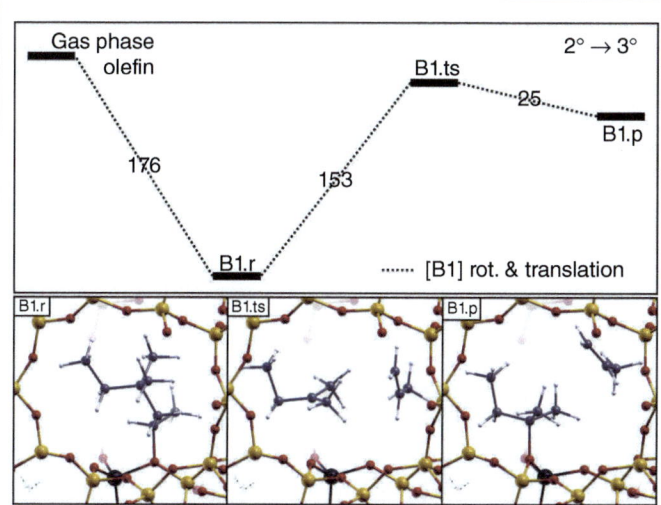

Figure 8i.2 DFT-calculated reaction intermediates and energy profile (kJ mol^{-1}) for the B1-type β-scission of a secondary-alcoxide (4,4-dimethyl-2-hexoxide) in a H-ZSM-5 channel. See Figure 8i.3 for the notation for different β-C—C type reactions. (Mazar et al. 2013 [50]. Reproduced with permission of American Chemical Society.)

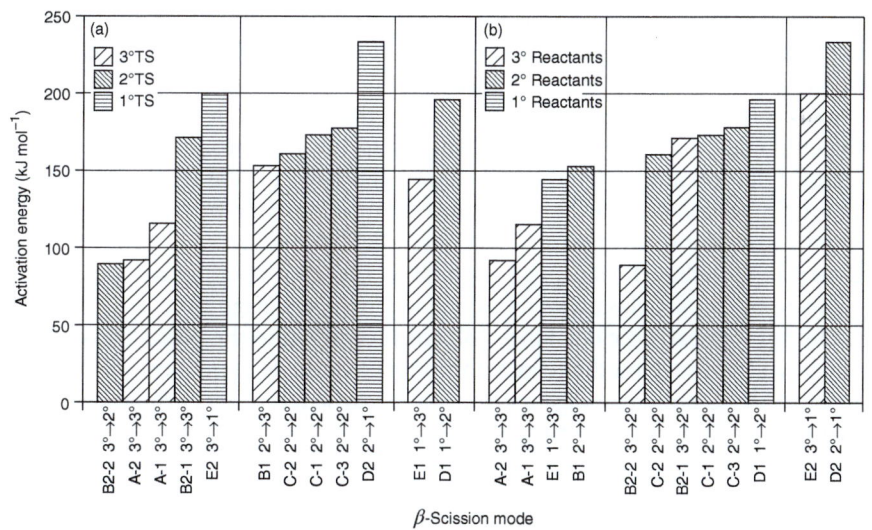

Figure 8i.3 Computed activation energies for the β-scission of C_8 and C_6 isomers in H-ZSM-5. (a) β-Scission elementary steps sharing the same substitution order of the carbocationic carbon atom in the reactant state (i.e., C$^+$) have been grouped together. (b) β-Scission elementary steps sharing the same substitution order of the carbocationic carbon atom in the transition state (i.e., β C) have been grouped. (Mazar et al. 2013 [50]. Reproduced with permission of American Chemical Society.)

(Continued)

Insert 8: (Continued)

Aliphatic carbenium ions | Alicyclic carbenium ions

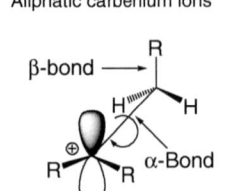
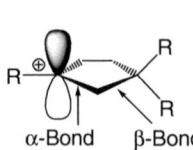

Figure 8i.4 The role of orbital orientation for the easiness of β-scissions of alkyl carbenium ions. After [14, 51].

Low activation energy β-C—C cleavage through tertiary carbenium ion intermediates is only possible for hydrocarbons with skeletons larger than seven carbon atoms. In this reaction, proton-activated isomerization of the alkene to a branched alkene has to occur which gives a tertiary carbenium cation. Subsequent β-C—C cleavage again gives a tertiary carbenium. A steep increase occurs in the relative rate for the C—C bond cleavage reaction as compared to skeletal isomerization reaction for alkenes with less than seven skeletal carbon atoms. This has a very important implication in the selectivity of the hydroisomerization and hydrocracking reaction of linear alkenes as a function of their hydrocarbon chain length [51], that we will discuss in 9.3.3.3.

There is also a large difference in the reactivity of β-C—C bond cleavage in aliphatic carbenium ions versus cleavage of alkyl site chains of planar C_5 ring naphthenic carbenium ions. The alkyl chains of the naphthenic molecule are remarkably robust due to the planarity of the naphthenic ring. This prevents hyperconjugation (overlap) of the empty $2p_z$ atomic orbital of the positively charged carbon atom with the σ C—C skeletal bond orbital that is to be broken. In an aliphatic molecule, this orbital overlap is possible because the C—C—C bond angle along the chain is less than 180°. Since β-C—C ring opening of the cyclopentadienyl cation is not possible, this ring-opening reaction requires catalysis by Pt particles as we discussed in Chapter 8, page 310.

Only proton catalytic cracking of alkanes according to the Haag–Dessau mechanism will produce methane as a coproduct. Methane formation will not occur by the β-C—C bond cleavage mechanism because it requires primary cation formation from intermediate carbenium ions.

9.3.2.6 The Hydride Transfer Reaction

In acid catalysis, the hydride transfer reaction (see Figure 9.3b) between reactant alkane and intermediate carbenium ion propagates the catalytic reaction cycle involving carbenium ions. It competes with termination of the reaction cycle by proton backdonation from the carbenium ion to the negatively charged solid acid lattice with formation of an olefin.

Due to the differences in energies of primary, secondary, and tertiary carbenium ions, the preferred hydride donating molecule is a tertiary alkane molecule or an alkene with more than two carbon atoms. The alkene can readily transfer a hydride ion since the cation that is formed is a relatively stable allyl intermediate.

The transition state of the hydride transfer reaction is similar to the transition state of the reverse reaction of proton-activated direct C—C bond cleavage. The hydride transfer reaction is thermodynamically neutral or exothermic. The proton activated C—C bond cleavage is an endothermic reaction. Therefore the activation barrier for hydride transfer is substantially lower than the barrier for direct proton-activated C—C bond cleavage. The intrinsic activation energy of C—C bond cleavage is of the order of 170 kJ mol^{-1}, whereas the activation energies for hydride transfer from a tertiary carbon of a reactant hydrocarbon can vary between 100 and 20 kJ mol^{-1} [52]. The respective activation energies are mainly determined by the activation energies to desorb the carbenium ions, which are sensitive to the primary, secondary, or tertiary nature of the respective intermediate carbenium ions (see Figure 5i.2).

As we will see, the catalytic cracking reaction is rapidly deactivated by oligimerization reactions that form carbonaceous residue.

In Section 9.3.2.6, we will discuss the mechanism of the *n*-alkane hydroisomerization reaction, which is catalyzed by bifunctional acidic zeolites promoted with Pt or Pd. This is a stable reaction that is operated in excess hydrogen.

n-Alkane isomerization is also possible when catalyzed by protonic zeolites in the absence of hydrogen. The reaction is initiated by the proton-activated formation of carbonium ions, which decompose to *n*-carbenium ions. These *n*-carbenium ions can isomerize to *i*-carbenium ions and the reaction is propagated by hydride transfer reaction steps. The elementary hydride transfer reaction that produces the *i*-alkane from the *i*-carbenium ion and a secondary carbenium ion from *n*-alkane has a relatively high barrier on the order of 70 kJ mol^{-1}, because a tertiary carbenium ion is replaced by a secondary, less stable cation. The reaction deactivates rapidly due to competitive alkene formation by proton backdonation of carbenium ions to the zeolite lattice. Deactivating low activation energy oligomerization reactions then produce carbonaceaous residue.

In the hydroisomerization reaction, the presence of hydrogen and a hydrogenation active catalyst reduces the alkene concentration, which suppresses the deactivating oligomerization reaction.

The hydride transfer reaction is a critical step in catalytic reactions that produce aromatics such as the catalytic cracking process and the conversion of methanol to aromatics. In this conversion, hydride transfer from olefin to primary or secondary cations is essential for the formation of aromatics. These reaction cycles will be discussed in detail in the following sections. Hydride transfer is also the propagating reaction step in the alkylation reaction of isobutane with propene or butene, which produces branched alkane molecules for jet fuel. We will discuss this reaction next.

9.3.2.6.1 Alkylation

The alkylation reaction is an important refinery process reaction that produces high octane gasoline from light molecules. The alkylation reaction converts isobutane with *n*-butene or propene into a branched C_8 or C_7 molecule. In practice, the reaction is catalyzed by oleum (neat H_2SO_4) or HF. Replacing the liquid acids by a heterogeneous catalyst is an important catalytic challenge.

We will discuss the physical and organic chemistry of this reaction in the liquid phase and we will compare the current efficiency for this reaction when catalyzed by a solid acid.

The catalytic reaction cycles proposed by Schmerling [53, 54] that constitute this reaction are shown in Figure 9.6 for the alkylation reactions with butene.

The key elementary reaction steps that compete are the hydride transfer between isobutane and an intermediate carbenium ion and the olefin oligomerization that leads to the deactivation of the catalytic system.

As illustrated in Figure 9.6a, the reaction is initiated by n-alkene protonation to give a secondary carbenium ion. This is converted into a n-alkane by hydride transfer from isobutane. This in turn gives an isobutyl cation and n-butane.

The propagation reaction consists of consecutive reaction steps in which the isobutyl cation adds to an alkene molecule. The resulting dimer cation is converted further in a second reaction with isobutane. It gives the C_8 product molecule and again generates an isobutyl cation intermediate. The propagation cycle is continued with these reactions. The reaction that competes with this alkylation propagation is the oligomerization of the olefins. It will deactivate the liquid acid catalyst by proton depletion through the formation of stable cationic aromatic species. In the superacid catalyzed refinery process, acid regeneration procedures are applied to recover the acid after reaction.

Application of solid acids would enable safer and more environmentally friendly process operations. Acidic zeolites are of interest, but their most significant drawback is their limited lifetime [49, 55–57]. The most long-lived zeolite material tested (zeolite β or La-promoted zeolite X and Y catalysts) had an effective life of less than 10 h.

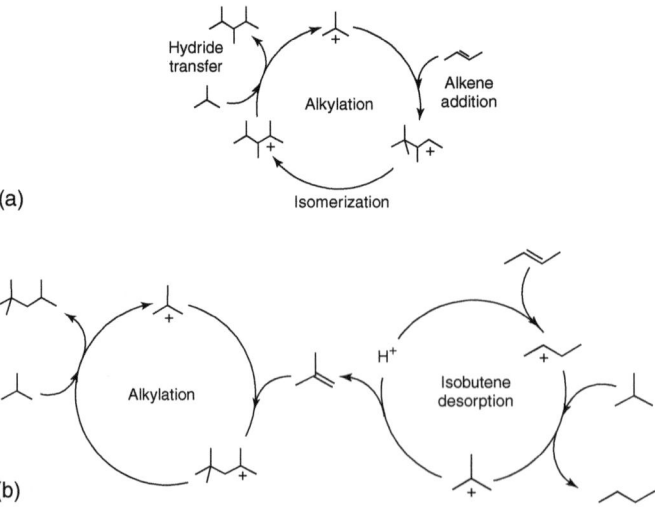

Figure 9.6 (a) Simplified alkylation cycle including the three key reaction steps. (b) Self-alkylation cycle, depicting the two steps: isobutene desorption and subsequent addition of isobutene to an isobutyl ion after hydride transfer to give 2,2,4-trimethylpentane. (Feller et al. 2004 [49]. Reproduced with permission of Elsevier.)

A classical experiment in which isobutane is reacted in D_2SO_4 [58] demonstrated that a tertiary isobutyl carbenium ion is formed as an intermediate. It was observed that only the nine H atoms of the methyl groups of isobutane will exchange with D^+, but that the hydrogen atom attached to the tertiary carbon will not exchange.

Reaction of isobutane is initiated by reaction of a proton of H_2SO_4 with the reactive tertiary H atom of isobutane to give a solvated isobutyl cation and H_2. In H/D exchange reactions with D_2SO_4, the isobutyl cation will initially backdonate a proton to give isobutene, which can protonate and deprotonate in consecutive reactions. This leads to proton exchange only with the hydrogen atoms attached to the primary carbon atoms. The tertiary carbon atom cannot take back a proton, since it can only accept a hydride ion.

Interestingly, active acidic heterogeneous catalysts such as a La-promoted zeolite X or Y [49] produce substantially more butane than would be expected based on their proton content. Apparently, intermediate isobutyl will also backdonate its proton to produce highly reactive isobutene, that can dimerize by reacting with an isobutyl cation (see Figure 9.6b).

The formation of butane indicates that proton backdonation competes with hydride transfer. Clearly, the proton donation affinity of the zeolite protons is not high enough to compete efficiently with the hydride transfer reaction.

The activation energy of the hydride transfer reaction is determined largely by the activation energy of the adsorbed secondary alcoxy intermediate that is formed upon the reaction of the isobutyl cation with n-butene. The tertiary isobutyl cation formed from isobutane will only weakly interact with the solid. The activation energy of the hydride transfer reaction will be on the order of 70 kJ mol^{-1}. A weaker O—C bond of the alcoxy intermediate will be favorable since it will lower the activation energy of hydride transfer. This is why high proton donation affinity of the acid is important. This will also prevent proton backdonation of the isobutyl to give isobutene that will also participate in the oligomerization reaction.

The activation energy of the competing oligomerization reaction that leads to catalyst deactivation is less sensitive to the proton donation affinity of the zeolite.

The neat liquid acids show better performance than the heterogeneous solid acid catalyst for several reasons:

- High proton donation affinity of liquid superacid results not only from a high proton concentration (low pH), but also from high reactivity of the protons. The acid molecules are partially ionized in the polar medium. The strong acidity is due to $H_3SO_4^+$ or H_2F^+ cations [59, 60]. That have substantially higher proton donation affinity than zeolitic protons, that are attracted to a neutral site.
- The temperature of operation (0–30 °C) of the liquid acid process is low compared to the temperature of 80 °C necessary for the zeolite-catalyzed reaction. The higher temperature for the zeolite reaction is required to desorb the reaction product from the solid catalyst. It however favors the competitive oligomerization reaction of the olefins that have higher activation energy.
- In the liquid system, undesirable oligomerization will also occur but does not result in the micropore blocking that affects the zeolite. Deactivation mainly

occurs because protons are consumed by the formation of stable aromatic cations, which can be observed from the red coloring of the system. This reaction relates to the paring reaction of Figure 12i.2.
- The concentration of free alkene in the liquid acid is lower than that in the micropores of the solid acid. The alkene becomes readily protonated into the superacid, which reduces the concentration of free alkene.

9.3.3 Catalytic Reaction Cycles and Kinetics

9.3.3.1 Physical Chemistry of Zeolite Catalysis

The protons of the zeolite are located in the nanopores of a small crystal particle, which can only be reached by transport processes from the crystallite exterior to its interior.

The overall catalytic reaction cycle including adsorption and diffusion that takes place in the zeolite particle is schematically illustrated in Figure 9.7.

In Figure 9.7, the time scales for the various elementary reaction steps and physical processes are compared.

It is important to realize that the rates of desorption can compete with the rates of elementary reaction steps. We will see in the following subsections that differences in catalyst performance among zeolites of various structures are often dominated by variations in the adsorption properties of the zeolitic materials rather than by changes in the reaction rates of elementary reaction steps.

As we discussed in Section 3.4, the rates of intra-zeolite crystal particle diffusion tend to be substantially faster than the elementary reaction rates, unless steric constraints inhibit mobility. Single file diffusion, which occurs in one-dimensional micropores and prevents molecules from passing each other, may also slow down reactions.

Generally, as long as crystal particles remain small enough, intra-zeolite crystal diffusion can be ignored. Extrinsic kinetics is instead mainly affected by inter-particle diffusion effects, which we will not be considered here.

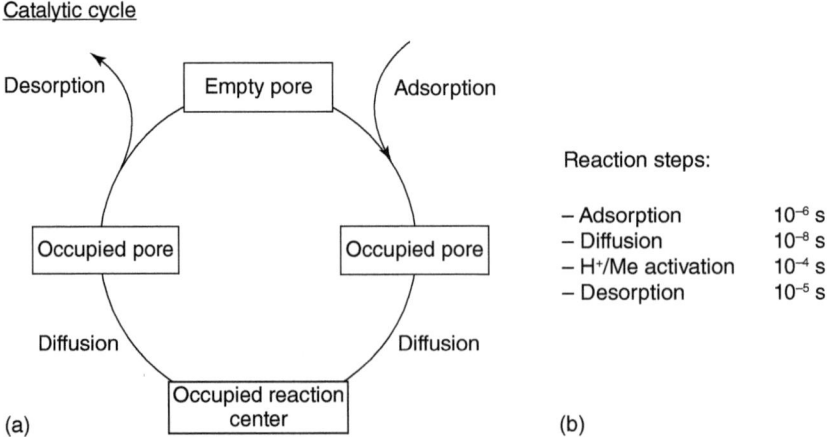

Figure 9.7 The catalytic cycle of a reaction with the reactive center located in the zeolite nanopore and representative time scales of the elementary reactions steps that are part of the zeolite catalytic reaction.

The catalytic reaction cycles and mechanisms of three important catalytic processes will be discussed in the next subsections:

- Catalytic cracking
- Hydroisomerization and hydrocracking
- Methanol conversion and methanol alkylation reactions.

The activity, selectivity, and stability of these reactions are sensitive not only to the composition of the zeolite framework but also, and especially, to the zeolite lattice structure. We will see that structural effects not only affect elementary reaction rates because of steric mismatch; even in the absence of a steric constraint, the dimension and shape of the zeolite micropores will have a large effect on the adsorption of reaction intermediates, sometimes with significant consequences for overall kinetics.

In catalytic cracking, the small size of zeolite micropores suppresses the formation of large deactivating aromatic molecules. In the hydrocracking reaction, the shape of the microchannels and cavities will bias formation of intermediates with a shape that provides an optimum fit. The product pattern is found to be dependent on the zeolite micropore structure and connectivity.

The selectivity of the conversion of methanol to ethylene versus aromatics, as in the Methanol To Olefin (MTO) process (see Chapter 2, page 27), is also strongly structure dependent. Zeolitic materials with the smaller micropores will only produce ethylene or propylene (H-SAPO-35) whereas zeolites with wider pores such as H-ZSM-5 will produce mainly aromatics.

When there are no steric constraints, differences in the apparent rates of zeolite-catalyzed reactions are often dominated by variations in the adsorption constants of the reaction intermediates. Product distribution will relate to the equilibrium constants of reaction intermediates formed in the zeolite micropores.

We have chosen to focus on the three processes mentioned above because they illustrate the complex interplay between catalytic kinetic networks and zeolite structure and composition.

In the catalytic cracking process, the catalyst reaction center is the proton, but the reaction network is a complex combination of many reaction steps. Reactions occur between nearly freely moving cationic intermediates in the nanopores and nanocavities of the zeolite.

Reaction initiation requires a demanding activation step through a carbonium ion transition state. Once carbenium ions are formed, propagation occurs readily by hydride transfer steps, by which alkanes are converted into reactive carbenium cation intermediates. The stereochemistry of bimolecular hydride transfer processes can be controlled by using zeolites with smaller micropores, which will suppress bimolecular reaction. For instance, in the catalytic cracking process, the use of a zeolite with smaller micropores reduces aromatics formation, but favors light alkene formation.

The apparent activation energy of the direct cracking of small alkanes, which goes through carbonium cation type transition states strongly depends on the hydrocarbon chain length due to differences in adsorption energies.

In hydroisomerization, we will find that the differences in the activity of zeolites also tends to be dominated by adsorption effects. These effects may determine the differences in apparent acidity of zeolites when their pore dimension is varied.

The lower temperature of the reactions in hydroisomerization and hydrocracking catalysis may cause a high occupation of the zeolite micropores by reaction intermediates. Selective adsorption effects then may dominate reaction selectivity. This will be the case especially when heavier molecules are processed. Methods for calculating adsorption isotherms of complex reaction mixtures as a function of catalyst loading by in zeolite nanopores have proven highly successful for explaining selectivity differences due to selective adsorption [61].

The final reaction that we will discuss is the conversion of methanol to higher hydrocarbons. After an initiation period, catalysis actually occurs by an *in situ* generated cationic intermediate that acts as an organocatalyst (see also Section 4.6.1). The catalytic cycle propagates through a complex series of consecutive alkylation and the ring-opening reactions of this cyclic hydrocarbon organocatalyst.

9.3.3.2 Catalytic Cracking

The catalytic cracking reaction network of an alkane is illustrated in Figure 9i.1. The activation of alkanes is initiated by Haag–Dessau direct cracking, but once alkenes are formed other reactions take over. The consecutive reactions shown in Figure 9i.1a that maintain the catalytic reaction chain are the hydride transfer and β-C—C bond cleavage reactions as well as the oligomerization reactions. This sequence of events is common to many catalytic processes: reaction is initiated by a slow reaction that forms a particular intermediate, but once this intermediate is present at high enough concentration, it becomes part of a catalytic reaction cycle that has a high turnover. In the particular case of alkane activation, the initiating reaction is carbonium ion formation.

Haag–Dessau cracking of alkanes can produce methane, but the β-C—C bond cleavage of alkenes will not produce methane. Therefore, an experimental indication of the relative importance of the Haag–Dessau activation versus the β-C—C bond cleavage reaction is the amount of methane that is formed as a coproduct.

The catalytic cycle is propagated by the hydride transfer reaction, when reagent alkane or an olefin transfers a hydride ion to an intermediate carbenium ion. The complexity of the reaction cycles that evolve for isobutane cracking is illustrated in Figure 9i.1b.

Aromatics will be produced by hydride transfer from alkenes (through intermediate allyl formation) to carbenium cations combined with proton backdonation to the solid. This heterolytic dehydrogenation process of combined H-transfer and proton backdonation gives dienes that can undergo consecutive ring closure reactions.

The very different rates of the reactions that participate in the catalytic reaction cycle are illustrated in Figure 9i.2. The β-C—C bond cleavage reaction of an alkene has a rate that is an order of magnitude faster than direct cracking of an alkane.

Insert 9: The Catalytic Cracking Reaction

I. **The Catalytic Reaction Cycles** Catalytic reaction cycles of catalytic cracking are illustrated in Figure 9i.1. Figure 9i.1a shows a schematic of the important reactions steps and Figure 9i.1b shows the complex cycle of isobutane conversion.

The very different rates for the main types of reactions that participate in the catalytic cracking process are compared in Figure 9i.2.

II. **Relations with Proton Concentration, Proton Reactivity, and Adsorption** Figure 9i.3 illustrates that direct cracking of hexane is linear in proton concentration at a low Al/Si ratios. Figure 9i.4 illustrates the differences in proton activity for ZSM-5 type materials, which contain Fe and Ga instead of Al in the zeolite framework. Figure 9i.5 illustrates the linear dependence of the apparent activation energy of n-alkane conversion as a function of alkane length. A comparison is made with experimentally measured adsorption energies in Figure 9i.5b.

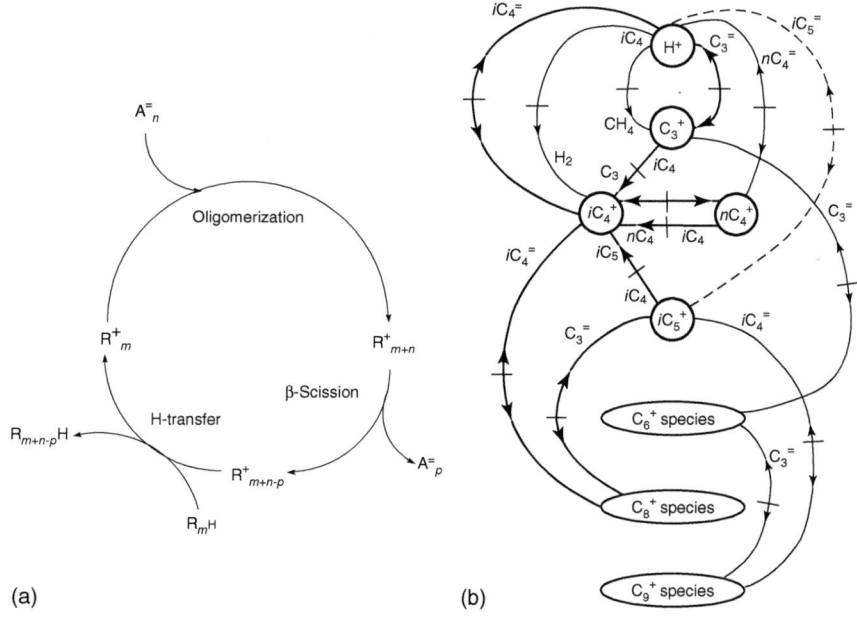

Figure 9i.1 (a) Oligomeric cracking route for an alkane molecule ($R_m H$) involving the oligomerization of an alkene with n carbon atoms ($A_n^=$) with a carbenium ion with m carbon atoms (R_m^+) and subsequent β-scission and hydride transfer. (b) Catalytic cycles for isobutane cracking catalyzed by the zeolite USY at 773 K and 11% conversion in a plug flow reactor. Illustrated examples notation: $iC_4^=$, isobutylene; C_3^+ propyl cation; nC_4, n-butane; C_6^+ species, hexyl cations. (Kotrel et al. 2000 [30]. Reproduced with permission of Elsevier.)

(Continued)

Insert 9: (Continued)

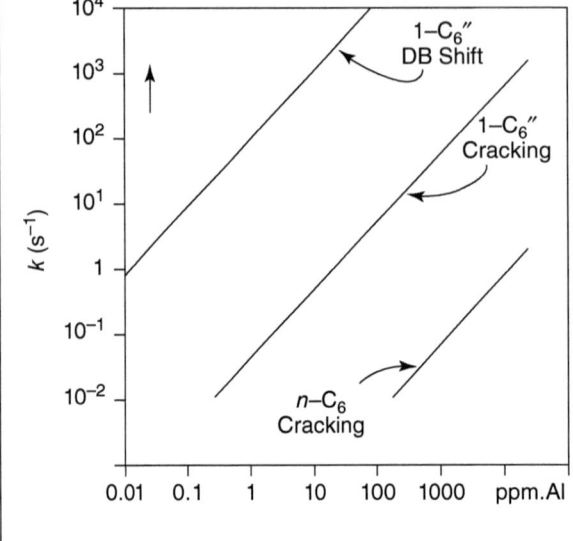

Figure 9i.2 Comparison of the rate constants at 454 °C, of hexane cracking, β-C—C hexene cracking and hexene double bond (DB) shift reactions. (Haag et al. 1984 [62]. Reproduced with permission of Nature Publishing Group.)

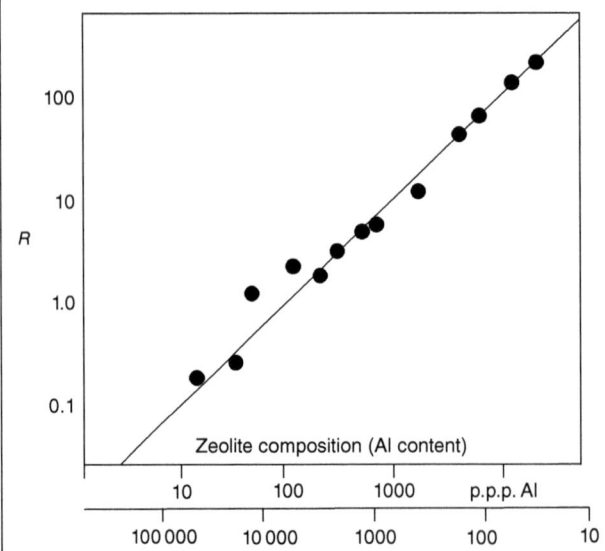

Figure 9i.3 The hexane cracking activity R plotted against the aluminum content in H-ZSM-5. (Haag et al. 1984 [62]. Reproduced with permission of Nature Publishing Group.)

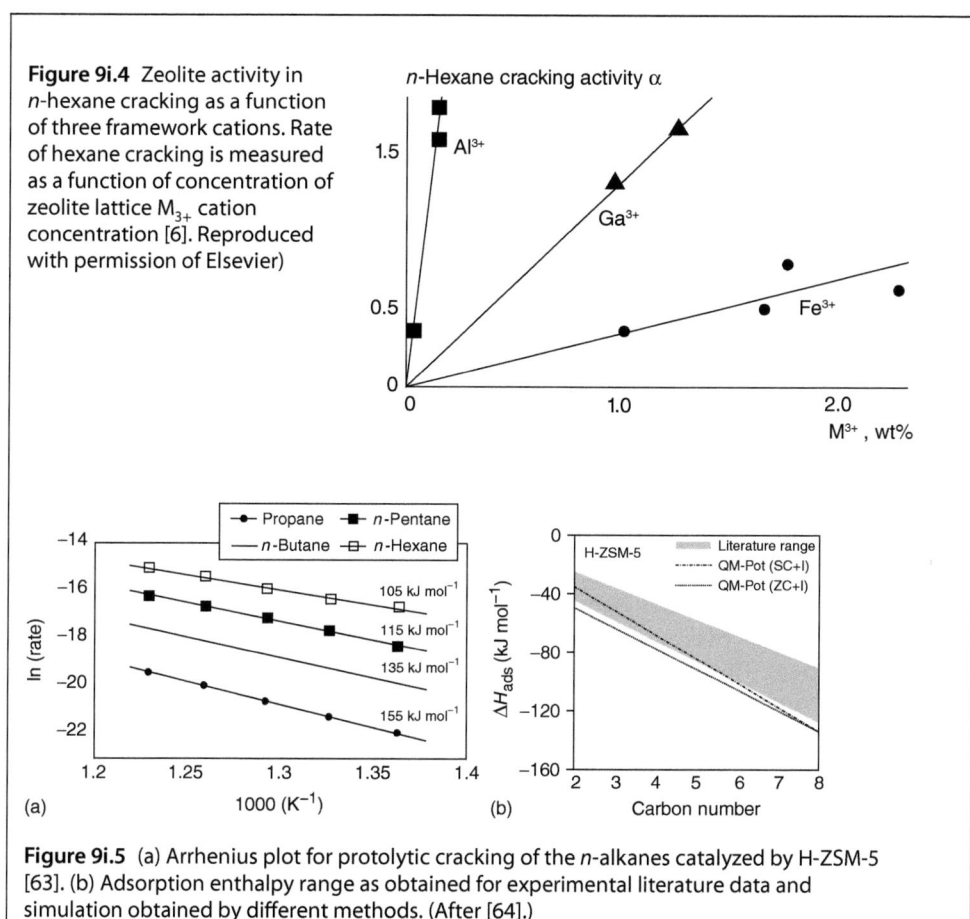

Figure 9i.4 Zeolite activity in n-hexane cracking as a function of three framework cations. Rate of hexane cracking is measured as a function of concentration of zeolite lattice M_{3+} cation concentration [6]. Reproduced with permission of Elsevier)

Figure 9i.5 (a) Arrhenius plot for protolytic cracking of the n-alkanes catalyzed by H-ZSM-5 [63]. (b) Adsorption enthalpy range as obtained for experimental literature data and simulation obtained by different methods. (After [64].)

Figure 9i.3 illustrates an important experiment that confirms that zeolite proton donation affinity does not change as long as Al/Si ratio is less than 10%. The turn over frequency (TOF) of direct cracking of hexane normalized per proton catalyzed by H-ZSM-5 remains independent of Al/Si ratio. Direct Haag–Dessau cracking dominates as long as hexane conversion remains low.

The decrease in proton donation affinity of a zeolite when Al^{3+} is substituted by Fe^{3+} or Ga^{3+} in the framework of ZSM-5 is illustrated in Figure 9i.4. From the slopes of the curves of the rates of the cracking reaction of hexane one deduces that proton donation affinity has the order $Al^{3+} > Ga^{3+} > Fe^{3+}$. The proton donation affinities decrease because the weaker covalent Fe—O and Ga—O interactions increase the relative OH bond energies of the zeolitic proton attached to the O atom (as expected from bond order conservation).

At the high temperature of the cracking process, the micropore occupation by alkanes in the zeolites is low. Then, the monomolecular reaction rate expression (Chapter 3, Insert 6, page 92) can be used. It follows that the apparent activation energy E_{app} decreases linearly with the adsorption energy of the hydrocarbon:

$$E_{app} = E^{\#} + E_{ads} \tag{9.3}$$

Note that in this expression the intrinsic activation energy $E_{\#}$ is positive, but E_{ads}, the adsorption energy of hydrocarbon is negative. The cracking data for n-alkanes in Figure 9i.5a shows a nearly linear decrease in apparent activation energy with increasing n-alkane length. According to Figure 9i.5b, the adsorption energy of these molecules also increases linearly. This implies that the intrinsic activation energy for cracking these molecules is independent of chain length and has the approximately constant value of 160 kJ mol^{-1}. The differences in apparent activation energies are seen to be dominated by their differences in adsorption energies.

The adsorption energies and entropies of alkanes adsorbed in a zeolite are discussed in Insert 10 as background for later chapters.

Insert 10: The Adsorption Energy and Entropy of Linear Hydrocarbons Adsorbed in the Zeolite Micropore

The adsorption energy of hydrocarbons arises from the dispersive van der Waals interactions of the hydrocarbon atoms with the polarizable oxygen atoms of the zeolite nanopore. This is linearly dependent on the contact area of the alkane chain with the nanopore. It strongly depends on the dimensions of the zeolite nanopore, because this determines match of the zeolite nanopore dimension and size of the hydrocarbon. For longer hydrocarbons, this may result in adsorption energies as high as 100–120 kJ mol^{-1}. This happens when the nanopore dimensions become comparable to the diameter of the molecule. Figure 10i.1 illustrates the linear increase of adsorption

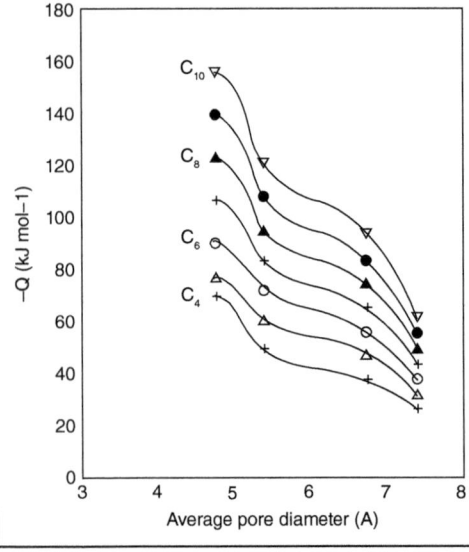

Figure 10i.1 A comparison of simulated adsorption energies $-Q$ of linear hydrocarbons as a function of hydrocarbon chain length and average siliceous zeolite nanopore dimension. (Bates et al. 1998 [65]. Reproduced with permission of Elsevier.)

energy with hydrocarbon chain length and also its variation with micropore dimension. The data shown in this figure are based on simulations with statistical mechanical Monte Carlo methods that can calculate adsorption free energies as a function of partial pressure and temperature. The interaction between zeolite and molecule is deduced form atomistic force field parameters [65].

The adsorption entropy decreases when molecules adsorb from the gas phase into the zeolite micropore. In linearly shaped micropores, the rotational motion of the adsorbed molecule will decrease as micropore diameter decreases. Differences in adsorption entropies and energies are compared in Table 10i.1 for the adsorption of n-pentane and n-hexane in the zeolite Ferrierite (TON) and Mordenite (MOR) that have one-dimensional linear channels. Mordenite has a 12-ring diameter microchannel and ferrierite (the TON structure) has a 10-ring diameter microchannel. From a comparison of adsorption equilibrium constants K_{ads} with adsorption energies ΔH_{ads} one deduces that the adsorption entropy of the molecules is smaller in the smaller channel. This contrasts with the adsorption energy, which increases with decreasing microchannel diameter.

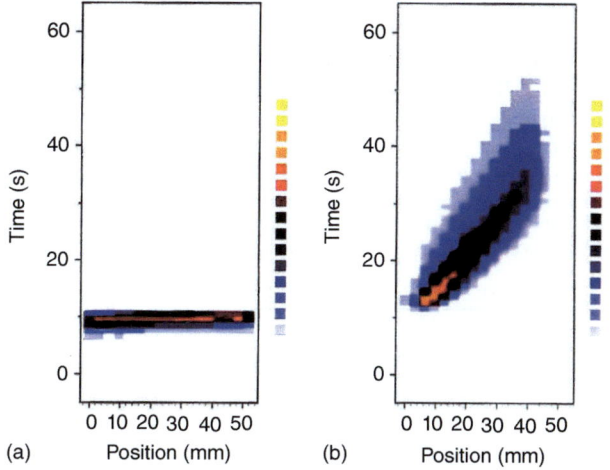

Figure 10i.2 Comparison of the residence times of a pulse of radiochemically labeled hexane into H_2 for the silica surface with only large micropores and in a zeolite with nano-sized pores. (a) A profile of the ^{11}C n-hexane labeled pulse as a function of time and position on silica at 200 °C is shown. (b) The residence time profile of a pulse of ^{11}C hexane over H-mordenite at 200 °C is presented. (Anderson et al. 1998 [66]. Reproduced with permission.)

The differences in adsorption free energy lead to very different nanopore adsorbate concentration dependences as a function of nanopore dimensions at different temperatures. This is illustrated by the gas chromatographic experiments with the injection of pulses of radioactively labeled ^{11}C hexane in Figure 10i.2. The slope of the residence time, measured by the radiochemical signal of a labeled reaction pulse as a function of position along the reactor bed is inversely proportional to the adsorption equilibrium constant of the hexane molecule. The residence time slope also increases when concentration of free adsorption sites reduces due to the presence of adsorbed molecules (see Figure 11i.2). While there is no measurable delay of the pulse signal for silica, it is large

(Continued)

Insert 10: (Continued)

for the nanoporous zeolite material. This means that the surface coverage of hexane on silica at experimental conditions is negligible.

The large adsorption equilibrium constants of hydrocarbons in zeolites will at low temperature lead to substantial nanopore occupation, which has important implications for the interpretation of zeolite-catalyzed reaction kinetics.

Table 10i.1 Energies of adsorption ΔH_{ads} and adsorption equilibrium constants K_{ads}.

	ΔH_{ads} [kJ mol^{-1}]		K_{ads} ($T = 513$ K) [mmol g Pa^{-1}]	
	Simulation	Literature	Simulation	Literature
n-Pentane/TON	63.6	71 [68] 62.1 [69]	4.8×10^{-6}	4.6×10^{-6} [68] 6.53×10^{-6} [69]
n-Pentane/MOR	61.5	59 [68] 55.7 [69]	4.8×10^{-5}	6.8×10^{-5} [68] 8.6×10^{-5} [69]
n-Hexane/TON	76.3	82 [68] 75.0 [69] 72.3 [70]	1.25×10^{-5}	8.0×10^{-6} [68] 1.99×10^{-5} [69] 7.8×10^{-6} [70]
n-Hexane/MOR	69.5	69 [68] 67.1 [69] 69 [70]	1.26×10^{-4}	1.9×10^{-4} [68] 5.5×10^{-4} [69] 1.3×10^{-4} [70]

A comparison between two zeolites with linear nanopores. Zeolite mordenite (MOR) has 12-ring channels, zeolite ferrierite (TON) has 10-ring channels.
Source: After [65].

9.3.3.3 Bifunctional Catalysis

9.3.3.3.1 Introduction

Bifunctional acidic catalysts are used in hydrocarbon conversion reactions such as isomerization and cracking reactions. A transition metal or another active compound (like the sulfides) is used to activate C—H bonds and dehydrogenate alkanes or hydrogenate intermediate alkenes. Protonation and deprotonation takes the place of the hydride transfer reactions in maintaining the catalytic reaction cycle (see Figure 9.3c). The main practical effect of the intermediate conversion of alkanes to alkenes is that the reaction can be run at a lower temperature, since proton activation of alkenes occurs more readily than that of the alkanes. This will favour isomerization from lineair to branched product. The stability of the catalyst is also substantially enhanced, because rapid hydrogenation reduces the state concentration of alkenes in the zeolite nanopores, so that carbonaceous residue formation is suppressed.

Close contact between the hydrogenation site and protonic site is required to prevent the deactivation of these catalysts by the non-selective oligomerization processes. Therefore, the metal or sulfide particles that are conventionally used for hydrogenation or dehydrogenation must be small and located inside the micropores close to the proton reaction center. A catalyst promoted by transition metal particles can also non-selectively produce methane that is an undesirable coproduct of the metal catalyzed hydrogenolysis reaction. In the next subsection we will discuss hydroisomerization in which the hydrocarbon skeleton is maintained. In the following subsection we will discuss hydrocracking by which the hydrocarbon is cracked into lighter molecules.

9.3.3.3.2 Hydroisomerization

The mechanism of the n-alkane hydroisomerization process and key reaction steps have been introduced before in Section 2.2.2.1.3 (Figure 2.6, page 26). The catalyst consists of a protonic zeolite promoted with a hydrogenation/dehydrogenation catalyst. The reaction is performed at high H_2/alkane ratio and a metal loading sufficient to maintain fast equilibration of alkane with alkene. The reaction conditions are chosen so that the steady state concentration of alkene is approximately 10_{-6} that alkane. The reaction rate of proton-activated alkene isomerization competes with alkane desorption from the zeolite micropore.

One would expect that the absolute reaction orders in alkane and H_2 partial pressures would be equal and of opposite sign. This would be the case if alkane and alkene concentrations within the nanopore are equilibrated with the gas phase. However, this is generally not observed, as seen in Table 11i.1, which illustrates measured and simulated reaction orders for hexane conversion.

The hexane concentration in the micropores is orders of magnitude higher than the concentration of hexene. At high pressure the micropore becomes saturated with hexane and the reaction becomes zero-order in hexane partial pressure but of order -1 in H_2 partial pressure (see Table 11i.1). The hydrogen concentration remains proportional in pressure. This implies that hexane is locally in equilibrium with hexene within the highly occupied micropores. However, under these conditions hexene does not desorb out of the micropores so that no equilibrium exists between the reacting molecules in the micropores and in the gas phase.

The lumped kinetics expression (Eq. (11i.1)) for hexane hydroisomerization illustrates that this difference between expected and observed reaction orders is also due to the competition of hexane and hexene for adsorption to the proton. Experimental validation of the high micropore occupation can be obtained from chromatographic transient experiments, in which the residence time of an isotope-labeled molecule is measured. As shown in Figure 11i.2, at atmospheric condition this residence time decreases to 30% of that in an empty micropore. This implies that the micropore has a hydrocarbon pore filling of 70%.

The significance of the effects of adsorption is also illustrated by experiments with mixtures of pentane and hexane (see Table 11i.3). In the narrow micropores of zeolitic mordenite, the adsorption constant of hexane is considerably larger than that of the shorter molecule pentane. This difference is substantially smaller in the wide pore zeolite Y catalyst. As a consequence, in mordenite the stronger adsorbing hexane suppresses the reaction of pentane, but in faujasite the relative rates are little affected compared to the rates of the individual components.

Insert 11: The Role of Adsorption in Kinetics and Selectivity of the Hydroisomerization Reaction

Data on the kinetics of the hydroisomerization reaction are given in Table 11i.1. Figure 11i.1 illustrates how the lumped sum equation (Eq. (11i.1)) for hydroisomerization can be used to extract information on the relative importance of adsorption versus proton activation of the reaction. The data in Table 11i.2 have been determined using a similar plot. Differences in the adsorption constants of hydrocarbons may lead to very different competitive affects as a function of micropore structure as shown in Table 11i.3.

Table 11i.1 (a) Experimentally measured and simulated orders of reaction at atmospheric pressure. (b) Simulated coverages at standard conditions ($H_2/C_6 = 30$, $p = 1$ atm, $T = 513.15$ K).

(a)

	Mordenite		ZSM-5	
	n-C_6 order	H_2 order	n-C_6 order	H_2 order
Sims	0.19	−0.12	0.21	−0.13
Expt	0.13	~ 0	0.51	−0.25

(b)

T (K)	2.0 wt% Pt/H-mordenite			0.5 wt% Pt/H-ZSM-5		
	$\theta_{n\text{-hexane}}$	$\theta_{n\text{-alkoxy}}$	TOF (h^{-1})	$\theta_{n\text{-hexane}}$	$\theta_{n\text{-alkoxy}}$	TOF (h^{-1})
493	0.77	0.98	1.41 (3.5)	0.86	0.98	7.45 (20.3)
513	0.62	0.91	4.54 (9.7)	0.72	0.90	23.83 (54.0)
533	0.44	0.73	11.32 (24.9)	0.53	0.72	57.13 (113.6)

Source: Van De Runstraat et al. 1997 [71]. Reproduced with permission of Elsevier.

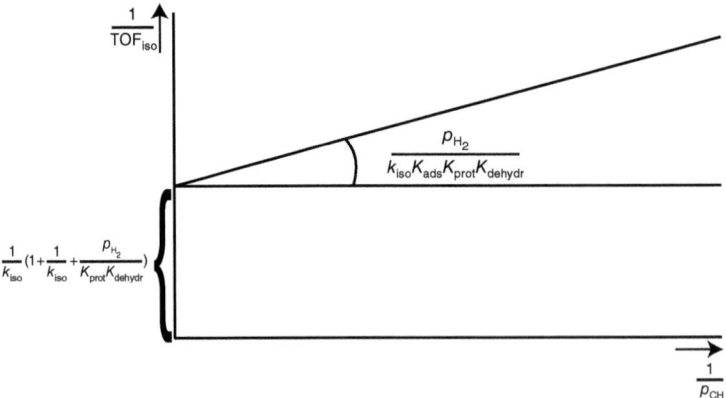

Figure 11i.1 Illustration of the plot of $(p_{CH})^{-1}$ against TOF^{-1}, that enables extraction of the elementary rate constant k_{isom} from experimental data, when the adsorption isotherm for the alkane is known. (de Gauw et al. 2002 [73]. Reproduced with permission of Elsevier.)

Table 11i.2 Results of activity measurements of the hydroisomerization of hexane ($T = 240\,°C$, $p_{nC_6} = 779\,Pa$).

	TOF (s^{-1})	K_{ads,nC_6} (Pa^{-1})	k_{iso} (s^{-1})
H-Beta	5.0×10^{-3}	6.4×10^{-5}	1.7×10^{-2}
H-MOR	1.1×10^{-2}	3.3×10^{-4}	2.7×10^{-2}
H-ZSM-5	4.1×10^{-3}	3.3×10^{-5}	2.8×10^{-2}
H-ZSM-22	1.6×10^{-3}	1.4×10^{-5}	1.7×10^{-2}

TOF are normalized per proton.
Source: de Gauw *et al.* 2002 [73]. Reproduced with permission of Elsevier

Table 11i.3 Comparison of the results of Pt–H-mordenite and Pd—H-faujasite catalyzed hydroisomerization at 514 K.

Total pressure (MPa)	C_5/C_6 vol ratio	H_2/HC	Mordenite		Faujasite	
			kC_5	kC_6	kC_5	kC_6
0.79	Pure C_5	10/1	0.241	—	0.187	—
2.17	Pure C_5	10/1	0.071	—	0.046	—
0.79	Pure C_6	10/1	—	0.337	—	0.431
2.17	Pure C_6	10/1	—	0.100	—	0.047
0.79	50/50	10/1	0.137	0.519	0.095	0.208
2.17	50/50	10/1	0.032	0.125	0.024	0.017

In contrast to wide pore faujasite the apparent rate constants in mixtures catalyzed by narrow pore mordenite show non-linear concentration dependence.
After [72].

Equation 11i.1 gives the lumped sum Langmuir–Hinshelwood expression for the TOF of hexane hydroisomerization, when alkane–alkene/H_2 equilibration catalyzed by the metal particle is considered fast compared to the proton-catalyzed reactions. The difference between this Eq. (11i.1) and the simpler Langmuir–Hinshelwood equation (presented earlier in Chapter 3, Eq 3.14) is that alkane and alkene competition for adsorption to the proton is included.

$$\text{TOF} = \frac{k_{iso} K_{ads} K_{dehydr} K_{prot} \frac{p_{nC_6}}{p_{H_2}}}{1 + K_{ads} p_{nC_6} + K_{ads} K_{dehydr} \frac{p_{nC_6}}{p_{H_2}} + K_{ads} K_{dehydr} K_{prot} \frac{p_{nC_6}}{p_{H_2}}} \tag{11i.1}$$

K_{ads}, K_{dehydr}, and K_{prot} are the equilibrium constants of, respectively, adsorption, dehydrogenation, and protonation of *n*-hexane adsorption and alkene protonation; k_{iso} is the rate constant of conversion of the intermediate *n*-alcoxy into isoalcoxy (no distinction is made between the various isomers); p_{H_2} is the hydrogen pressure; and p_{nC_6} is the *n*-hexane pressure.

(Continued)

Insert 11: (Continued)

Figure 11i.2 illustrates the high nanopore occupation of mordenite exposed to hexane at atmospheric conditions. Residence time measurement for a radiochemically ^{11}C-labeled hexane probe molecule are shown. A pulse injected into a H_2 flow has a residence time three times longer than a pulse injected into a hexane/H_2 mixture. The longer residence time reflects the difference in available sites for adsorption. In this experiment the micropore coverage with hexane is 30% at conditions used.

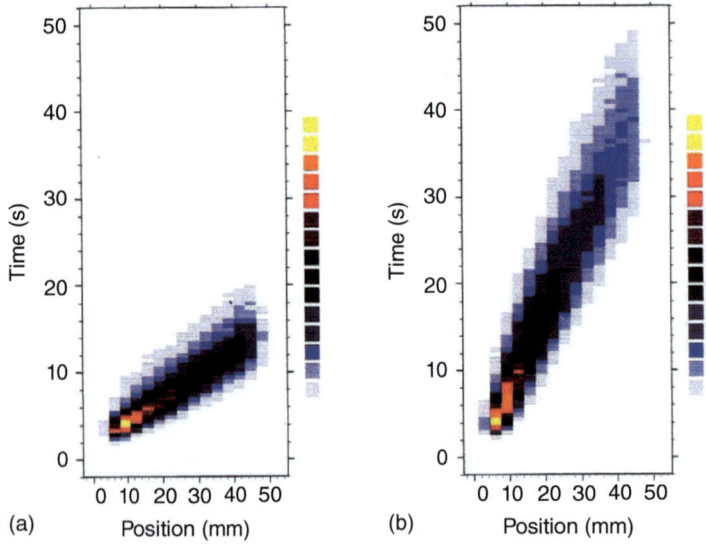

Figure 11i.2 Comparison of the residence time of ^{11}C labeled hexane pulse in an empty nanopore and a pore partially filled with hexane. The progress of the pulse through the catalyst bed can be followed as a function of time. (a) ^{11}C-n-hexane on platinum/H-mordenite 230 °C; 150 ml min^{-1} H_2 + n-hexane flow. (b) ^{11}C-n-hexane on platinum/H-mordenite 230 °C; 150 ml min^{-1} H_2 only flow. (Anderson *et al.* 1997 [67]. Reproduced with permission of Elsevier.)

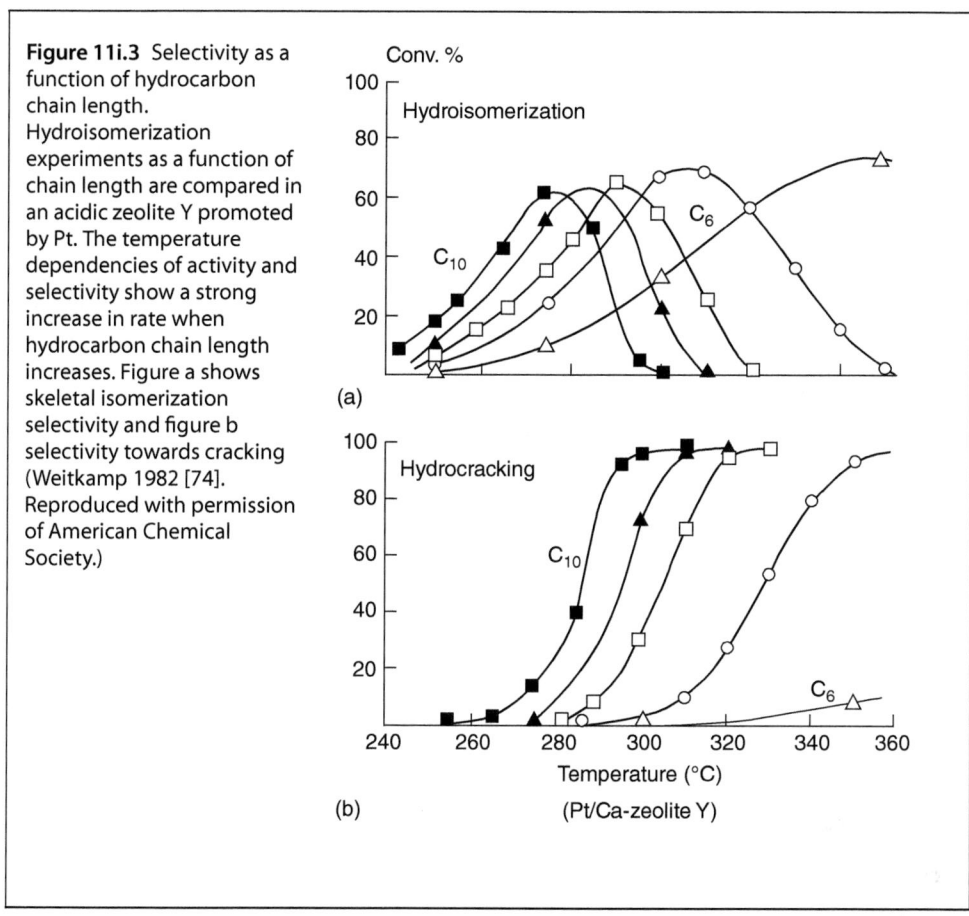

Figure 11i.3 Selectivity as a function of hydrocarbon chain length. Hydroisomerization experiments as a function of chain length are compared in an acidic zeolite Y promoted by Pt. The temperature dependencies of activity and selectivity show a strong increase in rate when hydrocarbon chain length increases. Figure a shows skeletal isomerization selectivity and figure b selectivity towards cracking (Weitkamp 1982 [74]. Reproduced with permission of American Chemical Society.)

When zeolites of different structures are compared it appears that differences in their reactivity are often dominated by differences in adsorption energies. For the hydroisomerization of n-hexane, measured intrinsic elementary reaction rate constants of proton-activated isomerization of the alkene, k_{iso} are compared with measured overall turnover frequencies for four zeolites in Table 11i.2. The intrinsic rate constants have been determined by the procedure illustrated in Figure 11i.1.

One notes the relatively small variation in the elementary rate constant of k_{iso} versus the large differences found for the TOFs, which measure the overall reaction rate. k_{iso} is the elementary reaction rate constant for the transformation of adsorbed n-alcoxy intermediate to i-alcoxy. The variation in the equilibrium constants for the adsorption of hexane to the different zeolites is also substantially larger than the corresponding variation of k_{iso}. This is a strong factor in determining the variation in TOFs. The variation in TOF between the different zeolites results largely from variations in hexane occupation in the zeolite micropores.

We will discuss in detail the difference in the activity of ferrierite (TON) and mordenite (MOR) caused by their analogous micropore structure. It is remarkable that the TOF is lower for the ferrierite zeolite with smaller one-dimensional micropores than for mordenite, but the adsorption energy in ferrierite is higher. At the temperature of reaction, the lower adsorption enthalpy in the wider pore zeolite mordenite is compensated by the smaller decrease in adsorption entropy. The adsorption equilibrium constant in narrow pore ferrierite is smaller, because the lower value of the free energy of adsorption is dominated by the larger loss in adsorption entropy. In the larger micropores the mobility of a molecule is less constrained than in the smaller pores.

A maximum rate of reaction requires an optimum micropore occupation of the catalyst. In the case of hydroisomerization of hexane this is determined by the micropore dimension that maximizes the free energy of adsorption at reaction temperature. The adsorption entropy differences dominate the difference in the free energies of adsorption and, therefore, also dominate the difference in respective TOFs between the two subject zeolites.

Similarly, as found for the catalytic cracking of linear alkanes, the reaction rate of the hydroisomerization reaction also depends on the length of the linear hydrocarbon, as does its selectivity. This is illustrated by experimental studies of the hydroconversion of linear alkanes that vary in length between C_6 and C_{10}, which are presented in Figure 11i.3. In this case, n-alkanes are catalyzed by the wide pore acidic CaY zeolite catalyst that is activated by Pt.

The yield of isomers that maintain the carbon skeleton number is compared with the yield of cracked product. For the longer hydrocarbons one observes a substantial increase in the rate of reaction, so that high yields are obtained at lower temperatures. The apparent activation energies decrease with hydrocarbon chain length, because of the increased hydrocarbon adsorption energies. Selectivity for skeletal isomerization depends on the relative rates of skeletal isomerization versus that of C—C bond cleavage. The latter is a consecutive reaction that it is responsible for the maximum on the skeletal isomerization yield curve.

The onset of the cracking reaction is also at a lower temperature. Although the temperature of the reaction is lower for the longer hydrocarbons, the selectivity of the reaction is decreased because the increased rate of branching, especially for molecules larger than C_7, allows for β-C—C cracking reactions that have a low activation energy. We will discuss this in the next subsection.

9.3.3.3.3 Mechanism and Selectivity of the Hydrocracking Reaction

The product patterns of n-alkane or naphthene hydrocracking show regular but different patterns. As illustrated in Figure 12i.1a, hydrocracking of linear alkanes with 8–15 skeletal carbon atoms gives a bell-shaped pattern of cracked products when molar concentration is plotted as a function of their carbon number. In the case of hydrocracking of ring molecules such as naphthenes, the M-shaped curve of Figure 12i.1b is seen when the molecule is large enough and has a skeleton atom number of at least 10 carbon atoms. These hydrocracking patterns are symmetric when hydrogenation and dehydrogenation of the respective intermediate alkenes or alkanes is fast compared to the rates of the proton-catalyzed reactions, and steric constraints due to zeolite microchannel dimensions are absent. The hydrocracking patterns of monocyclic or di-cyclic hydrocarbons are of interest because they provide a route to the conversion of undesirable polynuclear aromatic hydrocarbons into branched alkanes useful as components of high octane gasoline.

Insert 12: The Mechanism of Hydrocracking

Figure 12i.1 compares the product pattern of a linear alkane with that of a cyclic hydrocarbon. Figure 12i.2 gives the mechanistic scheme of the hydrocarbon cracking reaction of linear alkanes and naphthenes.

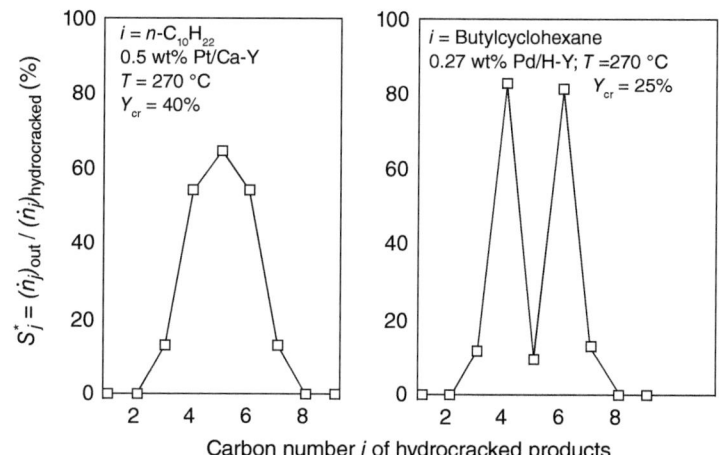

Figure 12i.1 Carbon number distributions of the hydrocracked products from n-decane ((a) bell-type curve) and butylcyclohexane ((b) M-type curve) on bifunctional zeolite catalysts. (Weitkamp 2012 [51]. Reproduced with permission of John Wiley and Sons.)

(Continued)

Insert 12: (Continued)

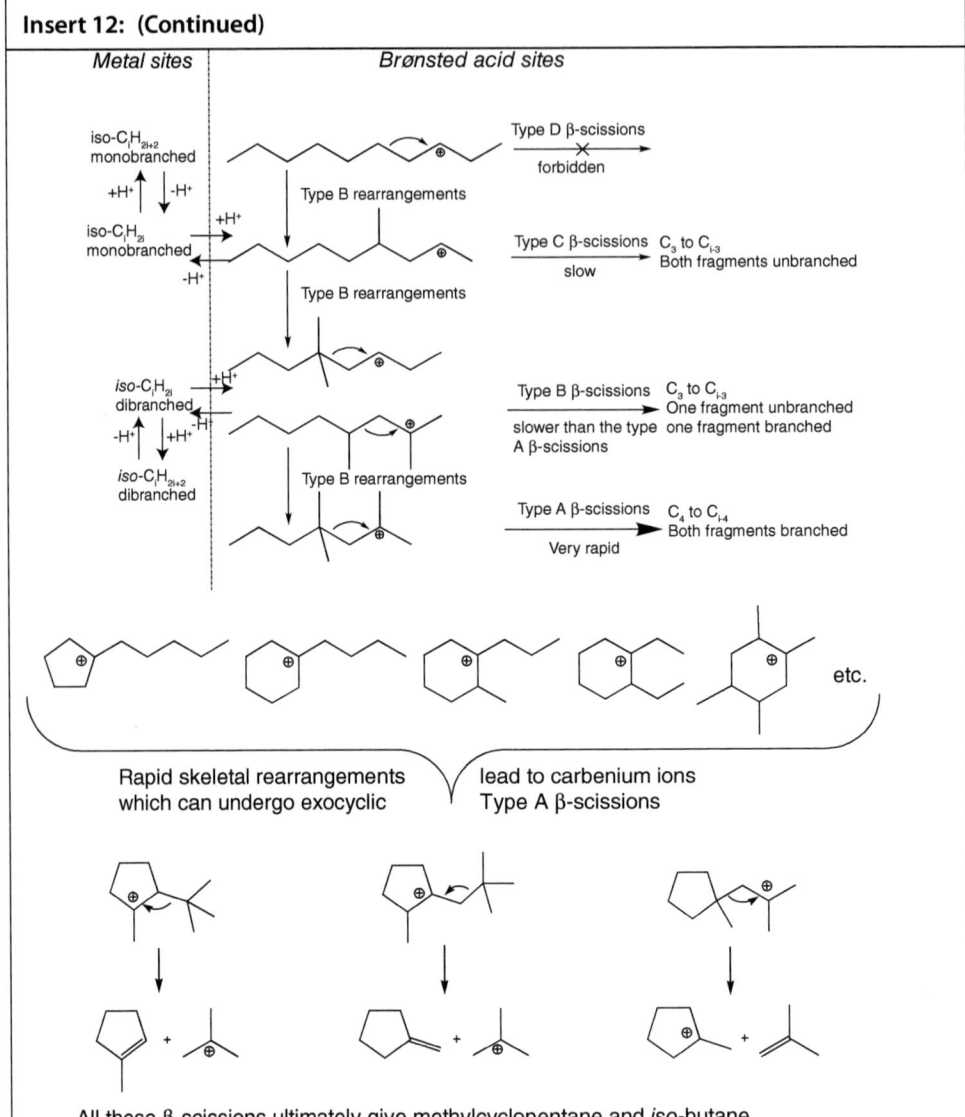

Figure 12i.2 (a) The proposed pathway of ideal hydrocracking of a long-chain n-alkane n-C_iH_{2i+2} on a bifunctional catalyst. (b) The paring reaction: hydrocracking of C_{10} naphthenes with any arbitrary structure (five examples are shown) involves a series of skeletal rearrangements until a carbon skeleton is reached which allows an exocyclic type A β-scission. In all cases, hydrocracking leads to isobutane and methylcyclopentane. (Weitkamp 2012 [51]. Reproduced with permission of John Wiley and Sons.)

9.3 Mechanism of Reactions Catalyzed by Zeolite Protons

The mechanistic explanation for the bell-shaped product pattern of alkane cracking is schematically illustrated in Figure 12i.2a.

The different types of β-scissions are denoted from A to D. They differ in rate because the β C—C cracking reactions involve intermediates with different degrees of branching. These intermediates are connected through two different mechanistic paths. In type A branching such as methyl shift reactions, the number of branches remains the same. In type B branching, the number of branches in a molecule increases or decreases (through the PCP mechanism, see Insert 7). The alkanes must be longer than C_7 in order to form tertiary carbenium ions, so that both the fragments formed by β-C—C cleavage are branched. The bell-shaped product pattern arises from the isomerization of the linear alkenes to the branched molecules that biases formation of reaction products with a carbon skeleton that is approximately half of that of the reactant linear alkane.

The role of the metal in establishing the alkane–alkene equilibrium is also indicated. The cracking and isomerization reactions are atom rearrangement reactions of the carbenium ions generated by protonation of the alkenes.

While primary carbenium ions adsorb with strong alcoxy bonds to the zeolite framework, secondary and tertiary carbenium ions interact only weakly with the framework. At reaction temperature they can be considered as moving freely through the zeolite micropores, analogous to the movement of carbenium ions in superacids.

The proton-catalyzed reactions will not form C1 and C_2 product hydrocarbons because their formation would involve intermediated primary carbenium ions. Non-selective methane formation will occur by the undesired hydrogenolysis reaction catalyzed by transition metals.

The hydrocracking pattern of monocyclic or di-cyclic hydrocarbons is qualitatively very different from that of the *n*-alkanes (Figure 12i.1b).

The M-shaped pattern of the hydrocracking of naphthene is indicative of the paring reaction (Figure 12i.2b) [75]. We will encounter the same reaction in the next subsection on methanol conversion that we also discussed in Chapter 4 (see Figure 4.10, page 138). The paring reaction is the recombination of methyl substituents around a cyclic hydrocarbon cation to give isobutyl and methyl cyclopentane. Because of the paring reaction, butylcyclohexane and decaline give very similar product patterns.

The paring reaction is based on the following three physical organic reaction principles:

- Contraction of a six-ring cation with a tertiary carbon atom to a methyl cyclopentyl cation is kinetically favorable.
- In the contraction of C_6 to C_5 ring cation, methyl substituents recombine with the methyl attached to the tertiary carbon atom through CPC intermediates.
- Exocyclic propylene or isobutylene formation becomes possible by tertiary–tertiary or tertiary–secondary β-C—C cleavage reactions (see Figure 8i.4) with larger substituents attached to the C_5 ring due to hyperconjugation. The unsubstituted C_5 ring will not form for this reason.

This generates the M-shaped product distribution profile because C_4 and C_6 are now the dominant products.

9.3.3.3.4 Hydrocracking of Hydrocarbons: Adsorption Effects

As already indicated in the previous section, the hydrocracking pattern may depend strongly on the micropore structure of the zeolite. When heavy hydrocarbons are converted, the micropore occupation will be high. This may contribute to the selectivity of reaction because the packing of molecules in the micropore may lead to the preferential adsorption of particular molecules. The product pattern becomes controlled by internal equilibration reactions of reaction intermediates, which are strongly affected by interaction with the zeolite wall and co-adsorbed molecules.

Three adsorption properties of long-chain molecules adsorbed at high concentration are of interest [76–78]:

- At high micropore occupation, the adsorption of small molecules is preferred over large molecules owing to entropic considerations.
- Linear molecules tend to stretch in narrow pores. This makes them larger and, hence, at high pore filling their adsorption is suppressed compared to adsorption of smaller molecules.
- In small micropores, the adsorption of branched molecules is suppressed.

This explains the product ratios for the hydrocracking reaction of n-C_{15} for dimethyl butane/n-hexane catalyzed by different zeolites shown in Figure 9.8. A maximum in selectivity toward the bulky branched molecule is found for the zeolite AFI with intermediate pore size.

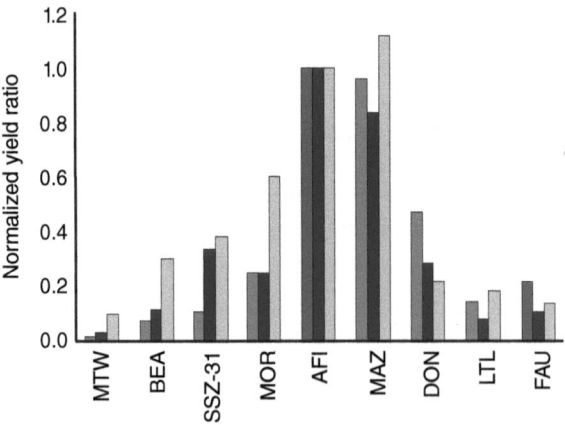

Figure 9.8 The effect of pore size on the hydroisomerization selectivity (ratio of 2,3-dimethylbutane to n-hexane) of the C_6 hydrocracking products formed during n-C_{15} hydroconversion (increasing pore sizes from MTW to FAU). When the pores are small (as with MTW type zeolites), repulsive adsorbent-adsorbate van der Waals interactions impede dimethyl butane (DMB) formation; when pores increase in size, those impeding interactions disappear and inter-adsorbent interactions favor formation of the better-packing DMB; when the pore size increases above 0.74 nm, differences in packing efficiencies disappear because the adsorbates no longer need to line up head-to-tail but can pack in an increasingly more random, liquid-like fashion. (Smit 2008 [35]. Reproduced with permission by American Chemical Society.)

When the pores are small, as in the MTW-type structure, repulsive adsorbate–zeolite wall interactions discourage adsorption of the bulky molecule in favor of adsorption of the linear molecule. When the pores increase in size, these unfavorable interactions disappear. Because of the high loading of the micropores, additional interactions with other adsorbed molecules become important. The effective volume of the bulky but more compact molecules is lower, since the linear molecules must align themselves along the channel. The more favorable entropy of the compact molecules also enables their relatively higher concentration. In contrast, when the pores become very large, the linear molecules can rotate freely and the entropy gain of the bulky molecules disappears.

The product patterns are similar to the relative concentrations of reaction intermediates found in simulations of the adsorption isotherms of mixtures of reaction intermediates at high micropore filling [35].

As seen in the hydroisomerization reaction, the reaction intermediates in the micropores equilibrate among themselves, but are not in equilibrium with the gas or liquid phase. The rate-controlling steps of the reaction are not the proton-activated elementary reaction steps but instead, the rates of product molecule desorption.

9.3.3.4 Methanol Aufbau Chemistry: Alkylation by Methanol

There are two main routes for the formation of chemical or liquid fuels from synthesis gas (see Sections 2.2.3.1; 2.2.3.2). The Fischer–Tropsch process uses catalysts such as Fe, Co, or Ru to convert synthesis gas and produces primarily linear hydrocarbons. The alternative process is based on two reaction steps: synthesis gas is converted to methanol and hydrocarbons are formed in a separate reaction by the dehydration of methanol. This reaction will be discussed here.

This endothermic process is operated at temperatures around 400 °C. It produces aromatics as a dominant product when a 12-ring channel H-ZSM-5 is used but with H-SAPO-34, which has cavities connected through the smaller eight rings, it preferentially produces ethylene or propylene (the MTO process).

There are two competing views on the mechanism of the reaction:

A. C—C bond formation occurs through the intermediate formation of a methoxonium ion or framework-adsorbed ylide anion. Stevens rearrangement of the methoxonium ion gives ethyl ether (Figure 9.9a). The ylide mechanism proposes intermediate methoxy formation and subsequent C—C bond formation through the ylide intermediate (Figure 9.9b) [79, 81]:
B. Catalysis by an organocatalytic intermediate that is formed during the initiation period of the reaction. This intermediate can be formed by oligomerization and ring closure reactions of ethylene or propylene.

Quantum-chemical estimates of the activation energies involved in the Stevens rearrangement or ylide reactions in acidic zeolite cavities give values of over 200 kJ mol^{-1} [82, 83]. This is too high for catalysis at the temperature of the reaction. However, this reaction may be part of the initiation reaction to form the organocatalytic intermediate of proposal B.

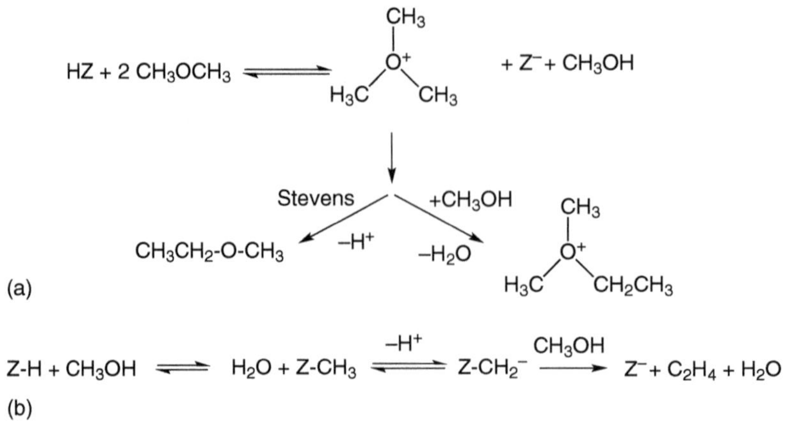

9 Solid Acid Catalysis, Theory and Reaction Mechanisms

$$HZ + 2\,CH_3OCH_3 \rightleftharpoons (CH_3)_3O^+ + Z^- + CH_3OH$$

Stevens ↙ ↓ +CH$_3$OH
−H$^+$ −H$_2$O

$CH_3CH_2\text{-}O\text{-}CH_3$ ← (CH$_3$)$_2$O$^+$CH$_2$CH$_3$

(a)

$$Z\text{-}H + CH_3OH \rightleftharpoons H_2O + Z\text{-}CH_3 \xrightleftharpoons[\ ]{-H^+} Z\text{-}CH_2^- \xrightarrow{CH_3OH} Z^- + C_2H_4 + H_2O$$

(b)

Figure 9.9 Intermediates proposed for C—C bond formation from methanol. (a) The oxonium intermediate mechanisms. (b) The framework ylide mechanism. (After [79, 80].)

Reaction through the initial formation of an organocatalyst intermediate is called the hydrocarbon pool mechanism [84, 85]. It catalyzes C—C bond formation at a lower temperature and proceeds by alkylation of an organocatalytic intermediate by methanol. C—C bond formation of product molecules occurs in consecutive reactions through the paring reaction (see also Figure 4.10) or the dual-cycle reaction of Figure 9.10.

Isotope-labeled experiments [86] of $^{13}CH_3OH$ and ethylene mixtures show no appearance of doubly labeled olefins, which excludes the recombination of two methanol molecules at comparable rates. Alkylation reactions of ethylene and consecutive reactions are more favorable. By co-feeding ^{13}C-labeled methanol with unlabeled benzene, isotope-labeled experiments have demonstrated the incorporation of the ^{13}C label in the benzene ring and the ^{12}C carbon in the product [87, 88]. This experiment supports the validity of the paring mechanism.

Consistent with the hydrocarbon pool proposal, the methanol conversion reaction initially shows low reactivity but after an initiation period conversion starts to proceed with high TOF. The organocatalytic intermediate is thought to be formed during the initiation period. Olefin from methanol impurities in the feed may be responsible for its formation or possibly the slow rearrangement steps according to mechanism A. Coke formation shortens the catalyst life, so the process is operated in a way that is similar to the fluid catalytic cracking process, with regeneration of the catalyst by steam as part of the process operation.

In the paring reaction mechanism, alkylation of the benzene ring by methanol gives a cation that transforms into a substituted cyclopentyl cation by ring contraction. In this step, a new σ C—C bond is formed. Olefin or propylene product formation occurs by consecutive β-C—C cleavage. Later, the C$_6$ aromatic ring is restored.

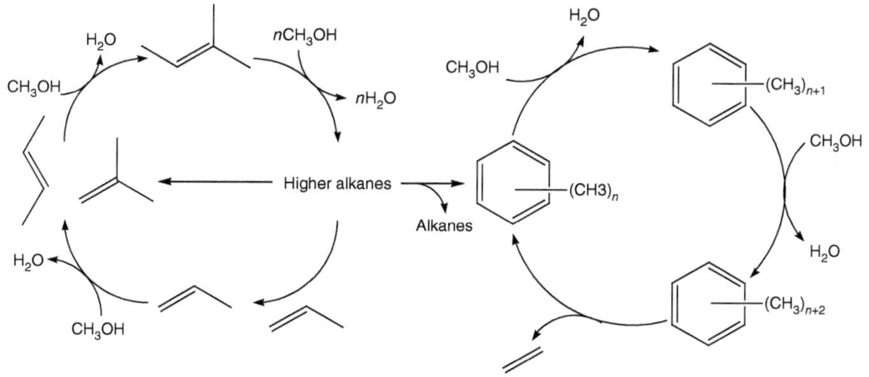

Figure 9.10 Suggested dual-cycle concept for the conversion of methanol over H-ZSM-5. (Olsbye et al. 2012 [89]. Reproduced with permission of John Wiley and Sons.)

Quantum-chemical studies of the elementary reaction steps according to the paring reaction mechanism predict activation energies with a maximum value on the order of 140 kJ mol^{-1} that are more consistent with the reaction temperature [83].

Figure 9.10 shows two coupled reaction cycles for the methanol to ethylene or propylene reactions catalyzed by the aromatic intermediate that acts as catalyst. Side chain alkylation of methylated aromatic rings to produce ethylene competes with isobutyl and propylene formation according to the paring reaction. The side chain mechanism is supported by the large amount of methylated benzyl molecules formed along with the products, especially by H-ZSM-5 [89, 90]. In the side chain reaction, methanol reacts with the methyl substituent benzene to yield an ethyl benzene intermediate, that when activated by a proton decomposes to benzene and ethylene.

The catalytically active intermediates are cationic so that the proton donation affinity of the zeolite protons is important. As illustrated in Figure 9.10, the two reaction cycles are coupled as an alkene cycle and an arene cycle. The reaction is autocatalytic in ethylene and propylene [91].

We will now discuss the alkylation reaction of the aromatic ring by methanol, in more detail.

Insert 13: The Alkylation Reaction of Methanol with Aromatics

In Figure 13i.1, the energetics scheme is given as a function cavity shape and size, as deduced from embedded zeolite cluster DFT calculations [83]. These figures illustrate the large effect of cavity size on the stabilization of cation transition states. In Figure 13i.2, transition state geometries are compared for different zeolite cavities.

(Continued)

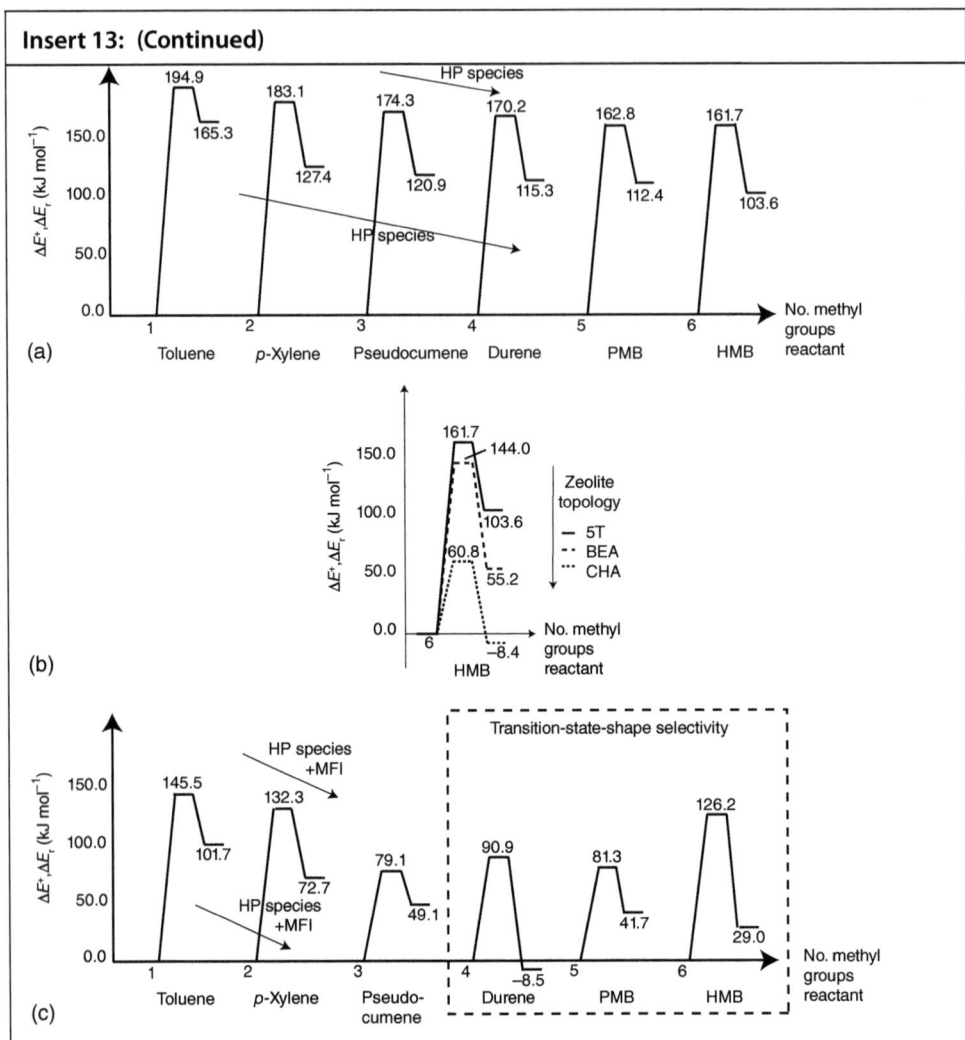

Figure 13i.1 Barrier heights $\Delta E^{\ddagger}$ and reaction energies ΔE_r in kJ mol^{-1} for (a) geminal methylation of different polymethylbenzenes in the 5T cluster, (b) geminal methylation of hexamethylbenzene in the zeolite topologies BEA and CHA, and (c) geminal methylation of different polymethylbenzenes in the space-limiting MFI structure. HP is the hydrocarbon pool species, the organo-catalyst; PMB is pentamethylbenzene; HMB is hexamethylbenzene. The MFI structure is identical to that of ZSM-5. (Lesthaeghe et al. 2007 [92]. Reproduced with permission of John Wiley and Sons.)

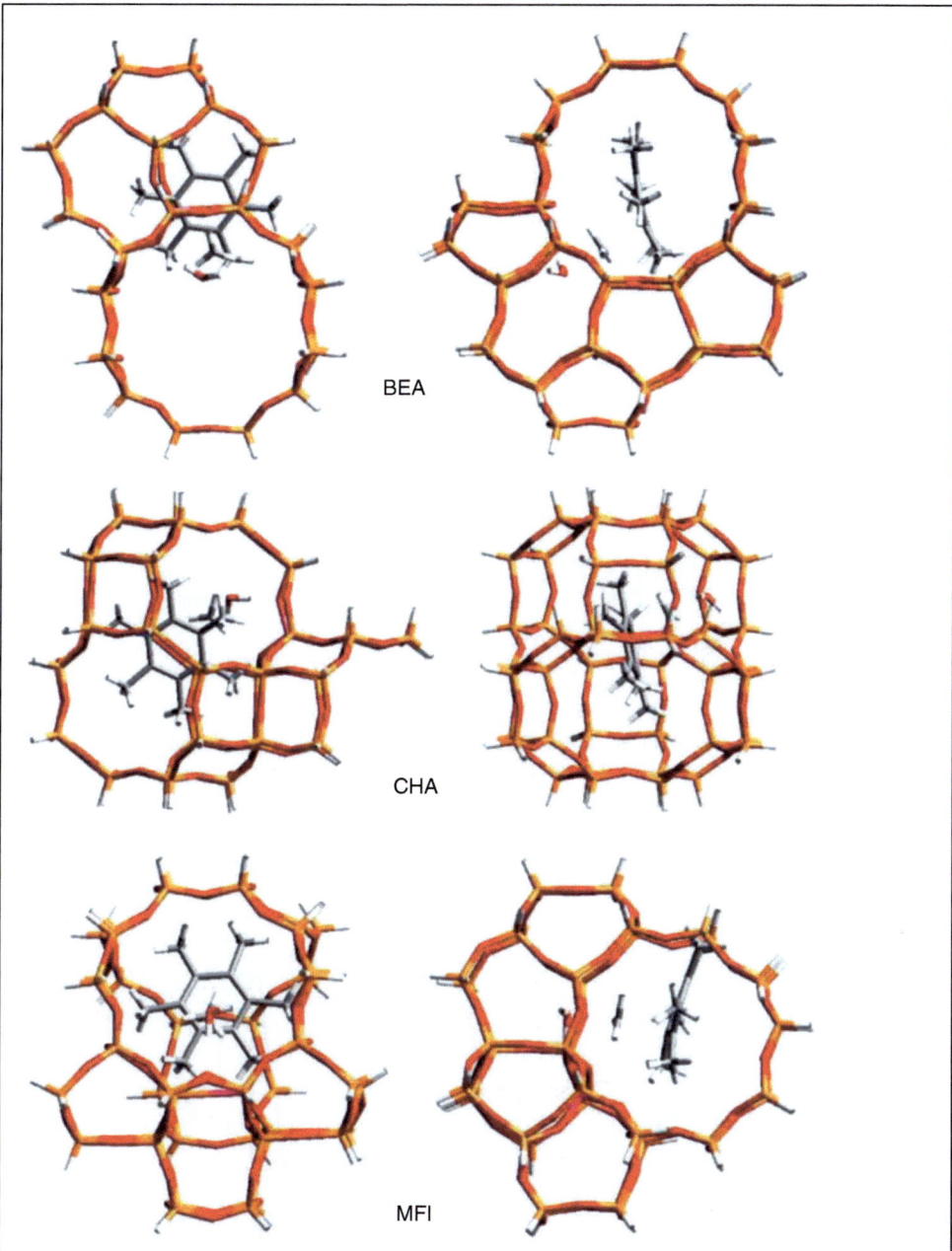

Figure 13i.2 Transition-state geometries for the formation of heptamethylbenzenium in the BEA, CHA, and MFI zeolite topologies. (Lesthaeghe *et al.* 2007 [92]. Reproduced with permission of John Wiley and Sons.)

(Continued)

Insert 13: (Continued)

In Figures 13i.3 and 13i.4, the mechanistic scheme and reaction energy diagram are shown for shape-selective alkylation of toluene with methanol catalyzed by the zeolite mordenite. These figures illustrate the importance of the pre-transition state orientation of molecules. The linear pore structure of mordenite biases the formation of *para*-xylene.

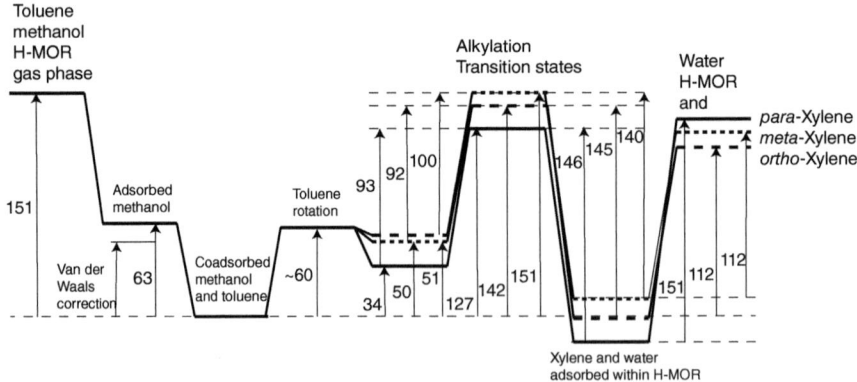

Figure 13i.3 Mechanism of the alkylation reaction of toluene by methanol catalyzed by an acidic zeolite. (Vos *et al.* 2001 [93]. Reproduced with permission of American Chemical Society.)

Figure 13i.4 Zeolite transition-state selectivity. Toluene alkylation with methanol catalyzed by H-MOR showing the energies of the key reaction intermediates. The reaction energy diagrams for *ortho-*, *meta-*, and *para*-xylene are compared. (Vos *et al.* 2001 [93]. Reproduced with permission of American Chemical Society.)

In Figure 13i.1a, the activation energies of methanol alkylation of aromatic molecules with different substituents are compared, as calculated from small zeolitic cluster calculations. As expected, the aromatic molecules become more reactive when they have more substituents. Based on embedded cluster calculations, Figure 13i.1b illustrates the large effect that cavities of different sizes may have on the energy of activation, even in the absence of steric constraint.

While the large cavity of zeolite-β shows only a small difference in activation barrier compared to that of the cluster, the smaller cavity of the chabasite zeolite shows a dramatic reduction in activation barrier. This reduction is due to the stabilizing interaction with the highly polarizable cavity lattice oxygen atoms.

In the ZSM-5 catalyst, steric effects also become important. Figure 13i.1c shows that for molecules without steric constraints, in the presence of the zeolite cavity the activation energy is reduced as compared with the cluster, but when steric constraints become important (such as for durene and hexametal benzene) the activation energy increases. These are the largest molecules that can be accommodated by the H-ZSM-5 micropore structure. This effect on transition state activity is called transition state selectivity.

Transition state selectivity is due to a mismatch of the transition state structure and the micropore size or shape. Shape mismatch selectivity requires reagents to prearrange molecules in the proper orientation to produce stereochemically desired molecules. We will illustrate this for the stereoselective alkylation of toluene by methanol to preferentially produce linear *para*-xylene. This reaction is selective when catalyzed by a zeolite such as mordenite that has a structure with one dimensional pores, which biases the formation of linear molecules.

The stereochemically relevant energy, or the energy cost to prearrange reactants geometrically so they can react with the properly oriented transition state, must be minimized. The activation energies for alkylation in the *ortho*, *para*, or *meta* positions show only very small differences, as illustrated by the respective reaction energy diagrams shown in Figures 13i.3 and 13i.4.

The mechanism of alkylation follows a concerted mechanism. Methanol adsorbed to a zeolitic proton transfers a CH_3^+ species directly to the aromatic ring of toluene. After the addition of the methyl cation to the toluene molecule, the protonated intermediate backdonates a proton to the solid and xylene is formed.

The reaction energy diagram shows that there is an energy cost to pre-orient the methanol and toluene molecules in a pre-transition state configuration that is suitable for the desired substitution of the aromatic ring. The difference in the overall activation energies of the reaction is comparable to the differences in pre-orientation energy.

The pre-orientation of reactants in a stereochemically desirable orientation is a feature that the zeolites share with enzymes, as we will shortly discuss in the next section, where we will compare the mechanistic principles of solid acid catalysis to those of enzymes.

9.4 Acid Catalysis and Hydride Transfer by Enzyme Catalysts

The TOF of a protolytic enzyme is at least 10^4 times faster than the TOF per proton of a solid acid zeolite. This is consistent with its biological function, which requires high reactivity at room temperature. It contrasts with the solid acid catalysts whose operational temperatures are typically around 500 K.

While the mechanisms of catalytic action of proton activating enzymes as lysozyme or chymotrypsine exhibit many similarities with the mechanism of zeolite-catalyzed reactions, the main reasons for the higher reactivity of the enzymes are as follows [94–96]:

- The catalytic reaction proceeds through a multi-point contact of the reaction center with the reacting substrate. In zeolites the protonation site is the only location where these is chemical control. The zeolitic lattice only interacts through van der Waals electrostatic interactions.
- There is optimum match of the reactive groups of the catalyst center with stereochemistry of the reactive molecule through cooperative action and the synergy of well-positioned basic and acidic reactive sites. The organic reacting groups of the enzyme can form labile intermediate covalent bonds that provide a balanced stabilizing force to the reaction intermediates.
- The high local electrostatic field of the enzyme's catalytic center contrasts with the low polarity at the center of the zeolitic proton site. In the enzyme, the high local electrostatic field that results from ionized peptide substituents stabilizes the separation of charge.
- The enzyme peptide backbone is flexible, and can adapt to changes in the shape of the reaction intermediate as the reaction proceeds.

Another major difference between the enzyme and the zeolite is that the enzyme cavity has been uniquely shaped for one particular reaction. As a result, enzyme catalysis is substantially more specific than zeolite catalysis.

In contrast to the inorganic systems used in heterogeneous catalysis, the catalytic enzymes are flexible systems that can achieve conformational adjustments of the peptide framework assisted by changes far from the reactive center. The clefts in the enzyme where the reactive center is located may alter the shape when molecules enter the enzyme. This optimizes the system for reaction with the reagent, whereas the more open structure of the free site will facilitate adsorption. This has been called an "induced fit" by Koshland [94], in contrast to the classical Fischer lock-and-key model of molecular recognition where the site is proposed to be rigid. The rigid site may lead to blocking and hence poisoning of the reaction when the fit of the reagent or product is too exact. A less precise fit is necessary to release product molecules.

In zeolites, the stereochemical fit is less precise and the framework is inflexible. This less-than-exact fit results in a compromise between stereochemical control and the release rate of product molecules. This scenario favors the overall rate of reaction, but it is not necessarily optimal.

References

1 Pauling, L. (1939) *Nature of the Chemical Bond*, Cornell University Press.
2 Chizallet, C., Costentin, G., Lauron-Pernot, H., Krafft, J.M., Bazin, P., Saussey, J., Delbecq, F., Sautet, P., and Che, M. (2006) Role of hydroxyl groups in the basic reactivity of MgO: a theoretical and experimental study. *Oil Gas Sci. Technol. - Rev. l'IFP*, **61**, 479–488. doi: 10.2516/ogst:2006023a.
3 Digne, M., Sautet, P., Raybaud, P., Euzen, P., and Toulhoat, H. (2004) Use of DFT to achieve a rational understanding of acid-basic properties of gamma-alumina surfaces. *J. Catal.*, **226**, 54–68.
4 Digne, M., Sautet, P., Raybaud, P., Euzen, P., and Toulhoat, H. (2002) Hydroxyl groups on gamma-alumina surfaces: a DFT study. *J. Catal.*, **211**, 1–5.
5 van Santen, R.A., Tranca, I., and Hensen, E.J.M. (2015) Theory of surface chemistry and reactivity of reducible oxides. *Catal. Today*, **244**, 63–84, http://linkinghub.elsevier.com/retrieve/pii/S092058611400488X (accessed 22 September 2016).
6 Post, M.F.M., Huizinga, T., Emeis, C.A., Nanne, J.M. and Stork, W.H.J. (1989) An infrared and catalytic study of isomorphous substitution in pentasil zeolites, *Stud. Surf. Sci. and Catal.*, **46**, 365–375.
7 Kazansky, V.B., Serykh, A.I., Semmer-Herledan, V., and Fraissard, J. (2003) Intensities of OH IR stretching bands as a measure of the intrinsic acidity of bridging hydroxyl groups in zeolites. *Phys. Chem. Chem. Phys.*, **5**, 966–969. doi: 10.1039/b212331m.
8 Kustov, L.M., Kazanskii, V.B., Beran, S., Kubelkova, L., and Jiru, P. (1987) Adsorption of carbon monoxide on ZSM-5 zeolites: infrared spectroscopic study and quantum-chemical calculations. *J. Phys. Chem.*, **91**, 5247–5251. doi: 10.1021/j100304a023.
9 Zecchina, A., Buzzoni, R., Bordiga, S., Geobaldo, F., Scarano, D., Ricchiardi, G., and Spoto, G. (1995) Host- guest interactions in zeolite cavities. *Stud. Surf. Sci. Catal.*, **97**, 213–222.
10 Lercher, J.A., Gründling, C., and Eder-Mirth, G. (1996) Infrared studies of the surface acidity of oxides and zeolites using adsorbed probe molecules. *Catal. Today*, **27**, 353–376, http://linkinghub.elsevier.com/retrieve/pii/0920586195002480 (accessed 6 February 2015).
11 Kotrla, J. and Kubelková, L. (1995) Activation of reactants by hydroxyl groups of solid acids. An FTIR study. *Stud. Surf. Sci. Catal.*, **94**, 509–516.
12 Pelmenschikov, A.G., van Wolput, J.H.M.C., Jaenchen, J., and van Santen, R.A. (1995) (A,B,C) triplet of infrared OH bands of zeolitic H-complexes. *J. Phys. Chem.*, **99**, 3612–3617. doi: 10.1021/j100011a031.
13 van Santen, R.A. (1994) Fano coupling in perturbed solid acid hydroxyls. *Recl. Trav. Chim. Pays-Bas*, **113**, 423–425, http://www.scopus.com/inward/record.url?eid=2-s2.0-43949154324&partnerID=40&md5=4cf801d9b77a3980a601a169bab0d03b (accessed 22 September 2016).
14 Brouwer, D.M. and Hogeveen, H. (1972) *Progress in Physical Organic Chemistry*, vol. **9**, p. 179, John Wiley & Sons.

15 Teunissen, E.H., Van Santen, R.A., Jansen, A.P.J., and Van Duijneveldt, F.B. (1993) Ammonium in zeolites: coordination and solvation effects. *J. Phys. Chem.*, **97**, 203–210. doi: 10.1021/j100103a035.

16 Haag, W.O. and Dessau, R.M. (1984) Duality of mechanism in acid catalyzed paraffin cracking. Proceedings of the 8th International Congress on Catalysis, vol. 2, p. 305.

17 Catlow, C.R.A., van Santen, R.A., and Smit, B. (2004) *Computer Modelling of Microporous Materials*, Elsevier.

18 Teunissen, E.H., Jansen, A.P.J., van Santen, R.A., Orlando, R., and Dovesi, R. (1994) Adsorption energies of NH3 and NH+4 in zeolites corrected for the long-range electrostatic potential of the crystal. *J. Chem. Phys.*, **101**, 5865. doi: 10.1063/1.467303.

19 Teunissen, E.H., Roetti, C., Pisani, C., de Man, A.J.M., Jansen, A.P.J., Orlando, R., van Santen, R.A., and Dovesi, R. (1999) Proton transfer in zeolites: a comparison between cluster and crystal calculations. *Model. Simul. Mater. Sci. Eng.*, **2**, 921–932.

20 Hensen, E.J.M., Pidko, E.A., Rane, N., and Van Santen, R.A. (2007) Water-promoted hydrocarbon activation catalyzed by binuclear gallium sites in ZSM-5 zeolite. *Angew. Chem. Int. Ed.*, **46**, 7273–7276.

21 Teunissen, E.H., Van Duijneveldt, F.B., and Van Santen, R.A. (1992) Interaction of ammonia with a zeolitic proton: ab initio quantum-chemical cluster calculations. *J. Phys. Chem.*, **96**, 366–371. doi: 10.1021/j100180a068.

22 Blaszkowski, S.R. and van Santen, R.A. (1995) Density functional theory calculations of the activation of methanol by a broensted zeolitic proton. *J. Phys. Chem.*, **99**, 11728–11738. doi: 10.1021/j100030a017.

23 Bonn, M., Bakker, H.J., Kleyn, A.W., and van Santen, R.A. (1996) Dynamics of infrared photodissociation of methanol clusters in zeolites and in solution. *J. Phys. Chem.*, **100**, 15301–15304.

24 Kramer, G.J., van Santen, R.A., Emeis, C.A., and Nowak, A.K. (1993) Understanding the acid behaviour of zeolites from theory and experiment. *Nature*, **363**, 529–531. doi: 10.1038/363529a0.

25 Esteves, P.M., Nascimento, M.A.C., and Mota, C.J.A. (1999) Reactivity of alkanes on zeolites: a theoretical ab initio study of the H/H exchange. *J. Phys. Chem. B*, **103**, 10417–10420. doi: 10.1021/jp990555k.

26 Beck, L.W., Xu, T., Nicholas, J.B., and Haw, J.F. (1995) Kinetic NMR and density functional study of benzene H/D exchange in zeolites, the most simple aromatic substitution. *J. Am. Chem. Soc.*, **117**, 11594–11595. doi: 10.1021/ja00151a031.

27 Almutairi, S.M.T., Mezari, B., Filonenko, G.A., Magusin, P.C.M.M., Rigutto, M.S., Pidko, E.A., and Hensen, E.J.M. (2013) Influence of extraframework aluminum on the Brønsted acidity and catalytic reactivity of faujasite zeolite. *ChemCatChem*, **5**, 452–466. doi: 10.1002/cctc.201200612.

28 Guisnet, M., Avendano, F., Bearez, C., and Chevalier, F. (1985) Evidence that butane disproportionation demands adjacent acid sites. *J. Chem. Soc., Chem. Commun.*, **6**, 336. doi: 10.1039/c39850000336.

29 Olah, G.A., Lecture, N., and Angeles, L. (1995) My search for carbocations and their role in chemistry (Nobel Lecture). *Angew. Chem. Int. Ed. Engl.*, **34**, 1393–1405. doi: 10.1002/anie.199513931.

30 Kotrel, S., Knözinger, H., and Gates, B.C. (2000) The Haag–Dessau mechanism of protolytic cracking of alkanes. *Microporous Mesoporous Mater.*, **35–36**, 11–20, http://linkinghub.elsevier.com/retrieve/pii/S1387181199002048 (accessed 7 February 2015).

31 Franklin, J.L. (1953) Calculation of the heats of formation of gaseous free radicals and ions. *J. Chem. Phys.*, **21**, 2029, http://scitation.aip.org/content/aip/journal/jcp/21/11/10.1063/1.1698737 (Accessed 7 February 2015).

32 Franklin, J.L. and Field, F.H. (1953) The energies of strained carbonium ions. *J. Chem. Phys.*, **21**, 550. doi: 10.1063/1.1698944.

33 Rozanska, X. et al. (2002) A periodic structure density functional theory study of propylene chemisorption in acidic chabazite: effect of zeolite structure relaxation. *J.Phys. Chem. B*, **106**, 3248–3254.

34 Blaszkowski, S.R. and van Santen, R.A. (1997) Theoretical study of the mechanism of surface methoxy and dimethyl ether formation from methanol catalyzed by zeolitic protons. *J. Phys. Chem. B*, **101**, 2292–2305. doi: 10.1021/jp962006+.

35 Smit, B. and Maessen, T.L.M. Molecular Simulations of Zeolites; Adsorption, Diffusion and Shape Selectiviy, *Chem. Rev.*, **108**(10), 4125–4184.

36 Houžvička, J. and Ponec, V. (1997) Skeletal isomerization of n-butene. *Catal. Rev.*, **39**, 319–344. doi: 10.1080/01614949708007099.

37 Blaszkowski, S.R., Nascimento, M.A.C., and van Santen, R.A. (1996) Activation of C–H and C–C bonds by an acidic zeolite: a density functional study. *J. Phys. Chem.*, **100**, 3463–3472. doi: 10.1021/jp9523231.

38 Angyan, J.G., Parsons, D., and Jeanvoine, Y. (2001) in *Theoretical Aspects of Heterogeneous Catalysis* (ed. M.A.C. Nascimento), Kluwer, p. 101.

39 Tranca, D.C., Hansen, N., Swisher, J.A., Smit, B., and Keil, F.J. (2012) Combined density functional theory and Monte Carlo analysis of monomolecular cracking of light alkanes over H-ZSM-5. *J. Phys. Chem. C*, **116**, 23408–23417.

40 Gounder, R. and Iglesia, E. (2009) Catalytic consequences of spatial constraints and acid site location for monomolecular alkane activation on zeolites. *J. Am. Chem. Soc.*, **131**, 1958–1971, http://www.ncbi.nlm.nih.gov/pubmed/19146372 (accessed 7 February 2015).

41 Redfern, P.C., Zapol, P., Sternberg, M., Adiga, S.P., Zygmunt, S.A., and Curtiss, L.A. (2006) Quantum chemical study of mechanisms for oxidative dehydrogenation of propane on vanadium oxide. *J. Phys. Chem. B*, **110**, 8363–8371, http://www.ncbi.nlm.nih.gov/pubmed/16623521 (accessed 9 March 2014).

42 Pine, H. (1977) *Base catalyzed Reactions of Hydrocarbons and Related Compounds*, Academic Press.

43 Gates, B.C., Katzer, J.R., and Schuit, G.C.A. (1979) *Chemistry of Catalytic Processes*, Wiley-Interscience.

44 Demuth, T., Rozanska, X., Benco, L., Hafner, J., van Santen, R.A., and Toulhoat, H. (2003) Catalytic isomerization of 2-pentene in H-ZSM-22—A

DFT investigation. *J. Catal.*, **214**, 68–77, http://linkinghub.elsevier.com/retrieve/pii/S002195170200074X (accessed 7 February 2015).

45 Meng, X., Xu, C., Li, L., and Gao, J. (2010) Kinetic study of catalytic pyrolysis of C4 hydrocarbons on a modified ZSM-5 zeolite catalyst. *Energy Fuels*, **24**, 6233–6238. doi: 10.1021/ef100943u.

46 Geus, J.W. and van Dillen, A.J. (1999) in *Environmental Catalysis* (eds F.J.J.C. Janssen and R.A. van Santen), Imperial College Press, pp. 151–169.

47 de Jong, K.P., Mooiweer, H.H., Buglass, J.G., and Maarsen, P.K. (1997) Activation and deactivation of the zeolite Ferrierite for olefin conversions. *Stud. Surf. Sci. Catal.*, **111**, 127–138.

48 Andy, P. (1998) Skeletal isomerization of n-Butenes II. Composition, mode of formation, and influence of coke deposits on the reaction mechanism. *J. Catal.*, **173**, 322–332, http://linkinghub.elsevier.com/retrieve/pii/S002195179791945X (accessed 7 February 2015).

49 Feller, A., Guzman, A., Zuazo, I., and Lercher, J.A. (2004) On the mechanism of catalyzed isobutane/butene alkylation by zeolites. *J. Catal.*, **224**, 80–93.

50 Mazar, M.N., Al-Hashimi, S., Cococcioni, M., and Bhan, A. (2013) β-scission of olefins on acidic zeolites: a periodic PBE-D study in H-ZSM-5. *J. Phys. Chem. C*, **117**, 23609–23620. doi: 10.1021/jp403504n.

51 Weitkamp, J. (2012) Catalytic hydrocracking-mechanisms and versatility of the process. *ChemCatChem*, **4**, 292–306. doi: 10.1002/cctc.201100315.

52 Yaluris, G. (1995) Isobutane cracking over Y-zeolites I. Development of a kinetic-model. *J. Catal.*, **153**, 54–64. doi: 10.1006/jcat.1995.1107.

53 Schmerling, L. (1945) The mechanism of the alkylation of paraffins. *J. Am. Chem. Soc.*, **67**, 1778–1783. doi: 10.1021/ja01226a048.

54 Schmerling, L. (1946) The mechanism of the alkylation of paraffins. II. Alkylation of isobutane with propene, 1-Butene and 2-Butene. *J. Am. Chem. Soc.*, **68**, 275–281. doi: 10.1021/ja01206a038.

55 Weitkamp, J. and Traa, Y. (1999) Isobutane/butene alkylation on solid catalysts. Where do we stand? *Catal. Today*, **49**, 193–199. doi: 10.1021/ja01156a075.

56 Yates, J.G. (1996) Effects of temperature and pressure on gas–solid fluidization. *Chem. Eng. Sci.*, **51**, 167–205.

57 Corma, A. and Martínez, A. (1993) Chemistry, catalysts, and processes for isoparaffin–olefin alkylation: actual situation and future trends. *Catal. Rev.*, **35**, 483–570.

58 Otvos, J.W., Stevenson, D.P., Wagner, C.D., and Beeck, O. (1951) The behavior of isobutane in concentrated sulfuric acid 1. *J. Am. Chem. Soc.*, **73**, 5741–5746. doi: 10.1021/ja01156a075.

59 Kazansky, V.B. (2000) Solvation of protons and the strength of superacids. *Top. Catal.*, **12**, 55–60. doi: 10.1023/A:1027264317915.

60 Kazansky, V.B. (2001) Solvation effects in catalytic transformations of olefins in sulfuric acid. *Catal. Rev.*, **43**, 199–232. doi: 10.1081/CR-100107477.

61 Smit, B. and Maesen, T.L.M. (2008) Molecular simulations of zeolites: adsorption, diffusion, and shape selectivity. *Chem. Rev.*, **108**, 4125–4184.

62 Haag, W.O., Lago, R.M., and Weisz, P.B. (1984) The active site of acidic aluminosilicate catalysts. *Nature*, **309**, 589–591.

63 Narbeshuber, T.F., Vinek, H., and Lercher, J.A. (1995) Monomolecular conversion of light alkanes over H-ZSM-5. *J. Catal.*, **157**, 388–395, http://linkinghub.elsevier.com/retrieve/pii/S0021951785713048 (accessed 22 September 2016).

64 De Moor, B.A., Reyniers, M.-F., Gobin, O.C., Lercher, J.A., and Marin, G.B. (2011) Adsorption of C2–C8 n-alkanes in zeolites †. *J. Phys. Chem. C*, **115**, 1204–1219. doi: 10.1021/jp106536m.

65 Bates, S.P. and Van Santen, R.A. (1998) The molecular basis of zeolite catalysis: a review of theoretical simulations. *Adv. Catal.*, **42**, 1–114.

66 Anderson, B.G., van Santen, R.A., and de Jong, A.M. (1999) Positrons as in situ probes of transient phenomena. *Top. Catal.*, **8**, 125–131.

67 Anderson, B.G., van Santen, R.A., and van Ijzendoorn, L.J. (1997) Positron emission imaging in catalysis. *Appl. Catal., A*, **160**, 125–138.

68 Eder, F. (1996) PhD thesis, Thermodynamic Siting of alkane adsorption in molecular sieves TU Twente.

69 Denayer, J.F., Baron, G.V., Martens, J.A., and Jacobs, P.A. (1998) Chromatographic study of adsorption of n-alkanes on zeolites at high temperatures. *J. Phys. Chem. B*, **102**, 3077–3081.

70 Noordhoek, N.J., van Ijzendoorn, L.J., Anderson, B.G., de Gauw, F.J., van Santen, R.A., and de Voigt, M.J. (1998) Mass transfer of alkanes in zeolite packed-bed reactors studied with positron emission profiling. 2. Modeling. *Ind. Eng. Chem. Res.*, **37**, 825–833.

71 Van De Runstraat, A., Van Grondelle, J., and Van Santen, R.A. (1997) Microkinetics modeling of the hydroisomerization of n-hexane. *Ind. Eng. Chem. Res.*, **36**, 3116–3125, http://www.scopus.com/inward/record.url?eid=2-s2.0-0031197359&partnerID=40&md5=4a7d567d0eff0cfe0bc0fbf09fef0169 (accessed 22 September 2016).

72 Spivey, J.J. and Bryant, P.A. (1982) Hydroisomerization of n-C5 and n-C6 mixtures on zeolite catalysts. *Ind. Eng. Chem. Process Des. Dev.*, **21**, 750–760. doi: 10.1021/i200019a034.

73 de Gauw, F.J.M.M., van Grondelle, J., and van Santen, R.A. (2002) The intrinsic kinetics of n-hexane hydroisomerization catalyzed by platinum-loaded solid-acid catalysts. *J. Catal.*, **206**, 295–304. doi: 10.1006/jcat.2001.3479.

74 Weitkamp, J. (1982) Isomerization of long-chain n-alkanes on a Pt/CaY zeolite catalyst. *Ind. Eng. Chem. Prod. Res. Dev.*, **21**, 550–558. doi: 10.1021/i300008a008.

75 Egan, C.J., Langlois, G.E., White, R.J., and Egan, J. (1962) Selective hydrocracking of C9 - to C12 -alkylcyclohexanes on acidic catalysts. Evidence for the paring reaction. *J. Am. Chem. Soc.*, **84**, 1204–1212. doi: 10.1021/ja00866a028.

76 van Santen, R.A. and Neurock, M. (2006) *Molecular Heterogenous Catalysis*, Wiley-VCH Verlag GmbH.

77 Schenk, M., Calero, S., Maesen, T.L.M., Van Benthem, L.L., Verbeek, M.G., and Smit, B. (2002) Understanding zeolite catalysis: Inverse shape selectivity revised. *Angew. Chem. Int. Ed.*, **41**, 2499–2502.

78 Talbot, J. (1997) Analysis of adsorption selectivity in a one-dimensional model system. *AIChE J.*, **43**, 2471–2478. doi: 10.1002/aic.690431010.

79 Olah, G.A., Doggweiler, H., Felberg, J.D., Frohlich, S., Grdina, M.J., Karpeles, R., Keumi, T., Inaba, S., Ip, W.M., Lammertsma, K. et al. (1984) Onium Ylide chemistry. 1. Bifunctional acid–base-catalyzed conversion of heterosubstituted methanes into ethylene and derived hydrocarbons. The onium ylide mechanism of the C1 to C2 conversion. *J. Am. Chem. Soc.*, **106**, 2143–2149. doi: 10.1021/ja00319a039.

80 Hutchings, G.J., Gottschalk, F., Hall, M.V.M., and Hunter, R. (1987) Hydrocarbon formation from methylating agents over the zeolite catalyst ZSM-5. Comments on the mechanism of carbon? Carbon bond and methane formation. *J. Chem. Soc., Faraday Trans. 1*, **83**, 571.

81 van den Berg, J.P., Wolthuizen, J.P., and van Hooff, J.H.C. (1980) in *Proceedings 5th International Zeolite Conference* (ed. L.V. Rees), Heyden, London, pp. 649–660.

82 Blaszkowski, S.R. and Van Santen, R.A. (1997) Theoretical study of C–C bond formation in the methanol-to- gasoline process. *J. Am. Chem. Soc.*, **119**, 5020–5027.

83 Lesthaeghe, D., Van Speybroeck, V., Marin, G.B., and Waroquier, M. (2006) Understanding the failure of direct C–C coupling in the zeolite-catalyzed methanol-to-olefin process. *Angew. Chem. Int. Ed.*, **45**, 1714–1719.

84 Svelle, S., Olsbye, U., Joensen, F., and Bjørgen, M. (2007) Conversion of methanol to alkenes over medium- and large-pore acidic zeolites: steric manipulation of the reaction intermediates governs the ethene/propene product selectivity. *J. Phys. Chem. C*, **111**, 17981–17984.

85 Dahl, I.M. and Kolboe, S. (1994) On the reaction mechanism for hydrocarbon formation from methanol over SAPO-34: I. Isotopic labeling studies of the co-reaction of ethene and methanol. *J. Catal.*, **149**, 458–464.

86 Dessau, R.M. (1982) On the mechanism of methanol conversion to hydrocarbons over HZSM-5. *J. Catal.*, **78**, 136–141, http://www.sciencedirect.com/science/article/pii/0021951782902925 (accessed 29 March 2015).

87 Westgård Erichsen, M., Svelle, S., and Olsbye, U. (2013) H-SAPO-5 as methanol-to-olefins (MTO) model catalyst: towards elucidating the effects of acid strength. *J. Catal.*, **298**, 94–101, http://www.sciencedirect.com/science/article/pii/S0021951712003557 (accessed 29 March 2015).

88 Bjørgen, M. (2004) The methanol-to-hydrocarbons reaction: insight into the reaction mechanism from [12C]benzene and [13C]methanol coreactions over zeolite H-beta. *J. Catal.*, **221**, 1–10, http://www.sciencedirect.com/science/article/pii/S0021951703002847 (accessed 29 March 2015).

89 Olsbye, U., Svelle, S., Bjrgen, M., Beato, P., Janssens, T.V.W., Joensen, F., Bordiga, S., and Lillerud, K.P. (2012) Conversion of methanol to hydrocarbons: how zeolite cavity and pore size controls product selectivity. *Angew. Chem. Int. Ed.*, **51**, 5810–5831.

90 Mole, T., Brett, G., and Seddon, D. (1983) Conversion of methanol to hydrocarbons over ZSM-5 zeolite: an examination of the role of aromatic hydrocarbons using 13carbon- and deuterium-labeled feeds. *J. Catal.*, **84**, 435–445, http://www.sciencedirect.com/science/article/pii/0021951783900143 (accessed 29 March 2015).

91 Bjørgen, M., Joensen, F., Lillerud, K.-P., Olsbye, U., and Svelle, S. (2009) The mechanisms of ethene and propene formation from methanol over high silica H-ZSM-5 and H-beta. *Catal. Today*, **142**, 90–97, http://www.sciencedirect.com/science/article/pii/S0920586109000340 (accessed 10 March 2015).

92 Lesthaeghe, D., De Sterck, B., Van Speybroeck, V., Marin, G.B., and Waroquier, M. (2007) Zeolite shape-selectivity in the gem-methylation of aromatic hydrocarbons. *Angew. Chem. Int. Ed.*, **46**, 1311–1314.

93 Vos, A.M., Rozanska, X., Schoonheydt, R.A., van Santen, R.A., Hutschka, F., and Hafner, J. (2001) A theoretical study of the alkylation reaction of toluene with methanol catalyzed by acidic mordenite. *J. Am. Chem. Soc.*, **123**, 2799–2809.

94 Koshland, D.E. (1959) in *The Enzymes* (eds P.D. Boyer and H. Hardy), Academic Press, New York, p. 305.

95 Rose, I.A. (1997) Restructuring the active site of fumarase for the fumarate to malate reaction. *Biochemistry*, **36**, 12346–12354.

96 Rose, I.A. (1998) How fumarase recycles after the malate to Fumarate reaction. Insights into the reaction mechanism. *Biochemistry*, **37**, 17651–17658.

10

Zeolitic Non-Redox and Redox Catalysis, Lewis Acid Catalysis

10.1 Introduction

In the zeolite catalyst, Lewis acidity can originate from cations located in both the micropores and in the zeolite framework. In this chapter, we will discuss the catalytic reactivity of such reducible and non-reducible cations or oxycationic clusters.

The first section will focus on Lewis acidic systems based on extra-framework cations. The Lewis acidic cations generate an electrostatic field that depends on their size, charge, and local chemical environment.

We will start with a short section describing their physical characterization, followed by a discussion of molecular recognition by these systems. The cations are often present as oxycations, which are useful in C—H bond activation as well as in selective oxidation.

The final two subsections will present a description of catalysis by Lewis acidic cations that are part of a zeolite framework. These catalysts, called *single-site catalysts*, are important for the production of fine and commodity chemicals. We will also discuss their application to convert biomolecules such as the sugars. These processes rely on selective oxidation, dehydration, and hydride transfer reactions. This catalysis has mechanistic similarities with the enzymes, which we will also describe.

For additional reading, refer to the book, *Design and Applications of Single-Site Heterogeneous Catalysts*, Thomas [1].

10.2 Non-Reducible Cations; The Electrostatic Field

The main chemical interaction between an isolated, non-reducible cation and an adsorbate is electrostatic. The electrostatic field around a cation can be characterized experimentally by using a molecule such as CO or methane (see Insert 1) as a probe. The interaction with the probe molecule increases with the cation charge and decreases with the cation radius.

The adsorption energy of CO to an Na$^+$ cation is 30 kJ mol^{-1} [2], which is similar to adsorption on a zeolitic proton [2]. The larger radius of the Na$^+$ cation compensates for the smaller effective charge on the zeolite proton. On the Ca^{2+} cation which has a higher charge, the adsorption energy of CO is 50 kJ mol^{-1} [3].

For a polarizable, non-polar probe molecule the electrostatic interaction energy is proportional to

$$E_{\text{Pol}} = -\frac{1}{2}\alpha_{\text{P}} \cdot \frac{q^2}{R^4} \tag{10.1}$$

α_{P} is the polarizability of the probe molecule, q is the charge of the cation, and R is the distance between the cation center and adsorbate.

In contrast with the case of adsorption to a transition metal, the vibrational frequencies of CO increase with interaction strength. The molecular orbitals rehybridize due their polarization in the electrostatic field of the cation. The CO molecule interacts preferentially with the C atom directed toward the cation. The electrostatic field at the C atom location is stronger than at the O atom, which causes the orbital energies of the C and O atoms to become more similar. This increases the bonding nature of the occupied σ-type molecular orbitals, since in CO the 5σ orbital localized on the C atom is largely antibonding (see Chapter 7, Figure 7i.1). As a result, the energy of the C—O chemical bond increases. This is illustrated in both Figure 1i.1 and Table 1i.1.

Insert 1: Infrared Adsorption Spectra of Probe Molecules in Contact with Cations

An experimental infrared spectrum of CO adsorbed to a zeolite cation is shown in Figure 1i.1. This figure shows the upwards shifted frequency of CO when in contact with the cation through the C atom, but a downwards shift when adsorb through the O atom. The band at 2140 cm^{-1} is that of free CO.

Table 1i.1 compares the downward frequency shift of the zeolite OH bands in contact with CO to the upwards frequency shift of CO when adsorbed to the Na cations in different adsorption environments. The high frequency band of CO has also been studied on sites comparable to those of Figure 1i.1 on ZSM-5 ion-exchanged with different alkali cations. This band shifts from 2188 cm^{-1} for Li to 2157 cm^{-1} for Cs [6], illustrating the decreasing interaction energy with CO as cation charge increases.

Figure 1i.2 shows the infrared spectra of methane in contact with a zeolite Zn^{2+} cation. For this non-polar molecule, infrared active transitions of the free molecule are not possible. When adsorbed to a cation polarization of the molecule. The charged cation induces absorption intensity.

10.2 Non-Reducible Cations; The Electrostatic Field | 431

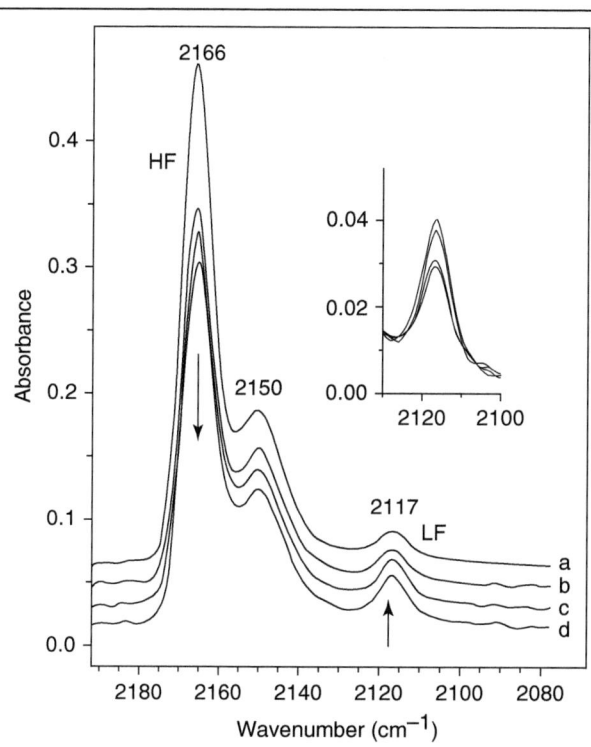

Figure 1i.1 FTIR spectra of CO (ca. 2 Torr) adsorbed on K-ZSM-5 at varying temperatures: (a) 130 K; (b) 152 K; (c) 166 K; (d) 171 K. For clarity, the spectra have been arbitrarily offset on the vertical scale. The inset shows an expanded view of the low-frequency band: spectra (a)–(d) from bottom to top. The band at 2166 cm^{-1} is due to CO adsorbed through its carbon atom, the band at 2117 cm^{-1} is due to CO adsorbed through its O atom. The energy difference between the two states of adsorption is only 3.2 kJ mol^{-1}. The assignment of the band to 2150 cm^{-1} is not clear, but is possibly due to K cations adsorbed at less accessible adsorption sites or at occluded liquid free CO. (Manoilova et al. 2001 [4]. Reproduced with permission of Elsevier.)

Table 1i.1 Frequency shifts (cm^{-1}) caused by the interaction of CO with OH-Groups and L-sites (L, Lewis acid).

Material	v_{OH}	Δv_{OH} after CO adsorption (downward shift)	v_{CO}	Δv_{CO} (upward shift) from CO gas (v_{CO} = 2143 cm^{-1})
Na-ZSM-5 (silanols)	3747	90		
Na-ZSM-5 (extraframework)	3673	138		
Na-ZSM-5 (Na$^+$)			2178	35
Na-ZSM-5 (strong L-sites in extra framework species)			2230	87

Source: Bordiga et al. 1992 [5]. Reproduced with permission of Elsevier

(Continued)

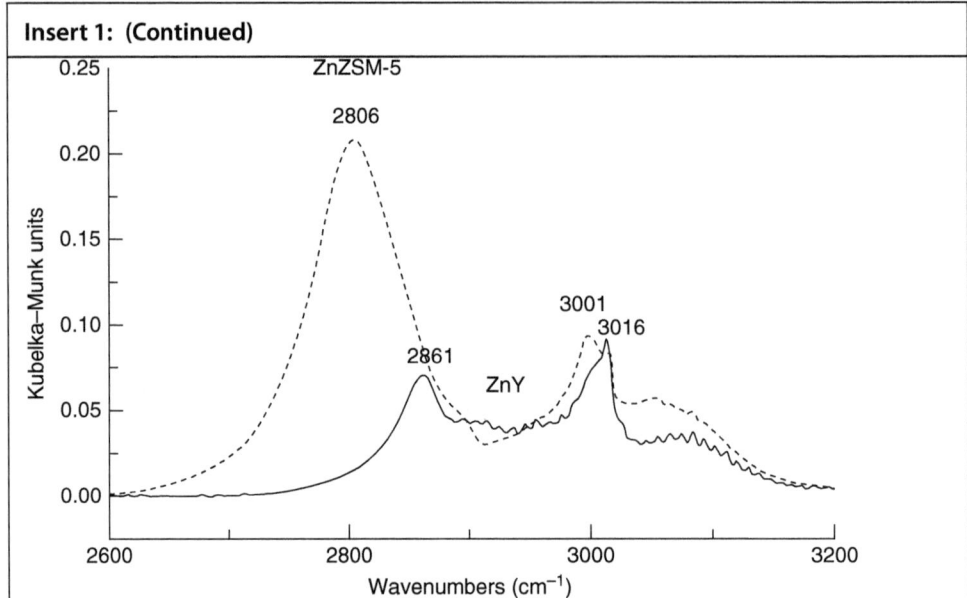

Figure 1i.2 Methane adsorption by a zinc-modified zeolite. Comparison between a low Al/Si framework ratio in ZSM-5 and a high Al/Si ratio in zeolite Y. The bands around 3000 cm^{-1} are from zeolitic protons, the bands at 2861 and 2806 cm^{-1} are due respectively to weakly and strongly interacting methane with a Zn^{2+} cation in ZnY the Zn^{2+} cation is partly shielded by the zeolite lattice O atoms. (Kazansky et al. 2004 [7]. Reproduced with permission of Elsevier.)

In contrast to CO, the methane molecule has no infrared active modes in the gas phase because it is apolar. It will become infrared active only in an external electrostatic field. In Figure 1i.2, the adsorption spectra for methane adsorbed to a Zn^{2+} ion located in the micropore of a low Al concentration zeolite ZSM-5 is compared with the adsorption spectra for a high Al concentration zeolite Y. The IR intensities around 3000 cm^{-1} are the frequencies of zeolitic protons left after ion exchange in the zeolite. The high-intensity peak at 2800 cm^{-1} and the less-disturbed peak at the higher frequency of 2861 cm^{-1} are due to methane adsorbed to the Zn^{2+} ions. The oxygen atoms around the Zn^{2+} cation in zeolite Y are more basic (with higher Al content) than those of the ZSM-5 lattice (with lower Al content). The electrostatic field due to the Zn^{2+} cations is dampened more by the negative charges in the surrounding O atoms in zeolite Y than in ZSM-5. Due to the higher effective charge of the Zn cation in ZSM-5, not only does the intensity of the methane infrared adsorption increase but the C—H bond frequency is also more weakened compared to the lower interaction case.

10.3 Catalysis with Non-Framework Non-Reducible Cations

10.3.1 Alkali and Earth Alkali Ions

According to the classical theory of zeolite reactivity, cation-exchanged zeolites should be considered solid electrolytes. The high electrostatic charge of the cations will stabilize the separation of charge in the zeolite cavity [8]. In this section, we will discuss systems exhibiting such a high internal local electrostatic field.

Two important factors determine the reactivity of these systems:

A. Molecular recognition by the structural fit between the reacting substrate molecules and cations located at the extra-framework cation sites.
B. Basicity of the zeolite framework as the result of the concentration of trivalent cations such as Al^{3+} in the zeolite framework, the concentration and type and location of extra-framework cations.

The relative positioning and size of the cations situated in a particular zeolite cavity leads to reactivity that is controlled by molecular recognition similar to that described by the Fischer lock and key concept initially introduced for enzymes. We will illustrate molecular recognition by examining the photochemical reaction of an alkene with O_2 [9]. The basicity effect and electrolytic behavior, exhibited by the decomposition of N_2O_4 will be discussed next.

Molecular recognition is responsable for the photochemical oxidation of dimethyl butene by an earth alkali-exchanged zeolite Y [10]. The photo-oxidation of the bulky molecule dimethyl butene with O_2 is initiated by a charge transfer from the occupied alkene π bond to an unoccupied O_2 π^* orbital, which generates a reactive O_2^- intermediate. The O_2^- anion initiates a radical oxidation reaction chain, which leads to oxygenated products.

The photo-excitation process requires precise orientation of the reacting molecule because the electron transfer between the two molecules requires orbital overlap between the adsorbed O_2 π^* and alkene π orbitals. Photo oxidation is sensitive to the size of the cations only takes place efficiently in the zeolite exchanged with Ca^{2+} cations. The cations adsorbed in the faujasite structure supercavity provide adsorption positions for the O_2 and dimethyl butene molecules (see Figure 2i.1).

Insert 2: Molecular Recognition by Cation-Exchanged Zeolites

Photo-Oxidation

Figure 2i.1 shows the relative positions of butene and O_2 adsorbed to Ca^{2+} exchanged zeolite Y that initiates photo-oxidation. Table 2i.2 compares the calculated photo-excitation probabilities for molecules adsorbed to cations in the presence and absence of a zeolite cavity, but with the same positioning as in the zeolite.

(Continued)

Insert 2: (Continued)

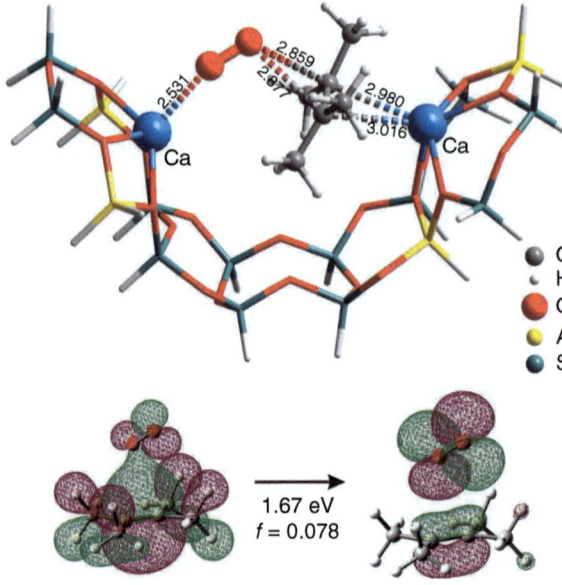

Figure 2i.1 Dimethylbutene and oxygen co-adsorption on a cluster that is part of the Y-zeolite supercage. The HOMO–LUMO transition of the butene–O_2 complex is shown. (Pidko et al. 2006 [10], reproduced with permission of American Chemical Society.)

N_2O_4 Dissociation

Figure 2i.2 presents the relative positioning of NO^- and NO_3^{-1} fragments in different alkali-exchanged zeolites. Table 2i.2 gives the energies of dissociative adsorption of N_2O_4. Figure 2i.3 gives the basicity of the respective zeolite lattices as determined spectroscopically.

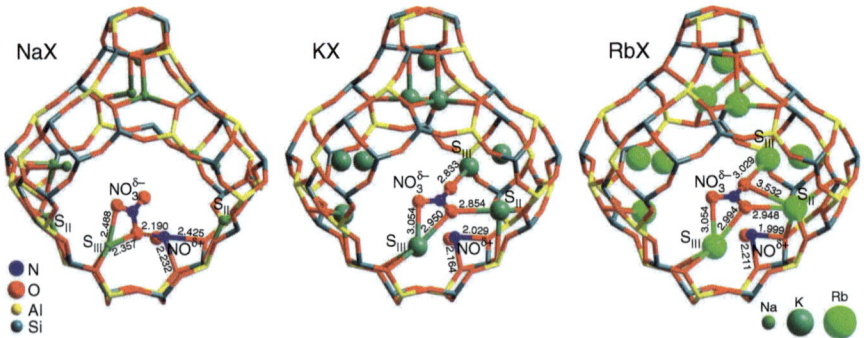

Figure 2i.2 Structures of the $NO_3^-\cdots NO^+$ ion pair formed by N_2O_4 disproportionation in alkali-exchanged zeolite X. (Schoonheydt et al. 2012 [15]. Reproduced with permission of Royal Society of Chemistry.)

Table 2i.1 The DMB/O_2 complex adsorbed in a NaY zeolite exhibiting a charge-transfer absorption band with onset at ~700 nm (DMB is dimethylbutene).

	R(DMB-O_2), Å			E, eV (nm)	f	Without cations	
	C_1—O_1/C_1—O_2	C_2—O_1/C_2—O_2	H—O			E, eV (nm)	f
DMB and O_2 by Mg	3.807/4.534	3.957/4.322	2.657/3.123	0.7953 (1560)	0.0071	1.2490 (992)	0.0047
DMB and O_2 by Ca	2.860/3.616	2.877/3.429	3.170/3.080	1.6636 (745)	0.0776	1.8992 (653)	0.0572
DMB and O_2 by Sr	3.783/4.623	3.754/4.447	3.370/4.027 3.372/3.495	1.2119 (1023)	0.0086	1.4056 (882)	0.0071
DMB-O_2 optimized	3.299/4.004	3.234/3.566	3.519/2.581	—	—	1.6570 (748)	0.0384

R measures bond distance, E is the excitation energy, and f is the calculated UV adsorption intensity. The excitation energies and adsorption intensities are compared in calculations with and without the presence of the zeolite framework and cations. This illustrates that only the relative positions of the molecules matter.

Source: (Pidko *et al.* 2006 [10]. Reproduced with permission of American Chemical Society.)

Insert 2: (Continued)

Table 2i.2 Calculated adsorption energies of N_2O_4 and comparison of experimental and calculated NO^+ stretching frequencies.

	Na^+-X	Na^+-Y	K^+-X	K^+-Y	Rb^+-X	Rb^+-Y
ΔE (kJ mol^{-1}) [11, 12]	−69	−30	−91	−40	−82	−50
$\nu(NO^+)$ (cm^{-1}); DFT [11, 12]	2017	2007	1991	1986	1972	1967
$\nu(NO^+)$ (cm^{-1}); experiment [13, 14]	1976	2090	1927	1973	1903	1968

Source: Schoonheydt et al. 2012 [15]. Reproduced with permission of Royal Society of Chemistry.

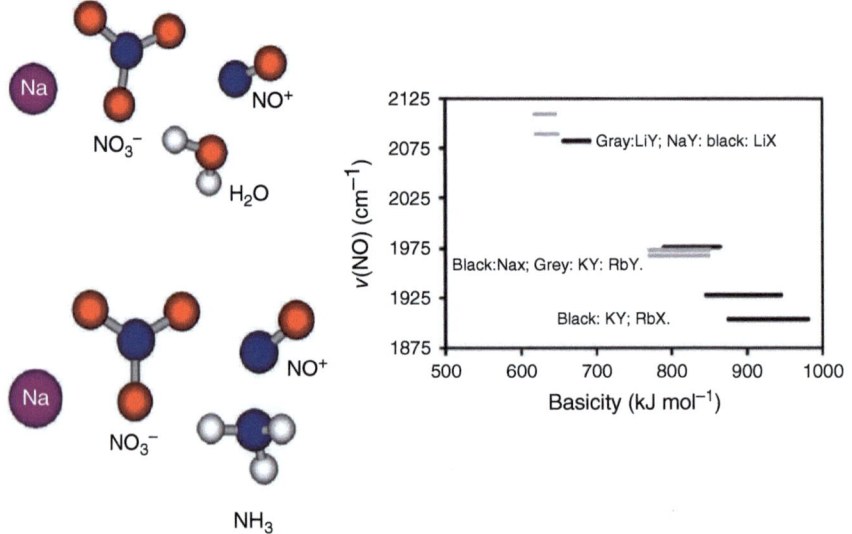

Figure 2i.3 (Left) Configurations of the complexes of N_2O_2 with H_2O and Na^+ (upper part) and with Na^+ and NH_3 (lower part). (Right) NO^+ stretching frequencies (cm^{-1}) as a function of basicity. (Schoonheydt et al. 2012 [15]. Reproduced with permission of Royal Society of Chemistry.)

The smaller Mg cations leave a separation between the two molecules, with no overlap of the π bond of butene with O_2. The larger Ca ions leave less distance between the two adsorbed molecules so that there is a substantial overlap, and the calculated adsorption intensity is a factor of 10 larger (see Table 2i.1). The essential role of the cations is twofold: their high cationic charge provides a large electrostatic field that apolar molecules will polarize and adsorb, and their size and relative position will orient the molecules in the proper positions.

The reactivity of N_2O_4 adsorbed to alkali-exchanged X and Y zeolites can be measured by the exothermicity of the charge separation reaction when it decomposes into NO_3^- and NO^+. The common feature of both zeolites is the high Al content of their frameworks. In zeolite Y the Al/Si ratio is around 0.3, and in zeolite X it equals 1. This difference in Al/Si ratio makes the oxygen ions of the zeolite X framework more basic.

N_2O_4 decomposition is the reaction originally studied by Rabo and Gajda [8] to analyze the solid electrolyte properties of ion-exchanged zeolites. This reaction is endothermic in the gas phase, but becomes highly exothermic in the zeolite.

The surprising observation is that this reaction has a higher exothermicity for the larger cations, which according to the Eq. (10.1) should produce a smaller electrostatic field stabilization (see Table 2i.2). Again, the relative orientation and the size of the cavity formed by the ion-exchanged cations determines reactivity [see Fig. 2i.2]. The NO_3^- anion interacts strongly with the alkali cations and NO^+ adsorbs to the framework oxygen atoms [15].

The vibrational frequencies of NO^+ can be used to measure the Lewis basicity of the oxygen framework atoms which can be derived by comparing the NO^+ frequencies in model calculations of dissociated N_2O_4 in contact with an alkali cation in which the zeolite framework is replaced by oxygen and ammonia (Figure 2i.3). A measure of the basicity of H_2O or NH_3 is their proton affinity. The NO^+ vibrational frequencies obtained from the zeolite calculations can then be calibrated with a basicity energy.

The basicity of the zeolite framework depends not only on the framework Al/Si ratio, but also on the cation size. Basicity tends to increase with increasing size of the alkali cation, but it is divided into subcategories for different framework compositions and cation sizes. The basicity of the framework oxygen atoms is between the basicity of H_2O and ammonia. The strong dependence of oxygen framework basicity on exchanged cation size also implies that the proton donation affinity zeolites will be sensitive to exchange concentration and the type of cation. The basicity of framework oxygen atoms increases when the protons are partially substituted by inorganic cations, which decreases the proton donation affinity. A lower Al/Si ration in the zeolite framework reduces the framework basicity and hence increases proton donation affinity.

10.3.2 Non Redox Oxycationic Clusters

Gallium- or zinc-exchanged zeolites can activate light alkanes and catalyze dehydrogenation to light alkenes. When Ga or Zn is used instead of a transition metal in bifunctional catalysts, these catalysts convert light hydrocarbons to higher hydrocarbons.

The reaction is in essence an "Aufbau" reaction of light alkane molecules to aromatics. Zeolites containing Ga are commercially applied to this reaction in the UOP Cyclar process which converts liquefied petroleum gas (LPG) directly into a liquid aromatic product in a single processing step.

The initial C—H bond cleavage reaction catalyzed by the oxycationic complex is a heterolytic reaction that generates an alkyl group adsorbed to Ga or Zn and a proton that is accepted by an O atom [see Fig. 2i.3]. The alkyl intermediate is converted into alkene by the β-CH decomposition of this complex. At the high temperature of the reaction, zeolitic protons oligomerize the alkenes and aromatics formed by hydride-transfer reactions. Zn or Ga are beneficial for the dehydrogenation of alkane because they selectively catalyze dehydrogenation and prevent competitive C—C bond activation (which would occur for the activation of alkanes by zeolitic protons).

C—H bond activation only becomes possible when it is assisted by reaction with an oxygen atom that is part of a cationic cluster containing Ga or Zn [16]. The bare cations are not catalytically active because the homolytic oxidative addition reaction by dissociative C—H bond cleavage with the Ga^+ ion changes its valency to Ga^{3+}. This reaction is highly endothermic. In addition, a change in the valency of the Zn^{2+} would pose a high energy demand.

Insert 3: Heterolytic C—H Bond by Ga_xO_y Complexes

Figure 3i.1 illustrates the different states of a Ga cation in the zeolite micropore. Figure 3i.2 compares activation energies of C—H bond cleavage and H_2 formation. Heterolytic C—H bond dissociation is schematically indicated. Figure 3i.3 shows the determination of the optimum Ga_xO_y complex according to the Sabatier principle.

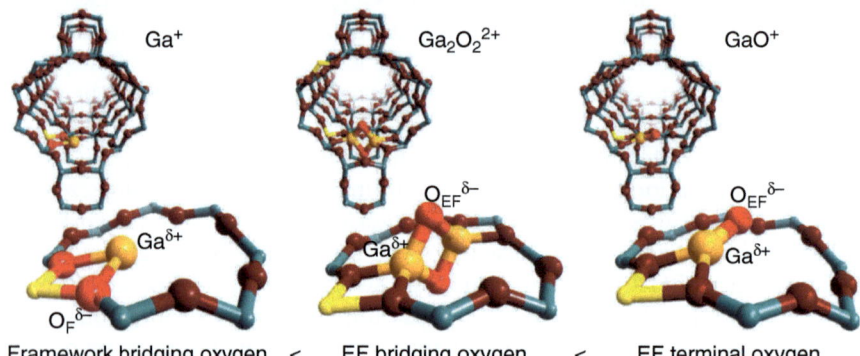

Framework bridging oxygen < EF bridging oxygen < EF terminal oxygen

Figure 3i.1 Different cationic states of Ga in the zeolite mordenite. (After [17].)

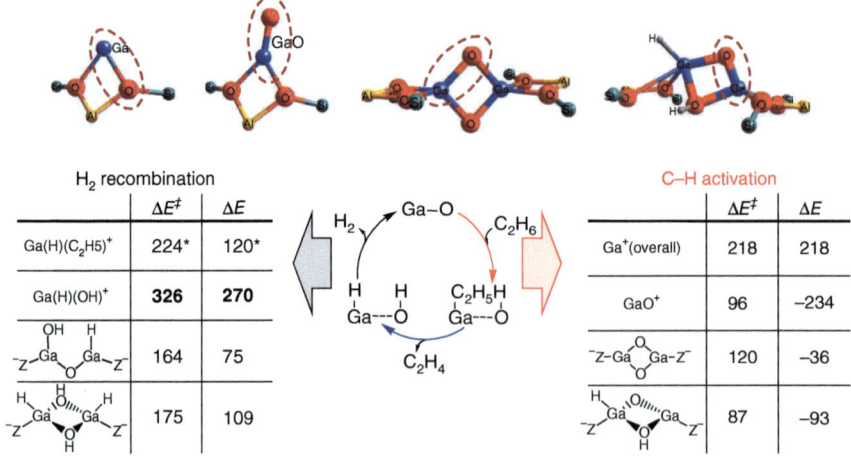

*One-step H_2 and C_2H_4 desorption for Ga^+

Figure 3i.2 C—H bond activation and H_2 formation by the Ga cation and different oxycationic complexes of Ga. $\Delta E\pm$ activation energy (kJ/mol), ΔE reaction energy (kJ/mol). $\Delta E^{\ddagger}$ activation energy (kJ/mol), ΔE reaction energy (kJ/mol). (Pidko et al. 2009 [17]. Reproduced with permission of Royal Society of Chemistry.)

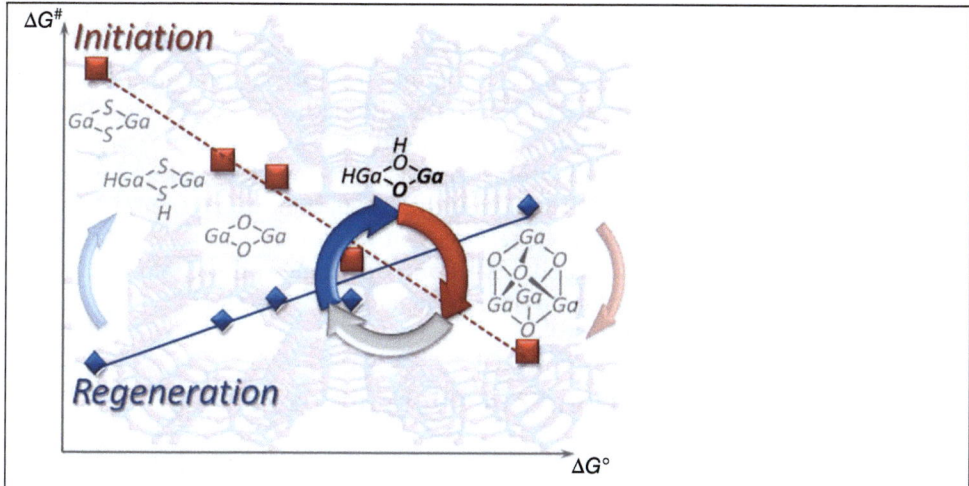

Figure 3i.3 The optimum oxycationic complex according to the Sabatier principle procedure. (Pidko and Van Santen 2009 [18]. Reproduced with permission of American Chemical Society.)

In the oxycationic cluster, the Ga cation is in its Ga^{3+} valency state and its valency does not change during the reaction. When ethane is activated, the negatively charged ethyl adsorbs to the Ga^{3+} cation and H^+ adsorbs to the O ion. Figure 3i.2 compares the activation energies of C—H bond activation as well as H_2 formation for ethane on clusters located in mordenite micropores for four different states of Ga: Ga^+, GaO^+, $Ga_2O_2^{2+}$, and $HGa_2O_2^{2+}$. There is a large decrease in the activation energy for C—H bond activation when activation by Ga^+ and GaO^+ are compared.

β-C—H bond cleavage of the formally negatively charged ethyl group generates ethylene and an adsorbed hydride. In this reaction, the Ga—C bond with ethyl is replaced by a GaH bond. The activation energy of approximately 170 kJ mol^{-1} for this reaction varies little with $Ga_xO_y^{n+}$ complex changes, as long as the valency of Ga remains 3+. Other oxycationic complexes that contain a Zn^{2+} ion have a very similar activation energy for this reaction. The reaction cycle is closed by heterolytic recombination where H^+ attaches to O and the H^- ion generated by β-C—H bond cleavage attaches to Ga^{3+} to give H_2.

The activation energy for C—H bond activation increases strongly when the coordination number of the O atom that accepts an H atom increases. This is to be expected based on the bond order conservation rule. The O—H bond that is attached to the O atom weakens with the increasing stability of the O atom. This weakened O—H bond interaction increases the activation energy of C—H bond activation, but decreases the activation energy for H_2 recombination.

The O—H bond energy is a reactivity performance parameter of the oxycation complex. As indicated in Figure 3i.3, there is an optimum value of O—H bond energy where the three elementary rate constants compete and there is no single rate-controlling step. This gives the reactivity performance descriptor the value of the maximum rate of the reaction.

The catalyst deactivates by reduction of the Ga oxycationic complex. H_2O formation is a reductive elimination reaction in which Ga^{3+} reduces to Ga^+. This competes with the dehydrogenation reaction, so a small amount of H_2O is added to the feed to increase the stability of the catalyst [19].

When oxycationic complexes are present in the zeolite micropores, there are essentially two kinds of oxygen atoms: the extra-framework oxygen atoms that are part of the zeolite framework and those that are part of the oxycationic clusters. This section illustrates the high reactivity of the extra-framework oxygen that participates in heterolytic bond activation. The framework oxygen atoms have low reactivity and, therefore, do not participate in the reaction.

10.4 Catalysis by Non-Framework Redox Complexes

We will now discuss redox catalysis by oxycationic clusters located in the zeolite micropores.

The two reactions that we will discuss are the reduction of NO by hydrocarbons and the decomposition of N_2O by promoted zeolites, both relevant to diesel engine exhaust catalysis and the Panov reaction. In exhaust catalysis, the use of zeolites is advantageous because the longer residence times of the molecules in the zeolite micropores increase the relative rate of consecutive reactions that lead to total N_2O combustion. In the Panov reaction [20, 21], catalyzed by extra-framework Fe in ZSM-5, phenol is selectively produced by the oxidation of benzene by N_2O. This reaction has been studied as a way to mitigate the production of N_2O, a greenhouse gas, which is a coproduct of nylon manufacturing. In the Panov system, the zeolite stabilizes the unique Fe^{2+} cationic state that is optimal for the reaction.

A great challenge is the activation of methane to methanol. Direct insertion of O into a CH bond is a reaction analogous to the benzene hydroxylation reaction, but so far has only been realized with low selectivity. We will also discuss methane activation by catalytic systems related to the Panov catalyst.

The redox chemistry of heterogeneous catalytic Panov-type reactions appears to be general. The section will be concluded with a comparison of the mechanism of CH hydroxylation by related homogeneous and enzyme-catalyzed oxidation reactions.

10.4.1 NO Reduction Catalysis: Selective Catalytic Reduction

Zeolites loaded with metal ions such as Cu, Co, or Pt and Pd have potential as catalysts for the reduction of NO at the oxidative condition found in diesel engine exhaust [22, 23].

This selective catalytic reduction (SCR) reaction of NO can be conducted using organic molecules or ammonia. When ammonia is used, the mechanism of the reduction reaction catalyzed by the zeolite cations is analogous to the mechanism for ammonia oxidation catalyzed by Pt^{2+} ions in zeolites, which we discussed in Section 8.3.1.3. A practical source of ammonia is urea that readily hydrolyzes to ammonia and CO_2.

Catalysis in the diesel exhaust occurs at lean burn conditions, which implies an excess of oxygen [see also Chapter 2, Figure 8i.1, page 47]. At these conditions, the metal centers will be cations or oxycations. Although these catalytically active centers will not readily dissociate NO, they can activate NO to give the auto redox products N_2O and NO_2. A reaction scheme for this reaction catalyzed by Cu ions as proposed by [24] is shown below:

$$[Cu-O-Cu]^{2+} + NO \rightarrow [Cu-[\,]-Cu]^{2+} + NO_2 \quad (10.2)$$

$$[Cu-[\,]-Cu]^{2+} + 2NO \rightarrow [Cu-O-Cu]^{2+} + N_2O \quad (10.3)$$

$$\text{net: } 3NO \rightarrow N_2O + NO_2$$

NO_2 and N_2O will be reduced in consecutive reactions.

Figure 10.1 illustrates how NO_2 can react with the hydrocarbons to give nitro- or nitrile organic compounds. As shown for the conversion of nitromethane [25], these compounds can be readily decomposed to give ammonia. Reaction intermediates for this reaction are shown in formule (4.1)–(4.3):

$$-CH_2-NO_2 \rightarrow -CH=NO(OH) \rightarrow -C=N=O \quad (4.1)$$

$$-C=N=O \rightarrow -N=C=O \rightarrow -NH_2 \rightarrow NH_3 \quad (4.2)$$

$$-CH_2-NO \rightarrow -CH=N(OH) \rightarrow -C\equiv N \rightarrow -C=N=O \quad (4.3)$$

NH_3 can be a product formed by the tautomerization of nitromethane to the corresponding oxime followed by dehydration to a nitrile N-oxide (Eq. (4.1)). This nitrile isomerizes to an isocyanate before yielding a primary amine and NH_3 by hydrolysis (Eq. (4.2)). The possibility of forming NH_3 from the reaction of organo-nitrile N-oxides species was confirmed by Obuchi *et al.* [26]. The same authors proposed that the organo-nitrile N-oxides were formed from organo-nitroso compounds via enol and cyanide formation (Eq. (4.3)).

10.4.2 N_2O Decomposition Catalysis

The direct decomposition reaction of N_2O into N_2 and O_2 is exothermic ($\Delta H_r = -163\,kJ\,mol^{-1}$). When catalyzed by the surface of a reducible oxide, the reaction proceeds differently than when it is catalyzed by a reducible cation exchange in a zeolite.

The rate of N_2O decomposition to adsorbed O and gas phase N_2 usually has a low activation barrier and O_2 molecule formation is the rate-controlling step. This can occur by the recombination of atomic oxygen, as is common on reducible oxides, or by a consecutive Eley–Rideal reaction where N_2O reacts in a subsequent step with previously adsorbed O to give O_2 and N_2. The second mechanism is representative of N_2O decomposition on cation-exchanged zeolites. The difference between the two systems is that on the oxide surface oxygen atoms are generated independently but on a single cation N_2O will decompose on a reaction center where an oxygen atom produced from a first reaction with N_2O is already present.

When reducible oxides are used in the first case O inhibits reaction, while in the second case of catalysis by zeolitic cations there is little inhibition by O_2 [27] although water may have an inhibitive effect [28].

In oxidation catalysis there are two competing models of the chemical reaction, the Langmuir–Hinshelwood type reaction versus the Eley–Rideal type reaction.

In the Langmuir–Hinshelwood mechanism, chemical bond formation occurs by the recombination of two intermediates that are both adsorbed to the catalyst surface. The recombination of the two oxygen atoms to give O_2 is an example of this mechanism. In the Eley–Rideal mechanism, the new chemical bond is formed by reaction of one intermediate adsorbed to the surface

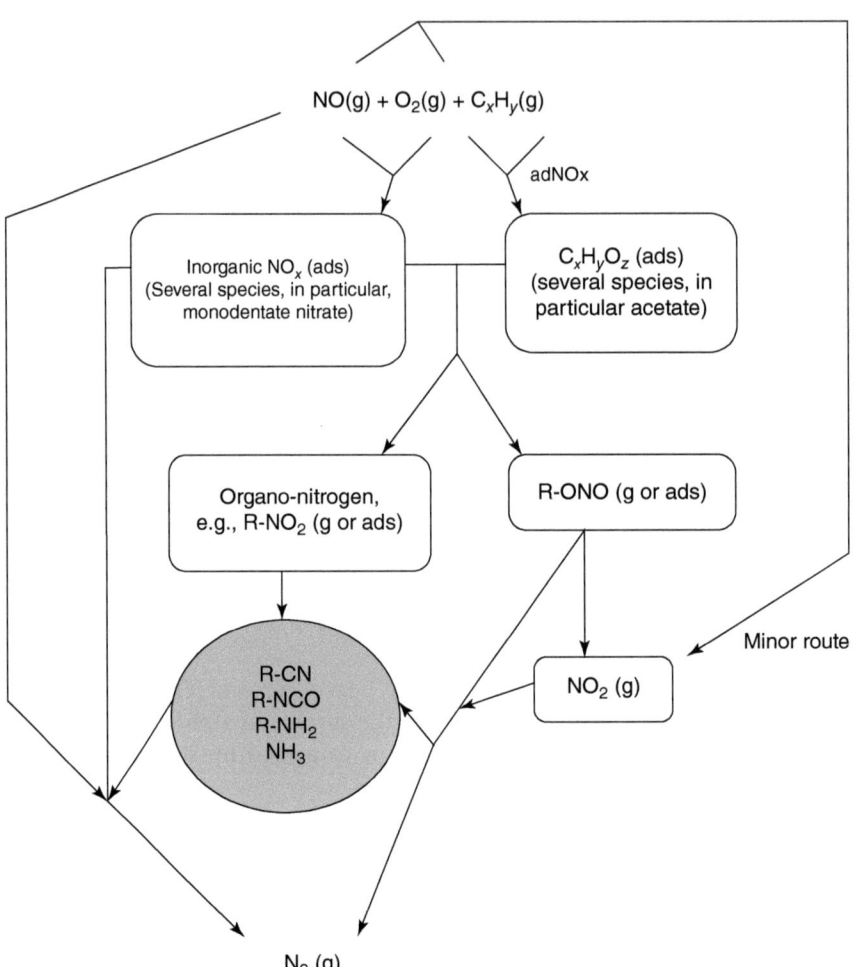

Figure 10.1 Simplified reaction scheme of the C_3H_6-SCR of NO over oxide catalysts giving the nature of the different species likely to be involved. It is proposed that the reduction to N_2 occurs through the reaction of oxidized and reduced (species in shaded circle) nitrogen compounds. (Beutel 1996 [24]. Reproduced with permission of American Chemical Society.)

with a molecule from the gas phase. The reaction of a surface O adatom and N_2O, which gives O_2 and N_2, is an example of an Eley–Rideal elementary reaction step.

10.4.3 Selective Oxidation of Benzene and Methane: The Panov Reaction

Fe cations exchanged in zeolites are catalysts for the Panov reaction that converts benzene into phenol with N_2O [21].

Insert 4: The Mechanism of the Panov Reaction

Figure 4i.1 schematically illustrates the mechanism of the N_2O–benzene reaction to phenol, which produces N_2 as a coproduct. The reaction center of an isolated Fe^{2+} cation gets a formal state of 4+ when an oxygen atom attaches. Figure 4i.2 gives a comparison of a computed reaction energy diagram of the reaction for an isolated cation and a dimeric Fe cationic cluster. In Figure 4i.3, the schematic for the deactivating phenolate forming reaction is shown.

Figure 4i.1 Catalytic cycle for benzene to phenol oxidation with N_2O over extra-framework iron (Fe_{EF}) sites. (Li et al. 2011 [29]. Reproduced with permission of Elsevier.)

(Continued)

Insert 4: (Continued)

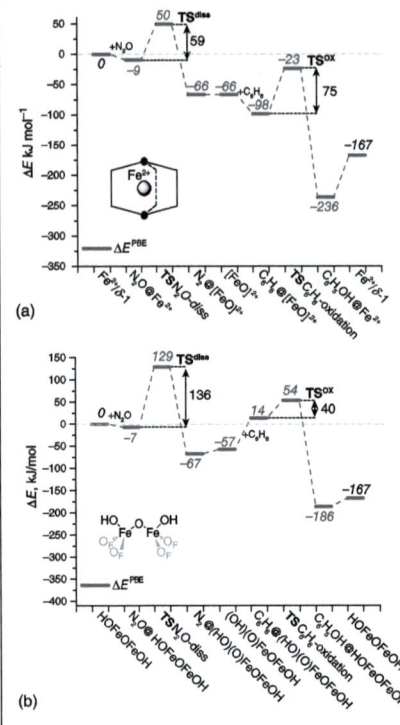

Figure 4i.2 (a) Reaction energy diagram for benzene oxidation to phenol by N_2O on mononuclear Fe^{2+} in ZSM-5. (b) Reaction energy diagram for benzene oxidation to phenol by 2O over a binuclear [HOFe(μ-O)FeOH]$^{2+}$ complex in ZSM-5. (After [30].)

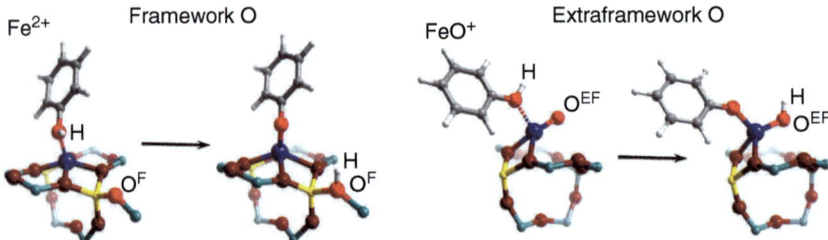

Figure 4i.3 Schematic representation of the deactivating phenolate-forming reaction step. When an O atom of phenol is co-adsorbed to a Fe ion, the proton of phenol is transferred to the O atom. The phenolate will induce consecutive oligomerization reaction steps. (After [30].)

In a zeolite, Fe cations are present as isolated cations, oxy cationic clusters or as oxide particles. The Fe cation can have a 2+ or 3+ charge.

There is strong experimental evidence that the selective oxidation of benzene is catalyzed by single-site Fe^{2+} cations [31]. DFT calculations indicate that the rate of reaction does not vary strongly when isolated cations or oxycations are compared (see Figure 4i.2). In this reaction the electrophilic oxygen atom from the decomposition of N_2O attaches to the Fe cation. When Fe^{2+} is used, its formal valency becomes 4+. The adsorbed oxygen atom inserts into the C—H bond of benzene by the oxene mechanism [see Fig. 4i.1].

The rate of phenol formation does not depend on the state of Fe. The isolated Fe^{2+} cation is preferred to avoid rapid deactivation of the catalytic system. When additional O atoms are present, as is the case for Fe oxycations, this atom co-adsorbed to the Fe ion readily reacts with the phenol molecule to give adsorbed phenolate and OH. This initiates the oligomerization of phenol. The resulting polyether will deactivate the catalyst. This deactivating proton transfer from phenol to O is schematically illustrated in Figure 4i.3. The reactivity of an extra-framework oxygen atom that is part of Fe oxycation is substantially higher than the reactivity of the oxygen atoms that are part of the zeolite framework. Only Fe^{2+} ions are present as isolated cations, is the basis to their unique reactivity.

The deactivation reaction cannot readily be prevented, since Fe oxycations will always be present. But bimodal porous zeolitic systems that also contain mesopores show substantially increased stability because access to the micropores by reactants can be maintained longer [30].

Methane conversion to methanol by the Fe-containing Panov catalyst [32] and by Cu-promoted ZSM-5 zeolites [33] has also been investigated.

The selectivity of this reaction is low. When the reaction is executed at low temperature and the catalyst is pre-oxidized with N_2O (as in Fe-promoted ZSM-5) or O_2 (as in Cu-promoted ZSM-5) methanol remains adsorbed to the Fe or Cu cations and can be recovered only by extraction with a suitable solvent.

Methanol is oxidized in the presence of an oxidant at higher temperatures and, therefore, there is no methanol selectivity. When a water solution is used and catalysts are activated by H_2O_2, a stable turnover of methane to methanol occurs, but with low selectivity [34].

The Cu catalyst is well understood. Its active state in ZSM-5 is a $[Cu-O-Cu]^{2+}$ intermediate. In the next section, we will discuss the similarity of this structure with the methane mono oxygenase (MMO) enzyme (see Figures 5i.4 and 5i.5) [36, 37].

The O atom in $[Cu-O-Cu]^{2+}$ is an electrophilic oxygen atom with radical properties: the formally 2^- oxygen ion is compensated by a 4+ charge! In Chapter 11, we will introduce related oxygen radical states that are attached to high valency Co^{4+} or Co^{5+} cations at the surface of Co_3O_4 electrocatalysts for the evolution of oxygen from water.

10.5 Related Homogeneous and Enzyme Oxidation Catalysts

Just as a Panov related system can convert methane to methanol by H_2O_2 in water, this reaction can be conducted using Fe^{2+} cations containing water solutions that are also activated by H_2O_2 and by biocatalysts. We will compare the mechanisms of these systems to show their remarkable similarity. We will begin with a short introduction of the systems, followed by a discussion of their mechanisms.

Insert 5: Methane Activation by Single and Dual Site Fe and Cu Catalysts

Figures 5i.1 and 5i.2 concern biocatalysis. The two competitive mechanisms of the methane to methanol reaction are presented. In Figures 5i.1a, b, the two corresponding reaction schemes are shown and, in Figures 5i.2a, b, the energetics for the hydrated Fe^{2+} cation are compared. Comparisson is also made with the reactivity of FeO_{2+} diatomic cations (see Figures 5i.2c, d). Figure 5i.3 gives the catalytic cycle for alkane hydroxylation by the cytochrome P450 system, and Figures 5i.4 and 5i.5 concern the MMO enzymes based respectively on Fe and Cu.

Figure 5i.1 Biocatalysis of methane oxidation. (a) The rebound mechanism. (b) The concerted oxene-insertion mechanism for hydroxylation proposed by Newcomb and coworkers. (Schröder et al. 2000 [38]. Reproduced with permission of Springer.)

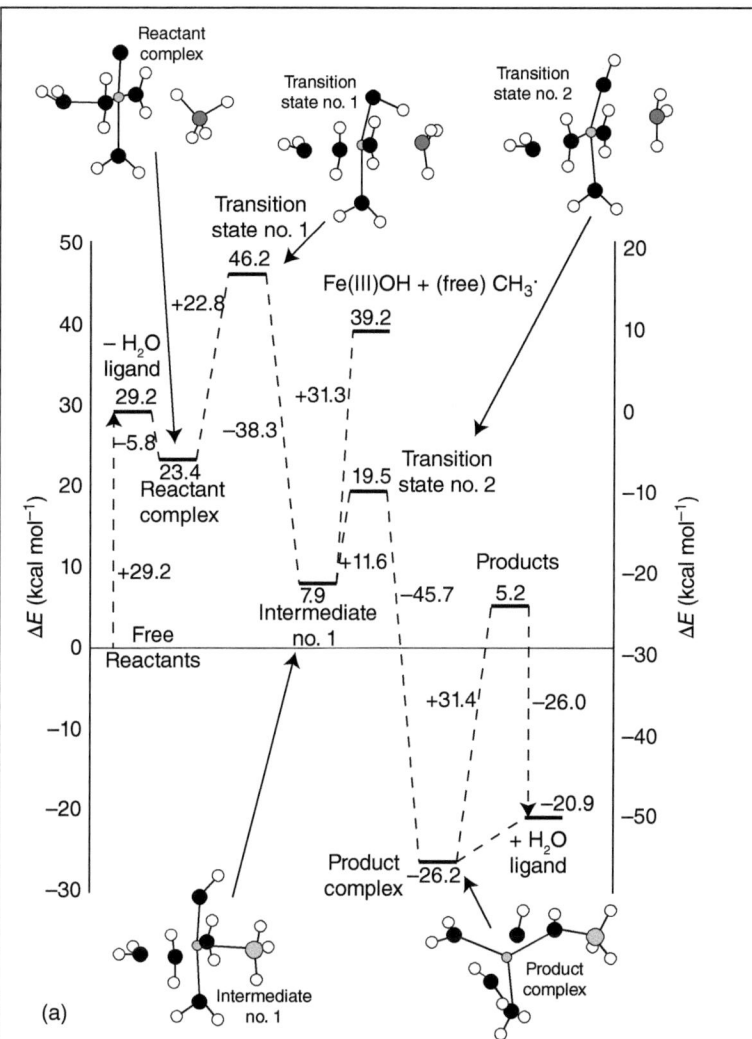

Figure 5i.2 Reactivity of FeO^{2+} complexes and their electronic structures. (a) Geometries and energy profiles (in kcal mol^{-1}) of the intermediate steps along the **methane coordination mechanism** of methane-to-methanol oxidation by the penta-aqua iron(IV) oxo species. The energy of the separated reactants is the offset for the left-hand energy axis. The energy axis on the right side serves to indicate the energy changes with respect to the separate tetra-aqua iron(IV) oxo complex and methane molecule, that is, after creating a vacant coordination site. (b) Geometries and energy profile (in kcal mol^{-1}) of the intermediate steps along the **oxygen-rebound mechanism** of the methane-to-methanol oxidation by the penta-aqua iron(IV) oxo species. The energy of the separated reactants is set to zero. The free energies are indicated by the dotted levels and the numbers between parentheses. (Ensing et al. 2004 [39]. Reproduced with permission of American Chemical Society.) (c) Potential energy diagrams along concerted H-atom abstraction pathways for FeO^{2+}. Relative energies are in kcal mol^{-1}. TS (direct) is a transition state along a direct H-atom abstraction pathway in the sextet state. (Yoshizawa et al. 1998 [40]. Reproduced with permission of Elsevier.) (d) Molecular orbital schemes for the ground state of molecular FeO^{2+}. (Schröder et al. 2000 [38]. Reproduced with permission of Springer.)

(Continued)

448 | *10 Zeolitic Non-Redox and Redox Catalysis, Lewis Acid Catalysis*

Insert 5: (Continued)

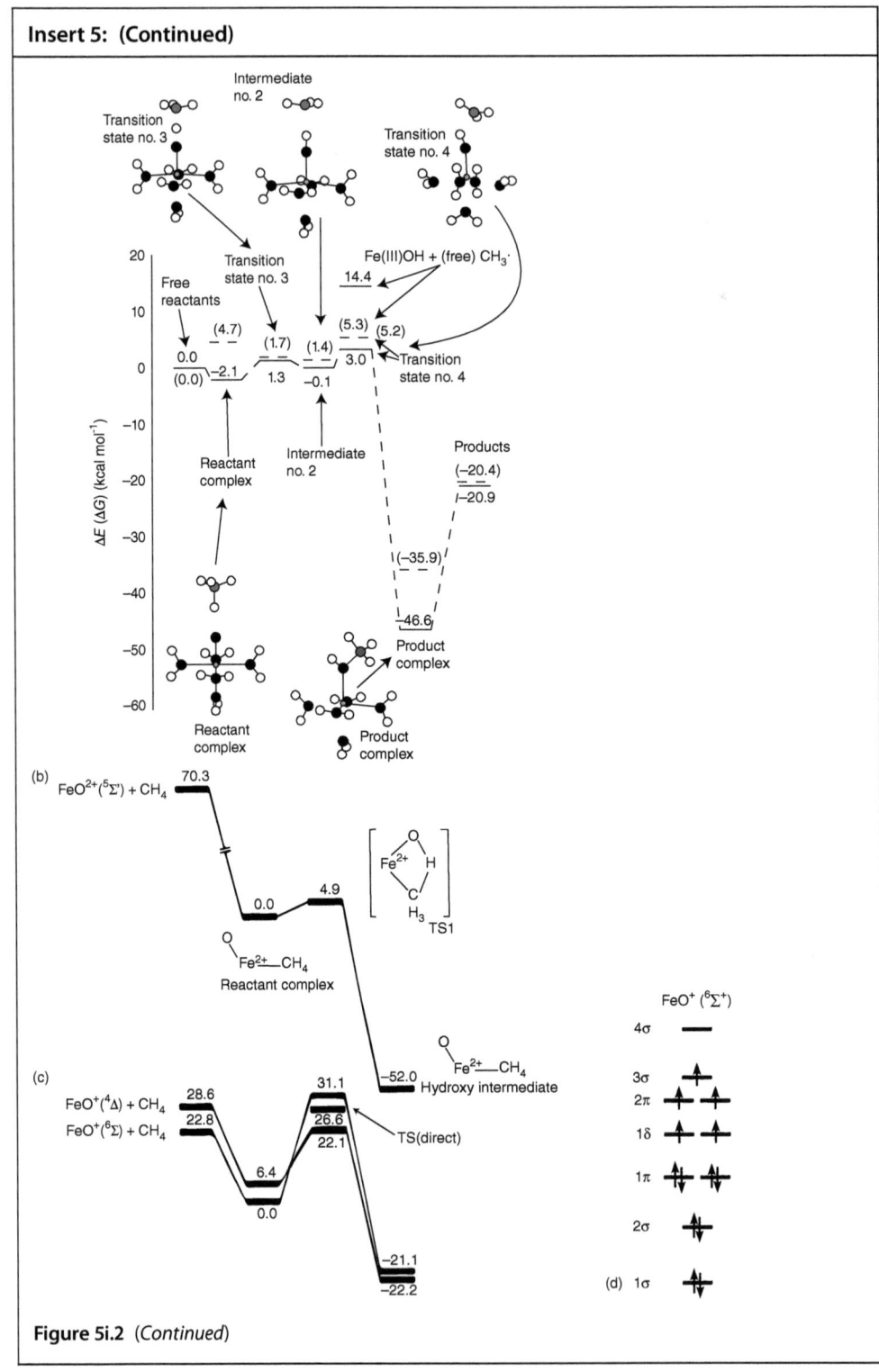

Figure 5i.2 *(Continued)*

Figure 5i.3 The catalytic cycle for cytochrome P450. The activation of O_2. (Groves and Nemo 1983 [41]. Reproduced with permission of American Chemical Society.)

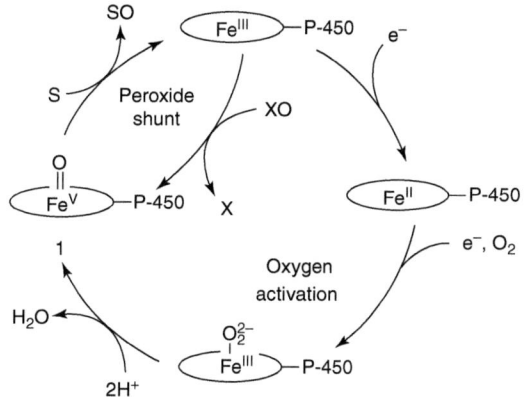

Figure 5i.4 (a) X-ray structure of the diferric form of the iron dimer complex of methane monooxygenase from *Methylococcus capsulatus*. (b) Suggested reaction sequence for hydroxylation in methane monooxygenase. (c) Potential energy surface for methanol formation from methane. Note the different scale for the negative *y*-axis. (Siegbahn and Crabtree 1997 [35]. Reproduced with permission of American Chemical Society.)

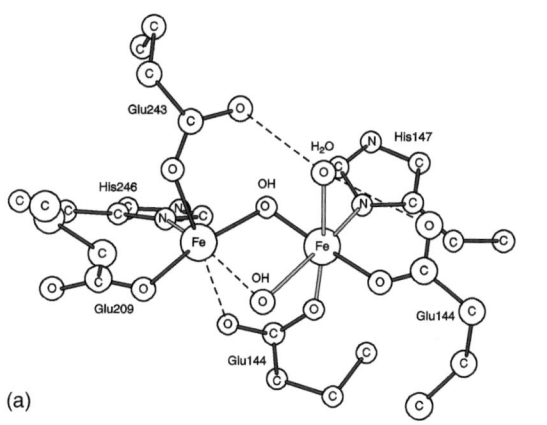

(a)

(Continued)

Insert 5: (Continued)

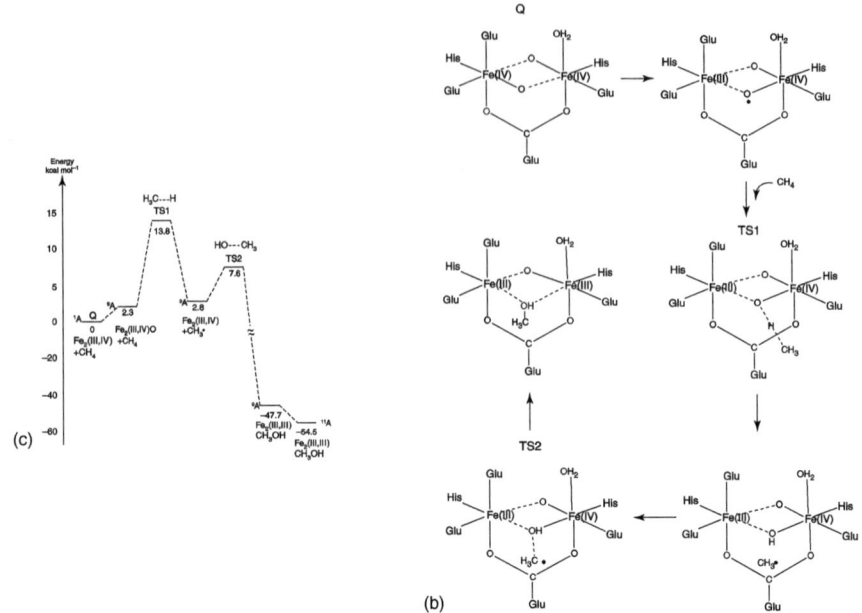

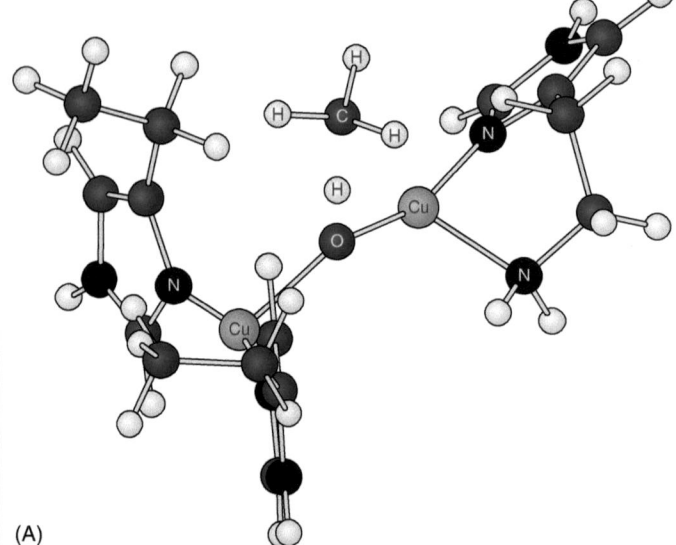

Figure 5i.5 (A) Cu^{II}_2O core modeled in the active site of pMMO: H-atom abstraction from methane at the transition state. (B) DFT-calculated structural models of the Cu_2O intermediate in ZSM-5 (a,c). (b) Shows the ring structure of ZSM-5 from which clusters have been constructed. (C) DFT-calculated reactivity of Cu^{II}_2O with methane (a), the HOMOs at the H-atom abstraction TS. The antibonding character of the CH_4 orbital fragment indicates initial polarization by the O atom. (b), and schematic of the Cu(I)-oxyl (c). (Solomon et al. 2011 [36]. Reproduced with permission of Royal Society of Chemistry.)

10.5 Related Homogeneous and Enzyme Oxidation Catalysts | 451

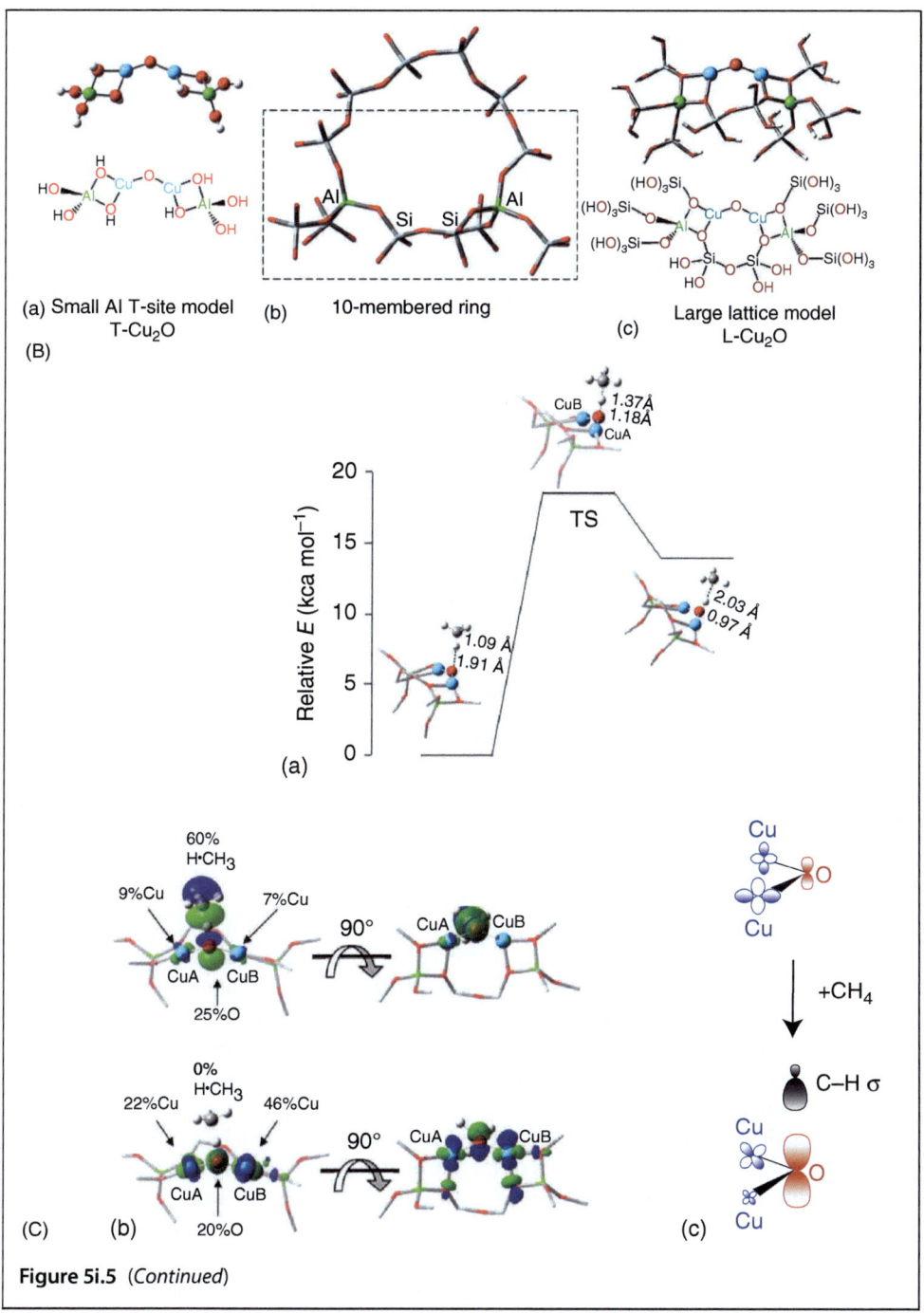

Figure 5i.5 (*Continued*)

The Fenton reagent is a solution of Fe^{2+} dissolved in H_2O with H_2O_2. The decomposition of H_2O_2 by Fe^{2+} is thought to give a hydrated FeO^{2+} intermediate, which can hydroxylate alkanes. There are also two enzymatic systems that can hydroxylate an alkane: the cytochrome P450 system with a porphyrin or a heme center (Figure 5i.3) and the non-heme MMO enzyme. The MMO systems contain dimer Fe cations (see Figure 5i.4) or Cu cations (Figure 5i.5).

Figure 5i.3 illustrates a catalytic cycle of the biocatalytic system. The hydroxylation reaction is part of an electrocatalytic system, driven by the electrochemical potential generated by the metabolism of the living system. It also provides the protons and electrons that are necessary to close the catalytic cycle. In the case of cytochrome P450, the oxygen donating center is generated by the acceptance of an O atom by the porphyrin Fe^{3+} cation. The overall oxidation process is achieved with O_2. When O_2 adsorbs to Fe^{3+}, two protons and electrons are used to decompose this compound to H_2O and an O atom that remains adsorbed and ready for the hydroxylation reaction. The porphyrin ligand donates one electron to the FeO^{3+} oxycation, which converts it to the reactive FeO^{2+} unit. This is similar to the reactive state of Fe in the Panov system.

In the dimer Fe-containing MMO enzyme, reaction with O_2 generates the Fe complex shown in Figure 5i.4. In the dimer complex, the reactive center is a FeO^{2+} cation; the other Fe cation can be considered a spectator of the reaction.

The $[Cu-O-Cu]^{2+}$ complex in ZSM-5 also activates methane as illustrated in Figure 5i.5a. This is the analog of the Cu-containing MMO enzyme.

There are two different activation routes for methane (see Figure 5i.1):

– The coordinative activation route
– The rebound mechanism

The coordination activation route with a FeO^{2+} cation is similar to the mechanism of activation of methane we discussed earlier for PdO (Section 7.4.2.3.1). The high charge on the cation enables methane to adsorb strongly (see Figure 5i.2c). The C—H bond cleaves in a heteropolar fashion with the formation of an $FeCH_3(OH)^{2+}$ intermediate. Methanol is produced by the reductive elimination of CH_3 and OH and the Fe cation reduces from a formal 4+ state to a 2+ state. This is the dominant mechanism for the reaction of a bare oxycation (Figure 5i.2c). This oxene mechanism is similar to the oxygen insertion reaction discussed for the hydroxylation of benzene in the Panov reaction. The comparative reaction energy scheme of this coordinative route in water is given in Figure 5i.2a.

C—H bond activation occurs in the rebound mechanism by a radical reaction of oxygen with a hydrogen atom of methane (see Figure 5i.2b). This is the dominant path for a system in which the Fe cation is surrounded by ligands. The presence of these ligands eliminates the possibility of direct interaction between the metal cation and the methane in the transition state. A methyl radical is created that then weakly interacts with the hydroxyl, and methanol is formed in a consecutive step (the rebound).

A slightly different reaction occurs in the Fe MMO system. After the methyl radical is formed, it ionizes and an electron is donated into the Fe—OH bond. In the final step, the methyl cation and OH^- recombine to give methanol [42].

The electrophilic nature of the FeO^{2+} cation becomes evident by inspecting its electronic structure, which is schematically illustrated in Figure 5i.2d. It is electronically related to O_2.

The FeO^{2+} unit forms bonding and antibonding π orbitals perpendicular to the Fe—O bond of two O 2p atomic orbitals and two Fe 3d atomic orbitals. Two d-atomic orbitals of Fe are non-bonding and a d_{z^2} atomic orbital forms a σ bonding and antibonding orbital with the O $2p_z$ orbital. The population of the antibonding σ^* (3σ in Figure 5i.2d) orbital is key to the reactivity of the system.

In FeO^{2+} this empty σ^* antibonding orbital has an energy low enough to accept an electron from the occupied methane orbitals. A $CH_3°$ radical as well as a $Fe(OH)^{2+}$ species are generated during this process. The activation energy of this reaction step is low as long as the FeO antibonding σ orbital is not occupied. While FeO^{2+} has an activation energy of only 5 kcal, FeO^+ with an electron occupied σ^* orbital has a barrier of at least 20 kcal mol^{-1}. The barriers become higher for the enzymes due to the presence of the ligands, but the chemical bonding principle that causes the barrier to be only low when the σ^* is unoccupied is similar.

The rather weak Fe—O bond and the electrophilic nature of O in the FeO^{2+} unit enables the porphyrins to act as versatile oxidation catalysts. They are also active catalysts for the epoxidation of unsaturated hydrocarbons. An upper bound to the Fe—O bond energy can be readily deduced from the (weak) exothermicity of the methane to methanol reaction (-122.1 kJ mol^{-1}). It implies that the Fe—O bond energy cannot be stronger than 320 kJ mol^{-1}.

For the Cu-based catalysts, there is also an analogy between the enzyme and inorganic systems. Spectroscopic and computational studies have shown that the reactive site of the Cu zeolite system in ZSM-5 is a dimer, just as in the enzyme [36]. The structure and energetics of the methane rebound mechanism of the ZSM-5 catalyst with the dimer Cu site are shown in Figure 5i.5. Calculations have been done on the cluster model of Figure 5i.5Bc that is part of the ZSM-5 ring pore structure. A low activation energy for CH bond cleavage requires a lowest unoccupied molecular orbital (LUMO) on the $[Cu_2O]^{2+}$ system of low energy. As is observerd Figures 5i.5Ca,b cleavage of the CH bond shifts the electron density of the oxygen atom to the Cu ions. This indicates polarizability of the Cu—O bond and that the electron density on O becomes reduced.

Recently, it has been discovered that the Cu trimer $(Cu_3(\mu\text{-}O_3))^{2+}$ complex located in the micropores of mordenite [43] also catalyzes methane to methanol conversion by a mechanism similar to the one exhibited for the dimer in ZSM-5.

10.6 Lewis Acid Catalysis by Non-Reducible Cations Located in the Zeolitic Framework

The substitution of zeolite framework cations by Lewis acidic reducible and non-reducible cations creates unique single-center catalysts. In this section, we will discuss the use of catalysis by non-reducible cations. Section 10.7 concerns catalysis with reducible cations located in the zeolite framework.

The systems described in these two sections find their main application in catalytic green chemistry reactions that can produce fine chemicals without harmful co-products. These reactions are designed to replace stoichiometric reactions that may create environmentally harmful waste.

Reactions that form oxygenated molecules are an important class of reactions. Since non-reducible cations cannot activate O_2, these systems use hydrogen peroxide or hydroperoxides. Reducible cations can activate O_2 and thus can catalyze reactions based on the use of O_2.

The Lewis acidic non-reducible systems are known to catalyze oxidation reactions with hydrogen peroxide that produce epoxides or lactones. The Lewis acidic cation activates the peroxide to insert an electrophilic oxygen atom.

Another class of reactions proceeds by hydride transfer in which the Lewis acid activates the HCOH group in alcohols or sugars for hydrogen transfer to keto groups. This can be accompanied by C—C bond cleavage or formation.

In their oxidic bulk structures, Lewis acidic cations such as Ti^{4+}, Zr^{4+}, and Sn^{4+} are six-coordinated to oxygen atoms (with coordination numbers even higher for Zr), but when they are incorporated into the framework of siliceous zeolites, they are four-coordinated. Their ionic radii (Ti = 0.68 Å; Zr = 0.80 Å; Sn = 0.71 Å) are larger than that of Si^{4+} (=0.54 Å), so that the zeolite lattices become strained. As a consequence, the M—O bond can become readily hydrolyzed when it reacts water. For example, the TiOSi bond is activated and hydrolyzed to give TiOH and SiOH groups. The Sn^{4+} ion located in the framework is also thought to be partially hydroxylated.

For additional reading on this class of systems we refer to Thomas [1].

10.6.1 Bayer–Villiger Oxidation

In the Bayer–Villiger reaction, a cyclohexanone is converted into a seven-ring lactone that is used to produce flavors and fragrances. This catalytic reaction replaces a process using Caro's peracid H_2SO_5 (see Figure 6i.1).

A zeolitic system that catalyzes this reaction with H_2O_2 contains Sn^{4+} in the framework of a siliceous zeolite β, which features wide pores [44]. In the next section, which deals with redox systems, we will introduce the reducible Mn and Co $AlPO_4$ redox systems that catalyze this reaction with O_2 by the *in situ* generation of peroxide.

Catalysis by the Sn^{4+} cation, most likely partially hydrolyzed, proceeds by deprotonation of H_2O_2 and formation of an intermediate hydroperoxide called the Criegee product. Deprotonation of H_2O_2 is assisted by the presence of the OH^- group co-adsorbed to Sn. The keto part of cyclohexanone connects to the Sn^{4+} ion (see Figure 6i.2) and is hence strongly activated. The OOH^- of the hydrogen peroxide initially reacts with the carbon atom of the keto group. In the steps that follow, the O atom from the peroxide is inserted into the activated C–C bond next to the keto group, H_2O desorbs and the proton is back donated. Proof of this mechanism has been provided by ^{18}O labeled studies of the keton [45], which demonstrates that the O atom from the hydrogen peroxide is inserted in the hydrocarbon ring.

Insert 6: Bayer–Villiger Oxidation

Figures 6i.1 illustrates the oxygen insertion reaction by H_2SO_5. Figure 6i.2 schematically presents the mechanism of the analogous heterogenous reaction of H_2O_2.

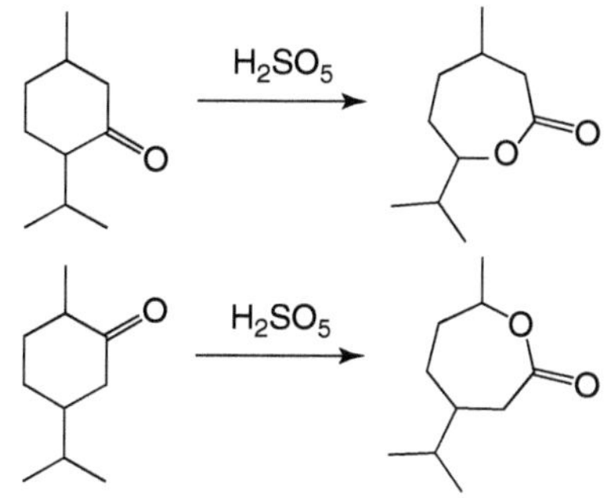

Figure 6i.1 Caro's acid, H_2SO_5, was used by Baeyer and Villiger to oxidize menthone and tetrahydrocarvone into products that are used in flavors and fragrances. (After [1].)

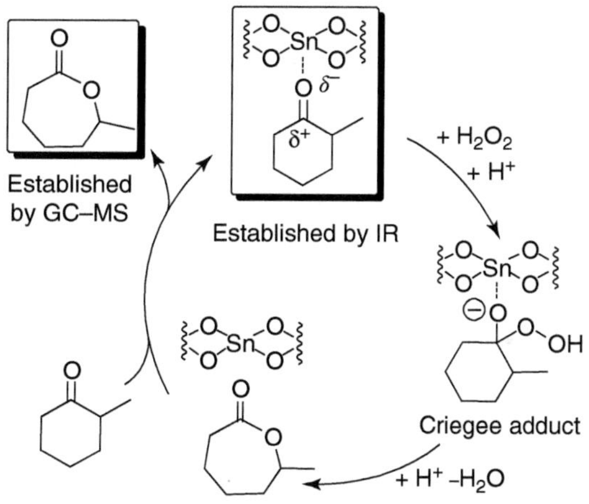

Figure 6i.2 Mechanism of the Baeyer–Villiger oxidation of 2-methylcyclohexanone catalyzed by Sn-β. (After [1, 45].)

Measurement of the C=O vibration frequency of hexanon adsorbed to different zeolite framework cations shows that the largest downward shift is for the Sn-doped zeolite. This indicates that the Sn^{4+} cation has a greater bond-weakening effect.

However, the Ti cation should have the highest Lewis acidity and the Zr cation the lowest based on their size differences, while the Sn cation should have an intermediate interaction. The stronger interaction of the Sn^{4+} cation is most likely due to its intermediate size, which exposes its positive charge more than the charge of the smaller Ti^{4+} cation, which remains largely screened by its tetrahedrally coordinated oxygen ions. The screening of the Zr ion is also less pronounced, but its large size lowers its interaction with the cation.

10.6.2 Meerwein–Ponndorf–Verley Reduction and Oppenauer Oxidation

In the Meerwein–Ponndorf–Verley (MPV) or Oppenauer oxidation (OP) reaction, a ketone is reduced by the oxidation of an alcohol through the combined hydride–proton transfer dehydrogenation of the alcohol and the addition of the hydride–proton pair to the ketone (see Figure 7i.1). This reaction has found important applications in organic chemistry.

The mechanism of this reaction, hydride transfer that results in hydrogenation or dehydrogenation, is of general importance to conversion reactions of oxygenated molecules. We discussed earlier similar hydride transfer reactions in solid acid catalysis (Chapter 9, page 391). The enzymatic catalytic conversion of glucose to fructose by D-xylose isomerase employs a closely related mechanism in converting the glucose six ring into a fructose five ring (see Figure 7i.2a) that also involves hydride transfer. Fructose is a sugar molecule that can be readily converted into chemically significant intermediates.

In sugar chemistry, such as in the transformation of glucose, the key elementary step for the isomerization reaction is internal hydride transfer involving the hydrogenation of a keto group of glucose by hydrogen from an internal alcohol group.

The invention of inorganic catalysts that can be used in this reaction makes the replacement of enzymatic catalysis possible, and the Sn-based zeolites again appear to be well suited to this role [13]. This reaction is important in the development of bio-based chemical processes.

10.6 Lewis Acid Catalysis by Non-Reducible Cations Located in the Zeolitic Framework

Insert 7: Meerwein–Ponndorf–Verley (MPV) and Oppenauer oxidation (OP) Inorganic Catalytic Systems for Sugar Conversion

Figure 7i.1 illustrates the mechanism of MPV and OP reactions, and Figure 7i.2 illustrates the application of this reaction to the isomerization of six-ring glucose into five-ring fructose. Figure 7i.3 shows the structure of the Sn site in zeolite beta and its reaction mechanism. Figure 7i.4 compares glucose isomerization by the biocatalyst and by the homogenous Cr-based systems.

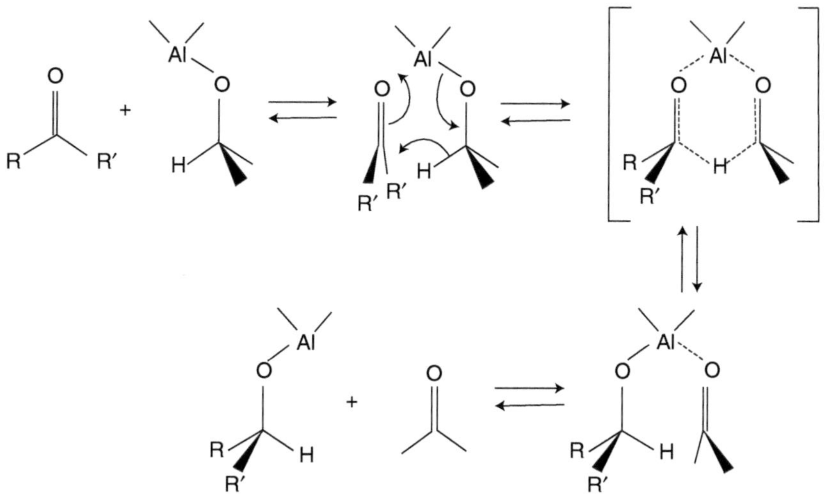

Figure 7i.1 The reaction intermediate of the Meerwein–Ponndorf–Verley reduction and Oppenauer oxidation reaction of a ketone with an alcohol. The alcohol deprotonates and the alcoholate adsorbs to the Al^{3+} cation. The ketone adsorbs to the same Al^{3+} cation and the hydride ion transfers. (de Graauw *et al.* 1994 [14]. Reproduced with permission of Thieme.)

(Continued)

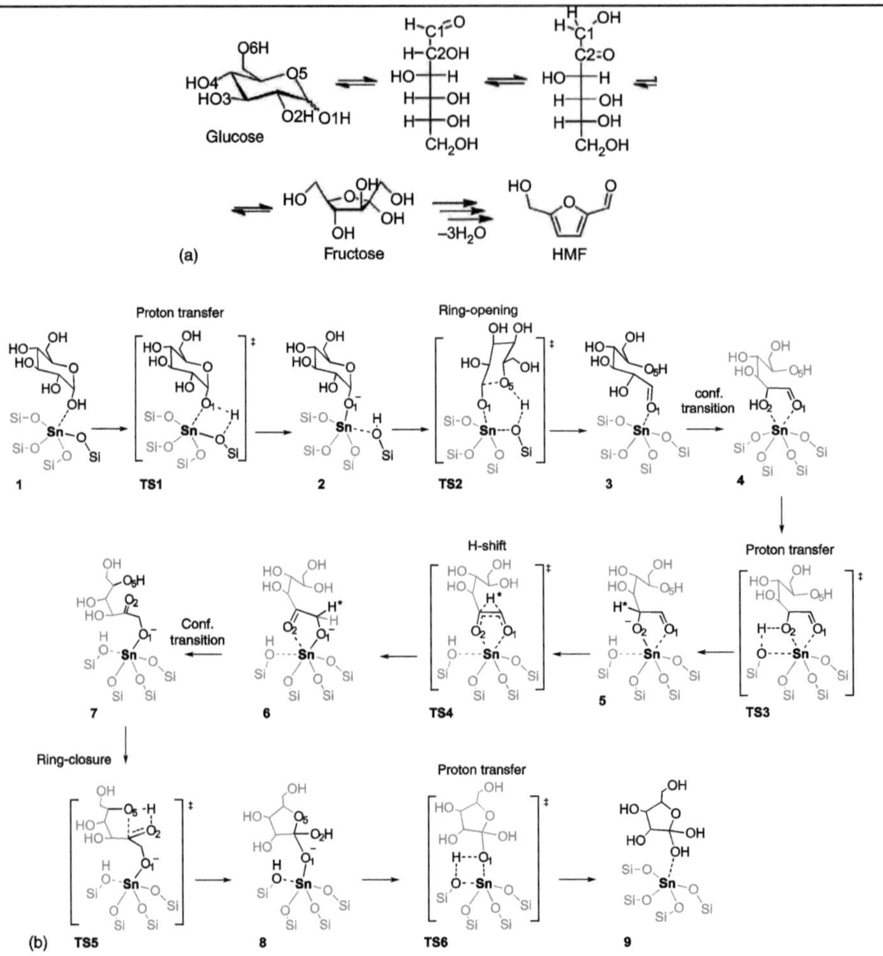

Figure 7i.2 (a) Isomerization of glucose to fructose as a first step of the dehydration reaction to HMF. (b) Mechanism of glucose isomerization to fructose catalyzed by a perfectly tetrahedral Sn^{IV} lattice site in Sn-BEA model. (Yang *et al.* 2013 [11]. Reproduced with permission of American Chemical Society.)

10.6 Lewis Acid Catalysis by Non-Reducible Cations Located in the Zeolitic Framework

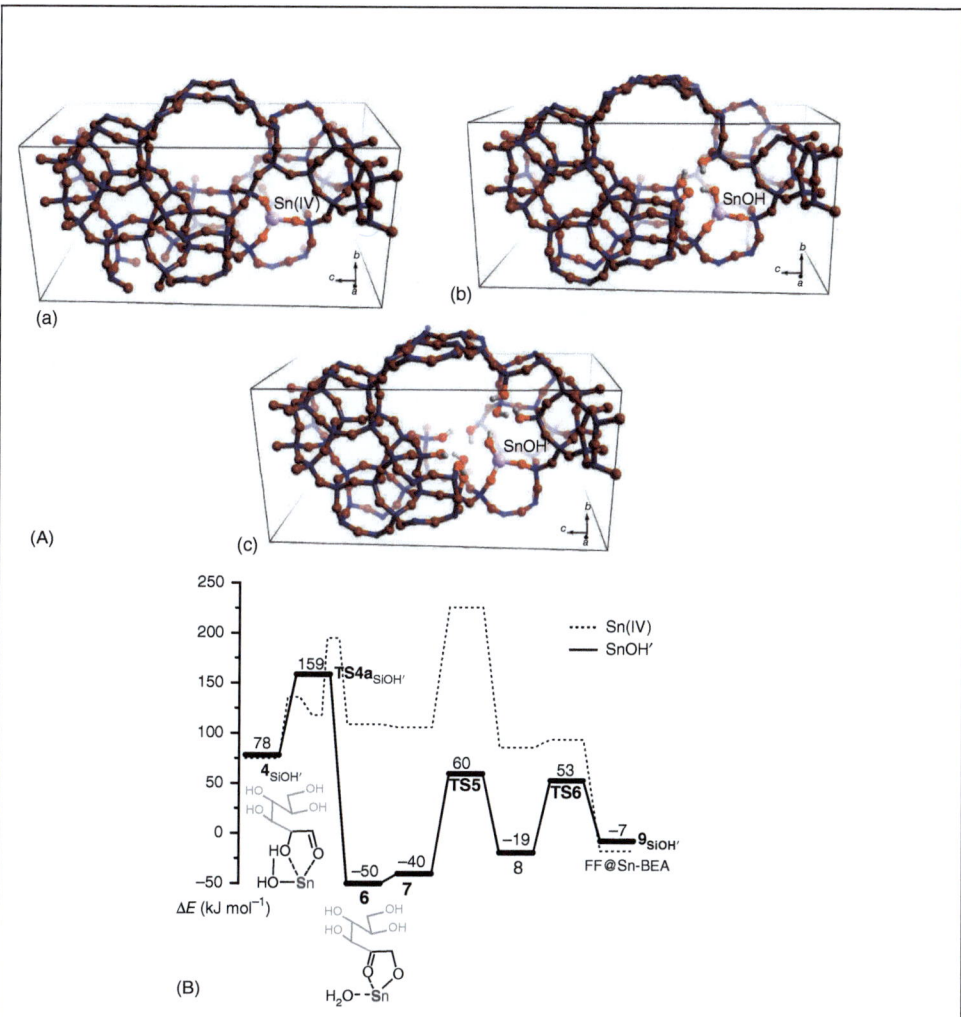

Figure 7i.3 (A) DFT-optimized structures of Sn-BEA: (a) model I containing a perfect Sn^{IV} lattice site, (b) model II with a partially hydrolyzed lattice SnOH site, and (c) model III with a SnOH' site in the vicinity of a large lattice defect formed by more extensive desilication of the framework. (B) DFT-computed reaction energy diagram for the key steps of glucose isomerization over the extended SnOH' defect site in Sn-BEA zeolite (model III). The energetics for the corresponding steps over the perfect Sn^{IV} site is given for comparison as a dashed line. (Adapted from Yang et al. 2013 [11, 12].)

(Continued)

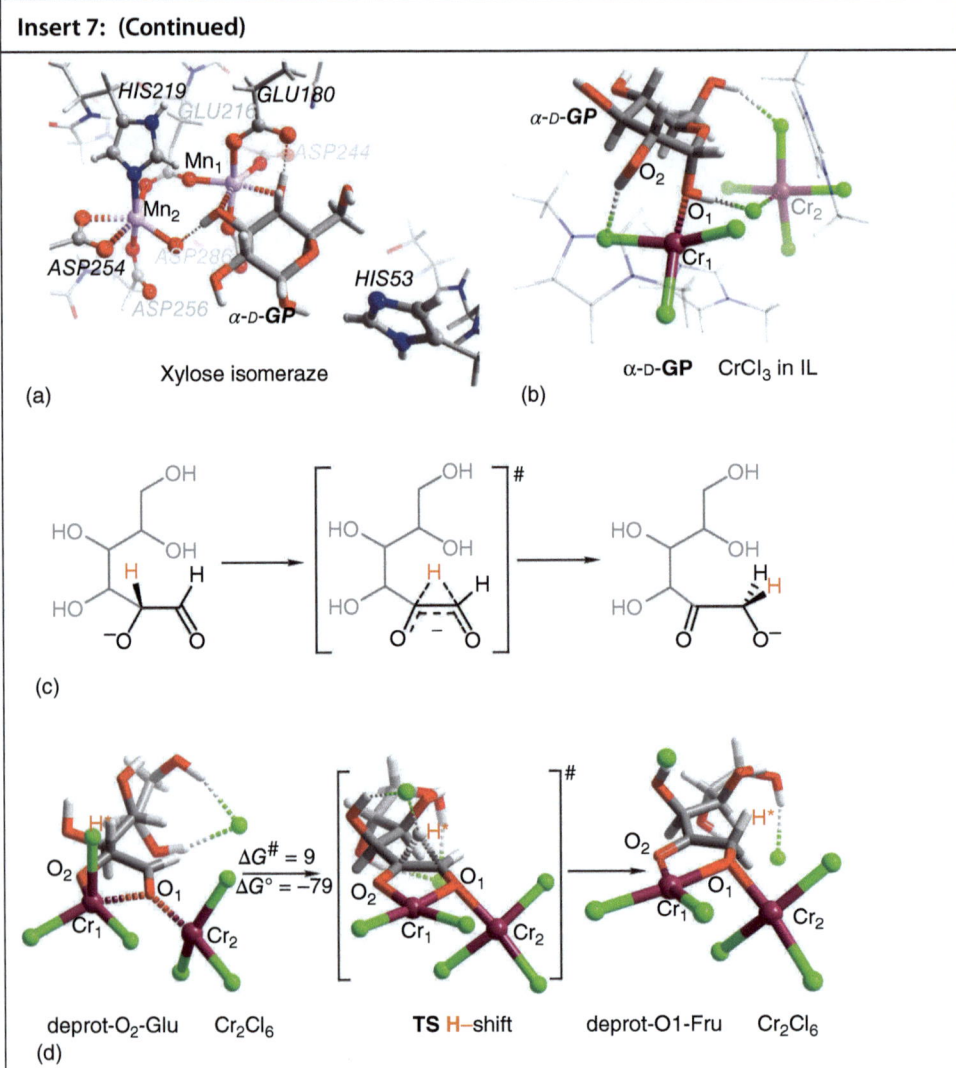

Figure 7i.4 (a) Atomic-resolution XRD structure-based calculation of glucose interacting with the binuclear active site of a xylose isomerase enzyme. (After [46].) (b) DFT-optimized structure of the initial coordination complex between glucose and chromium(III) chloride. (Garcia-Viloca et al. 2003 [47]. Reproduced with permission of John Wiley & Sons.) (c) Scheme illustrated showing the H shift that takes place during glucose isomerization. (d) DFT-computed binuclear Cr^{III} complexes with the deprotonated sugar intermediates shown in (c). DG values are given in kJ mol^{-1}. TS, transition state. (Pidko et al. 2012 [48]. Reproduced with permission of John Wiley and Sons.)

Sn^{4+} located in zeolite beta is again the preferred catalyst for this reaction. It can also be used to produce lactic acid (useful for biodegradable polymers) from fructose [49, 50].

These reactions require the following steps (see Figure 7i.2b): proton removal from a hydroxyl, stabilization of the negative charge generated between the deprotonated hydroxyl and keto group that receives the hydride and subsequent proton backdonation.

The unique activity of the Sn-substituted zeolite relates to the formation of "Sn–OH" substituents by hydrolysis of the Sn–O–Si bonds. The basic OH attached to Sn assists in the initial proton transfer from the ring-opened glucose hydroxyl. This is illustrated in Figure 7i.3, which shows the large reduction of activation energies by the hydrolyzed Sn system. Computational modeling has indicated that a site that contains a defect not only provides more space, but also enhances reactivity through the creation of a water–hydroxyl network. The co-adsorbed H_2O atoms form a network and act as proton shuttles that help to stabilize the charged anion [12].

10.6.3 Homogeneous and Biocatalyst Analogs

The mechanism of isomerization of glucose to fructose by the single center Sn^{4+} zeolite β is related to that of the enzyme xylose isomerase [51]. This enzyme contains a cofactor with two bivalent cations such as Mn^{2+}, Mg^{2+}, Fe^{2+}, or Co^{2+} cations that catalyze the hydride transfer reaction. The mechanistic similarity arises from the essential role of the positive cation charge in stabilizing the two negative polar parts of the OC—HCO$^-$ unit for efficient internal hydride transfer (see Figure 7i.4). Figures 7i.4a,b illustrate the similarity of the enzyme interaction with glucose and the structure deduced with DFT calculations when glucose interacts with two $CrCl_3$ complexes. These complexes when dissolved in ionic liquids are efficient isomerization catalysts. In Figure 7i.4c the hydride transfer step is schematically illustrated. Computed reaction intermediates of the hydride transfer reaction are shown in Figure 7i.4d. The low barrier of activation of the hydride transfer reaction is due to the interaction with the two Cr_{3+} ions that stablize the negatives charges on the oxygen atoms that arise when OH deprotonates and the hydride transfer generates an other unsaturated hydroxyl [52, 53].

10.6.4 Propylene Epoxidation

The classical process for producing propylene is by the intermediate production of chlorohydrine when hyprochlorite reacts with alkene. In the overall process, Cl_2 is converted into $CaCl_2$, and epoxide is formed from the chlorinated intermediate by reaction with $Ca(OH)_2$. The atom efficiency of this process is only 28% (Part II, Introduction page 168). Direct epoxidation of propylene with O_2 by Ag-promoted catalysts has only a low yield and the ready formation of the allyl intermediate from propylene leads to total combustion.

Propylene epoxidation with a high yield is possible by reaction with H_2O_2. The atom efficiency increases to 75% when the epoxide is made directly from hydrogen peroxide.

A selective catalyst for the epoxidation reaction of propylene can be created by substituting Ti^{4+} for Si^{4+} in the framework of siliceous ZSM-5 [53]. The strong Lewis acidity of Ti activates H_2O_2. The catalytic reaction cycle of propylene epoxidation is shown in Figure 8i.1. The framework Ti—OSi bond opens upon reaction with H_2O_2 to give Ti—OOH and Si—OH substituents. The electrophilic oxygen atom from OOH attached to Ti inserts into propylene. The reaction cycle closes by the formation of H_2O and closure of the Ti—OSi bond. As Figure 8i.2 illustrates, the O insertion step is promoted by coadsorption of methanol, which adsorbs through its OH group to the Ti cation. The OH proton hydrogen bridges with the hydroperoxide, which makes the oxygen atom that is inserted more electrophilic.

This Ti site is stable even in the presence of water because of the hydrophobicity of the microchannels of silicalite, the siliceous form of ZSM-5. The reaction is uniquely catalyzed by single-site Ti. That is tetrahedrally coordinated with the oxygen atoms of the rather rigid zeolite framework.

Insert 8: Mechanism of Propylene Epoxidation

The catalytic reaction cycle on Ti-silicate is shown in Figure 8i.1. Promotion of the rate of reaction by methanol co-adsorption is illustrated by Figure 8i.2 and the mechanism of the reaction catalyzed by Ti attached to the silica surface is shown in Figure 8i.3.

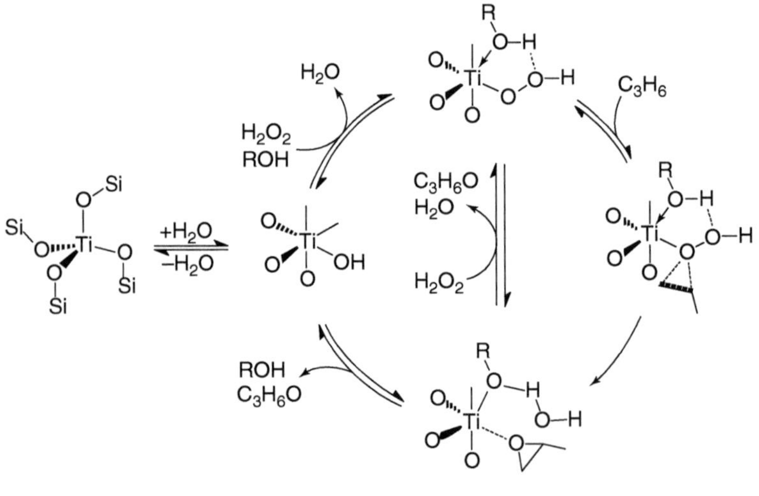

Figure 8i.1 Catalytic reaction cycle of the epoxidation reaction of propylene catalyzed by Ti silicate. (Clerici 1993 [53]. Reproduced with permission of Elsevier.)

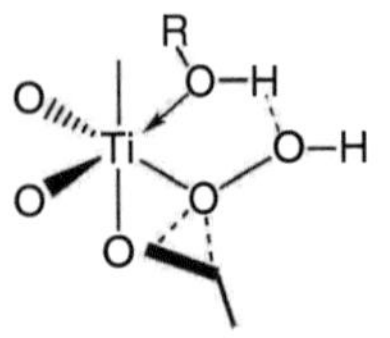

Figure 8i.2 Methanol co-adsorption on the Ti silicate site. (Clerici 1993 [53]. Reproduced with permission of Elsevier.)

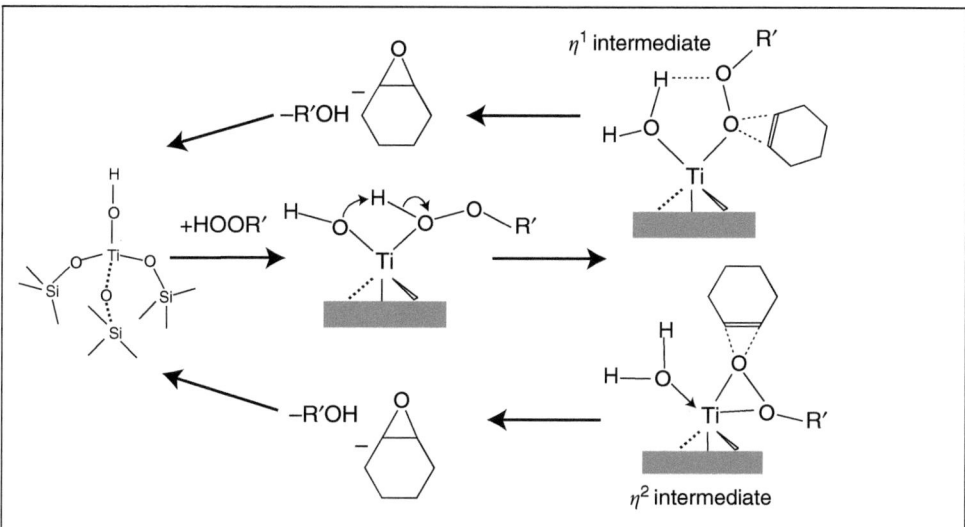

Figure 8i.3 Mechanism of propylene epoxidation on Ti silicate. (Thomas et al. 2002 [54]. Reproduced with permission of Royal Society of Chemistry.)

Instead of hydrogen peroxide, a hydroperoxide can also be used for this reaction, as in the Shell SMPO process of propylene epoxidation by ethylbenzene hydroperoxide [55]. A single site Ti complex attached to SiO_2 through free O atoms is the catalyst for this reaction (see Figure 8i.3). This Ti center is more reactive than the four-coordinated Ti site located in the zeolite framework. If this catalyst is instead used with hydrogen peroxide, the molecule will decompose into O_2 and H_2O.

Only Ti^{4+} attached to the silica surface through three Ti—O—Si surface bonds is stable. Ti complexes created during catalyst preparation that are attached to the surface with fewer Ti-O-Si bonds readily hydrolyze and Ti oxide particles agglomerate. The Ti complex is terminated by a hydroxyl or alcoxy group. The Ti-OH is replaced by Ti—OOR upon reaction with the hydroperoxide, which reacts with propylene to give the epoxide and an alcohol. Useful coproducts styrene or i-butene are formed upon dehydration of the alcohol products. When isobutyl hydroperoxide is used instead of ethylbenzene hydroperoxide dehydration of i-butyl alcohol that is formed in that case, gives i-butene as coproduct of the process.

10.7 Catalysis by Redox Cations located in the Zeolitic Framework: The Thomas Oxidation Catalysts

In contrast to non-reducible Lewis site catalyzed oxidation reactions that require activated oxygen in the form of H_2O_2, oxidation catalysis by reducible cations can be done with O_2. Thomas et al. developed zeolitic redox systems for reaction with O_2 [1], which are not based on siliceous zeolites but instead on analogous three-dimensional microporous systems of the same structure with $AlPO_4$ stoichiometry.

Non-acidic redox systems can be designed by replacing Al^{3+} with a reducible cation that has a three valent charge. These systems are useful catalysts amongst others for the selective oxidation reaction with O_2 that initiates radical chain chemistry. The catalysts are selective because of the unique constraints induced by the presence of the zeolite microchannels. In solution, the catalysts may deactivate by leaching of the reactive metal cations. Acidity in the $AlPO_4$ material can be introduced by the substitution of P^{5+} by Si^{4+} and a proton. Alternatively, acidity can be introduced by the substitution of Al^{3+} with a bivalent cation and a proton.

10.7.1 Bayer–Villiger Oxidation with Molecular Oxygen

Mn^{3+} or Co^{3+} substituted wide pore $AlPO_5$ systems catalyze lactone formation from six-ring cyclic molecules such as cyclohexanone through oxidation with oxygen. The Bayer–Villiger insertion reaction requires H_2O_2 or per acid to insert an O atom into the cyclohexanone ring (see INSERT 6). Therefore the oxidation reaction of cyclohexanone by O_2 has to be executed using a sacrificial molecule. A sacrificial benzaldehyde is used to produce an intermediate perbenzoic acid. The Mn^{3+} that is substituted for Al_{3+} in the $AlPO_4$ framework initiates a radical reaction [45]:

$$PhCHO + Mn^{III} \rightarrow PhCO^{\cdot} + H^+ + Mn^{II}$$
$$PhCO^{\cdot} + O_2 \rightarrow PhCOOO^{\cdot}$$
$$PhCOOO^{\cdot} + PhCHO \rightarrow PhCOOOH + PhCO^{\cdot}, \text{ etc.}$$

The Mn^{3+} cation is reduced by the reactive H atom of benzaldehyde, to Mn^{2+} and a proton is generated. The benzaldehyde radical reacts with O_2, which initiates a radical chain reaction. The peracid is formed by a bimolecular reaction between benzaldehyde and the oxygen radical. The peracid forms the lactone from a cycloketone in a process analogous to the reaction with Caro's acid (see Figure 6i.1). Mn^{3+} is regained by oxidation with O_2 with cogeneration of H_2O.

10.7.2 Zeolite Catalysts for Caprolactam Synthesis

Caprolactam is an important intermediate for nylon production. Caprolactam is a seven-ring ketone with an NH group next to the CO group as part of the carbon atom ring. Nylon is formed by ring opening of the strained peptide bond to give the nylon polymer (see Figure 9i.1). Caprolactam is made from a reaction between cyclohexanone and NH_2OH. In the classical industrial process, the oxime is converted into cyclohexanone by reaction with oleum, which is then converted into ammonium sulfate.

The catalytic process aims to produce caprolactam without a coproduct. Ti-silicate can be used to convert ammonia and hydrogen peroxide with cyclohexanone into the oxime, which can be converted to caprolactam in a gas phase reaction.

The Thomas catalytic system can achieve this reaction using O_2 [54]. Al^{3+} can be substituted with Co^{3+}, Mn^{3+}, or Mg^{2+} in the $AlPO_{4-5}$ structure. When Mg^{2+} is used, protons are present that induce Brønsted acidity to compensate for the charge. The reducible cation Co^{3+} or Mn^{3+} catalyzes oxidation of NH_3 to NH_2OH, which in turn reacts with hexanon to give the oxime. This is converted to caprolactam by the protonic centers (see Figure 9i.2).

Insert 9: Thomas Oxidation Catalysis

Figure 9i.1 gives the reaction scheme of the caprolactam and nylon reactions. The catalytic reaction invented by Thomas is illustrated by Figure 9i.2. Table 9i.1 and Figures 9i.3 and 9i.4 illustrate selective oxidation of linear alkanes.

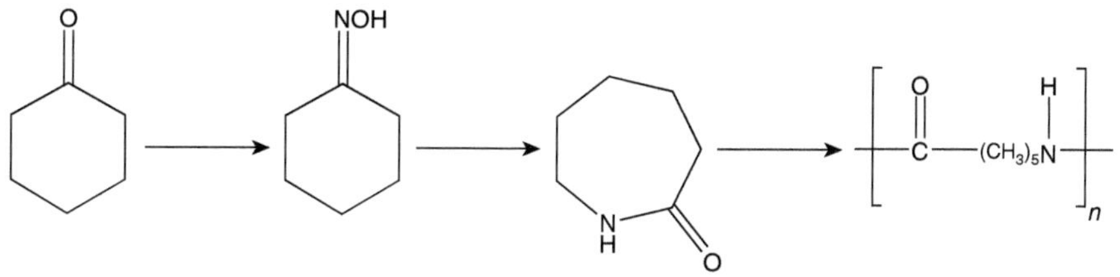

Figure 9i.1 Summary of sequence of conversions that produce ε-caprolactam (and its linear polymer, nylon 6) from cyclohexanone via its oxime. (After [1].)

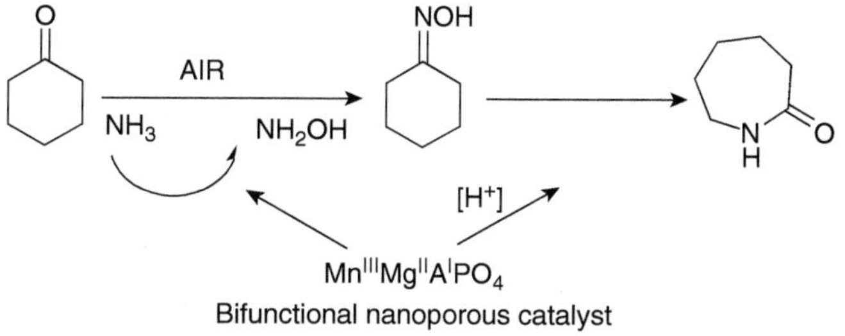

Figure 9i.2 Simplified illustration of the single-step, solvent-free, environmentally benign synthesis of ε-caprolactam using molecular O_2. (Thomas et al. 2005 [56]. Reproduced with permission of National Academy of Sciences.)

(Continued)

Insert 9: (Continued)

Table 9i.1 Oxidation of *n*-alkanes over MAPOs: primary selectivity.

Framework	Pore dimensions (nm)	Substrate	Metal	Primary sel. (%)
AlPO$_4$-18	0.38 × 0.38 nm	*n*-Pentane	Co	33
			Mn	39
		n-Hexane	Co	61
			Mn	66
		n-Octane	Co	60
			Mn	62
AlPO$_4$-11	0.39 × 0.63 nm	*n*-Hexane	Co	19
AlPO$_4$-36	0.65 × 0.75 nm	*n*-Pentane	Co	5
			Mn	0
		n-Hexane	Co	23
			Mn	0
		n-Octane	Co	12
			Mn	7
AlPO$_4$-5	0.73 × 0.73 nm	*n*-Hexane	Co	9

Source: Dugal 2000 [57]. Reproduced with permission of John Wiley and Sons.

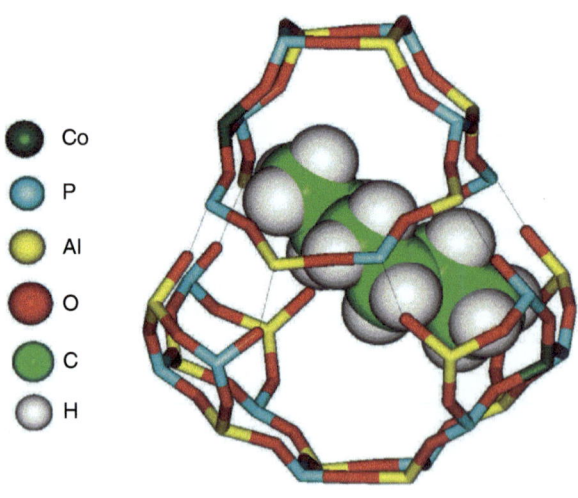

Figure 9i.3 Redox active centers in AlPO molecular-sieve catalysts effect region selective oxyfunctionalization of alkanes at terminal methyl groups. MAPO-18 (M=Co^{III}, Mn^{III}) is especially effective for this purpose. (Thomas *et al*. 2001 [58]. Reproduced with permission of American Chemical Society.)

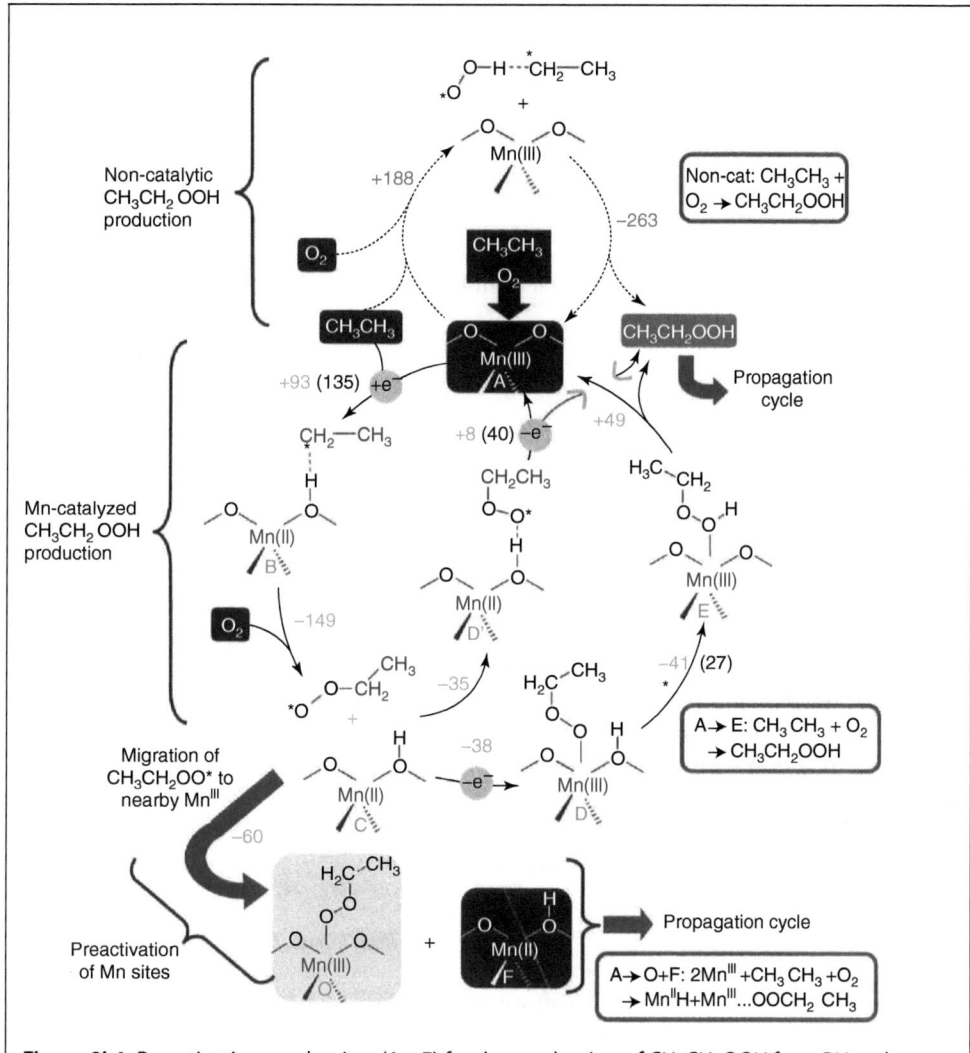

Figure 9i.4 Preactivation mechanism (A→E) for the production of CH_3CH_2OOH from RH and O_2 without (top) and with (middle) the assistance of Mn^{III}. A dark background indicates initial catalyst and reactant molecules while a light background indicates reaction intermediate or product molecule. (After [59].)

10.7.3 Alkane Oxidation

Radical chain oxidation reactions of alkanes can have superior selectivity when activated by reducible cations that are part of the AlPO framework, due to limitations imposed by the zeolitic micropores. Fe^{3+}, Co^{3+}, and Mn^{3+} cation substitution can especially be used to selectively produce alcohols, ketones, or acids from linear and cyclic alkanes while the skeletal carbon number is maintained.

Radical chain auto-oxidation reactions occur by a sequence of initiation, chain propagation, and chain termination reaction steps. The initiation step involves the formation of an OOH intermediate and hydrocarbon-free radical R·. Subsequently, a reactive ROO· species is formed by the reaction of an alkyl radical with O_2.

The reaction propagation step is:

$$ROO\cdot + RH \rightarrow ROOH + R\cdot$$

Chain termination occurs by a bimolecular reaction such as:

$$2R(H)OO\cdot \rightarrow ROH + (RH) - O + O_2$$

Reactions at mild temperatures are initiated by redox reactions with reducible cations. These reactions have a substantially lower activation energy than the non-catalytic initiation reaction, on the order of 100 kJ mol^{-1} versus 190 kJ mol^{-1} [59].

Direct alkane activation is possible by using reactions similar to those discussed for the activation of benzaldehyde:

$$R + Mn^{3+} \rightarrow R\cdot + Mn^{2+} + H^+_{ads}$$
$$ROO\cdot + Mn^{2+} + H^+ \rightarrow ROOH + Mn^{3+}$$

In small zeolitic micropores, the bimolecular chain termination reaction is suppressed because it inhibits the recombination of the bulky peroxy radicals. The reaction with linear alkanes will be initiated with the primary CH_3 groups, because the endpoints will be directed into the cavities where the activating cations are located (see Figure 9i.3). Products will be formed by decomposition of the peroxides by reactions such as:

$$ROOH + M^{2+} + H^+ \rightarrow RO + H_2O + M^{3+}$$

Table 9i.1 shows the selectivity for primary carbon atom oxidation from AlO_4 materials with different micropore structures. The production of adipic acid, a C_6 hexyl terminated by two end-on carboxylic groups, is unique and indicative of the role of microcavity inclusion. This is an important step in nylon production.

Figure 9i.4 illustrates the full complexity of the catalytic reaction cycle of radical chain catalysis induced by reducible cations located in the framework of $AlPO_4$ materials. It shows the interplay of CH activation, O_2 activation, and redox catalysis.

10.8 Summary of Zeolite Catalysis

Zeolite catalysis, as discussed in this chapter as well as the previous chapter, provides a rich variety of reactions that can be used to produce a wide range of products. The constraints of chemistry within the catalyst micropores results in unique reactivity.

We can distinguish the following elementary reaction steps:

- Proton transfer and proton-activated catalysis. This gives rise to Brønsted acid catalysis.
- Hydride transfer reactions. These are general and usually occur near activated groups such as C=C, keto-enol group combinations. These reactions lead to hydrogenation, dehydrogenation, C—C bond cleavage, and C—C bond formation.
- Heterolytic C—H bond activation by reactive oxidic cationic sites. Oxygen of a reactive (cationic) particle accepts a proton and a metal cation of the same particle binds an alkyl intermediate, inducing a variety of consecutive reactions.
- Charge separation and electron transfer reactions, within or between reactants. These are usually accompanied by steric selectivity and molecular recognition preference.
- Competitive adsorption into CH, C=C, and activated C–C bonds.
- Radical chain reactions quenched by the micropores.

In addition to these reactions, there are the specific effects that are due to zeolite micropore structure and dimension:

- Selective biased concentration of reaction intermediates. This relates to free energies of adsorption that can also depend on micropore occupation.
- Equilibration of intermediates within the micropores versus equilibration with the external reaction environment.
- Steric constraints that relate to the size mismatch between reaction intermediates and the micropores.

References

1 Thomas, J.M. (2012) *Design and Applications of Single-Site Heterogeneous Catalysts*, Imperial College Press.
2 Gribov, E.N., Cocina, D., Spoto, G., Bordiga, S., Ricchiardi, G., and Zecchina, A. (2006) Vibrational and thermodynamic properties of Ar, N_2, O_2, H_2 and CO adsorbed and condensed into (H,Na)-Y zeolite cages as studied by variable temperature IR spectroscopy. *Phys. Chem. Chem. Phys.*, **8**, 1186–1196.
3 Bonelli, B., Areán, C.O., Armandi, M., Delgado, M.R., and Garrone, E. (2008) Variable-temperature infrared spectroscopy studies on the thermodynamics of CO adsorption on the zeolite Ca—Y. *ChemPhysChem*, **9**, 1747–1751.
4 Manoilova, O.V., Peñarroya Mentruit, M., Turnes Palomino, G., Tsyganenko, A., and Otero Areán, C. (2001) Variable-temperature infrared spectrometry of carbon monoxide adsorbed on the zeolite K-ZSM-5. *Vib. Spectrosc.*, **26**, 107–111, http://www.sciencedirect.com/science/article/pii/S0924203101001047 (accessed 31 March 2015).
5 Bordiga, S., Platero, E.E., Areán, C.O., Lamberti, C., and Zecchina, A. (1992) Low temperature CO adsorption on Na-ZSM-5 zeolites: an FTIR investigation. *J. Catal.*, **137**, 179–185, http://www.sciencedirect.com/science/article/pii/002195179290147A (accessed 31 March 2015).

6 Zecchina, A., Bordiga, S., Lamberti, C., Spoto, G., Carnelli, L., and Arehn, C.O. (1994) Low temperature Fourier transform infrared study of the interaction of CO with cations in alkali-metal exchanged ZSM-5 zeolites. *J. Phys. Chem.*, **98**, 9577–9582.

7 Kazansky, V.B., Serykh, A.I., and Pidko, E.A. (2004) DRIFT study of molecular and dissociative adsorption of light paraffins by H-ZSM-5 zeolite modified with zinc ions: methane adsorption. *J. Catal.*, **225**, 369–373.

8 Rabo, J.A. and Gajda, G.J. (1989) Acid function in zeolites: recent progress. *Catal. Rev.*, **31**, 385–430.

9 Blatter, F., Sun, H., Vasenkov, S., and Frei, H. (1998) Photocatalyzed oxidation in zeolite cages. *Catal. Today*, **41**, 297–309.

10 Pidko, E.A. and Van Santen, R.A. (2006) Confined space-controlled olefin-oxygen charge transfer in zeolites. *J. Phys. Chem. B*, **110**, 2963–2967.

11 Yang, G., Pidko, E.A., and Hensen, E.J.M. (2013) Structure, stability, and Lewis acidity of mono and double Ti, Zr, and Sn framework substitutions in BEA zeolites: a periodic density functional theory study. *J. Phys. Chem. C*, **117**, 3976–3986. doi: 10.1021/jp310433r

12 Yang, G., Pidko, E.A., and Hensen, E.J.M. (2013) The mechanism of glucose isomerization to fructose over Sn-BEA zeolite: a periodic density functional theory study. *ChemSusChem*, **6**, 1688–1696.

13 Moliner, M., Román-Leshkov, Y., and Davis, M.E. (2010) Tin-containing zeolites are highly active catalysts for the isomerization of glucose in water. *Proc. Natl. Acad. Sci. U.S.A.*, **107**, 6164–6168.

14 de Graauw, C.F., Peters, J.A., van Bekkum, H., and Huskens, J. (1994) Meerwein–Ponndorf–Verley reductions and Oppenauer oxidations: an integrated approach. *Synthesis*, **10**, 1007–1017. doi: 10.1055/s-1994-25625

15 Schoonheydt, R.A., Geerlings, P., Pidko, E.A., and van Santen, R.A. (2012) The framework basicity of zeolites. *J. Mater. Chem.*, **22**, 18705.

16 Pidko, E.A., Hensen, E.J.M., and van Santen, R.A. (2012) Self-organization of extraframework cations in zeolites. *Proc. R. Soc. London, Ser. A*, **468**, 2070–2086. doi: 10.1098/rspa.2012.0057

17 Pidko, E.A., van Santen, R.A., and Hensen, E.J.M. (2009) Multinuclear gallium-oxide cations in high-silica zeolites. *Phys. Chem. Chem. Phys.*, **11**, 2893–2902.

18 Pidko, E.A. and Van Santen, R.A. (2009) Structure-reactivity relationship for catalytic activity of gallium oxide and sulfide clusters in zeolite. *J. Phys. Chem. C*, **113**, 4246–4249.

19 Hensen, E.J.M., Pidko, E.A., Rane, N., and Van Santen, R.A. (2007) Water-promoted hydrocarbon activation catalyzed by binuclear gallium sites in ZSM-5 zeolite. *Angew. Chem. Int. Ed.*, **46**, 7273–7276.

20 Panov, G.I. (2000) Advances in oxidation catalysis; oxidation of benzene to phenol by nutrous oxide. *CATTECH*, **4**, 18–31. doi: 10.1023/A:1011991110517

21 Panov, G.I., Sheveleva, G.A., Kharitonov, A.S., Romannikov, V.N., and Vostrikova, L.A. (1992) Oxidation of benzene to phenol by nitrous oxide over Fe-ZSM-5 zeolites. *Appl. Catal., A*, **82**, 31–36.

22 Iwamoto, M. and Hamada, H. (1991) Removal of nitrogen monoxide from exhaust gases through novel catalytic processes. *Catal. Today*, **10**, 57–71.

23 Burch, R., Breen, J.P., and Meunier, F.C. (2002) A review of the selective reduction of NOx with hydrocarbons under lean-burn conditions with non-zeolitic oxide and platinum group metal catalysts. *Appl. Catal., B*, **39**, 283–303.

24 Beutel, T., Sárkány, J., Lei, G.-D., Yan, J.Y., and Sachtler, W.M.H. (1996) Redox chemistry of Cu/ZSM-5. *J. Phys. Chem.*, **100**, 845–851. doi: 10.1021/jp952455u

25 Sata, T., Yamaguchi, T., Kawamura, K., and Matsusaki, K. (1997) Transport numbers of various anions relative to chloride ions in modified anion-exchange membranes during electrodialysis. *J. Chem. Soc., Faraday Trans.*, **93**, 457–462.

26 Obuchi, A., Wögerbauer, C., Köppel, R., and Baiker, A. (1998) Reactivity of nitrogen containing organic intermediates in the selective catalytic reduction of NO(x) with organic compounds: A model study with tert-butyl substituted nitrogen compounds. *Appl. Catal., B*, **19**, 9–22.

27 Kapteijn, F., Marban, G., Rodriguez-Mirasol, J., and Moulijn, J.A. (1997) Kinetic analysis of the decomposition of nitrous oxide over ZSM-5 catalysts. *J. Catal.*, **167**, 256–265.

28 Heyden, A., Bell, A.T., and Keil, F.J. (2005) Kinetic modeling of nitrous oxide decomposition on Fe-ZSM-5 based on parameters obtained from first-principles calculations. *J. Catal.*, **233**, 26–35.

29 Li, G., Pidko, E.A., Van Santen, R.A., Feng, Z., Li, C., and Hensen, E.J.M. (2011) Stability and reactivity of active sites for direct benzene oxidation to phenol in Fe/ZSM-5: a comprehensive periodic DFT study. *J. Catal.*, **284**, 194–206.

30 Li, Y., Feng, Z., van Santen, R.A., Hensen, E.J.M., and Li, C. (2008) Surface functionalization of SBA-15-ordered mesoporous silicas: oxidation of benzene to phenol by nitrous oxide. *J. Catal.*, **255**, 190–196. doi: 10.1002/cctc.201200612

31 Chernyavsky, V.S., Pirutko, L.V., Uriarte, A.K., Kharitonov, A.S., and Panov, G.I. (2007) On the involvement of radical oxygen species O^- in catalytic oxidation of benzene to phenol by nitrous oxide. *J. Catal.*, **245**, 466–469.

32 Sobolev, V.I., Dubkov, K.A., Panna, O.V., and Panov, G.I. (1995) Selective oxidation of methane to methanol on a FeZSM-5 surface. *Catal. Today*, **24**, 251–252.

33 Groothaert, M.H., Smeets, P.J., Sels, B.F., Jacobs, P.A., and Schoonheydt, R.A. (2005) Selective oxidation of methane by the bis(mu-oxo)dicopper core stabilized on ZSM-5 and mordenite zeolites. *J. Am. Chem. Soc.*, **127**, 1394–1395.

34 Hammond, C., Jenkins, R.L., Dimitratos, N., Lopez-Sanchez, J.A., Ab Rahim, M.H., Forde, M.M., Thetford, A., Murphy, D.M., Hagen, H., Stangland, E.E. *et al.* (2012) Catalytic and mechanistic insights of the low-temperature selective oxidation of methane over Cu-promoted Fe-ZSM-5. *Chem. Eur. J.*, **18**, 15735–15745.

35 Siegbahn, P.E.M. and Crabtree, R.H. (1997) Mechanism of C–H activation by diiron methane monooxygenases: quantum chemical studies. *J. Am. Chem. Soc.*, **119**, 3103–3113. doi: 10.1021/ja963939m

36 Solomon, E.I., Ginsbach, J.W., Heppner, D.E., Kieber-Emmons, M.T., Kjaergaard, C.H., Smeets, P.J., Tian, L., and Woertink, J.S. (2011) Copper dioxygen (bio)inorganic chemistry. *Faraday Discuss.*, **148**, 11–39; discussion 97–108.

37 Woertink, J.S., Smeets, P.J., Groothaert, M.H., Vance, M.A., Sels, B.F., Schoonheydt, R.A., and Solomon, E.I. (2009) A $[Cu_2O]^{2+}$ core in Cu-ZSM-5, the active site in the oxidation of methane to methanol. *Proc. Natl. Acad. Sci. U.S.A.*, **106**, 18908–18913.

38 Schröder, D., Schwarz, H., and Shaik, S. (2000) in *Metal-Oxo and Metal-Peroxo Species in Catalytic Oxidations* (ed. B. Meunier), Springer-Verlag, Berlin, pp. 91–123.

39 Ensing, B., Buda, F., Gribnau, M.C.M., and Baerends, E.J. (2004) Methane-to-methanol oxidation by the hydrated iron(IV) oxo species in aqueous solution: a combined DFT and car-parrinello molecular dynamics study. *J. Am. Chem. Soc.*, **126**, 4355–4365.

40 Yoshizawa, K., Shiota, Y., and Yamabe, T. (1998) Abstraction of the hydrogen atom of methane by iron-oxo species: the concerted reaction path is energetically more favorable. *Organometallics*, **17**, 2825–2831, http://www.scopus.com/inward/record.url?eid=2-s2.0-11644305451&partnerID=40&md5=5eeefeac3987cc6387465c16e70fcd19 (accessed 28 September 2016).

41 Groves, J.T. and Nemo, T.E. (1983) Epoxidation reactions catalyzed by iron porphyrins. Oxygen transfer from iodosylbenzene. *J. Am. Chem. Soc.*, **105**, 5786–5791. doi: 10.1021/ja00356a015

42 Siegbahn, P.E.M. (2001) O–O bond cleavage and alkane hydroxylation in methane monooxygenase. *J. Biol. Inorg. Chem.*, **6**, 27–45.

43 Grundner, S., Markovits, M.A.C., Li, G., Tromp, M., Pidko, E.A., Hensen, E.J.M., Jentys, A., Sanchez-Sanchez, M., and Lercher, J.A. (2015) Single-site trinuclear copper oxygen clusters in mordenite for selective conversion of methane to methanol. *Nat. Commun.*, **6**, 7546. doi: 10.1038/ncomms8546

44 Corma, A., Nemeth, L.T., Renz, M., and Valencia, S. (2001) Sn-zeolite beta as a heterogeneous chemoselective catalyst for Baeyer-Villiger oxidations. *Nature*, **412**, 423–425.

45 Corma, A. (2004) Attempts to fill the gap between enzymatic, homogeneous, and heterogeneous catalysis. *Catal. Rev. Sci. Eng.*, **46**, 369–417.

46 Fenn, T.D., Ringe, D., and Petsko, G.A. (2004) Xylose isomerase in substrate and inhibitor michaelis states: atomic resolution studies of a metal-mediated hydride shift. *Biochemistry*, **43**, 6464–6474. doi: 10.1021/bi049812o

47 Garcia-Viloca, M., Alhambra, C., Truhlar, D.G., and Gao, J. (2003) Hydride transfer catalyzed by xylose isomerase: mechanism and quantum effects. *J. Comput. Chem.*, **24**, 177–190. doi: 10.1002/jcc.10154

48 Pidko, E.A., Degirmenci, V., and Hensen, E.J.M. (2012) On the mechanism of Lewis acid catalyzed glucose transformations in ionic liquids. *ChemCatChem*, **4**, 1263–1271.

49 Holm, M.S., Saravanamurugan, S., and Taarning, E. (2010) Conversion of sugars to lactic acid derivatives using heterogeneous zeotype catalysts. *Science*, **328**, 602–605.

50 Taarning, E., Saravanamurugan, S., Holm, M.S., Xiong, J., West, R.M., and Christensen, C.H. (2009) Zeolite-catalyzed isomerization of triose sugars. *ChemSusChem*, **2**, 625–627.
51 Kovalevsky, A.Y., Hanson, L., Fisher, S.Z., Mustyakimov, M., Mason, S.A., Trevor Forsyth, V., Blakeley, M.P., Keen, D.A., Wagner, T., Carrell, H.L. *et al.* (2010) Metal ion roles and the movement of hydrogen during reaction catalyzed by D-xylose isomerase: a joint x-ray and neutron diffraction study. *Structure*, **18**, 688–699.
52 Pidko, E.A., Degirmenci, V., Van Santen, R.A., and Hensen, E.J.M. (2010) Glucose activation by transient Cr^{2+} dimers. *Angew. Chem. Int. Ed.*, **49**, 2530–2534.
53 Clerici, M. (1993) Epoxidation of lower olefins with hydrogen peroxide and titanium silicalite. *J. Catal.*, **140**, 71–83, http://www.sciencedirect.com/science/article/pii/S0021951783710699 (accessed 28 September 2016).
54 Thomas, J.M., Catlow, C.R.A., and Sankar, G. (2002) Determining the structure of active sites, transition states and intermediates in heterogeneously catalysed reactions. *Chem. Commun.*, **24**, 2921–2925.
55 Sheldon, R.A. (1982) New catalytic methods for selective oxidation. *J. Mol. Catal.*, **20**, 1–26.
56 Thomas, J.M. and Raja, R. (2005) Design of a "green" one-step catalytic production of epsilon-caprolactam (precursor of nylon-6). *Proc. Natl. Acad. Sci. U.S.A.*, **102**, 13732–13736.
57 Dugal, M., Sankar, G., Raja, R., and Thomas, J.M. (2000) Designing a heterogeneous catalyst for the production of adipic acid by aerial oxidation of cyclohexane. *Angew. Chem. Int. Ed.*, **39**, 2310.
58 Thomas, J.M., Raja, R., Sankar, G., and Bell, R.G. (2001) Molecular sieve catalysts for the regioselective and shape-selective oxyfunctionalization of alkanes in air. *Acc. Chem. Res.*, **34**, 191–200.
59 Gómez-Hortigüela, L., Corà, F., Sankar, G., Zicovich-Wilson, C.M., and Catlow, C.R.A. (2010) Catalytic reaction mechanism of Mn-doped nanoporous aluminophosphates for the aerobic oxidation of hydrocarbons. *Chem. Eur. J.*, **16**, 13638–13645. doi: 10.1002/chem.201090227

11

Reducible Solid State Catalysts

11.1 Introduction

The main topic of this chapter is the surface reactivity of reducible inorganice materials, primary transition metal oxides. In addition we will briefly discuss catalysis by sulfide catalysts. We will present selective oxidation and reduction reactions as well as hydrodesulfization (HDS) and related reactions, in addition to others that include electrocatalytic and photocatalytic water splitting. Also included is a section on the chloride supported systems, which are applied in Ziegler-Natta polymerization catalysis.

While most of the solid state catalysts considered consist of the surface of a transition metaloxide or sulfide, this chapter will also deal with single-site coordination compounds attached to non-reducible high surface area oxide supports as used for the metathesis reaction.

Some of these reactions have already been introduced earlier in Chapters 2 and 3. Here we provide a more in-depth discussion with a focus on the molecular aspects of the reactions.

The chemical binding aspects of transition metal oxides and their surfaces are discussed in detail in the next section 11.2. This will be followed by the later sections that discuss the mechanism of the different catalytic reactions.

11.2 Chemical Bonding of Transition Metal Oxides and Their Surfaces

11.2.1 Electronic Structure of the Metal Oxide Chemical Bond

The chemical bond in a metal oxide is polar and its energy consists of both an ionic or electrostatic contribution and a covalent contribution to the bond energy. An interesting consequence of the important contribution of the electrostatic energy to the chemical bond energy is the trend of the surface energy of the oxide compounds.

While for the transition metals the surface energies follow the trends in bulk cohesive energies (Chapter 7, Figure 1i.1) this is generally not the case for the

Modern Heterogeneous Catalysis: An Introduction, First Edition. Rutger A. van Santen.

transition metal oxides. The trends in surface energies of the ionic oxides are contrary to the trends for the heats of formation of the corresponding bulk oxides. This is illustrated by Figures 1i.2 in this chapter.

When an ionic solid is considered to consist of ions approximated as point charges equal to their valence, the electrostatic energy is given by the Madelung energy, which we defined earlier in Chapter 6 (Eqs. (6.1a)–(6.1c)).

The Madelung constant in the Madelung energy expression defines the electrostatic energy of an ionic solid as a function of its structure. It is the proportionality constant in the electrostatic energy expression of the solid. The electrostatic energy is also proportional to the product of ion charges and inversely proportional to the nearest neighbor distance of cation and anion.

The Madelung constant is the same for materials of the same structure. Table 1i.1 compares the Madelung constants for different structures and Table 1i.2 compares its value at the surface with its value in the bulk for different surfaces of the rutile and anatase structures of TiO_2. These are shown in Figures 1i.1 with the corresponding DFT calculated surface energies.

The ionic solids have low surface energies. This is because when their cation and anion charges balance the overall surface charge is zero. Therefore the electrostatic interaction between the surfaces becomes zero at large distances.

The low surface energy of an ionic solid is due to the small decrease in the electrostatic interaction at the surface. The decrease is so small because the electrostatic interactions are long-range and hence do not depend strongly only on the nearest neighbor interactions (see Table 1i.2). This is different from covalent interactions. The covalent interaction depends strongly only on nearest neighbor atom coordination. It relates to the distribution of electrons over the bonding, non-bonding, antibonding orbitals or chemical bond.

The electrostatic interaction due to the charges on cations and anions of the oxide contributes considerably to their heats of formation, especially for the ionic oxides of the metals located at the left part of a row in the periodic system.

In Chapter 7, we discussed the trends in adsorption energies of oxygen to transition metal surfaces. The adatom energies as well as the polarity of their respective M—O bonds increase for transition metals from right to left along a row of the periodic table. The same trend is found for the metal oxides. This trend relates to the decreasing ionization potential of the metal atoms, which is exhibited by corresponding decreases in the work functions of the corresponding metals (see Chapter 7, Figure 1i.4).

With a change in position of the metal along a row of the periodic system the relative contribution of the covalent interaction to the M—O chemical bond energy increases with the decrease in electrostatic interaction.

The trend in surface energies of the ionic oxides anti-parallel to the trend for the heats of formation of the corresponding bulk oxides is a consequence of the large electrostatic contribution to the bulk energy of formation.

The electrostatic contribution to the surface energies varies only slightly between the oxides, whereas the change in bulk energies is dominated by variation in the covalent as well as electrostatic energy contributions.

Insert 1: The Ionic Bond

Figure 1i.1 shows the surface structures and the respective surface energies of the anatase and rutile surfaces. Table 1i.1 gives the Madelung constants for some bulk oxides. Table 1i.2 compares the Madelung constants and the electrostatic energies of TiO_2 surfaces and bulk. Figure 1i.2 compares the changes in bulk and surface energies for several oxides.

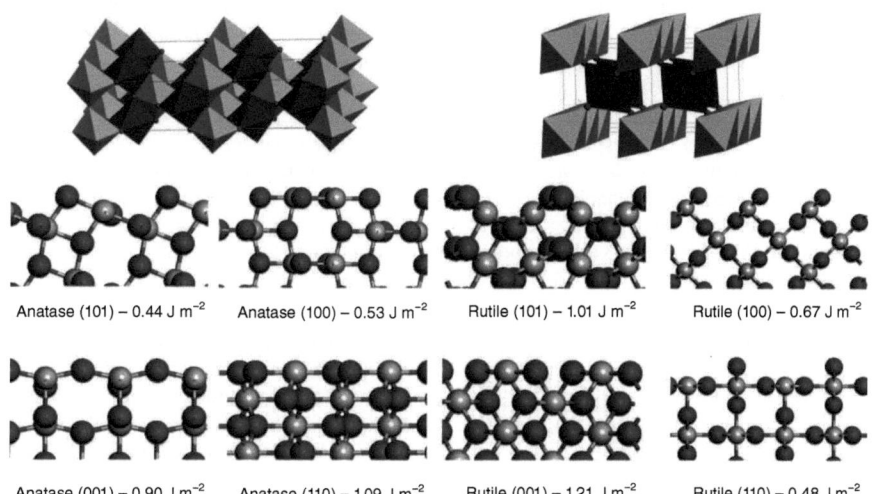

Anatase (101) – 0.44 J m^{-2} Anatase (100) – 0.53 J m^{-2} Rutile (101) – 1.01 J m^{-2} Rutile (100) – 0.67 J m^{-2}

Anatase (001) – 0.90 J m^{-2} Anatase (110) – 1.09 J m^{-2} Rutile (001) – 1.21 J m^{-2} Rutile (110) – 0.48 J m^{-2}

Figure 1i.1 The structures, the common surface terminations, and the corresponding surface free energies computed by DFT for the TiO_2 anatase bulk and TiO_2 rutile bulk. (After [1, 2].)

Table 1i.1 Madelung constants of several oxides with different crystal structures.

PdO	Cu_2O	TiO_2 (rutile)	TiO_2 (anatase)	SiO_2
1.605	4.44	4.82	4.80	4.44

Source: Greenwood 1968 [4]. Reproduced with permission of Elsevier.

Table 1i.2 Madelung potentials Φ and Madelung constants M for TiO_2 anatase (100) (a) surface and TiO_2 rutile (110) (b) surface.

(a)

l	$\Phi^{(5)}(Ti^{4+})$(e nm^{-1})	$\Phi^{(2)}(O^{2-})$(e nm^{-1})	$\Phi^{(3)}(O^{2-})$(e nm^{-1})	M
0	−28.66	17.29	19.92	4.45053
1	−29.56	19.66	19.66	4.76644
2	−29.57	19.63	19.63	4.76539
3	−29.59	19.61	19.61	4.76531

(Continued)

Insert 1: (Continued)

Table 1i.2 (Continued)

(b)

l	$\Phi^{(5)}(Ti^{4+})$(e nm^{-1})	$\Phi^{(6)}(Ti^{4+})$(e nm^{-1})	$\Phi^{(3)}(O^{2-})$(e nm^{-1})	$M^{(5)}$	$M^{(6)}$
0	−20.53	−24.16	26.87	4.61313	4.99642
1	−22.68	−22.53	26.41	4.77649	4.76231
2	−22.64	−22.67	26.43	4.77163	4.77254
3	−22.67	−22.67	26.36	4.77185	4.77182

"l" denotes the layer number with respect to the surface. Coordination numbers of titanium and oxygen in the surface layer are indicated as superscripts of Φ. The differences of the Madelung constant values compiled in this table with Table 1i.1 are due to small differences in structural parameter values.
Source: Woning and van Santen 1983 [5]. Reproduced with permission of Elsevier.

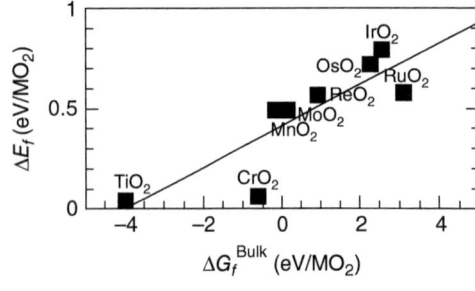

Figure 1i.2 DFT-calculated surface energies ΔE_f for rutile metal oxide (110) surfaces in eV/MO$_2$ versus calculated bulk heats of formation ΔG_f^{Bulk}. (Mowbray et eal. 2011 [3]. Reproduced with permission of American Chemical Society.)

The relatively low surface energy of the oxides versus that of the metals has an interesting consequence for the reducibility trend of small oxide particles as a function of particle size. When particles become smaller, the energy differences between the oxide and the metal increase. Since in contrast to the oxides the surface energy of the transition metals strongly depends on surface atom coordination number. This causes small oxidized particles to be relatively more resistant to reduction than the corresponding large particles. It is consistent with the observation that the reduction of large oxidized metal particles generally occurs at lower temperatures than the reduction of smaller oxide particles.

11.2.2 The Electronic Structure of the Transition Metal Oxides

The electronic structure of the chemical bond in the dominantly ionic oxide is very different than in a dominantly covalent oxide. In both cases, bonding and antibonding molecular M—O orbital fragments are formed, but the interaction of metal atom valence atomic orbitals and oxygen atom valence atomic orbitals are very different (see Insert 2).

Insert 2: The Electronic Structure of the Oxides

The PdO bulk structure and surface structure of the PdO(101) surface are given in Chapter 7 (Figures 10i.1 and 10i.3). In bulk PdO, the Pd^{2+} cation shares four oxygen atoms that are located in a planar square.

In Figure 2i.1, the partial density of states (PDOS) on the Pd ion is resolved into the contributions of the respective Pd d-atomic orbitals.

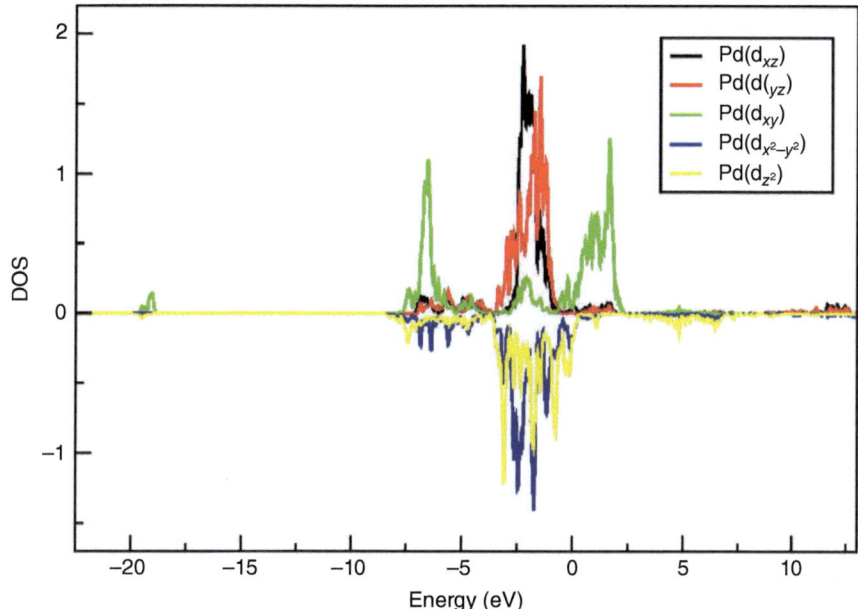

Figure 2i.1 Decomposition of the Pd 4d PDOS into its components. All the PDOS have been calculated at the DFT-GGA level using the VASP code. (van Santen *et al*. 2015 [6]. Reproduced with permission of Elsevier.)

The bond order density of the PdO bond are shown in Chapter 7, Figure 10i.3. The structures of bulk TiO_2 and its surfaces are presented in Figure 1i.1. In TiO_2, the Ti^{4+} ion is octahedrally coordinated with six oxygen atoms. The electronic structure of TiO_2 is shown in Figures 2i.2.

The electronic features PDOS and –COHP (crystal orbital Hamiltonian population) plotted in these figures are defined in Chapter 6, Insert 1.

(Continued)

Insert 2: (Continued)

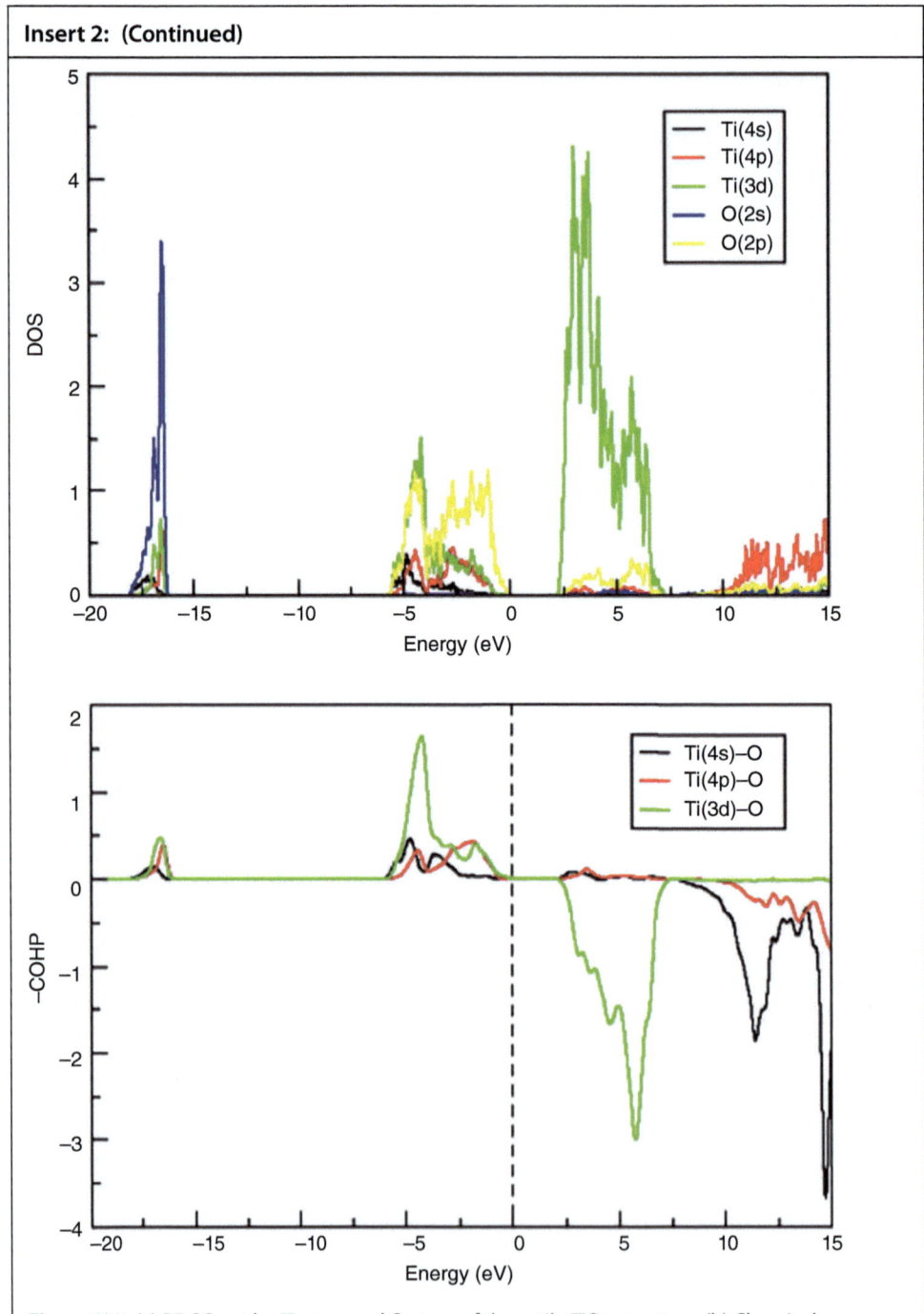

Figure 2i.2 (a) PDOS on the Ti atom and O atom of the rutile TiO_2 structure. (b) Chemical bonding analysis (COHP) of the free TiO bond. The PDOS has been computed at the DFT–GGA level using VASP code. The COHP analysis has been performed using the Lobster code based on VASP data. In both plots, the zero energy level was chosen to coincide with the Fermi level E_F. (van Santen *et al.* 2015 [6]. Reproduced with permission of Elsevier.)

The bonding or antibonding nature of respective orbital fragments follows from the sign of the bond order densities ρ_{ij}^h in the –COHP plots of PdO and TiO$_2$ respectively.

In PdO, we can distinguish three electron density regimes: the electron density between -3 and -8 eV (regime I), the electron density between -3 and $+1$ eV (regime II), and the electron density above $+1$ eV (regime III). Comparison with the valence electron distribution of Pd metal (Chapter 6, Figure 5i.2) shows that the d-valence electrons of Pd metal have a finite PDOS density only in regime II.

As follows from the COHP plots in Chapter 7, Figure 10i.3b, the electron density regime I in PdO is dominated by the bonding M—O molecular orbital fragments. The electron density in energy intervals II and III is antibonding. The three partial density regimes I, II, and III, correspond with the ligand field splitting as sketched in Figure 3i.1. The Pd $4d_z^2, d_x^2$ and the $4d_y^2$ atomic orbitals participate in bonding orbitals. The $4d_{x^2-y^2}$ atomic orbital participates in the antibonding orbitals below the Fermi-level and the $4d_{xy}$ atomic orbital in antibonding orbitals above the Fermi-level. In region III, the appearance of a substantial PDOS for the Pd d-valence electron density as well as O-2p atomic orbital density above the Fermi-level also implies that the d-valence electron band as well as the O 2p atomic orbitals are not completely occupied. Indeed, we would expect the Pd cation to have a 2+ charge and the O atom a 2– charge. Clearly, in the PdO compound the charges on the atoms are lower, because the atomic orbitals of O are not completely filled.

Different from PdO the antibonding molecular orbital regime dominated by the Ti 3d-valence atomic orbitals is located completely above the Fermi level. Similarly as in PdO, and in agreement with the ligand field splitting for an octahedrally coordinated metal cation, the PDOS spectrum above the Fermi-level is split into two regimes. At lower energies, the larger peak is mainly non-bonding and the higher smaller peak is antibonding.

The extreme difference in PDOS ratio for the Ti d-valence atomic orbitals and the O 2p atomic orbitals in the respective bonding valence band regime and antibonding conduction band regime is indicative of the high ionicity of the TiO$_2$ metal oxide chemical bond. The Ti d-valence electron density is nearly competely pushed above the Fermi level. The total electron population on the Ti atom is nearly zero (there is only a small PDOS contribution below the Fermi level). The charge on Ti is therefore slightly less than 4+. Below the Fermi-level the atomic orbitals of O are nearly completely occupied.

The narrow bandwidth of the respective molecular orbitals also indicates that the d-electrons are mainly localized on the metal cations.

As we already indicated valence electronic ligand field theory of coordination complexes [7] can be used to model the electronic structure of the metal cations of the transition metal oxides. It is a simplified model and considers only the local environment of the metal atom, while direct metal atom–metal atom interactions are ignored [8, 9]. The interaction between the metal and oxygen atom is viewed to involve only one doubly occupied σ-symmetric atomic orbital on the oxygen atom. Figure 3i.1 shows the ligand field molecular orbital correlation diagram for the Pd^{2+} cation in PdO.

11 Reducible Solid State Catalysts

Insert 3: Ligand Field Theory

Figures 3i.1 and 3i.2 illustrate the construction of ligand field splitting of the d-valence atomic orbitals on the metal cation for different coordination of the cations. Figure 3i.1 shows the resulting split of the d-valence electron energies in PdO. The Pd ion is planar and fourfold coordinated; Figure 3i.2 shows the d-valence electron energy distribution for a cation such as Mn^{2+} that is octahedrally coordinated. The low- and high-spin cases are compared. Within ligand field theory, the ligand field splitting is due to the bonding and antibonding interactions with the atomic orbitals mainly localized on the neighbouring anion atoms.

Crystal field theory provides a way to estimate the splitting of the d atomic orbital levels based on an electrostatic interaction model. This is illustrated by Figure 3i.3. Table 3i.1 summarizes the effect of crystal field splitting on chemical bond stabilization for low- and high-spin electron distributions and Figure 3i.4 illustrates this for solvation energies of metal cations. Water surrounds these cations octahedrally.

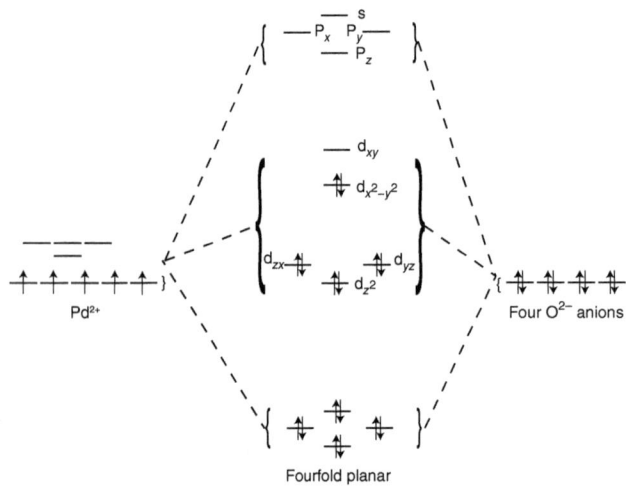

Figure 3i.1 Local ligand field electronic structure model of the chemical bond of a planar fourfold-coordinated Pd^{2+} ion in PdO. (van Santen et al. 2015 [6]. Reproduced with permission of Elsevier.)

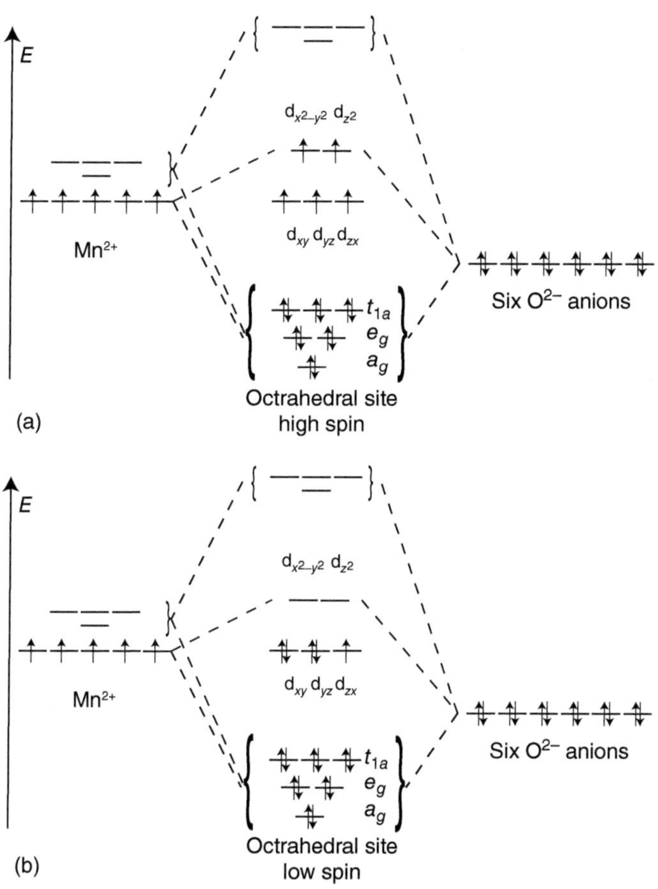

11.2 Chemical Bonding of Transition Metal Oxides and Their Surfaces

Figure 3i.2 Ligand field splitting in an octahedral environment with (a) high- and (b) low-spin d-electron distributions. (van Santen et al. 2015 [6]. Reproduced with permission of Elsevier.)

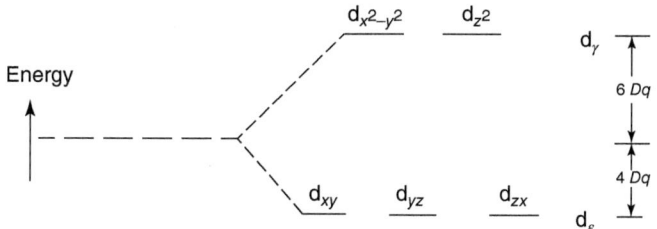

Figure 3i.3 Influence of an octahedral field on the position of the energy levels of the d orbitals according to crystal field theory [10].

(Continued)

Insert 3: (Continued)

Table 3i.1 Ligand field stabilizing energy (LFSE) for the d atomic orbitals in an octahedral field.

Number of d electrons	0	1	2	3	4	5	6	7	8	9	10
High-spin LFSE (in Dq)	0	4	8	12	6	0	4	8	12	6	0
Low-spin LFSE (in Dq)	0	4	8	12	16	20	24	18	12	6	0
Difference (in Dq)	0	0	0	0	10	20	20	10	0	0	0

Source: Greenwood 1968 [4]. Reproduced with permission of Elsevier.

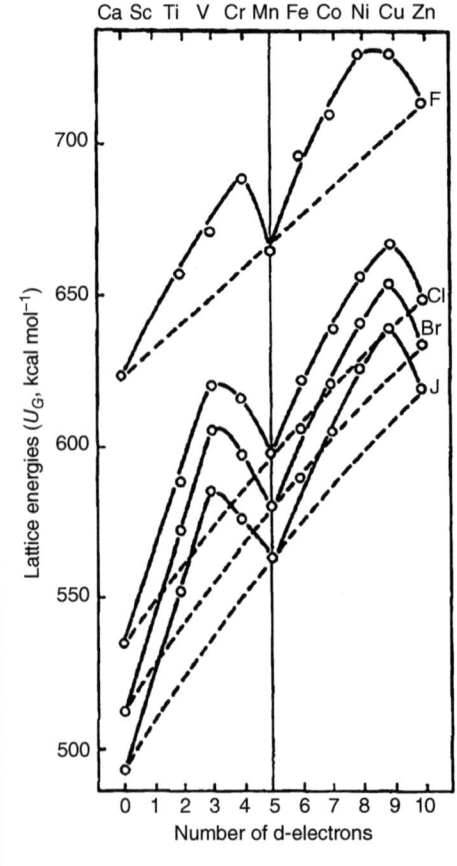

Figure 3i.4 Ligand field stabilization energy of dihalogenides from the first row of transition elements. (Waddington 1959 [11]. Reproduced with permission of Elsevier.)

Within the ligand field model, the five 4d atomic orbitals of Pd^{2+} hybridize with its one 5s and three 5p valence atomic orbitals. The Pd ion in PdO has a planar coordination with four oxygen atoms. This results in four bonding and four antibonding orbital combinations. Orbital overlap between the Pd d_{xz}, d_{xy} atomic orbitals, and O σ atomic orbitals is prevented by symmetry, and the Pd

d_{z^2} atomic orbital interacts only weakly with the surrounding O atoms because it is not directed toward them (the z-axis is perpendicular to the Pd-4O plane). These orbitals are therefore non-bonding. The two $d_{x^2-y^2}$ (directed between the O atoms) and d_{xy} (directed toward the O atoms) atomic orbitals are antibonding. The d_{xy} orbital, which mixes strongly with the O orbitals, is unoccupied. Within the Pauling valence bond hybridization model [12], the Pd valence atomic orbitals can be characterized as four d_{sp^2} hybridized orbitals, directed to the corners of a square.

The calculated orbital scheme used to construct Figures 3i.1 and 3i.2 does not completely agree with the simplified ligand field splitting model because the three 2p atomic orbitals of the O atom mix in the solid, contrary to the simplifying rule, with one σ atomic orbital on the O atom.

When a metal atom is octahedrally coordinated to six oxygen atoms, as in TiO_2, the symmetry of ligand field splitting is as illustrated in Figure 3i.2. Ligand field splitting predicts three non-bonding degenerate t_{2g} atomic orbitals and two antibonding degenerate e_g levels (for group theory notations for the symmetries of orbitals see [13]). In the electronic structure of TiO_2 shown in Figure 2i.2, these can be recognized as the unoccupied conduction band d-atomic orbitals on Ti in the +4 to +8 eV energy regime. Note that the intensity is 3/2-fold higher for the three t_{2g} orbitals than for the two e_g orbitals, and that the calculated antibonding character is substantially higher for the e_g orbitals than for the t_{2g} orbitals.

The schematic illustration of ligand field splitting of MnO (see Figure 3i.2) illustrates an important additional feature of 3d transition metal cations. When electron-electron repulsion is weak compared to the ligand field splitting energies, the lower energies will be doubly occupied. This gives a low total electronic spinstate (Figure 3i.2b). The large one-atomic repulsion of two electrons in the same atomic orbitals may be released by pushing an electron into an empty higher energy atomic orbital. This state has a lower energy when the orbitals energy difference is less than the intraorbital electron–electron repulsion energy. It will result in the high spin state sketched in Figure 3i.2a which prescribes single electrons in separate orbitals to have parallel spin.

An alternative to the covalent ligand field splitting model is the crystal field splitting model. This can be used to estimate the effect of low spin–high spin excitation on the bond energy.

The crystal field model is an electrostatic model, which calculates relative energies of the cation d-atomic orbitals based on their relative orientation with respect to the negative point charges of the anions. The electron in an orbital not directed toward the point charges has a lower energy than the average electrostatic energy, while an electron in an atomic orbital directed toward the point charges has a higher than average energy. This is illustrated in Figure 3i.3 for a cation with octahedral coordination. Orbitals below the average electrostatic energy are stabilized, orbitals above the average electrostatic energy are destabilized.

By counting the number of electrons in the up or down d atomic orbitals, we can estimate the crystal field splitting stabilization as a function of the number of d-electrons on the metal cation. This procedure is illustrated in Table 3i.1. The low spin case shows maximum crystal field splitting stabilization when the system has six electrons, while the high spin case shows a maximum for three and eight electron occupation and a minimum for a five electron occupation. The high spin case is representative for metal cations of the third row. Figure 3i.4 illustrates the double maxima in stabilization energy for third row cations of similar charge when the hydration energy is plotted along a row in the periodic table.

These crystal field stabilization estimates will be important when we discuss trends in the reactivity of third row metal oxides (see Sections 11.3.1 and 11.6.1).

Trends in the covalent contribution to the metal-oxide bond can be deduced from Figure 11.1, which shows calculated PDOS plots of the valence band (the occupied metal oxide orbitals) and conduction band (the unoccupied metal oxide orbitals) electron density distributions for different metal oxides with the rutile structure.

From top to bottom, the metals vary in position along a row in the periodic table from left to right. This implies a gradual increase in both the number of d-valence electrons and in transition metal atom ionization potential.

We recognize the unoccupied Ti d-atomic orbitals in TiO_2, which are part of the bonding conduction band orbitals, and the occupied O 2p atomic orbitals that are part of the antibonding valence electron band orbitals. The bandgap between valence and conduction band is dominated by the difference in electrostatic energy of an electron on the O atom and an electron on the Ti atom (the electron on Ti will have a repulsive interaction in the field of the negatively charged oxygen atoms). The degree of covalency of the chemical bond is low, as deduced from an estimate of the ratio of the PDOS of the oxygen atomic orbitals and metal valence atomic orbitals (see also Chapter 6, Insert 2).

When the d-electron count increases the valence electrons will also occupy the antibonding conduction band orbitals, which will weaken the M—O bond energy. The Fermi level will be positioned within the antibonding valence band regime. For VO_2, the bandgap that arises from the ionicity of the chemical bond is located within the occupied electron energy density regime. This bandgap tends to disappear for the metal oxides on the right side of the periodic system, indicating the increasing covalent character of the M—O bonds.

As we have discussed for PdO (see Insert 2), which has a mainly covalent M—O bonds, the d-valence band density separates into three regimes. Regions I and III, which correspond to the bonding and antibonding M—O orbitals, and region II, dominated by non-bonding and weakly antibonding d atomic orbitals.

In Figure 11.1, which allows for comparison of these electron density regimes of different oxides, the appearance of a region II density is already apparent for CrO_2. When one compares the oxides with transition metals from the left to right along a row of the periodic system, the oxides such as MoO_2, WO_2, or VO_2 are still highly ionic as follows from the low relative contributions of the O atomic orbital in the unoccupied region III electron density PDOS, and the high polarity of their M—O bonds. There is still a bandgap between the region I and region II electron energy densities. On the other hand, for metal oxides such as RuO_2 and IrO_2, the strong signature of the covalent interactions is made apparent by the features of mixed O 2p and transition metal atomic orbital densities in region III and the absence of a bandgap between the regions I and II electron densities.

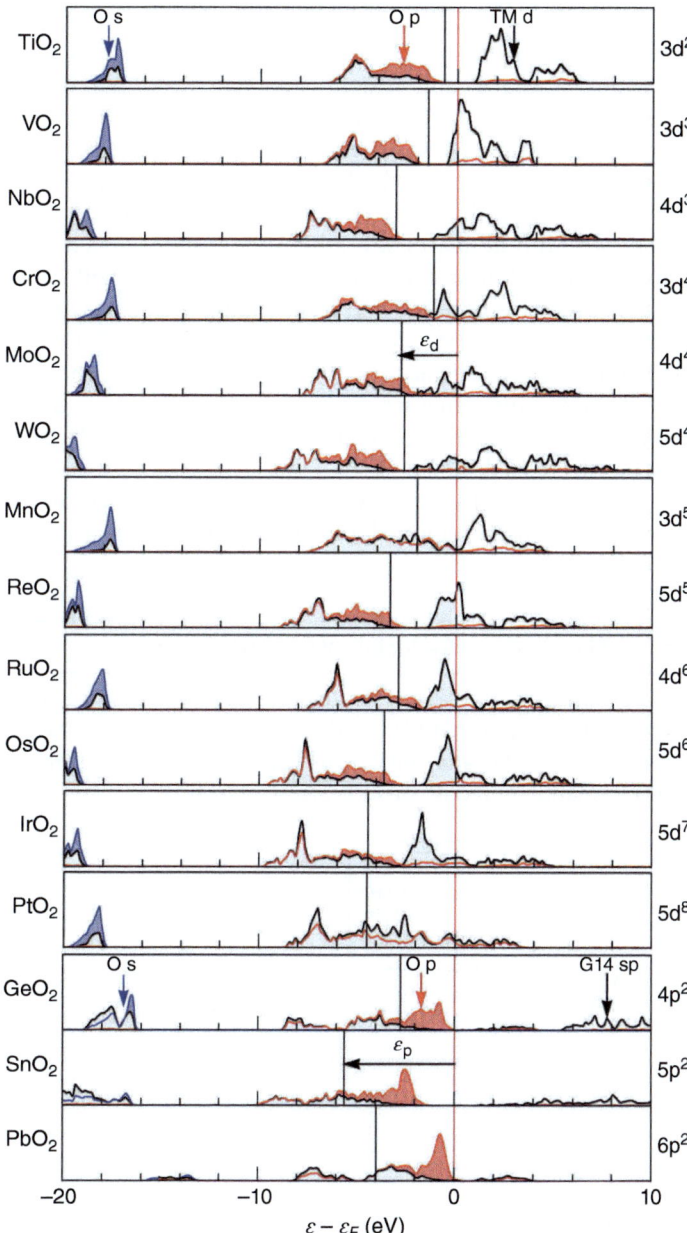

Figure 11.1 Density of states for bulk rutile metal oxides projected onto the transition-metal (TM) d orbitals (black solid line), oxygen p orbitals (red solid line), and oxygen s orbitals (blue solid line) versus energy, ε, in electron volts relative to the Fermi level, ε_F (thin red vertical line). Filling is denoted by the shaded regions. (Mowbray et al. 2011 [3]. Reproduced with permission of American Chemical Society.)

11.2.3 The Electronic Structure of the Transition Metal Oxide Surface

Two types of Pd and O ions are present on the PdO(101) surface, each with different reactivities. Cations and anions are four-coordinated and three-coordinated, respectively (see Figure 10i.1b). The adsorption energy of H_2 to a fourfold-coordinated surface Pd ion is endothermic ($+58\,kJ\,mol^{-1}$), but exothermic on the coordinatively unsaturated Pd ion ($-32\,kJ\,mol^{-1}$) [14].

In a covalent oxide such as PdO, there is a change in PDOS when the coordination number decreases. The atoms of the PdO(101) surface, which have three coordination instead of bulk four coordination, have a smaller difference in energy δ between the respective bonding and antibonding molecular orbital regimes than the bulk. The electronic structures of the atoms at the PdO(101) surface are shown in Figure 4i.1. The surface structure of PdO(101) and the bulk structure of PdO with its electronic structure can be found in Chapter 7, Insert 10. The smaller value of δ implies that the interaction energy between palladium and the oxygen atoms has decreased.

Less of the d-valence electron density on Pd is above the Fermi level due to the reduced energy difference of density regions I and III. This implies that the Pd cation charge at the surface is lower.

Insert 4: Chemical Bonding and Reactivity of Reducible Metal-Oxide Surfaces

For the structure of the Pd(101) surface, we refer to Chapter 7, Figure 10i.1b. Figure 4i.1 shows the electronic structures of three-coordinated Pd and O, respectively, at this surface. Figure 4i.2 gives the structure of bulk CeO_2 and three of its surfaces. Bandgap, surface energies, and oxygen vacancy formation energies are indicated (Table 4i.1).

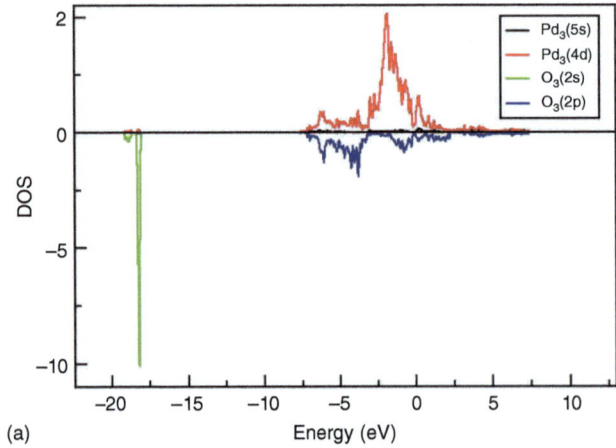

Figure 4i.1 (a) For the PdO(101) surface, the PDOSs associated with the most reactive Pd and O surface atoms (Pd_3-coordinated, O_3-coordinated) are given. (b) Chemical bonding analysis (COHP) for the Pd_3—O_3 bond. All the PDOS have been calculated at the DFT–GGA level using the VASP code, while the COHP has been obtained from a static calculation done in Siesta. (van Santen et al. 2015 [6]. Reproduced with permission of Elsevier.)

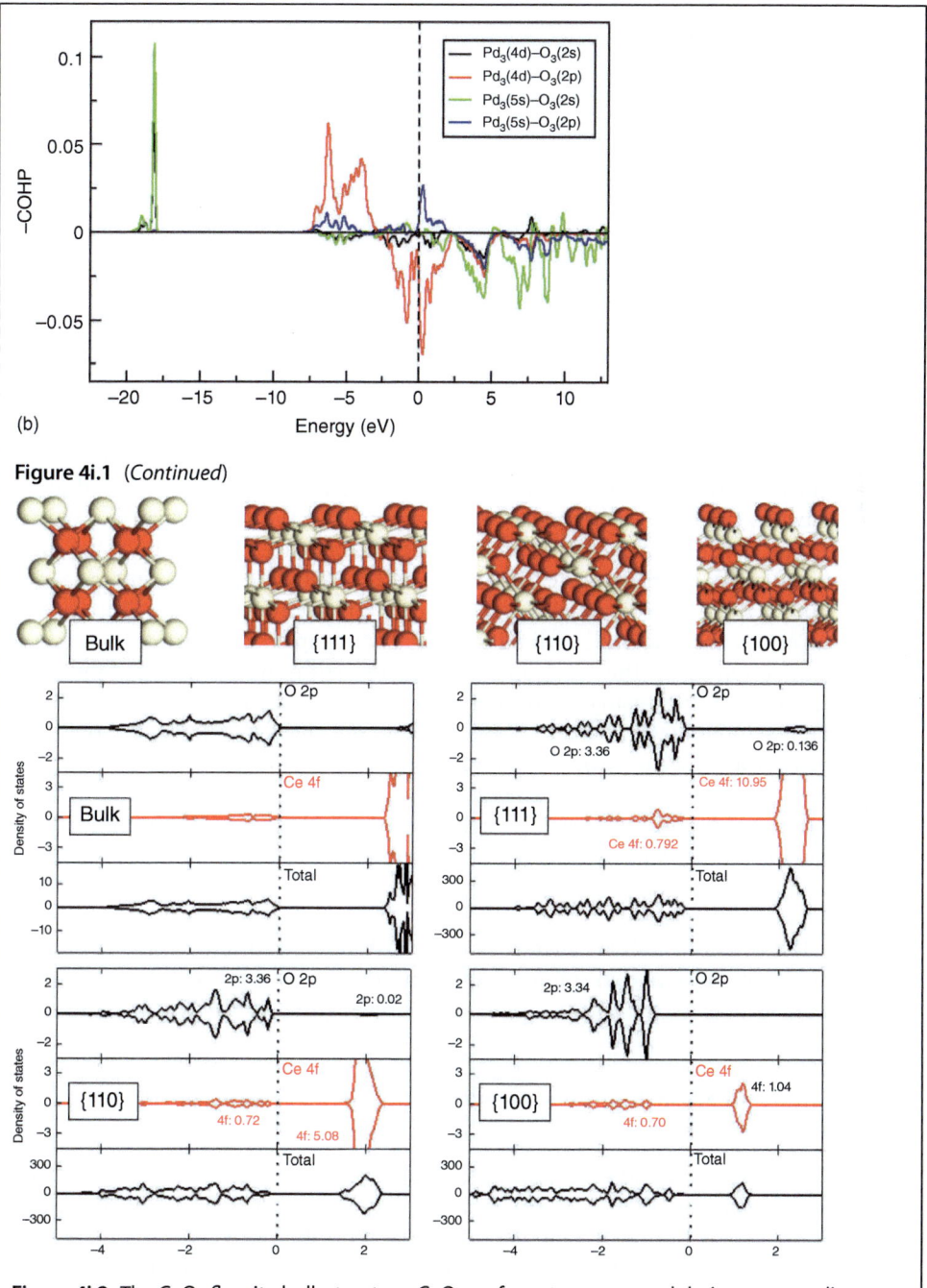

Figure 4i.1 (Continued)

Figure 4i.2 The CeO_2 fluorite bulk structure, CeO_2 surface structures, and their corresponding PDOS (DFT-VASP) [6]. Note that only the PDOS contributions of the Ce 4f and O_{2p} atomic orbitals are shown. From left to right, top to bottom, the spin up and spin down partial density of states of the CeO_2 bulk structure, CeO_2 (111) surface, CeO_2 (110) surface, and CeO_2 (100) surface.

(Continued)

> **Insert 4: (Continued)**
>
> **Table 4i.1** Electronic properties and energies of CeO_2.
>
> | | Bulk | (111) | (110) | (100) |
> |---|---|---|---|---|
> | Band gap (eV) | 2.35 | 2.02 | 1.67 | 1.69 |
> | Surface energies (J m^{-2}) | – | 0.68 | 1.01 | 1.41 |
> | Vacancy formation (eV) | – | 2.60 | 1.99 | 2.27 |
>
> Values from [6, 15, 16].

In Figure 4i.2, the three low-energy surfaces of CeO_2 are shown and compared with the bulk structure of CeO_2. Bulk CeO_2 adopts the fluorite structure, with the larger Ce atoms showing the face centered cubic (fcc) packing and eight-coordination with the O atoms that occupy the tetrahedral sites between the four Ce atoms in a cubic arrangement. The (111) surface contains seven-coordinated Ce atoms and three-coordinated O atoms, the (110) surface contains six-coordinated Ce atoms and three-coordinated O atoms, and the (100) surface has six-coordinated Ce atoms and two-coordinated O atoms.

CeO_2 is an ionic solid with a substantial bandgap energy between valence and conduction band. The bandgap energy decreases on the surfaces and diminishes further with increasing coordinative unsaturation of the surface ions. This difference in bandgap (see also Table 4i.1) energy suggests a lower energy cost to transfer an electron from the oxygen anion to a Ce cation, which reduces the oxygen vacancy formation energy of the surface with the more coordinatively unsaturated ions. It also reflects a decrease in the Madelung constant, which is a measure of the electrostatic field, at the surface of the ionic solid. Table 4i.1 indeed shows that the surface energies increase with decreasing coordination energy of the surface ions.

The dense CeO_2(111) surface is also most difficult to reduce. However CeO_2, surface reconstruction stabilizes the coordinatively most unsaturated (100) surface, so that its reduction temperature exceeds that of the (110) CeO_2 surface [17].

11.2.4 Trends in Adsorption Energies of O Adatoms to Transition Metal Oxide Surfaces

The energies for the adsorption of atomic oxygen to cations of reducible oxides are a strong function of cation charge and d-valence electron occupation. This is illustrated by the adsorption energy trends reproduced in Figure 11.2 [18].

Adsorption energies are compared for cubic monooxides and the mixed oxide perovskites whose structure is shown in Figure 11.3.

The perovskite structure allows for a wide range of compositions. The reactive metal cation (Ti) is octahedrally coordinated with O, and the other cations (Ba) are large and occupy the corners of a cube. They are coordinated with 12 oxygen atoms, which have a cubic fcc arrangement. When the charge on the large cation

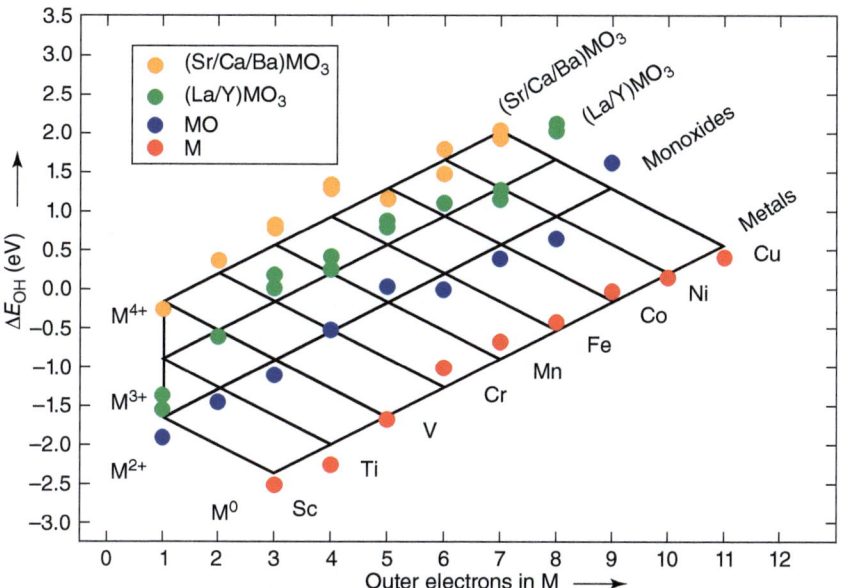

Figure 11.2 Adsorption energy grid for *O adsorbed on various classes of oxides as a function of the number of metal atom valence electrons. A comparison with adsorption of an oxygen atom to the transition metal surface is shown. (Calle-Vallejo et al. 2013 [18]. Reproduced with permission of Royal Society of Chemistry.)

Figure 11.3 Perovskite structure of compound $BaTiO_3$: grey (Ti), black (O), and white (Ba).

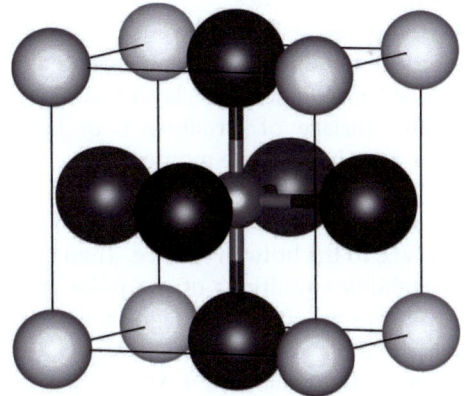

changes, the cation charge on the transition metal must adapt. This is the reason for the great variety of components with perovskite structure.

In the calculations used in Figure 11.2, the transition metal cation that binds the adsorbing oxygen atom is a truncated octahedral site, missing one of its oxygen corner atoms. We observe a high adsorption energy for metals with a low number of valence electrons that are located at the left of a row of the periodic system. The adsorption energy decreases for transition metals located at the right of the same row in the periodic system. This is similar to the trends found for oxygen adsorption to the transition metal surfaces (Chapter 7, Insert 4).

The metal atom-oxygen atom bond is highly polar. The low electron count metals have low nuclear charge and hence they have low ionization energy. The high polarity of the chemical bond is the reason for their strong bond energy.

The nuclear charge of the metal atom increases with increasing electron count so that ionization potential decreases and the chemical bond gets a stronger covalent contribution. Both the bonding and antibonding fragment molecular orbitals between M—O become occupied. The increasing d-electron occupation of the antibonding orbitals is an additional reason for the decrease in the strength of the M—O adsorbate bond.

Figure 11.2 indicates that, without exception, the bonding strength to a cation on an oxide surface is lower than the adsorption strength to the corresponding transition metal surface. The higher the charge on the cation becomes, the lower will be its interaction with the adsorbing oxygen atom, since electron donation to the oxygen atom becomes reduced.

11.2.5 Reconstruction of Polar Surfaces

We have seen several examples of surface reconstruction. It can have a large effect on the reactivity of an oxide surface. This is of particular importance for polar surfaces. Tasker [19] has provided a theory that gives insights into the electrostatic driving forces behind the reconstruction of these surfaces.

This is illustrated in Figure 5i.1a. Structure (a) in this figure is electrostatically stable since the sum of its positive and negative charges is zero in the surface layer. When the surfaces are separated by an infinite distance, their electrostatic interaction vanishes.

The polar surface of layered compounds, case b in Figure 5i.1a, is also stable because the dipole moments oriented in opposite directions and perpendicular to the basal plane in one layer cancel.

The surface of structure c in Figure 5i.1a is unstable because it has a non-vanishing dipole moment.

As illustrated in Figure 5i.1b, a vanishing dipole moment is generated from structure c in Figure 5i.1a when half of its surface charge transfers from the top surface to the bottom surface. Then the structure again becomes stable.

Stability conditions of the polar surface of compounds with the corundum structure are shown in Figures 5i.2 and 5i.3 [20]. In the bulk, each cation is octahedrally coordinated by oxygen. This is also the case for the cations on the polar (001) surfaces of γ-Al_2O_3, Cr_2O_3, V_2O_3, and γ-Fe_2O_3 as well as γ-Ga_2O_3. The Tasker type reconstructions are realized differently for each system. The specific kind of terminations for the (001) surface of the different materials is schematically illustrated as a function of oxygen chemical potential. Four atom composition type terminations are found: full oxygen, full metal, half metal, and metal-oxo. Figure 5i.3 shows that the ionic compounds only occur as a half metal surface. The more covalent compounds can also stabilize the full metal phase.

Insert 5: Reconstruction of Polar Surfaces

Figure 5i.1 gives the schematic representation of the stable reconstruction of the polar surface.

Figure 5i.2 shows the different kind of terminations that occur for the (001) surface of the different materials, and Figure 5i.3 schematically illustrates the surface reconstructions of this surface as a function of oxygen chemical potential.

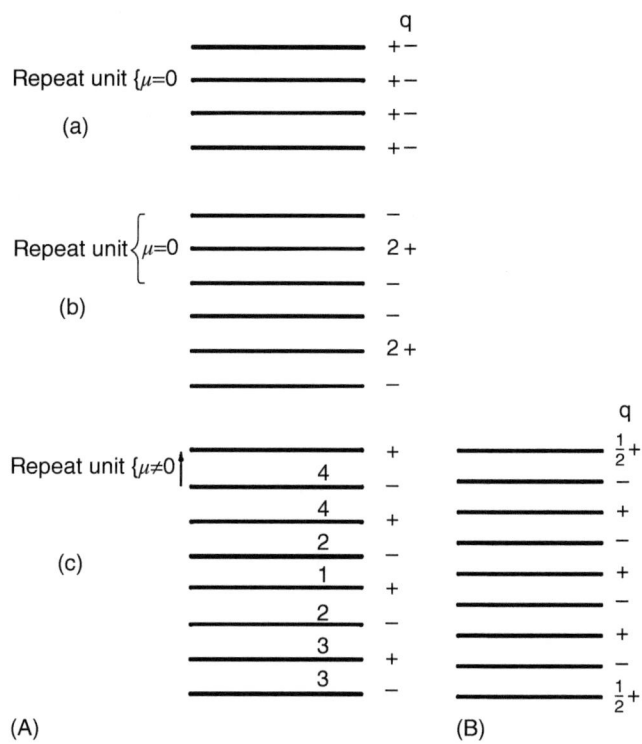

Figure 5i.1 The stability of ionic crystals. (Reproduced from [19].) (A) Distribution of charges q on planes for three stacking sequences parallel to the surface. (a) Type 1 with equal anions and cations on each plane. (b) Type 2 with charged planes but no net dipole moment perpendicular to the surface. (c) Type 3, charged planes and a dipole moment normal to the surface. (B) Distribution of charge for a stable crystal block with a type 3 stacking sequence. This can be achieved by vacancies or adatoms on opposite surfaces.

(Continued)

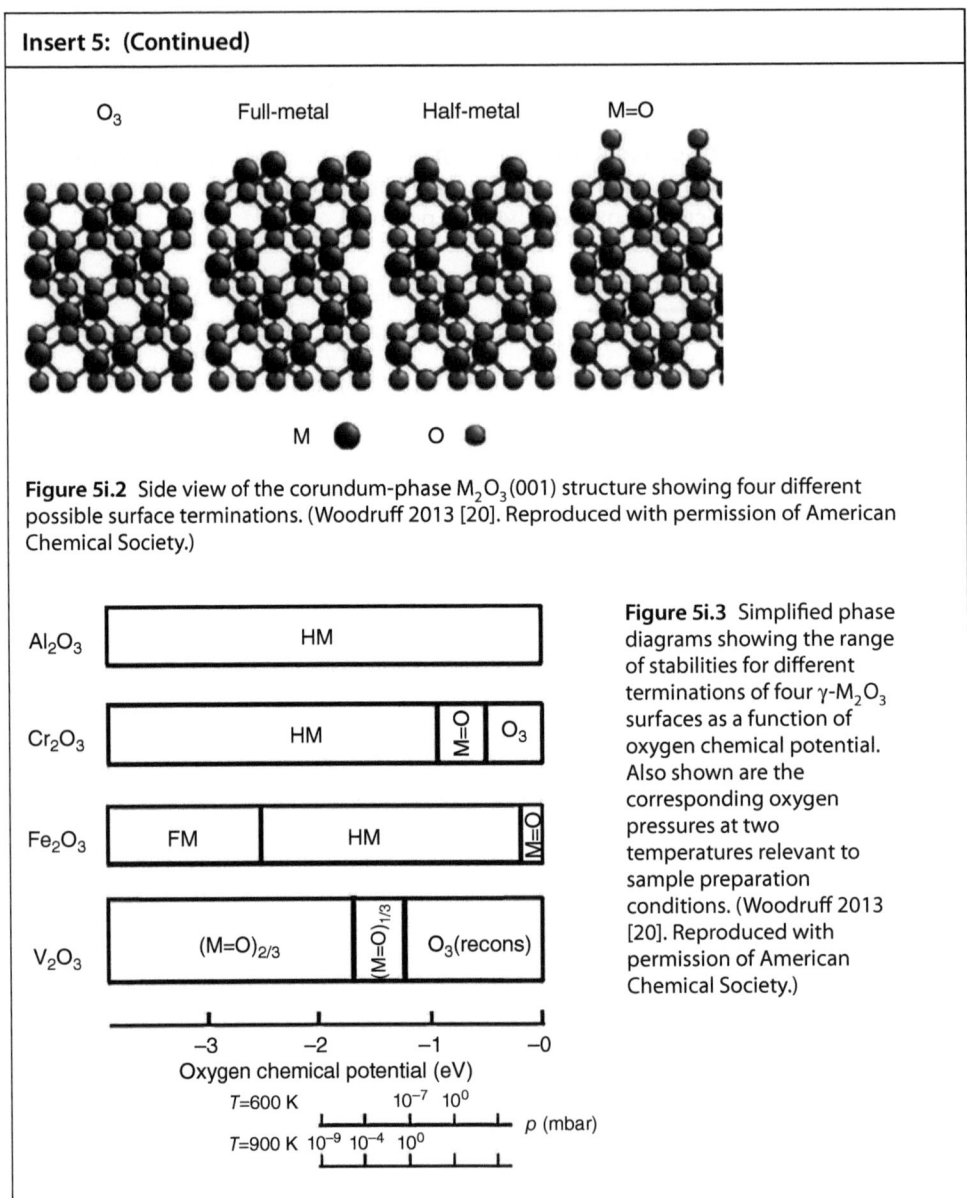

Figure 5i.2 Side view of the corundum-phase M_2O_3 (001) structure showing four different possible surface terminations. (Woodruff 2013 [20]. Reproduced with permission of American Chemical Society.)

Figure 5i.3 Simplified phase diagrams showing the range of stabilities for different terminations of four γ-M_2O_3 surfaces as a function of oxygen chemical potential. Also shown are the corresponding oxygen pressures at two temperatures relevant to sample preparation conditions. (Woodruff 2013 [20]. Reproduced with permission of American Chemical Society.)

11.3 Mechanism of Oxidation Catalysis by Group V, VI Metal Oxides

11.3.1 Reactivity Trends

Group V and VI reducible transition metal oxides such as V_2O_5 or MoO_3 are selective oxidation catalysts. A high selectivity and catalyst stability is achieved by the design of multicomponent materials, often with very complex structures containing V or Mo cations as important components. A major issue is the suppression of total combustion.

The component metal cations can have several different valence states; thus, there is a wide variation in the metal–oxygen atom stoichiometries and the redox energies for these compounds.

Catalysis by these materials takes place by the Mars–van Krevelen mechanism[21]. Oxidation catalysis can be viewed as a sequence of elementary surface oxidation and reduction steps, occurring at various dissimilar sites. The oxidation of substrate reagent molecules occurs by reaction with the surface oxygen atoms. This is a surface reduction step. Reoxidation of the surface occurs by dissociative adsorption of O_2, which can occur at different sites than the oxygen–reagent reactions. This is made possible by the rapid diffusion of oxygen atoms through lattice vacancies. Proof of the Mars–van Krevelen mechanism was exhibited by isotope-labeled experiments with $^{18}O_2$ in the gas phase, where product formation is followed as a function of time. In principle, the initial product oxygen can be ^{18}O or ^{16}O labeled. In the second case, the surface reaction occurs with the unlabeled bulk oxygen atoms. The Mars–van Krevelen mechanism occurs when product molecules are initially found that have the same label as the non-labeled bulk oxide.

The strength of the M—O bond, its polarity, and the redox energies of the reducible cations that compose the metal oxide are reactivity descriptors for the oxide catalyst. When the M—O bond is strong because of multiple σ and π symmetric chemical bonds, the charge on the reducible cation high, and the redox energy for cation reduction relatively low, substrate activation by O is nucleophilic and of the radical type.

The reactivity of the reducible oxides depends strongly on particle size. Large crystallites tend to be more reactive than small oxide clusters. As illustrated in Figure 6i.1 and Table 6i.1, the V=O bond is strong in small clusters, but substantially weaker in layered bulk Vanadium oxide. In the bulk system, the V=O group is in contact with a vanadyl group of the layer below. This contact enhances the reactivity of the upper vanadyl group since it stabilizes the vanadium cation. We will see in Section 11.3.2.1 that there is also a great difference between the reactivity of a single-site Mo center and a large MoO_3 cluster. Single-site Mo catalyzes the metathesis reaction without oxygen insertion, but the MoO_3 oxide particle is an active oxidation catalyst.

Reaction with a saturated organic molecule is initiated by the formation of OH and a radical intermediate. Butane activation is illustrated by Figure 6i.2. The basal (001) plane of V_2O_5 contains terminal V=O reactive vanadyl groups and V—O—V sites with bridging oxygen atoms. An alkane molecule such as butane is activated as a radical with the formation of a VOH species and alcoxy species that adsorbs to a terminal vanadyl (see Figure 6i.2) and the V_5 cation is reduced to V_4 in this step. This is the initial step of the butane oxidation toward maleic anhydride reaction catalyzed by vanadium phosphate. This reaction is selective because the stability of maleic anhydride makes it resistant to further oxidation.

The surface reactivity of the bulk oxides is determined to a large extent by differences in the reducibility of the oxides. Table 6i.2 compares the heats of formation for the reducible oxides as a function of metal cation redox state.

> **Insert 6: Reactivity of Reducible Oxides**
>
> Figure 6i.1 and Table 6i.2 compare bond energies of the vanadium vanadyl group as a function of cluster size. Figure 6i.2 shows calculated reaction intermediates of propane activation by V_2O_5. Figures 6i.3 and 6i.4 compare the reactivity trend of reducible oxides with the trend found for the adsorption energies of oxygen.
>
>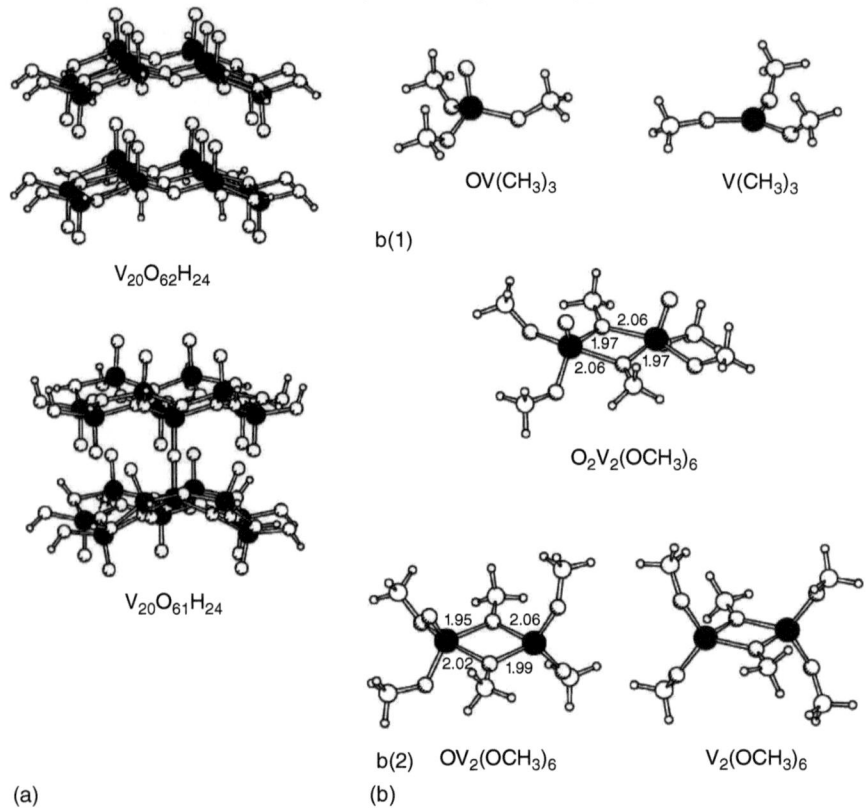
>
> **Figure 6i.1** (a) Two-layer $V_{20}O_{62}H_{24}$ cluster model for the (001) crystal surface of V_2O_5 and the relaxed structure after formation of a vanadyl O defect. (b) The $O=V(OCH_3)_3$ molecule (b1) and its dimer (b2). Relaxed structures after removal of one and two vanadyl O atoms are also shown. (After [22, 23].)

Table 6i.1 Calculated V=O bond dissociation energies for some complexes [22, 23].

	V=O bond dissociation energy (kJ mol^{-1})
$V_{20}O_{62}H_{24}$ (a1)	113 (286)[a]
$OV(OCH_3)_3$ (b1)	312
$[OV(OCH_3)_3]_2$ (b2)	210

a1, b1, and b2 refer, respectively, to the structures presented in Figure 6i.1.
a) Before structure relaxation.

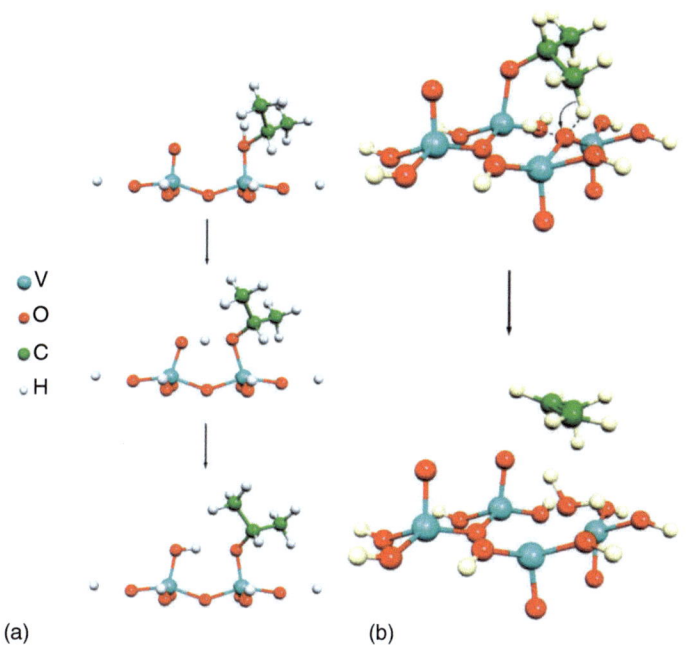

(a) (b)

Figure 6i.2 (a) Illustrations of the reactant, transition, and product states involved in the removal of a hydrogen atom from propane to form *i*-propoxide species. (Gilardoni *et al*. 2000 [24]. Reproduced with permission of American Chemical Society.) (b) Illustrations of the reactant and product states involved in the formation of propene and water from an adsorbed *i*-propoxide species.

(Continued)

Insert 6: (Continued)

Table 6i.2 Heats of formation of important catalytically active reducible oxides.

(a)

V	Energy (kcal mol^{-1})	Cr	Energy (kcal mol^{-1})	Mn	Energy (kcal mol^{-1})	Nb	Energy (kcal mol^{-1})
V_2O_5	−373	—	—	—	—	Nb_2O_5	−463.2
V_2O_4	−344	—	—	—	—	Nb_2O_4	−367.8
V_2O_3	−290	Cr_2O_3	−209	—	—	—	—
V_2O_2	−200	—	—	MnO_2	−124	—	—

(b)

Mo	Energy (kcal mol^{-1})	Ta	Energy (kcal mol^{-1})	W	Energy (kcal mol^{-1})
—	—	Ta_2O_5	−550	—	—
MoO_3	−180	—	—	WO_3	−201
MoO_2	−130	—	—	WO_2	−136

Source: After [22].

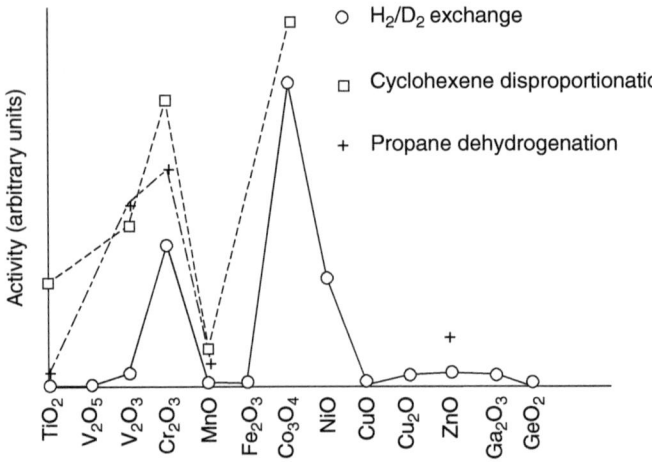

Figure 6i.3 Activity pattern for H—H, C—H and C—C bond activation as a function of metal-element position in the third row of the periodic system. (Dowden 1972 [25]. Reproduced with permission of Taylor & Francis.)

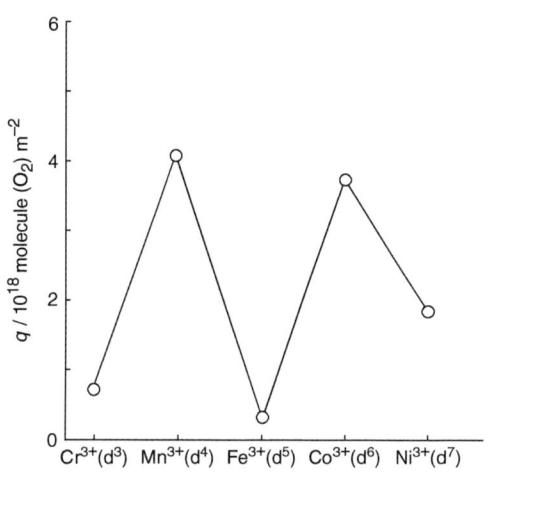

Figure 6i.4 The adsorption of O_2 at 25 °C and $P = 150$ mmHg on $LaMO_3$ (M = Cr, Mn, Fe, Co, Ni). (Kremenic et al. 1985 [26]. Reproduced with permission of Royal Society of Chemistry.)

The smaller energy difference of the redox states of the vanadium oxides makes them the most reactive oxidation catalyst. This is why V_2O_5 will activate strong CH bonds in an alkane, while MoO_3 or MoO_2 do not activate these bonds, but only activate alkenes. These Mo compounds are selective for O insertion into activated alkene intermediates as we will discuss in the next section. Nb_2O_5 has the lowest reactivity due to its higher redox energies and is thus not an effective oxidation catalyst.

A pronounced doubly peaked curve occurs on the plot of the reactivity for the oxides of the third row metals as a function of the metal's position along the row in the periodic table (Figure 6i.3). This is very different from the single volcano-type, bell-shaped dependence on d-valence electron count that we have seen for reactions catalyzed by transition metals (see Chapter 8, Figure 11i.1). We will introduce a similar double-peaked dependence on increasing metal cation d-electron occupation for related systems used in electrocatalysis later the Section 11.6.1.

The double-peaked reactivity pattern is also seen when the trend in the adsorption energy of oxygen is studied (see Figure 6i.4). The reason for the double-peaked dependence of the M—O adatom energy on d-valence electron count in these systems is the redistribution of the d-valence electrons when cations are in a high spin state. As we have seen in Insert 3, for octahedrally coordinated metal cations in the high spin state, crystal field splitting of the d-valence atomic orbitals causes a dip in the M—O interaction energy for metal cations with five electrons instead of the maximum in energy expected for the low spin case (see also Figure 3i.4). High magnetic spin states commonly occur in third row metal oxides with Mn^{2+} and Fe^{3+}.

Interestingly, the catalytic reactions that are the basis of the plot of Figure 6i.3 are not oxidation reactions but instead involve the activation of H_2, CH, or C=C bonds. The relative energies of surface reaction intermediates behave in parallel with the M—O bond energies.

In the following sections, we will provide a detailed description of the mechanism of propylene to acrolein oxidation, which illustrates the general mechanistic principles of the solid state oxide catalyzed reactions. Also radical reaction catalysis of methane oxidation will be discussed as well as electrocatalytic evolution of oxygen from water.

11.3.2 Selective Oxidation of Propylene and Propane

Acrolein and acrylamide are important base chemicals. Catalysis to produce these chemicals from propylene was developed around 1960. More recently, processes have been developed to produce these chemicals from propane. While the catalysts for selective propylene oxidation are based on Mo, selective propane oxidation requires activation by multicomponent catalysts that also contain vanadium.

11.3.2.1 Propylene to Acrolein Conversion

The composition of the active phase of the propylene oxidation catalyst is $Bi_2Mo_3O_{12}$. The reactivity of an oxygen atom attached to Bi^{3+} is very different from the reactivity for oxygen attached to Mo^{6+}. The reaction is initiated by radical reaction activation of propylene to give an allyl intermediate and OH. The molybdate does not participate in this reaction, which is initiated by a BiO moiety that is converted to BiOH. Deuterium-labeled propylene experiments indicate that the rate-controlling step of the reaction is CH bond cleavage of the CH_3 group [27–29].

The allyl radical reacts with a molybdate M—O bond to an oxygen-σ-allyl species (see [30]). Acrolein is formed with oxygen with equal probability at each of the two propylene-ending C atoms. One of these atoms was initially part of CH_3 and the other was part of CH_2. In the allyl intermediate the two end C atoms are equivalent.

Pure Bi_2O_3 gives 1,5-hexadiene as a product, which is the dimer of the allyl. Acrolein formation occurs from the σ-oxygen allyl species adsorbed to Mo. In this reaction step, an H atom is transferred from the allyl intermediate to an additional O atom attached to Mo.

Experiments with MoO_3 show that it can only convert activated molecules such as allyl iodide, which readily gives allyl, to acrolein [31]. In the process of acrolein formation, Mo^{6+} reduces to Mo^{4+}. The catalyst is restored by re-oxidation with O_2 according to the Mars–van Krevelen mechanism. Figure 7i.1 gives a schematic representation of the complex reaction network of this reaction as modeled with a specially designed appropriate reactive force field.

Insert 7: Mechanism of Acrolein Formation and Ammoxidation

Figure 7i.1 is the reaction network of acrolein oxidation catalysis. Figure 7i.2 shows the crystal structure of a catalyst that activates propane. The corresponding ammoxidation mechanism is shown in Figure 7i.3.

11.3 Mechanism of Oxidation Catalysis by Group V, VI Metal Oxides

Figure 7i.1 Proposed mechanism and energetics for propene oxidation over bismuth molybdate catalyst. The top energy is the ΔE from QM, the middle is $\Delta H_{0K} = \Delta E + \Delta ZPE$, and the bottom is ΔG_{593K}. All reported values are in kilocalories per mole. Results were derived from molecular dynamics simulations. (After [22, 32].)

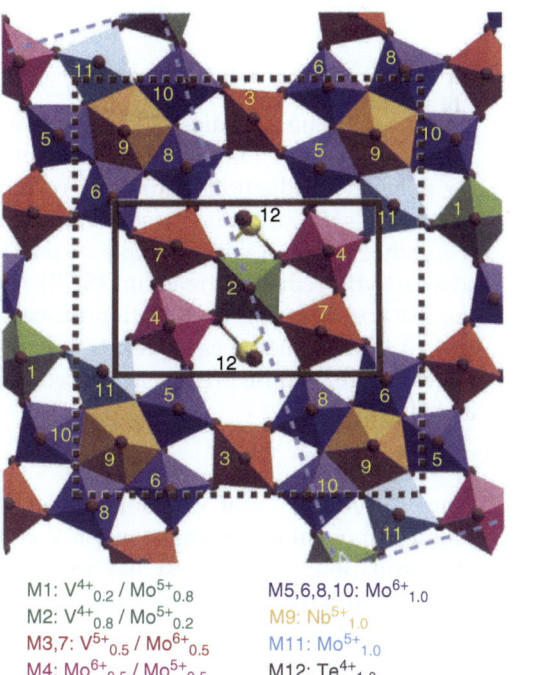

Figure 7i.2 Catalytically active center of $Mo_{7.5}V_{1.5}NbTe_{29}$ in [001] projection and schematic depiction of the active site. (Grasselli et al. 2003 [33]. Reproduced with permission of Springer.)

M1: $V^{4+}_{0.2}$ / $Mo^{5+}_{0.8}$
M2: $V^{4+}_{0.8}$ / $Mo^{5+}_{0.2}$
M3,7: $V^{5+}_{0.5}$ / $Mo^{6+}_{0.5}$
M4: $Mo^{6+}_{0.5}$ / $Mo^{5+}_{0.5}$
M5,6,8,10: $Mo^{6+}_{1.0}$
M9: $Nb^{5+}_{1.0}$
M11: $Mo^{5+}_{1.0}$
M12: $Te^{4+}_{1.0}$

(Continued)

Insert 7: (Continued)

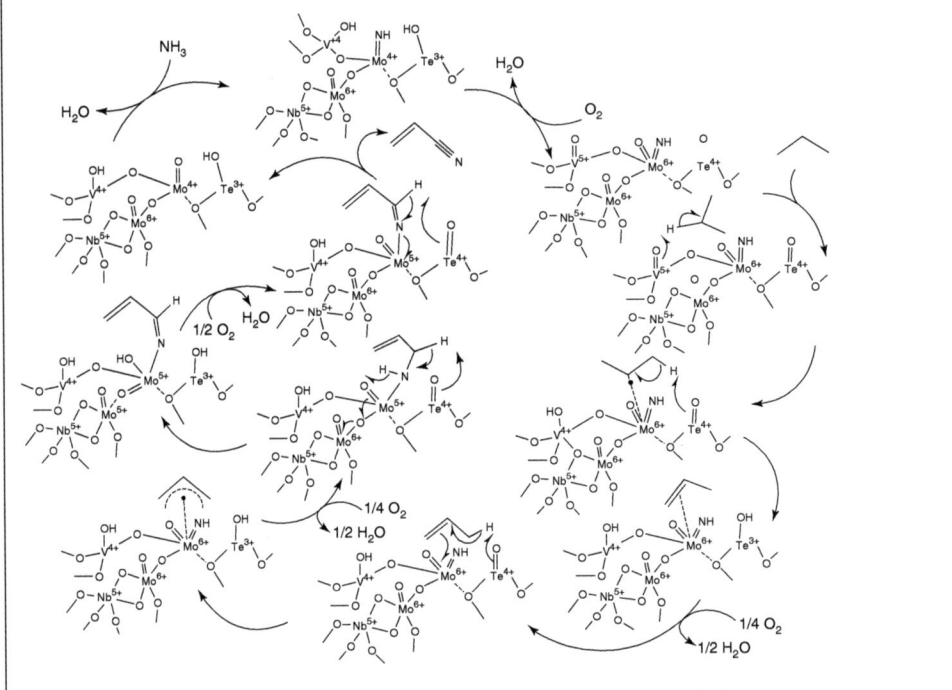

Figure 7i.3 Proposed propane ammoxidation mechanism over Mo–V–Nb–Te–Ox catalysts. (After [22, 33].)

When propene is oxidized in the presence of ammonia, acrylonitrile is formed instead of acrolein. Like acrolein, acrylonitrile is also an important intermediate for polymer production. Ammonia readily reacts with M=O bonds to give a reactive M=NH$_x$ intermediate that can be inserted into the allyl intermediate.

Another reaction with ammonia catalyzed by MoO$_3$ and V$_2$O$_5$ is the selective catalytic reduction reaction (SCR) with NO, which is useful for reducing the exhaust emission of oxidation processes with air (see also Sections 2.2.5.3 and 10.4.1). Ammonia easily reacts with M=O bonds to give a reactive M=NH$_x$ intermediate, and this reacts in consecutive reaction steps with NO to give N$_2$ and H$_2$O.

11.3.2.2 Propane Ammoxidation

Since the Mo—O unit will not activate the CH bond of an alkane, complex multicomponent oxides that contain V and Mo have been designed for propane ammoxidation. The catalyst has to combine propane dehydrogenation with propene ammoxidation. The catalyst system requires that both reactions are catalyzed at the same temperature. The corresponding reaction network is shown in Figure 6i.3 and the structure of the oxide that catalyzes the reaction is shown in Figure 6i.2. In this catalyst, the Nb^{5+} centers isolate the Mo and V centers. Total oxidation is suppressed by site isolation, which reduces excess availability of reactive oxygen atoms.

Surface enrichment of complexes that have weaker M—O bond strength may occur during a reaction so that the stoichiometry at the surface can differ from the stoichiometry in the crystalline bulk. The surface restructuring under formation of V- and Te-containing clusters anchored to the crystalline solid is thought to create structurally and electronically isolated active sites. The mobility of Te, especially in the presence of water vapor, may contribute to the development of site isolation under reaction conditions [34].

11.3.3 Methane Conversion to Higher Hydrocarbons

No practical process exists for the conversion of methane to methanol by selective oxidation (see also Section 10.4.3). While the oxidation reaction of methane to methanol is possible, high selectivity is not found at a high conversion rate, because the activation of methanol by oxygen is easier than the activation of methane [35].

The low temperature process yields methanol, while at higher temperatures ethylene or acetylene can be produced. Pyrolytic processes without oxygen can also be used to activate methane to produce acetylene or ethylene.

In oxidative processes, selective product formation must compete with total oxidation. Pyrolytic processes must compete with the deposition of carbon.

Catalytic systems that oxidatively convert methane to ethane and ethylene operate at a high temperature of 900 K [36, 37], with a maximum yield of 20%. One of the best-performing and well-investigated catalytic systems is the LiO/MgO system. In Section 11.3.3.2, we will see that its reaction mechanism is a radical reaction. This is a unique catalytic reaction. In contrast to this reaction, the direct oxidative conversion of methane by transition metals at high temperature will lead to total combustion unless short contact times on the order of 10^{-3} s are used [38]. Then synthesis gas is selectively produced.

In the pyrolytic Huels process, which has a high yield, methane is converted to acetylene in an electric arc plasma. In the electric arc process, thermodynamics prescribes a high temperature for acetylene production but again a short contact time of 10^{-3} s must be used to prevent the formation of thermodynamically favored carbon and aromatic oils [39].

Figure 11.4 illustrates the thermodynamics of ethylene and acetylene formation; the formation of carbon and aromatics is excluded. Note the onset of ethylene formation around 1000 K.

Thermodynamically, carbon is always more stable than methane. Benzene formation becomes thermodynamically possible at 800 K, and at higher temperatures acetylene and ethylene can be formed.

In the next subsection, the catalysis of direct methane conversion to benzene is discussed followed by a subsection on oxidative ethylene formation in the presence of oxygen.

11.3.3.1 Direct CH_4 Conversion to Aromatics

Since methane to benzene conversion is less endothermic than ethylene formation, bifunctional catalysts will catalyze this reaction at a lower temperature than

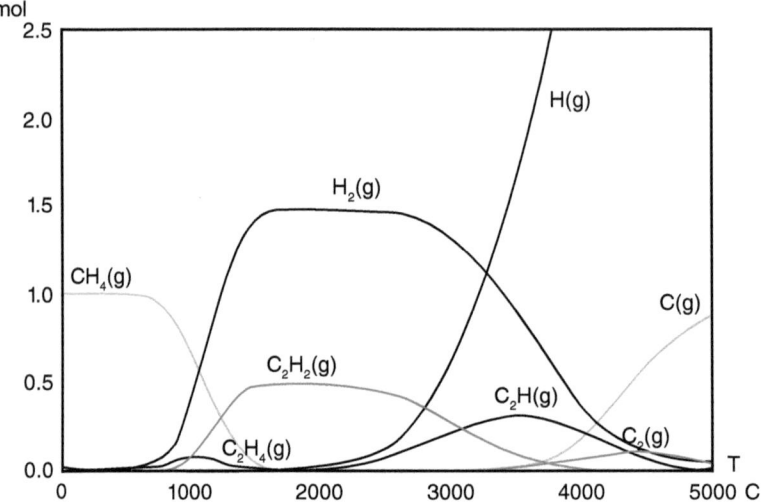

Figure 11.4 Simplified equilibrium diagram for methane where the solid phase of carbon is not included as a species. (After [40].)

necessary for ethylene formation. A zeolite such as HZSM-5 that is promoted by Mo, V, Fe, or Cr can catalyze conversion of methane to benzene [41–43]. Reaction temperature is 900 K. It has been demonstrated for Mo/ZSM-5 catalysts that monomeric Mo(=O)$_2$, 2+ intermediates species originally present in the micropores of the zeolite convert into small dimeric or trimeric M$_x$C$_y$ oligomers upon contact with methane. These carbide clusters activate methane and produce ethylene through the recombination of a CH and CH$_3$ intermediate in a reaction analogous to that of ethylene formation in the Fischer–Tropsch reaction [44]. The carbided oligomers have a positive charge and particles are stabilized in the zeolite micropores by the negatively charged sites of Al substituted in the zeolite framework. The acidic protons present in the zeolite catalyze the consecutive conversion of ethylene into aromatics. The catalyst becomes deactivated by carbonaceous residue formation and the agglomeration of carbide particles.

At higher temperatures, catalytic materials promoted with reducible cations can be used to catalyze non-oxidative ethylene conversion from methane. These catalysts initiate CH$_3$ radical intermediate formation and product formation occurs by gas phase radical reactions. At a temperature of 1400 K, conversion of 60% methane with 50% selectivity to ethylene has been reported [45, 46]. The lifetime of the catalysts is limited by non-selective naphthenic oil formation.

11.3.3.2 The Mechanism of Oxidative Methane Coupling, the LiO–MgO System

The oxidative methane coupling reaction to ethylene can be conducted at a temperature of 900 K. An active catalyst for this reaction is the LiO–MgO system. As in the non-oxidative process, ethylene is formed through intermediate radical formation [47, 48].

Two different reaction paths for methane activation to ethylene have been proposed that are analogous to those proposed for the low temperature redox systems that produce methanol (see Section 10.5):

- The harpooning reaction in which methane collides with an oxygen atom, OH and CH_3 radicals are generated and the cation is reduced [49].
- Reaction occurs through the initial coordination of methane to metal cations and oxygen anions, and the heteropolar dissociation of the CH bond. In this reaction, no redox is involved [50].

Substitution of a MgO unit by LiO at the MgO surface introduces a radical center into the solid. This radical center can initiate radical reaction with methane to form a LiOH and CH_3 group. The catalytic reaction cycle originally proposed according to this mechanism is shown in Figure 8i.1.

Calculations indicate that this reaction, which proceeds through the harpooning route, has a low activation barrier of 12 kJ mol^{-1}. This value is much lower than the experimentally observed activation energy that is at least 90 kJ mol^{-1} [52].

According to the heterolytic mechanism, the activation energy at an edge of the MgO particle is calculated as close to the experimentally observed value (Figure 8i.2b). Although radical desorption of the CH_3 from the $MgCH_3$ intermediate has a high barrier, it occurs with a low barrier when O_2 adsorption is included in the same reaction step. O_2 adsorption is very exothermic at this site and has a value of -191 kJ mol^{-1}. The overall energy change of this reaction given of which the chemical reaction step are given by Eq. (11.1) is 37 kJ mol^{-1}.

$$[Mg^{2+}O^{2-}]_{MgO} + H-CH_3 + O_2 \rightarrow (O_2^{*-})[HO^-Mg^{2+}]_{MgO} + *CH_3 \quad (11.1)$$

The adsorbed superoxo is proposed to re-oxidize the surface through intermediate OOH formation.

In this heterolytic mechanism of methane activation, Li promotion is not necessary for the reaction. The role of Li promotion has been shown to change the morphology of the MgO crystals, which creates more active steps and corner sites (see Chapter 9, Insert 1). The reaction can also be catalyzed by non-Li-promoted MgO, with an apparent activation energy comparable to the promoted system [51]. The much lower reactivity of the non-promoted system is due to its higher surface area and its reduced fraction of active edge and corner sites.

Insert 8: Activation of Methane by LiO/MgO

Figure 8i.1 shows the radical reaction mechanism as proposed originally by Lunsford. Figure 8i.2 illustrates the alternative reaction paths according to heterolytic mechanism.

(Continued)

Insert 8: (Continued)

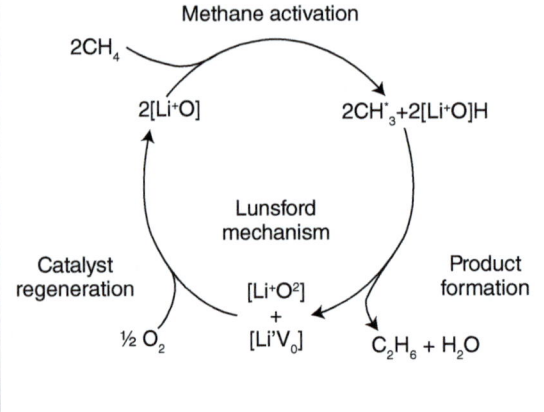

Figure 8i.1 The Lunsford mechanism for the oxidative coupling of methane. (Schwarz 2011 [48]. Reproduced with permission of John Wiley and Sons.)

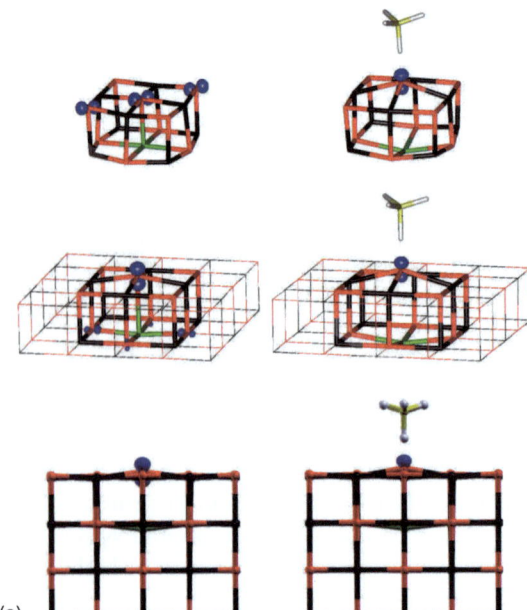

Figure 8i.2 (a) From top to bottom: cluster model, constraint cluster model embedded in a periodic array of point charges, and periodic model of an $Li^+O\cdot^-$ site on the MgO (001) terrace. Left, active site structure and right, transition structure for H abstraction from CH_4, O, red; Mg, black; Li, green; C, yellow; H, gray; spin density, blue. The activation energy for generation of $CH_3\cdot$ intermediate is 12.6 kJ mol^{-1}. (b) Reaction energy diagram for chemisorption of CH_4 onto corner/edge sites of a Mg_9O_9 cluster showing C—H bond addition on an $Mg^{2+}O^{2-}$ pair. B3LYP energies are given in kJ mol^{-1}. EC means encounter complex; TS, transition state; IN, intermediate; colors as in (a). (Kwapien et al. 2014 [51]. Reproduced with permission of John Wiley and Sons.)

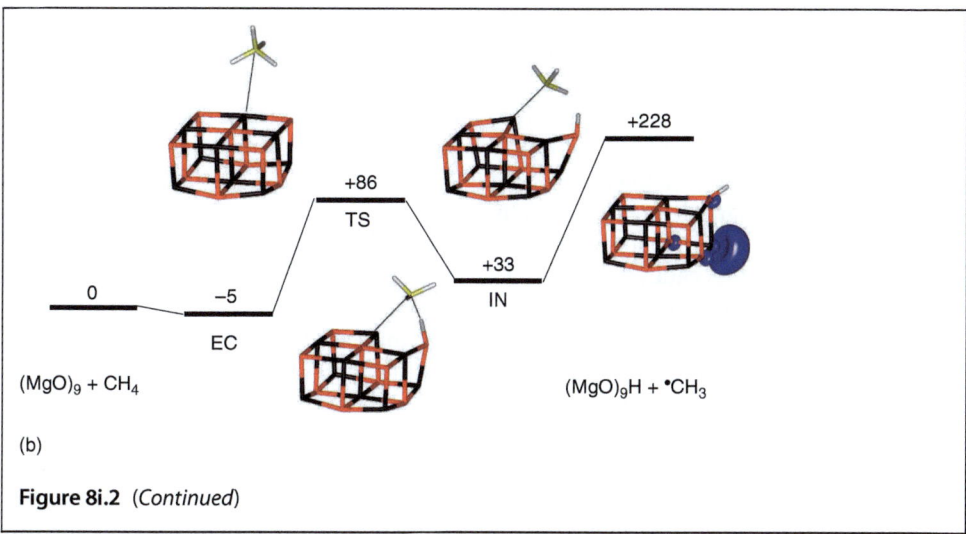

(b)

Figure 8i.2 (Continued)

11.4 Metathesis and Polymerization Catalysis: Surface Coordination Complexes

In this section, we will discuss two important related catalytic reaction systems: the metathesis or disproportionation reactions and olefin polymerization reactions.

These reactions can be catalyzed by homogeneous as well as heterogeneous catalytic systems. The heterogeneous catalysts are surface single-site organometallic or coordination complexes which are prepared by the immobilization of a molecular complex on a non-reducible support. The surface can be considered a large ligand in contact with the reaction center. For the organometallic complexes, we will see that differences in the coordination of the catalytically active reaction centers to the supporting material may substantially alter reactivity. In polymerization catalysis, local topology will also affect the stereoselectivity of the reaction.

11.4.1 Alkene Disproportionation and Metathesis

Selective catalysts of the disproportionation reaction, also called the *metathesis reaction* [53], are homogeneous as well as heterogeneous catalysts with single-site catalytic centers of reactive, highly charged cations such as Mo^{6+}, W^{6+}, Ta^{5+}, or Re^{7+}. Most of them are active in their high d_0 valence state.

An introduction to these processes and their mechanism was presented in Section 2.2.3.3. Figure 2.10 provides an illustration of the mechanism of this reaction. The key catalyst intermediate is a M=CR$_2$ metal carbene complex, which reacts with an olefin through a four atom metallocycle complex. The C=C bond of the incoming olefin cleaves, and a new C=C bond is formed by the recombination of one of the carbon atoms with carbene. This creates an alkene product and a new carbene fragment, left over from the other half of the reacting alkene.

The heterogeneous metathesis catalyst is an oxidic single-center metal cation attached to the surface oxygen atoms of a non-reducible oxidic support. Generally, such catalysts are prepared by reacting a hydroxylated surface with a carbonyl compound, which is subsequently calcined. It is essential that the catalyst reactive site is a single center and that the presence of oligomeric or large oxidic particles is prevented. In practice, usually only a few of the metal complexes present on the catalyst particle are catalytically reactive.

The original industrial application of this reaction was made using a $MoO_3/\gamma\text{-}Al_2O_3$ catalyst to catalyze the disproportionation of ethylene and propylene to butene-2 [52].

In this section, we will discuss the mechanism of this reaction in some detail. A recent computational and experimental study [54] of the metathesis reaction catalyzed by monomeric Mo oxide sites on different surfaces of $\gamma\text{-}Al_2O_3$ can be used to illustrate some of the mechanistic steps. Different adsorption modes of a $MoCH_2$ complex are shown in Figure 9i.1. The reactivity of these complexes can vary substantially. The metallocycle intermediates formed on the respective structures in Figure 9i.1 are shown in Figure 9i.2. In the metathesis reaction, the carbene monomer that is inserted is regenerated as part of the reaction cycle. The metallocycle intermediate of the propagation step can be considered an allyl intermediate, weakly interacting with a hydride ion. It is stabilized by the high positive charge and the symmetry of the empty d-valence electron orbitals of the d_0 cation. In subsequent reaction steps it cleaves the C=C bond. The cation is highly reactive because of its high positive charge and available unoccupied d-atomic orbitals.

Insert 9: Metathesis Reaction by Supported Mo Complexes

Figures 9i.1 and 9i.2 show computed reaction intermediates and their energies. Figure 9i.3 illustrates structural changes due to dehydration.

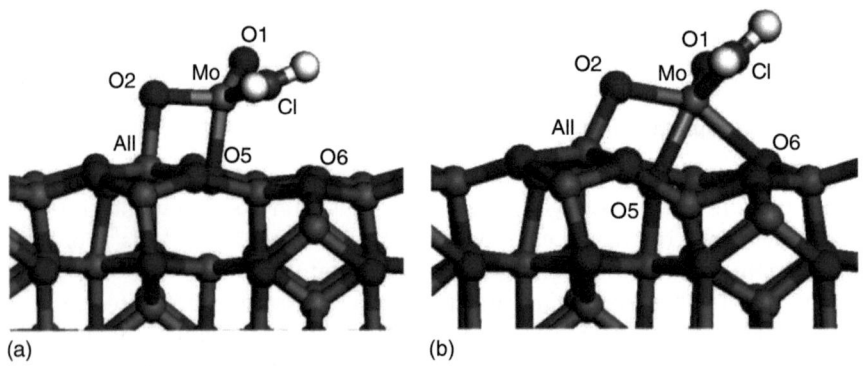

(a)   (b)

Figure 9i.1 A two- (a) and threefold (b) Mo-carbide complex intermediate immobilized on the γ-alumina (100) surface. (Handzlik and Sautet 2008 [54]. Reproduced with permission of Elsevier.)

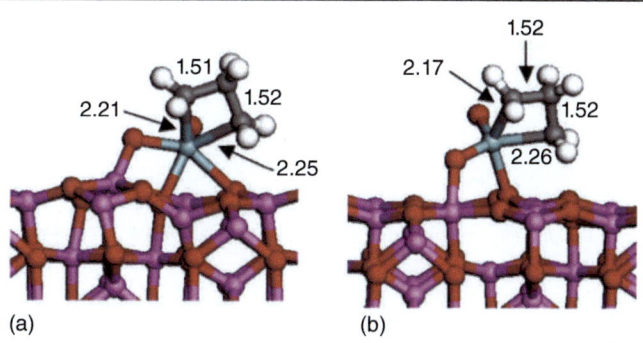

Figure 9i.2 The four-center metallocycle intermediates in ethene metathesis on the Mo-methylidene center formed on the respective Mo complexes of Figure 9i.1. (Handzlik and Sautet 2008 [54]. Reproduced with permission of Elsevier.)

Figure 9i.3 Dehydration of a Mo=CH_2^- containing complex attached to a γ-alumina. Schematic comparison of hydrated and non-hydrated structures (a) and (b) ($\delta E_{dehyr} = +161$ kJ mol^{-1}) on γ-alumina (100) surface and structures (c) and (d) ($\delta E_{dehydr} = +196$ kJ mol^{-1}) adsorbed to the γ-alumina (110) surface. (Handzlik and Sautet 2008 [54]. Reproduced with permission of Elsevier.)

As is schematically illustrated in Figure 9i.3, the Mo species can be adsorbed one-, two-, or threefold through oxygen atoms to the alumina surface. One can consider the surface as a ligand to the reactive complex.

The rate of metathesis by reaction of the $MoCH_2$ carbene intermediate with alkene depends on Mo coordination, and is strongly influenced by the O—Mo—O bond angles between Mo and the oxygen atoms of the support [55]. The great majority of the sites are not reactive because the molybdo–cyclobutane intermediate is too stable. Only a few sites provide the proper angle and coordination needed for high reactivity.

The Mo-methylidene species on the surfaces of γ-alumina can be active at room temperature as long as they are dehydrated. The structure illustrated in Figure 9i.3a is the best candidate for the metathesis active site at room temperature. More stable "dehydrated" Mo species on the more reactive (110) face such as structure d (Figure 9i.3) can be active only at high temperatures.

A true activation energy of 34 kJ mol^{-1} was derived from experimental kinetic data by applying the Langmuir–Hinshelwood kinetics [54]. This value is close to the activation barrier predicted for ethene metathesis on the Mo-methylidene (Figure 9i.3a) site.

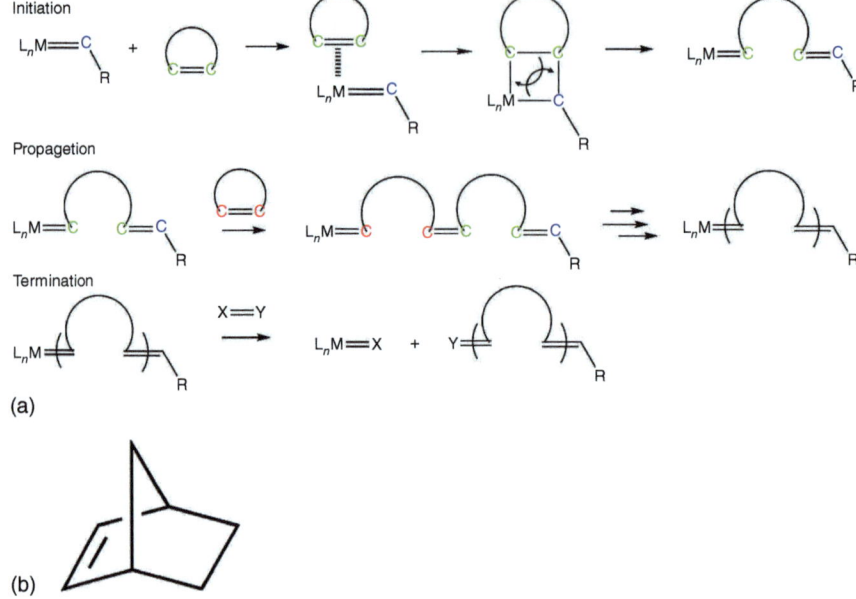

Figure 11.5 (a) Mechanism of ROMP reaction. (b) Norborene, which is often used for ROMP.(Sutthasupa et al. 2010 [55]. Reproduced with permission of Nature Publishing Group.)

Many reactions analogous to the original ethylene–propylene disproportionation reaction are now used in the petrochemical industry and for fine chemical production. These reaction systems are homogenous. Progress in developing these new processes has been due to the invention of highly selective organometallic complexes for these reactions [56, 57]. The mechanism of the homogenous and heterogeneous metathesis systems are similar. Their unique reactivity results from an ability to tune the metal cation ligands.

The ring opening metathesis polymerization (ROMP) catalysis of unsaturated cyclic alkenes is technologically important [56]. The mechanism of the ROMP reaction is illustrated in Figure 11.5.

Cyclic olefins are polymerized and the propagating chain remains attached as a carbene to a cation that has a high cation charge. The reaction is driven by the release of the ring strain energy of the cyclic monomer when it reacts. Strained cyclic olefins that are often used for this reaction are substituted and non-substituted norborene molecules (see Figure 11.5b).

The ROMP mechanism of polymerization is very different from the Ziegler–Natta polymerization of ethylene or propylene, which we will discuss next. Instead of intermediate carbene formation, the Ziegler–Natta polymerization is based on insertion into an alkyl intermediate (see Figure 2.11).

11.4.2 Polymerization of Propylene

Polymerization of light olefins can also be catalyzed both homogeneously and heterogeneously. Similar to the metathesis reaction, the catalyst originally discovered for this process was heterogeneous. More recently, highly active homogenous metallocene complexes (organometallic complexes with negatively charged cyclopentadienyl derived ligands) of group IV or V metals as Ti, V, or Cr have been discovered. They can be tuned to create extremely selective and active catalytic systems [58, 59]. The mechanism of these homogeneous systems is closely related to the mechanism of the heterogeneous systems.

We will discuss the mechanism of the heterogeneous Ziegler–Natta catalysts discovered in 1953–1954 for the polymerization of ethylene and propylene [60–62]. These catalysts are heterogeneous $TiCl_3$ catalysts or $TiCl_4$ immobilized on $MgCl_2$. The catalysts are often promoted with electrophilic organic molecules to induce improved stereoregularity of the polymer.

Insert 10: Mechanism of Ziegler–Natta Polymerization

Figure 10i.1 illustrates the transition state structure of the insertion reaction of the olefin into the growing alkyl chain. In Figure 10i.2a, the reaction centers of immobilized $TiCl_3$ or $TiCl_4$ are schematically illustrated. Figure 10i.2b illustrates stereochemical control at a Ti-Chloride site. Figure 10i.3 gives the notations used in Figure 10i.2b.

Figure 10i.1 The insertion reaction step according to the Cossée–Arlman mechanism. (After [63, 64].)

(Continued)

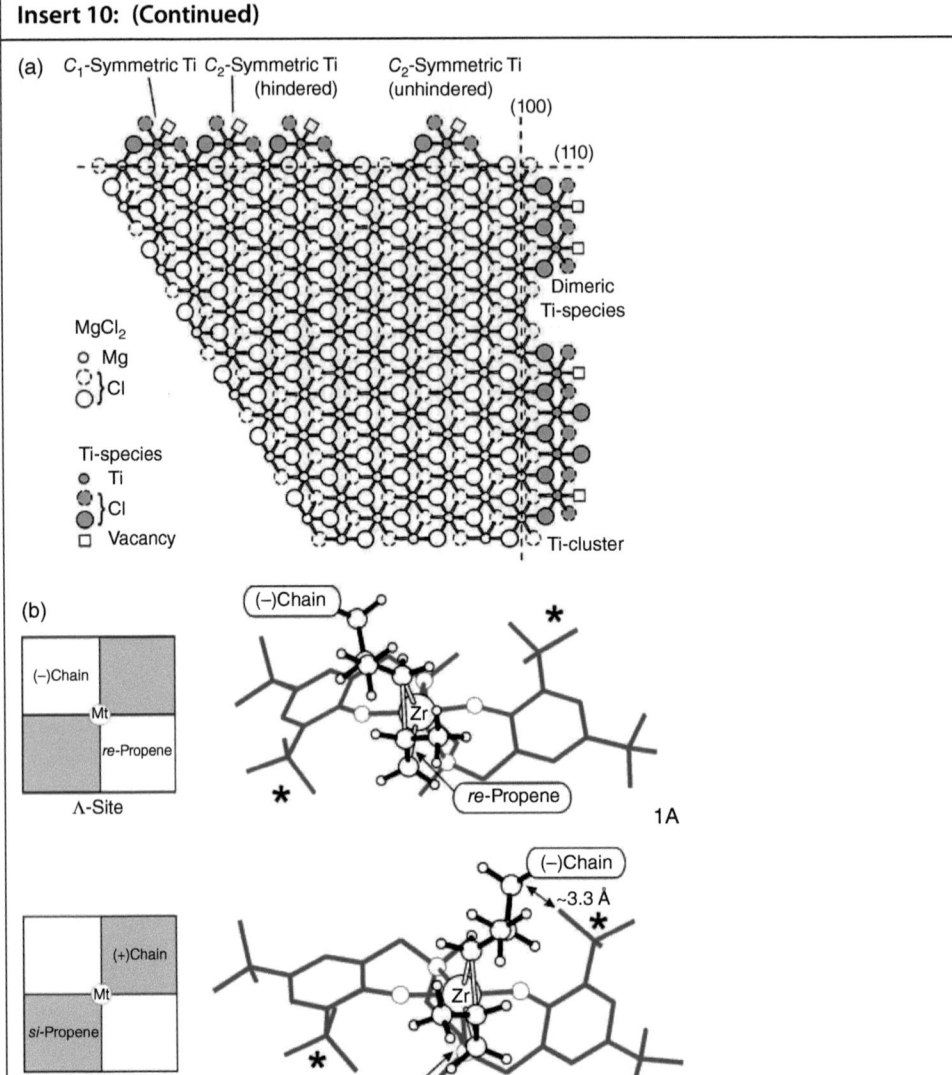

Figure 10i.2 (a) Structural layer of $MgCl_2$. Solid line Cl atoms represent atoms above the plane of the Mg atoms, and dashed Cl atoms represent atoms below the plane of the Mg atoms. (100) and (110) lateral cuts with five- and four-coordinated Mg atoms are indicated. Ti^{III}-chloride species with different steric environments are shown. (Corradini et al. 2004 [65]. Reproduced with permission of American Chemical Society.) (b) Low-energy transition states (right) for propene insertion into a titanium (primary chain) bond and quadrants representation (left) of the same systems. Gray quadrants correspond to crowded zones occupied by the Cl atoms marked by a star on the right.

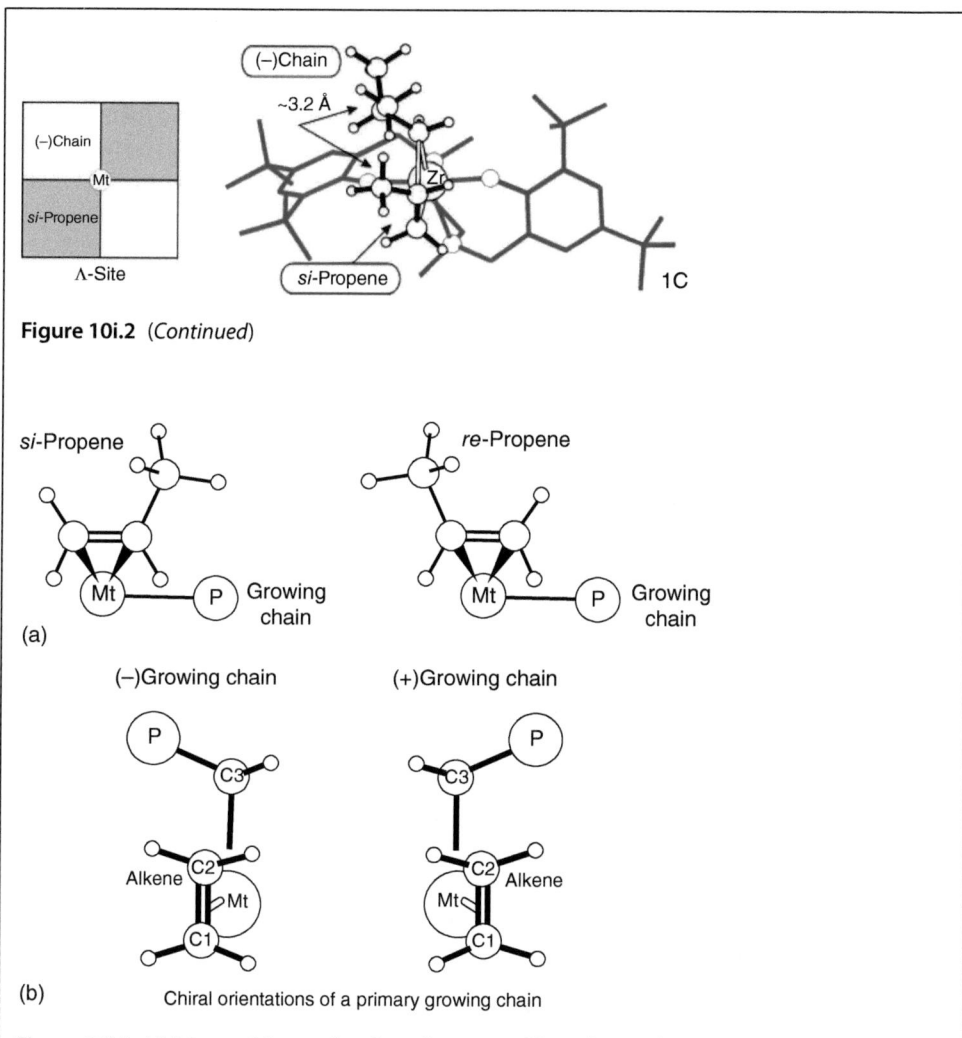

Figure 10i.2 (Continued)

Figure 10i.3 (a) Primary 1,2 coordination of propene. (Corradini *et al*. 2004 [65]. Reproduced with permission of American Chemical Society.) (b) Chiral orientations of a primary growing chain.

The polymerization reaction is initiated by alkylation of the catalyst Ti center. In metathesis catalysis, the alkyl is converted into the carbene that initiates the reaction. On Ti^{4+} or Ti^{3+} the alkyl itself initiates the growing polymer chain. Chain growth occurs by the insertion of a propylene or ethylene molecule, which adsorbs to a vacant Ti site next to the alkyl intermediate (see Figure 10i.1). This will produce a longer alkyl chain, which will undergo a subsequent insertion step with the olefin that adsorbs on the site that was vacated. The continuation of this process leads to a long hydrocarbon chain. The reaction that terminates the propagation reaction is the β-H abstraction of the growing alkyl chain, which must be a slow reaction step compared to the rate of propagation.

An interesting property of the polymers of propylene is that their methyl groups can be arranged in different ways on the hydrocarbon backbone: in identical positions (isotactic), in identical but alternating left or right orientation (syndiotactic), or in an irregular (atactic) arrangement. Control of the chirality of the reactive center is thought to determine the crystallinity and type of product.

However, homogeneous systems without a chiral center are also known to give a stereoregular polypropylene product. Propylene has a stereogenic center at carbon atom 2, since propylene itself is chiral. Stereochiral polymer production then is due to chain-end control and is dictated by the chirality of the last inserted monomer. In the next section, we will closely follow the description of the mechanism of the reaction of Corradini *et al.* [65].

Natta proposed that steric control is due to the structure of catalytic sites on the border of crystal layers of $TiCl_3$, [65]. Arlman and Cossee [63, 64] developed this idea further, based on the insertion model of the polymerization reaction (see Figure 10i.1).

If we cut a $TiCl_3$ layer, in which the Ti^{3+} cations are octahedrally coordinated, parallel to the line connecting two bridged Ti atoms, electroneutrality conditions dictate that each Ti atom at the surface is bonded to only five Cl atoms. Four of these are strongly bonded to the Ti atoms because they are bridged to other metal atoms. The fifth Cl atom, which is not bridged, may be replaced by an alkyl group. The free sixth octahedral position may coordinate an alkene. For these sites, the two positions that are accessible to the growing chain and to the monomer are different.

In the $MgCl_2$-supported catalysts, $MgCl_2$ has crystal structures somewhat similar to those of violet $TiCl_3$. This presents the possibility for an epitaxial coordination of $TiCl_4$ units (or $TiCl_3$ units after reduction) on the lateral faces of the $MgCl_2$ crystals. The epitactic placement of Ti_2Cl_6 and Ti_4Cl_{12} units on the (100) face of $MgCl_2$ is shown in Figure 10i.2a. The environment of the Ti atoms is chiral.

The heterogeneous systems control the local environment of the reactive center poorly, which puts them at a disadvantage when compared with the homogeneous organometallic catalysts. The homogeneous catalysts can be optimized by variations in their ligands. Complexes containing Ti, Zr, Cr, or V have been developed with high steric control and extremely high turnover numbers.

11.4.3 Ziegler–Natta Polymerization versus Metathesis Reaction

The key feature of an effective Ziegler–Natta olefin polymerization catalyst is the ability to suppress β-H transfer from the alkyl chain to the cationic metal center that terminates the chain growth reaction. The transition metals as Ti^{4+} or Zr^{4+} are useful in this respect due to their rather weak interaction with H and the preference of the d_0 state of the cation. The insertion reaction proceeds through a metallocycle intermediate, which is stabilized by the interaction with the empty d-atomic orbitals of proper symmetry (see Figure 10i.1).

A metathesis catalyst requires a more active reaction center. A metallocycle intermediate is also formed, but between carbene and olefin. The high reactivity of a cation such as Mo^{6+} allows cleavage of the C=C bond, and a new carbene species is formed that propagates the reaction. This makes the highly charged or reactive cations of the group VI or VII metals necessary for metathesis catalysis.

Single-site systems are also essential. The corresponding multisite oxide compounds have been discussed in Section 11.3.2. They are selective oxidation catalysts, but show very poor performance in metathesis or polymerization reactions.

11.5 Sulfide Catalysts

Hydrotreating reactions are hydrogenation reactions catalyzed mainly by Ni- or Co-promoted MoS$_2$ or WS$_2$ catalysts. They can hydrodesulfurize and hydrodenitrogenate organic compounds in coal- or oil-derived petrochemical feed streams or they can be used to deoxygenate biomass-derived process streams [66] (see also Section 2.2.4).

In aromatic molecules, the hydrotreating reaction occurs through the partial hydrogenation of unsaturated rings that contain a heteroatom and subsequent C—X bond cleavage. Reaction schemes for HDS and hydrodenitrogenation (HDN) are shown in Figure 11i.1. A challenge to HDS and HDN is to remove S or N atoms from polyaromatic molecules without saturation of the aromatic ring part of the molecules. This will reduce overall hydrogen consumption of the process and at the same time maintain a high octane number of the oil fraction. A catalyst is needed that can reduce the undesirable hydrogenation of aromatic rings not containing the heteromolecules, which occurs in parallel to H$_2$S and NH$_3$ formation.

MoS$_2$ and WS$_2$ are compounds that contain cations sandwiched between two layers of S atoms. Their basal planes are stabilized by hexagonally packed sulfur atoms. As a consequence, the sites with reactive metal cations will only be exposed at the edge of the basal surfaces (as we discussed for TiCl$_3$ in the previous section).

The activation of H$_2$ occurs heterolytically at sulfur vacancies located at these edges. At reaction conditions, some of the sulfur atoms at the crystal edge will be converted into sulfohydril (SH) groups and H adsorbed to the metal cation. H$_2$S can also heterolytically adsorb at a vacant sulfur edge and only sulfhydryl groups are generated at the step edge. The valence state of Mo or W in the bulk sulfides is 4+. The structure of a MoS$_2$ particle and its reactive edge structure is illustrated in Figure 11i.2.

Insert 11: Mechanism of Catalytic Hydrodesulfurization

Figure 11i.1 shows the reaction schemes of the HDS and HDN hydrotreating reactions. The structure of the catalytically reactive sites is shown in Figure 11i.2. Figures 11i.3 and 11i.5 display volcano plot relationships between the HDS and bulk sulfide reactivity indicators. Table i.1 shows the corresponding reactionorders of th and hydrogen of Figure 11i.5. Figure 11i.4 shows a reaction energy diagram for thiophyene disulferizaiton.

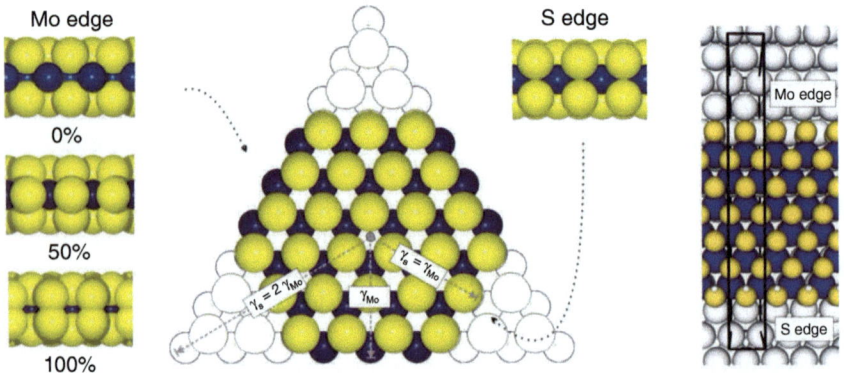

Figure 11i.1 Reaction schemes of HDN and HDS. (After [22].)

Figure 11i.2 Atomic ball model (top view) showing a hypothetical, bulk-truncated MoS$_2$ hexagon exposing the two types of low-index edges, the S edges, and Mo edges. The Mo atoms (dark) at the Mo edge are coordinated to only four S atoms (light). To the left, the stripped Mo edge is shown in a side view together with two more stable configurations with S adsorbed in positions predicted from theory: the 50% covered (monomer) and a 100% covered Mo edge (dimer). To the right, a side view of the S edge with a full coordination of six sulfurs per Mo atom is shown. Plotted on the MoS$_2$ hexagons are vectors with lengths corresponding to the edge free energies of Mo for the (1010) Mo edge and S edge. The envelope of tangent lines drawn at the end of each vector constructs a hexagon if S equals Mo. If S > 2 × Mo the result is a triangle (outlined shape) terminated exclusively by the Mo edge, or vice-versa for the S edge. Intermediate values result in clusters with a hexagonal symmetry. (Lauritsen et al. 2004 [67]. Reproduced with permission of Elsevier.)

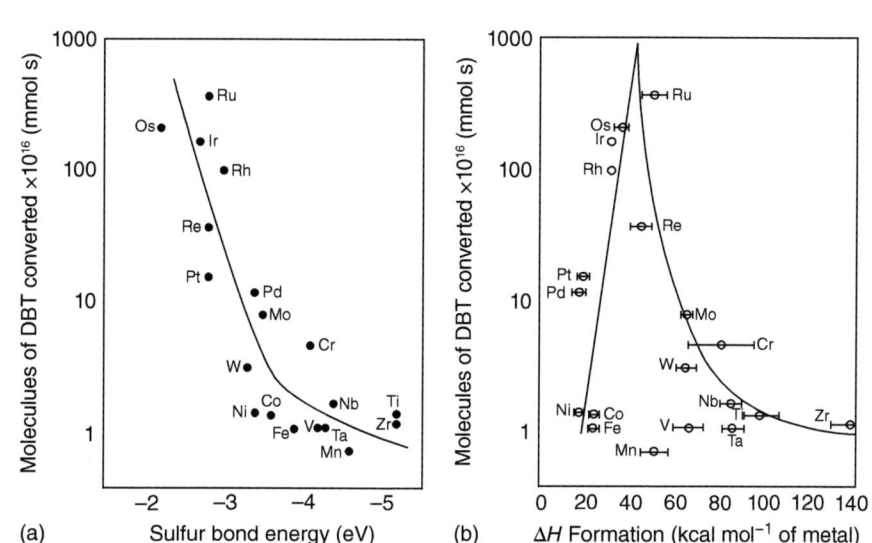

Figure 11i.3 (a) HDS activities versus calculated metal—sulfur bond energies. (b) HDS activities versus heats of formation normalized per mole of metal. Values for V_2S_3 and MnS have been added to the data set (closed circles). (Adapted from [68].)

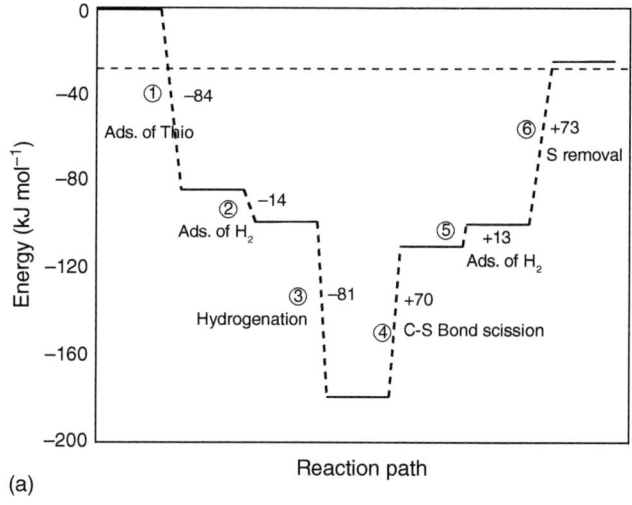

Figure 11i.4 A computed catalytic HDS cycle for thiophene desulfurization through a 2,5-dihydrothiophene intermediate on Ni_3S_2 clusters. (a) Diagram of energy changes for each reaction step. (Neurock and van Santen 1994 [69]. Reproduced with permission of American Chemical Society.) (b) Structures of successive reaction intermediates of the catalytic reaction cycle.

(Continued)

Insert 11: (Continued)

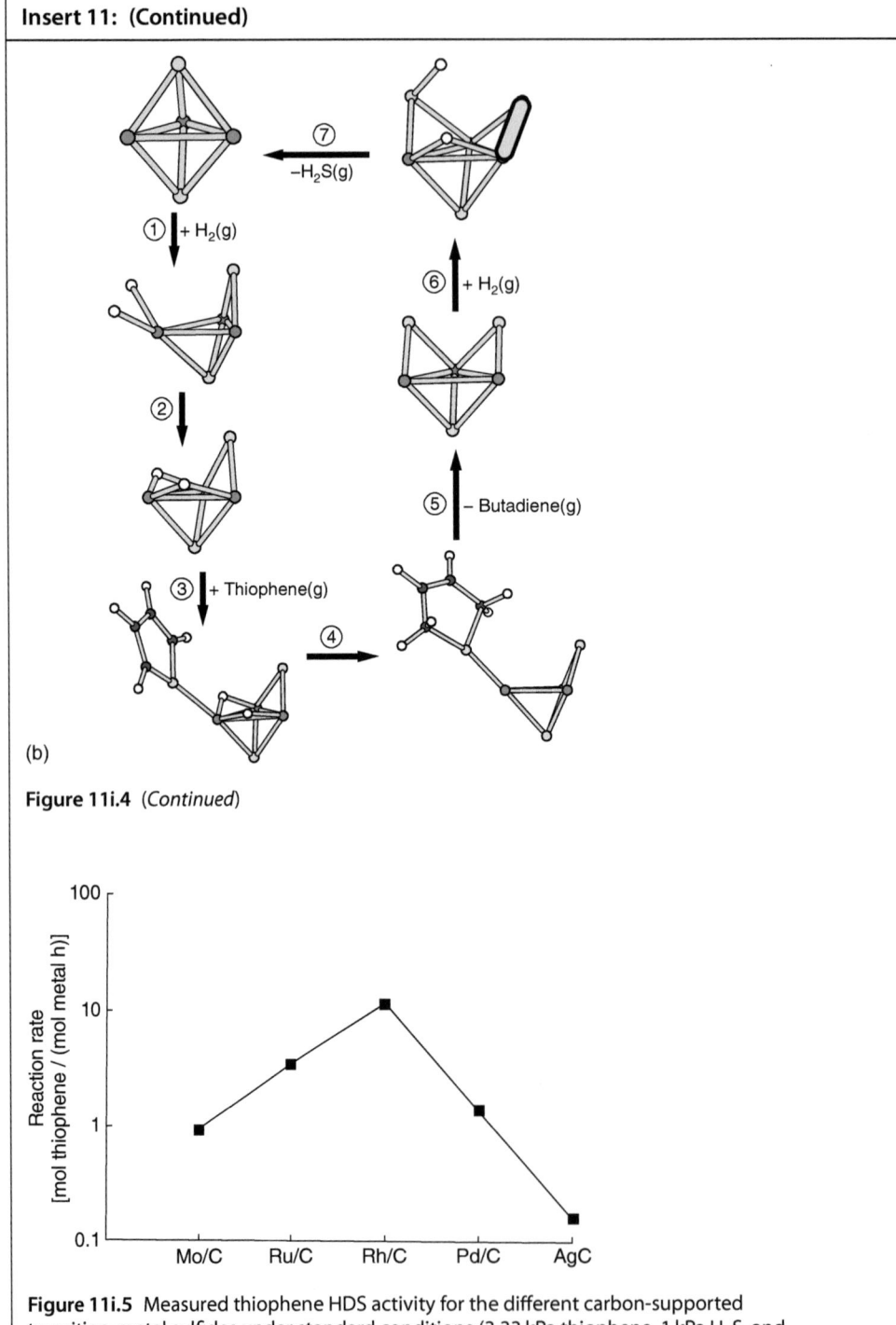

(b)

Figure 11i.4 (Continued)

Figure 11i.5 Measured thiophene HDS activity for the different carbon-supported transition-metal sulfides under standard conditions (3.33 kPa thiophene, 1 kPa H_2S, and $T = 573$ K). (Hensen et al. 2000 [71]. Reproduced with permission of Elsevier.)

Table 11i.1 Reaction orders of thiophene (n_T), H_2S (n_S), and H_2 (n_H) under different conditions.[a]

Catalyst	T = 573 K					T = 623 K		
	n_T[b]	n_T[c]	n_S	n_H[b]	n_H[c]	n_T	n_S	n_H[b]
Mo/C	0.40	0.50	−0.32	0.54	0.57	0.65	−0.34	0.74
Ru/C	0.28	0.39	−0.25	0.56	0.53	0.57	−0.27	0.93
Rh/C	0.21	0.31	−0.83	0.71	0.93	0.53	−0.59	1.03
Pd/C	0.50	0.65	−1.04	0.77	0.99	0.77	−0.97	1.42
CoMo/C	0.10	0.12	−0.46	0.61	0.78	0.28	−0.30	0.92

a) 95% Confidence interval for $n_T \pm 0.05$, $n_S \pm 0.07$, $n_H \pm 0.02$.
b) Inlet H_2S partial pressure: 0 kPa.
c) Inlet H_2S partial pressure: 1 kPa.
Source: Hensen et al. 2000 [71]. Reproduced with permission of Elsevier.

The structures of the sulfide edge structures have become understood mainly through computational studies on MoS_2 [67]. The sulfur atoms at the edge will rearrange, and one can distinguish both metal cation terminated surface edges and sulfur terminated surface edges. At representative reaction conditions, the Mo atoms at the Mo edge are coordinately saturated, so that the Mo atom at the surface edge is six-coordinated as it is in the bulk. At the sulfur edge, the Mo atoms are 50% sulfided and tetrahedrally coordinated.

Unpromoted MoS_2 or WS_2 catalysts have low reactivity mainly because the rate of reaction is limited by the diminished surface vacancy concentration due to H_2S adsorption. This is consistent with the interpretation of a plot of the activity of HDS as a function of the M—S bond energies (see Figure 11i.3a). The sulfide with weakest M—S bond energy has the maximum activity for thiophene desulfurization. Thiophene HDS is often used as a model reaction. The non-volcano dependence of rate on M—S bond strength is different from that observed in the volcano plots constructed by Chianelli et al. from the measurement of the dibenzothiophene desulfurization activities of bulk sulfides as a function of the bulk energies of sulfide formation (Figure 11i.3b). The plots in Figure 11i.3a,b are consistent, when the particular dependence of the M—S bond energy on metal-sulfide structure is taken into account.

As predicted by the Sabatier principle, the two main reaction steps that compete in the catalytic reaction cycle are the activation of reactant molecule and product desorption. A density functional theory (DFT) study of the reaction energy diagram of thiophene desulfurization by a Ni_3S_2 cluster is shown in Figure 11i.4. In this reaction, Ni_3S_2 particles have been shown to be highly reactive when positioned in the micropores of a zeolite [70]. There is an increasing exothermicity of the reaction when molecules adsorb and hydrogen is added to the thiophene molecule. One of the most endothermic steps is the desorption of H_2S.

Figure 11i.5 and Table 11i.1 show the actual volcano-type dependence of the thiophene HDS as a function of metal position of the sulfide along a row in the periodic system. According to the Sabatier principle a change in the order of reactions is expected at the volcano maximum. At the maximum one notes the relatively small negative order in H_2S, which indicates weaker adsorption, and also the relatively low positive order of thiophene, which implies a relatively high rate of thiophene activation. The volcano curve in this case is the consequence of the competition of the two reaction steps and may therefore be expected to strongly depend on reaction conditions or the type of molecule that is converted.

Highly active catalysts can be produced by promotion of MoS_2 or WS_2 with Co (in the case of HDS) or with Ni (in the case of HDN, where the inhibition by ammonia is less since ammonia binds less strongly to Ni than Co). The Co^{2+} and Ni^{2+} ions substitute for Mo^{4+} at the crystal edges. This can only be done in a charge-neutral way if a $(MoS)^{2+}$ unit is replaced by Ni^{2+} or Co^{2+}. This creates surface vacancies, which is the reason for the increased rate of HDS.

11.6 Electrocatalysis: The Oxygen Evolution Reaction (OER)

In the next subsection, periodic trends in the reactivity of reducible oxides for the electrocatalytic oxygen evolution reaction (OER) will be discussed initially based on the thermodynamics approach to estimate the reaction's overpotential. While this is a useful initial approach to understand differences in reactivity, information on current densities cannot be deduced in this way. Instead, kinetics studies are required.

The kinetics of elementary reaction steps of the reactions will be presented in a second subsection, using the results of DFT calculations on the highly reactive RuO_2 system as an example. As with chemocatalysis, electrocatalysis is also strongly dependent on surface structure. This dependence of reactivity on surface structure will be explored by comparing the turnover frequencies of the OER reaction on different surfaces of Co_3O_4.

The section will conclude with a comparison of the O—O bond formation reaction catalyzed by a cubic $Co_4O_4^{y+}$ and the cube-like $Mn_4O_4Ca^{x+}$ clusters that are part of the biological photosynthesis systems. Similarities and differences between the inorganic and enzyme systems will be discussed.

11.6.1 Trends in OER Reactivity

The Sabatier optimum of electrocatalytic oxygen evolution at the oxide anode as a function of the M—O bond energy of transition metal oxides is shown in Figure 12i.1. The overpotential as calculated using the thermodynamics method of Section 3.3.2.4 is plotted as a function of $\Delta G_O - \Delta G_{HO}$ [72].

The difference $\Delta G_{HO} - \Delta G_O$ is the deprotonation energy of adsorbed OH. Following the scaling law (Section 7.2.7) the bond energy of OH, ΔG_{HO}, scales with $1/2\,\Delta G_O$. Therefore, Figure 12i.1 is essentially a plot of electrocatalytic overpotential as a function of the adsorption energy of an oxygen atom to the metal cation of the respective oxides. One notes the maximum in reactivity for Co_3O_4 and RuO_2, which corresponds to a theoretical prediction of the minimum of the overpotential of 0.4 V.

At the left of the volcano plot for weak M—O bonds, the activation of water limits the reaction rate, but O—O bond formation or desorption of O_2 may be rate-controlling when the M—O bond is strong. These are the competing elementary reactions that determine the electrocatalytic OER optimum, which we will consider in detail in Section 11.6.2.

Insert 12: Volcano Plots of Electrochemical Reactivity

See Figures 12i.1 and 12i.2.

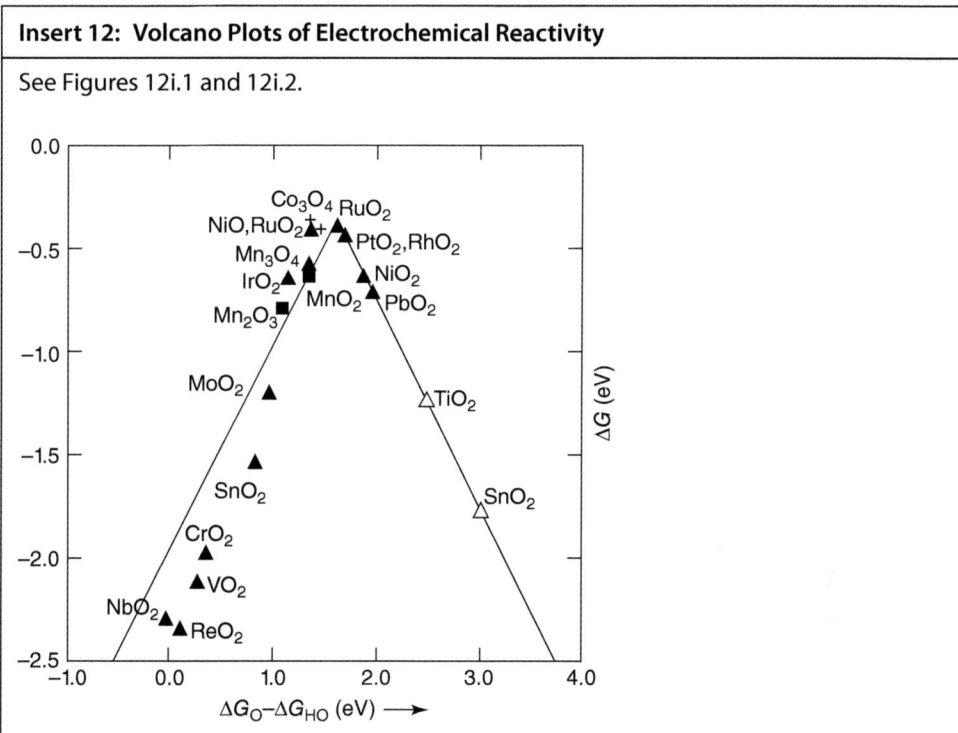

Figure 12i.1 Activity trends toward oxygen evolution for rutile oxides (△), Co_3O_4 (+), and Mn_xO_y (■). Open and solid symbols represent the adsorption energies on clean surfaces and on high-coverage surfaces, respectively. The negative value of the theoretical overpotential is plotted as a function of the standard free energy of $\Delta G_{O^*} - \Delta G_{HO^*}$. This is the difference of the free energies of adsorption of the O atom and that of the hydroxyl respectively. (Man et al. 2011 [72]. Reproduced with permission of John Wiley and Sons.)

(Continued)

Insert 12: (Continued)

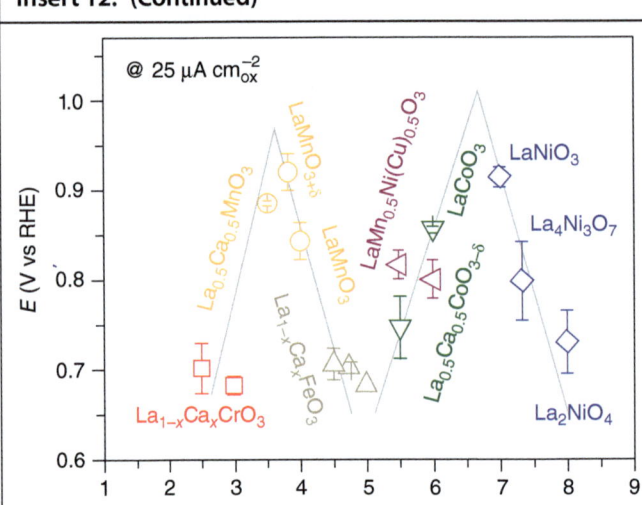

Figure 12i.2 Oxygen reduction reaction activity of perovskite (ABO_3) transition-metal-oxide catalysts. Data symbols vary with type of B ions: red (Cr), orange (Mn), gray (Fe), green (Co), blue (Ni), purple (mixed compounds), where $x = 0$ and 0.5 for Cr, and 0, 0.25, and 0.5 for Fe. Welters et al. 1994 [70]. Reproduced with permission of Elsevier.

In contrast to the plot as a function of M—O bond interaction energy, the plot of the overpotential of electrocatalytic water decomposition on a series of perovskites as a function of the number of d-valence electrons on the reactive transition metal cation shows a double volcano curve. The positions of the peak maxima are very similar to those in Figure 6i.3 for catalytic conversion on reducible oxides with the same cations. Again, we find the minimum in reactivity for cations with five d-electrons and maxima in reactivity at four and six electrons. As in Section 11.3.1, the maxima of activity agree with the crystal field theory prediction for the high spin state of an octahedrally coordinated site of Insert 3. For an electron count of less than five electrons the e_g orbitals have to remain empty at the maximum M—O interaction, and for cations with more than five electrons the M—O bond energy is at a maximum when all the t_{1u} orbitals are doubly occupied.

Therefore, the double-peaked structure of Figure 12i.2 follows the bond energies of adsorbed oxygen. This parallel behavior indicates that the reactivity of the perovskites is related to the trend in activation of H_2O. Their relatively high overpotential implies that their activity is low because the adatom M—O bond on these systems is too weak.

11.6.2 Reaction Mechanism of OER Reaction

In order for O_2 to evolve, the water molecule must deprotonate. The state of the surface may be expected to change as a function of applied potential. The degree of surface hydroxylation will change and the surface oxidation state also changes with increasing O atom coverage.

The change in surface state, will affect its reactivity. The activation energies of the elementary reaction steps will change when the surface composition alters.

A computational study of the OER reaction of the active $RuO_2(110)$ surface illustrates the relationship between surface reactivity and surface state [73].

Three important stages can be distinguished in the state of the surface as a function of potential.

At low potential, coordinatively unsaturated Ru ions become covered with OH and a proton moves to a bridging RuO_2 surface atom. At higher potential the surface deprotonates, and above a potential of 1.58 V the surface oxidizes further to become covered with only atomic O. No vacant Ru sites are present at this point (Figure 13i.1). This is the most reactive surface phase and the potential at which this phase is established determines the minimum potential to run the process. An important question is whether the mechanism of O_2 formation occurs by the recombination of two surface O atoms or by the reaction of a water molecule with adsorbed O through an O—OH intermediate.

There are two successive elementary reaction steps of H_2O that lead to O—O bond formation:

$$H_2O \rightarrow O_t + 2H^+ + 2e^- \tag{11.2}$$

$$H_2O + O_t \rightarrow O_2 + 2H^+ + 2e^- \tag{11.3}$$

O_t is the terminal O atom generated by the first H_2O dissociation.

An important issue is whether the mechanism of O_2 formation occurs by the recombination of two surface O atoms or by the reaction of a water molecule with adsorbed O through an O—OH intermediate.

Figure 13i.2 shows that the mechanism involving the direct O—O recombination of two O_t atoms requires an overall higher activation energy than the O—OH recombination step. This is due to the excessively strong binding of O_t.

At potentials higher than 1.58 V (overpotential of 0.35 V), the deprotonation rate increases further and hence the activation energy of the O—OH formation step, Eq. (11.3), decreases (Figure 13i.3).

Insert 13: OER of RuO_2 Surface

Figure 13i.1 shows the surface state as a function of potential. (After [73].) Figure 13i.2 presents the reaction energy diagram of O_2 formation on the oxygen-terminated RuO_2 surface. Figure 13i.3 illustrates the change in activation energy for the two reactions as a function of potential. Eley–Rideal type reactions of a H_2O molecule with surface, which also cause a change in redox state, are strongly affected by an electrostatic field. The surface recombination reaction of two O atoms is not very sensitive to potential. Figure 13i.4 shows the M—O bond energy necessary for a reactive OER surface for a particular potential regime.

(Continued)

Insert 13: (Continued)

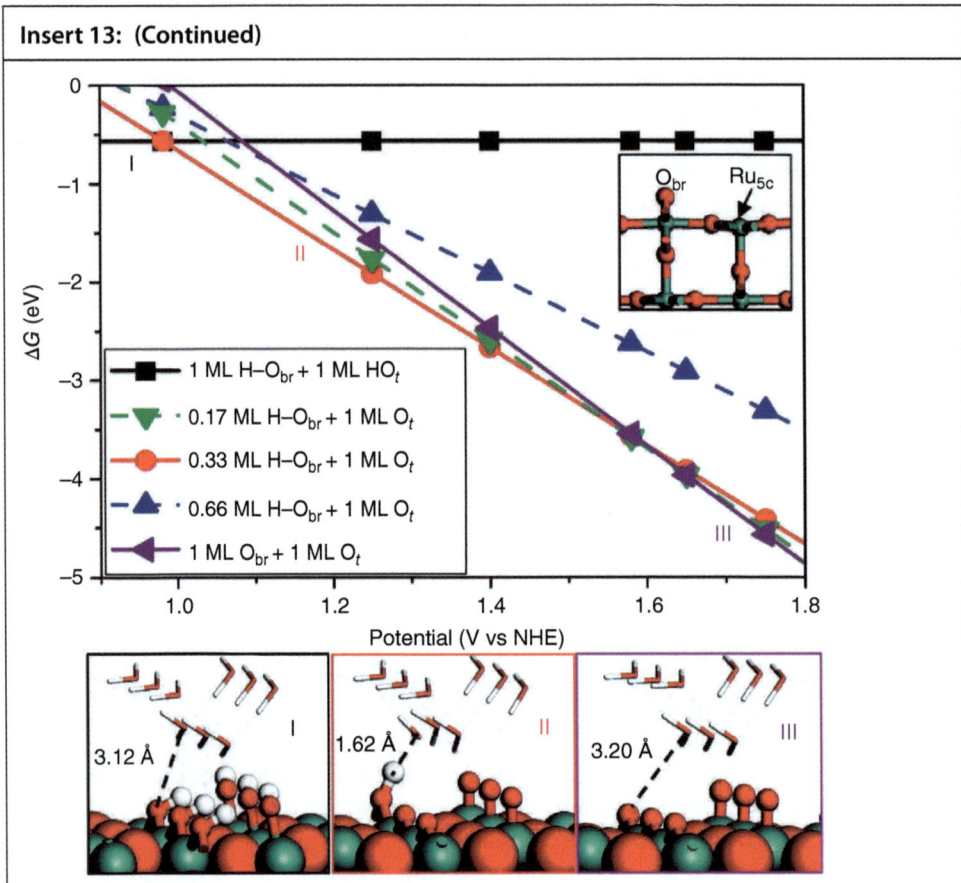

Figure 13i.1 The surface phase diagram of the RuO$_2$ (110) surface and the optimized structures of the three stable surface phases. Phase I: the OH-terminated phase (1 ML H—O$_{br}$ + 1 ML H—O$_t$); phase II: the 0.33 ML H—O$_{br}$/O$_t$ mixed phase (0.33 ML H—O$_{br}$ + 1 ML O$_t$); and phase III: the O-terminated phase (1 ML O$_{br}$ + 1 ML O$_t$). The insert show the surface structures of a bare RuO$_2$ (110) surface (upright corner). At the bottom from left to right the succesive surface phases I, II and III are shown. O, Red ball; H, white ball; Ru, green ball. (Fang and Liu 2010 [73]. Reproduced with permission of American Chemical Society.)

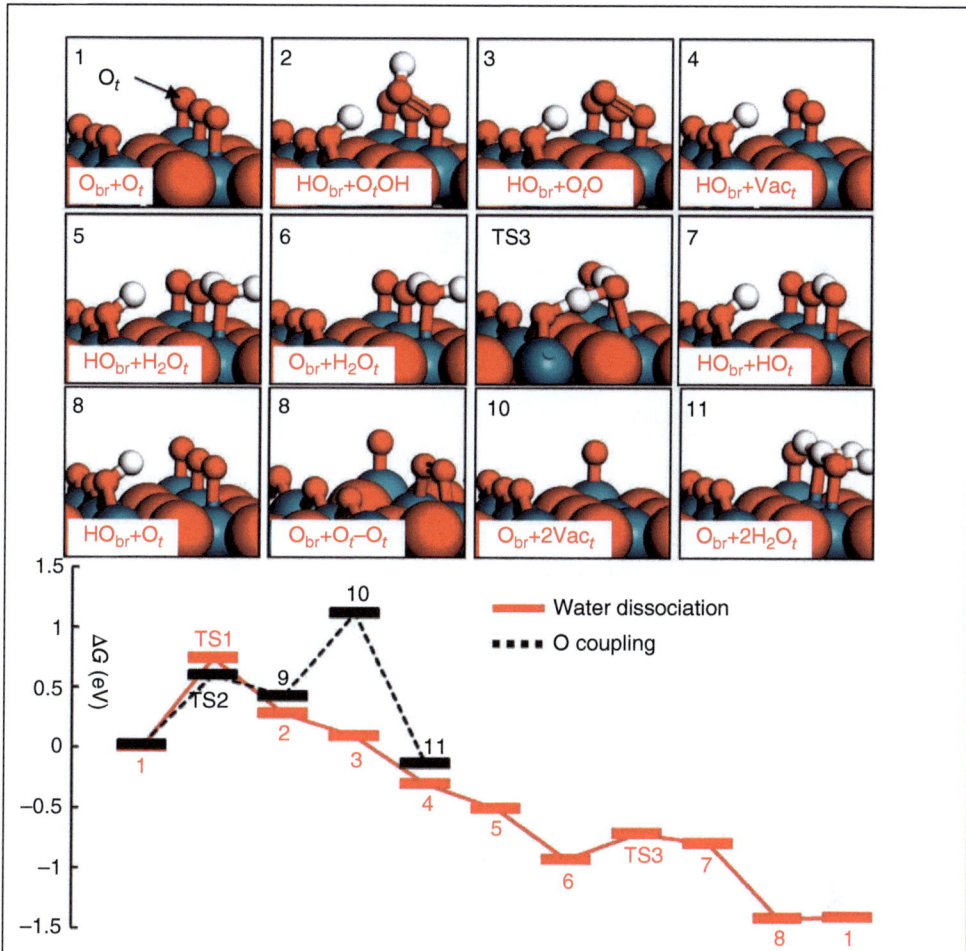

Figure 13i.2 The optimized structures of intermediate states and the free energy profile for OER on the O-terminated phase of the RuO_2 (110) surface at 1.58 V. O_{br}, O_t, and V_{act} are the bridging O, the terminal O on Ru_{5c} and the vacant Ru_{5c} site, respectively. For the optimized structures, the first H_2O layer is omitted for clarity. O, Red ball; H, white ball; Ru, green ball. (Fang and Liu 2010 [73]. Reproduced with permission of American Chemical Society.)

(Continued)

Insert 13: (Continued)

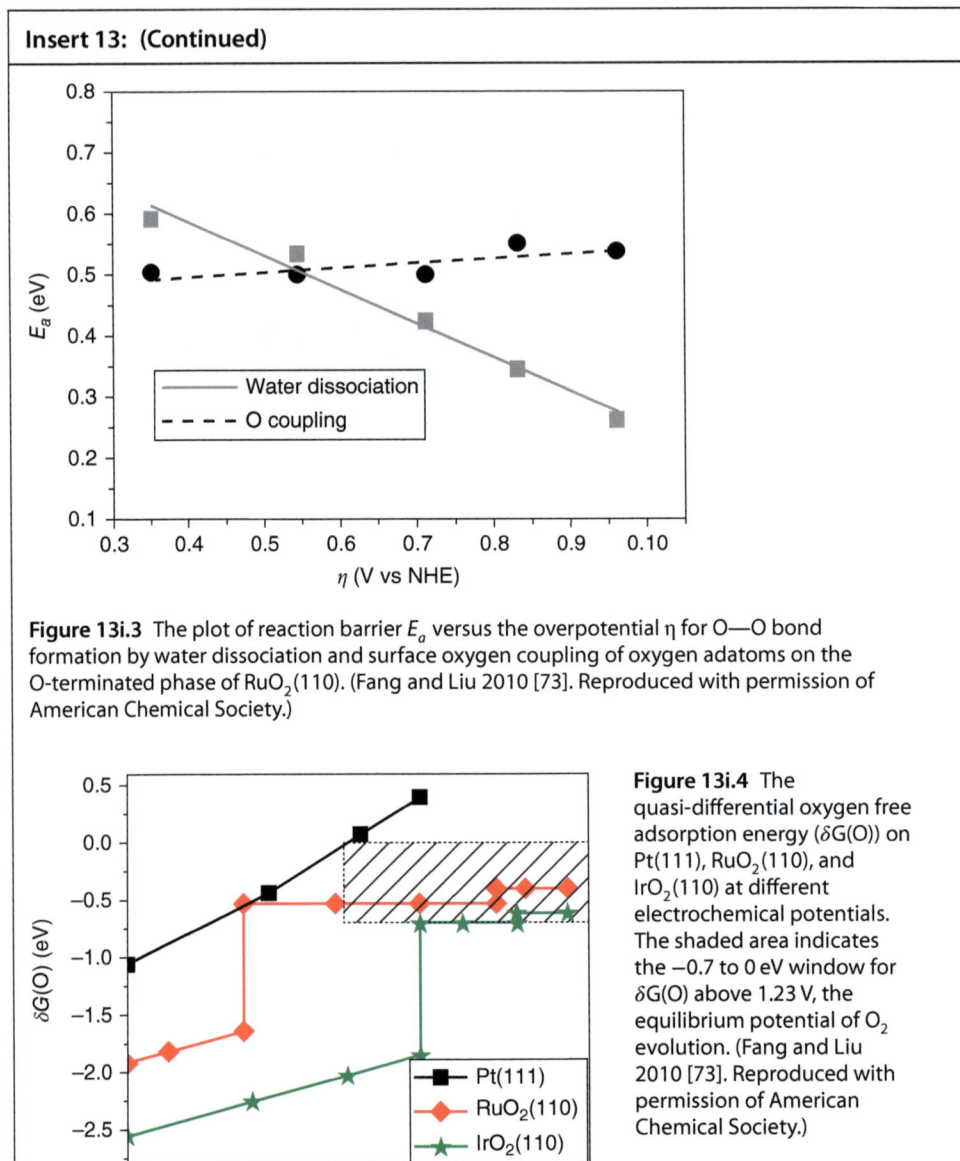

Figure 13i.3 The plot of reaction barrier E_a versus the overpotential η for O—O bond formation by water dissociation and surface oxygen coupling of oxygen adatoms on the O-terminated phase of $RuO_2(110)$. (Fang and Liu 2010 [73]. Reproduced with permission of American Chemical Society.)

Figure 13i.4 The quasi-differential oxygen free adsorption energy ($\delta G(O)$) on Pt(111), $RuO_2(110)$, and $IrO_2(110)$ at different electrochemical potentials. The shaded area indicates the −0.7 to 0 eV window for $\delta G(O)$ above 1.23 V, the equilibrium potential of O_2 evolution. (Fang and Liu 2010 [73]. Reproduced with permission of American Chemical Society.)

In order to form an O—O chemical bond according to Eq. (11.3), proton vacancies are necessary on the surface oxygen atoms to accept protons from the reacting H_2O. The anode potential must increase to create proton vacancies. The electrode potential should be high enough for free oxygen atoms to be generated from H_2O, but not so high that the surface oxygen atom basicity becomes so low that proton acceptance from H_2O is prevented.

The overpotential of reaction depends first on the electrode potential necessary to prepare the surface in its reactive state and secondly on the activation barrier of the rate-controlling elementary reaction step.

When the free energy of the M—O bond is zero or positive, O—O bond formation is thermodynamically favorable, but surface coverage with oxygen will be low. The free energy of M—O bond formation should be between 0 and $-70\,kJ\,mol^{-1}$ in order to achieve a low overpotential (see Figure 13i.4) [73].

Figure 13i.4 illustrates the potential dependence of the metal-oxygen bond. The high potential increases oxygen coverage, which weakens the individual M—O bond energies. The favorable overpotential of RuO_2 is due to its optimum M—O bond energy. A similar conclusion was reached from the thermodynamic volcano plots in Figure 12i.1. The activity of the system may be expected to be higher at lower overpotential when the electrochemical potential necessary to form the deprotonated oxidic overlayer is lowered by a slight increase of the M—O bond energy.

Similar results have been found in a computational study by García-Mota et al. [74] on the reactivity of the Co_3O_4 surface. Again, the minimum overpotential determines the surface state for efficient O—OH bond formation.

We will now explore the structure sensitivity of the reactions and the chemical performance descriptors that determine the turnover frequency (TOF) as well as overpotential for different surfaces. This discussion is based on DFT calculations on the three Co_3O_4 (001), (311), and (011) surfaces respectively [75]. The surface oxygen atoms become increasingly more coordinatively unsaturated in this order. This is one of the reasons for the larger differences in OER activity between these surfaces.

The M—O bond energy is strongly influenced by the coordination number of the surface oxygen atom with the surface metal ions and the valence state of the cations to which the oxygen atom is bonded. The O atom can be attached to a single metal cation as a terminal O or adsorbed in a two- or threefold bridge site. On the Co oxide surface, the terminal oxygen atom is stable when adsorbed to a Co^{5+} ion, but when the Co ion has a lower valence state, only higher coordinations are stable.

Generally, the higher the O coordination number, the stronger the surface chemical bond. The electrochemical potential needed to convert surface hydroxyls to reactive O centers will be lower on sites with a stronger M_x—O surface bond. Then the O—H and O—OH bonds will be weaker. In contrast, the activation energy for O—OH bond formation on the more highly coordinated oxygen atoms will be higher than on the coordinatively less saturated surface oxygen atoms. When the site is initially hydroxylated, the higher oxygen coordination site will have reduced overpotential for dihydroxylation compared to an end-on adsorbed hydroxyl and the activation energy for O—O bond formation will be higher. As we will see, these counteracting effects may ultimately produce a higher overpotential on the coordinatively less saturated surface.

As we discussed earlier for RuO_2, the apparent activation energy of O—OH bond formation from H_2O and surface O atoms depends not only on the O—O bond formation energy but also on the basicity of the surface. H_2O must deprotonate in order to react as OH with surface O. This is illustrated in Figure 14i.1. At the potential of reaction a bridged OH is deprotonated at the Co site on a surface where OH⁻ and H_2O are also initially co-adsorbed. Deprotonation occurs in a complex series of proton reshuffling reactions.

The deprotonation energy is very different for the different surfaces. For example, energy for the (100) surface is 0.68 kJ mol⁻¹, while for the (311) surface is 0.51 kJ mol⁻¹.

Figures 14i.2 show the catalytic reaction cycles for the oxygen evolution reaction at two different Co-O sites. Simulated TOFs of the OER reaction as a function of electropotential are shown in Figure 14i.3. The strong dependence of the TOF on potential and the large differences in performance of the different surfaces are notable.

Sites with terminal and bridging oxygen atoms exhibit differing reactivities. The sites with terminal oxygen require a higher potential to form O_2. This is because reactive oxygen atoms that have these weaker M—O bonds become stabilized only at a higher potential. However, the terminal surface oxygen atoms will react with water to form O—O bonds with rate constants of lower activation energies. When this reaction is rate-controlling it will increase the overall TOF, compared to the TOF of other surfaces with less reactive highly coordinated O atoms that reach the reactive surface state at the lower over potential

The TOF of a surface saturates at a finite value when the maximum coverage with reactive O is reached at a particular potential. At an even higher potential the TOF may actually decrease because no surface vacancies for H_2O decomposition are available.

Insert 14: OER by Co_3O_4 Surfaces

Figure 14i.1 shows the OER reaction schemes for H_2O with a terminal versus bridging O atom. (After [75].) Figure 14i.1 illustrates that the apparent activation energy of O—OH bond formation will not only depend on the O—O bond formation energy, but also on the basicity of the surface, since H_2O must deprotonate in order to react as OH with surface O. Figure 14i.2 displays the electrocatalytic cycles on the two surfaces. The calculated TOF versus overpotential curves are shown for three surfaces in Figure 14i.3. At higher overpotential η, reactions with terminal oxygen atoms also become competitive. At the end of the insert, the equations to calculate the TDFs are summarized.

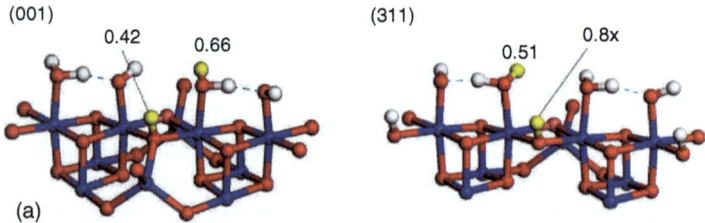

(a)

Figure 14i.1 (a) Structures of the active site prior to water addition to a bridging oxo on the (001) and (311) surfaces. The numbers indicate the overpotential at which the respective protons (yellow) are removed. (b) Comparison of the structures and energetics (eV at η = 0) of the oxidation of the single-Co site on the (110) and (001) surfaces prior to water addition. The lower oxidation potential of the (110) site is due to the stabilization of Co(V) in the oxidized state by basic η-OH.

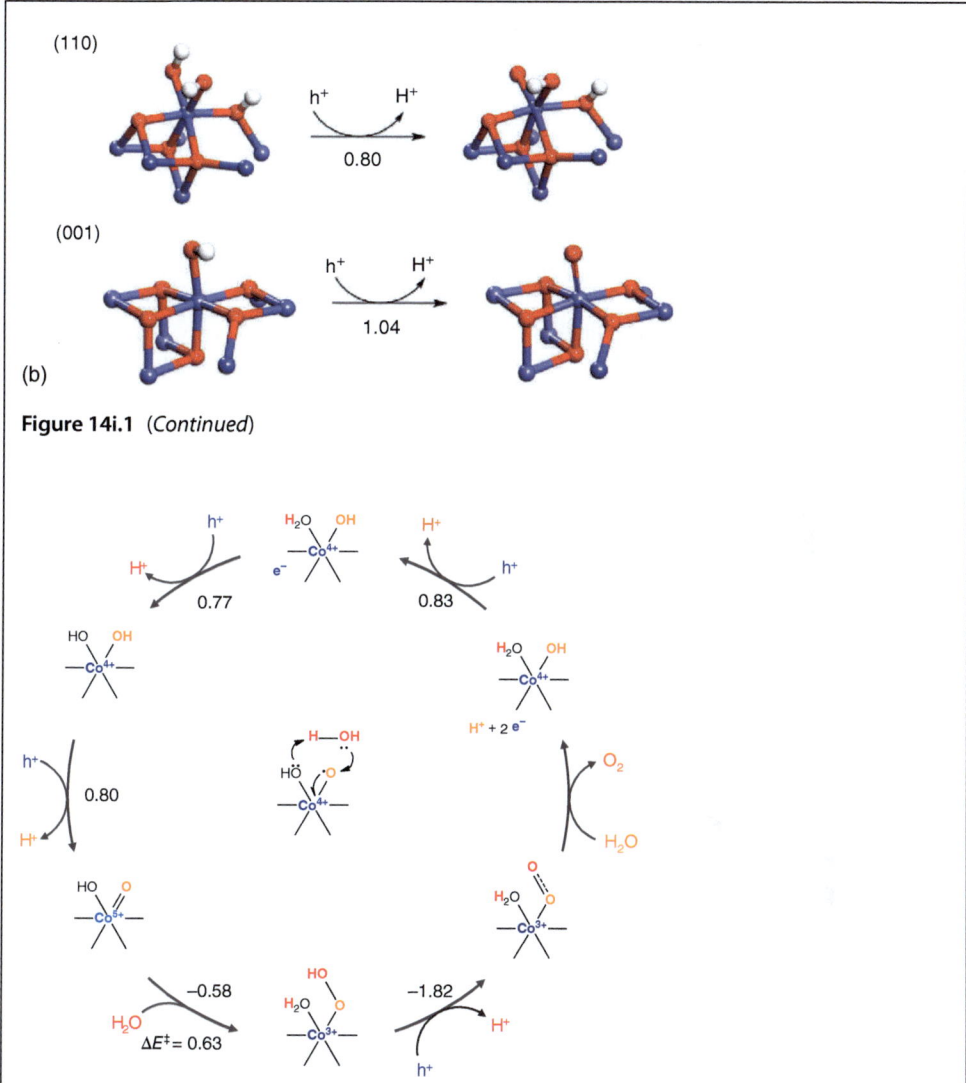

Figure 14i.1 (Continued)

Figure 14i.2 Catalytic cycles and energetics (eV at $\eta = 0$) of the OER for the single-Co site on the (110) surface and the dual-Co site on the (311) surface. H$^+$ and e$^-$ labels in the structures indicate protons and electrons located at a nearby site that is not shown in the structure.

(Continued)

Insert 14: (Continued)

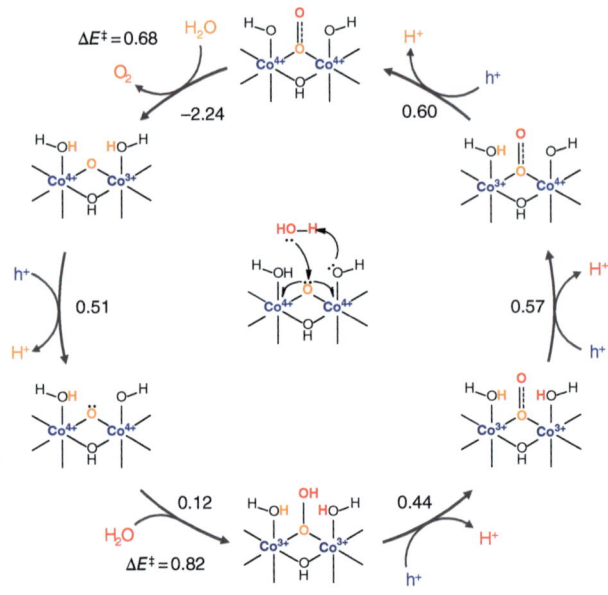

Figure 14i.2 (Continued)

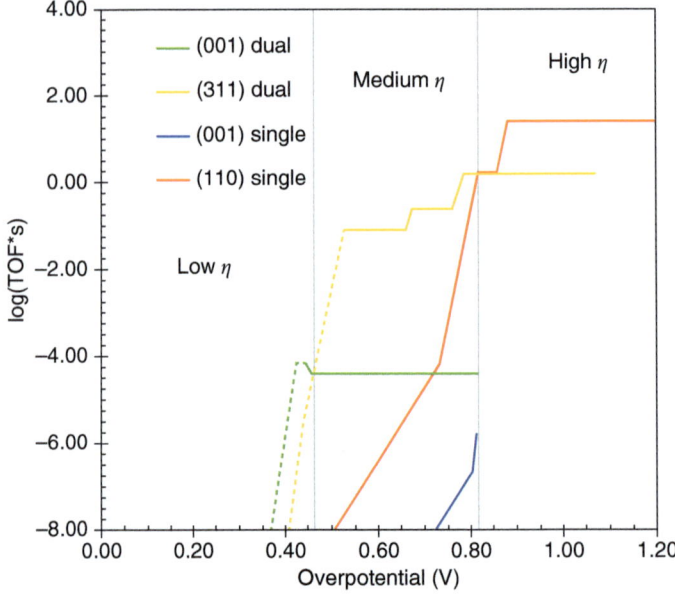

Figure 14i.3 Oxygen evolution Co_3O_4 TOF of the OER on the dual-Co sites on the (001) and (311) surfaces and the single-Co sites on the (001) and (110) surfaces as a function of overpotential. The three overpotential regimes are indicated. Solid curves show the regions where H_2O addition is rate-limiting. Dashed curves show the regions where O_2 release is rate-limiting.

> The equations used to derive the curves of Figure 14i.3 are as follows:
>
> $$\text{TOF} = \frac{1}{\frac{1}{\text{TOF}_1} + \frac{1}{\text{TOF}_2}} \tag{14i.1}$$
>
> $$\text{TOF}_n = \frac{k_B T}{h} \sum_j \exp\left(-\frac{\Delta G_j^\ddagger}{k_B T}\right) \theta_j \tag{14i.2}$$
>
> In Eq. (14i.2), $\Delta G_j^\ddagger$ is the free energy difference between a transition state and the resting state j preceding it while θ_j is the probability that the active site is in resting state j, θ_j is computed from
>
> $$\theta_j = \left[\sum_i \exp\left(\frac{\Delta G_{ij} - e\eta \Delta n_{ij}}{k_B T}\right)\right]^{-1} \tag{14i.3}$$
>
> The term ΔG_{ij} is the free energy difference between resting states j and i at an overpotential of zero (1.23 V vs SHE) and is computed according to the method of Nørskov et al. [76].
>
> At most overpotentials, one term in the sums in Eqs. (14i.1) and (14i.2) will be much larger than the others and the TOF can be approximated by a single term that corresponds to the resting state and transition state with the lowest free energies at that overpotential.

The (001) surface becomes reactive at the lowest potential, but the TOF of the (001) surface is very low, and the reaction is rate limited by O_2 desorption. At slightly higher potential, it is surpassed by the activity of the (311) surface, but the rate of reaction is still initially limited by the rate of O_2 desorption. The (311) surface has a higher overpotential compared to the (001) surface because the (311) surface oxygen has lower reactivity. On the other hand, when the overpotential increases a surface state generates on the (311) surface which increases the rate of O—OH bond formation. This results in a substantially higher TOF once the potential has induced the (311) surface into this reactive state. At higher potential, the least reactive surface (110) dominates, with reactive terminal rather than bridging O atoms and the TOF at the higher overpotential is maximum.

11.6.3 Summary Mechanism of OER Reaction

The overpotential and TOF of a catalyst active for OER depend on:

- The electropotential necessary to stabilize a surface containing surface O atoms of the desired reactivity. The potential affects the strength of the M—O bond and the degree of protonation of the surface oxide layer.
- The activation energy of the rate-controlling reaction step. The rate of the elementary reaction step determines the TOF. The surface state (determined by the electropotential) governs which elementary reaction step is rate-controlling.

The reaction mechanism of the OER contains the following reaction steps:

1. **The Activation of Water to Give OH_{ads} or O_{ads}** On the oxide surface this requires a vacant site for water adsorption and a basic surface oxygen atom

which accepts a proton. The reaction requires an M—O bond energy that is strong enough to bind to a cation but weak enough to accept a proton generated by heterolytic water splitting.

2. Formation of the O—O bond that can occur through two reaction paths:

   a. ***Reaction of Adsorbed O with an H_2O Molecule*** The O—OH bond formation reaction can occur by a Langmuir–Hinshelwood type reaction through the recombination of adsorbed OH and adsorbed O or by an Eley–Rideal type mechanism through the reaction of a surface oxygen atom with a water molecule. The Eley–Rideal mechanism occurs on RuO_2 and Co_3O_4.

   We will discuss the OER catalyzed by TiO_2 in Section 11.7.2. In this case, O—OH formation is of the Langmuir–Hinshelwood type. A terminal-adsorbed OH radical reacts with a bridging lattice O atom. The O—OH reaction is favored for intermediate M—O bond energies.

   b. ***Recombination of Two Surface Oxygen Atoms*** Recombination of two surface oxygen atoms will be the preferred pathway to O_2 when the M—O interaction becomes weaker. This may happen on metal oxides at higher potentials. As we will see in the next section, a spin selection rule may also determine whether O—O or O—OH recombination will be preferred. On Co oxide clusters the O—OH reaction with H_2O from the water phase is favored, whereas on Mn oxide clusters direct O—O bond formation occurs.

3. ***Desorption of the O_2 Molecule*** The desorption of O_2 may be facilitated by the adsorption of H_2O in the vacant position it generates. Since this adsorption also favors H_2O decomposition, O_2 desorption may compete with H_2O activation even when the M—O bond interaction is small.

An analysis of the overpotential of the OER reactions based on thermodynamics and a reaction scheme following the formation of O—OH intermediate has been given in Section 3.3.2.4. According to this analysis based on the simplified reaction scheme of four reaction steps, the overpotential of the OER reaction has a theoretical minimum of approximately 0.4 V. A zero overpotential requires the free energy differences of the four reaction steps to be the same, which would give the ideal equal energy distant reaction energy diagram in Figure 3.15. However, this cannot be realised because the difference in adsorption energies of M—OH and M—OOH cannot be varied independently due to scaling relations. This determines the differences in the free energies of the transformation of M—OH to M—O and M—O to the M—OOH intermediate respectively. Scaling relations indicate that the M—O bond change of M—OH and M—OOH scale linearly when M is varied (Section 7.2.7).

We will return to this issue at the end of the next section.

11.6.4 Comparison with the OER in Enzyme Catalysis

Small oxide clusters supported by semiconducting materials have undiscovered that photocatalytically decompose water [77]. We will discuss these in the next section on photocatalysis. The enzymes that are part of the biological photosystems and are involved in the water splittting reaction contain related oxide clusters as cofactors. This makes the comparison of inorganic systems with enzyme catalysis interesting.

We will compare the mechanism of the OER reaction involving a small cubic Co oxide cluster to the mechanism for the same reaction with a Mn_4Ca cluster, which is the reactive center in the oxygen evolution enzyme of photosystem II. Photosystem II is the reactive center that is part of the biological photosynthesis system.

The OER reaction in photosystem II is driven by the electrochemical potential that evolves when light is absorbed.

We will start with a summary of the computational results on the Mn oxide cluster. The structure of the cubic oxidic Mn_4Ca system in contact with the peptide ligands of the enzyme is shown in Figure 15i.1a. It consists of a cube formed by three Mn cations and Ca cation bridged by the fourth Mn cation.

Insert 15: The OER Reaction Catalyzed by Small Oxide Clusters

In this insert, Figure 15i.1 summarizes the mechanistic results for the Mn oxide cluster reactivity of photosystem II. Figure 15i.2 summarizes the computational results of OER-catalyzed Co clusters. The reaction with the Co_4 clusters is compared in figure 15i.2A, B. Figure 15i. 2 C, D is the schematic representation of O—O recombination energies of the Mn clusters compared with that of the Co cluster [79]. Figure 15i.3 compares the thermodynamics predictions for the bulk oxides surface and the enzyme.

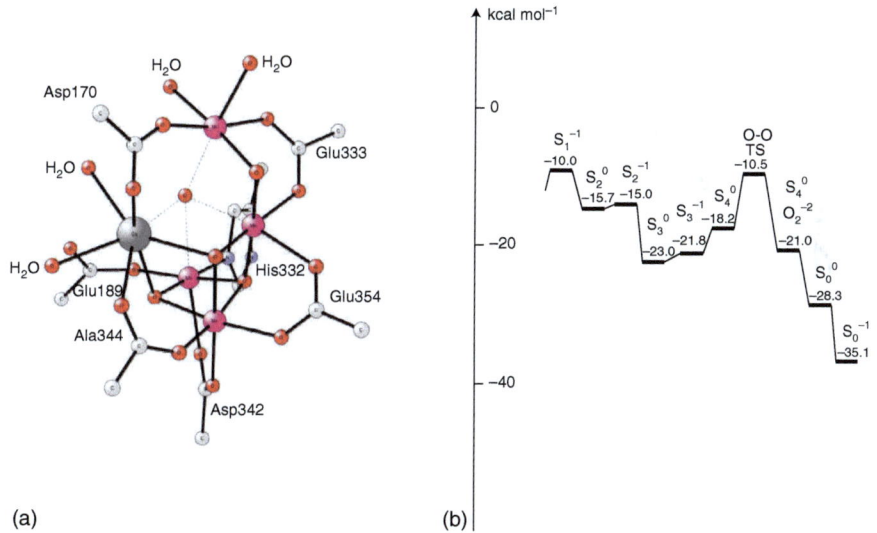

(a)   (b)

Figure 15i.1 Photosystem II simulations of the Mn5Ca system. (a) Oxygen-evolving complex embedded in the protein skeleton of PSII. (b) Full-energy diagram for water oxidation from S_0 back to S_0 with a full potential membrane gradient structures are shown in (C). (c) Schematic picture of the different S transitions. The structures have been optimized, but only the most important atoms are shown. An asterisk marks the atom that has been oxidized in that transition. (Blomberg et al. 2014 [78]. Reproduced with permission of American Chemical Society.)

(Continued)

534 | *11 Reducible Solid State Catalysts*

Insert 15: (Continued)

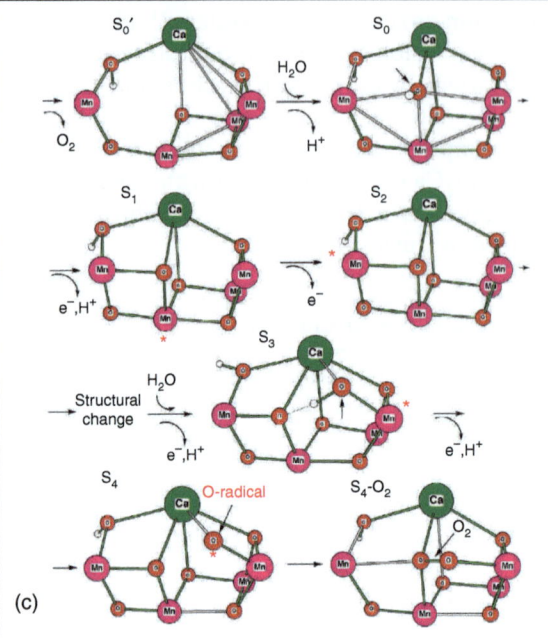

Figure 15i.1 (Continued)

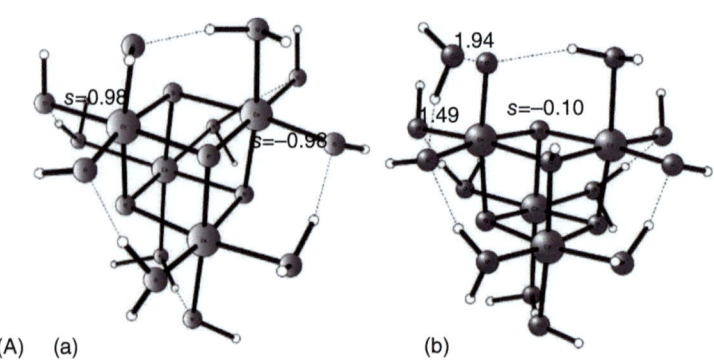

Figure 15i.2 The reactivity of the Co_4 clusters. Optimized structures and energies (Figure A–C) (A) (a) Optimized structure for the neutral Co_4 model where two of the metals are Co(IV). (b) Optimized TS structure with an overall barrier of 31.6 kcal mol^{-1} (Li and Siegbahn 2013 [79]. Reproduced with permission of American Chemical Society.)

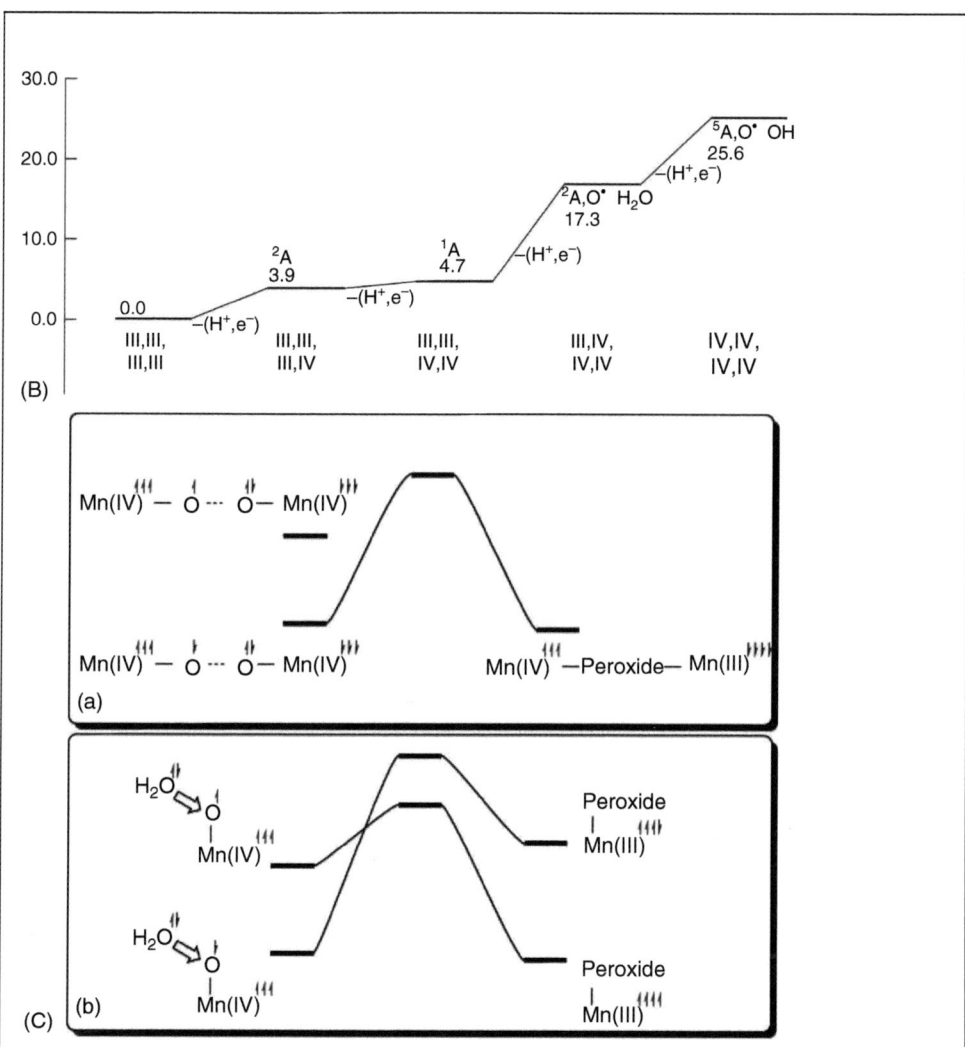

Figure 15i.2 B: Energy levels for Co_4. Each structure is characterized by its formal oxidation state (at the x-axis), its spin state, and by the number of oxygen radicals (O·). Only states that are reactive for O—O bond formation are shown in the diagram.
C: Schematic comparison mechanisms for O—O bond formation in the Mn oxide cluster of PSII; the direct O—O recombination (DC) mechanism in (a) and the water attack mechanism in (b).

(Continued)

Insert 15: (Continued)

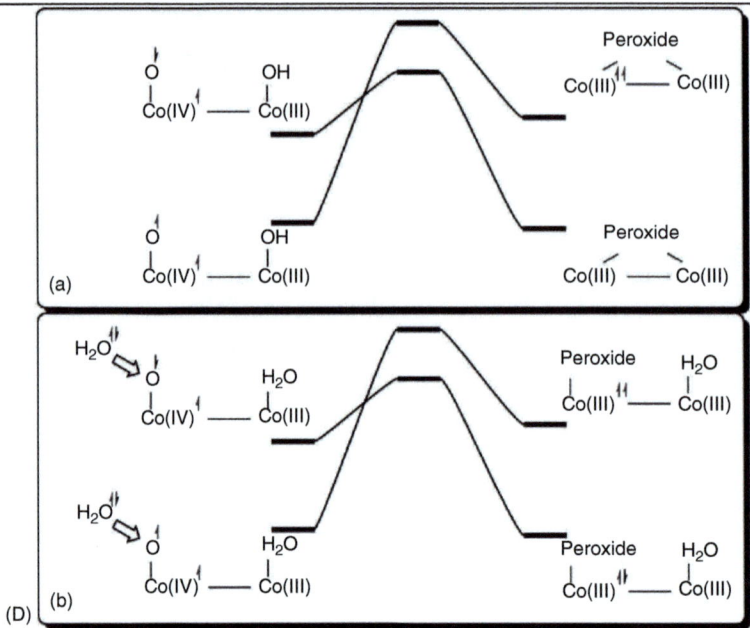

Figure 15i.2 D: Schematic mechanism for O—O bond formation for the Co_4 cluster at the formal (III, III, IV, IV) oxidation level; the DC mechanism in (a) and the water attack mechanism in (b).

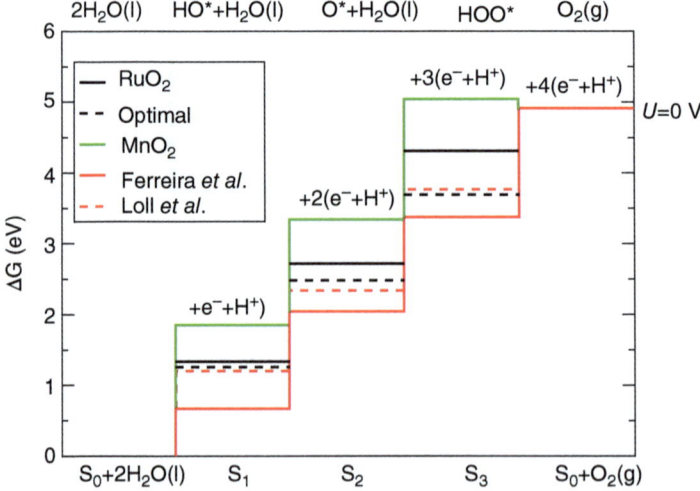

Figure 15i.3 Relative free energies for intermediates along the oxygen evolution reaction for RuO_2 (solid black line), the optimal catalyst (dashed black line), MnO_2 (green line), and energies from two different X-ray structures of the Mn cluster found in enzymes (red line represents the Ferreira et al. [80] structure and red dashed line represents the Loll et al. [81] structure). (After [82].)

11.6 Electrocatalysis: The Oxygen Evolution Reaction (OER)

A considerable volume of literature exists on the structure and mechanism of the OER reaction catalyzed by this center, which is well summarized in [78, 79] that we will follow here. Figures 15i.1b,c illustrates the reaction energies and mechanism. In the enzyme system the electropotential is determined by the biological system and hence the overall energy change is exothermic. The rate-controlling step is the formation of the O_2 molecule.

Figure 15i.1c shows that in the first reaction step with H_2O an OH species is formed that is twofold coordinated to Mn and also bridges with a Ca^{2+} cation. In a next step, the system is oxidized so that a reactive O atom is generated at the same position. Then, a reaction occurs with a second H_2O molecule, and the OH group that is generated by this reaction locates between Ca^{2+} and another Mn cation. After a further oxidation, an O radical is formed that reacts with the first O atom to give O_2.

In contrast to electrocatalysis on the Co oxide surface, O_2 is formed by recombination of the O anion and O radical instead of by a recombination of OH and O. As we will see, this difference in reactivity relates to the spin state changes that occur within the respective Mn or Co cations of the respective oxidic clusters.

The cubic Co cluster is related to the state of Co in the amorphous Co oxide systems used in photocatalytic applications. The chemistry of the cubic cluster is especially interesting because it reacts in a manner similar to the Co_3O_4 surfaces. This is very different from the Mn_4Ca cluster reaction discussed above.

Figure 15i.2a gives the structure of the partially hydrated cubic Co_4O_x cluster, which is used in the calculations by Li and Siegbahn [79]. The corresponding reaction energy diagram for the minimum reaction energy path at zero potential is shown in Figure 15i.2b. Because now there is no external potential, the reaction is endothermic. The charge state of the Co ions and the spin state of the Co atoms in the cluster are indicated in this figure. Schematic illustrations of the O—O bond forming reactions on the respective clusters are given in Figure 15i.2c,d.

The important difference between these two cases is that direct coupling (DC) occurs in the Mn cluster without change in spin state, whereas O—O bond formation in the Co cluster requires a change in spin state. This is due in essence to the Mn^{3+} cation's preference for a high spin state and the Co^{4+} and Co^{3+} preference for low-spin coupling states. In Mn, antiparallel spin coupling occurs with an O radical, which is not possible on Co along either reaction path. The computed activation energy for the photosystem II reactive center is only 44 kJ mol^{-1}, but a higher value of 125 kJ mol^{-1} is found for the Co cluster. Although this value is higher than the activation energy for Co_3O_4 surfaces or the comparable experimental amorphous system (90 kJ mol^{-1}), the cluster reactivity is subject to the same spin considerations as the Co_3O_4 surface.

It is interesting to use the thermodynamics approach described in Section 3.2.2 to compare the reactivity of the reducible oxide surfaces and the enzyme Mn_4Ca oxide cluster. This comparison is shown in Figure 15i.3. It can be seen that one of the materials with lowest overpotential, RuO_2, is still far from the ideal equal distance free energy profile (see Section 3.2.2), and, as expected, MnO_2 is even farther removed. However, the free energies of the resting states of the Mn_4Ca cluster model structure of Loll et al. [81] are very close to ideal thermodynamics. The overpotential of reaction is mainly determined by the activation energy of the rate-controlling O—O bond formation step [83]. The OH—OOH Bond-Order-Conservation rule correlation that prevents equal distance free energies in the electrocatalytic OER reaction energy diagram is resolved in the biological system by the alternative O—O recombination route found for the MnCa oxide cluster.

11.7 Photocatalytic Water Splitting

11.7.1 Device Considerations

Solar light capture and its conversion into electricity is a renewable source of energy. An important practical issue in this process is storing this electric energy when there is no daylight. Unless a worldwide electric grid is available to compensate during times when solar light cannot be captured, electricity will have to be stored in batteries or converted into a chemical energy carrier such as H_2. This requires combination of a photovoltaic (PV) system that generates electricity with an electrolysis process that produces hydrogen.

The maximum PV photoefficieny is 20% and that of commercial electrolyzers is 65%, which gives a maximum overall efficiency for H_2 generation of 13%. It is a question of substantial interest whether a process that is analogous to the photosynthesis process for converting solar light into H_2 and O_2 is feasible and whether it is more efficient than separate electricity generation and water electrolysis.

In this section, we will introduce the basic principles of a photocatalytic device for the water splitting reaction.

Figure 16i.1 illustrates the fundamental chemical and physio-chemical steps in this process. Light absorption in a semiconductor can generate an electron and leaves a positive charge, the hole, behind. The electron must be used to reduce H^+ from water and the positive charge must be used to oxidize water to O_2. Hence, high efficiency requires the effective separation of electron and hole and the suppression of electron–hole recombination. The electron and hole have to separate and diffuse to the surface of the semiconductor where current can be collected. A potential gradient is necessary for the separation of the hole and the electron. This gradient is often provided by doping the photoabsorber to separate it into an n-type part that attracts the holes, and a p-type part that attracts the negative charge. The state of the surface in contact with a solvent may create a charge on the surface (we refer to the ζ potential discussed in Chapter 4, Insert 1), that may lead to additional potential gradients but nearer to the surface.

Imperfections in a crystal lattice are often the sites of electron–hole recombination. The surface of the semiconductor can be considered a large defect that will catalyze the recombination of electrons and holes. Special precautions to prevent electron–hole recombination must be taken when high surface area particles are used.

A disadvantage in the use of small semiconductor particles is also that their total light absorption capacity is decreased compared to that of a large particle, which reduces the electric performance capacity of the system.

The minimum potential that the photocatalytic device must generate is 1.23 V, which is the thermodynamic potential for splitting water into hydrogen and oxygen. In practice, the minimum potential needed is around 1.70 V due to overpotential losses on the electrocatalyst particles.

Photoactive material generates electric potential by light absorption and the excitation of electrons. The difference in energy between the occupied valence band and non-electron-occupied conduction band, or the bandgap, must be at least equal to the energy needed to decompose water. In addition, the energies of valence band must match the donating energy levels of water and the conduction band levels must match the electron accepting water levels that lead to H_2 generation. The relationship between semiconductor bandgap, valence band, and conduction band energy positions with respect to the proton reduction and O^{2-} redox potentials for photoactive materials are schematically shown in Figure 16i.2.

Matching with the bandgap is never satisfactory, but matching of the valence or conduction band position individually is often possible. A proposed solution is the use of the so-called Z-scheme, which involves the tandem use of two semiconductors coupled through a redox system (see Figure 16i.3). The redox system transmits electrons from the oxygen generation half of the system to the hydrogen generation part. The application of two semiconductors that adsorb light at different frequencies also utilizes a broader part of the solar energy spectrum.

For H_2 generation, the electrons that are generated by light must be collected by a catalytic particle situated at the surface of the semiconductor, which reduces H^+ to H_2. This electrocatalytic reaction is the hydrogen evolution reaction (HER) we discussed in Section 8.2.5. The positive charge must be collected on a catalytic particle that catalyzes the OER reaction, discussed in the previous Section 11.6. The main overpotential cost to the overall water splitting reaction is caused by the catalytic OER reaction.

A protective layer is needed on the surface to prevent recombination of the holes and electrons, and to prevent possible corrosion of the photoabsorbing material. Catalytically active particles on the semiconductor surface can act as the protective layer for reducing electron–hole recombination. The use of electrons and holes by the catalytic reaction will reduce their concentration at the surface and suppress recombination.

Hence, a material that has proper bandgap and energy band position relations, but suffers from fast electron–hole recombination may improve in photo-efficiency by use as a water splitting catalyst. Special precautions may be necessary to prevent recombination of H_2 or O_2. Coating of the catalytically active particles with Cr has been proposed to suppress this back reaction [86].

The theoretically maximum obtainable efficiency of a solar light water splitting device is 7% for a single semiconductor device, but 17% for a tandem device [87]. In principle, the tandem concept competes with the non-integrated combination which has an efficiency of 13%.

The device concept for the photovoltaic tandem system combined with water decomposition is shown in Figure 16i.4. The semiconductor combination is used for the generation of photovoltaic current and potential. Even when valence and conduction bands match the energy requirement of the water decomposition reaction, electrocatalytic reactions by these materials may have high overpotentials. Efficient catalysis requires activation by promoting active HER catalyst and OER catalyst particles. The semiconductor portion of the device may require a protective layer to prevent corrosion, which can be induced also by photoactivation.

The interphase between the semiconductor, the protective layer, and the catalytically active particles must be chosen carefully because an imperfect match between valence and conduction band or electron and hole trapping by lattice imperfections may cause additional energy losses.

The main difficulty in realizing this type of device is the choice of semiconductor material. For instance, a small bandgap Si is suitable for the anodic O_2 evolution reaction, but a large bandgap material is necessary for cathodic H_2 evolution. Despite extensive screening programs [89] no satisfactory large bandgap material has been identified, which is efficient and provides the necessary voltage.

The discovery of an OER material that produces O_2 by water splitting with a lower overpotential than that in currently known catalysts is also necessary.

11.7.2 Mechanism for Photoactivation of Water

In the previous section, we discussed photocatalytic systems where photoadsorption generates the electropotential for an electrocatalytic reaction with H_2O. Photoexcitation and H_2O activation are considered decoupled in these systems.

Insert 16: Photocatalytic Device Issues and Mechanism

The first four figures concern the (photo)physics of the photocatalytic water splitting system. The final two figures relate to the photo-activation processes of water.

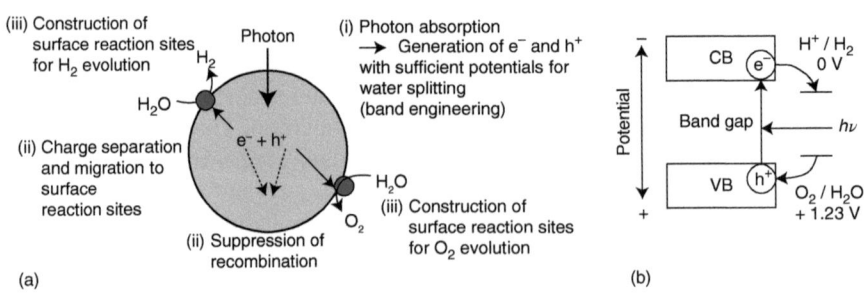

Figure 16i.1 (a) Main processes in photocatalytic water splitting. Kudo and Miseki 2009 [84]. Reproduced with permission of Royal Society of Chemistry. (b) Energetics of water splitting using semiconductor photocatalysts.

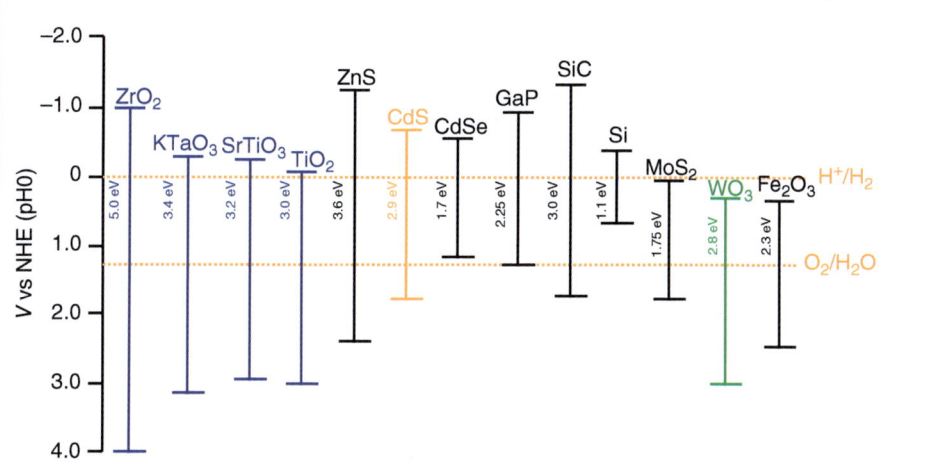

Figure 16i.2 Relationship between band structure of semiconductor and redox potentials of water splitting. (After [84, 85].)

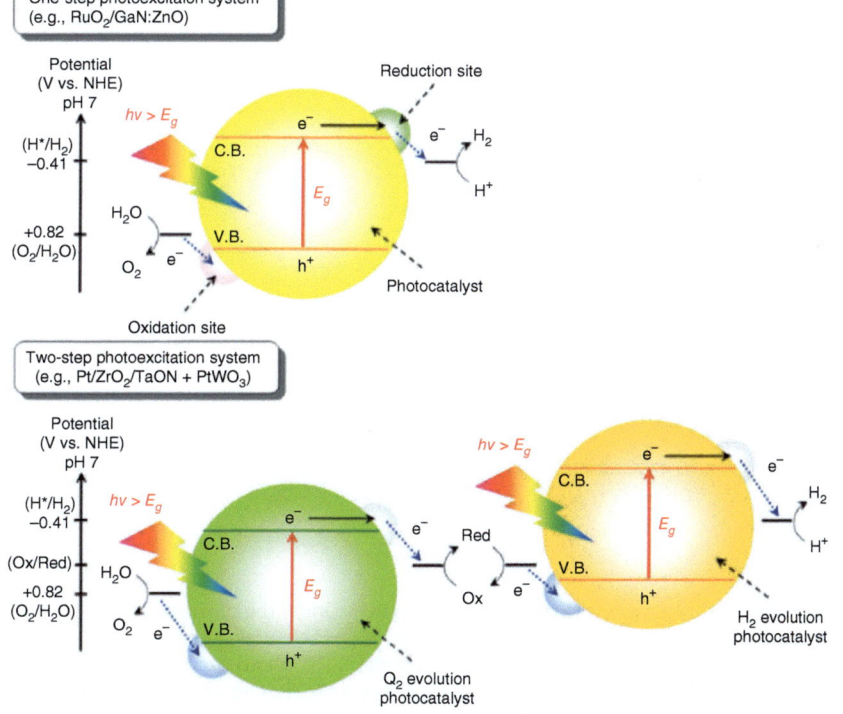

Figure 16i.3 Schematic energy diagrams of photocatalytic water splitting by one-step and two-step photoexcitation systems. C.B., conduction band; V.B., valence band; and E_g, band gap. (Maeda and Domen 2010 [86]. Reproduced with permission of American Chemical Society.)

(Continued)

Insert 16: (Continued)

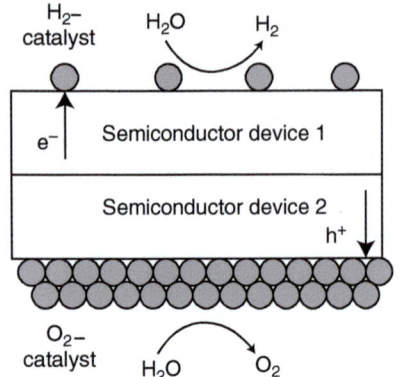

Figure 16i.4 The tandem device with catalyst particles for HER and OER. (After [88].)

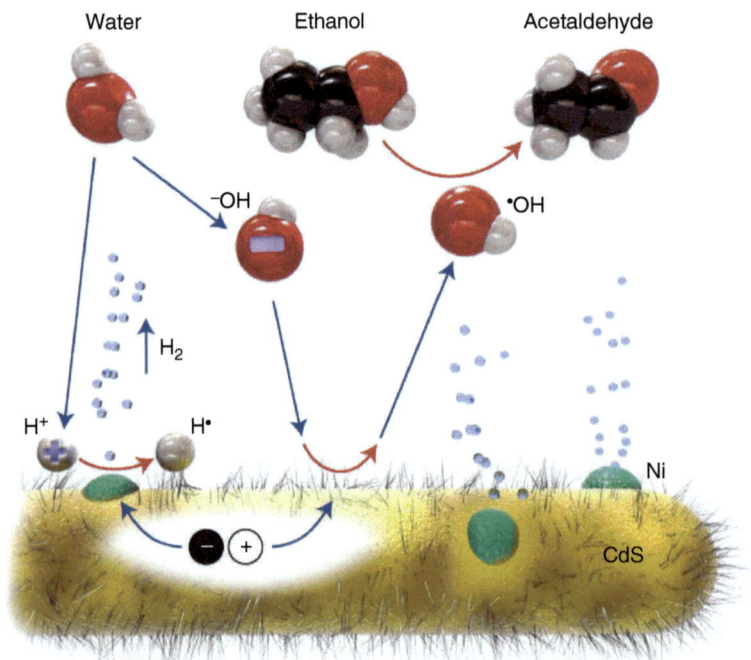

Figure 16i.5 Schematic of the photocatalytic generation of H_2 under the proposed hole-shuttle mechanism. Nickel forms 2–8 nm nanoparticles on the cysteine-stabilized CdS nanorods where water is reduced to hydrogen on illumination. The photo-excited holes oxidize hydroxyl anions that – as a radical – carry away the positive charges and in turn oxidize ethanol to acetaldehyde. The blue arrows denote movement of the species; the red arrows denote a redox reaction. (Simon *et al.* 2014 [90]. Reproduced with permission of Nature Publishing Group.)

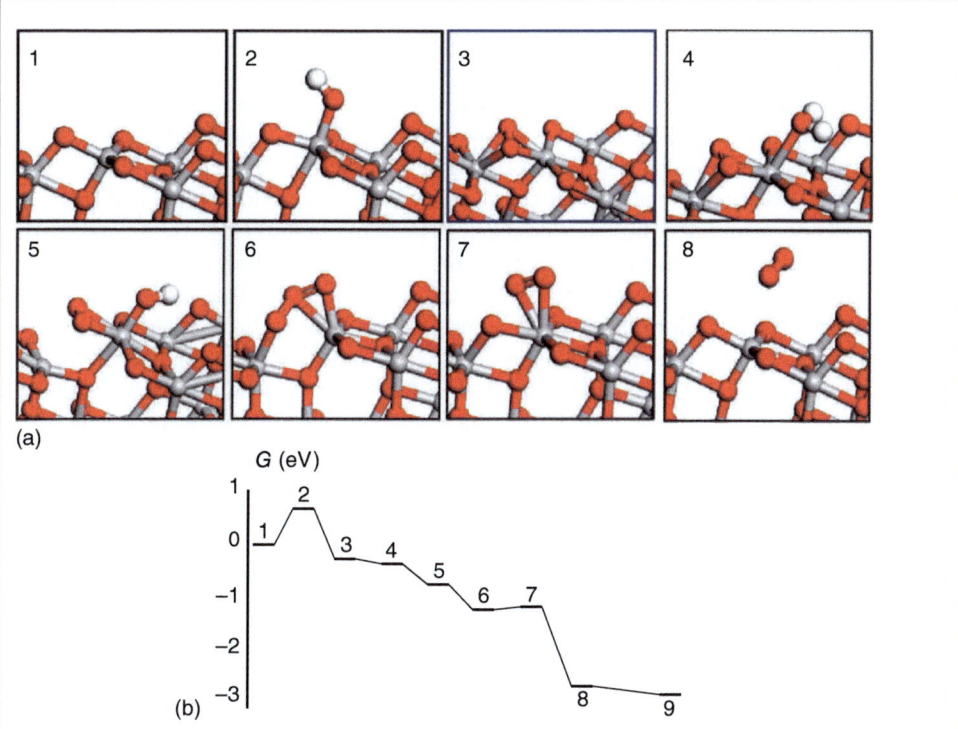

Figure 16i.6 (a) Optimized structures of the intermediate states in OER catalyzed by the anatase (101) surface. The reaction is initiated by the rate-controlling step that forms OH from dissociation of water. The OH intermediate inserts into the TiOTi bond. A second water molecule adsorbs and dissociates. The O_2 molecule evolves and the TiOTi bond is restored. (Li *et al.* 2010 [98]. Reproduced with permission of American Chemical Society.) (b) Free energy (G, eV) profile of OER on (101) surface at an overpotential of 0.7 V (1.93 V vs SHE).

Some of the photocatalytic systems are also active oxidation catalysts. Co-oxidation can even be exploited to improve the photon yield for H_2 evolution, as has been found for the CdS system [90]. This is illustrated in Figure 16i.5. The photo-generated electrons are collected on a Ni particle where they reduce protons to give H_2. The holes are located on the sulfur atoms. The potential energy of the holes is large enough to generate OH· radicals by reaction with adsorbed OH^- anions from the water medium. Formation of these OH^- radicals prevents recombination of the hole with an excited electron, which has a substantial beneficial effect on the photo-efficiency of H_2 generation as long as the radicals can be reacted away. In the example shown in Figure 16i.5 the OH· radicals react readily with ethanol to produce an aldehyde or acid.

Such OH· radical formation is well known for photocatalysis by TiO_2 [91]. At anaerobic conditions, it is used to catalyze a reaction for alcohol similar to the CDS system. At aerobic conditions, it can also be used to oxidize alkane molecules, because peroxide radical intermediates will also be formed with O_2.

It has also been demonstrated for TiO$_2$ that the electron–hole recombination is linked to the efficiency of water splitting [92].

There has been a long controversy over the actual mechanism of H$_2$ and O$_2$ formation by the TiO$_2$ system, which was the first system to show the water splitting reaction [93]. It was initially proposed that a hole located on the TiO$_2$ oxygen atom would generate an OH· radical from OH$^-$, which would recombine to give hydrogen peroxide. In consecutive radical reaction steps, the hydrogen peroxide would decompose to oxygen [94]. While initial OH$^-$ radical formation does occur, a thermodynamic objection to the proposal of consecutive O$_2$ formation via intermediate hydrogen peroxide formation is that the overall energy cost for this reaction path would be more than double the experimentally measured potential. The energy cost to split water is 1.23 V, and the electrochemical potential of OH· radical formation from OH$^-$ is 2.8 V. Water splitting from TiO$_2$ is only found at a potential of 3 V, which is consistent only with the formation of one OH radical as the intermediate [95]. Quantum-chemical calculations suggest that initial OH radical formation occurs from water by photo excited TiO$_2$ [96].

Convincing spectroscopy and isotope labeling experiments by Nakamura and Nakato [97] have proven that no hydrogen peroxide is formed in the photocatalytic reaction. The primary activation step is similar to hydroxyl radical formation by electron transfer of OH$^-$ to a hole on an oxygen atom, but the radical species adsorbs to a Ti cation. In this process, a TiO bond is broken and an adsorbed oxygen radical is generated:

$$[Ti - O - Ti]_s + h^+ + H_2O \rightarrow [Ti - O - HO - Ti]_s + H^+ \tag{11.4}$$

OOH is formed in a consecutive nucleophilic reaction step, which decomposes to O$_2$ as illustrated by the quantum-chemical calculations of Figure 16i.6.

Experiments with isotope-labeled water prove that one oxygen atom derives from the lattice, the other from water. The OER reaction can be considered to follow a Mars–van Krevelen-type mechanism.

DFT calculations confirm the high barrier of OH formation on the TiO$_2$ surface. Figure 16i.6a shows the reaction intermediates on the (101) surface of TiO$_2$ anatase and Figure 16i.6b shows the corresponding reaction energies [98].

The nucleophilic Nakamura mechanism of photoactivated water splitting by TiO$_2$ is closely related to the electrocatalytic reaction catalyzed by transition metal oxides, where it is also proposed that a lattice oxygen atom reacts with a water molecule to form intermediated adsorbed OOH. We refer to Section 11.6.4 for a summary and comparison of the mechanism and electrocatalysis of the OER reaction catalyzed by oxides of different compositions.

TiO$_2$ is not an efficient OER catalyst because of its overpotential of 1.72 V for the OER reaction. It is also not efficient for PV generation due to its unsuitable bandgap. However, useful applications of the photocatalytic activity of TiO$_2$ have been found in its application to the oxidation of environmentally harmful molecules such as chlorinated hydrocarbons and related compounds in water [95].

References

1 Perron, H., Domain, C., Roques, J., Drot, R., Simoni, E., and Catalette, H. (2007) Optimisation of accurate rutile TiO2 (110), (100), (101) and (001) surface models from periodic DFT calculations. *Theor. Chem. Acc.*, **117**, 565–574, http://link.springer.com/10.1007/s00214-006-0189-y (accessed 24 September 2016).

2 Lazzeri, M., Vittadini, A., and Selloni, A. (2001) Structure and energetics of stoichiometric TiO2 anatase surfaces. *Phys. Rev. B*, **63**, 1555409.

3 Mowbray, D.J., Martinez, J.I., Calle-Vallejo, F., Rossmeisl, J., Thygesen, K.S., Jacobsen, K.W., and Nørskov, J.K. (2011) Trends in metal oxide stability for nanorods, nanotubes, and surfaces. *J. Phys. Chem.*, **115**, 2244.

4 Greenwood, N.N. (1968) *Ionic Crystals, Lattice Defects and Nonstoichiometry*, Butterworth & Co., London.

5 Woning, J. and van Santen, R.A. (1983) Electrostatic potential calculations on crystalline TiO2: the surface reducibility of rutile and anatase. *Chem. Phys. Lett.*, **101**, 541–547, http://linkinghub.elsevier.com/retrieve/pii/0009261483870304 (accessed 14 March 2014).

6 van Santen, R.A., Tranca, I., and Hensen, E.J.M. (2015) Theory of surface chemistry and reactivity of reducible oxides. *Catal. Today*, **244**, 63–84, http://linkinghub.elsevier.com/retrieve/pii/S092058611400488X (accessed 24 September 2016).

7 Ballhausen, C.J. (1962) *Introduction to Ligand Field Theory*, McGraw Hill.

8 Hoffmann, R. (1988) *Solids and Surfaces*, Wiley-VCH Verlag GmbH, Weinheim.

9 Albright, T.A., Burdett, J.K., and Whangbo, M.-H. (2013) *Orbital Interactions in Chemistry*, John Wiley & Sons, Inc., Hoboken, NJ.

10 Mingos, D.M.P. (1998) *Essential Trends in Inorganic Chemistry*, Oxford University Press.

11 Waddington, T.C. (1959) Lattice energies and their significance in inorganic chemistry. *Adv. Inorg. Chem. Radiochem.*, **1**, 157–221.

12 Pauling, L. (1939) *Nature of the Chemical Bond*, Cornell University Press.

13 Cotton, F.A. (1990) *Chemical Applications of Group Theory*, 3rd edn, John Wiley & Sons, Inc., New York.

14 Wales, D.J. and Jenkins, S.J. (2009) Mechanisms for H 2 reduction on the PdO {101} surface and the Pd {100}-(5 × surface oxide 5) R 27-O. *J. Phys. Chem. C*, **113**, 16757–16765.

15 Nolan, M., Grigoleit, S., Sayle, D.C., Parker, S.C., and Watson, G.W. (2005) Density functional theory studies of the structure and electronic structure of pure and defective low index surfaces of ceriax. *Surf. Sci.*, **576**, 217–229.

16 Pacchioni, G. (2008) Modeling doped and defective oxides in catalysis with density functional theory methods: room for improvements. *J. Chem. Phys.*, **128**, 182505.

17 Wu, Z., Li, M., and Overbury, S.H. (2012) On the structure dependence of CO oxidation over CeO2 nanocrystals with well-defined surface planes. *J. Catal.*, **285**, 61–73.

18 Calle-Vallejo, F., Inoglu, N.G., Su, H.-Y., Martinez, J.I., Man, I.C., Koper, M.T.M., Kitchin, J., and Rossmeisl, J. (2013) Number of outer electrons as descriptor for adsorption processes on transition metals and their oxides. *Chem. Sci.*, **4**, 1245.

19 Tasker, P.W. (1979) The stability of ionic crystal surfaces. *J. Phys. C: Solid State Phys.*, **12**, 4977.

20 Woodruff, D.P. (2013) Quantitative structural studies of corundum and rocksalt oxide surfaces. *Chem. Rev.*, **113**, 3863–3886.

21 Mars, P. and van Krevelen, D.W. (1954) Oxidations carried out by means of vanadium oxide catalysts. *Chem. Eng. Sci.*, **3**, 41.

22 Prins, R. (2012) in *Hydrodesulfurization in Catalysis* (eds. Beller, M., Renken, A., and van Santen, R.A.), Chapter 18, Wiley–VCH, 390–395.

23 Sauer, J. and Dobler, J. (2004) Structure and reactivity of V(2)O(5): bulk solid, nanosized clusters, species supported on silica and alumina, cluster cations and anions. *Dalton Trans.*, **19**, 3116–3121, http://www.ncbi.nlm.nih.gov/pubmed/15452641 (accessed 9 March 2014).

24 Gilardoni, F., Bell, A.T., Chakraborty, A., and Boulet, P. (2000) Density functional theory calculations of the oxidative dehydrogenation of propane on the (010) surface of V2O5. *J. Phys. Chem. B*, **104**, 12250–12255.

25 Dowden, D.A. (1972) Crystal and ligand field models of solid catalysts. *Catal. Rev.*, **5**, 1–32. doi: 10.1080/01614947208076863

26 Kremenic, G., Nieto, J.M.L., Tascón, J.M.D., and Tejuca, L.G. (1985) Chemisorption and catalysis on LaMo3 oxides. *J. Chem. Soc., Faraday Trans.*, **1**, 939–949.

27 Adams, C. and Jennings, T.J. (1963) Investigation of the mechanism of catalytic oxidation of propylene to acrolein and acrylonitrile. *J. Catal.*, **2**, 63–68, http://linkinghub.elsevier.com/retrieve/pii/0021951763901404 (accessed 24 September 2016).

28 Sachtler, W.M.H. (2010) Mechanism of the catalytic oxidations of propene to acrolein and of isomeric butenes to butadiene. *Recl. Trav. Chim. Pays-Bas*, **82**, 243–245. doi: 10.1002/recl.19630820305

29 Voge, H. (1963) Mechanism of propylene oxidation over cuprous oxide. *J. Catal.*, **2**, 58–62, http://linkinghub.elsevier.com/retrieve/pii/0021951763901398 (accessed 24 September 2016).

30 Hodnett, B.K. (2000) *Heterogeneous Catalytic Oxidation: Fundamental and Technological Aspects of the Selective and Total Oxidation of Organic Compounds*, John Wiley & Sons, Ltd, Chichester.

31 Grzybowska, B., Haber, J., and Janas, J. (1977) Interaction of allyl iodide with molybdate catalysts for the selective oxidation of hydrocarbons. *J. Catal.*, **49**, 150–163, http://linkinghub.elsevier.com/retrieve/pii/0021951777902512 (accessed 24 September 2016).

32 Goddard, W.A., Chenoweth, K., Pudar, S., Duin, A.C.T., and Cheng, M.-J. (2008) Structures, mechanisms, and kinetics of selective ammoxidation and oxidation of propane over multi-metal oxide catalysts. *Top. Catal.*, **50**, 2–18, http://link.springer.com/10.1007/s11244-008-9096-x (accessed 9 March 2014).

33 Grasselli, R.K., Burrington, J.D., Buttrey, D.J., DeSanto, P., Lugmair, C.G., Volpe, A.F., and Weingand, T. (2003) Multifunctionality of active centers in

(amm)oxidation catalysts: from Bi-Mo-Ox to Mo-V-Nb-(Te, Sb)-Ox. *Top. Catal.*, **23**, 5–22.

34 Sanfiz, A.C., Hansen, T.W., Teschner, D., Schnörch, P., Girgsdies, F., Trunschke, A., Schlögl, R., Looi, M.H., and Hamid, S.B.A. (2010) Dynamics of the MoVTeNb oxide M1 phase in propane oxidation. *J. Phys. Chem. C*, **114**, 1912–1921.

35 Hargreaves, J.S.J., Hutchings, G.J., Joyner, R.W., and Taylor, S.H. (1996) Methane partial oxidation to methanol over Ga2O3 based catalysts: use of the CH4/D2 exchange reaction as a design tool. *Chem. Commun.*, **4**, 523.

36 Keller, G.E. and Bhasin, M.M. (1982) Synthesis of ethylene via oxidative coupling of methane I. Determination of active catalysts. *J. Catal.*, **73**, 9–19, http://www.sciencedirect.com/science/article/pii/0021951782900756 (accessed 24 September 2016).

37 Lunsford, J.H. (1995) The catalytic oxidative coupling of methane. *Angew. Chem. Int. Ed. Engl.*, **34**, 970–980. doi: 10.1002/anie.199509701

38 Hickman, D.A. and Schmidt, L.D. (1993) Production of syngas by direct catalytic oxidation of methane. *Science*, **259**, 343–346, http://www.ncbi.nlm.nih.gov/pubmed/17832347 (accessed 20 January 2015).

39 Gladisch, H. (1962) How Huels makes acythylene by DR Arc, *Hydrocarbon Process. Pet. Refin.*, **41**, 159–164.

40 Fincke, J.R., Anderson, R.P., Hyde, T., Wright, R., Bewley, R., Haggard, D.C., and Swank, W.D. (2000) Thermal Conversion of Methane to Acetylene. Bechtel BWXT Idaho, LLC.

41 Wang, L., Tao, L., Xie, M., Xu, G., Huang, J., and Xu, Y. (1993) Dehydrogenation and aromatization of methane under non-oxidizing conditions. *Catal. Lett.*, **21**, 35–41.

42 Weckhuysen, B.M., Wang, D., Rosynek, M.P., and Lunsford, J.H. (1998) Conversion of methane to benzene over transition metal ion ZSM-5 Zeolites I. Catalyst characterization by X-ray photoelectron spectroscopy. *J. Catal.*, **175**, 347–351, http://linkinghub.elsevier.com/retrieve/pii/S0021951798920115 (accessed 24 September 2016).

43 Weckhuysen, B.M., Wang, D.J., Rosynek, M.P., and Lunsford, J.H. (1997) Catalytic conversion of methane into aromatic hydrocarbons over iron oxide loaded ZSM-5 zeolites. *Angew. Chem. Int. Ed. Engl.*, **36**, 2374–2376, ISI:A1997YH94400028..

44 Gao, J., Zheng, Y., Jehng, J.-M., Tang, Y., Wachs, I.E., and Podkolzin, S.G. (2015) Identification of molybdenum oxide nanostructures on zeolites for natural gas conversion. *Science*, **348**, 686–690. doi: 10.1126/science.aaa7048

45 Guo, X., Fang, G., Li, G., Ma, H., Fan, H., Yu, L., Ma, C., Wu, X., Deng, D., Wei, M. *et al.* (2014) Direct, nonoxidative conversion of methane to ethylene, aromatics, and hydrogen. *Science*, **344**, 616–619, http://www.ncbi.nlm.nih.gov/pubmed/24812398 (accessed 24 September 2016).

46 Van Der Zwet, G.P., Hendriks, P.A.J.M., and Van Santen, R.A. (1989) Pyrolysis of methane and the role of surface area. *Catal. Today*, **4**, 365–369.

47 Hutchings, G.J., Scurrell, M.S., and Woodhouse, J.R. (1988) The role of gas phase reaction in the selective oxidation of methane. *J. Chem. Soc., Chem. Commun.*, **10**, 253.

48 Schwarz, H. (2011) Chemistry with methane: concepts rather than recipes. *Angew. Chem. Int. Ed.*, **50**, 10096–10115.

49 Ito, T., Lin, C.H., and Lunsford, J.H. (1987) Oxidative dimerization of methane over lithium-promoted zinc oxide. *J. Am. Chem. Soc.*, **32**, 242–248.

50 Tosoni, S. and Sauer, J. (2010) Accurate quantum chemical energies for the interaction of hydrocarbons with oxide surfaces: CH(4)/MgO(001). *Phys. Chem. Chem. Phys.*, **12**, 14330–14340.

51 Kwapien, K., Paier, J., Sauer, J., Geske, M., Zavyalova, U., Horn, R., Schwach, P., Trunschke, A., and Schlögl, R. (2014) Sites for methane activation on lithium-doped magnesium oxide surfaces. *Angew. Chem. Int. Ed.*, **53**, 8774–8778.

52 Heckelsberg, L.F., Banks, R.L., and Bailey, G.C. (1969) Olefin disproportionation catalysts. *Ind. Eng. Chem. Prod. Res. Dev.*, **8**, 259–261. doi: 10.1021/i360031a009

53 Irvin, K.J. and Mol, J.C. (1997) *Olefin Metathesis and Metathesis Polymerization*, Academic Press.

54 Handzlik, J. and Sautet, P. (2008) Active sites of olefin metathesis on molybdena-alumina system: a periodic DFT study. *J. Catal.*, **256**, 1–14.

55 Sutthasupa, S., Shiotsuki, M., and Sanda, F. (2010) Recent advances in ring-opening metathesis polymerization, and application to synthesis of functional materials. *Polym. J.*, **42**, 905–915.

56 Grubbs, R.H. and Tumas, W. (1989) Polymer synthesis and organotransition metal chemistry. *Science*, **243**, 907–915.

57 Schrock, R.R. (1986) High-oxidation-state molybdenum and tungsten alkylidyne complexes. *Acc. Chem. Res.*, **19**, 342–348. doi: 10.1021/ar00131a003

58 Kaminsky, W. (1998) Highly active metallocene catalysts for olefin polymerization. *J. Chem. Soc., Dalton Trans.*, **9**, 1413–1418

59 Busico, V. (2013) *Polymerization in Catalysis*, Chapter 12 (eds M. Beller, A. Renken, and R.A. van Santen), 261–285, Wiley-VCH Verlag GmbH.

60 Ziegler, K., Holzkamp, E., Breil, H., and Martin, H. (1955) Das Mülheimer Normaldruck-Polyäthylen-Verfahren. *Angew. Chem.*, **67**, 541–547.

61 Natta, G. and Pasquon, I. (1959) The kinetics of the stereospecific polymerization of alfa-olefins. *Adv. Catal.*, **11**, 1–66.

62 Moore, E.P. Jr., (1996) Polypropylene handbook: polymerization, characterization, properties, processing, applications, in *Polypropylene Handbook* (ed. E.P. Moore), Hanser Publishers, Munich.

63 Cossee, P. (1964) Ziegler–Natta catalysis. I. Mechanism of polymerization of α-olefins with Ziegler–Natta catalysts. *J. Catal.*, **3**, 80–88, http://www.sciencedirect.com/science/article/pii/0021951764900958 (accessed 24 September 2016).

64 Arlman, E. (1964) Ziegler–Natta catalysis. II. Surface structure of layer-lattice transition metal chlorides. *J. Catal.*, **3**, 89–98, http://www.sciencedirect.com/science/article/pii/002195176490096X (accessed 24 September 2016).

65 Corradini, P., Guerra, G., and Cavallo, L. (2004) Do new century catalysts unravel the mechanism of stereocontrol of old Ziegler Ziegler–Natta catalysis Natta catalysts? *Acc. Chem. Res.*, **37**, 231–241.

66 Topsøe, H., Clausen, B.S., and Massoth, F.E. (1996) *Hydrotreating Catalysis*, Springer.
67 Lauritsen, J.V., Bollinger, M.V., Lægsgaard, E., Jacobsen, K.W., Nørskov, J.K., Clausen, B.S., Topsøe, H., and Besenbacher, F. (2004) Atomic-scale insight into structure and morphology changes of MoS 2 nanoclusters in hydrotreating catalysts. *J. Catal.*, **221**, 510–522.
68 Topsøe, H., Clausen, B.S., Mamoth, F.S. (1996) *Hydrotreating Catalysis*, Vol. **11**, Series Catalysis Science and Technology, Springer.
69 Neurock, M. and van Santen, R.A. (1994) Theory of carbon–sulfur bond activation by small metal sulfide particles. *J. Am. Chem. Soc.*, **116**, 4427–4439.
70 Welters, W.J.J., Vorbeck, G., Zandbergen, H.W., Dehaan, J.W., Debeer, V.H.J., and Vansanten, R.A. (1994) HDS activity and characterization of zeolite-supported nickel sulfide catalysts. *J. Catal.*, **150**, 155–169, http://linkinghub.elsevier.com/retrieve/pii/S0021951784713327 (accessed 24 September 2016).
71 Hensen, E.J.M., Brans, H.J.A., Lardinois, G.M.H.J., de Beer, V.H.J., Rob van Veen, J.A., and van Santen, R.A. (2000) Periodic trends in hydrotreating catalysis: thiophene hydrodesulfurization over carbon-supported 4d transition metal sulfides. *J. Catal.*, **192**, 98–107, http://www.sciencedirect.com/science/article/pii/S0021951700928240 (accessed 24 September 2016).
72 Man, I.C., Su, H.-Y., Calle-Vallejo, F., Hansen, H.A., Martínez, J.I., Inoglu, N.G., Kitchin, J., Jaramillo, T.F., Nørskov, J.K., and Rossmeisl, J. (2011) Universality in oxygen evolution electrocatalysis on oxide surfaces. *ChemCatChem*, **3**, 1159–1165. doi: 10.1002/cctc.201000397
73 Fang, Y.H. and Liu, Z.P. (2010) Mechanism and tafel lines of electro-oxidation of water to oxygen on $RuO_2(110)$. *J. Am. Chem. Soc.*, **132**, 18214–18222.
74 García-Mota, M., Bajdich, M., Viswanathan, V., Vojvodic, A., Bell, A.T., and Nørskov, J.K. (2012) Importance of correlation in determining electrocatalytic oxygen evolution activity on cobalt oxides. *J. Phys. Chem. C*, **116**, 21077–21082.
75 Plaisance, C.P. and van Santen, R.A. (2015) Structure sensitivity of the oxygen evolution reaction catalyzed by cobalt(II,III) oxide. *J. Am. Chem. Soc.*, **137**, 14660–14672. doi: 10.1021/jacs.5b07779
76 Nørskov, J.K., Rossmeisl, J., Logadottir, A., Lindqvist, L., Kitchin, J.R., Bligaard, T., and Jónsson, H. (2004) Origin of the overpotential for oxygen reduction at a fuel-cell cathode. *J. Phys. Chem. B*, **108**, 17886–17892.
77 Nocera, D.G. (2012) The artificial leaf. *Acc. Chem. Res.*, **45**, 767–776.
78 Blomberg, M.R.A., Borowski, T., Himo, F., Liao, R.-Z., and Siegbahn, P.E.M. (2014) Quantum chemical studies of mechanisms for metalloenzymes. *Chem. Rev.*, **114**, 3601–3658, http://www.ncbi.nlm.nih.gov/pubmed/24410477 (accessed 24 September 2016).
79 Li, X. and Siegbahn, P.E.M. (2013) Water oxidation mechanism for synthetic Co-oxides with small nuclearity. *J. Am. Chem. Soc.*, **135**, 13804–13813.
80 Ferreira, K.N., Iverson, T.M., Maghlaoui, K., Barber, J., and Iwata, S. (2004) Architecture of the photosynthetic oxygen-evolving center. *Science*, **303**, 1831–1838.

81 Loll, B., Kern, J., Saenger, W., Zouni, A., and Biesiadka, J. (2005) Towards complete cofactor arrangement in the 3.0 A resolution structure of photosystem II. *Nature*, **438**, 1040–1044.

82 Rossmeisl, J., Dimitrievski, K., Siegbahn, P., and Nørskov, J.K. (2007) Comparing electrochemical and biological water splitting. *J. Phys. Chem. C*, **111**, 18821–18823.

83 McAlpin, J.G., Stich, T.A., Casey, W.H., and Britt, R.D. (2012) Comparison of cobalt and manganese in the chemistry of water oxidation. *Coord. Chem. Rev.*, **256**, 2445–2452.

84 Kudo, A. and Miseki, Y. (2009) Heterogeneous photocatalyst materials for water splitting. *Chem. Soc. Rev.*, **38**, 253–278.

85 Serpone, N. and Pelizetti, E. (1989) *Photocatalysis: Fundamentals and Applications*, John Wiley & Sons, Inc., New York.

86 Maeda, K. and Domen, K. (2010) Photocatalytic water splitting: recent progress and future challenges. *J. Phys. Chem. Lett.*, **1**, 2655–2661. doi: 10.1021/jz1007966

87 Laursen, A.B., Kegnæs, S., Dahl, S., and Chorkendorff, I. (2012) Molybdenum sulfides—efficient and viable materials for electro-and photoelectrocatalytic hydrogen evolution. *Energy Environ. Sci.*, **5**, 5577.

88 Kaizer, B., Jaegermann, W., Flechter, S., and Lewerens, H.J. (2011) Direct photoelectrochemical conversion of sun light into hydrogen for chemical energy storage. *Bunsen Mag.*, **13**, 104.

89 Castelli, I.E., Landis, D.D., Thygesen, K.S., Dahl, S., Chorkendorff, I., Jaramillo, T.F., and Jacobsen, K.W. (2012) New cubic perovskites for one- and two-photon water splitting using the computational materials repository. *Energy Environ. Sci.*, **5**, 9034.

90 Simon, T., Bouchonville, N., Berr, M.J., Vaneski, A., Adrović, A., Volbers, D., Wyrwich, R., Döblinger, M., Susha, A.S., Rogach, A.L. et al. (2014) Redox shuttle mechanism enhances photocatalytic H2 generation on Ni-decorated CdS nanorods. *Nat. Mater.*, **13**, 1013–1018. doi: 10.1038/nmat4049

91 Fox, M.A. and Dulay, M.T. (1993) Heterogeneous photocatalysis. *Chem. Rev.*, **93**, 341–357. doi: 10.1021/cr00017a016

92 Tang, J., Durrant, J.R., and Klug, D.R. (2008) Mechanism of photocatalytic water splitting in TiO2. Reaction of water with photoholes, importance of charge carrier dynamics, and evidence for four-hole chemistry. *J. Am. Chem. Soc.*, **130**, 13885–13891.

93 Fujishima, A. and Honda, K. (1972) Electrochemical photolysis of water at a semiconductor electrode. *Nature*, **238**, 37–38.

94 Wilson, R.H. (1980) Observation and analysis of surface states on TiO2 electrodes in aqueous electrolytes. *J. Electrochem. Soc.*, **127**, 228.

95 Juodkazis, K., Juodkazyte, J., Jelmakas, E., Kalinauskas, P., Valsiūnas, I., Mecinskas, P., and Juodkazis, S. (2010) Photoelectrolysis of water: solar hydrogen-achievements and perspectives. *Opt. Express*, **18** (Suppl. 2), A147–A160.

96 Kazaryan, A., van Santen, R., and Baerends, E.J. (2015) Light-induced water splitting by titanium-tetrahydroxide: a computational study. *Phys. Chem. Chem. Phys.*, **17**, 20308–20321. doi: 10.1039/C5CP01812A

97 Nakamura, R. and Nakato, Y. (2004) Primary intermediates of oxygen photoevolution reaction on TiO_2 (rutile) particles, revealed by in Situ FTIR absorption and photoluminescence measurements. *J. Am. Chem. Soc.*, **126**, 1290–1298.

98 Li, Y.F., Liu, Z.P., Liu, L., and Gao, W. (2010) Mechanism and activity of photocatalytic oxygen evolution on titania anatase in aqueous surroundings. *J. Am. Chem. Soc.*, **132**, 13008–13015.

Index

a

ABC band 358
accessible free valence
   CH_3 and NH_3 chemisorption, the agostic interaction 235
   chemisorptions, molecular fragments 234
acetaldehyde, ethylene to 326
acetylene 309
acid catalysis 420
   alkane activation 378
   alkane isomerization 383
   alkene protonation 374
   β-C–C bond cleavage 388
   n-butene isomerization 386
   hydride transfer reaction 390
   alkylation 391
acrolein conversion 500
   and ammoxidation 500
   propylene to 500
acrylamide 500
acrylonitrile 502
activation and deactivation 123
activation entropy 270
   elementary surface reaction 81
active acidic heterogeneous catalysts 393
adatom binding energy 220
adatom bond, chemisorbed C 221
adatom co-assisted activation
   hydrogen activated dissociation 277
   oxygen assisted X–H bond cleavage 277
adatom energies 476
adsorbate chemical bond
   bonding electron density 263
   nonbonding, antibonding regime 263
adsorbate σ-bond activation
   activation entropy 270
   methane activation
      bond cleavage, CH_4 267
      CH bonds activation 265
      Rh terrace $vs.$ Rh step-edge 267
   oxidative addition, reductive elimination model 268
   umbrella effect 269
adsorption energies
   C atom 82
   reaction 209
adsorption free energy 401
advanced kinetics, mean field approximation breakdown
   catalytic self-organizing systems
      exploding reactions 153
      heterogeneous catalytic self-organizing systems 154
   heterogeneous catalytic reactions, kinetics 145
   kinetic Monte Carlo method, RuO_2 catalyzed oxidation 146
   single molecule spectroscopy 149
aldehyde $vs.$ olefinic bond 309
alkali and earth alkali ions 433
alkane activation 378
alkane adsorption 266
alkane–alkene equilibrium 411
alkane cracking 411
alkane oxidation 467

Index

alkene disproportionation and metathesis   507
alkene isomerization   383
alkene protonation   374, 392
alkylation   391, 414
alloys and promoting effects   133
α–γ C–C bond activation   312
ammonia adsorption   364
ammonia synthesis   19
   enzyme catalysis   297
   heterogeneous catalytic reaction   293
   molecular basis   293
   synthesis gas conversion   297
ammoxidation   500
Anderson–Schulz–Flory (ASF) distribution   304, 305
antibonding surface ad atom orbitals   229
apparent activation energy   75
Arrhenius reaction rate expression   9
Aufbau reaction   437
autocatalytic generation, surface vacancies   156
automotive exhaust catalysis; NO reduction by CO   45
average difference, bonding vs. antibonding d-valence electron energies   218

b

Bayer–Villiger insertion reaction   464
Bayer–Villiger oxidation   454
   molecular oxygen   464
Berzelius' catalysis law   8
β-C–C bond cleavage reaction   387, 396
BET adsorption isotherm   65
bifunctional catalysis
   adsorption effects   412
   hydrocracking reaction   408
biocatalytic system, catalytic cycle   452
biomass refinery   37
Boltzmann constant (k)   69
bond order conservation   83
bond order overlap population density (BOOPD)   173, 174

bond polarity   180
Born–Haber cycle analysis   243
Born–Haber process   323
bracket notation   174
Brønsted acid catalysis   469
   alkane to aromatics   22
   catalytic cracking   22
   hydroisomerization   26
Brønsted acidic protons   126
Brønsted acidity   345, 346, 351
Brønsted basicity   346, 350
Brønsted–Evans–Polanyi (BEP)   79
   bond order conservation   83
   extrapolation   301
Brønsted–Evans–Polanyi (BEP) extrapolation   320
bulk Rh atom   213
butane activation   495
n-butene isomerization   386

c

calcination (oxidation)   127
caprolactam synthesis   464
carbenium ions   372, 374, 388, 411
carbide mechanism   298, 300
carbon atom adsorption, surface molecular complex   221
carbonium cation   378
carbonium ions   372
catalysis, basic laws of
   Berzelius' catalysis law   7
   Ostwald's catalysis law   8
   Sabatier's catalysis law   10
catalyst and process   16
catalyst characterization
   Langmuir adsorption isotherm   64
   pore volume measurement   65
   porosity   66
   temperature-programmed reactivity measurements   67
catalytic cracking   396
   process   18
catalytic platforming process   25
catalytic reaction cycles and kinetics   395
   bifunctional catalysis   402
      adsorption effects   412

hydrocracking reaction   409
hydroisomerization   403
catalytic cracking   396
zeolite catalysis   394
catalytic self-organizing systems
  exploding reactions   154
  heterogeneous catalytic
    self-organizing systems   154
cationic organo-catalysis   137
C atom 2p atomic orbitals   221
CCR platforming process   25
CeO_2
  electronic properties and energies   490
  substantial bandgap energy   490
charge separation   181
C–H bond activation   315, 439, 452
CH_3 and NH_3 chemisorption, the
    agostic interaction   235
CH_4 conversion, to aromatics   503
chemical bonding analysis (COHP)   480, 488
chemical bonding and reactivity,
    transition metal surfaces
  accessible free valence
    chemisorptions, molecular
      fragments   233
  σ-symmetric vs. π-symmetric
      orbital interactions   233
  chemisorptions, atoms and molecules   214
  quantum chemistry   209
  surface chemical bond
    electronic structure   210
  surface chemical reactivity   209
chemical bonding, transition metals
  electronic structure   190, 475
  relative stability   199
chemisorptions
  carbon atom adsorption, surface
      molecular complex
    adatom bond, chemisorbed C   221
    COHP distributions   226
  H adatom, the covalent surface bond
    electronic structure, atop and
      threefold coordinated H   214

trends, periodic table metal
    position   219
chloric acid   7
coal liquification process   30
cobalt carbonyl complexes   124
Co carbonyl complex $HCo(CO)_4$ 33
CO chemisorption   249
CO insertion mechanism   298, 300
CO oxidation reaction   119, 128
Co_3O_4 surfaces   528
CO_2 production rate vs. temperature   160
compound formation   123
computational quantum-chemical
    studies   169
Cossée–Arlman mechanism   511
covalency, NaCl chemical bond   182
cracking or isomerization reactions   19
crystal field splitting model   485
crystal field theory   482
crystal overlap Hamiltonian population
    densities (COHPs)   190
Cu-based catalysts   453
CuO/ZnO-based catalyst   21
cyclic olefins   510

d

Deacon process   7
decreased surface reactivity and site
    blocking   275
degenerate $1\pi_g$ bonding molecular
    orbitals   185
degenerate $1\pi_u$ antibonding orbitals   185
dehydrogenation   309
delocalization, d-valence electrons   191
density functional theory   61
deprotonation energy   364, 528
descriptors partial density of states
    (PDOS)   174
diatomic molecules dissociation
  π-bonds, non-shared bonding
    principle   274
dibenzothiophene desulfurization   519
diesel engines; NO reduction by NH_3 46
diesel exhaust catalysis   440

diesel exhaust treatment 47
diffusion
　concentration profiles 106
　effectiveness factor 107
dimethyl butane (DMB) formation 412
direct alkane activation 468
direct coupling (DC) 537
disproportionation reaction, single site catalysis 32
Döbereiner of catalytic oxidation 5
donative and backdonative interactions 249
double-peaked reactivity pattern 499
d-valence electron bandwidth 198
D-xylose isomerase 456

e

effectiveness factor, monomolecular reaction 107
Einstein relation 106
electrocatalysis 499, 520
　H_2 evolution 316
　surface reactivity descriptor 94
electrocatalytic water decomposition 522
electron–electron repulsion energies 213
electron microscopy 68
electronic basis, lattice stability differences 200
electronic structures
　adsorbed O atom 229
　bulk NaH and NaCl 186
　chemical bond 174
　HCl and NaCl 179
　N_2, O_2, F_2 183
electron occupation, molecular orbital (N_i) 175
electrons redistribution, bonding and antibonding orbitals 213
electro-oxidation, Pt surfaces 122
electrophilic surface oxygen 284, 286
electrostatic energy 475, 486
electrostatic field 432

electrostatic interactions 258
　energy 430
　model 482
elementary catalytic reaction kinetics
　apparent activation energy 74
　fundamental catalytic kinetics equation 77
　Michaelis–Menten kinetics 76
　site number conservation law 72
elementary kinetics
　lumped kinetics expressions
　　catalytic reaction 71
　　elementary reaction, rate constant 68
　Sabatier volcano curve 84
elementary reaction, rate constant 69
Eley–Rideal mechanism 532
Eley–Rideal reaction 441, 442
Eley–Rideal vs. Langmuir–Hinshelwood kinetics 78
endothermic reaction 92
energies and electronic features for N_2, O_2, F_2 and Cl_2 184
energy gap, valence and conduction band electrons 189
enzyme catalysts 296
　acid catalysis and hydride transfer 420
enzyme oxidation catalytic systems 446
ethylene 325
　acetaldehyde 326
　adsorption and hydrogenation of 308
　epoxidation 333
　vinyl acetate 329
ethylene epoxidation 284
　process 42
ethylene epoxide (EO) vs. acetaldehyde (AA) 334
ethylene–propylene disproportionation reaction 510
exothermic reaction 93
extended X-ray absorption fine structure (EXAFS) spectroscopy 68
Eyring transition state rate expression 69

f

face-centered cubic (fcc) 199
Fe-containing MMO enzyme 452
Fermi level 320, 486, 488
Fischer–Tropsch reaction 305
Fischer lock-and key model 420
Fischer–Tropsch catalysis 29, 120
Fischer–Tropsch process 413
Fischer–Tropsch reaction 297, 300, 301, 304, 504
fructose 456
fuel cell 6
fundamental catalytic kinetics equation 77

g

gallium-or zinc-exchanged zeolites 437
gas phase kinetics 80
globally-synchronized behaviour, diffusion 159
global oscillations and pattern formation 163
group V, VI metal oxides 494

h

Haag–Dessau cracking mechanism 372, 390
  β-C–C bond cleavage reaction 396
Haber–Bosch ammonia synthesis 5, 10
Haber–Bosch process 293
H adatom, the covalent surface bond
  electronic structure, atop and threefold coordinated H 214
  hydrogen chemisorption 215
  trends, periodic table metal position 219
Hamiltonian matrix element (H_{ij}) 175
harpooning reaction 505
H/D exchange reactions 393
Henry's law 64
heterogeneous catalysis
  chemical compounds 3
  early 19th century discoveries 5
  later 19th century discoveries 7
  reactions 19, 293

heterogeneous catalytic self-organizing systems
  different surface phases alteration, spiral wave formation
    CO oxidation, Pt(110) surface 155
  self-organizing heterogeneous catalytic systems 164
heterogeneous metathesis catalyst 508
heterogeneous Ziegler–Natta catalysts 511
heterolytic mechanism 505
heterolytic water splitting 532
H_2 evolution, electrochemical volcano curve 316
hexagonal close packed (hcp) 199
hexamethylbenzene 416
Heyrovsky reaction 319
homoatomic molecules, N_2, O_2, F_2 183
homogeneous and biocatalyst analogs 461
homogeneous catalytic systems 446
H_2+O_2 recombination 164
Horiuti–Polanyi reaction 332
H-SAPO-34, structure of 28
hydride–proton transfer dehydrogenation 456
hydride transfer reaction, alkylation 391
hydrocarbons
  conversion reactions 345
  hydrocracking of 412
hydrocarbons, hydroconversion of
  acetylene 309
  ethylene 306
  hydrogenolysis and isomerization 310
  NH_3 and CH_4 to HCN 314
hydrocarbons, zeolite micropore wall 111
hydrocarbon transformation reactions
  biomass refinery 37
  Brønsted acid catalysis
    alkane to aromatics 22
    catalytic cracking 22
    hydroisomerization 26
    hydrodesulfurization 36

hydrocarbon transformation reactions (*contd.*)
  oligomerization, polymerization catalysis
    Aufbau reaction   27
    Fischer–Tropsch catalysis   29
hydroconversion processes   22
hydrocracking catalysis   396
hydrocracking process   23
hydrocracking reaction   409
hydrodenitrogenation (HDN)   515
hydrodeoxygenation (HDO)   37
hydrodesulfurization   19, 36
hydrodesulphurization (HDS)   515–517, 520
hydrodynamics   59
hydroformylation reaction scheme   34
hydrogenation reactions   309
  ammonia synthesis
    enzyme catalysis   296
    gas conversion   297, 300, 301, 304, 305
    heterogeneous catalytic reaction   293, 296
    electrocatalysis; H_2 evolution   316, 317, 319, 320
  enzyme catalysis   297
  hydroconversion of hydrocarbons
    acetylene   309
    NH_3 and CH_4 to HCN   314
    ethylene   306
    hydrogenolysis and isomerization   310, 312
  transition metal catalysts
    ammonia synthesis   19
    Olefin hydrogenation   21
    synthesis gas, methanol   21
hydrogen evolution reaction (HER)   539, 540
hydrogen–deuterium (H/D) exchange reactions   367
hydrogenolysis and isomerization   310
hydrogenolysis reaction   135, 312
hydroisomerization   23, 26, 307, 372, 391, 396, 403, 412, 413
hydrotreating process   37
hydrotreating reactions   515

hydroxylated MgO surface   350, 351
HZSM-5 zeolite   165

i
INEOS MTO process scheme   29
infrared adsorption spectra   430
infrared and Raman spectroscopies   68
intraorbital electron-electron repulsion energy   485
intrinsic activation energy   400

k
kinetic expressions, R_{CH4} 74
kinetic Monte Carlo method (KMC)
  catalytic self-organization   160
  RuO_2 catalyzed oxidation   149

l
Langmuir–Hinshelwood–Hougen–Watson (LHHW) expression   78
Langmuir–Hinshelwood model   509
Langmuir–Hinshelwood type reaction   442, 532
Langmuirian view   134
Lewis acid catalysis   429, *see also* zeolitic non-redox and redox catalysis 429
  by non-reducible cations   453
    Bayer–Villiger oxidation   454
    homogeneous and biocatalyst analogs   461
    Meerwein–Pondorf–Verley (MPV) reduction   456
    Oppenauer oxidation (OP) reaction   456
    propylene epoxidation   461
Lewis acidic non-reducible cations   453, 454
Lewis acidic reducible cations   453
Lewis acidity   346, 349, 429
Lewis basicity   346, 349, 357, 437
ligand field stabilizing energy (LFSE)   484
LiO-MgO system, oxidative methane coupling   504
liquefied petroleum gas (LPG)   437
local ligand field electronic structure model   482

Lorentzian adsorption peak 357
low activation barrier, π bond 273
low or high d-valence electron occupation 204

m

Madelung constant 189
Madelung energy 476, 477
maleic anhydride reaction 495
Mars–van Krevelen mechanism 495, 544
Mars–van Krevelen surface reaction 333, 337
mechanical relations, energy and entropy 70
Meerwein–Pondorf–Verley (MPV) reduction 456
metal (oxide) surface 284
metathesis and polymerization catalysis
   alkene disproportionation and 507
   of propylene 511
   surface coordination complexes 507
   Ziegler–Natta polymerization vs. metathesis reaction 514
metathesis catalysis 513
metathesis reaction,
   single site catalysis 32
   vs. Ziegler–Natta polymerization 514
methane activation, PdO 281
methane conversion 445
   hydrocarbons 503
methane mono oxygenase (MMO) 445, 452
methane oxidation 322
   biocatalysis of 446
methanol
   to gasoline process 28
   to olefin process 28
methanol Aufbau chemistry
   alkylation by methanol 413
methanol conversion 27
methyl cyclopentane 104
methyl phenyl ketone (MPK), hydrogenation 43
Michaelis–Menten kinetics 76, 153

micro-cavity structure, zeolite faujasite 17
microkinetics equations
   ODE 88
   2R ordinary differential equations 88
microkinetics simulations 296, 301, 304, 315
microporous zeolites 22
mixed oxide surface state 139
M–O bond
   interaction energy 522
   polarity 228
M–O chemical bond energy 476, 486, 527
molecular heterogeneous catalysis
   atom utilization 168
   surface chemical reactivity of catalysts 167
molecular orbitals 174
molecular recognition,
   cation-exchanged zeolites 433
molybdo–cyclobutane intermediate 509
monocyclic/di-cyclic hydrocarbons 411
Mordenite zeolite 306
most abundant reaction intermediate (MARI) 306

n

n-alkane
   hydroisomerization reaction 391
   isomerization 391
   naphthene hydrocracking 408
NH_3 oxidation
   to NO and N_2 323
nitrogenase enzyme 296
N_2O_4 decomposition 437
N_2O decomposition catalysis 441
N_2O_4 dissociation 434
non-acidic redox systems 464
non-framework redox systems, catalysis by
   NO reduction catalysis 440
   N_2O decomposition catalysis 441
   Panov reaction 443

non-monomolecular reactions   102
non redox oxycationic clusters   440
non-reducible cations
   alkali and earth alkali ions   433
   electrostatic field   429
non-reducible oxidic systems   167
non-selective hydrogenolysis reactions   128
NO reduction catalysis   440
NO reduction, H_2 or NH_3 164
nuclear magnetic resonance spectroscopy (NMR)   68
nucleophilic surface oxygen   284, 286

o

O adatoms   490
   perovskite structure   490
OER, see oxygen evolution reaction (OER)   520
OER ideal catalyst   95
OH bond energy   439
OH bond frequency   351, 357
Olefin hydrogenation   21
olefins, protonation of   375
oligimerization reactions   391, 413
oligomerization
   ethylene   33
     polymerization catalysis
       Aufbau reaction   27
       Fischer–Tropsch catalysis   29
       surface coordination complex catalyst   35
oligomerization reactions   393, 396
O-2p atomic orbitals   481
operando spectroscopic techniques   169
Oppenauer oxidation (OP) reaction   456
ordinary differential equations (ODE)   72, 88
Ostwald reaction   323
   process   39
Ostwald's catalysis law   9
overlap of atomic orbitals (S_{ij})   175
oxametallocycle intermediate   334
oxene mechanism   452
oxidation and reduction reactions
   NH_3 to NOx oxidation   39
   steam reforming   38
oxidation catalysis   495
   methane conversion
     CH_4 to aromatics   503
     LiO-MgO system   504
   propylene and propane   500
     ammoxidation   502
     to acrolein conversion   500
   reactivity trends   495
oxidation reactions   321
   selective reaction   337
   synthesis gas from methane   321
oxidative addition   268
oxide overlayers   281
   ethylene epoxidation   284
   methane activation by PdO   281
oxygen assisted X–H bond cleavage   277
oxygen evolution reaction (OER)
   electrocatalysis   520
   reaction mechanism   522
   RuO_2 surface   523
   summary mechanism   531
   volcano plots and trends in   520
oxygen-reagent reactions   495

p

Panov reaction   440, 443, 445
   mechanism   443
partial density of states (PDOS)   173, 479
   bandwidth   204
   distributions, atomic orbitals   189
partition function (Q)   70
Pauling charge excess   346
Pauli principle   174
Pauli repulsion   236
PDOS, bulk hcp Ru   191
PDOS orbital density, see energy gap, valence and conduction band electron   189
pentamethylbenzene   416
photocatalysis   532
photocatalytic water splitting   538
   device considerations   538
   photoactivation of water   540

photo-excitation 433, 540
photo-oxidation 433
photosystem II simulations 533, 537
physical chemistry, elementary kinetics
   catalyst characterization
      Langmuir adsorption isotherm 63
      pore volume measurement 65
      porosity 66
      spectroscopic techniques 68
      temperature-programmed
         reactivity measurements 67
   diffusion
      concentration profiles 106
      zeolitic micropores 110
π-bonded ethylene 309
π orbitals, atop-adsorbed CO 257
plug flow micro reactor 63
polarity (POL) value 189
polar surface reconstruction 492
polymerization chain growth, olefin 35
positron emission profiling (PEP) 307
potential energy saddle point diagram 69
primary carbenium ion *vs.* secondary ion 377
product fragment NH_2 272
propane ammoxidation 502
propane dehydrogenation 502
propylene 500, 514
   to acrolein conversion 500
   epoxidation 461
   polymerization of 511
propylene epoxidation
   process 43
   lewis acid catalysis 43
propylene epoxide 43
protonated cyclopropane (PCP) 384, 386, 411
proton-catalyzed alkane activation 378
protonic zeolites 345
proton-oxygen interaction energy 371
proton transfer reaction 364
pyrolytic Huels process 503

q

quantum chemical concepts
   BOOPD 174

chemical bonding, transition metals
   electronic structure 190
   relative stability 199
   diatomic molecules with π bonds 182
   electronic structure, molecules and solids 186
   PDOS 174
   quantum-chemical bond 173
quantum-chemical accuracy 364
quantum-chemical calculations 72
quantum-chemical reactivity 301
quantum chemistry, chemical bond 167
quasi-differential oxygen free adsorption energy 526

r

radical chain auto-oxidation reactions 468
rapid equilibration, gas phase CO with CO 87
rate control 89
reaction energy diagram, catalytic reaction cycle
   methanation reaction 91
   transition state energies 90
reaction mechanism, catalytic reaction 168
reactivity 127
   transition metal surface atoms 274
redox catalysis 440
redox system 505, 539
reducible metal-oxide surfaces, chemical bonding and reactivity 488
reducible solid state catalysts
   electrocatalysis, OER 520
   metathesis and polymerization catalysis 507
   oxidation catalysis by group V, VI metal oxides 494
   photocatalytic water splitting 538
   sulfide catalysts 515
   transition metal oxides 475
relation, d valence band energy shift 244
relative stability of the Co 121

reoxidation 495
Rh catalysts, CO oxidation 128
ring opening metathesis polymerization (ROMP) 510

S

Sabatier principle 519
Sabatier's catalysis law 10
Sabatier volcano curve
  reaction energy diagram, catalytic reaction cycle
    succession, energy reaction 90
    transition state energies 91
  simulation 84
scaling laws
  adsorbate $X-H_x$ 259
  BEP reaction 262
  free valence, $X-H_x$ 261
  offset 262
  surface edge atom 262
secondary vs. primary carbenium ion 377
selective catalytic reduction (SCR) reaction 440, 502
selective oxidation, alkanes and olefins
  butane to maleic acid oxidation 39
  ethylene epoxidation; Ag catalysis 41
  propane conversion, acrolein and acrylonitrile 41
self-organization, Ertl system and CO oxidation 155
shell SHOP process 18
$3\sigma_g$ and $3\sigma_u$ orbitals 185
σ-symmetric adsorbate orbitals 233
single molecule spectroscopy 149
single-nanoparticle catalysis, single-turnover resolution 151
single-site catalysts 429
site number conservation law 73
site-specific catalytic activity, single-shaped nanocrystals 152
small oxide clusters 533
SMPO process 44
solid acid catalysis 345
  enzyme catalysts 420
  surface acidity and basicity 345
  zeolite protons 371
spectroscopic techniques
  electron microscopy 68
  infrared and Raman spectroscopies 68
  XPS, EXAFS 68
spin component, d-band width and d-band center spin 197
stability 128
stability and electronic structure, transition metals 192
steady-state isotopic transient kinetic analysis (SSITKA) 153
steady-state surface coverage 99
steam reforming 321
step-edge sites 133
stereochiral polymer production 514
stereoregular polypropylene 514
strong metal atom adsorbate bonding 119
structure dependence, transition metal catalysts 129
structure sensitivity, transition metal particles
  heterogeneous catalytic reactions, particle and strucrure dependence 129
  site generation 133
sugar chemistry 456
sulfide catalysts 515
sulfohydril (SH) groups 515
superacids anionic catalysis 16
supported small metal particles
  reactivity 128
  stability 128
  support material, nature of catalyst preparation 125
  common oxide and carbon supports 126
surface acidity and basicity 345
  Pauling charge excess 345
  zeolitic proton 351
    H/D exchange reactions 367
    OH bond, vibrational spectroscopy of 352
    proton transfer reaction 364

zeolite acidic OH chemical bond 358
surface atom coordination number 244
surface chemical bond
  CO chemisorption 249
  donative and backdonative interactions 249
  oxygen adatom, the polar surface bond 228
  relation, d valence band energy shift 243
surface coordination complexes, metathesis and polymerization catalysis 507
surface distribution, Brønsted basic OH groups 126
surface energy 210
surface reconstruction 119
synthesis gas conversion, methane and liquid hydrocarbons 297
synthesis gas from methane 321
  ethylene, selective oxidation of 325
    acetaldehyde 326
    epoxidation 333
    vinyl acetate 329
    methane oxidation 322
    NH_3 oxidation 323
    steam reforming 321
synthesis gas, methanol 21

t

Tafel reaction 319
Taylor view 135
Temkin kinetic expression 296
temperature dependence of catalytic reaction rate 96
temperature-programmed reduction (TPR) 67
thermodynamics 371
thiophene hydrodesulphurization 519
Thomas catalytic system 465
Thomas oxidation catalysts 463, 465
  alkane oxidation 467
  Bayer–Villiger oxidation 464
  caprolactam synthesis 464

3d transition metal cations 485
threefold-adsorbed ammonia 236
three-way catalyst (TWC) control system 46
Ti 3d-valence atomic orbitals 481
Ti-silicalite (TS-1) 43
transient kinetics
  TOF 105
  transient behavior, CO methanation 105
transition metal bulk cohesive energy 210
transition metal catalyzed reactions 293
  hydrogenation reactions 293
  oxidation reactions 321
transition metal oxide surfaces
  electronic structure 478, 488
  O adatoms, adsorption energies of 490
  polar surface reconstruction 492
transition metal surface atom 265
transition state selectivity 419
transition states, elementary surface reactions
  adsorbate σ-bond activation 265
transition state theory 70
turn over frequency (TOF) 105, 315, 321, 399, 527

u

umbrella effect 269

v

valence electron structure, bulk Ru 190
vanadium vanadyl group 496
van der Waals interaction 377
vibrational spectroscopy, of OH bond 352
vinyl acetate 329
Volcano plot, Sabatier principle 11
Volmer reaction 319

w

Wacker reaction 326
water-hydroxyl network 461

working catalyst, state of
   activation and deactivation 123
   catalytically reactive particle 117
   compound formation 123
   high surface area oxides of low
      intrinsic reactivity 118
   supported small metal particles
      reactivity 128
      stability 128
      support material, nature of 125
   surface reconstruction 119
working zeolitic catalysts 136

x

X-ray photo emission spectroscopy (XPS) 68

z

Z-scheme 539
zeolite acidic OH chemical bond 358
zeolite Brønsted acidity 345
zeolite catalyst 429
zeolite micropores 400, 408
zeolite protons 357, 358, 371
   acid catalysis, elementary reactions in 374
   acid catalyzed reactions and mechanism 371
   catalytic reaction cycles and kinetics 394
   H/D exchange reactions 367
   OH bond vibrational spectroscopy 352
   proton transfer reaction 364
   quantum-chemistry 358
zeolite reactivity 433
zeolites 17
   hydrocarbon conversion reactions 345
zeolitic non-redox and redox catalysis 429
   homogeneous and enzyme oxidation 446
   non-framework redox systems 440
   non-reducible cations 429
      catalysis with non-framework 433
      Lewis acid catalysis by 453
   Thomas oxidation catalysts 463
Ziegler–Natta catalysts 36
Ziegler–Natta polymerization 510, 511
   vs. metathesis reaction 514
ZSM-5, structure of 27